The Structure, Dynamics and Equilibrium Properties of Colloidal Systems

NATO ASI Series

Advanced Science Institutes Series

A Series presenting the results of activities sponsored by the NATO Science Committee, which aims at the dissemination of advanced scientific and technological knowledge, with a view to strengthening links between scientific communities.

The Series is published by an international board of publishers in conjunction with the NATO Scientific Affairs Division

A Life Sciences **B Physics**	Plenum Publishing Corporation London and New York
C Mathematical and Physical Sciences **D Behavioural and Social Sciences** **E Applied Sciences**	Kluwer Academic Publishers Dordrecht, Boston and London
F Computer and Systems Sciences **G Ecological Sciences** **H Cell Biology**	Springer-Verlag Berlin, Heidelberg, New York, London, Paris and Tokyo

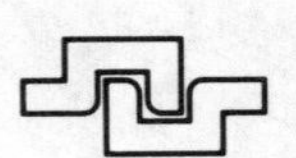

Series C: Mathematical and Physical Sciences - Vol. 324

The Structure, Dynamics and Equilibrium Properties of Colloidal Systems

edited by

D. M. Bloor

and

E. Wyn-Jones

Department of Chemistry and Applied Chemistry,
University of Salford,
Salford, U.K.

Kluwer Academic Publishers

Dordrecht / Boston / London

Published in cooperation with NATO Scientific Affairs Division

Proceedings of the NATO Advanced Study Institute on
Properties of Colloidal Systems
Aberystwyth, Wales, U.K.
September 10–23, 1989

Library of Congress Cataloging-in-Publication Data

NATO Advanced Study Institute on the Properties of Colloidal Systems (1989 : Aberystwyth, Wales)
The structure, dynamics, and equilibrium properties of colloidal systems : proceedings of a NATO Advanced Study Institute on the Properties of Colloidal Systems, held at the University College of Wales, Aberystwyth, September 10-23, 1989 / edited by D.M. Bloor and E. Wyn-Jones.
p. cm. -- (NATO ASI series. Series C, Mathematical and physical sciences ; 324)
Includes index.
ISBN 0-7923-0993-6 (alk. paper)
1. Colloids--Congresses. 2. Surface chemistry--Congresses.
I. Bloor, D. (David), 1937- . II. Wyn-Jones, E. (Evan)
III. Title. IV. Series: NATO ASI series. Series C, Mathematical and physical sciences ; no. 324.
QD549.N2854 1989
541.3'45--dc20 90-48011

ISBN 0–7923–0993–6

Published by Kluwer Academic Publishers,
P.O. Box 17, 3300 AA Dordrecht, The Netherlands.

Kluwer Academic Publishers incorporates the publishing programmes of
D. Reidel, Martinus Nijhoff, Dr W. Junk and MTP Press.

Sold and distributed in the U.S.A. and Canada
by Kluwer Academic Publishers,
101 Philip Drive, Norwell, MA 02061, U.S.A.

In all other countries, sold and distributed
by Kluwer Academic Publishers Group,
P.O. Box 322, 3300 AH Dordrecht, The Netherlands.

Printed on acid-free paper

Printed in the Netherlands

CONTENTS

PREFACE

The papers in this volume are as a result of contributions given at the NATO Advanced Study Institute held at Llandinam Building, University College of Wales, Aberystwyth, 10 - 23 September 1989. The Institute considered the physical and chemical properties of a variety of colloidal systems ranging from simple micellar solutions to concentrated colloidal dispersions. The purpose of the NATO Advanced Study Institute was to create a forum so that research scientists working in different areas concerned with colloid science could interact. The emphasis of the contributions were on the interpretation of the different experimental and theoretical approach to give information on the structure, dynamics and equilibrium properties of these systems. The application of several different techniques in colloid science have been described; new developments and perspectives have been covered by several authors. The present volume reviews the current state of the art in this area and it is hoped that it will be used as an incentive for further studies particularly with reference to new areas of research.

In the organisation of the scientific programme for the NATO meeting we would like to acknowledge the assistance of Professors J. Lyklema, D.G. Hall and J. Holzwarth. We wish to thank Miss Mandy Rudd for all the secretarial assistance in setting up the meeting and for the invaluable assistance in preparing the manuscripts. In connection with the proceedings we would also like to thank Miss Sandra Fahy for assistance. The help of Paul Jones and Mrs G. Wyn-Jones during the meeting is also gratefully acknowledged. We would also like to express our deepest gratitude to the NATO Science Division for the award of the grant which enabled the meeting to be held. Last but not least we are grateful for financial assistance from Unilever Ltd, B.P., I.C.I., and Harcross Chemicals.

THE TIME-RESOLVED FLUORESCENCE QUENCHING METHOD FOR THE STUDY OF MICELLAR SYSTEMS AND MICROEMULSIONS : PRINCIPLE AND LIMITATIONS OF THE METHOD

J. LANG
Institut Charles Sadron (CRM-EAHP), CNRS-ULP
6, rue Boussingault
67083 Strasbourg Cédex
France

ABSTRACT. The application of time-resolved fluorescence quenching method to the determination of the size of micelles and oil-in-water or water-in-oil microdroplets is described. It is also shown that this method gives information on dynamic processes occurring in micellar solutions and microemulsions. In this account the scope and limitations of the method are discussed with special emphasis placed on the assumptions used to interpret the fluorescence decay data and on the selection of appropriate probe and quencher molecules. Recent developments are also briefly presented.

1. Introduction

The time-resolved fluorescence quenching (TRFQ) method is now widely used for the determination of the size of micelles and of water-in-oil (w/o) or oil-in-water (o/w) microdroplets. This method can also be used to investigate some kinetic processes occurring in micellar solutions and in microemulsions. As a result, one of the advantages of this method is that both structural and dynamic characteristics of such systems can be investigated.

Extensive reviews of the results obtained with the TRFQ method have been reported recently [1-3]. The purpose of this article is to give a brief description of the principle and of the scope and limitations of this method so that a foundation is laid for those readers who wish to rapidly acquaint themselves with the method. Emphasis is placed on the nature of the structural and dynamic information than can be derived from these studies together with the main assumptions involved in the interpretation of the experimental data. Many original references are also quoted for the reader who wishes to pursue the technique in more detail. Finally recent developments are also mentioned.

For the sake of simplicity, the term micelle will be used to designate normal or inverse micelles and o/w or w/o microdroplets. It is only when circumstances demand that the type of aggregate will be specified. In addition the terms probe and quencher will be often referred to as reactants in the following treatment.

D. M. Bloor and E. Wyn-Jones (eds.),
The Structure, Dynamics and Equilibrium Properties of Colloidal Systems, 1–38.

2. Principle of the method

The principle of the TRFQ method is to dissolve luminescent (mostly fluorescent) probe and quencher molecules in a micelle and to measure and analyze the luminescence intensity decay of the probe. From this analysis, which assumes a given distribution of the quencher amonst the micelles, information about the size of the micelles and the dynamics of the system are obtained.

In Figure 1 the fluorescence decay curves of a fluorescent probe in the presence of quenchers in a <u>homogeneous</u> medium (Fig.1A) (pure solvent in which the fluorescent probe and the quencher molecules are uniformly distributed) is shown together with the fluorescence decay found in a <u>heterogeneous</u> medium (Fig.1B) such as micellar solutions or w/o or o/w microemulsions where appropriate probe and quencher molecules are dissolved in the micelles. It is seen that in the case of the homogeneous solutions the fluorescence decay decreases linearly

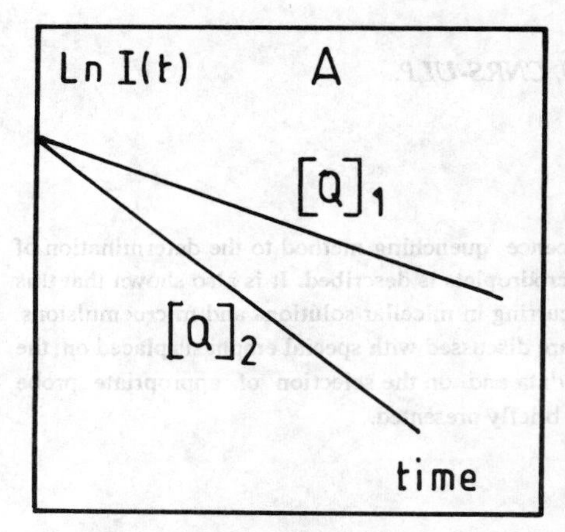

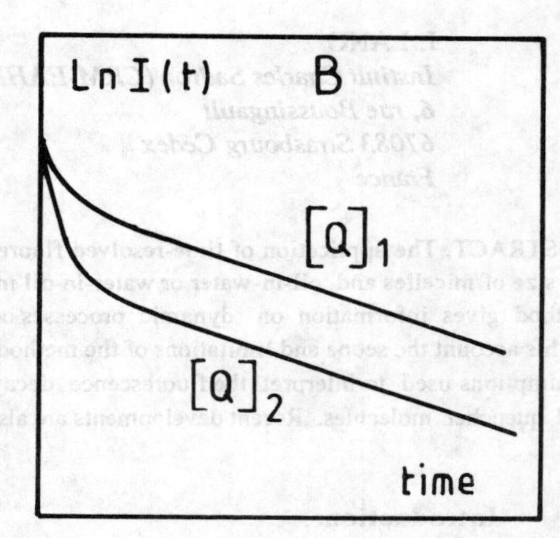

Figure 1.: Fluorescence decay curves of a fluorescent probe in the presence of a quencher in the homogeneous case (A) and in the heterogeneous case where the solution contains micelles (B). The quencher concentration increases from $[Q]_1$ to $[Q]_2$. In the heterogeneous case it is assumed that there is no intermicellar exchange of the reactants.

with time, with a slope which increases with the quencher concentration, whereas in the case of the heterogeneous solution two regimes are observed in the decay curves. A fast decay at short time followed by a linear decay at long time. The behaviour observed in the heterogeneous system, arises from the fact that the probe and quencher are compartmentalized in restricted parts of the solution, namely the micelles. Indeed the statistical distribution of the probe and quencher molecules among the micelles leads to four types of micelles which are schematically illustrated in Fig. 2A. One can distinguish : (i) empty micelles, (ii) micelles with quencher but without probe, (iii) micelles with probe but without quencher and (iv) micelles with

probe and quencher. Only the two last types of micelles will be responsible for the emitted fluorescence. In comparison to what happens in the homogeneous case where all the probe molecules have the same probability to be quenched by the quencher molecules, in the heterogeneous case only the probe molecules inside a micelle which also contains at least one quencher molecule, can be quenched. This type of micelle leads to the fast decrease observed at the beginning of the fluorescence decay curve in Fig. 1B and whose amplitude increases with the quencher concentration. On the other hand the linear decrease, at long times, is due to the micelles which contain only probe molecules. The slope of this decay at long times is not affected by the increase in quencher concentration as long as the reactants are not exchanged between the micelles during the lifetime of the probe. This is the case for the examples shown in Fig. 1B. The reactants are then said to be frozen or immobile in the micelles during the lifetime of the probe.

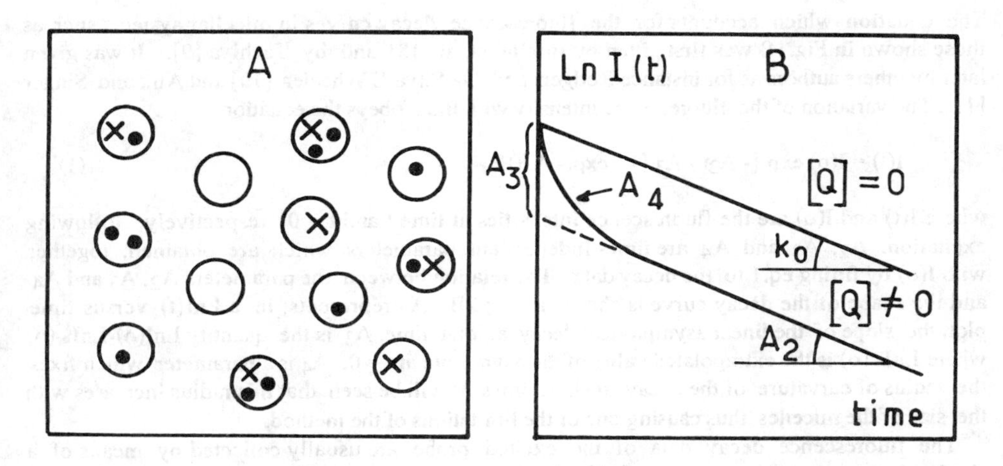

Figure 2.:(A) : Schematic representation of the distribution of probe (x) and quencher (•) molecules among micelles. (B) :Variations of the fluorescence intensity with time of probe molecules solubilized in the micelles, in the absence and presence of quencher molecules.

This argument assumes that in the heterogeneous medium there are no reactants in appreciable amount in the bulk, otherwise the situation will be that of a heterogeneous and a homogeneous medium. It will be seen that in some systems quencher molecules are partly dissolved into the bulk and this will affect the fluorescence decay curves presented in Fig. 1B. However, in all the theoretical work done so far for micellar systems it has always been assumed that no quenching of the probe occurs in the bulk, although exchange of the quencher through the bulk has been considered. Quenching of the probe both in the bulk and in the micelles would lead to a situation in which an interpretation of the time resolved fluorescence decay would be extremely difficult if not impossible. Great care must therefore be taken in the choice of the reactants and especially of the probe molecule in order to have them principally solubilized into the micelles. The way these choices are made will be discussed below.

Before proceeding with the description of the TRFQ method it is worth pointing out that the first measurement of a micellar aggregation number with a fluorescence method was done by steady-state fluorescence in 1974 by Dorrance and Hunter [4]. This work has been followed by several others, for example those of Turro and Yekta [5] and Koglin et al. [6]. The steady-state fluorescence method is very easy to use but it is practically restricted to micelles of rather small aggregation number (< 70 for direct micelles) and to immobile probe and quencher [7]. For larger micelles the steady-state method underestimates the aggregation number which can be corrected if the lifetime and the intramicellar quenching rate constant of the probe are known. However this needs fluorescence decay measurements. Therefore most of the micellar size determinations are now undertaken with the TRFQ method.

3. Analysis of the fluorescence decay curves

The equation which accounts for the fluorescence decay curves in micellar systems such as those shown in Fig.1B was first given by Infelta et al. [8] and by Tachiya [9]. It was given later by others authors as for instance Rodgers and Da Silva E Wheeler [10] and Atik and Singer [11]. The variation of the fluorescence intensity with time obeys the equation:

$$I(t) = I(o) \exp \{- A_2 t - A_3 [1 - \exp(- A_4 t)]\} \quad (1)$$

where I(t) and I(o) are the fluorescence intensities at time t and t = 0, respectively, following excitation. A_2, A_3 and A_4 are time- independent parameters which are obtained, together with I(o) by fitting eq.1 to the decay data. The relation between the parameters A_2, A_3 and A_4 and the shape of the decay curve is shown in Fig.2B. A_2 represents, in a LnI(t) versus time plot, the slope of the linear asymptotical decay at long time. A_3 is the quantity $\mathrm{LnI(o)} - \mathrm{LnI}_\infty(o)$ where $\mathrm{LnI}_\infty(o)$ is the extrapolated value of the asymptote at t = 0. A_4 is a parameter which fixes the radius of curvature of the decay at short times. It will be seen that this radius increases with the size of the micelles thus causing one of the limitations of the method.

The fluorescence decay data of the excited probe are usually collected by means of a single-photon counting apparatus [12]. The fitting of eq.1 to the data is currently done with a nonlinear weighted least squares procedure which includes deconvolution of the data by the profile of the exciting pulse. It must be noticed, however, that the deconvolution is necessary only for very small micelles where intramicellar quenching of the probe is very fast and gives a decay at short time which is affected by the width of the exciting pulse.

3.1. CASE WHERE THE PROBE AND THE QUENCHER DISTRIBUTIONS ARE FROZEN ON THE PROBE FLUORESCENCE TIME SCALE

It has been shown [13] that in this case the expressions for A_2, A_3 and A_4 are :

$$A_2 = k_0, \quad A_3 = [Q]/[M], \quad A_4 = k_q \tag{2}$$

The same expressions were established to interpret the results of intramicellar photoredox reaction [14] and can be deduced from the decay equation given by Infelta et al. [8] and by Tachiya [9] assuming that the quencher molecules are exclusively solubilized into the micelles.

In eq.2, k_0 represents the fluorescence decay rate constant of the probe in the micelles without quencher ($k_0^{-1} = \tau_0$ is the probe fluorescence lifetime), k_q is the pseudo-first-order rate constant for intramicellar quenching when only one quencher is present in the micelle, [Q] is the total quencher concentration and [M] the total micelle concentration in the solution. It appears therefore that the TRFQ method gives the concentration of the micelles [M] in the solution but not a direct measurement of the size of the micelles. However, structural parameters can be deduced from [M] under fairly reasonable assumptions. For instance, the mean surfactant aggregation number, N, in a micelle is given by :

$$N = \frac{C - C_f}{[M]} = \frac{(C - C_f)A_3}{[Q]} \tag{3}$$

where C is the total surfactant concentration and C_f the free surfactant concentration i.e. the surfactant concentration in the bulk which does not participate to the formation of the micelles. In the case of direct micelles C_f is usually taken equal to the critical micelle concentration (cmc). For large values of C the error made assuming C_f = cmc is negligible. In solutions of inverse micelles or w/o microemulsions C_f is usually very low and is therefore neglected compared to C.

In Fig.2B the variation of the fluorescence intensity versus the time is shown for two identical micellar solutions, one with and one without quencher. The slopes of these two curves at long time are equal. This indicates that there is no exchange of reactants between the micelles during the lifetime of the probe. Equation 2 can then be readily used for the interpretation of the fluorescence decay data obtained with the solution containing the quencher. It will be shown that when intermicellar exchange of the reactants occurs in the solution, these two slopes are not equal. Expressions other than those given by eq.2 are then valid for A_2, A_3, and A_4.

It must be emphasized that the comparison of the values of the slopes of the decay curves obtained with and without quencher in the solution must be made for each system investigated. This is necessary in order to check if eq.2 is appropriate for the interpretation of the fluorescence decay data.

Several assumptions have been made for the determination of eqs.1 and 2, which are listed below:

a) The micelles are assumed to be monodisperse. As a consequence only one intramicellar quenching rate constant, k_q, has been considered in the kinetic equations.

b) There is no limit for probe and quencher solubility in the micelles.

c) Only one probe is excited at one time in the same micelle.

d) The number of quenchers in a micelle is independent of the presence of a probe and vice versa.

e) The kinetics of intramicellar quenching is first-order. This can be assumed for small micelles only, where diffusion process and Fick's law cannot be applied to the reactants. For large micelles, like infinite cylindrical micelles, a second-order kinetics must be considered.

f) The rate of intramicellar quenching is proportional to the number of quenchers, x, in the micelle. This assumption is sometimes referred to as a statement that the intramicellar quenching rate constant is proportional to x or that k_q must be replaced by xk_q when x quenchers are present in a micelle, k_q being the quenching rate constant when only one quencher is in the micelle.

g) The distribution of the quenchers among the micelles corresponds to a Poisson distribution.

In order to fulfil assumptions b, c and d one has to ensure that the molar concentration ratio [probes]/[micelles] is kept much below 1 (usually a ratio 0.01 is employed) and that the molar concentration ratio [quenchers]/[micelles] is close to 1. Also in order to fulfil assumption c, a small exciting pulse intensity is used.

Tachiya [15], following a treatment proposed by Hunter [16], has treated theoretically the case where there is a limit to the number of solubilized molecules in the micelles. His treatment includes also exchange of solubilizates between micelles. However, the analysis of the decay data is then much more complicated and has not yet been carried out for real systems.

3.2. DISTRIBUTION OF THE REACTANTS AMONG MICELLES

It is now generally considered that the distribution of the quenchers among the micelles corresponds to a Poisson distribution. Other distributions have also been examined as for instance the geometric distribution. The difference between these two distributions comes from the value which is taken for the exit rate constant of the quencher from the micelle. For the Poisson distribution this rate constant is assumed to be proportional to the number of quenchers in the micelle, whereas for the geometric distribution this rate constant is assumed to be independent of the number of quenchers in the micelles [17]. The association of the quencher to the micelles can be described by a series of equations of the type :

$$MQ_{x-1} + Q \underset{(x)k_{-1}}{\overset{k_1}{\rightleftarrows}} MQ_x \qquad (4)$$

where M and Q stand for the micelles and the quencher molecules, respectively, and x is the number of quencher molecules in the micelle MQ_x. Therefore the kinetic equations at equilibrium can be written as follows:

(i) for the Poisson distribution:

$$k_1 [MQ_{x-1}][Q] = xk_{-1} [MQ_x] \tag{5}$$

and

(ii) for the geometric distribution:

$$k_1 [MQ_{x-1}][Q] = k_{-1} [MQ_x] \tag{6}$$

where k_1 and k_{-1} are the entry and exit rate constants of one quencher molecule to/from a micelle, respectively.

From the series of eqs.5 and 6 it is easy to calculate the probability of occurrence of a micelle containing x quenchers. One finds :

(i) for the Poisson distribution :

$$P_x = \frac{\bar{R}^x e^{-\bar{R}}}{x!} \tag{7}$$

and

(ii) for the geometric distribution:

$$G_x = \frac{\bar{R}^x}{(1+\bar{R})^{x+1}} \tag{8}$$

where $\bar{R} = \sum_{x=1}^{\infty} x\,[MQ_x] / \sum_{x=0}^{\infty} [MQ_x] = [Q] / [M]$ is the mean occupancy number of the micelles.

Figure 3 shows that there is an appreciable difference between the Poisson and the geometric distribution especially for $\bar{R} > 1$. It is important to know that, depending on the statistical distribution adopted, very different expressions of $LnI(o)-LnI_\infty(o)$ and therefore very different values of [M] can be found [17]. This is why various experiments have been carried out with the aim of finding out which statistics best account for the fluorescence decay data.

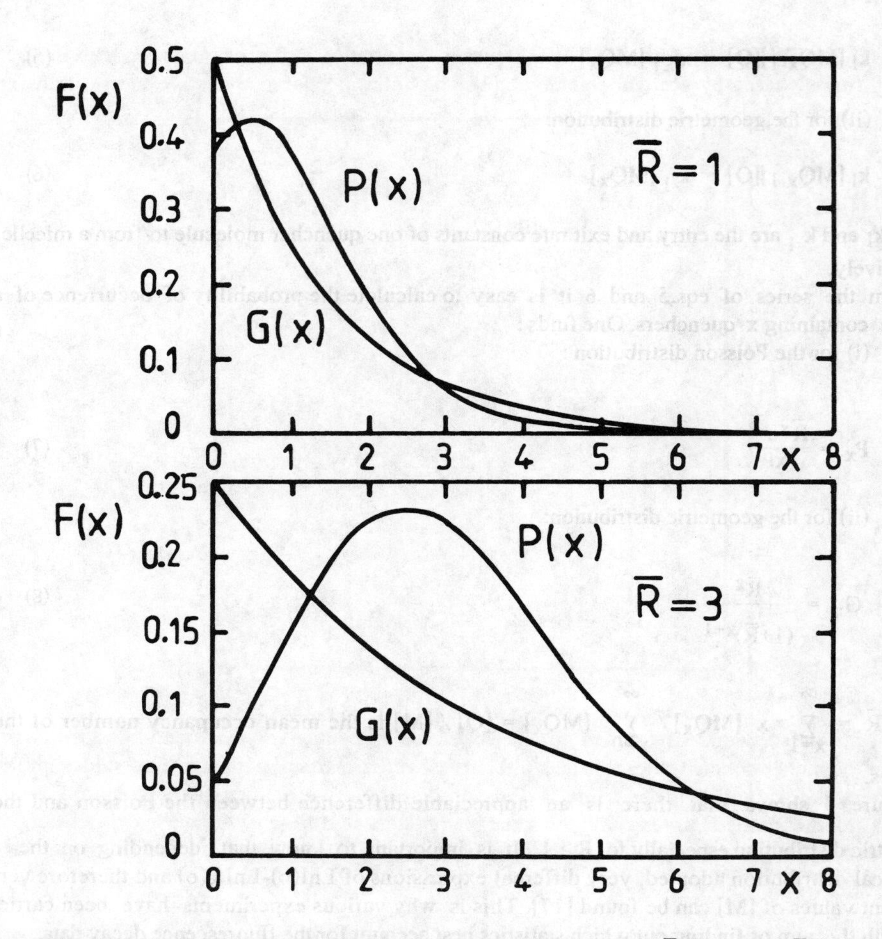

Figure 3.: Distribution of quencher molecules among micelles for $\bar{R}$ = 1 and $\bar{R}$ = 3. P(x) = Poisson distribution, G(x) = geometric distribution. F(x) is the fraction of micelles containing x quencher molecules.

Figure 4 represents the fluorescence decay data obtained by Atik and Thomas [17] for a sodium bis(2-ethylhexyl)sulfosuccinate (AOT) - water-heptane w/o microemulsion. The results show that a much better fit is found assuming the Poisson distribution, rather than the geometric distribution.

The results reported in Figure 5 further support the argument in favour of the Poisson distribution compared to the geometric distribution. Indeed eq.2, based on a Poisson distribution of the quencher among micelles, indicates that $LnI(o)-LnI_\infty(o) = A_3$ varies linearly with [Q] and this is what is found experimentally in Fig.5A, whereas the geometric distribution predicts a curvature for this variation which is not found. Alternatively $LnD^T(\infty)-LnD^T(o)$, where $D^T(\infty)$ and $D^T(o)$ are the donor triplet concentration at time $t = \infty$ and $t = 0$,

respectively, varies linearly with [A], the acceptor concentration in the example reported in Fig.5 which is from a work by Rottenberger et al. [18] on intramicellar irreversible energy transfer

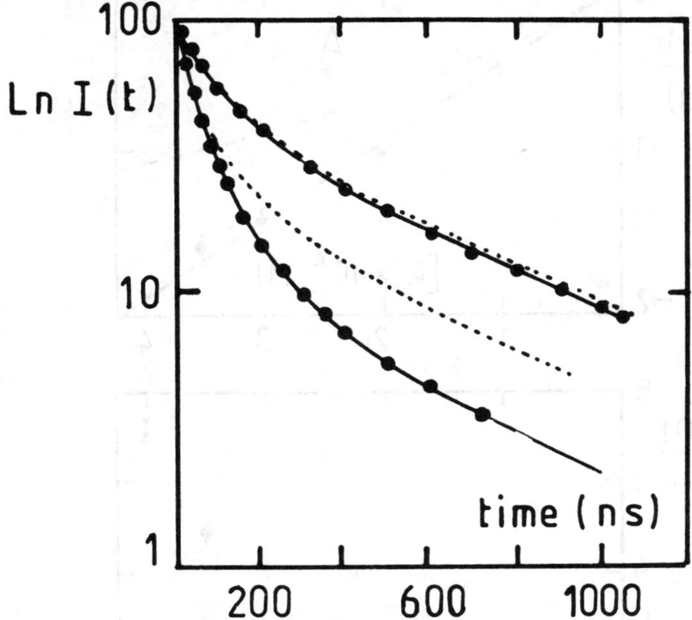

Figure 4.:Fluorescence decay curves for a water-AOT-heptane w/o microemulsion with two different quencher concentrations. The solid curves were obtained with identical values of the parameters k_0, k_q and [M] assuming a Poisson distribution of the quenchers among the micelles and the dashed curves with the same parameters assuming a geometric distribution of the quenchers among the micelles. Only the Poisson distribution gives a good fit to the data [after reference 17]

reaction. In fact, assuming a geometric distribution, it is the quantity $I(o)/I_\infty(o)$ (or $D^T(o)/D^T(\infty)$) which should vary linearly with [Q] (or [A]). The results reported in Fig.5B show that the experimental variation of $D^T(o)/D^T(\infty)$ versus [A] is not a straight line. This result confirms that the geometric distribution fails to account for the data.

Other results have been found in favour of the Poisson distribution. For instance Miller et al. [19] have studied the kinetics of excimer formation of pyrene in aqueous sodium dodecylsulphate (SDS) solutions and shown that the distribution of pyrene among micelles can be described by a Poisson distribution. Recently, Siemiarczuk and Ware [20] analyzing, by the maximum entropy method, the fluorescence decay of pyrene in SDS solutions quenched by Cu^{2+}, were able to directly recover the Poisson distribution of pyrene amongst micelles.

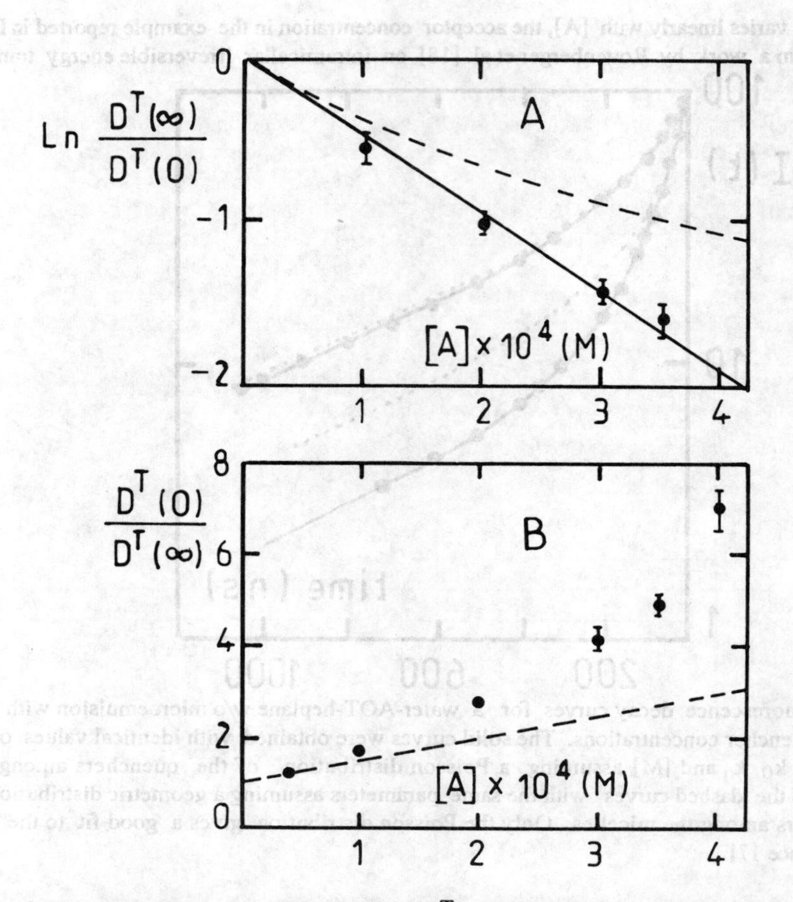

Figure 5.:Variation of the triplet concentration D^T of a donor D (N-methylphenothiazine) after completion of the irreversible energy transfer between D and an acceptor A (trans-stilbene). The micelles in which the energy transfer takes place are normal micelles of cetyltrimethylammonium bromide in water. The solid line corresponds to a Poisson distribution and the dashed lines to a geometric distribution. Only the Poisson distribution gives a good fit to the data [after reference 18].

Thus, enough evidence seems to have been obtained which confirm the use of the Poisson distribution of the quencher amongst the micelles for the analysis of the fluorescence decay data.

It must be emphasised that since the probe concentration is chosen to be very much lower than the concentration of micelles one can assume that, as far as the probe is concerned, there are only micelles without or with one probe molecule. A rapid calculation shows that , for a [probe]/[micelle] molar concentration of 0.01, and assuming Poisson statistics for the probe among the micelles, these two types of micelles represent 99.99% of the total number of micelles, and the number of micelles with one probe molecule is 200 times larger than the number of micelles with more than one probe molecule. Therefore the distribution of the probe among the micelles has not been considered in the determination of eqs.1 and 2.

3.3. EXAMPLES OF PROBE AND QUENCHER MOLECULES WHICH DO NOT EXCHANGE THROUGH THE CONTINUOUS PHASE

Table 1 gives examples of probe and quencher molecules which do not exchange through the bulk on the probe fluorescence time scale, and which are used for the study of micellar solutions

TABLE 1. Typical probe and quencher molecules used in TRFQ studies of micellar systems and microemulsions, which do not give a detectable exchange via the continuous phase on the fluorescence time scale of pyrene or $Ru(bpy)_3^{2+}$. P,Q and E stand for fluorescent probe, quencher and excimer, respectively.

For normal micelles and o/w microdroplets		
Anionic Surfactants	Cationic Surfactants	Nonionic Surfactants
Pyrene (P,E)	Pyrene (P,E)	Pyrene (P,E)
DPyrCl (Q)	HPyrCl (Q)	HPyrCl (Q)
TPyrCl (Q)		
HPyrCl (Q)		
$Ru(bpy)_3^{2+}$ (P)		
$Ru(bpy)_3^{2+}$ - $(C_{17})_2$(P)		
1-methyl-1'-tetra-decyl-4,4'-bipyridinium^{2+}(P)		
For w/o microdroplets		
Anionic Surfactants	Cationic Surfactants	
$Ru(bpy)_3^{2+}$ (P)	$Ru(bpy)_3^{2+}$ (P)	
$Fe(CN)_6^{3-}$ (Q)	Methylviologen^{2+} (Q)	

or microemulsions. Pyrene is a fluorescent probe which is only very slightly soluble in water ($7x10^{-7}$ M at 25ºC). It solubilizes in the hydrophobic part of the micelles and is therefore used almost exclusively for the study of micelles and o/w microemulsions. Pyrene has been studied either together with a quencher or alone using excimer formation. Good quenchers of pyrene in anionic micelles are dodecyl-, tetradecyl- or hexadecylpyridinium chlorides (DPyrCl, TPyrCl and HPyrCl, respectively). The long alkyl chain and the positive charge of these molecules makes them suitable to bind to anionic micelles, their alkyl chain being incorporated into the micelles. For cationic and non ionic micelles only HPyrCl which has a long alkyl chain can safely be used as a quencher which does not exchange through the water phase. Indeed there is in this case no electrostatic attraction between the micelle and the quencher. There is even a repulsion in the case of cationic micelles and only a large hydrophobic attractive interaction can maintain the quencher inside the micelle. Ruthenium tri(bipyridyl)chloride ($Ru(bpy)_3^{2+}$) and 1-methyl-1'-tetradecyl-4-4'-bipyridinium chloride which possesses two positive charges, have also been used as fluorescent probe for the study of normal micelles and o/w microemulsions made of anionic surfactants. The modified $Ru(bpy)_3^{2+}$, $Ru(bpy)_3^{2+}$-$(C_{17})_2$, where one of the bypyridine is dialkylated by heptadecyl chains can also be used as a fluorescent probe in anionic micellar systems where the micelles are not to small and cannot be notably perturbated by the heptadecyl chains. In the case of w/o microemulsions the $Ru(bpy)_3^{2+}$ ion can be safely used as a fluorescent probe since it is not soluble in the organic bulk. As efficient quenchers, the most often used are the $Fe(CN)_6^{3-}$ and the methylviologen^{2+}(MV) ions for w/o microemulsions based on anionic and cationic surfactants, respectively. However, it must be pointed out that the formation of ion pairs, for instance between cationic probes and anionic surfactants which could migrate through the bulk, cannot be entirely excluded.

3.4. CASE WHERE INTERMICELLAR EXCHANGE OF THE REACTANTS OCCURS ON THE PROBE FLUORESCENCE TIME SCALE

Three types of exchange processes which are schematically represented in Figure 6 have been investigated.

Path 1 represents the exchange of the reactants by collision and transient merging of two micelles followed by the splitting of the transient dimer. During the merging of the two micelles all kinds of intermicellar exchanges can occur : exchange of the water molecules, the surfactant ions, the counterions and the reactants. In particular, a probe and a quencher molecule initially in two different micelles can be found in the same micelle at the end of this process. If the fluorescence probe was excited at time t = 0 its lifetime will be shortened by the presence of the quencher molecule in the new micelle formed. The consequence of this redistribution of the reactants during the probe lifetime is that the asymptote of the fluorescence decay curve will no longer be parallel to the fluorescence decay obtained for the solution without quencher. The slope of this asymptote will increase with the quencher concentration.

This type of exchange has been found principally for microdroplets in w/o microemulsions made of ionic surfactants [17,21-24] and for nonionic micelles [25]. Indeed the microdroplets, considered as a whole, are uncharged in w/o microemulsions and interdroplets collisions are possible. In the case of o/w microdroplets and normal micelles one can imagine that the high charge density carried by the micelles renders the collisions between micelles very much improbable, unless high ionic strength is used. However, as it will be indicated below, exchange by collisions between direct ionic micelles of quenchers bound to the micelles, has also been observed.

Path 2 represents the exchange of the quencher via the continuous phase. It has been already mentioned that the fluorescent probe must be chosen to be entirely solubilized in the micelles or to have a solubility in the continuous phase so low that its exit rate constant is very small

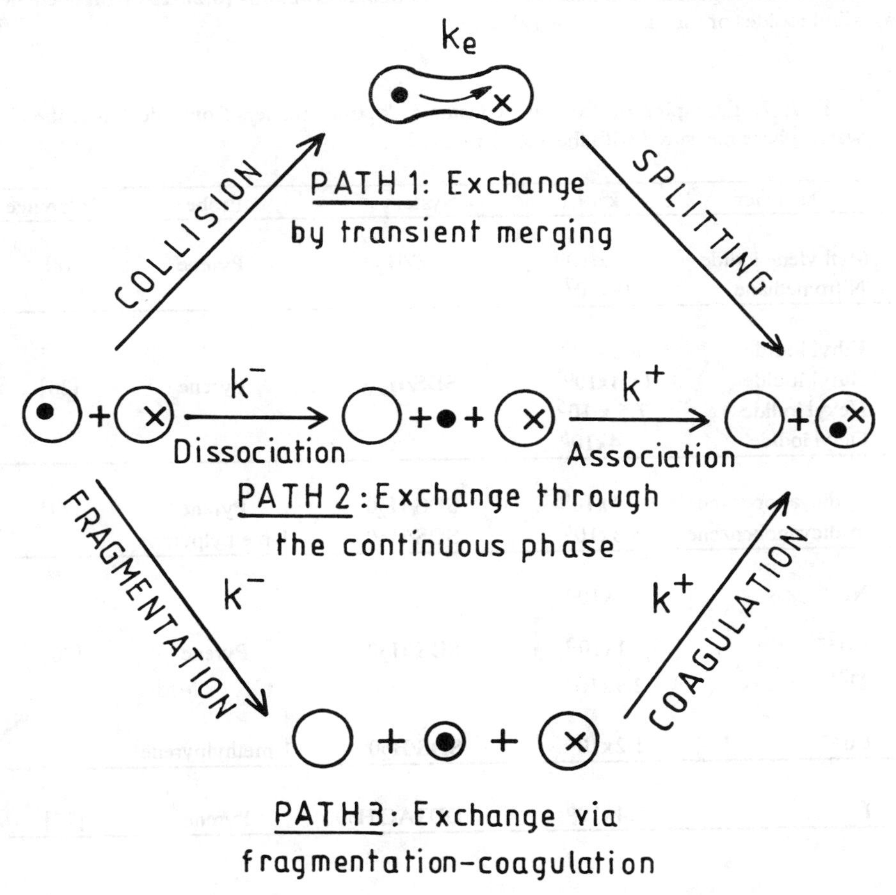

Figure 6.: General scheme for intermicellar exchanges of the reactants. These exchanges are illustrated by the migration of a quencher (o) from a micelle without probe to a micelle containing one probe (X).

compared to its fluorescence decay rate constant. However, as far as the quencher is concerned, various types have been employed, in particular quenchers with a relatively high solubility in the water continuous phase were used. This solubility is responsible for their exchange from one micelle to another micelle via the continuous phase. Again, this type of exchange ensures that an excited probe, which is initially in a micelle without quencher, has a finite probability to be in contact with a quencher coming from another micelle through the continuous phase. When such exchanges occur on the probe fluorescence time scale, the lifetime of the probe, measured in the presence of quencher ($1/A_2$ in eq.1), is therefore lower than τ_0.

This type of exchange has been observed in direct micellar solutions [8,26-29] with quenchers such as alkyl iodides or metal ions (see Table 2).

TABLE 2. Examples of exit rate constants, k^-, of quenchers from micelles to the water phase measured with the TRFQ method

Quencher	$k^- s^{-1}$	System	Probe	Reference
Methylene iodide	9.5×10^6	SDS/H_20	Pyrene	[8]
Nitromethane	(6×10^7)			
Ethyl iodide	8.3×10^6			
Butyl iodide	$(1.4 \times 10^6$	SDS/H_20	Pyrene	[28]
Hexyl iodide	7.5×10^5			
Octyl iodide	4×10^5			
m-dicyanobenzene	5×10^6	SDS/H_20	Pyrene	[29]
m-dicyanobenzene	6.3×10^6	SDS/H_20	1-methylpyrene	
Ni^{2+}	1×10^5 }			
Co^{2+}	1×10^5 }	SDS/H_20	Pyrene	[26]
Tl^{2+}	2.9×10^6			
Cu^{2+}	1.2×10^6	SDS/H_20	1-methylpyrene	
I^-	2.4×10^6	$DDTAC/H_20$	Pyrene	[27]

Path 3 illustrates the exchange of a quencher via the fragmentation- coagulation process. In this process a quencher leaves a micelles embedded in a micellar fragment which migrates through the continuous phase and associates to another micelle. In this process, an excited probe molecule inside a micelle without quencher has also a finite probability to be visited by a quencher. This process again will decrease the lifetime τ_0 of the probe in the micelles without quencher at t = 0.

This type of exchange has been found in mixed surfactant+alcohol micelles and is related to the relatively high micellar polydispersity in the system [30]. Indeed the presence of micellar fragments means that the system is polydisperse.

Thus, one of the effects of these three exchange processes is to reduce the probe lifetime compared to τ_0. This is schematically presented in Figure 7B. However, it must be emphasized that these exchange processes can be detected only if the rate constant k_e or k^-, are large enough compared to k_0, the fluorescence decay rate constant of the probe molecule in the micelles without quenchers. The smaller the value of k_0 (that is the larger is the probe fluorescence lifetime) results in smaller values of k_e and k^- which can be measured by the TRFQ method (for numerical values of k_e and k^- see references given in [1-3]). When exchange of reactants between micelles occurs, eq.1 still applies but then the expressions for A_2, A_3 and A_4 are different from those given by eq.2. However, they have been derived on the same assumptions than those made in the case of immobile reactants and presented above in section 3.1.

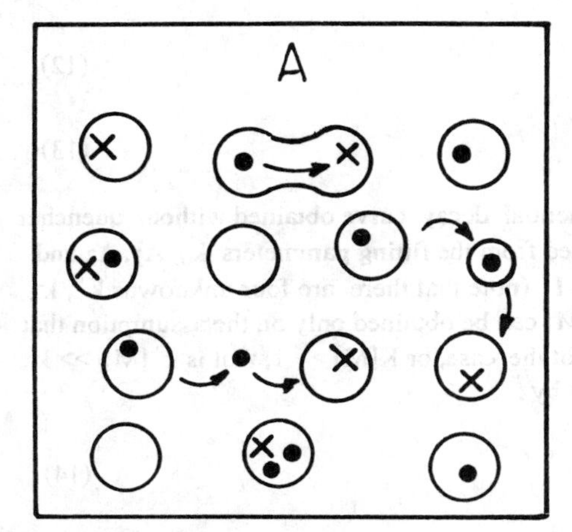

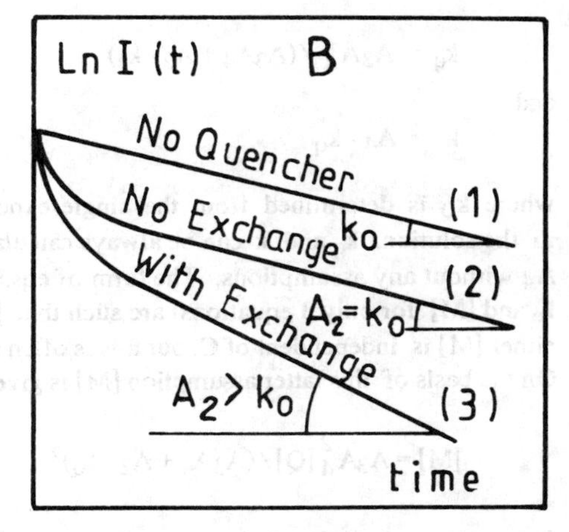

Figure 7.:(A) : Schematic representation of the distribution of probe (X) and quencher (•) molecules among micelles and of the intermicellar exchanges of the reactants shown in Fig.6. (B) : Variations of the fluorescence intensity with time of a probe solubilized in the micelle : (1) = when only probes and no quencher are in the micelles, (2) = when probes and quenchers are in the micelles but without intermicellar exchange of the reactants and (3) = when probes and quenchers are in the micelles and intermicellar exchange of the reactants occurs on the probe fluorescence time scale.

If the exchange occurs via the dissociation-association process (path 2) the expressions for A_2, A_3 and A_4 are given by [8-9]:

$$A_2 = k_0 + \frac{k_q k^+[Q]}{A_4(1+K[M])} \qquad (9)$$

$$A_3 = \left(\frac{k_q}{A_4}\right)^2 \frac{K[Q]}{1+K[M]} \qquad (10)$$

$$A_4 = k_q + k^- \qquad (11)$$

where k^+ is the second-order rate constant for the association of a quencher to a micelle, k^- is the first-order rate constant for the dissociation of a quencher from a micelle (see Fig.6) and $K = k^+/k^-$. Eqs.9-11 can be solved to obtain :

$$k_q = A_3 A_4^2 / (A_3 A_4 + A_2 - k_0) \qquad (12)$$

and

$$k^- = A_4 - k_q \qquad (13)$$

where k_0 is determined from the single-exponential decay curve obtained without quencher in the solution. k_q and k^- can be always calculated from the fitting parameters k_0, A_2, A_3 and A_4 without any assumptions. The form of eqs.9-11 (note that there are four unknowns k^+, k^-, k_e and [M] for only 3 equations) are such that [M] can be obtained only on the assumption that either [M] is independent of C, but this is often not the case, or $K[M] >> 1$, that is $k^+[M] >> k^-$. On the basis of this latter assumption [M] is given by :

$$[M] = A_3 A_4^2 [Q] / (A_3 A_4 + A_2 - k_0)^2 \qquad (14)$$

It must be noticed that if the assumption $K[M] >> 1$ is not valid the true [M] values are lower than that given by eq.14.

If the exchange occurs via the fragmentation-coagulation process the expression for A_2, A_3 and A_4 are also given by eqs.9-11, since the mathematics are identical for exchanges occurring via paths 2 or 3. In this case k^+ represents the second-order rate constant for the coagulation of a fragment to a full micelle and k^- is the first-order rate constant for the detachment of a fragment from a full micelle. k_q and k^- are also given by eqs.12 and 13, respectively, and [M] is given by eq.14 also on the assumption that [M] is independent of C or $K[M] >> 1$.

The treatment for the case of exchange of quencher between water pools via interdroplets collisions (path 1) has been given by Atik and Thomas [17,21]. The following expressions of A_2, A_3 and A_4 have been found :

$$A_2 = k_0 + \frac{k_q k_e}{A_4}[Q] \tag{15}$$

$$A_3 = \left(\frac{k_q}{A_4}\right)^2 \frac{[Q]}{[M]} \tag{16}$$

$$A_4 = k_q + k_e[M] \tag{17}$$

where k_e represents the second-order rate constant associated with collisions giving rise to migration of the reactants between micelles (see Fig.6). Note that k_e is an overall rate constant which includes the diffusion, the fusion and the separation of the two droplets. In the present case, k_0 is also obtained from an independent transient fluorescence experiment without quencher in the solution. From eqs.15- 17, k_q, k_e and [M] are obtained as a function of the fitting quantities k_0, A_2, A_3 and A_4:

$$k_q = A_3A_4^2/(A_3A_4 + A_2 - k_0) \tag{18}$$

$$k_e = A_4(A_2 - k_0)/(k_q[Q]) \tag{19}$$

$$[M] = A_3A_4^2[Q]/(A_3A_4 + A_2 - k_0)^2 \tag{20}$$

Note that eq.18 and 20 are identical to eq.12 and 14, respectively, but here [M] is obtained without any assumptions.

Almgren et al. [31,32] have shown that eqs.15-17 are not valid if exchange of the excited probe by intermicellar collisions is taken into account. However, eqs.18 and 20 are not modified, and only eq.19, which gives k_e, is then different. However the analysis of the fluorescence decay data is not straightforward and eqs.18-20 are currently used. Moreover it has been found, in the study of AOT-water- alkanes w/o microemulsions [24], that the k_e-values calculated with eq.19 are slightly lower, by less than 20%, compared to those obtained if the analysis proposed by Almgren et al. is used.

The expressions of A_2, A_3 and A_4 have also been given when the exchanges using paths 1 and 2 occur simultaneously [33,34] :

$$A_2 = k_0 + \frac{k_q(k^+ + k_eK[M])[Q]}{A_4(1+K[M])} \tag{21}$$

$$A_3 = \left(\frac{k_q}{A_4}\right)^2 \frac{K[Q]}{1+K[M]} \tag{22}$$

$$A_4 = k_q + k^- + k_e[M] \tag{23}$$

In fact these equations were given prior to eqs.15-17. Of course, eqs.15-17 follow directly by setting $k^- = 0$ in eqs.21-23.

Eqs.21-23 have been used for the study of the quenching of fluorescent probes by single or multivalent ions in normal ionic micellar solutions [26,27,33]. In these studies the rate constants k^+, k^- and k_e were determined, k_e being the rate constant for the transfer of the ions (quenchers) bound to the micelles, from the host micelle to another micelle during collision between the two micelles. However, since eqs.21-23 contain five unknowns, these equations can only be used by making some assumptions and if [M] is known from another experiment. Eqs.21-23 were also employed for instance for the study of the quenching of fluorescent probes by neutral molecules in SDS solutions [35], where the quencher was assumed to transfer directly between two colliding micelles. In eqs.21-23, k_e then represents the rate constant for a quencher to transfer between two micelles.

Equations 9,15 and 21 indicate that A_2 is larger than k_0. Thus, mathematics show what has been emphasised already several times: i.e. when intermicellar exchange of reactants occurs on the probe fluorescence time scale, the slope of the decay curve at long times in the presence of quencher (A_2), in a LnI(t) versus time plot, is larger than the slope of the decay curve found without quencher (k_0). Note that eqs.9,15 and 21, also show that A_2 increases as [Q] increases.

It must be noticed that the rate constant k_e in the collision- splitting process and the rate constants k^- and k^+ in the fragmentation-coagulation process are parameters which characterize the micelles alone, whereas the rate constants k^- and k^+ in the dissociation-association process are parameters which characterize principally the quencher molecule, in a given type of system (direct or inverse micelles or o/w or w/o microemulsions). Indeed, in this last process the values of k^- and k^+ depend principally on the partition coefficient of the quencher between the continuous and the micellar phases. This partition coefficient depends on the nature of the quencher, for example on the number of hydrophobic groups or electric charges that the quencher carries.

3.5. EXCIMER FORMATION

The case of excimer formation is more complicated than the case of quenching since the respective decays of the monomer and excimer are kinetically coupled through the dissociation reaction of the excimer. In principle it is possible to measure the fluorescence of the excited monomer as well as that of the excimer as a function of time. The equations which govern these variations have been given by Infelta and Grätzel [36] assuming that the pyrene distribution among micelles is frozen during the pyrene lifetime. However, generally only the fluorescence decay curve of the monomer is used with the assumption that the excimer dissociation rate constant is small and much lower than the excimer decay rate constant. With these assumptions the equation given by Infelta and Grätzel for the fluorescence decay of the monomer becomes identical to eq.1. Equation 2 and eqs.9-23 are then also valid, but in these equations k_q is replaced by k_E the pseudo-first-order rate constant for intramicellar excimer formation, and [Q] by [P] the total pyrene concentration. The rate constants k^-, k^+ and k_e are then associated to the intermicellar exchanges of the probe molecule (pyrene). The fluorescence decay rate constant k_0 for pyrene is measured from the single exponential decay curve obtained with a molar concentration ratio [Pyrene]/[M] no larger than 0.01 to avoid excimer formation.

It must be added that eq.1 is valid only for studies made at moderate temperature. Indeed, as the temperature increases the excimer dissociation rate constant k_{-E} increases and the assumption, that k_{-E} is small and much lower than the excimer dissociation rate constant, may no longer hold.

3.6. THE INTRAMICELLAR QUENCHING RATE CONSTANT

Various parameters influence the value of the intramicellar quenching rate constant kq as for instance:

(i) - the size of the micelle : the larger the micelle size the lower is the value of k_q, since as the size of the micelle increases the mean distance between probe and quencher increases

(ii) - the microviscosity of the micelle where the reactants are located : the larger is this microviscosity the lower is the mobility of the reactants and the value of k_q

(iii) - the possible existence of preferential sites for the reactants due to specific interactions between the reactants and the micelle. For example attractive or repulsive interactions between the internal charged surface of a water pool and ionic reactants : the k_q-value will depend on the respective electric charges of the surfactant and reactant molecules

(iv) - the shape of the micelle : for a given micellar aggregation number the k_q-value may differ from a spherical to a cylindrical micelle of the same aggregation number

(v) - the intrinsic rate constant of the quenching of the probe molecule by the quencher molecule: k_q depends on the nature of the probe and quencher.

The parameter which is the most easily modified by the experimentalist is the size of the micelle. It is also the most interesting to change in the studies of w/o or o/w microemulsions since one has often to cover a large domain of compositions in the phase diagram. This can lead to very large variations of the size of the microdroplets.

Figure 8 shows the variation of the fluorescence intensity with time, according to eq.1 where A_2, A_3 and A_4 are given by eq.2 (no intermicellar exchange of the reactants), for various values of the ratio $r = k_q/k_0$. The rate constant k_0 has been chosen equal to 2.5×10^6 s^{-1} (τ_0 = 400 ns). Note that the curves extend to three times the value of τ_0. It is seen that the asymptote is reached only for $r > 1$ (recall that the decays at long time should all fall on the same line in Fig.8 since it is assumed that there is no intermicellar exchange of the reactants). These simulated decay curves show that the important parameter is not k_q but the ratio r. Usually, if this ratio is below 1 (for too large micelles or too low fluorescence probe lifetime) the fit of eq.1 to the decay data gives incorrect values of the fitting parameters A_2, A_3 and A_4 since then the asymptote of the decay curve is not reached. For instance, Fig.8 shows that for r = 0.01 the decay at long time is very close to k_0. Indeed the asymptote of the curve with r = 0.01 seems almost parallel to the one obtained for the curve with r = 100, which is equal to k_0. Thus, fitting eq.1 to the decay curve obtained with r = 0.01, will give almost the correct value for k_0 but will give a value for A_3 much too low, say about 0.01, compared to the value 1 used to draw the curve. Of course, in such a case, the values found for k_q and r will give an indication whether one can have any confidence in the fitted parameters. Moreover, when r is below 1 the slope of the decay curve at long time might be found larger than k_0 (see for example the curve obtained with r = 0.5) and this can be interpreted as due to intermicellar exchange of the reactants, although such exchange does not occur in the system during the lifetime of the probe.

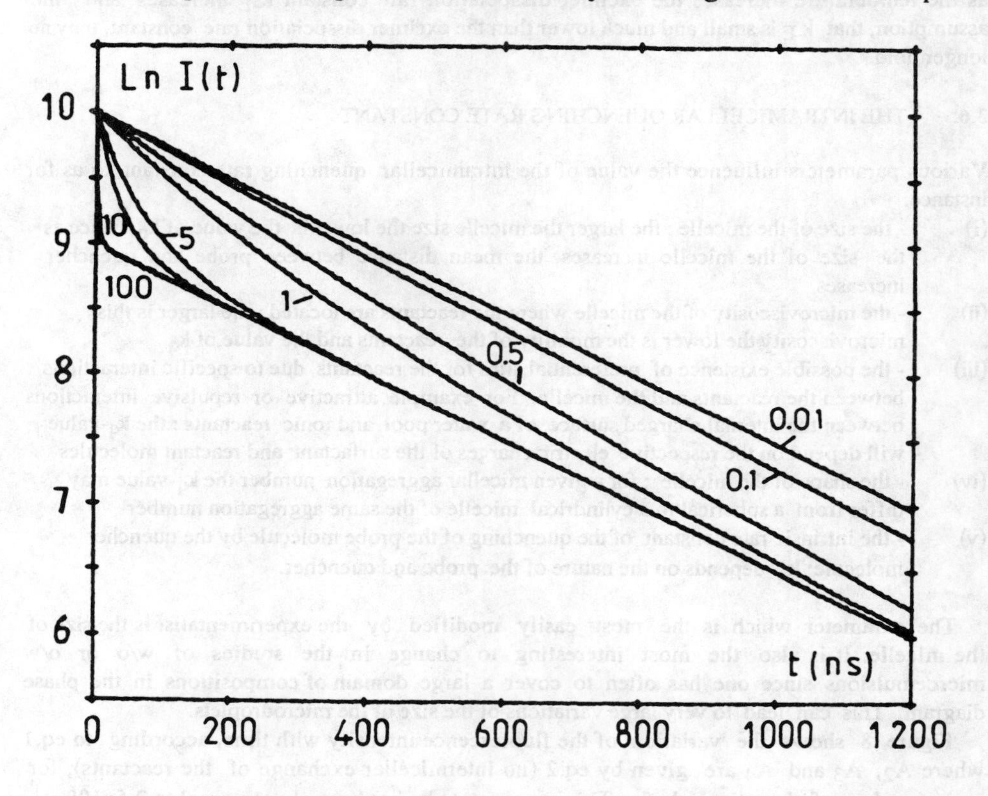

Figure 8.:Plots of fluorescence decay curves according to eqs.1 and 2, for k_0 = $2.5 \times 10^6 s^{-1}$, A_3 = 1 and various values of $k_q = rk_0$. The value of r is indicated on each curve.

Thus great care must be taken in the procedure involving the accumulation of the fluorescence decay data. As a result caution must be exercised when large micelles are investigated and/or when exchange processes characterized by large k^- or k_e values ($A_2 >> k_0$) are occurring in the system under investigation.

First, one must take a probe with a lifetime as long as possible. This is why pyrene is the probe that is most often used in direct micellar solutions and o/w microemulsions studies (τ_0 for pyrene = 350-400 ns, depending on the nature and concentration of additives in the micelles). In w/o microemulsions $Ru(bpy)_3^{2+}$ is used since τ_0 = 600- 800 ns, depending on the nature of the surfactant and on the size of the water pool. It must be emphasised, however, that if one uses probes with increasing lifetimes then it might turn out that some exchange processes, not detectable with short living probe molecules because of the low values of the rate constants k^- and k_e, can appear on a longer time scale. Next, one has to accumulate the fluorescence decay curves over as long a time scale as possible. In some experiments the decay has been extended to seven times the value of τ_0. The accumulation of the decay data must then be made during a very long period of time. Indeed it may need several hours to obtain a good signal/noise ratio at the longest decay time. One must also compare the value of the slope of the asymptote, A_2, as one first fits eq.1 to all the decay data and then only to the first three quarters of the decay data. If, under these two conditions, A_2 is not constant, within the accuracy in the determination of A_2 (2-3%), it might be that the asymptote of the fluorescence decay curve calculated with all the decay data has not been reached. As indicated previously, this can lead to erroneous interpretations of the fluorescence decay data. Thus, as the micelles become too large there is a limitation in the TRFQ method when the asymptote of the decay curve is not obtained. Surfactant aggregation number, N, up to 200 for normal micelles and 1500 for w/o microdroplets are currently determined with an accuracy better than 10%. Larger aggregation numbers have been measured for particular systems only by taking great care in the accumulation of the fluorescence intensity decay data. Usually for w/o microdroplets larger values of N have been determined than for o/w microdroplets and normal micelles. This comes mostly from the longer lifetime of $Ru(bpy)_3^{2+}$ compared to the lifetime of pyrene. It must also be mentioned that probe with lifetimes longer than the lifetime of $Ru(bpy)_3^{2+}$ have been synthesized [37]. These however have not yet been systematically used for the investigation of the size and dynamics of w/o microdroplets. One of the reasons is that these probes are very large molecules and might perturb the micelle unless the water pool is itself very large.

A last remark must be made which concerns very small micelles. If very small micelles are to be investigated very large values of k_q can lead to static quenching and part of the information which appears at the beginning of the decay curve can then be lost. For example during the analysis the procedures for evaluating the parameter A_3 sometimes lead to artificially low values - even zero in some cases. Fortunately, this is a favourable case where the steady state fluorescence method can be used for the determination of [M] and thus of the surfactant aggregation number and other structural parameters which characterize the micelle.

Instructive discussions have been presented on the effects of the relative values of k_q, k_0, k^- and k_e on the shape and analysis of the fluorescence intensity decay curves [38,39].

3.7. STRUCTURAL PARAMETERS

Once the aggregation number N is known from [M] (eq.3), other parameters concerning the size of the micelles can be obtained by means of some appropriate working model.

In the case of w/o microdroplets for instance, if it is assumed that all the water is inside the micelles, that the water cores are spheres separated from the continuous oil phase by a monomolecular film of N surfactant ions, and that the droplet size is monodisperse (an assumption already made for the determination of [M] and N). The radius of the water pool, R_w, and the mean spherical surface area, σ_s, per surfactant at the surface of the water pool then follow from simple geometrical considerations:

$$R_w = \left[\frac{3N}{4\pi}(\omega\, v_w + v_{ci} + v_{hg})\right]^{1/3} \tag{24}$$

and

$$\sigma_s = \frac{4\pi R_w^2}{N} \tag{25}$$

In eq.24 ω represents the molar concentration ratio $[H_2O]/[\text{surfactant}]$ in the solution, v_w the molecular volume of a water molecule (29.9 $Å^3$ at 25°C), v_{ci} the apparent molecular volume of the counterion of the surfactant and v_{hg} the apparent volume of the ionic head group of the surfactant which may be considered as being inside the water pool. Note that most of the time v_{hg} has not been taken into account in the calculation of R_w. Note also that in eq.24 no distinction between free and bound counterions is made. However, this equation gives R_w to a very good approximation.

If some additives are solubilized in the water core the expression for R_w is :

$$R_w = \left[\frac{3N}{4\pi}(\omega\, v_w + v_{ci} + v_{hg} + \omega_a v_a)\right]^{1/3} \tag{26}$$

and σ_s is still given by eq.25. In eq.26 ω_a represents the molar concentration ratio [additive]/[surfactant] in the solution and v_a is the apparent molecular volume of the additive in the water pool.

The additive can be partitioned between the water pool, the interfacial layer and the oil-rich continuous phase. This is for instance the case for medium chain length alcohol molecules. In this case, in eq. 26 $\omega_a v_a$ must be replaced by $f\omega_a v_a$ where f represents the fraction of additives in the water pool, and σ_s is also given by eq.25. It is worth remarking that if additives are present in the interfacial layer σ_s still represents the mean surface per surfactant at the surface of the water core, but then includes the surface occupied by the additives.

The overall droplet radius R_W can be taken as :

$$R_M = R_W + 1 \tag{27}$$

where l is the length of the fully extended hydrophobic part of the surfactant ion if v_{hg} is included in the calculation of R_W, or the length of the fully extended surfactant ion if v_{hg} is not included in the calculation of R_W.

Equations analogous to eqs.24-27 can be derived for the case of normal micelles and o/w microdroplets.

4. Recent developments of the TRFQ method

4.1. EFFECT OF POLYDISPERSITY IN SIZE OF THE MICELLES ON THE ANALYSIS OF THE FLUORESCENCE DECAY CURVES

In the previous section the analysis of the fluorescence decay data has been made assuming that the micelles were monodisperse. Recently theoretical developments have been carried out which take into account the polydispersity in size of the micelles [40-44]. It has been shown that for polydisperse micelles the measured value of N (designated by $<N>_Q$ and called quenching-average aggregation number [40]) should decrease with an increase of the quencher concentration [40-42]. Such a decrease has indeed been observed in several systems [40,41,44-49]. Almgren and Löfroth [40] have proposed a treatment in which the fluorescence decay curve is the sum of all the contributions of the individual micelles in which each micelle contains s surfactant ions, x quenchers and z fluorescent probes. These authors assume that :

a) The Poisson distribution of the probes and quenchers among micelles remains valid for each subset of micelles of a particular size (as in the case of monodisperse micelles), but the mean number of probes and quenchers in a micelle is proportional to the aggregation number s, i.e. to the size of the micelle.

b) The number of probes in a micelle is independent of the presence of a quencher and vice versa (as in the monodisperse case).

c) The intramicellar quenching rate constant is proportional to the number of quenchers in the micelles, inversely proportional to the micellar size and independent of z. Thus, contrary to the monodisperse case where only one value of k_q is considered, in the polydisperse case various values of k_q are considered due to the micellar polydispersity.

d) The probe and quencher distributions are frozen during the lifetime of the probe. This assumption implies that the proposed treatment is valid only in the cases where intermicellar exchanges of the reactants are not detected on the fluorescence time scale of the probe molecule.

The above assumptions lead to the following expression for the decay of the fluorescence in a polydisperse system after pulse excitation at time t = 0 [40] :

$$I(t) = I(o)\exp(-k_0 t)\frac{\sum_s sA(s)\exp\{-s\eta[1-\exp(-K_q t/s)]\}}{\sum_s sA(s)} \tag{28}$$

where A(s) represents the concentration of micelles with s surfactant molecules, K_q is the intramicellar quenching rate constant which plays the role of a second-order rate constant, and η is given by :

$$\eta = \frac{[Q]}{C\text{-cmc}} \tag{29}$$

which assumes that all the quencher molecules are inside the micelles.

The apparent quenching-average aggregation number is defined as [40]:

$$\langle N\rangle_Q = \frac{1}{\eta}\,\mathrm{Ln}\,\frac{I(o)}{I_\infty(o)} \tag{30}$$

which is a relationship identical to eq.3 where A_3 is obtained with eq.1 from the quantity $\mathrm{Ln}[I(o)/I_\infty(o)]$. In eq. 30 $\mathrm{Ln}I_\infty(o)$ is, like in section 3, the extrapolation at time t = 0 of the fluorescence decay curve at long time (with slope $A_2 = k_0$) in the LnI(t) versus time plot.

From eqs.28-30, $\langle N\rangle_Q$ is finally given by [40] :

$$\langle N\rangle_Q = \frac{1}{\eta}\,\frac{\sum_s sA(s)}{\sum_s s\exp(-\eta s)A(s)} \tag{31}$$

Notice that at low quencher concentration ($\eta \to 0$), $\langle N\rangle_Q$ turns out to be the weight-average aggregation number $\langle N\rangle_w$.

A more general relationship between $\langle N\rangle_Q$ and η, involving a polynomial of up to order 3 in η, has also been given for unsymmetrical distributions [42,50] :

$$\langle N\rangle_Q = \langle N\rangle_w - \frac{\sigma^2\eta}{2} + \frac{\Lambda\eta^2}{6} - \frac{E\eta^3}{24} \tag{32}$$

where σ, Λ and E are the second, third and fourth cumulants of the weight distribution sA(s). σ is the root-mean-square deviation and Λ and E are called the raw skewness and raw excess or kurtosis, respectively. σ^2 and E must be positive, and Λ can take either sign. The skewness and kurtosis are best standardised with respect to a Gaussian(symmetrical) distribution [42] :

$$\lambda = \frac{\Lambda}{2\sigma^3} \qquad \varepsilon = \frac{1}{8}\left[\frac{E}{\sigma^4} - 3\right] \tag{33}$$

A positive value of λ indicates positive skewness and positive ε indicates that the weight distribution sA(s) is more peaked about $\langle N\rangle_w$ than in a Gaussian distribution.

An analytical expression for $\langle N\rangle_Q$ as a function of η (eq.31) has been given [40] for the Gaussian and exponential distributions. For the Gaussian distribution A(s) is equal to :

$$A(s) = A(s_N) \exp[-(s-s_N)^2/2\sigma^2] \tag{34}$$

where $s_N = \langle N\rangle_N$ is the number-average aggregation number, σ the root- mean-square deviation of s and $A(s_N)$ the number of micelles at the maximum of the distribution, and $\langle N\rangle_Q$ is given by :

$$\langle N\rangle_Q = \langle N\rangle_N + \sigma^2 \left[\frac{1}{\langle N\rangle_N} - \frac{\eta}{2}\right] = \langle N\rangle_w - \frac{\sigma^2}{2}\eta \tag{35}$$

For the exponential distribution A(s) is equal to :

$$A(s) = A(s_0) \exp[-(s-s_0/\sigma)] \qquad \text{for } s \geq s_0$$

$$A(s) = 0 \qquad \text{for } s < s_0 \tag{36}$$

where s_0 is the starting value of s for the distribution, σ a measure of the decay length of the distribution, and $\langle N\rangle_Q$ is given by :

$$\langle N\rangle_Q = s_0 + \frac{2}{\eta}\mathrm{Ln}(1+\eta\sigma) - \frac{1}{\eta}\mathrm{Ln}\left(1 + \frac{\eta s_0 \sigma}{s_0 + \sigma}\right) \tag{37}$$

The fit of eqs.35 and 37 to the experimental variation of $\langle N\rangle_Q$ versus η should tell if the Gaussian or the exponential distribution can account for the data and should give the parameters which characterize the micelle distribution function for the investigated system. Note that eq.32 can also be used to obtain some information about the micellar distribution. Equations 32, 35 and 37 have been used in the study of several polydisperse systems [40-42,45-50]. Other distribution functions, as for instance the zeroth-order logarithmic normal distribution (ZOLD) [51], or the Schulz, the Stevenson-Heller- Wallach and negatively skewed distributions [52] can also be used, but then the distribution function must directly be introduced into eq.31 since no analytical relationship for $\langle N\rangle_Q$ as a function of η has been given so far for these distributions. This has been done for instance with the ZOLD and other distribution functions [49].

Figure 9 summarizes the expected variations of $<N>_Q$ versus η for a monodisperse system, and for two polydisperse systems, one having a symmetrical and the other a non-symmetrical micellar size distribution. These variations can be easily deduced from eq.32. Figure 9 shows that the surfactant aggregation number should be independent of the quencher concentration if the micelles are monodisperse in size (σ, Λ and E equal to zero in eq.32) although, in practice, this is never the case. However for small globular micelles and w/o microdroplets the polydispersity is found often to be very low. In these circumstances it has been shown [40] that then it may be assumed that a number- average surfactant aggregation number is obtained. Figure 9 also shows that for symmetrical distribution (Λ and E equal to zero in eq.32), as for instance the Gaussian distribution, a linear decrease of $<N>_Q$ versus the quencher concentration should be found whereas in the case of a non-symmetrical distribution, e.g. the ZOLD or the exponential distribution, $<N>_Q$ should decrease non linearly as [Q] increases.

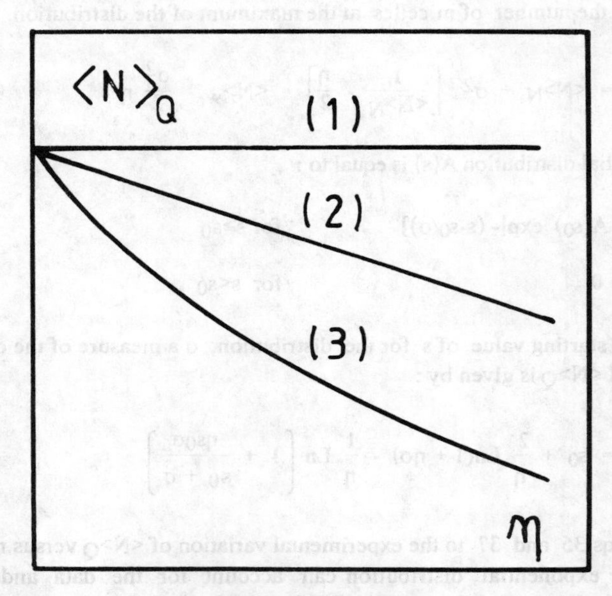

Figure 9.:Plots of the variation of the quenching-average surfactant aggregation number $<N>_Q$, as a function of the molar concentration ratio of quenchers to surfactants in the micelles, η. (1) = The micelles are monodisperse : $<N>_Q$ is independent of η . (2) = The micelles are polydisperse in size and the distribution is symmetrical:$<N>_Q$ decreases linearly as η increases. (3) = The micelles are polydisperse in size and the distribution is non-symmetrical : $<N>_Q$ decreases non-linearly as η increases.

Finally it must be emphasized that the theory of Almgren and Löfroth [40] is based strictly on the cases where the micellar aggregation number s is proportional to the size of the micelle. This is introduced in their calculations by an intramicellar quenching rate constant inversely proportional to s. However, in the case of mixed micelles, for instance made of surfactant and alcohol molecules, the proportionality between s and the size of the micelle is no longer valid if the composition of the mixed micelles varies with their size. Therefore some corrections to the proposed theory to take into account more complex systems may appear useful in the future.

4.2. THE TRFQ METHOD FOR THE STUDY OF CYLINDRICAL MICELLES

The main difference in the theoretical treatment of the fluorescence decay curves found with small and large aggregates is that in the case of small aggregates the intramicellar quenching rate constant k_q can be assumed to be pseudo-first-order, whereas in the case of large aggregate k_q is a second-order rate constant. Indeed the distance which separates a probe and a quencher molecule in very large aggregates, means that the intramicellar quenching will depend on the mutual diffusion of the probe and quencher molecules.

The expression of the intramicellar quenching rate constant k_q has been derived for cylindrical micelles [53]. It has been shown that k_q scales as N^{-1} or $N^{-1.77}$ depending on whether the reaction between probe and quencher is reaction-controlled or diffusion-controlled, respectively, as compared to $N^{-0.87}$ and $N^{-1.07}$ for spherical micelles.

The expression for the fluorescence intensity decay I(t), in the presence of a quencher in an infinite cylindrical micelle, without intermicellar exchange of reactants, has also been derived [44,54]:

$$I(t) = I(o)\ \exp\{-k_0t - B_1C_0[\exp(B_2t)\mathrm{erfc}(B_2t)^{1/2} - 1 + 2(B_2t/\pi)^{1/2}]\} \qquad (38)$$

with $B_1 = 2h^{-1}$, $B_2 = h^2D_m$ and $h = 2k_qR/3D_m$, where R is the radius of the cylindrical micelle, D_m the probe and quencher diffusion coefficients in the cylinder and C_0 the number of quenchers per unit length of cylinder.

This expression for I(t) differs very much from eq.1. In particular, eq.(38) never shows a truly exponential behaviour at long times. Nevertheless, due to statistical noise, an experimental decay curve which corresponds in reality to the situation described by eq.38, can be fitted fairly well by eq.1. Such fittings always lead to $A_2 > k_0$. Therefore a system with very long micelles may be confused with a system of small micelles with intermicellar exchange of the reactants. However this uncertainty may be partially avoided by using a so-called global analysis where the least square procedure is applied to a series of decay curves obtained for a micellar system at increasing quencher concentration [32,47,54-60].

Almgren et al. [44] have also discussed the case of polydisperse systems of finite rod-like micelles and gave the first values of k_q and D_m for cylindrical micelles. These authors have also examined the case of infinite rod-like micelles with exchange of the quencher through the continuous phase [44]. The analysis of the decay data then becomes very difficult.

Numerical simulations performed by Van der Auweraer et al. [54] give useful indications concerning the characteristics of surfactant aggregates, quenchers and probes to which eq.1 or alternately eq.38 applies. The conclusion of these calculations, which agree with that of Almgren et al. [44] is that probes with long lifetime and efficient quenchers with long residence times in the micelles should be always used. The ratio [Q]/[M] should be as high as possible, without perturbing the size of the micelle. These simulations have also shown that eq.1 can be used as long as the length L of the micelle is such that $L/2 < L_{Diff}$ where the diffusion length, $L_{Diff} = (\pi D_m/k_0)^{1/2}$, is approximately the length over which a quencher diffuses during the lifetime of the probe, by a motion intermediate between diffusion in one and two dimensions. Equation 38 should be used whenever $L/2 > L_{Diff}$. The above condition applies when a micelle of length L contains one excited probe and one quencher. The case where the same micelle contains two quenchers in addition to the probe has also been discussed [54]. A criterion for the choice between eq.1 and 38 for the analysis of the decay data has also been proposed. If the value of A_4 and A_2 obtained from the fit of eq.1 to the decay data is such that $A_4 < A_2$ then eq.38 has to be used.

4.3. THE TRFQ METHOD FOR THE STUDY OF VESICLES

The TRFQ method has also been used for the study of vesicles on the assumption that there is no intermicellar exchanges of the reactants on the probe fluorescence time scale [61]. As in the case of infinite cylindrical micelles the quenching kinetics was taken as second-order since the vesicles studied were large aggregates. A fluorescence decay equation based on a three-dimensional diffusion of the reactants in the vesicles has been given for the interpretation of the transient fluorescence experiments:

$$I(t) = I(o)\exp[-(k_{app}t + k_{Do}t^{1/2})] \tag{39}$$

with

$$k_{app} = k_0 + k_v[Q_a]$$

and

$$k_{Do} = k_D[Q_a] \tag{40}$$

where k_v and k_D are rate constants which are a function of the reactant diffusion coefficients and reaction encounter distance. I(t), I(o) and k_0 have the usual meaning and $[Q_a]$ is the quencher concentration based on the volume available to the quenchers (total aggregate volume). The term $k_{Do}t^{1/2}$ comes from the so-called depletion effect which occurs in the vesicles and originating when the diffusion of the quencher is slow compared to the lifetime of the probe. Note that at long times where the term $t^{1/2}$ becomes less important, eq.39 reduces to the well known Stern-Volmer equation.

The authors also indicate that it would probably make little difference, within the accuracy of the TRFQ method, if two dimensional diffusion of the reactants in the vesicles is assumed.

Another equation for the fluorescence decay of the probe in the presence of quenchers has been proposed for the case where the solution contains both normal micelles and vesicles [61]. In this case the proposed fluorescence decay equation is given by the following equation which is simply a combination of eqs. 1 and 39.

$$I(t) = I(o)\{f_m\exp[-A_2t - A_3(1 - \exp(-A_4t))] + (1 - f_m)\exp[-(k_{app}t + k_{Do}t^{1/2})]\} \tag{41}$$

where A_2, A_3 and A_4 are given by eq.2 (no intermicellar exchange of the reactant), k_{app} and k_{Do} by eq.40, f_m is the fraction of surfactant in micellar form and $(1-f_m)$ the fraction of surfactant in vesicles (or in other large aggregates like liposomes which might be present in this solution).

From eq.41 one should, in principle, obtain the fraction of surfactant in the micelles and in the vesicles, the size of the micelles and the various rate constants associated with the quenching processes in the micelles and in the vesicles. However eq.41 contains six adjustable parameters. Here k_0 is fixed, and is measured from an independent experiment without quencher in the solution. It has been shown [61] from simulation calculations that it is very difficult due to the statistical noise in the experiments to determine these six parameters with a least-squares procedure, Thus, it has been concluded that only qualitative information can be obtained from eq.41. In fact a test has been proposed to determine whether the system contains both small micelles and large aggregates. By means of equation 41 it can be shown that the presence of vesicles or liposomes in the solution is revealed by a dependence of the fluorescence decay at short times on the quencher concentration. This can be confirmed by simulating fluorescence decays using equation 41 and fitting the resulting data using equation 1. Therefore, it has been proposed to fit equation 1 to the decay curves obtained with various quencher concentrations [Q], and to verify how the parameter A_4, which characterizes the decay curves at short times, varies with [Q]. A large increase of A_4 as [Q] increases would reveal the presence of a significant number of large aggregates in the system.

The authors also emphasised that if intermicellar exchanges of the reactants occur on the probe fluorescence time scale in solutions containing only micelles then A_4 remains constant and it is the parameter A_2 which increases as [Q] increases (see eqs.9,11,15,17,21 and 23). This allows one to distinguish between normal micellar solutions with intermicellar exchanges of reactants and mixture of micelles and vesicles without intermicellar exchanges of reactants. However the case of intermicellar exchanges of reactants in systems containing both micelles and vesicles would lead to an unworkable decay equation and as such has not been examined. Fortunately in aqueous solutions made of ionic surfactants the exchange of reactants solubilized inside the aggregates via collisions between aggregates is, in principle, unlikely. On the other hand, the exchange via the water phase is under the control of the experimentalist who must choose probe and quencher molecules which are insoluble in the water phase in order to avoid this type of exchange. Exchange by fragmentation will appear only for polydisperse systems.

In summary, the proposed analysis permits one to detect qualitatively the simultaneous presence of micelles and vesicles in a given system. Even if a large decrease of the experimental noise is achieved a quantitative analysis of the fluorescence decay data with eq.41 will be very difficult due to the large number of parameters appearing in this equation. A quantitative analysis might be possible if some of these parameters are obtained from independent measurements.

4.4. ANALYSIS OF THE TRFQ DATA WITH FRACTAL MODELING

In fractal modeling the distribution of quenchers around an excited probe molecule in a micelle is considered fractal and the diffusion of the quencher toward the excited probe is equivalent to a random walk from site to site [62-65]. In the viscous environment of a micelle or a vesicle, quenchers close to the excited probe have a better chance to quench the probe than f from the excited probe.. This results in a short- and long-time behaviour of the fluorescence decay profile which is no longer a single exponential. Also the exploration of the quencher within the micelles or the vesicles can be characterized as compact or non-compact which are respectively related to the short- and long-time behaviour.

From these considerations the equation which describes the fluorescence decay of the probe in the presence of quenchers in micelles or vesicles is given by :

$$I(t) = I(o) \exp \left\{ - (k_0 + k_1 [Q])t - \frac{k_2[Q]}{f} t^f \right\} \quad (42)$$

with $f = d_s/2$, where k_0 has the usual meaning, k_1 is a diffusion- controlled second-order rate constant, k_2 is a constant, d_s is the so-called fractal dimension and [Q] the quencher concentration. In eq.42 the term $k_1[Q]t$ is associated to the non-compact exploration and $(k_0 + k_1[Q])t$ is equivalent to the term which appears in the Stern-Volmer equation. The term $k_2[Q]t^f/f$ is associated to the compact exploration. The authors of this model also mention that in polydisperse systems it is possible that the exploration is compact in small aggregates and non-compact is very large ones. Thus eq.42 can also account for the fluorescence decay obtained with polydisperse systems.

Equation 42, in which a fractal dimension appears, has been used for the interpretation of the fluorescence decays obtained in various systems like w/o microemulsions or solutions containing vesicles [62- 65]. From the fitting of eq.42 to the decay data various parameters could be obtained like k_1, k_2 and d_s but the relationships between these parameters and the real size of the micelles and the rate constants characterizing exchange processes occurring in micellar solutions or microemulsions has yet not been established.

4.5. OTHER STUDIES WITH THE TRFQ METHOD

Studies of liquid crystal phases have been made by Johansson and Söderman [66], Fletcher [67] and Liang and Thomas [68]. The decay of $Ru(bpy)_3^{2+}$ adsorbed on clays containing quenching impurities has also been studied [69] and the parameters of the decay function have been correlated to the chemical composition of the clays and to their swelling properties. The size of nonionic [70] and anionic [71] micelles at the solid-liquid interface has been also investigated, whereas an instructive discussion of the characterization of surface by excited state, which includes TRFQ decay measurements, has been made by Thomas [72]. Finally the recent study of Fletcher concerning the comparison of the size of normal micelles in glycerol and water must be mentioned [73].

5. Conclusive remarks

We have shown that the TRFQ method gives the surfactant aggregation number of micelles and microdroplets and also some other characteristics of these aggregates, for example their radius (assuming they form spheres). Other geometrical models like ellipsoids or disks have also been explored and the results will be found in the literature. This method can also give information on dynamic processes involving the micelles as a whole (rate of collision between micelles with transient merging of the micelles and exchange of reactants, principally in w/o microemulsions and nonionic direct micelles), or micellar fragments (rates of detachment and coagulation of a fragment from/to a full micelle, in polydisperse systems). Dynamic processes involving the micelles and free or micelle-bound quencher molecules can also been investigated. Such processes occur via the dissociation/association of a quencher from/to a micelle or, upon collision between two micelles, via the transfer of a quencher ion from the surface of a charged micelle to another charged micelle or via the hopping of a neutral quencher from the surface of a micelle to another micelle. These last processes mainly characterize the quencher molecule and not so much the micelle. The effect of the number of hydrophobic groups or electric charges carried by the quencher on the rate constants which characterize these exchanges in a given micellar system, have been investigated.

We have also shown that the TRFQ method can give information on the distribution in size of the micelles. However the treatment has been made so far only for the case where there is no intermicellar exchange of the reactants. Although polydispersity and exchange processes can occur simultaneously, the difficulties in the analysis of the fluorescence decay are in this case very important and such analysis has not been done so far for real systems. Notice that the fragmentation-coagulation process has been studied using equations established assuming the micelles are monodisperse, although this process implies a certain polydispersity [30]. Thus the values of the rate constants obtained in these cases are only approximate but are probably of the right order of magnitude. Recall that it can always be verified if the polydispersity is large or not by measuring the quenching-average surfactant aggregation number as a function of the quencher concentration.

The TRFQ method can also be use to obtain information on the rate of diffusion of reactants in large cylindrical micelles and on the extent of the amount of vesicles present in solutions containing a mixture of micelles and vesicles.

Finally, one must emphasize the great advantage of the TRFQ method compared to scattering methods. In the latter method the size of the micelles can only be accurately obtained at low micelle concentrations where the interaction between micelles is low. The TRFQ method permits the determination of the size of the micelles at almost any concentration of micelles in the system, since the fluorescence decay of the probe is, in principle, independent of the interactions between micelles. As a result it has been shown that in some systems the size of the micelles depends on their concentration [24]. Moreover the values of the rate constants k_e, k^- and k^+ can be measured at any micellar concentration, provided that they are large enough to be detected on the time scale of the fluorescence of the probe. It must be pointed out that k_e could be directly correlated to the attractive interactions between micelles as evidenced by other methods. This has been done in several studies, as for example the one reported in the present volume [74] which illustrates also that the sizes of the micelles obtained from the TRFQ method are in excellent agreement with those measured by quasielastic light scattering at low micelle concentrations. The TRFQ method has been employed for a very large variety of studies where the effect of a multitude of parameters such as the structure of the surfactant and of the oil; the nature and concentration of added electrolytes or other additives; the temperature; the structure and concentration of cosurfactants; the nature of the surfactant counterion; on the micellar size and dynamics have been investigated [1-3].

6. References

[1] Zana, R. (1987) "Luminescence probing methods", in R. Zana (ed.), Surfactant Solutions: New Methods of Investigation, Marcel Dekker, New-York, pp 241-294.

[2] Grieser, F. and Drummond, C.J. (1988) "The physicochemical properties of self-assembled surfactant aggregates as determined by some molecular spectroscopic probe techniques", J. Phys. Chem. 92, 5580-5593.

[3] Zana, R. and Lang, J. (1990) "Recent developments in fluorescence probing of micellar solutions and microemulsions", to appear in Colloids Surf.

[4] Dorrance, R.C. and Hunter, T.F. (1974) "Absorption and emission studies of solubilization in micelles. Part 2. - Determination of aggregation number and solubilisate diffusion in cationic micelles", J. Chem. Soc. Faraday Trans.1, 70, 1572-1580.

[5] Turro, N.J. and Yekta, A. (1978) "Luminescent probes for detergent solutions. A simple procedure for determination of the mean aggregation number of micelles", J. Am. Chem. Soc. 100, 5951-5952.

[6] Koglin, P.K.F., Miller, D.J., Steinwandel, J. and Hauser, M. (1981) "Determination of micelle aggregation numbers by energy transfer", J. Phys. Chem. 85, 2363-2366.

[7] Infelta, P.P. (1979) "Fluorescence quenching in micellar solutions and its application to the determination of aggregation numbers", Chem. Phys. Lett. 61, 88-91.

[8] Infelta, P.P., Grätzel, M. and Thomas, J.K. (1974) "Luminescence decay of hydrophobic molecules solubilized in aqueous micellar systems. A kinetic model", J. Phys. Chem. 78, 190-195.

[9] Tachiya, M. (1975) "Application of a generating function to reaction kinetics in micelles. Kinetics of quenching of luminescent probes in micelles", Chem. Phys. Lett. 33, 289-292.

[10] Rodgers, M.A.J. and Da Silva E Wheeler, M.F. (1978) "Quenching of fluorescence from pyrene in micellar solutions by cationic quenchers", Chem. Phys. Lett. 53, 165-169.

[11] Atik, S.S. and Singer, L.A. (1978) "Nitroxyl radical quenching of the pyrene fluorescence in micellar environments. Development of a kinetic model for steady-state and transient experiments", Chem. Phys. Lett. 59, 519-524.

[12] Pfeffer, G., Lami, H., Laustriat, G. and Coche, A. (1963) "Détermination des constantes de temps de scintillateurs", C.R. Hebd. Séances Acad. Sci. 257, 434-437.

[13] Atik, S.S., Nam, M. and Singer, L. (1979) "Transient studies on intramicellar excimer formation. A useful probe of the host micelle", Chem. Phys. Lett. 67, 75-80.

[14] Maestri, M., Infelta, P.P. and Grätzel, M. (1978) "Kinetics of fast light-induced redox processes in micellar systems : Intramicellar electron transfer", J. Chem. Phys. 69, 1522-1526.

[15] Tachiya, M. (1980) "Kinetics of quenching of luminescent probes in micellar systems. II", J. Chem. Phys. 76, 340-348.

[16] Hunter, T.F. (1980) "The distribution of solubilisate molecules in micellar assemblies", Chem. Phys. Lett. 75, 152-155.

[17] Atik, S.S. and Thomas, J.K. (1981) "Transport of photoproduced ions in water in oil microemulsions : movement of ions from one water pool to another", J. Am. Chem. Soc. 103, 3543-3550.

[18] Rothenberger, G., Infelta, P.P. and Grätzel, M. (1979) "Kinetic and statistical features of triplet energy transfer processes in micellar assemblies", J. Phys. Chem. 83, 1871-1876.

[19] Miller, D.J., Klein, U.K.A. and Hauser, M. (1980) "Occupation numbers in micellar solubilisation - An excimer study", Ber. Bunsenges. Phys. Chem. 84, 1135-1140.

[20] Siemiarczuk, A. and Ware, W.R. (1989) "A novel approach to analysis of pyrene fluorescence decays in sodium dodecylsulfate micelles in the presence of Cu^{2+} ions based on the maximum entropy method", Chem. Phys. Lett. 160, 285-290.

[21] Atik, S.S. and Thomas, J.K. (1981) "Transport of ions between water pools in alkanes", Chem. Phys. Lett. 79, 351-354.

[22] Fletcher, P.D.I. and Robinson, B.H. (1981) "Dynamic processes in water-in-oil microemulsions", Ber. Bunsenges. Phys. Chem. 85, 863-867.

[23] Fletcher, P.D.I., Howe, A.M. and Robinson, B.H. (1987) "The kinetics of solubilisate exchange between water droplets of a water-in-oil microemulsion", J. Chem. Soc., Faraday Trans.1, 83, 985-1006.

[24] Lang, J., Jada, A. and Malliaris, A. (1988) "Structure and dynamics of water-in-oil droplets stabilized by sodium bis(2- ethylhexyl)sulfosuccinate", J. Phys. Chem. 92, 1946-1953 and references therein.

[25] Zana, R. and Weill, C. (1985) "Effect of temperature on the aggregation behaviour of nonionic surfactants in aqueous solutions", J. Physique Lett. 46, L-953-L-960.

[26] Dederen, J.C., Van der Auweraer, M. and De Schryver, F.C. (1981) "Fluorescence quenching of solubilized pyrene and pyrene derivatives by metal ions in SDS micelles", J. Phys. Chem. 85, 1198-1202.

[27] Grieser, F. (1981) "The dynamic behaviour of I- in aqueous dodecyltrimethyl-ammonium chloride solutions. A model for counter-ion movement in ionic micellar systems", Chem. Phys. Lett. 83, 59-64.

[28] Löfroth, J.-E. and Almgren, M. (1982) "Quenching of pyrene fluorescence by alkyl iodides in sodium dodecyl sulfate micelles", J. Phys. Chem. 86, 1636-1641.

[29] Croonen, Y., Geladé, E., Van der Zegel, M., Van der Auweraer, M., Vandendriessche, H., De Schryver, F.C. and Almgren, M. (1983) "Influence of salt, detergent concentration, and temperature on the fluorescence quenching of 1-methylpyrene in sodium dodecyl sulfate with m-dicyanobenzene", J. Phys. Chem. 87, 1426 -1431.

[30] Malliaris, A., Lang, J., Sturm, J. and Zana, R. (1987) "Intermicellar migration of reactants : effect of additions of alcohols, oils and electrolytes", J. Phys. Chem. 91, 1475-1481.

[31] Almgren, M., Löfroth, J.-E. and Van Stam, J. (1986) "Fluorescence decay kinetics in monodisperse confinements with exchange of probes and quenchers", J. Phys. Chem. 90, 4431-4437.

[32] Almgren, M., Van Stam, J., Swarup, S. and Löfroth, J.-E. (1986) "Structure and transport in the microemulsion phase of the system Triton X-100-toluene-water", Langmuir, 2, 432-438.

[33] Dederen, J.C. and Van der Auweraer, M. and De Schryver, F.C. (1979) "Quenching of 1-methylpyrene by Cu^{2+} in sodium dodecylsulfate. A more general kinetic model", Chem. Phys. Lett. 68, 451-454.

[34] Grieser, F. and Tausch-Treml, R. (1980) "Quenching of pyrene fluorescence by single and multivalent metal ions in micellar solutions", J. Am. Chem. Soc. 102, 7258-7264.

[35] Van der Auweraer, M., Dederen, C., Palmans-Windels, C. and De Schryver, F.C. (1982) "Fluorescence quenching by neutral molecules in sodium dodecyl sulfate micelles", J. Am. Chem. Soc. 104, 1800-1804.

[36] Infelta, P.P. and Grätzel, M. (1979) "Statistics of solubilizate distribution and its application to pyrene fluorescence in micellar systems. A concise kinetic model", J. Chem. Phys. 70, 179-186.

[37] Krisnagopal Mandal, Hauenstein, B.L., Jr., Demas, J.N. and DeGraff, B.A. (1983) "Interactions of ruthenium (II) photosensitizers with Triton X-100", J. Phys. Chem. 87, 328-331.

[38] Yekta, A., Aikawa, M. and Turro, N.J. (1979) "Photoluminescence methods for evaluation of solubilization parameters and dynamics of micellar aggregates. Limiting cases which allow estimation of partition coefficients, aggregation numbers, entrance and exit rates", Chem. Phys. Lett. 63, 543-548.

[39] Verbeeck, A. and De Schryver, F.C. (1987) "Fluorescence quenching in inverse micellar systems : possibilities and limitations", Langmuir 3, 494-500.

[40] Almgren, M. and Löfroth, J.-E. (1982) "Effects of polydispersity on fluorescence quenching in micelles", J. Phys. Chem. 76, 2734- 2743.

[41] Löfroth, J.-E. and Almgren, M. (1984) "Fluorescence quenching aggregation numbers in a non-ionic micelle solution", in K.L. Mittal and B. Lindmann (Eds.), Surfactants in Solutions, Plenum Press, New-York, pp 627-643.

[42] Warr, G.G. and Grieser, F. (1986) "Determination of micelle size and polydispersity by fluorescence quenching", J. Chem. Soc., Faraday Trans.1, 82, 1813-1828.

[43] Chen, J.-M., Su, T.-M. and Mou, C.Y. (1986) "Size of sodium dodecyl sulfate micelles in concentrated salt solutions", J. Phys. Chem. 90, 2418-2421.

[44] Almgren, M., Alsins, J., Mukhtar, E. and Van Stam, J. (1988) "Fluorescence quenching dynamics in rodlike micelles", J. Phys. Chem. 92, 4479-4483.

[45] Warr, G.G., Grieser, F. and Evans, D.F. (1986) "Determination of micelle size and polydispersity by fluorescence quenching", J. Chem. Soc., Faraday Trans.1, 82, 1829-1838.

[46] Warr, G.G., Drummond, C.J., Grieser, F., Ninham, B.W. and Evans, D.F. (1986) "Aqueous solution properties of nonionic n-dodecyl b-D-maltoside micelles", J. Phys. Chem. 90, 4581-4586.

[47] Almgren, M., Alsins, J., Van Stam, J. and Mukhtar, E. (1988) "The micellar sphere-to-rod transition in CTAC-NaClO3. A fluorescence quenching study", Prog. Colloid Polym. Sci. 76, 68- 74.

[48] Warr, G.G., Magid, L.J., Caponetti, E. and Martin, C.A. (1988) "Micellar growth and overlap in aqueous solutions of hexadecyl- octyldimethylammonium bromide : a fluorescence quenching and small-angle neutron scattering study", Langmuir 4, 813-817.

[49] Lang, J. (1990) "Surfactant aggregation number and polydispersity of SDS + 1-pentanol mixed micelles in brine by time-resolved fluorescence quenching", J. Phys. Chem., to appear.

[50] Brown, W., Rymdén, R., Van Stam, J., Almgren, M. and Svensk, G. (1989) "Static and dynamic properties of nonionic amphiphile micelles : Triton X-100 in aqueous solution", J. Phys. Chem. 93, 2512-2519.

[51] Espenscheid, W.F., Kerker, M. and Matijević, E. (1964) "Logarithmic distribution functions for colloidal particles", J. Phys. Chem. 68, 3093-3097.

[52] Yan, Y.D. and Clarke, J.H.R. (1989) "In situ determination of particle size distributions in colloids", Adv. Colloid Interface Sci. 29, 277-318.

[53] Van der Auweraer, M. and De Schryver, F.C. (1987) "On the intramicellar fluorescence quenching rate constant in cylindrical micelles", Chem. Phys. 111, 105-112.

[54] Van der Auweraer, M., Reekmans, S., Boens, N. and De Schryver, F.C. (1989) "The intramicellar quenching in cylindrical micelles. II", Chem. Phys. 132, 91-113.

[55] Löfroth, J.-E. (1985) "Deconvolution of single photon counting data with a reference method and global analysis", Eur. Biophys. J.13, 45-58.

[56] Löfroth, J.-E. (1986) "Time-resolved emission spectra, decay- associated spectra, and species-associated spectra", J. Phys. Chem. 90, 1160-1168.

[57] Boens, N., Malliaris, A., Van der Auweraer, M., Luo, H. and De Schryver, F.C. (1988) "Simultaneous analysis of single-photon timing data with a reference method : application to a Poisson distribution of decay rates", Chem. Phys. 121, 199-209.

[58] Boens, N., Luo, H., Van der Auweraer, M., Reekmans, S., De Schryver, F.C. and Malliaris, A. (1988) "Simultaneous analysis of fluorescence decay curves for the one-step determination of the mean aggregation number of aqueous micelles", Chem. Phys. Lett. 146, 337-342.

[59] Luo, H., Boens, N., Van der Auweraer, M., De Schryver, F.C. and Malliaris, A. (1989) "Simultaneous analysis of time-resolved fluorescence quenching data in aqueous micellar systems in the presence and absence of added alcohol", J. Phys. Chem. 93, 3244-3250.

[60] Boens, N., Janssens, L.D., and De Schryver, F.C. (1989) "Simultaneous analysis of single-photon timing data for the one- step determination of activation energies, frequency factors and quenching rate constants. Application to tryptophan photophysics", Biophys. Chem. 33, 77-90.

[61] Miller, D.D. and Evans, D.F. (1989) "Fluorescence quenching in double-chained surfactants. 1. Theory of quenching in micelles and vesicles", J. Phys. Chem. 93, 323-333.

[62] Lianos, P. and Modes, S. (1987) "Fractal modeling of luminescence quenching in microemulsions", J. Phys. Chem. 91, 6088-6089.

[63] Duportail, G. and Lianos, P. (1988) "Fractal modeling of pyrene excimer quenching in phospholipid vesicles" Chem. Phys. Lett. 149, 73-78.

[64] Lianos, P. (1988) "Luminescence quenching in organized assemblies treated as media for noninteger dimensions", J. Chem. Phys. 89, 5237-5241.

[65] Modes, S. and Lianos, P. (1989) "Luminescence probe study of the conditions affecting colloidal semiconductor growth in reversemicelles and water-in-oil microemulsions", J. Phys. Chem. 93, 5854-5859.

[66] Johansson, L.-B.-0. and Söderman, O. (1987) "The cubic phase (I_1) in the dodecyltrimethylammonium chloride/water systems. A fluorescence quenching study", J. Phys. Chem. 91, 5275-5278.

[67] Fletcher, P.D.I. (1988) "Time-resolved fluorescence study of the structure and dynamics of the cubic I_1 lyotropic mesophase of dodecyltrimethylammonium chloride", Mol. Cryst. Liq. Cryst.154, 323-333.

[68] Liang, P. and Thomas, J.K. (1988) "Photophysical studies in liquid crystal solutions and liquid crystal foams", J. Colloid Interface Sci. 124, 358-364.

[69] Habti, A., Keravis, D., Levitz, P. and Van Damme, H. (1984) "Influence of surface heterogeneity on the luminescence decay of probe molecules in heterogeneous systems. $Ru(bpy)_3^{2+}$ on clays", J.Chem. Soc. Faraday Trans.2, 80, 63-67.

[70] Levitz, P. and Van Damme, H. (1986) "Fluorescence decay study of the adsorption of nonionic surfactants at the solid-liquid interface. 2. Influence of polar chain length", J. Phys. Chem. 90, 1302-1310.

[71] Chandar, P., Somasundaram, P. and Turro, N.J. (1987) "Fluorescence probe studies on the structure of the absorbed layer of dodecyl sulfate at the alumina-water interface", J. Colloid Interface Sci. 117, 31-46.

[72] Thomas, J.K. (1987) "Characterization of surfaces by excited states", J. Phys. Chem. 91, 267-276.

[73] Fletcher, P. and Gilbert, P. (1989) "Structure and dynamic properties of decylammonium chloride micelles in water and glycerol", J. Chem. Soc., Faraday Trans.1, 85, 147-156.

[74] Lang, J., Zana, R. and Lalem, N. (1990) "Droplet size and dynamics in water-in-oil microemulsions. Correlations between results from time-resolved fluorescence quenching, quasielastic light scattering, electrical conductivity and water solubility measurements", this volume.

SOLUBILIZATION OF ALCOHOLS AND ALKANEDIOLS IN AQUEOUS SURFACTANT SOLUTIONS

H. HØILAND and A.M. BLOKHUS,
Department of Chemistry,
University of Bergen,
N-5007 Bergen,
Norway.

ABSTRACT. The aqueous solubility of otherwise sparingly soluble substances may be significantly increased by adding surfactant, i.e. they become solubilized. At low surfactant concentrations the solubilities of alcohols and diols exhibit a linear correlation between the amount dissolved and the amount of surfactant added. However, the solubility of primary alcohols reaches a maximum at a sodium dodecyl sulfate (SDS) concentration of about 0.05 m. Closer inspection of the separating phase suggests that this also changes around the maximum. Below the surfactant concentration of maximum solubilization the separating phase is L_1, above it seems to be a D-phase, and this obviously influences the solubilities Viscosity and conductance measurements suggest that this practically coincides with a change in micellar structure, from small spherical micelles to large rod-like structures. In the region of small micelles there is a distinct break in the slope when compressibilities are plotted versus alcohol concentration at constant SDS concentration. The partial molar compressibilities of the alcohols suddenly increase. This suggests that the alcohol moves towards a more hydrophobic environment, presumably it becomes solubilized in the micellar interior.

1. Introduction

An important property of micellar solutions is their ability to solubilize compounds that are otherwise sparingly soluble in water. Micellar solubilization is important for many technical applications such as detergency, pharmaceuticals, polymerization processes, and enhanced oil recovery [1-5]. The solubilization process can be treated within the framework of the phase separation model. The micelles are considered a pseudo-phase, and the solubilization process can be treated as a distribution equilibrium between two phases. Several studies have been carried out in recent years [6-19] demonstrating how the extent of solubilization depends on the chemical characteristics of the surfactant and the solubilizate.

The localization of the solubilizate within the micelles may also vary depending on the solubilizate. Pure hydrophobic compounds such as hydrocarbons are thought to be solubilized in the hydrocarbon core of the micelles while amphiphilic molecules, such as alcohols are thought to be oriented in the same way as the surfactant molecules [20]. Aromatic compounds seem to be localized at the micellar surface, but higher concentration favour location in the micellar interior [20-22].

Solubilization may influence the micellar size and shape. Solubilizates like alkanes may penetrate the micellar core and produce a swelling of the micelles [20,23], and polar solubilizates may induce changes from small spherical to large rod-like micelles [24-26]. Different types of

D. M. Bloor and E. Wyn-Jones (eds.),
The Structure, Dynamics and Equilibrium Properties of Colloidal Systems, 39–48.

liquid crystalline phases will also form as a result of adding alcohols. In this paper we shall examine the solubilization process with respect to the amount that can be solubilized, the localization of the solubilizate within the micelles, and the effects on the micellar size.

2. Experimental

2.1 METHODS

The solubilities of alkanes and alcohols in aqueous micellar solutions were determined by density and turbidity measurements. In both types of measurements, the solubility equilibria were attained in thermostatted water baths controlled to ± 0.05 K. The methods have been described in detail elsewhere [16].

Viscosities of micellar solutions were measured by Ostwald viscosimeters with an accuracy of ± 0.1%.

The electrical conductivities of the solutions were measured by a Wayne-Kerr autobalance precision bridge. They were kept in a dilution cell fitted with platinized electrodes.

Densities were measured by a Paar density meter with an accuracy of $\pm 5 \times 10^{-6}$ g cm^{-3}.

Isentropic compressibilities were determined by measuring the speed of sound through the solutions. The accuracy was ± 0.05 ms^{-1}.

2.2 MATERIALS

Water was distilled and filtered. It had a conductivity of about 1 µS. Sodium dodecyl sulfate was from BDH, "specially pure". The alcohols and diols were from Fluka at the best quality obtainable. All chemicals were used without further purification.

3. Results and Discussion

Ekwall *et al* [27] pioneered work on surfactant-water-alcohol phase diagrams. In addition to an aqueous micellar phase, termed L_1, one generally finds a lamellar D-phase, a hexagonal phase E, and an inverted micellar phase, L_2. In the following we shall concentrate on the L_1 phase.

The distribution equilibrium of the solubilizate between the micellar pseudophase and the surrounding aqueous phase can be described by a distribution coefficient:

$$K = \frac{x_A^{mic}}{x_A^{aq}} \tag{1}$$

Here x_A^{mic} and x_A^{aq} are the mole fractions of solubilizate in the micellar and aqueous states, respectively. By introducing moles and molalities this expression can be rearranged to read:

$$m_A = \frac{K\,M_w\,m_A^{aq}}{1 - K\,M_w\,m_A^{aq}}\,(m_s - m_s^{cmc}) + m_A^{aq} \tag{2}$$

Here m_s is the surfactant molality, m_s^{cmc} the surfactant molality at the c.m.c., m_s is the solubilizate molality at saturation, and M_w the molar mass of water. Eq. (2) shows that a linear relationship is expected between the maximum amount that can be solubilized and the surfactant molality. The intercept is equal to the alcohol solubility in pure water.

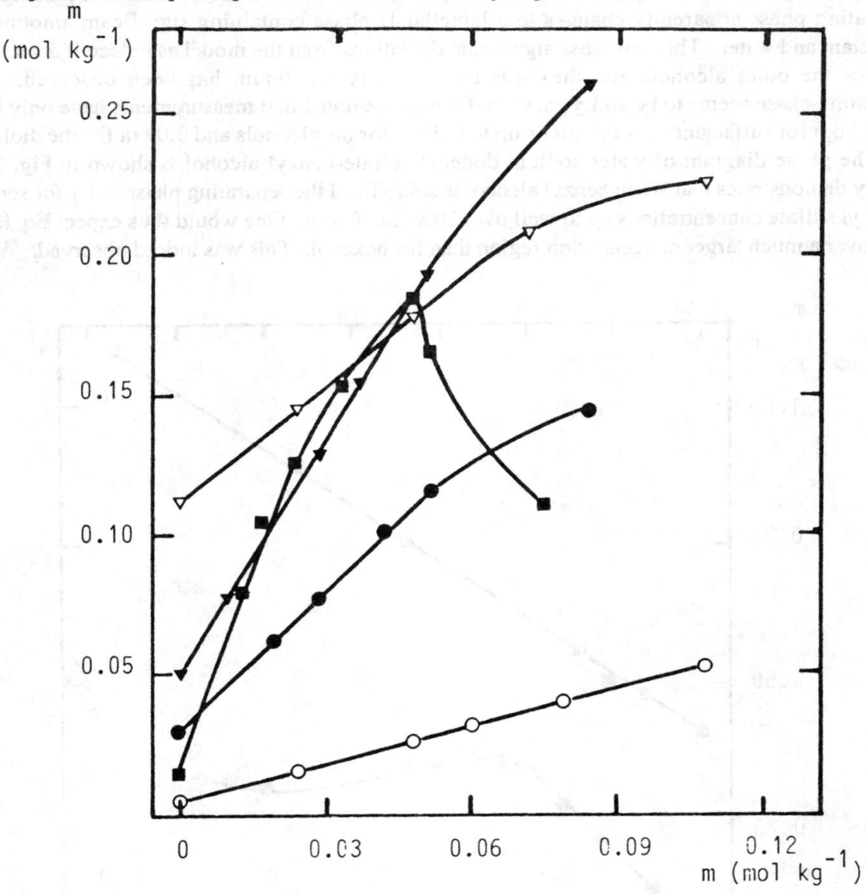

Fig. 1. The solubility of 1-heptanol (■), 2-heptanol (●), 4-heptanol (▼), 1,8-octanediol (▽), and 1,10-decanediol (O) as a function of the sodium dodecyl sulfate molality.

Figs. 1 and 2 show experimental data for alcohols and diols. The model, Eq. (2), generally holds, but only at very low surfactant concentrations. The intercept agrees excellently with literature data on the aqueous solubility of alcohols [28,29]. 1-Hexanol and 1-heptanol differ from the other compounds by the fact that their solubilities reach a maximum at a certain SDS concentration, about 0.05 m. Activity coefficients have been neglected in the derivation of Eq. (2), but this can not explain a maximum in the plot. However, the separating phase will be of importance. Implicit in Eq. (1) it has been assumed that on reaching the solubility limit, the separating phase is pure alcohol. Closer inspection reveals that for the linear part of the diagram, the separating phase is an L_2 phase; i.e. almost pure alcohol. When the maximum is reached, the separating phase apparently changes to a lamellar D-phase containing significant amounts of surfactant and water. This can cause significant deviations from the model as indeed it does.

For the other alcohols and the diols no solubility maximum has been observed. The separating phase seems to be an L_2 phase, but it must be noted that measurements have only been carried out for surfactant concentrations up to 0.06 m for the alcohols and 0.09 m for the diols.

The phase diagram of water-sodium dodecyl sulfate-benzyl alcohol is shown in Fig. 3. It clearly demonstrates that when benzyl alcohol is solubilized the separating phase is L_1 for sodium dodecyl sulfate concentrations up to well over 20%, i.e. 0.7 m. One would thus expect Eq. (2) to hold over a much larger concentration region than for hexanol. This was indeed observed. We

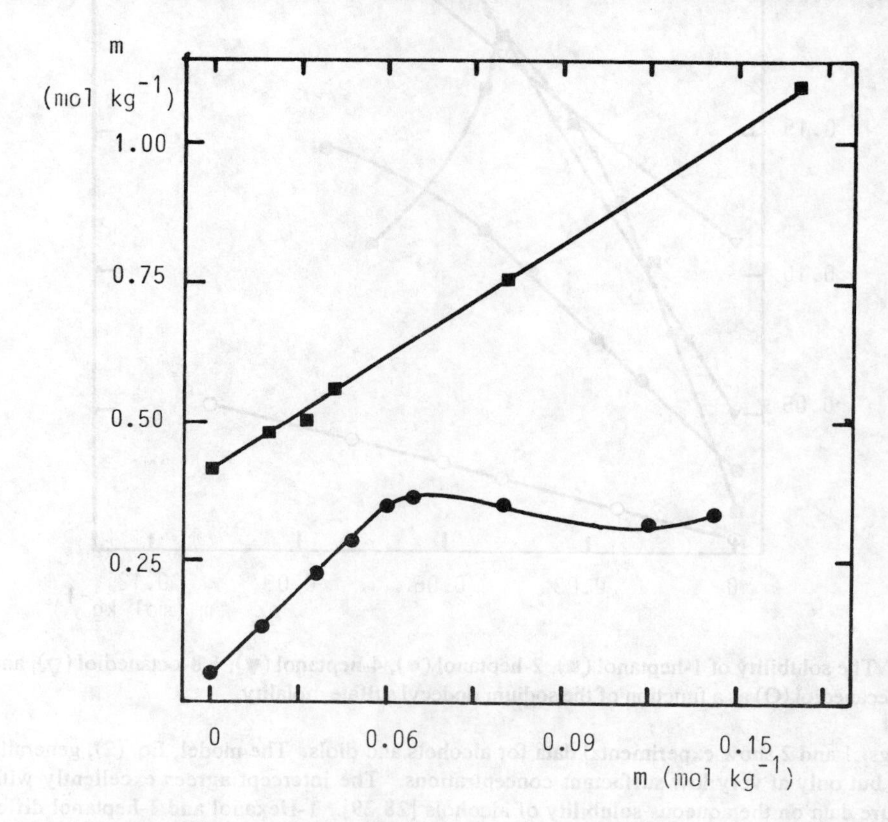

Fig. 2. The solubility of 1-hexanol (●) and benzyl alcohol (■) as a function of the sodium dodecyl sulfate molality.

found the predicted linearity between the amount of benzyl alcohol that could be dissolved and the amount of surfactant added to be linear up to an SDS concentration of 0.5 m, at least.

The distribution coefficients are presented in table 1. These are related to the standard molar

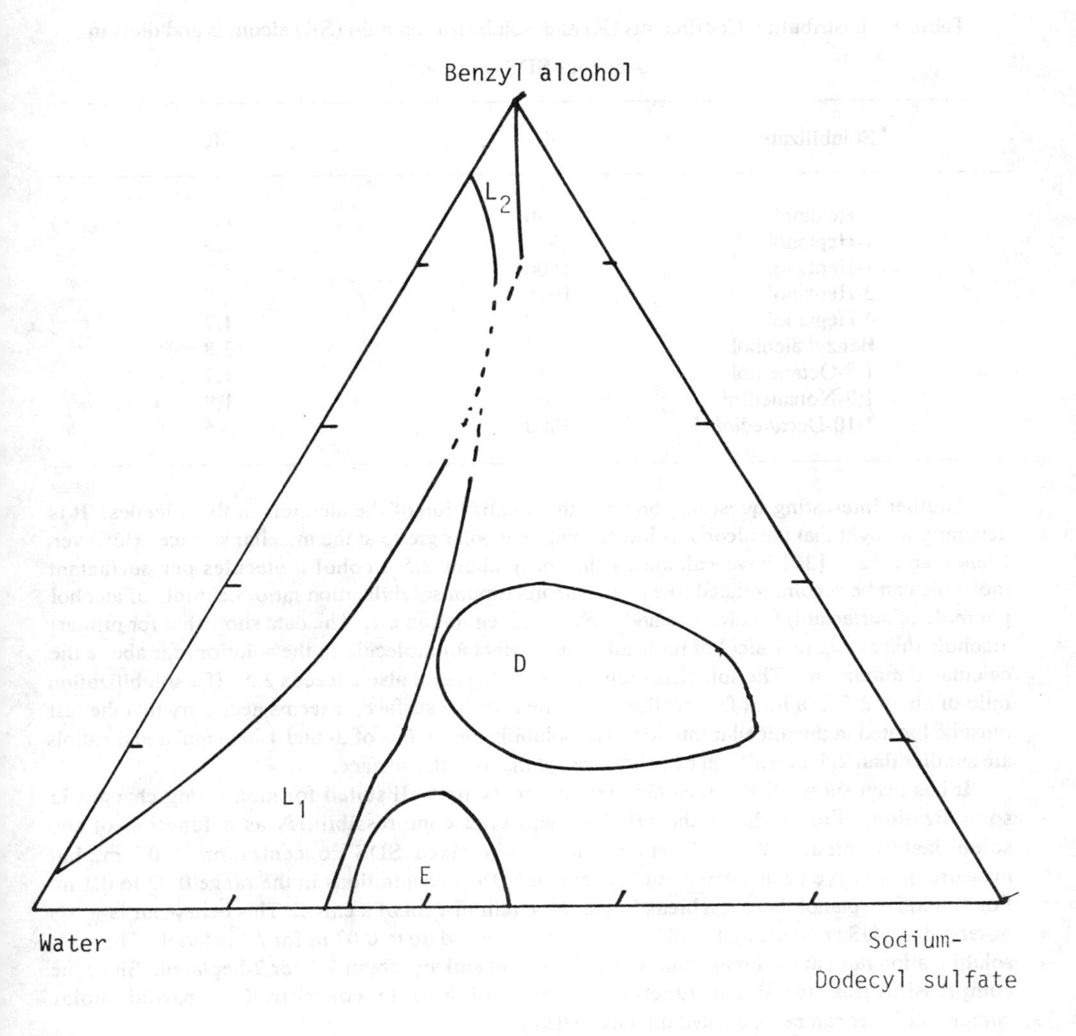

Fig. 3. The phase diagram of water-sodium dodecyl sulfate-benzyl alcohol at 298.15K. (It is not clear if the L_1 and L_2 phases are connected).

Gibbs' energy by the normal equation, $\Delta G^o{}_m$ = -RT 1n K. The data for 1-hexanol and 1-heptanol agrees well with the data of Abuin and Lissi [14]. The increment of $\Delta G^o{}_m$ per CH_2 group becomes -2.66 kJ mol^{-1}. The data show that the transfer of the alkane part of the alcohols water to micelles is, in principle, analogous to the transfer of an alkyl group from water to a nonpolar solvent.

Table 1. Distribution Coefficients (K) and Solubilization ratio (SR) alcohols and diols in Aqueous SDS solutions

Solubilizate	K	SR
1-Hexanol	760	7.1
1-Heptanol	2650	6.3
2-Heptanol	1500	3.6
3-Heptanol	1010	2.0
4-Heptanol	930	1.7
Benzyl alcohol	173	3.8
1,8-Octanediol	311	1.7
1,9-Nonanediol	743	1.9
1,10-Decanediol	3800	0.5

Another interesting question concerns the localization of the alcohols in the micelles. It is generally thought that the alcohol is located with the polar group at the micellar surface. However, Lianos and Zana [30] have calculated that only about 2.5 alcohol molecules per surfactant molecule can be accommodated like that. The maximum solubilization ratio, i.e. mole of alcohol per mole of surfactant, for alcohols and diols are given in Table 1. The data show that for primary alcohols there is up to 7 alcohol molecules per surfactant molecule in the solution, far above the calculated maximum. The solubilization ratio of 2-heptanol also exceeds 2.5. If a solubilization ratio of about 2.5 is a limit for localization at the micellar surface, it seems necessary that the rest must be located in the micellar interior. The solubilization ratios of 3- and 4-heptanol and the diols are smaller than 2.5, and all can be solubilized at the micellar surface.

It has been shown that ultrasonic measurements are well suited for monitoring changes in solubilization. Fig. 4 shows the relative isentropic compressibilities as a function of the solubilizate content. We only show data at one fixed SDS concentration, 0.05 m, but measurements have been carried out for several SDS concentrations in the range 0.02 to 0.1 m. For 1- and 2-heptanol there is a break in the slope reminiscent of a c.m.c. This behaviour is observed at SDS contents up to 0.05 m for 1-heptanol and up to 0.07 m for 2-heptanol. The solubilization ratio at the break point is 1.5 for 1-heptanol and about 2.1 for 2-heptanol. Since the compressibilities are linear functions of the solubilizate concentration, partial molar compressibilities can be calculated directly by [31]:

$$K_{s,2} = \frac{w_1 + w_2}{\rho}\left(\frac{\partial \kappa}{\partial n_2}\right)_{n1} + \kappa V_2$$

Here w_1 is the mass of the solvent, water and SDS, w_2 the mass of solute, alcohol or diol, ρ and κ the solution density and compressibility, respectively, and V_2 the partial molar volume of the solubilizate. The results are presented in Table 2. It shows that the compressibility increases after the break, suggesting that the alcohol moves to a more hydrophobic environment. The partial molar compressibilities after the break come close to the values in pure hydrocarbon [32]. The simplest explanation for this phenomenon is that at a certain concentration the 1- and 2-heptanol

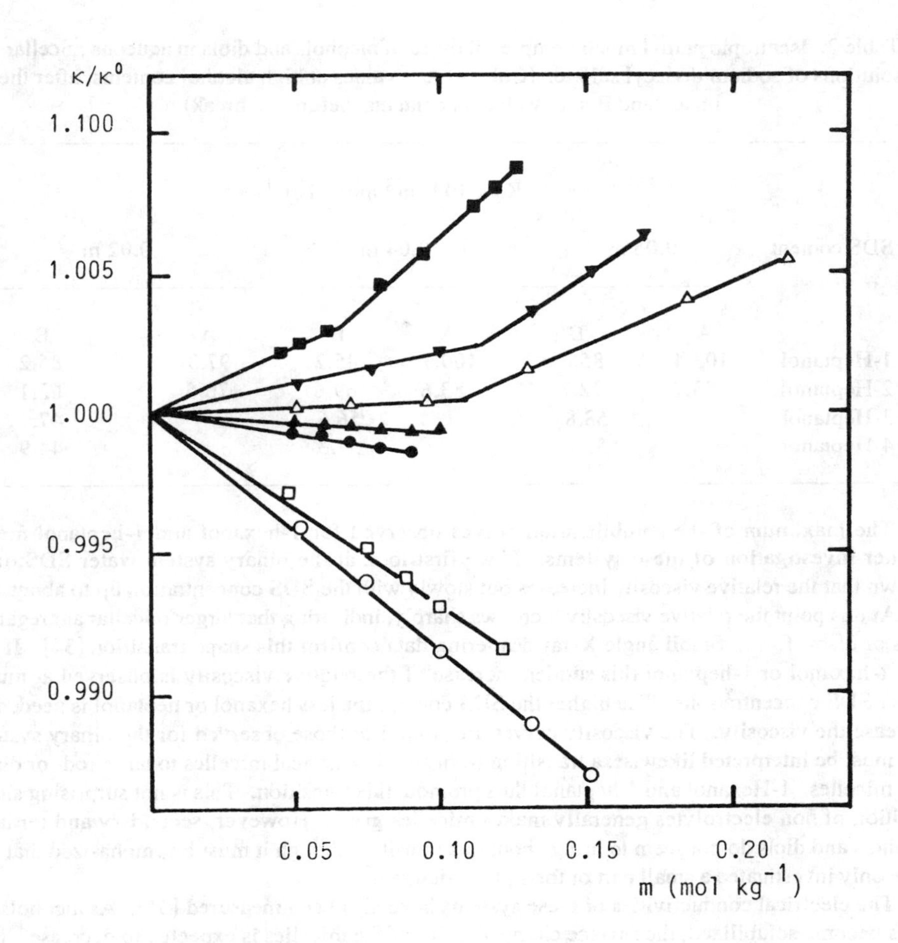

Fig. 4. The relative isentropic compressibilities of 1-hexanol (Δ), 1-heptanol (■), 2-heptanol (▼), 3-heptanol (▲), 4-heptanol (●), 1,8-octanediol (O), and 1,9-nonanediol (□) as a function of the alcohol or diol molality in 0.05 m sodium dodecyl sulfate aqueous solution.

becomes solubilized in the micellar interior. This has also been concluded from fluorescence probe studies of the solubilization of primary alcohols [30]. Structurally it is easily seen that 1-heptanol can be solubilized parallel to the surfactant molecules of the micelle with the polar hydroxyl group in aqueous surroundings. From this position it should be possible for the alcohol to penetrate to the interior of the micelle. Sterically, one would expect that 2-heptanol is less likely to solubilize in the micellar interior. The compressibility data suggest that it still does. For 3- and 4-heptanol and the α,ω-alkanediols steric difficulties seem even greater, and the data show no evidence for solubilization in the micellar interior. This agrees well with small angle X-ray scattering data on 1,8-octanediol in sodium octanoate where it seems that the diol resides in the polar region of the micelles [33].

Table 2. Isentropic partial molar compressibilities of alcohols and diols in aqueous micellar solutions of sodium dodecyl sulfate. (Column A is values at high alcohol contents, after the break, and B at low alcohol contents, before the break)

	$K_2 \times 10^4$ cm^3 mol^{-1} bar^{-1}					
SDS content	0.05 m		0.04 m		0.02 m	
	A	B	A	B	A	B
1-Heptanol	103.4	85.4	100.1	85.2	97.3	85.2
2-Heptanol	93.7	72.7	83.6	69.6	76.5	62.1
3-Heptanol		58.8		56.6		47.7
4-Heptanol		57.2		54.3		44.9

The maximum of the solubilization curves observed for 1-hexanol and 1-heptanol merit further investigation of these systems. If we first look at the binary system water-SDS, it is known that the relative viscosity increases but slowly with the SDS concentration up to about 1.1 m. At this point the relative viscosity increases sharply, indicating that larger micellar aggregates, rods or discs, form. Small angle X-ray scattering data confirm this shape transition [34]. If we add 1-hexanol or 1-heptanol this sudden increase of the relative viscosity is observed at much lower SDS concentrations. The higher the SDS content the less hexanol or heptanol is needed to increase the viscosity. The viscosity curves are similar to those observed for the binary system and must be interpreted likewise: a transition from small, spherical micelles to large rod- or disc-like micelles. 1-Hexanol and 1-heptanol thus promote this transition. This is not surprising since addition of non electrolytes generally makes micelles grow. However, secondary and tertiary alcohols and diols do not seem to bring about this transition though it must be emphasized that we have only investigated a small part of their phase diagrams.

The electrical conductivities of these systems have also been measured [35]. As alcohols or diols become solubilized, the surface charge density of the micelles is expected to decrease. The result is that associated counter ions will be discharged from the micellar surface, and the electrical conductivity should increase. This has also been observed. However, for 1-hexanol and 1-heptanol the conductivity reaches a maximum at the point where the micelles transform from small spherical to large rod- or disc-like aggregates, defined as the point where the viscosity starts increasing. The degree of counter ion association has been reported to increase when rod-like micelles are formed [36], and thus confirms the conclusions reached from viscosity measurements. The viscosity and conductivity measurements also show that the most hydrophobic alcohol is the most efficient in promoting a micellar shape transition.

Based on the data one may divide the L_1 phase of the system water-SDS-1-hexanol and water-SDS-1-heptanol into various regions. At low surfactant contents the solubility is a linear function of the SDS content. Here the micelles remain small and presumably spherical as alcohol is added. The L_1 phase is in equilibrium with the L_2 phase. At higher SDS concentrations alcohol will bring about a shape transition to large rod- or disc-like micelles. We observe large deviations from linearity between alcohol solubility and SDS concentration. The L_1 phase is in equilibrium with the D phase. (The exact point where the equilibrium shifts has not been established). Even in the region of small spherical micelles there exist two regions as seen by measuring the isentropic compressibilities. At a certain point the compressibility increases, suggesting that the alcohol moves to a more hydrophobic environment. Apparently the localization of the solubilizate shifts from the micellar surface to the micellar interior.

Acknowledgement. Part of this work has been supported by VISTA, a research cooperation between the Norwegian Academy of Science and Den Norske Stats Oljeselskap A/S (Statoil).

4. References

1. McBain, M.E.L. and Hutchinson, E. (1955) 'Solubilization and Related Phenomena', Academic Press, New York.
2. Elworthy, P.H., Florence, A.T., and MacFarlane, C.B. (1968) 'Solubilization by Surface Active Agents', Chapman & Hall, London.
3. Wennerstrom, H. and Lindman, B. (1979) Phys. Rep. **52**, 1.
4. Mukerjee, P. (1979) 'Solution Chemistry of Surfactants Vol 1' in K.L. Mittal (ed.), Plenum Press, New York, pp. 153-174.
5. Mittal, K.L. and Mukerjee, P. (1977) 'Micellization, Solubilization, and Microemulsions' Vol 1 in K.L. Mittal (ed) Plenum Press, New York, pp 1-21.
6. Manabe, M., Shirahama, K., and Koda, M. (1976) Bull. Chem. Soc. Japan **49**, 2904.
7. Hayase, K. and Hayano, S. (1977) Bull. Chem. Soc. Japan **50**,83.
8. Gettins, J., Hall, D., Jobling, P.L., Rassing, J., and Wyn-Jones, E. (1978), J. Chem.Soc. Faraday Trans. II **74**, 1957.
9. Almgren, M., Grieser, F., and Thomas, J.K. (1979) J. Chem. Soc. Faraday Trans. I **75**, 1674.
10. Yiv, S., Zana, R., Ulbricht, W., and Hoffman, H. (1981) J. Colloid Interface Sci. **80**, 224.
11. Kaneshina, S., Kamaya, H., and Ueda, I. (1981), J. Colloid Interface Sci. **83**, 589.
12. Treiner, C. (1982) J. Colloid Interface Sci. **90**, 444.
13. Stilbs, P. (1982) J. Colloid Interface Sci. **87**, 385.
14. Abuin, E.B. and Lissi, E.A. (1983) J. Colloid Interface Sci. **95**, 198.
15. Spink, C.H. and Colgan, S. (1984) J. Colloid Interface Sci. **97**, 41.
16. Høiland, H., Ljosland, E., and Backlund, S. (1984), J. Colloid Interface Sci. **101**, 467.
17. Høiland, H., Blockhus, A.M., Kvammen, O., and Backlund, S. (1985) J. Colloid Interface Sci. **107**, 576.
18. Blockhus, A.M., Høiland, H., and Backlund, S. (1986) J. Colloid Interface Sci. **114**, 9.
19. Gao, Z., Wasylishen, E., and Kwak, C.T. (1989) J. Phys. Chem. **93**, 2190.
20. Mukerjee, P. and Cardinal, J.R., (1978) J. Phys. Chem., **82**, 1620.
21. Eriksson, J.C. and Gillberg, G. (1966) Acta Chem Scand. **20**, 2019.
22. Nagarajan, R., Chaiko, M.A., and Ruckenstein, E. (1984) J. Phys. Chem. **88**, 2915.
23. Almgran, M. and Swarup, S. (1983) J. Colloid Interface Sci. **91**, 256.

24. Larsen, J.W., Magid, L.J., and Payton, V. (1973), Tetrahedron Lett., **29**, 2663.
25. Tominaga, T., Stem, T.B., and Evans, D.F. (1980) Bull. Chem. Soc. Japan **53**, 795.
26. Lindblom, G., Lindman, B., and Mandell, L.J. (1971) J. Colloid Interface Sci. **35**, 519.
27. Ekwall, P. (1975) 'Advances in Liquid Crystals' Vol. 1 in G.H. Brown (ed.) Academic Press, New York. pp. 1-142.
28. Kinoshita, K., Ishikawa, H., and Shinoda, K. (1958) Bull. Chem. Soc. Japan **31**, 1081.
29. Amidon, G.L., Yalkowski, S.H., and Leung, S. (1974) J. Pharm. Sci. **63**, 1858.
30. Lianos, P. and Zana, R. (1984) J. Colloid Interface Sci. **101**, 587.
31. Vikholm, I., Douheret, G., Backund, S., and Høiland, H. (1987) J. Colloid Interface Sci., **116**, 582.
32. Vikingstad, E. (1979) J. Colloid Interface Sci. **72**, 75.
33. Frimann, R. and Rosenholm, J. (1982) Progr. Colloid Polymer Sci. **260**, 545.
34. Reiss-Husson, F. and Luzzati, V. (1964) J. Phys. Chem., **68**, 3504.
35. Backlund, S., Rundt, K., Veggeland, S., and Høiland, H. (1987) Progr. Colloid Polymer Sci. **74**, 93.
36. Kamenka, N., Fabre, H., Chorro, M., and Lindman, B. (1977) J. Chim. Phys. **74**, 510.

THERMODYNAMICS OF SOLUBILIZATION INTO SURFACTANT MICELLES

Y. MOROI
Department of Chemistry
Faculty of Science
Kyushu University 33
Higashi-ku, Fukuoka 812
JAPAN

ABSTRACT. Solubilization into surfactant micelles is examined from the viewpoint of the phase rule. Treating micelles as a separate phase, into which bulk phase solubilizate can partition, turns out to be inconsistent with the rule. On the contrary, the treatment of micelles as chemical species can perfectly elucidate solubilization in a manner consistent with the phase rule. The maximum additive concentration (MAC) of 4-n-alkylbenzoic acids (C_n-BA: n-0, 1,2,3,4,5,6,7,8) in aqueous solutions of dodecylsulfonic acid was determined spectrophotometrically at 25^oC with surfactant concentration varied. From the association constants (K_1) the free energy change of solubilization per methylene group was estimated to be -2.59 kJ mol^{-1}. The free energy change per methylene group from solid state to solubilized state was evaluated to be 0.67 kJ mol^{-1}. These values suggest that an interface of tension does not seem to exist in the micellar interior and that partial crystallinity of the alkyl chain is still remaining in the solubilized state.

1. Introduction

Increased aqueous solubility of otherwise slightly soluble organic substances brought about by the presence of surfactant micelles is well known as solubilization [1,2], a phenomenon that plays a very important role in industrial and biological processes. However, its quantitative treatment has not been fully established yet. Most theoretical treatments have been semi-empirical and model the partitioning as an equilibrium of solubilizates between a solubilized micellar phase and intermicellar bulk phase [3,4], in spite of the fact that micelles are most often not a separate phase at all [5]. In the preceding papers [6-9] the author made clear the advantages of modelling solubilization into micelles from a thermodynamic point of view using stepwise association equilibria between micelles with average surfactant aggregation number and solubilizate molecules.

This paper stresses the advantage of a model in which micelles are regarded as chemical species not only for solubilization but also for micellization [10]. It also inquires experimentally whether micelles can be treated as a separate phase or whether an interfacial tension does exist in the micellar interior [11]. This is because as for solubilization in particular a pressure increase of the micelle inside due to Laplace pressure has often been employed to elucidate a diminished free energy change of transfer per methylene group from aqueous bulk into micellar interior compared with the free energy change from aqueous bulk into liquid hydrocarbon [12,13].

D. M. Bloor and E. Wyn-Jones (eds.),
The Structure, Dynamics and Equilibrium Properties of Colloidal Systems, 49–62.

2. Examination of Solubilization by Phase Rule

Hitherto, many published papers concerning solubilization have been based on the phase separation model of micellization [12-14]. There, the solubilization was treated from the viewpoint of partition of solubilizate molecules between micellar phase and intermicellar bulk phase. If micelles are regarded as a phase, there exists three phases in the system in the case where an excess solubilizate phase is present (micellar phase, intermicellar bulk phase, and solubilizate phase). The total number of components of the system is three (solvent, surfactant, and solubilizate), and the system becomes divariant. In other words, we cannot change surfactant concentration at constant temperature and pressure in spite of the real fact that the maximum additive concentration (MAC) changes with total surfactant concentration. Even if it is admitted that an increase in MAC with surfactant concentration above critical micelle concentration (CMC) is due to an increase in total micellar phase, the concentration of solubilizate in the micellar phase should remain constant, because the concentration is an intensive property of the system and is homogeneous throughout the micellar phase. That is, the increase in micellar phase does not lead to an increase in solubilizate concentration in it. On the contrary, in the system where excess solubilizate phase does not co-exist, the number of degrees of freedom becomes three, increasing by one. Then, the surfactant concentration is a unique variable which can determine every intensive property of the system at constant temperature and pressure. In other words, solubilizate monomer concentration in the intermicellar bulk phase is determined automatically at a certain surfactant concentration irrespective of total solubilizate concentration in the system. This means that the concentration of solubilizate molecules in the bulk phase too will be determined by the surfactant concentration. This is not only totally incorrect but also quite contrary to experimental evidence that the concentration of solubilizate is determined only by the amount added to the system. From the above discussion, it becomes quite clear that the phase separation model of micelles and the partition model of solubilization are quite contrary to experimentally observed facts.

The above contradiction can be easily solved by treating micelles as chemical species. That is, micelle formation can be expressed by the following association equilibrium between surfactant monomers (S) and micelles (M):

$$m\,S \underset{}{\overset{K_m}{\rightleftharpoons}} M \tag{1}$$

where K_m is the equilibrium constant of micelle formation and the mono-dispersity of micelles of aggregation number m is assumed in the absence of solubilizate in order to remove the difficulty arising from their polydispersity. Even in the case of the polydispersity the present discussion remains essentially the same [7]. The stepwise association equilibria between micelles and solubilizates (R) are presented by:

$$M + R \overset{K_1}{\rightleftharpoons} MR_1,\ MR_1 + R \overset{K_2}{\rightleftharpoons} MR_2, \ldots MR_{n-1} + R \overset{K_n}{\rightleftharpoons} MR_n \tag{2}$$

where MR_i designates the micelles associated with i molecules of solubilizate, K_i is the stepwise association constant between MR_{i-1} and a monomer molecule of solubilizate, and n is an arbitrary number. The total number of components of this system is n+4 including solvent molecules (solvent, S, R, M, MR_1,....MR_n), and the number of phases is two (micellar solution

phase and solubilizate phase). The solubilizate phase is inevitable in the case where the MAC is a point of discussion. The n+1 equilibrium equations for a micellar system including eq. (1) reduce the number of degrees of freedom by n+1, resulting in three degrees of freedom. Hence, at constant temperature and pressure, only one other intensive variable can be selected to specify the thermodynamic system. The total surfactant concentration is generally selected for the variable. From eqs. (1) and (2), we have the following equations for the total micellar concentration ($[M_t]$), the total equivalent concentration of solubilizate ($[R_t]$) or MAC, and the average number of solubilizate molecules per micelle ($\bar{R}$):

$$[M_t] = K_m[S]^m\{1 + \sum_{i=1}^{n} (\prod_{j=1}^{i} K_j)[R]^i\} \quad (3)$$

$$[R_t] = [R] + K_m[S]^m \sum_{i=1}^{n} i (\prod_{j=1}^{i} K_j)[R]^i \quad (4)$$

$$\bar{R} = ([R_t] - [R])/[M_t] \quad (5)$$

In the case where excess solubilizate phase is absent, the degrees of freedom increase by one, resulting in four degrees of freedom. Hence, at constant temperature and pressure, two other intensive variables can be selected to prescribe the thermodynamic system. Three sets of combinations of two intensive variables were examined to derive other intensive properties in the preceding paper [6]. The difference between the presence and the absence of excess solubilizate phase is that the concentration of solubilizate ($[R]$) is automatically determined as solubility at a specified temperature and pressure resulting in the decrease of freedom by one. Consequently a total surfactant concentration becomes only one variable remaining at constant temperature and pressure in the former case. As can be seen above, the thermodynamical discussion which regards micelles as chemical species can perfectly explain real experimental facts, and micelles as chemical species turns out to be correct. In addition, an important point is that a solubilization system is specified by both association constants and monomeric solubilizate concentration, as is clear from eq. (4). That is, higher association constant does not always lead to larger solubilization. On the contrary, in general, a solubilizate whose aqueous solubility is smaller results in a smaller total equivalent concentration in spite of its higher association constant [8,14].

The next step is to develop the above equations under reasonable assumptions. When the concentration of solubilizate is less than a few times the micellar concentration, an incorporation of the solubilizates into micelles can be assumed not to change intrinsic properties of original micelles. In other words, the discussion here concerns such small extent of solubilization that micelles keep their original physicochemical properties. In this case, the following mathematical calculation is available. Let us suppose that the solubilization of ($[R_t]$-$[R]$) molecules into $[M_t]$ micelles is equivalent to placing ($[R_t]$-$[R]$) balls into $[M_t]$ cells randomly, where both balls and cells are independent and indistinguishable. Then the probability P(i) that a specified cell contains exactly i balls is given by [15]

$$P(i) = \binom{r}{i} (1/q^i)(1-1/q)^{r-i} \quad (6)$$

where $r = [R_t]$-$[R]$ and $q = [M_t]$. The magnitude of r and q are of the order of Avogadro's number and i is very small in number compared with them, less than ten at maximum for the usual case. This equation is a special case of the so-called binominal distribution. Using the following approximation, which is reasonable for the present conditions,

$$1 - 1/q = \exp(-1/q) \tag{7}$$

and the numerical conditions as to r, q and i values, we finally obtain [7]

$$P(i) = \bar{R}^{i} \exp(-\bar{R})/i! \tag{8}$$

which is the Poisson distribution. The total surfactant concentration which is an intensive variable to specify the solubilization system must determine every intensive variable of the system. The total equilibrium concentration of solubilizate $[R_t]$ must, of course, be determined only by the total surfactant concentration. As the concentration of micelles associated with i solubilizate molecules is

$$[MR_i] = [M_t]\, P(i) \tag{9}$$

the association constant K_i is given by

$$K_i = [MR_i]/([R]\,[MR_{i-1}]) = K_1/i \tag{10}$$

where the Poisson distribution is assumed. Introduction of eqs. (9) and (10) into eq. (4) gives:

$$([R_t] - [R])/[R] = K_1[M_t] \tag{11}$$

where the summation is over an infinite n.

3. Solubilization of 4-n-Alkylbenzoic Acids into Dodecylsulfonic Acid Micelles

The maximum additive concentration is determined under condition that an excess solubilizate phase coexists in the system. Then, it is important to make clear the physicochemical properties of the solubilizate phase, the solid phase of p-n-alkylbenzoic acids in the present case. The melting point and the aqueous solubility are criteria of stability of the solid state. The solubilities in the present study were determined as the extrapolated MAC of the acids at zero surfactant concentration. More than 99% of the dissolved acids are in an undissociated state from their acidity constants [16] and the solution pH less than 3. In Fig. 1 are shown the melting points and the aqueous solubilities of p-n-alkylbenzoic acids plotted against the carbon number of the alkylchain. The melting points of the acids with odd carbon number are much higher than those with even number up to number 4. This indicates an increased stability of acids with odd carbon number compared with those with even number, which is reflected in their solubilities; the solubility of the former is less than that of the latter. This even-odd irregularity of physico-chemical properties is often the case as far as properties of the solid state are concerned [17]. From five carbon atoms upwards the difference in these two properties disappears. Especially the solubilities can be roughly fitted to a straight line, although the melting point at five carbon atoms deviates downwards a bit from a smooth curve of even carbon acids. It can be roughly said that the physicochemical properties of organics with an attached alkyl chain regularly change with increasing carbon number in the alkyl chain above a certain value.

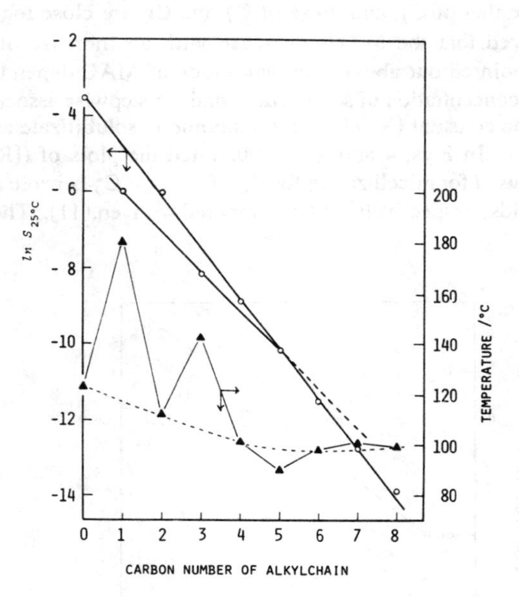

Figure 1. Melting points and aqueous solubilities of p-n-alkylbenzoic acids plotted against carbon number of the alkyl chain.

From the slope of the solubility plots the free energy change of transfer from solid state into the aqueous bulk phase per methylene group is estimated to be 3.30 and 2.54 kJ mol^{-1} for alkyl chains with even and odd carbon numbers less than five, respectively, where the following equation was employed assuming no ionization:

$$\Delta G^{o\,d}_{CH_2} = -RT\,\Delta \ln S/\Delta n \qquad (12)$$

The smaller value for the odd carbon acids indicates that the enhanced stability of acids with odd carbon number mentioned above is caused by a more stabilized arrangement of the benzoic acid groups in the crystalline state. These values are almost 70% or less of those for dissolution of ionic surfactants [18].

We turn now to the MAC variations with the acid chain lengths and with surfactant concentrations. In Figs. 2 and 3 are illustrated the variations for C_0-, C_1-, C_2-, C_3-benzoic acids and C_4-, C_5-, C_6-, C_7-, C_8-benzoic acids, respectively. As with the case of solubilization into surfactant micelles, the plots of MAC against the surfactant concentrations can be divided into two straight lines. The surfactant concentration and the MAC at an intersection of the two lines can be considered to be the CMC of surfactant of the system and the monomer concentration of the solubilizates at the CMC, respectively. These two values are used to determine the first stepwise association constant between monomeric solubilizate and vacant micelle. Both the order of MAC and its difference in magnitude above the CMC do not always regularly decrease with an increase in the carbon number of the alkylchain of the acids. The order of C_1 and C_2 is reversed, and the MAC of C_0 is far above that of C_1, and those of C_7 and C_8 are close together, although the tendency can be observed that the MACs decrease with an increase of hydrophobicity of solubilizates. As was pointed out above, the magnitude of MAC depends on a combination between the monomeric concentration of solubilizates and the stepwise association constants. The first stepwise association constant (K_1) between monomeric solubilizate and vacant micelle is estimated using eq. (11). In Figs. 4 and 5 are illustrated the plots of $([R_t]-[R])/[R]$ against surfactant concentration used for micellization for C_0-, C_1-, C_2-, C_3-benzoic acids and C_4-, C_5-, C_6-, C_7-, C_8-benzoic acids, respectively, to be compared with eq.(11). The slope of the line

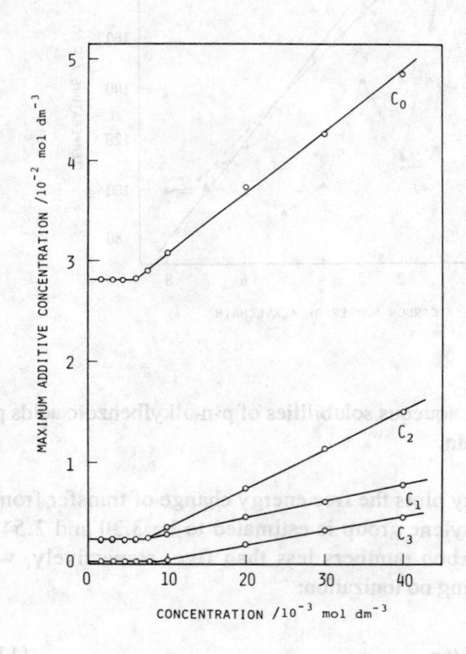

Figure 2. Solubility changes of C_0-, C_1-, C_2-, and C_3-benzoic acids with dodecylsulfonic acid concentrations.

obtained by the plots gives K_1/N, where $[M_t] = (C\text{-}CMC)/N$, N is an aggregation number of micelles, and C is the total surfactant concentration. The excellent linearities are obtained for every solubilizate acid up to the highest surfactant concentration of the present experiment, indicating a constant composition of micelles due to a definite concentration of monomeric solubilizate as mentioned above. The values of K_1 are evaluated using N=61[19]. The value of

$-RT \ln K_1$ is the standard free energy change for the first stepwise association, and the plots of the change with carbon number of the alkyl chain of the solubilizates are shown in Fig. 6. A good linearity is also obtained in this case. The incremental free energy change of methylene group for solubilization is evaluated to be -2.59 kJ mol^{-1}. Then, the solubilizates are found to be stabilized by their incorporation into surfactant micelles, although the free energy decrease by methylene group is much less than that from aqueous bulk into liquid hydrocarbon.

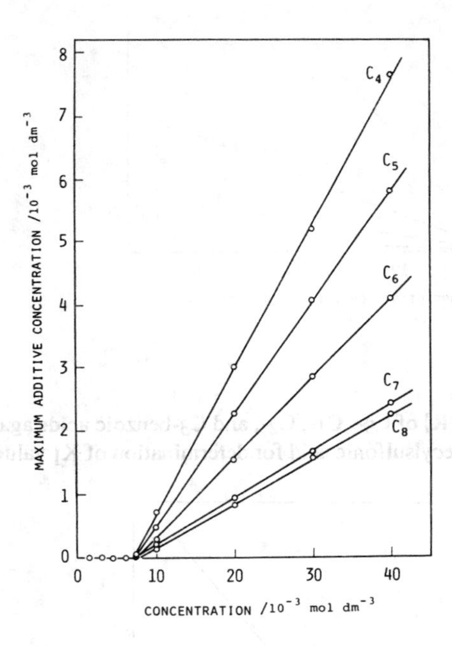

Figure 3. Solubility changes of C_4-, C_5-, C_6-, C_7-, and C_8-benzoic acids with dodecylsulfonic acid concentrations.

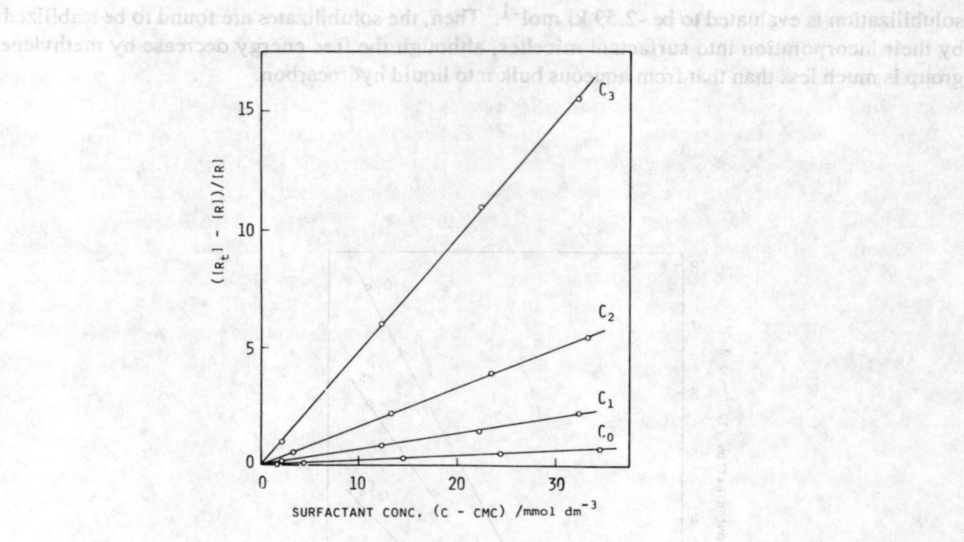

Figure 4. Plots of $([R_t]-[R])/[R]$ of C_0-, C_1-, C_2-, and C_3-benzoic acids against micellar concentration (C-CMC) of dodecylsulfonic acid for determination of K_1 values.

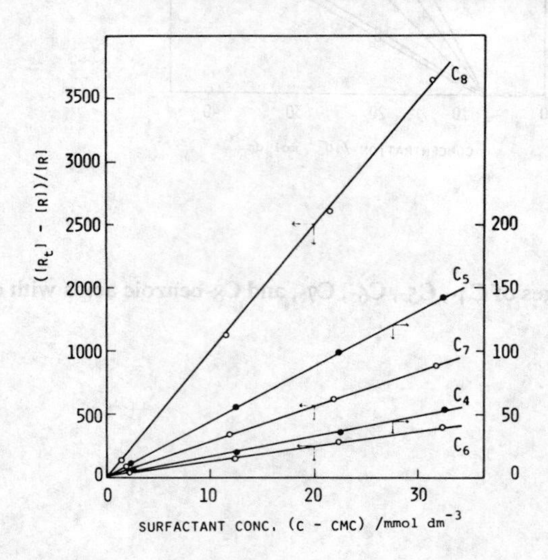

Figure 5. Plots of $([R_t]-[R])/[R]$ of C_4-, C_5-, C_6-, C_7-, and C_8-benzoic acids against micellar concentration (C-CMC) of dodecylsulfonic acid for determination of K_1 values.

4. Examination of the Laplace Pressure of the Micellar Interior

The attached alkyl chains of solubilizates range from C_0 to C_8 and the full length of solubilizate of C_8 is almost equal to that of the surfactant molecule. Then, if a plane of tension is located inside the micelle, the longer alkyl chains could go inwards enough to reach the micellar core region where the pressure is increased by the Laplace pressure. The expected plots then would become less steep above a certain carbon number in Fig.6, because the alkyl chain must intrude into the inner micellar region against the increased pressure. However, the experimental evidence does not show this. The question here is whether the aggregation number of the micelle used for the present calculation can be applicable to every solubilizate system. The total monomeric concentration ([R]+CMC) remains almost the same except for C_1-acid, which means that a decrease in aggregation number due to mixed micelle formation of the surfactant with solubilizates is unlikely for the solubilizates with shorter alkyl chains. One possibility is an increase in the aggregation number in the system of solubilizates with shorter alkyl chains. However, the CMCs do not decrease much from that of the surfactant only, 8.7×10^{-3} mol dm^{-3}, and the mole fractions of C_1- and C_3-acids in the micelle are both less than 0.15. Therefore, such an increase in the aggregation that brings about a distinct upward deviation of plots for small carbon numbers in Fig. 6 is highly improbable. The plots of C_0- and C_2-acids whose mole fraction in the micelle is as high as 0.38 and 0.28, respectively, are also in a line with those of smaller mole fractions. This fact depends on the normalized value of K_1, which is the first stepwise association constant between monomeric solubilizate and empty micelle.

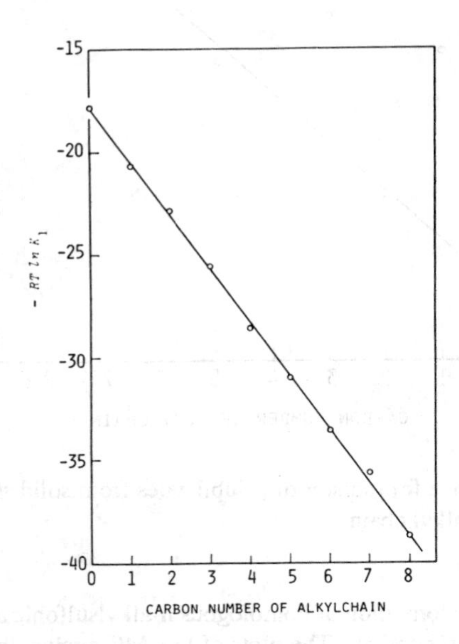

Figure 6. Standard free energy change for the first stepwise association constant against carbon number of the alkyl chain.

Let us now consider the equation

$$\Delta G^{o,s} = - RT \ln ([R] \times K_1) \quad (13)$$

$\Delta G^{o,s}$ represents the free energy change when a solid solubilizate is incorporated into micelles. The slope of the plots of $\Delta G^{o,s}$ against n indicates, therefore, the incremental free energy change per methylene group for the process (Fig. 7). The incremental change of 0.67 kJ mol^{-1} obtained by the linear regression analysis is only 15% of the incremental enthalpy change per methylene group on the melting of n-alkyl-carboxylic acids (4.3 kJ mol^{-1}) [20]. The value of 0.67 kJ mol^{-1} is almost equal to 0.71 kJ mol^{-1} from the solubility data (Fig. 8). Such a small change strongly suggests that the solubilized alkyl chains are still very similar in physicochemical property to their solid states with a hydrophilic group anchored to the micellar surface. This is quite contrary to the general concept that alkyl chains in micelles are liquid-like in character. This partial crystallinity has often been ascribed to pressure increase inside the micelle due to the Laplace pressure. If the pressure increase were the case, the following calculation could be possible.

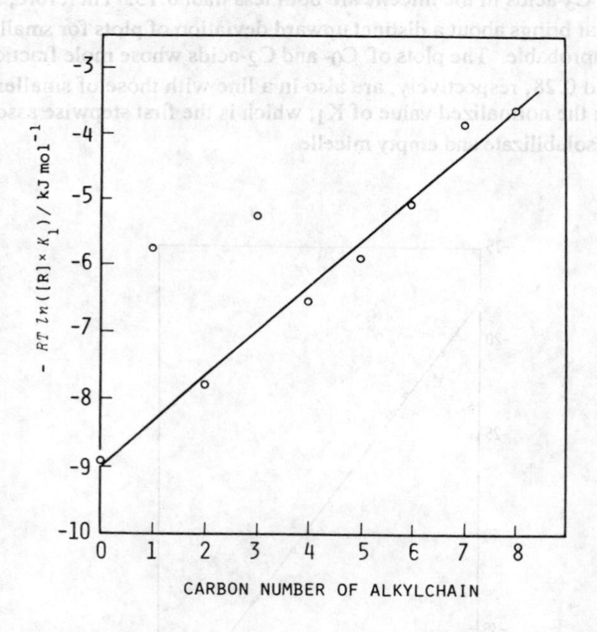

Figure 7. Free energy change for transfer of solubilizates from solid state to solubilized state against carbon number of the alkyl chain.

Let us consider the micellar formation of homologous n-alkylsulfonic acids whose chain length ranges from 10 to 22 in carbon number. The plots of ln CMC against the carbon number give a straight line [21] whose slope is -0.612. The incremental free energy change per methylene group for a transfer from aqueous bulk to micelle is evaluated to be -2.83 kJ mol^{-1}, where the degree of counterion association to micelle was assumed to be 0.75. The important point here is that the value remains constant irrespective to the chain length of the surfactants. In other words, it is independent of the radius of micelles. On the other hand, the free energy change for a transfer of a methylene group from aqueous phase to organic liquid phase is -3.51 kJ mol^{-1} [22]. When their difference of 0.68 kJ mol^{-1} was ascribed only to the pressure increase ($2\gamma/z$)

$$V_{CH_2} \times 2\gamma/z = 0.68 \text{ kJ mol}^{-1} \quad (14)$$

the interfacial tension(γ) inside the micelles increases with increasing alkyl chain of the surfactant molecules as shown in Fig. 9, where the following parameters for partial molecular volume of methylene group (V_{CH_2}) and radius (z) of the inner hydrophobic sphere of micelle made of alkyl chain with carbon number n are assumed [23];

$$V_{CH_2} = 26.9 \text{ Å}^3 \text{ and } z = 1.5 + 1.265 \times n \text{ Å} \quad (15)$$

The problem is then whether the increase of interfacial tension is reasonable or not. If an imaginary interfacial tension inside the micelle were equivalent to an interfacial tension between water and liquid hydrocarbon where surfactants are at their saturation adsorption, the interfacial tension and the maximum adsorption would remain the same independent of the chain length of the surfactants[24]. In addition, an increase in interfacial tension up to 60 mN m^{-1} is quite unbelievable judging from the reasonable values of the above two parameters, because an interfacial tension between water and liquid hydrocarbon is around 50 mN m^{-1} and because the interfacial tension between liquid hydrocarbon and aqueous micellar solution around CMC is less than 15 mN m^{-1} (Table 1). Indeed, factors such as curvature, head group repulsion, and entropic

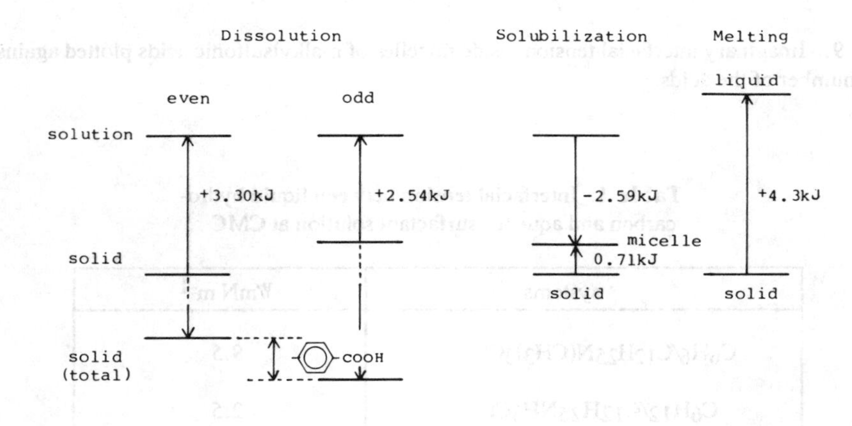

Figure 8. Standard free energy change per methylene group.

effects must be taken into account as for micellization of surfactants with different kinds of hydrophilic head groups including counterions, but the present discussion is made for micellization of homologous surfactants with identical hydrophilic head groups and therefore almost the same physicochemical factors mentioned above can be expected for the homologous micelles, except for the micellar radius. This fact can be seen from mixed micelle formation for homologous surfactants [25].

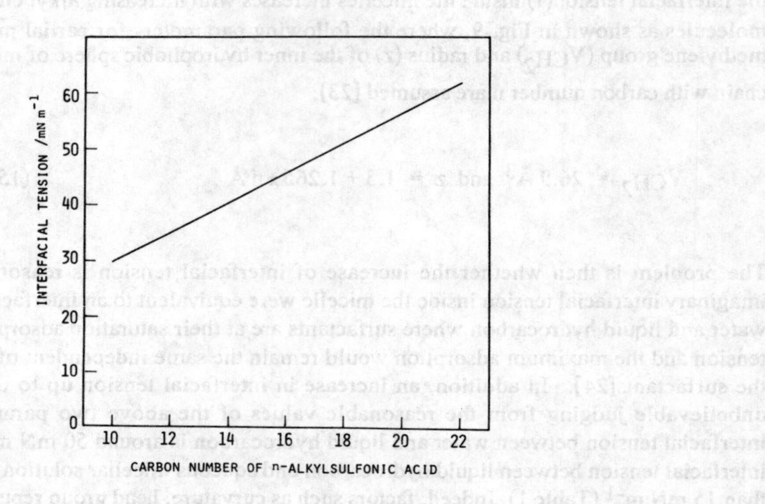

Figure 9. Imaginary interfacial tension inside micelles of n-alkylsulfonic acids plotted against carbon number of the acids.

Table 1. Interfacial tension between liquid hydrocarbon and aqueous surfactant solution at CMC

systems	γ/mN m^{-1}
$C_6H_6/C_{12}H_{25}N(CH_3)_3Cl$	9.5
$C_6H_{12}/C_{12}H_{25}NH_3Cl$	2.5
$C_6H_{12}/C_{12}H_{25}N(CH_3)_3Cl$	11.0
$C_6H_{12}/C_{12}H_{25}SO_3Na$	7.4

Taking the above calculations into consideration, the existence of a plane of interfacial tension inside a micelle is quite unlikely. The decrease in the free energy change for the transfer of a methylene group results from a decrease in mobility of solubilizate or surfactant molecule inside micelles due to an anchorage of head groups of these molecules at the micellar surface. In fact, solubilizate molecules whose head group is bulky and difficult to move about micellar surface like 10-alkylphenothiazines have less free energy change [10]. Even for non-polar solubilizates like methane, ethane, and propane, their movement becomes difficult and constrained in micelles where one end of a long alkyl chain of surfactants is loosely fixed at the micellar surface.

5. References

1. McBain, M.E.L. and Hutchinson, E. (1955) Solubilization and Related Phenomena, Academic Press, New York.
2. Elworthy, P.H., Florence, A.T., and Macfarlane, C.B. (1968) Solubilization by Surface-Active Agents and Its Application in Chemistry and Biological Sciences, Chapman & Hall, London.
3. Bunton, C.A. and Sepulveda, L. (1979) 'Hydrophobic and Coulombic Interactions in the Micellar Binding of Phenols Phenoxide Ions', J. Phys. Chem. 83, 680-682.
4. Prapaitrakul, W. and King, Jr. A. D. (1985) 'The Solubility of Gases in Aqueous Solutions of Decyltrimethyl- and Cetyltrimethyl ammonium Bromide', J. Colloid Interface Sci. 106, 186-193.
5. Moroi, Y., Sugii, R., and Matuura, R. (1984) 'Examination of Micelle Formation by Phase Rule', J. Colloid Interface Sci. 98, 184-191.
6. Moroi, Y. (1980) 'Distribution of Solubilizates among Micelles and Kinetics of Micelle Catalyzed Reactions', J. Phys. Chem. 84, 2186-2190.
7. Moroi, Y., Sato, K., and Matuura, R. (1982) 'Solubilization of Phenothiazine in Aqueous Surfactant Micelles', J. Phys. Chem. 86, 2463-2468.
8. Moroi, Y., Noma, H., and Matuura, R. (1983) 'Solubilization of N-Alkylphenothiazine in Aqueous Anionic Surfactant Micelles', J. Phys. Chem. 87, 872-876.
9. Moroi, Y., Sato, K., Noma, H., and Matuura, R. (1984) 'Solubilization of Phenothiazine and its N-Alkyl Derivatives into Anionic Surfactant Micelles', in K.L. Mittal and B. Lindman (eds.), Surfactants in Solution, Plenum Publishing Corporation, New York, 963-979.
10. Moroi, Y. and Matuura, R. (1988) 'Thermodynamics of Solubilization into Surfactant Micelles: Effect of Hydrophobicity of Both Solubilizate and Surfactant Molecules', J. Colloid Interface Sci. 125, 456-462.
11. Moroi, Y. and Matuura, R. (1988) 'Solubilization of 4-n-Alkylbenzoic Acids into Dodecylsulfonic Acid Micelle: Examination of Laplace Pressure of Micellar Interior', J. Colloid Interface Sci. 125, 463-471.
12. Matheson, I.B.C. and King, Jr. A.D. (1978) 'Solubility of Gases in Micellar Solutions', J. Colloid Interface Sci. 66, 464-469.
13. Ownby, D.W. and King, Jr. A.D. (1984) 'The Solubility of Propane in Micellar Solutions of Sodium Octyl Sulfate and Sodium Dodecyl Sulfate at 25, 35, and 45°C', J. Colloid Interface Sci. 101, 271-276.
14. Ogino, K., Abe, M., and Takeshita, N. (1976) 'A Study of the Solubilization of Polar Oily Materials by Sodium Dodecyl Sulfate', Bull. Chem. Soc. Jpn. 49, 3679-3683.

15. Fellar, W. (1967) An Introduction to Probability Theory and Its Applications, 3rd ed. Vol. 1, Wiley, New York.
16. Kortum, G., Vogel, W., and Andrussow, K. (1961) Dissociation Constants of Organic Acids in Aqueous Solution', Butterworths, London.
17. Bailey, A.E. (1950) Melting and Solidification of Fats, Interscience, New York.
18. Moroi, Y., Sugii, R., Akine, C., and Matuura, R. (1985) 'Anionic Surfactants with Methylviologen and Cupric Ions as Divalent Cationic Gegenion (II): Effect of Alkylchain Length on Solubility and Micelle Formation', J. Colloid Interface Sci. 108, 180-188.
19. Moroi, Y., Humphrey-Baker, R., and Gratzel, M. (1987) 'Determination of Micellar Aggregation Number of Alkylsulfonic Acids by Fluorescence Quenching Method', J. Colloid Interface Sci. 119, 588-591.
20. Garner, W.E., Madden, F.C., and Rushbrooke, J.E. (1926) 'Alternation in the Heats of Crystallization of Normal Monobasic Fatty Acids. Part II', J. Chem. Soc. 2491-2502; Garner, W.E. and King, A.M. (1929) 'Alternation in the Heats of Crystallization of the Normal Monobasic Fatty Acids. Part IV', J. Chem. Soc. 1849-1861.
21. Saito, M., Moroi, Y., and Matuura, R. (1980) 'Micelle Formation of Long-Alkyl-Chain Sulfonic Acids in Aqueous Solutions', J. Colloid Interface Sci. 76, 256-258.
22. Mukerjee, P. (1977) 'Size Distribution of Micelles: Monomer-Micelle Equilibrium, Treatment of Experimental Molecular Weight Data, the Sphere-to- Rod Transition and a General Association Model' in K.L. Mittal (eds.), Micellization, Solubilization, and Microemulsions, Plenum Press, New York and London, pp. 171-194.
23. Leibner, J.E. and Jacobus, J. (1977) 'Charged Micelle Shape and Size', J. Phys. Chem. 81, 130-135.
24. Vader, F. van V. (1960) 'Adsorption of Detergents at the Liquid-Liquid Interface. I', Trans. Faraday Soc. 56, 1067-1077.
25. Lange, von H. and Beck, K.H. (1973), 'Zur Mizellbildung in Mischlösungen Homologer und Nichthomologer Tenside', Kolloid-Z.u.Z. Polymere, 251, 424-431.

APPLICATION OF THE DIFFERENTIAL CONDUCTIVITY METHOD: THE EFFECT OF CHAIN-LENGTH OF HOMOLOGOUS SURFACTANTS ON THE PARTITION COEFFICIENT OF ALKANOLS BETWEEN BULK WATER AND MICELLES

M. MANABE, H. KAWAMURA, Y. YAMASHITA, AND S. TOKUNAGA,
Niihama National College of Technology,
Niihama,
Ehime 792,
Japan.

ABSTRACT. The partition coefficient (K) of alkanols (C_4-C_7) between bulk water and micelles of an homologous series of sodium alkyl sulphates (C_9-C_{11}) has been determined. The results show that for a given alkanol, K tends to increase with increasing chain length of surfactant and that the increase is more appreciable for longer chain alkanols. However, the dependence is very small, appearing within the determination uncertainty. In the determination of K, a new method named "differential conductivity method" was applied to the present systems consisting of surfactants with differing cmc values over a wide region.

1. Introduction

In order to represent the equilibrium in solubilization systems, the partition coefficient (K) of solubilizate between the bulk water and micelles has been used. In recent years K for various types of solubilizates has been determined in some micellar systems [1-12]. The authors have mainly paid attention to the properties of the solubilizates, compared to those of surfactants. Against that, one of the present purposes is to study the effect of the chain-length of an homologous series of surfactants on K for a given alkanol.

In the determination of K, a variety of methods were adopted. In a previous study, a novel method named "differential conductivity method" was proposed by which, K for an homologous series of alkanols was determined in micellar solutions of sodium dodecyl sulphate [11]. The other present purpose is to examine the effectiveness and limitation of the method, by applying it to the partition systems consisting of the homologous surfactants with widely differing cmc values.

2. Experimental

Sodium undecyl sulphate (SUS), sodium decyl sulphate (SDeS), sodium nonyl sulphate (SNS) were synthesized and purified by the same procedure described elsewhere [13]. The surface tension-concentration curve for a solution of each surfactant gave no minimum around its critical micelle concentration. Commercially available 1-alkanols (carbon number n_a = 4,5,6,7) were

D. M. Bloor and E. Wyn-Jones (eds.),
The Structure, Dynamics and Equilibrium Properties of Colloidal Systems, 63–69.

used after fractional distillation. Water purified by ion exchange and distillation was used as the solvent. Its specific conductivity was lower than 2 μScm^{-1} at 298.15 K. The conductivity measurements were carried out following an identical procedure using the same apparatus as described in a previous study [11]. The conductivity of the surfactant solution at a given concentration was measured as a function of concentration of alkanol, where the alkanol solution prepared with the same surfactant solution was added incremently to the surfactant solution. The measurements were made at 298.15 K.

3. Results and Discussion

The specific conductivity, κ, of SDS solution is shown in Fig. 1 as a function of its concentration, C_s. It is well known, the break point from the two straight lines provides the critical micelle concentration, cmc. The equivalent conductivity, or molar conductivity, Λ, expressed as

$$\Lambda = 1000\ \kappa/C_s \tag{1}$$

is also plotted against the square root of C_s. Equally well known, Λ, declines at first slowly

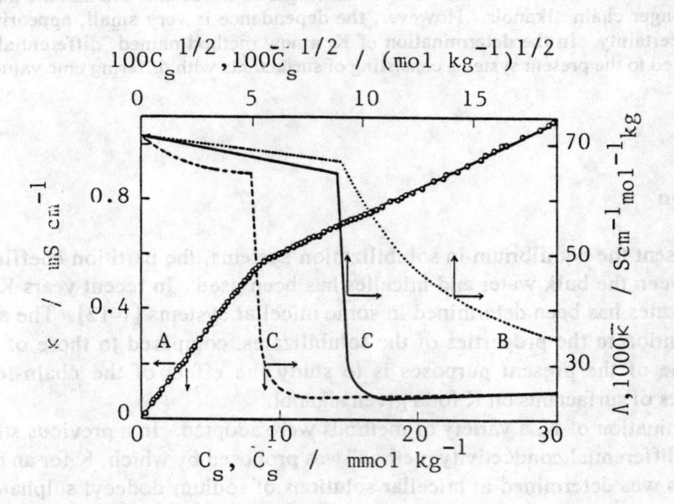

Figure 1. The relation between conductivity and concentration of SDS for its solution at 298.15 K. A(κ), B(Λ), C($\bar{\kappa}$).

according to the Kohlrausch's square root law, and then more rapidly after the cmc, causing a break in the curve. Another conductivity, named "differential conductivity" and denoted by $\bar{\kappa}$, defined in a previous paper [11] is

$$\bar{\kappa} = d\kappa/dC_s = (\kappa_2-\kappa_1)/(C_{s2}-C_{s1}) \tag{2}$$

which is a differential quantity at a concentration C_s or a slope between a pair of nearest neighbour points of the κ vs C_s plot at the mean concentration $\bar{C}_s$ (= $(C_{s1}+C_{s2})/2$). The relation between $\bar{\kappa}$ and $\bar{C}_s$ or $\bar{C}_s^{1/2}$ is shown in Fig. 1, too. In contrast to the pattern of the curve of Λ, the plot of $\bar{\kappa}$ vs $\bar{C}_s^{1/2}$ has a characteristic profile that after a linear slight decrease as in the case of Λ, $\bar{\kappa}$ drops suddenly in a very narrow concentration region and then attains an almost constant value at around 12 mmol kg^{-1}, while Λ vs $C_s^{1\backslash 2}$ plot above the break point is non-linear. If two linear relations of Λ and $\bar{\kappa}$ in Fig. 1 are expressed as

$$\Lambda = \Lambda_0 + \beta\, C_s^{1/2} \tag{3}$$

$$\bar{\kappa} = \bar{\kappa}_0 + \gamma\, \bar{C}_s^{1/2} \tag{4}$$

$\Lambda_0=\kappa_0$ and $\beta=(2/3)\gamma$ is obtained by taking the definitions of Λ and κ into account. Such a sudden decrease in $\bar{\kappa}$ was utilized for the determination of cmc in mixed surfactant systems [14,15]. The C_s at the break point is taken to be the cmc. The transitional concentration region is called the cmc region.

Comparing the cmc values determined from respective plots, the value is lower in the order of $\bar{\kappa}$-plot (8.0 mmol kg^{-1}), Λ-plot (8.1 mmol kg^{-1}), and κ-plot (8.3 mmol kg^{-1}). The κ-plot value is close to the C_s at the half height point of the $\bar{\kappa}$-plot in the cmc region. It should be noted that the cmc values determined by various methods [16] lie within the cmc region appearing in the $\bar{\kappa}$-plot. It seems that at the lowest cmc corresponding to that in $\bar{\kappa}$-plot, small aggregates (premicelles or micellar embryo) start to form, consisting of completely ionized surfactant ion. With increasing C_s, the aggregates grow and the degree of ionization of the aggregates decrease until $\bar{\kappa}$ attains an almost constant value.

The sigmoid pattern of the $\bar{\kappa}$-plot is consistent with the cooperativity of micelle formation, in line with the phase separation model [17]. If κ is taken as a function of concentration of surfactant in monomeric form (C_s^f) and in micellar form (C_s^m), the total differential is represented as

$$d\kappa = (d\kappa/dC_s^f)\, dC_s^f + (d\kappa/dC_s^m)\, dC_s^m \tag{5}$$

where

$$C_s = C_s^f + C_s^m \tag{6}$$

Then, its integral form is

$$\kappa = \bar{\kappa}^f C_s^f + \bar{\kappa}^m C_s^m \tag{7}$$

where $\bar{\kappa}^f = d\kappa/dC_s^f$ and $\bar{\kappa}^m = d\kappa/dC_s^m$. Mathematically, the definition of $\bar{\kappa}$ is identical with the "partial molar quantity", whereas Λ which represents the slope between the origin and a point in the κ-C_s plot corresponds to the "apparent molar quantity" in the thermodynamics. If it is possible to estimate the numerical values of the $\bar{\kappa}$'s for respective ionic species, the additivity as Eq. (7) allows us to determine their concentrations from the conductivity data. In this respect, the differential conductivity is believed to have wider application than the traditional equivalent conductivity.

The most striking characteristic of $\bar{\kappa}$ for ionic surfactant solutions is the independence of $\bar{\kappa}$ above the cmc on C_s, in contrast to the monotonical decrease in Λ, even though both $\bar{\kappa}$ and Λ have a small dependence on C_s below the cmc, obeying the square root law. On the basis of the phase separation model of micelle formation, $\bar{\kappa}$ below the cmc and $\bar{\kappa}$ above cmc (Fig. 1) can be taken to be $\bar{\kappa}^f$ and $\bar{\kappa}^m$ as defined in Eq. (7), respectively. If the concentration dependence of $\bar{\kappa}$ is not to be ignored in the following calculation, the linear relationship of $\bar{\kappa}$ with $C_s^{1/2}$ below and above cmc in Fig. 1 is available.

When small amounts of an alkanol is successively added to a micellar solution at a certain concentration of the surfactant, κ changes in the manner shown in Fig. 2 as a function of C_a, the total alkanol concentration. When C_s is just above the cmc, κ tends to decrease most remarkably. The tendency becomes less pronounced with increasing C_s, and eventually κ tends to increase at high C_s. When C_s is even higher, the increment in the resistance of the micellar solution on the addition of alkanol is too small to detect. Therefore, for short chain surfactants with a high cmc,

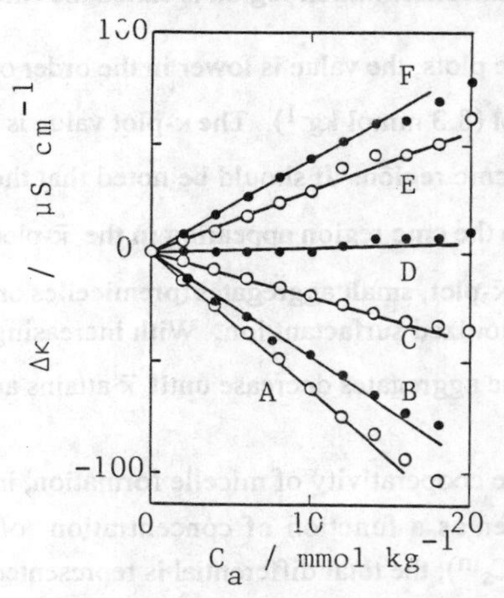

Figure 2. The conductivity change of SDeS solution on the addition of pentanol. C_s/mmol kg^{-1}; A (40.40), B (50.94), C (69.93), D (100.2), E (149.8), F (200.1).

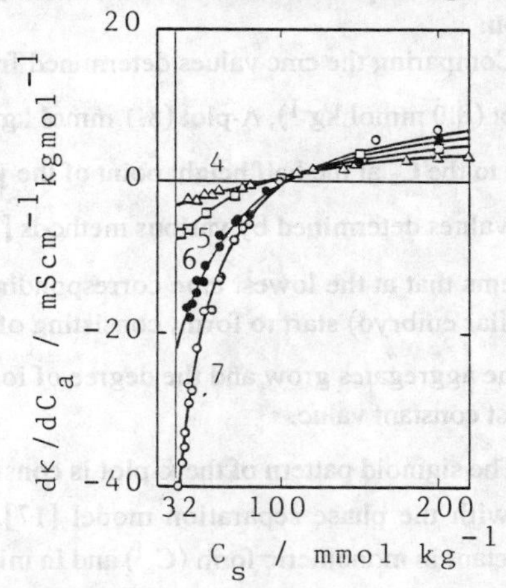

Figure 3. The relation between $d\kappa/dC_a$ and concentration of SDeS. The number indicates n_a.

the detectable C_s region is limited. According to the differential conductivity method for the determination of partition coefficient of an alkanol between the bulk water and micelles [11], the slope of the straight line at a limiting dilution of alkanol ($d\kappa/dC_a$) is used as described below. $d\kappa/dC_a$ is also a differential conductivity. The decrease of κ is explained to be due to the transfer of monomerically dissolving surfactant (completely ionized) to a micellar one (definitely ionized), resulting in the decrease in the concentration of free ions. The increase in κ, i.e., increase in the ion concentration, is explained by the enhancement of the degree of ionisation of the micelles, the alkanol incorporated in the micelles diluting the surface charge density of the micelles. Both effects are competitive.

The limiting slope, $d\kappa/dC_a$, is plotted against C_s (Fig. 3). For each alkanol, $d\kappa/dC_a$ tends to increase from a negative value at the cmc to positive ones at high C_s's. Further, the increasing tendency is more pronounced for a longer chain alkanol. The equation representing the relation between $d\kappa/dC_a$ and C_s can be derived as follows. When Eq. (7) is differentiated with respect to C_a at a given C_s the following equation is obtained,

$$d\kappa/dC_a = \bar{\kappa}^f (dC_s^f/dC_a) + \bar{\kappa}^m (dC_s^m/dC_a) + C_s^f (d\bar{\kappa}^f/dC_s) + C_s^m (d\bar{\kappa}^m/dC_s) \quad (8)$$

Here, we define the quantities,

$$K_x = x_a^m/x_a^f; \quad j = dC_a^m/dC_a; \quad k = dC_s^f/dC_a^f \quad (9)$$

K_x is the partition coefficient of alkanol expressed in mole fraction of alkanol in the micellar phase (x_a^m) and in the intermicellar bulk phase (x_a^f), j is the fraction of solubilized alkanol to the total amount of added alkanol, and k refers to the effect of the intermicellar concentration of alkanol (C_a^f) on the intermicellar concentration of surfactant (C_s^f) in a micellar solution. At the cmc, k corresponds to the well known quantity, $dcmc/dC_a$, the rate of cmc-decrease on the addition of

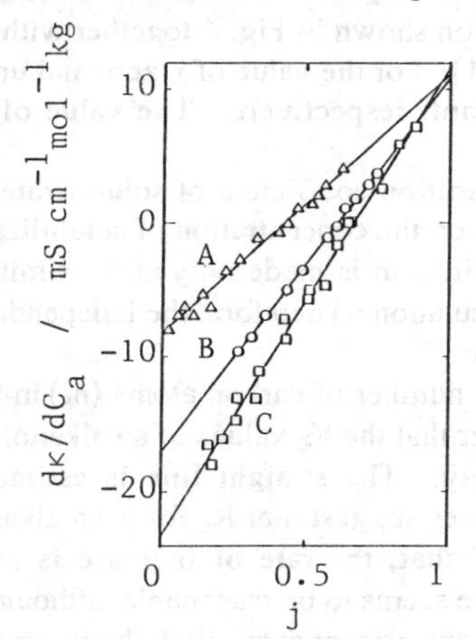

Figure 4. The relation between $d\kappa/dC_a$ and j for hexanol. Surfactant: A (SDS) [11], B (SUS), C (SDeS).

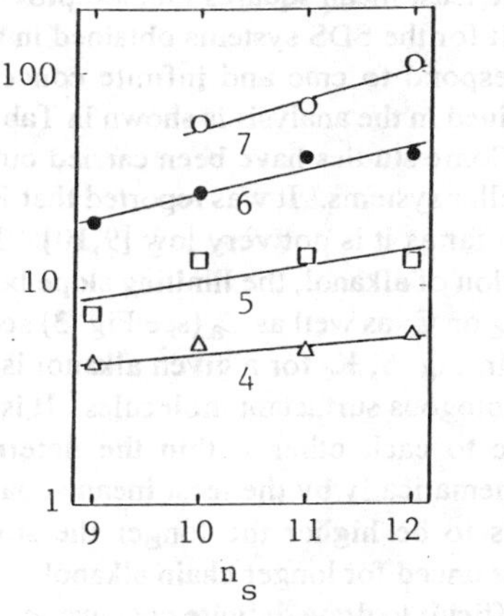

Figure 5. The relation between K_c and n_s. The number indicates n_a.

alkanol [13,18]. In order to consider the effect of solubilized alkanol on the degree of ionization of micelles, α, the following equation may be used,

$$\bar{\kappa}^m = \alpha\, \bar{\kappa}^{m*} \quad (10)$$

where $\bar{\kappa}^{m*}$ refers to the hypothetical value of $\bar{\kappa}^{m}$ for the completely ionized micelles. Taking these quantities in Eqs. (9) and (10) into account and making the approximation of C_a being extremely low, we obtain

$$d\kappa/dC_a = (\bar{\kappa}^f - \bar{\kappa}^m)\ k + (\bar{\kappa}^{m*}(d\alpha/dx_a{}^m) - (\bar{\kappa}^f - \bar{\kappa}^m)\ k)\ j \qquad (11)$$

where under the condition of C_a being extremely low, j can be approximated as

$$j = K_c/(1+K_c\, C_s{}^m) \qquad (12)$$

K_c is a partition coefficient in molality units and correlated with K_x as $K_c=K_x/n_w$ (n_w: the mole number of water in a unit kg). The derivation of the equations shown above has been described in more detail in a previous paper [11].

Eq. (11) suggests that if k and $d\alpha/dx_a{}^m$ (the effect of surface charge density dilution on α) are independent of C_s as well as C_a, $d\kappa/dC_a$ is a linear function of j. Accordingly, taking K_c in Eq. (12) as a parameter, a regression analysis for the linear relation between $d\kappa/dC_a$ and j gives the most suitable value of K_c. For the calculation, $C_s{}^m$ is approximated as C_s-cmc. For the numerical value of $\bar{\kappa}^{m*}$, the value of $\bar{\kappa}^f$ at the cmc (Fig. 1) is adopted. The regression analysis of the least mean squares method provides the linear relation shown in Fig. 4 together with the result for the SDS systems obtained in a previous study [11]. For the value of j, zero and unity correspond to cmc and infinite concentration of surfactant, respectively. The value of K_c obtained in the analysis is shown in Table 1.

Some studies have been carried out recently on the partition coefficient of solubilizates in micellar systems. It was reported that K_c strongly depends on the concentration of solubilizates in so far as it is not very low [9,10]. The present determination is made only at the limiting dilution of alkanol, the limiting slope being used in the calculation. Therefore, the independence of K_c on C_s as well as C_a (see Fig. 3) seems to be reasonable.

In Fig. 5, K_c for a given alkanol is plotted against the number of carbon atoms (n_s) in the homologous surfactant molecules. It is possible to consider that the K_c values of an alkanol are close to each other within the determination uncertainty. The straight line is estimated mathematically by the least mean squares method. The lines suggest that K_c for each alkanol tends to be higher the longer the surfactant chain, and that, the rate of increase is more pronounced for longer chain alkanols. The increasing slope seems to be reasonable, although it is difficult to draw definite conclusions from the slope with the present data. Both the increasing slope which is slight but not to be ignored (Fig. 5) and the drop of $\bar{\kappa}$ which appears in a narrow but a definite concentration region (Fig. 1) may be attributed to the definite size of micelles. These findings reflect the limitation and validity of the phase separation model of micelle formation.

Table 1. The value of K_c of alkanols in sodium alkylsulfate micellar systems at 298.15 K.

	K_c / kg mol^{-1}			
n_a	4	5	6	7
SNS	4.4	7.4	19.7	
SDeS	5.3	13.4	27.0	56.7
SUS	5.0	13.4	39.8	68.7
SDS*	5.9	13.0	40.5	108

n_a Carbon number in alkanol molecule
* From Ref. [11]

4. References

[1] Dougherty, S.J.; Berg, J.C. (1974) *J. Colloid Interface Sci.*, **48**, 110.
[2] Manabe, M.; Shirhama, K; Koda, M. (1976) *Bull. Chem. Soc. Jpn.*, **49**, 2904.
[3] Hayase, K.; Hayano, S. (1977), *Bull. Chem. Soc. Jpn.*, **50**, 83.
[4] Almgran, M.; Grieser, F.; Thomas, J.K. (1979), *J.C.S. Faraday I*, **75**, 1674.
[5] Goto, A.; Nihei, M.; Endo, F. (1980), *J. Phys. Chem.*, **84**, 2268.
[6] Kaneshina, S.; Kamaya, H.; Ueda, I. (1981), *J. Colloid Interface Sci.*, **83**, 589.
[7] Abuin, E.B.; Lissi, E.A. (1983), *J. Colloid Interface Sci*, **95**, 198.
[8] Hoiland, H.; Ljosland, E.; Backlund, S. (1984), *J. Colloid Interface Sci.*, **101**, 467.
[9] Christian, S.D.; Tucker, E.E.; Smith, G.A.; Bushong, D.S. (1986), *J. Colloid Interface Sci.*, **113**, 439.
[10] Nguyen, C.M.; Scamehorn, J.F.; Christian, S.D. (1988), *Colloid and Surfaces*, **30**, 335.
[11] Manabe, M.; Kawamura, H.; Yamasita, A.; Tokunaga, S. (1987), *J. Colloid Interface Sci.*, **115**, 147.
[12] Kawamura, H.; Manabe, M.; Miyamoto, Y.; Fujita, Y.; Tokunaga, S. (1989), *J. Phys. Chem.*, **93** , 5536.
[13] Manabe, M.; Tanizaki, Y.; Watanabe, H.; Tokunaga, S. (1983) Memoirs of the Niihama Technical College (Science and Engineering) **19**, 50.
[14] Mysels, K.J.; Otter, R.J. (1961), *J. Colloid Sci*, **16**, 462.
[15] Mukerjee, P.; Yang, A.Y.S., (1976), *J. Phys. Chem.*, **80**, 1388.
[16] Mukerjee, P.; Mysels, K.J. "Critical Micelle Concentration of Aqueous Surfactant Systems" NSRDS-NBS **36**, Washington, D.C., (1971).
[17] Shinoda, K.; Nakagawa, T.; Tamamushi, B.; Isemura, T. "Colloidal Surfactants" Academic Press, New York, (1963), p. 25.
[18] Manabe, M.; Koda, M. (1978) *Bull. Chem. Soc. Jpn.*, **51**, 1599.

MIXED MICELLE THEORY AS AN AID TO SURFACTANT FORMULATION

J.H. CLINT
Colloid Science Branch
BP Research Centre
Chertsey Road
Sunbury-on-Thames
MIDDLESEX TW16 7LN

ABSTRACT. In practical applications, surfactants are usually formulated as mixtures which often provide superior performance to that obtainable by single surfactants. The phase separation model of micelle formation can be extended to mixed surfactants which can behave as ideal or non-ideal mixtures. In the latter case, surfactant interactions can be treated by introducing separate non-ideality parameters for mixed micelle formation and for mixing at liquid interfaces, both of which can be calculated from experimental data. The theory is applied to examples of real practical interest including foaming, detergency and enhanced oil recovery where attractive interactions between the components leads to synergism. Other systems are illustrated where a repulsive interaction leads to antagonism.

1. Introduction

The use of surface active agents (surfactants) in formulations for products or for processes, is now very widespread. In many cases the development of these formulations has been achieved by trial and error or by a series of successive optimisations of performance. This product development has resulted in the use of surfactant mixtures capable of performances unobtainable by single or pure surfactants. Application of theory to such systems, especially mixed micelle theory, is fairly recent and this forms the basis of this article.

Soap was the first detergent, dating from Roman times, and it is still one of the commonest. The introduction of synthetic detergents in the 1930's saw the start of a revolution in both domestic and industrial cleaning products. To illustrate how this has now grown, world wide annual consumption of surfactants in 1987 in household and personal care products exceeded 8 million tonnes, of which soap still accounted for well over half.

In addition to these domestic uses of surfactants, there are large industrial uses other than for detergency totalling a further 7 million tonnes. These include uses in dyestuffs, fibres, mineral processing, oil field chemicals, paints, pesticides, pharmaceuticals, and plastics. Each of these is a major industry and the total usage of surfactants in formulations represents a significant factor in the economy.

D. M. Bloor and E. Wyn-Jones (eds.),
The Structure, Dynamics and Equilibrium Properties of Colloidal Systems, 71–84.

2. Practical Surfactant Systems

Whenever surfactants are used on a commercial scale they are invariably mixtures. Commercial surfactants are mixtures anyway for a number of reasons. First they are usually made from feedstocks which have mixed chain lengths. Second, depending on their method of synthesis, they are often mixtures of isomers. Third, when for example a sulphate is made from a fatty alcohol, incomplete reaction leads to a product with typically no more than 80% conversion to the sulphate. Whatever the origin of the mixture, it is usually not feasible to purify the surfactant to any great extent since the cost would soon become prohibitive.

In addition it is found that in most practical applications that well chosen mixtures of surfactants can be made to perform better than single surfactants. An example would be in emulsion formation where a mixture of two different emulsifiers gives a more stable emulsion than would a surfactant with properties intermediate between those of the two surfactants in the mixture. Another example is in household detergents where mixtures of anionic and nonionic surfactants are used to formulate a product with superior properties to one containing only anionic or only nonionic surfactants. Yet another example is in surfactant enhanced oil recovery where the complex multiple requirements of low interfacial tension, low adsorption on the reservoir rock and correct phase behaviour to ensure uniform progression of the surfactant front through the oil field, can usually only be met by using a well optimised mixture of surfactants.

In each of these 3 examples the reasons for choosing a mixture rather than a pure surfactant are different. However the fact remains that mixtures usually outperform single component systems, and this synergism between components is at the very heart of most well formulated surfactant systems. It would clearly be advantageous to understand the ways in which surfactants interact in mixtures and how this in turn controls the performance of the system. It is the intention of the remainder of this article to show how relatively simple theories of mixed micelle formation have been developed, how these can predict the distribution of components between monomers, micelles and interfaces and how this understanding can be used to quantify the notion of synergism in mixtures.

3. Ideal Mixed Micelle Theory

In most formulated products the surfactant concentration may range from a few % to 100% so that surfactant phase behaviour, especially the formation of liquid crystalline phases, is important in controlling product properties and stability. In contrast, usage concentration is usually fairly low, and it is the micellar behaviour of the surfactants which is then of most interest. The number of molecules per micelle is dependent on many factors including the chain length of the hydrophobic tail, the nature of the polar head group, especially whether it is charged, and the composition (especially ionic) of the aqueous solution. It can be as few as 2 or 3 when the chain length is as short as six carbon atoms, but an increase of chain length to only seven gives an aggregation number in excess of 25 [1]. For surfactants of more commercial importance, with a chain length of 12 to 18, the aggregation number for spherical micelles of anionic surfactants is more usually in the range 50 to 100 [2].

Because the aggregation number is quite large, it is found that the thermodynamics of micelle formation can be treated using a phase separation model which treats the micellar material as a separate phase from that of the dissolved surfactant monomers [3]. Consider a mixture of two pure surfactants, 1 & 2, where α is the mole fraction of component 1. If we adopt the phase separation model for micellisation, treat a mixed micelle as an ideal mixture of its components and assume that the activity coefficient for free surfactant monomers is equal to unity, then it can be shown [4] that the monomer concentration of component 1 is given by

$$C_1^m = x_1 \cdot CMC_1 \tag{1}$$

where x_1 is the mole fraction of component 1 in the mixed micelles, and CMC_1 is the critical micelle concentration of pure component 1. A similar equation can be written for component 2 and then from a simple mass balance equation, x_1 can be eliminated to give

$$\frac{1}{CMC} = \frac{\alpha}{CMC_1} + \frac{(1-\alpha)}{CMC_2} \tag{2}$$

where CMC is the critical micelle concentration of the mixture. Equation 2 shows that the cmc of any ideal mixture of two surfactants can be calculated simply from the cmc values of the two pure components.

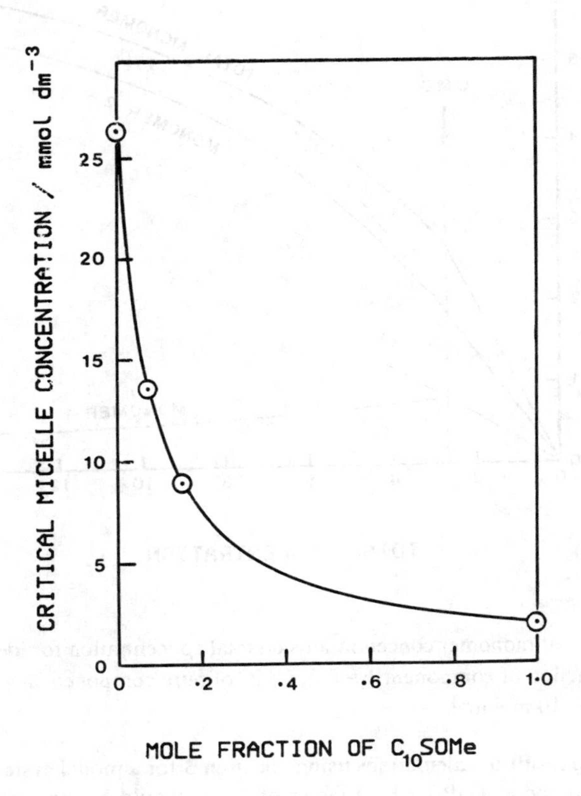

Figure 1. Variation of CMC with mole fraction of mixtures of C_8SOMe and C_{10}SOMe. Circles - experimental points; line - theoretical relation based on equation 2.

An example of a system which obeys equation 2 is shown in Figure 1 where the cmc of a mixture of two homologues is plotted against the mole fraction of component 1. The continuous line is that calculated using equation 2 and its passing accurately through the points for two different mixtures indicates that they are behaving ideally.

As well as cmc values, it is possible [4] to calculate monomer concentrations at any total concentration. If we write an equation for component 2 analogous to equation 1 and use a mass balance equation to eliminate $C_2{}^m$ we obtain a quadratic equation in $C_1{}^m$ whose solution is

$$C_1{}^m = \frac{-(C-\Delta) + [(C-\Delta)^2 + 4\alpha C\Delta]^{1/2}}{2[(CMC_2/CMC_1) - 1]} \tag{3}$$

where $\Delta = CMC_2 - CMC_1$.

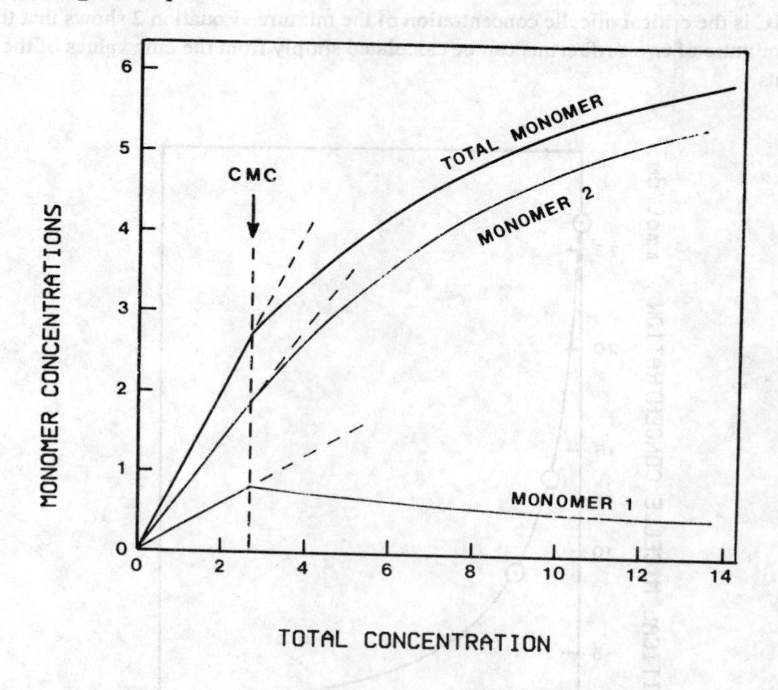

Figure 2. Dependence of monomer concentrations on total concentration for ideal mixture of two surfactants. Mole fraction of component 1 = 0.3; CMC of pure component 1 = 1 mol m^{-3}; CMC of pure component 2 = 10 mol m^{-3}.

Figure 2 shows the result of calculations using equation 3 for a model system where the cmc values for the pure components differ by a factor of 10, as would be the case for homologues differing in chain length by 2 carbon atoms. The most striking feature is that monomer concentrations are by no means constant as total concentration is increased above the mixed cmc. If an adsorbed layer at an interface is in equilibrium with these monomer concentrations then the composition of that layer would be expected to vary with total concentration above the cmc. This will be shown to be the case in the next section.

3.1 SURFACE TENSION OF MIXED SURFACTANT SOLUTIONS

Combination of the Langmuir isotherm and the Gibbs equation leads to a multi-component form of the Szyskowski equation for the surface tension of a mixed surfactant solution:

$$\pi = -RT.\left[\sum_i \Gamma_i K_i C_i^m / \sum_i K_i C_i^m\right].\ln(1 + \sum_i K_i C_i^m) \qquad (4)$$

where the K_i are Langmuir adsorption constants and the Γ_i are saturated surface concentrations for each pure component and both of these quantities can be calculated from the γ vs. lnC curves for pure components. Therefore, for a two component system, if we know the values of C_1^m from equations such as 3 then we are able to calculate the expected variation of surface tension with total concentration for any mixture using a combination of equations 3 and 4. This has been done for two homologous surfactants in Figure 3. The lines show the surface tensions predicted by equation 4 and the experimental points fall remarkably close to these lines, again indicating that such a mixture behaves ideally.

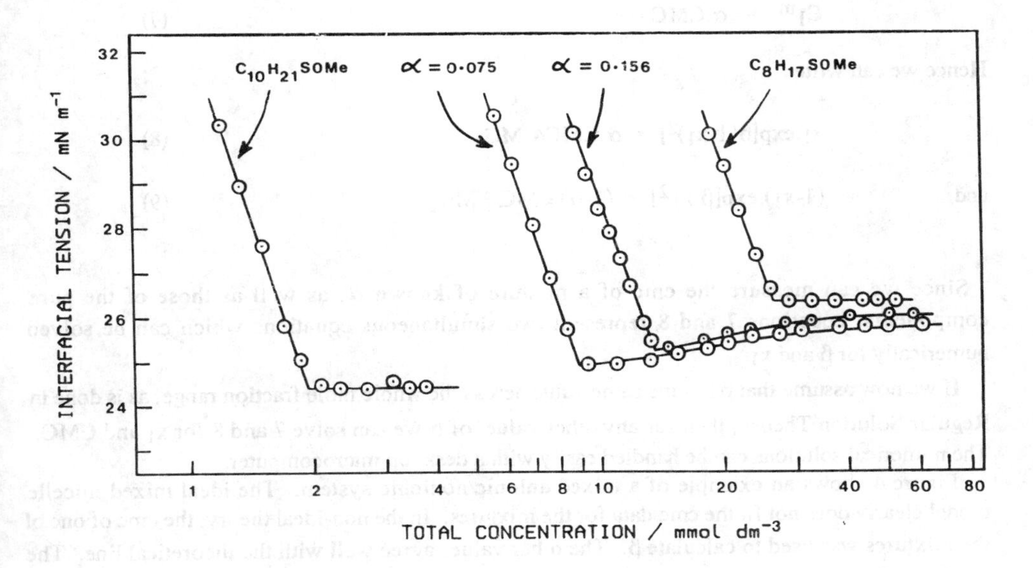

Figure 3. Surface tension plotted against logarithm of concentration for pure n-decyl methyl sulphoxide, pure n-octyl methyl sulphoxide and mixtures of the two with mole fractions 0.075 and 0.156 of the n-decyl component. Circles - experimental points; lines for the two mixtures - theoretical curves based on equations 3 and 4.

4. Non-Ideal Mixed Micelle Theory

Very few pairs of surfactants fit the ideal mixed micelle model. The first extension of this theory to account for non-ideal behaviour was derived by Corkill [5]. He argued that since the micelle core has been shown to be essentially liquid like, then Regular Solution Theory, which has had considerable success in treating non-ideal liquid mixtures, should be useful for mixed surfactant micelles. Corkill introduced activity coefficients for components 1 and 2 within the micelles thus making equation 1 become

$$C_1^m = x_1.f_1.CMC_1 \quad (5)$$

where the activity coefficient, f_1, is given by the Regular Solution Theory as

$$f_1 = \exp[\beta(1-x_1)^2] \quad (6)$$

with similar expressions for component 2. β is a non-ideality parameter.

At the mixed cmc,

$$C_1^m = \alpha.CMC \quad (7)$$

Hence we can write

$$x_1.\exp[\beta(1-x_1)^2] = \alpha.CMC/CMC_1 \quad (8)$$

and

$$(1-x_1).\exp[\beta\, x_1^2] = (1-\alpha).CMC/CMC_2 \quad (9)$$

Since we can measure the cmc of a mixture of known α, as well as those of the pure components, equations 7 and 8 represent two simultaneous equations which can be solved numerically for β and x_1.

If we now assume that β has the same value across the whole mole fraction range, as is done in Regular Solution Theory, then for any other value of α we can solve 7 and 8 for x_1 and CMC. The numerical solutions can be handled easily with a desk top microcomputer.

Figure 4 shows an example of a mixed anionic/nonionic system. The ideal mixed micelle model clearly does not fit the cmc data for the mixtures. In the non-ideal theory, the cmc of one of the mixtures was used to calculate β. The other values agree well with the theoretical line. The small negative value of β = -1.87 indicates a slight attractive interaction between the two ingredients.

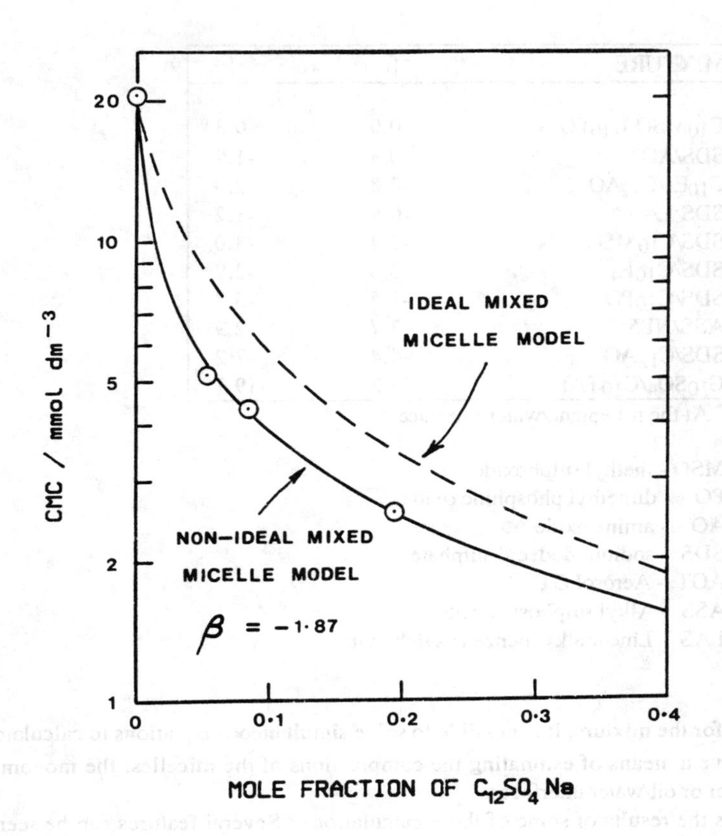

Figure 4. CMC data for mixtures of C_8SOMe and $C_{12}SO_4Na$. Circles - experimental points; line - theoretical relation from solution of equations 8 and 9.

5. Non-Ideality in the Interface

In the same way that non-ideality in the micelle can be accounted for by introduction of the β parameter, it is possible to treat non-ideality in the mixed monolayer at an interface in equilibrium with the bulk micellar solution. Holland [6] introduced a non-ideality parameter β^s for the mixed monolayer which has a mole fraction $x_1{}^s$ of component 1 in it. The complete treatment requires the γ vs. lnC data for each pure component and for one mixture of known composition. The bulk non-ideal theory is first used to calculate β and x_1. Then, using the area per molecule and maximum surface pressure for each pure component (obtained from the γ - lnC curves) and the

TABLE 1. Interaction parameters in mixed micelles and monolayers

MIXTURE	β	$β^s$
$C_{10}MSO/C_{10}PO$	0.0	-0.3
SDS/AOT	-0.9	-1.9*
$C_{10}E_4/C_{12}AO$	-0.8	-2.0
SDS/LAS	-0.9	-1.2
$SDS/C_{10}MSO$	-2.4	-3.0
$SDS/C_{10}E_4$	-3.6	-2.9
$SDS/C_{10}PO$	-3.5	-3.7
ASS/NP5	-3.7	-2.9
$SDS/C_{12}AO$	-4.4	-7.2
$C_{10}SO_4/C_{10}TAB$	-13.2	-19.7

* At the n-heptane/water interface.

MSO - methyl sulphoxide
PO - dimethyl phosphine oxide
AO - amine oxide
SDS - sodium dodecyl sulphate
AOT - Aerosol OT
ASS - Alkyl sulphosuccinate
LAS - Linear alkyl benzene sulphonate

surface pressure for the mixture, it is possible to solve simultaneous equations to calculate $β^s$ and $x_1{}^s$. We thus have a means of estimating the compositions of the micelles, the monomers and either the air/water or oil/water interfaces.

Table 1 shows the results of some of these calculations. Several features can be seen in the data which are worth comment.

(1) Deviations from ideality tend to be negative, i.e. indicating net attractive interactions between the components.

(2) When the surfactant head groups of the two components are very similar in nature, the non-ideality is very small. An example of this behaviour is the mixture of a methyl sulphoxide and a phosphine oxide, both of which are so-called semipolar compounds.

(3) Anionic/nonionic mixtures show considerable non-ideality for two reasons. First the non-ionic component acts so as to partially shield the repulsion between negatively charged anionic head groups. Second, there is attraction between the components anyway due to ion dipole interactions.

(4) The greatest non-ideality is shown by anionic/cationic mixtures where strongly attractive coulombic forces dominate the interaction.

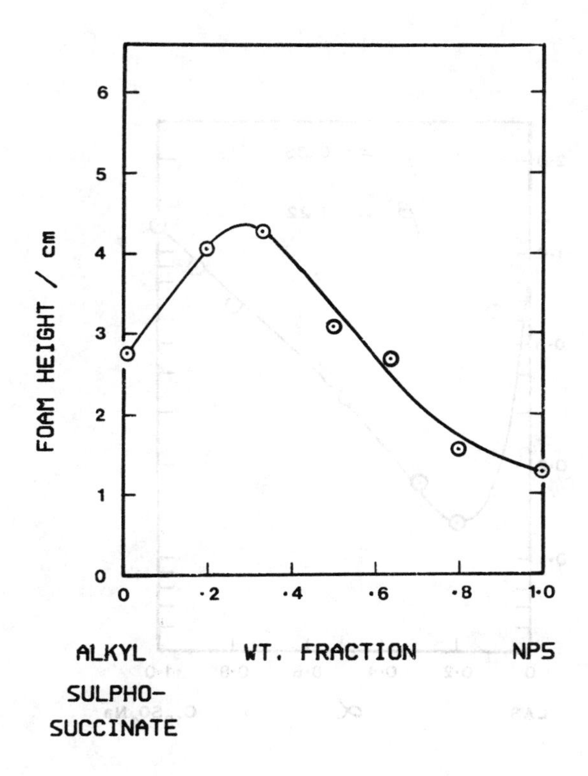

Figure 5. Variation of foam height with composition for mixtures of alkyl sulpho-succinate and nonyl phenol ethoxylate (5EO). (From von Rybinski and Schwuger, [8]).

In general the non-ideality in the interface is greater than that in the mixed micelles. Some authors use this criterion to define true synergism between the components. For example, Hua and Rosen [7] maintain that synergism in surface tension reduction effectiveness requires $|\beta^s| > |\beta|$ but examples can be found in the literature where this is not the case.

Figure 5 [8] shows the variation of foam height with composition for mixtures of alkyl sulpho-succinate and nonyl phenol ethoxylate (5EO). The results indicate a clear synergism in foaming behaviour whereas the values in Table 1 shows that $|\beta| > |\beta^s|$. One inference may be that foam stability is concerned with more than simply equilibrium surface tension reduction.

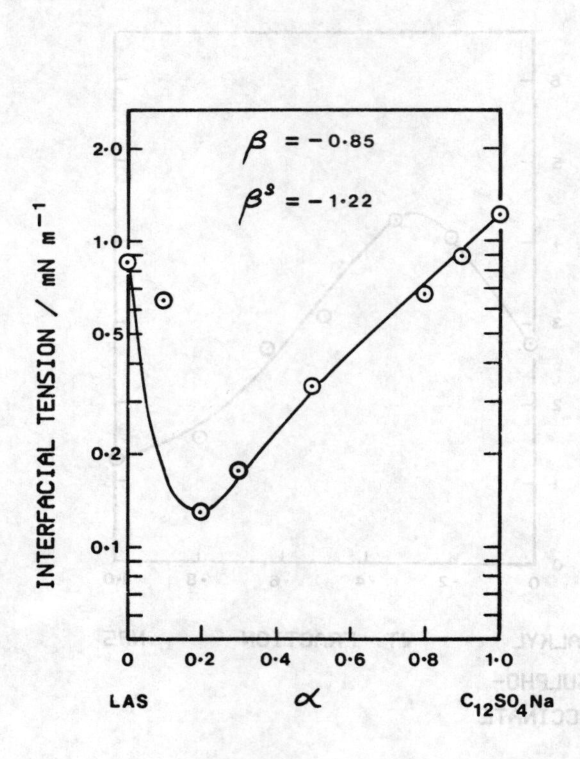

Figure 6. Tension of the olive oil/water interface for aqueous mixtures of linear alkylbenzene sulphonate (LAS) and sodium dodecyl sulphate (SDS) plotted against mole fraction of SDS. (Schwuger, [9]). Circles - experimental points; line - theoretical curve using the method of Holland [6].

Another example of synergistic behaviour is shown in Figure 6 (Schwuger [9]) where interfacial tension at the olive oil/water interface is plotted against mole fraction for mixtures of LAS and SDS. The points are measured tension values and the line is the theoretical curve obtained with $\beta = -0.85$ and $\beta^s = -1.22$ showing a good fit with experiment. Minimum interfacial tension is obtained at a mole fraction of SDS equal to 0.2. Sets of data for soil removal from wool and for dishwashing using mixtures of these two components also show the best cleaning results at this mole fraction (Figure 7).

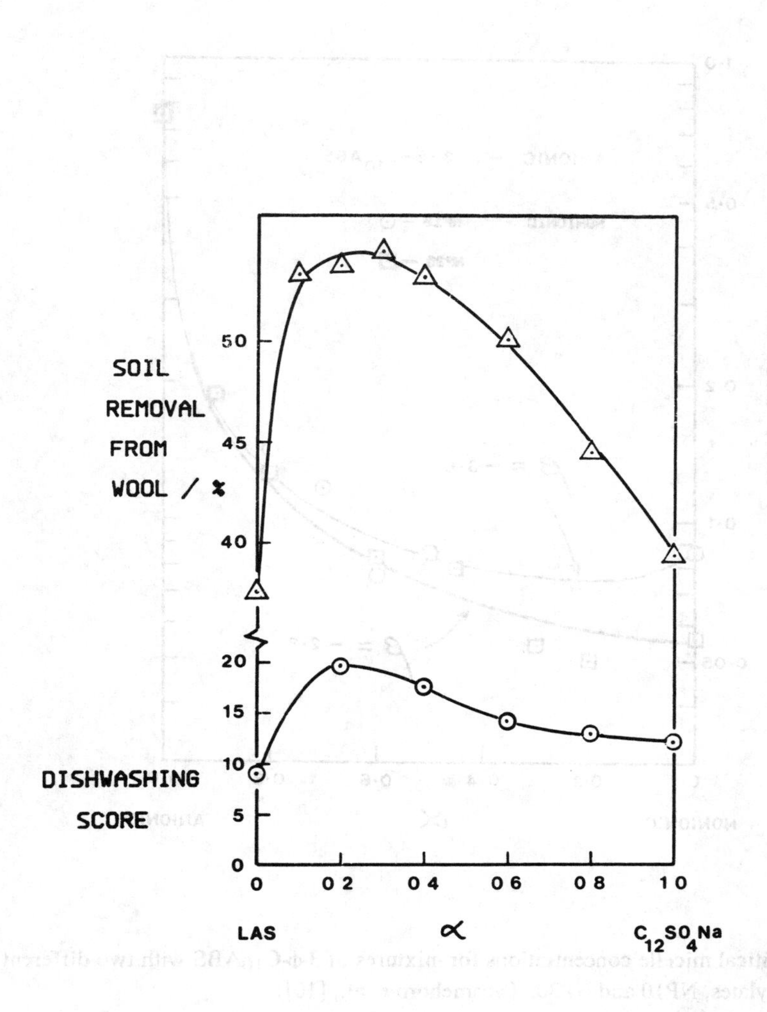

Figure 7. Cleaning performance for mixtures of linear alkylbenzene sulphonate (LAS) and sodium dodecyl sulphate (SDS) as a function of mole fraction of SDS. (Schwuger, [9]).

A final example of synergism between surfactants is concerned with adsorption at the solid/liquid interface. Figure 8 shows the CMC data for an anionic surfactant, 3-Φ-C_{10}ABS mixed with a nonionic, in one case NP10 and in another, NP30 (Scamehorn et al).[10] Best fits with the experimental data are obtained using $\beta = -2.8$ for NP30 and -3.6 for NP10, indicating stronger attractive interactions with the nonionic having a shorter EO chain. The effect which

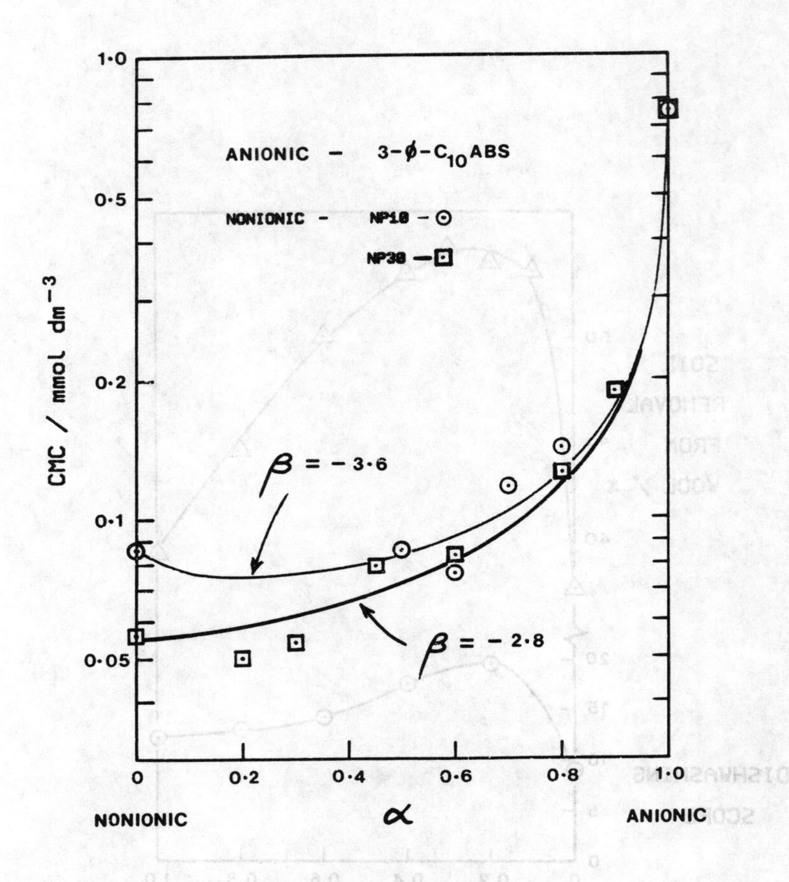

Figure 8. Critical micelle concentrations for mixtures of 3-ϕ-C_{10}ABS with two different nonyl phenol ethoxylates, NP10 and NP30. (Scamehorn et al., [10]).

these interactions have on adsorption on kaolinite is shown in Figure 9 (Scamehorn et al) [11]. The greater synergism in the case of mixtures with NP10 produces adsorption greater than 40 μmol/g, a figure which would rule out use of such mixtures for enhanced oil recovery on economic grounds, kaolinite being an important component of reservoir sandstones.

The series of examples above illustrate synergism in mixtures at the air/liquid, liquid/liquid and solid/liquid interfaces. Use of the non-ideal mixed micelle theory helps considerably by quantifying the degree of interaction which is responsible for the synergism. In all cases above the interaction is attractive. This is by no means always the case, and some very interesting systems arise in mixtures showing repulsive interactions as explained in the next section.

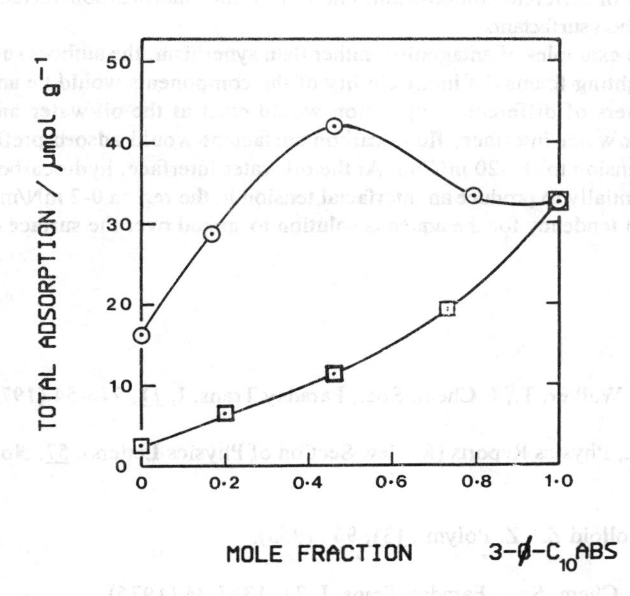

Figure 9. Adsorption of surfactants on kaolinite for mixtures of 3-ϕ-C_{10}ABS with two different nonionic surfactants. (Systems and symbols same as in Figure 8). (Scamehorn et al., [11]).

6. Positive Deviations from Ideality

In regular solution theory the parameter β is related to molecular interactions in the mixed micelle by

$$\beta = (W_{11} + W_{22} - 2W_{12})/kT \qquad (10)$$

where W_{11} and W_{22} are the energies of interaction between molecules in the pure micelle and W_{12} is the interaction energy between the two species in the mixed micelles. The entropy of mixing is assumed to be that expected for ideal mixing.

In situations where β is positive, there is antagonism between the components and the most common examples are mixtures of hydrocarbon and fluorocarbon surfactants. Shinoda and Nomura [12] studied mixtures of $C_7F_{15}COONa$ and $C_{10}H_{21}SO_4Na$. Application of mixed micelle theory with $\beta = 1.07$ provides a very good fit with the experimental CMC data.

An interesting case arises when the positive deviations become large. In regular solution theory the limit for β is +2, beyond which systems become immiscible over some of their composition range. Shinoda and Nomura [12] quote an example of $C_8F_{17}COONH_4$ and $C_{12}H_{25}SO_4NH_4$. They calculate $\beta = 2.2$ for their mixed CMC data and imply that "pseudo phase separation" is occuring in the micelles. This is interpreted as indicating the co-existence of two types of micelles of different composition, one rich in the fluorocarbon surfactant and the other rich in hydrocarbon surfactant.

Although these are examples of antagonism rather than synergism, the authors suggest that in application to fire-fighting foams the immiscibility of the components would be an advantage. Two mixed monolayers of different composition would exist at the oil/water and water/air interfaces. At the air/water interface, fluorocarbon surfactant would adsorb preferentially to depress the surface tension to 15-20 mN/m. At the oil/water interface, hydrocarbon surfactant would adsorb preferentially to produce an interfacial tension in the region 0-2 mN/m. The result would be a very great tendency for the aqueous solution to spread over the surface of (burning) oil.

7. References

1. Clint, J.H. & Walker, T., J. Chem. Soc., Faraday Trans. I, 71, 946-54 (1975).

2. Tiddy, G.J.T., Physics Reports (Review Section of Physics Letters), 57, No.1, 1-46 (1980).

3. Lange, H., Kolloid Z. - Z. Polym., 131, 96 (1953).

4. Clint, J.H., J. Chem. Soc., Faraday Trans. I, 71, 1327-34 (1975).

5. Corkill, J.M., Internal report of the Procter and Gamble Co., Cincinnati, Ohio, USA (1974). Dr Corkill's untimely death in 1974 prevented him from publishing it. The theory was subsequently published by a colleague [See Rubingh, D.N., in 'Solution Chemistry of Surfactants' (ed. Mittal, K.L.), New York: Plenum Press, 1979, Vol. 1, 337].

6. Holland, P.M., Colloids and Surfaces, 19, 171-183 (1986).

7. Hua, X.Y. and Rosen, M.J., J. Colloid Interface Sci., 125, 730-2 (1988).

8. von Rybinski, W. and Schwuger, M.J., Langmuir, 2, 639-643 (1986).

9. Schwuger, M.J., ACS Symposium Series, M.J. Rosen Ed., 253, 1 (1984).

10. Scamehorn, J.F., Schechter, R.S. and Wade, W.H., J. Dispersion Sci. Technol., 3, 261-278 (1982).

11. Scamehorn, J.F., Schechter, R.S. and Wade, W.H., J. Colloid Interface Sci., 85, 494 (1982).

12. Shinoda, K. and Nomura, T., J. Phys. Chem., 84, 365-369 (1980).

PARTITIONING OF CATIONIC SURFACTANTS BETWEEN HEPTANE AND WATER

J.M.WATES
Akzo Chemicals Ltd
Hollingworth Road
Littleborough
Lancashire, OL15 OBA.

ABSTRACT. This work is concerned with partitioning of cationic surfactants between water and a solvent. The vast majority of commercially interesting cationic surfactants are simple or derivatised amines and quaternary ammonium compounds having a mixed alkyl group based on a naturally occurring fatty acid. Many of the systems in which these products are used consist of several phases in contact which often include water and an immiscible organic phase. It is very important to know how the materials are distributed between the liquid phases as the efficiency of a product in a real system depends on it being present in sufficient quantities in the correct phase. Examples of cases where partition plays an important role in determining performance of a product include the use of corrosion inhibitors in oil pipelines, biocide and herbicide action, and the addition of agents to improve adhesion of bitumen to road surfaces.

1. Cationic Surfactants [1]

A cationic surfactant consists of amphiphilic molecules in which the hydrophilic head group is either reversibly or permanently positively charged. The vast majority of commercially-interesting cationic surfactants are simple or derivatised amines and quaternary ammonium compounds having an alkyl group based on a naturally-occurring fatty acid.

e.g. $R - NH_2$
$R - (CH_3)_2$
$R - N(CH_2CH_2OH)_2$
$R - NH(CH_2)_3NH_2$
R - N O (ring)
$R - N(CH_3)_3^+ Cl^-$

where R = coco (mainly $C_{12/14}$), tallow (mainly $C_{16/18}$, partially unsaturated) or hydrogenated tallow (mainly $C_{16/18}$, saturated). Single chain length materials are prepared by distillation of either the alkyl feed stock or the final product.

D. M. Bloor and E. Wyn-Jones (eds.),
The Structure, Dynamics and Equilibrium Properties of Colloidal Systems, 85–93.

2. Partition coefficient [2]

Many systems encountered in applications of cationic surfactants consist of two or more phases which include water and an immiscible organic liquid. It is important to know how the surfactants are distributed between the liquid phases since the efficiency of a commercial product depends on it being present in sufficient quantities in the correct phase. Moreover in certain applications the transport of material from one phase to another is necessary before the product can reach the part of the system in which it is active.

e.g.
- use of corrosion inhibitors in oil/water mixtures in oil field pipelines.
- use of adhesion agents to improve binding between bitumen and aggregate in road construction.
- application of biocides and herbicides in water-based sprays.

The distribution of the solute between the liquid phases may be defined by the partition coefficient P.

$$P = \frac{\text{concentration of solute in organic phase}}{\text{concentration of solute in aqueous phase}}$$

By this definition, a partition coefficient greater than 1 indicates that the solute is hydrophobic, preferentially entering the organic phase, whereas a hydrophilic solute has a partition coefficient less than 1.

A widely-used concept in comparing the relative affinities of nonionic surfactants for organic solvents and water, particularly in applications involving emulsification, is that of hydrophile- lipophile balance (HLB); surfactants are assigned values ranging from 0 (100% hydrophobic) to 20 (100% hydrophilic) [3]. However, the HLB scale is empirical and values are not directly related to a measurable physical property; attempts to include ionic surfactants in the scale have met with only limited success. In contrast, the partition coefficient of a solute is a well-defined thermodynamic parameter and thus is a better tool when attempting to quantify hydrophilicities of different types of compounds.

3. Measurement of partition coefficients

Partition coefficients are usually measured by shaking together the solute and the solvents and analysing one or both phases for solute content. This method was used to measure the partition coefficients of a range of aliphatic amines between heptane and 0.1M aqueous sodium hydroxide solution; an alkaline aqueous phase was used to inhibit protonation of the amine so that the partition of the neutral compound could be obtained.

The heptane phase was analysed for amine content by gas-liquid chromatography (g.l.c.). This technique has several advantages; it is sufficiently sensitive (detection limit < 10ng solute) that solutions which are almost ideally-dilute can be studied, and it enables the investigation of the partition of individual components of a mixed commercial product, eliminating the need for lengthy preparation and purification of the separate compounds. Figure 1 shows a typical chromatogram for ARMEEN C®, a primary amine commercial product having an alkyl group based on coco fatty acid.

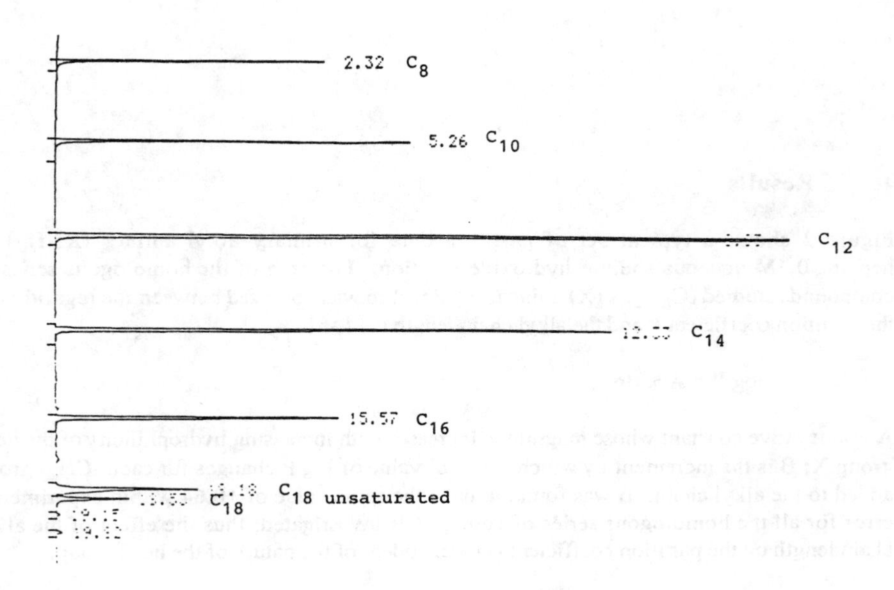

Figure 1. **Chromatogram showing typical composition of coco-alkyl primary amine (ARMEEN C ®)**

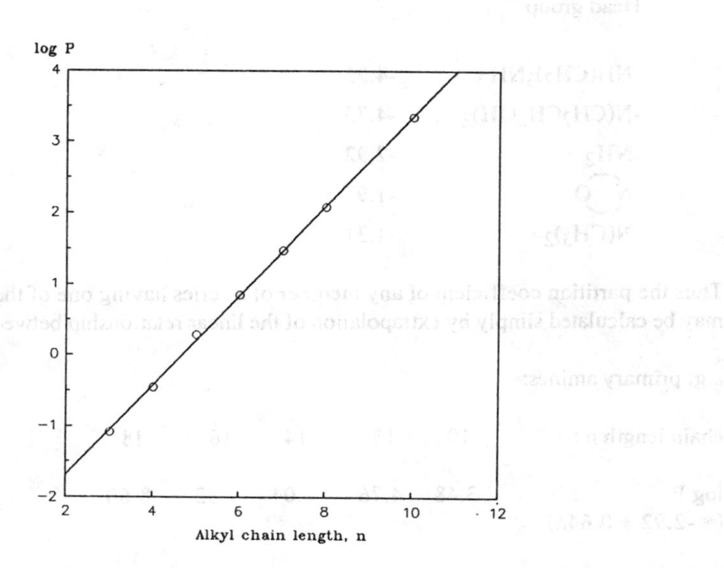

Figure 2. **Typical partition data for primary alkyl amines in heptane/0.1 M aq NaOH.**

4. Results

Figure 2 shows a typical set of partition data for primary alkyl amines (RNH_2) in heptane/0.1M aqueous sodium hydroxide solution. For each of the homologous series of compounds studied ($C_nH_{2n+1}X$) a linear relationship was observed between the logarithm of the partition coefficient P and the alkyl chain length n [4, 5].

$$\log P = A + Bn$$

A is a negative constant whose magnitude increases with increasing hydrophilicity of the head group X; B is the increment by which the total value of log P changes for each $-CH_2-$ group added to the alkyl chain. B was found to have the same value of +0.64 within experimental error for all the homologous series of compounds investigated; thus the effect of the alkyl chain length on the partition coefficient is independent of the nature of the head group.

5. Calculation of partition coefficients from experimental results

Values of A, the contribution of the hydrophilic head group to log P, are given below in decreasing order of hydrophilicity.

Head group	A
$-NH(CH_2)_3NH_2$	-4.95
$-N(CH_2CH_2OH)_2$	-4.73
$-NH_2$	-2.92
-N O (morpholino ring)	-1.97
$-N(CH_3)_2$	-1.31

Thus the partition coefficient of any member of a series having one of the above head groups may be calculated simply by extrapolation of the linear relationship between log P and n.

e.g. primary amines:

chain length n :	10	12	14	16	18
log P : (= -2.92 + 0.64n)	3.48	4.76	6.04	7.32	8.60

It is also possible to predict the partition of a mixed commercial product of known composition if it is assumed that every component partitions independently of the others in the mixture. The distribution of the individual amines between the heptane and aqueous phases is calculated and the total amount of material in each solvent obtained by addition, taking into consideration the relative proportions of the components in the original product.

6. Effect of pH on partition

It has been shown that positively-charged species (protonated amines and quaternary ammonium compounds) partition almost completely into the aqueous phase of a heptane/water mixture [6]. The fraction of an amine which becomes protonated in aqueous solution is a function of the ambient pH; thus the overall partitioning behaviour of the amine also depends on pH. The amounts of protonated (A_w) and unprotonated (B_w) amine in the aqueous phase of a sample are given by:-

$$A_w = \frac{C(v_1 + v_2)}{v_1\ (1/R + 1)} \qquad B_w = \frac{C(v_1 + v_2)}{(Pv_2 + v_1)\ (R + 1)}$$

where

$$R = \frac{\text{amount of protonated amine}}{\text{amount of unprotonated amine}}\ ;\ \log R = Pk_a = pH - \log\ (P\frac{v_2}{v_1} + 1)$$

C = total concentration of amine

v_1, v_2 = volume of aqueous and solvent phases respectively

P = partition coefficient of unprotonated amine

The amount of amine in the heptane phase is found by difference (= $C(v_1 + v_2) - (A_w + B_w)$). Thus from knowledge of the partition coefficients and the pKa values of the neutral amines, it is a simple procedure to calculate the partition behaviour of a pure or mixed product at any pH. The results of carrying out such a calculation for the coco-alkyl primary amine product ARMEEN C ® are illustrated in Figure 3.

7. Ethoxylated amines

Ethoxylated amines are probably the most versatile group of cationic surfactants, since both their alkyl portion and head group may be varied to modify the properties of the product. They are made by the reaction of a primary amine with a variable amount of ethylene oxide which yields a product having the general formula

$$R-N\begin{matrix} \diagup (CH_2CH_2O)_xH \\ \diagdown (CH_2CH_2O)_yH \end{matrix} \qquad \text{where } (x + y) \geq 2$$

This method of preparation results in random addition of ethylene oxide groups to the amine molecules which leads to a statistical distribution in the degree of ethoxylation in the product molecules. The chromatogram in Figure 4 illustrates the composition of the product formed by reacting coco-alkyl primary amine with ethylene oxide in the ratio 1:5 (ETHOMEEN C/15 ®); the peaks correspond to components having different alkyl chain lengths and degrees of ethoxylation such that $2 \leq (x + y) \geq 10$. For the sake of clarity only the predominant C_{12}-alkyl components are labelled.

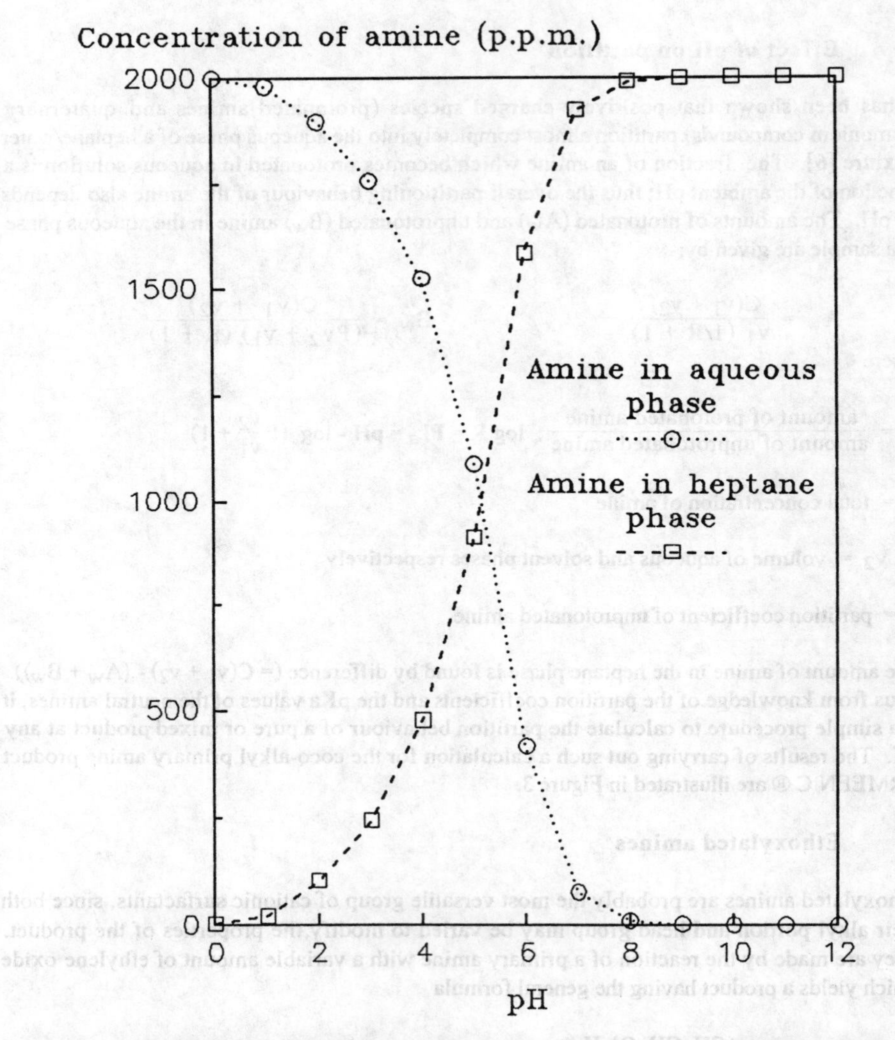

Figure 3. Calculated dependence on pH of the partitioning of ARMEN C ® between heptane and water.

8. Partitioning of ethoxylated amines

The effect of the ethylene oxide content of a compound on its partition was measured by studying the distribution of the individual C_{12}-alkyl components of ETHOMEEN C/15 ® between heptane and 0.1M aqueous sodium hydroxide. It was found that the incremental

contribution of one -(CH_2CH_2O)- unit to log P was -0.47; as might be expected, increasing the degree of ethoxylation makes the amine molecules more hydrophilic and thus decreases the extent of partition into the heptane phase.

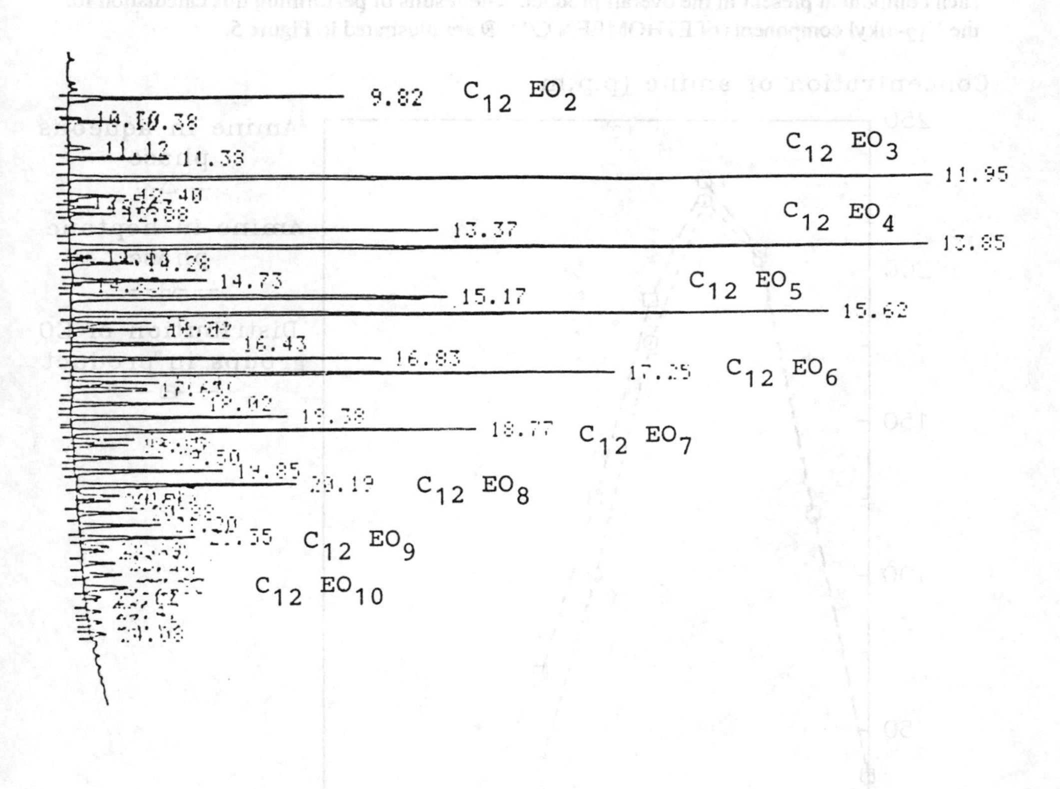

Figure 4. **Chromatogram showing typical composition of the product formed by reacting coco-alkyl primary amine with 5 moles of ethylene oxide (ETHOMEEN C/15 ®)**

In principle the overall partition of a commercial ethoxylated amine product can be determined by adding together the amounts of each component measured by means of g.l.c. in the heptane phase. However in practice the complexity of the chromatograms for the multi-

component mixtures make such a procedure time-consuming and the results inaccurate; thus it proved more convenient to calculate the partition behaviour from theoretical considerations as follows:

The composition of the product was estimated from the initial ratio of primary amine:ethylene oxide and the alkyl chain length distribution of the amine; it was assumed that random addition of ethylene oxide to the amine molecules resulted in a modified Poisson distribution in the degree of ethoxylation of the head groups [7]. The partition coefficients of the various components were obtained by extrapolation of the linear relationships between log P and alkyl chain length or degree of ethoxylation. The distribution of product between the heptane and aqueous phases was then calculated, allowance being made for the percentage of each component present in the overall product. The results of performing this calculation for the C_{12}-alkyl components of ETHOMEEN C/15 ® are illustrated in Figure 5.

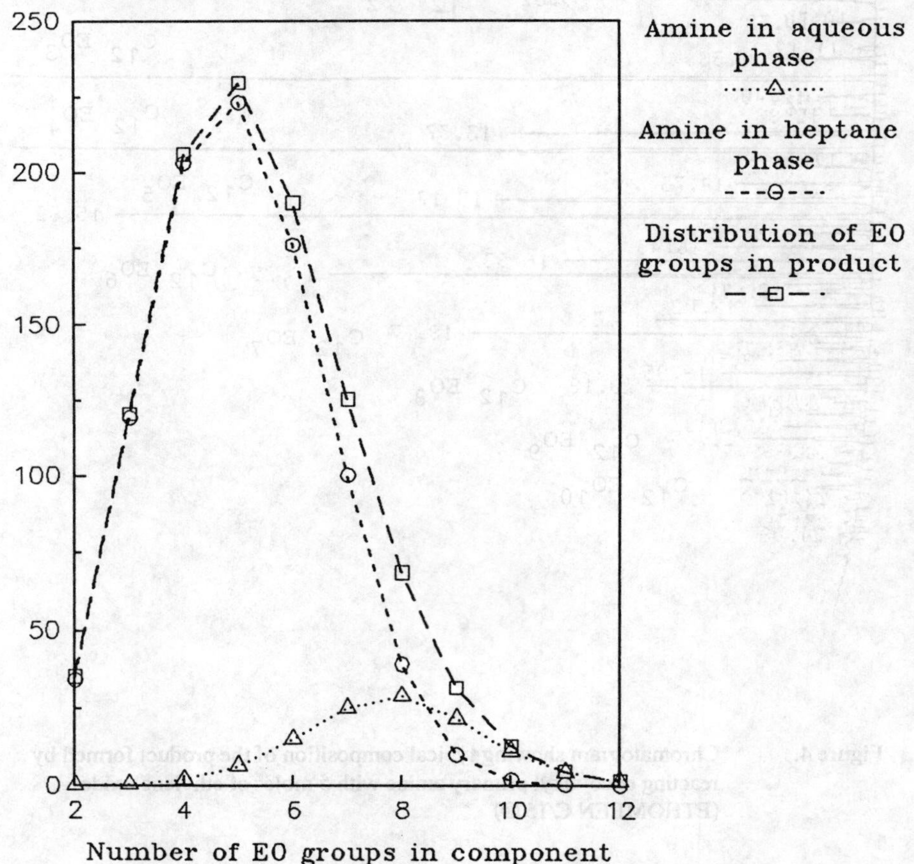

Figure 5. Calculated partition behaviour of C_{12}-alkyl components of ETHOMEEN C/15 ® in heptane/0.1 M NaOH.

9. Limitations of partition calculation

The procedures described above for calculating the partition behaviour of cationic surfactants have been found to work well at very low concentrations of amine; the results of the calculation agree well with experimental observation.

However, at higher concentrations the calculation consistently underestimates the amount of amine in the aqueous phase. This error is a result of micellisation of the surfactant in the aqueous phase, which introduces an equilibrium competing with partition:

monomer (heptane) $\rightleftarrows$ monomer (aqueous) $\rightleftarrows$ micelle (aqueous)

The free energies of micellisation and partition into a hydrocarbon are comparable in value for -CH_2- groups in the alkyl chain (approximately -3.4 and -3.7 kJ/mol respectively), while for a polar head group the former process is energetically more favourable [8]. Thus micellisation is a significant factor in determining the overall partition of cationic surfactants; unfortunately, attempts to allow for the effect of micellisation in the calculations have as yet proved unsuccessful.

Acknowledgement

The methods for analysing amine solutions in heptane by means of g.l.c. were developed by G.E.Varcoe, Analytical Department, Akzo Chemicals Ltd., Littleborough.

References

1. "Cationic Surfactants", ed. E. Jungermann, Marcel Dekker, Inc., New York, 1977.

2. A. Leo, C. Hansch and D. Elkins, Chem. Rev., 71 (6), 1971, 525-616.

3. W.C. Griffin, Offic. Dig. Federation Paint and Varnish Production Clubs, 28, 1956, 446-55.

4. A.H. Beckett and A.C. Moffat, J. Pharm. Pharmac., 21 (Supplement), 1969, 144S-150S.

5. S.S. Davis, T. Higuchi and J.H. Rytting, J. Pharm. Pharmac., 24 (Supplement), 1972, 30P-46P.

6. P. Dallat, J.C. Colleter, J.J. Girand and B. Penicant, C.R. Acad. Sc. Paris, 284, 1977, 125-8.

7. S.A. Miller, B. Bann and R.D. Thrower, J. Chem. Soc., 1950, 3623-8.

8. P. Mukerjee, Adv. Coll. Int. Sci., 1, 1967, 241-75.

ADSORPTION OF SODIUM DODECYL SULFATE AND 1-BUTANOL ON SOLID ALUMINIUM OXIDE

H. HØILAND, A.M. BLOKHUS, T. LIND, AND A. SKAUGE
Department of Chemistry
University of Bergen
Allegt 41
N-5007 Bergen
NORWAY

ABSTRACT. The adsorption of sodium dodecyl sulfate (SDS) and 1-butanol on solid aluminium oxide has been studied at 298.15 K by the batch method. It appears that butanol does not adsorb on its own, but when SDS is present, it will. Though SDS adsorption is preferred, one can adsorb more butanol by increasing its bulk concentration. By comparing the adsorbed layer of SDS and butanol with a lamellar lyotropic liquid crystalline phase, a D-phase, one can estimate the areas occupied by each component on the solid surface. These appear to be somewhat larger than in a D-phase. Adsorption of SDS in Winsor III systems has also been studied. It appears that the adsorption is highest from the middle phase and lowest from the upper phase microemulsion.

1. Introduction

Adsorption of surfactants at the liquid/solid interface is of interest especially for enhanced oil recovery processes. During surfactant flooding it is important to keep control of the chemical loss. This depends on interactive forces between a variety of components; surfactant, cosurfactant, inorganic salts, oil, and reservoir rock. Though there are several processes by which chemicals may be lost, adsorption is a key factor, and adsorption isotherms will be fundamental information for the understanding of the mechanisms involved.

Surfactant adsorption isotherms on mineral oxide particles are generally S-shaped, and they have been described in terms of four distinct regions [1-5]. In the dilute region where the surface coverage is low, the adsorption isotherm obeys Henry's law. As the surfactant concentration increases a distinct deviation from ideal behaviour is observed as the adsorption density increases markedly. At even higher surfactant concentrations the adsorption isotherm levels off, and it finally reaches a point after which the adsorption density remains practically constant. The plateau is generally reached at a surfactant concentration equal to the critical micelle concentration, c.m.c., though it is possible to saturate the solid surface at lower surfactant concentrations by keeping a very high liquid to solid ratio.

A characteristic feature of surfactant adsorption is that lateral forces of interaction between the adsorbed molecules are of significance. Their participation is considered to be a major factor accounting for the high adsorption density observed. These forces of interaction are thought to be the same forces that are responsible for micelle formation. As a result the adsorbed surfactant may aggregate as bilayered structures, often termed hemimicelles [5-8]. From studies of adsorption of

D. M. Bloor and E. Wyn-Jones (eds.),
The Structure, Dynamics and Equilibrium Properties of Colloidal Systems, 95–102.

homologous series of surfactants the free energy of adsorption per CH_2 group has been determined, and it compares well with the corresponding value for micelle formation [9,10].

In this paper we present data for the adsorption of sodium dodecyl sulfate (SDS) on aluminium oxide with and without added 1-butanol. Adsorption from the surfactant rich phase of Winsor two and three phase systems has also been studied. In describing the packing of molecules on the solid surface, a direct parallel has been drawn to the geometries found in lyotropic liquid crystalline phases with a lamellar geometry (generally termed a D-phase).

2. Experimental

2.1. MATERIALS

Sodium dodecyl sulphate was obtained from BDH, "specially pure". Sodium chloride was from Merck, p.a. quality. 1-butanol and cyclohexane was obtained from Fluka, puriss quality. C^{14}-labelled 1-butanol was from Du-Pont. SDS was dried in vacuo at 50°C and NaCl at 120°C. The compounds were used without further purification.

The adsorbent (Al_2O_3) was α-alumina obtained from Ventron. The particle size was given as 40 µm. BET measurements showed a specific surface area of 103 m^2/g. The aluminium oxide was dried at 105°C before use.

2.2. METHODS

The adsorption measurements were carried out at 25°C from aqueous solutions at constant ionic strength, obtained by adding 0.3 mol dm^{-3} NaCl to the water. The adsorption isotherms were obtained by using a batch method. Approximately 2.5g Al_2O_3 was mixed with 25 ml surfactant solution in a centrifuge tube with screw-top. Mixing was carried out by rotating the tubes in a water bath at constant temperature for 24 hours.

After mixing the solid/liquid suspension was centrifuged and the equilibrium solution was removed for further analysis. The equilibrium concentration of SDS was determined by titration with Hyamin 1622, a cationic surfactant, in a two-phase system of water and chloroform as described by Reid et al. [11]. The equilibrium concentration of 1-butanol was determined by means of a liquid scintillation technique [12].

2.3. WINSOR III SYSTEMS

By mixing water, NaCl, SDS, 1-butanol, and cyclohexane one obtains a system that consists of two or three phases as described by Winsor [13]. In this work we mixed equal amount of salt water and cyclohexane, i.e. originally 44.997% by weight of each. The salt water contained 5.0% NaCl. The original amounts of SDS and 1-butanol were also equal, 5.003% by weight. This resulted in a Winsor II-system, i.e. the surfactant was present in the lower, aqueous phase. By increasing the butanol content to 5.567% keeping the relative amounts of the other components constant, a three phase system, Winsor III, was obtained. The middle phase contains most of the surfactant. By increasing the butanol content further to 6.00 and 7.00% the system becomes a Winsor II+, i.e. the surfactant resides in the oleic phase. By separating the phases one can carry out adsorption studies on the phase containing the SDS. This was done as described above.

3. Results

Figure 1 shows the adsorption isotherms of SDS on Al_2O_3. When 1-butanol is added, the saturation value of the SDS adsorption density is lowered. The equilibrium concentration of SDS at which the saturation value is reached also decreases. This just reflects the effect of butanol on the c.m.c. of SDS.

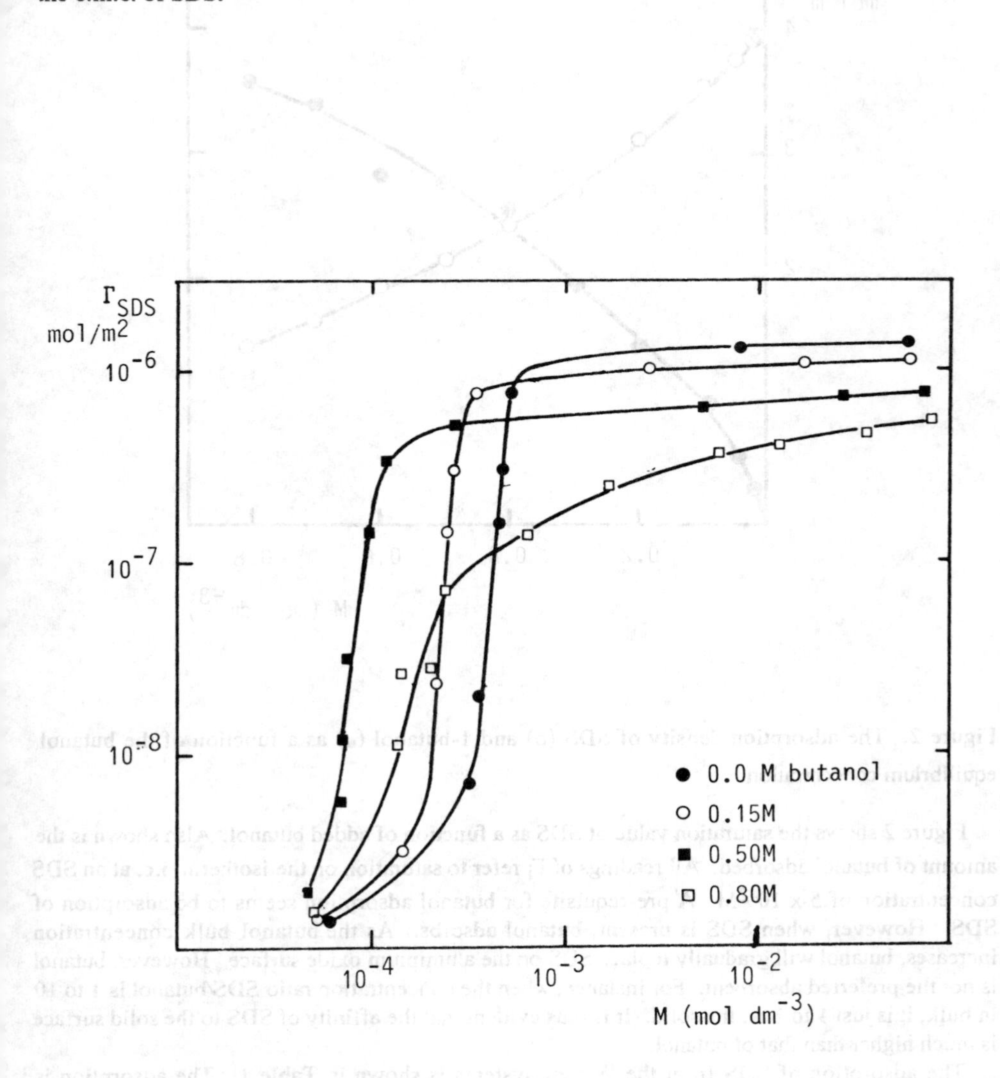

Figure 1. Adsorption isotherms of SDS on Al_2O_3 with varying contents of 1-butanol as a function of the SDS equilibrium concentration.

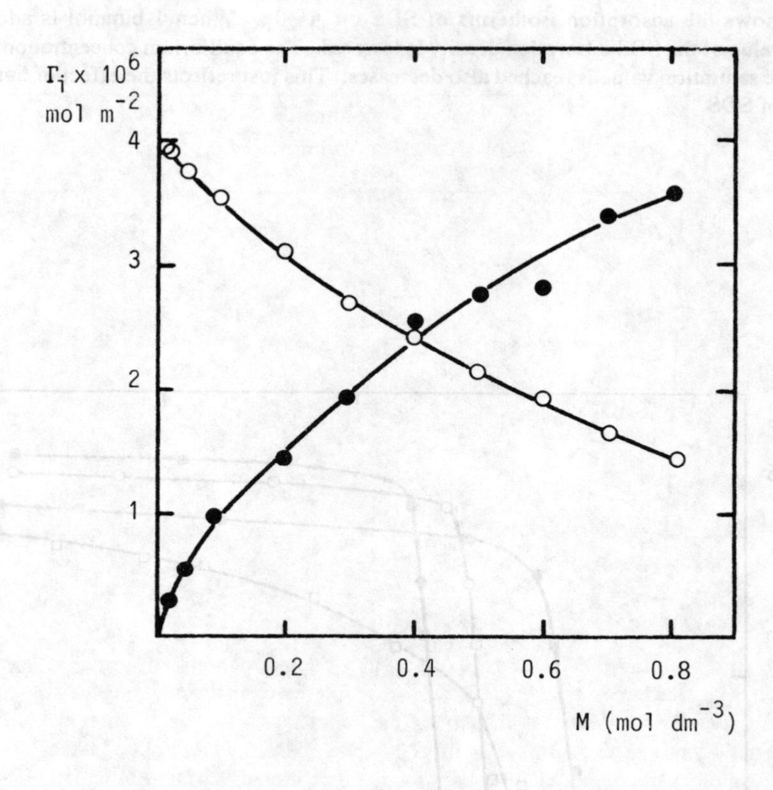

Figure 2. The adsorption density of SDS (o) and 1-butanol (●) as a function of the butanol equilibrium concentration.

Figure 2 shows the saturation value of SDS as a function of added butanol. Also shown is the amount of butanol adsorbed. All readings of Γ_i refer to saturation on the isotherm, i.e. at an SDS concentration of 5 x 10^{-2}M. A pre-requisite for butanol adsorption seems to be adsorption of SDS. However, when SDS is present, butanol adsorbs. As the butanol bulk concentration increases, butanol will gradually replace SDS on the aluminium oxide surface. However, butanol is not the preferred absorbent. For instance, when the concentration ratio SDS/butanol is 1 to 10 in bulk, it is just 1 to 1 on the solid. It is thus evident that the affinity of SDS to the solid surface is much higher than that of butanol.

The adsorption of SDS from the Winsor systems is shown in Table 1. The adsorption is highest from the middle phase microemulsion, and lowest from the upper phase microemulsion, apparently decreasing with the butanol content.

Table 1. Concentration of SDS before and after adsorption, and the adsorption density of SDS of the surfactant-rich phase of two and three phase systems, Winsor systems. The system contains equal weights of cyclohexane and salt water. (The water contains 5.0% NaCl).

	content mass %		concentration $mol\ dm^{-3}$ before	after	Γ_i $mol\ m^{-2}$
SDS	5.003	II-	0.192	0.097	2.3
Butanol	5.003				
SDS	5.002	III	0.172	0.060	2.7
Butanol	5.567				
SDS	5.000	II+	0.174	0.096	1.9
Butanol	6.000				
SDS	5.000	II+	0.194	0.136	1.4
Butanol	7.000				

3. Discussion

Normally when discussing mechanisms of adsorption complementary information is obtained from zeta potential measurements. In our case we are interested in the competitive adsorption of an anionic surfactant and a polar cosurfactant. In the bulk phase mixtures of water alcohol and SDS will eventually produce a lyotropic liquid crystalline phase, a lamellar D-phase. The structure of a lyotropic liquid crystalline D-phase is built up of a repetition of surfactant/alcohol bilayers and intervening water layers. It thus seems sensible to draw a parallel between the surfactant/alcohol adsorbed layers on solid and the surfactant/alcohol layers of a D-phase.

The surface charge density of the interface of the lamellar D-phase depends on the total area ascribed to the anionic sulfate and the nonionic hydroxylic groups. If the repetition distance between the bilayers is d, one obtains the following equation [14,15]:

$$1/\rho = (d\, A^{tot} N_A w_s)/2M_s \qquad (1)$$

Here ρ is the density, N_A is the Avogadro's number, w_s is the weight fraction of surfactant, and M_s its molar mass. Experimental values of ρ and d can be found in the literature for the system water-sodium octanoate-decanol [16]. The total area, A^{tot}, can thus be found from Eq. (1).For mixed surfactant-cosurfactant layers A^{tot} is the sum of the area required by the surfactant and the alcohol molecules. It is reasonable to assume that this total area will depend on the molar ratio alcohol to surfactant. If so, the following equation should hold:

$$A^{tot} = A^s + (n_{ROH}/n_s)A^{ROH} \qquad (2)$$

This was observed to hold for the system water-sodium octanoate-decanol, and the area ascribed to the surfactant and alcohol could be calculated. The result was 0.30 nm^2 for sodium octanoate and 0.22 nm^2 for decanol [14,15].

If we assume that the total area of the solid aluminium oxide surface is covered by a monolayer of SDS and butanol, we will have a mixed layer of SDS and alcohol that can be compared to a D-phase layer. We can define a total area as:

$$A^{tot} = 1/N_A \Gamma_i \qquad (3)$$

Γ_i is the adsorption density of species i, and N_A is Avogadro's number.

The adsorption density can be calculated from:

$$\Gamma_i = \Delta c_i \; V^o/m\,a \qquad (4)$$

Here Δc_i is the concentration difference due to adsorption, V^o the dilution volume, m the mass of the solid material, and a its specific area. The total area is built up by SDS and butanol, and we can test an equation similar to Eq. (2). The ratio n_{ROH}/n_{SDS} will refer to the ratio of butanol to SDS of the adsorbed phase. The result is seen in Fig. 3.

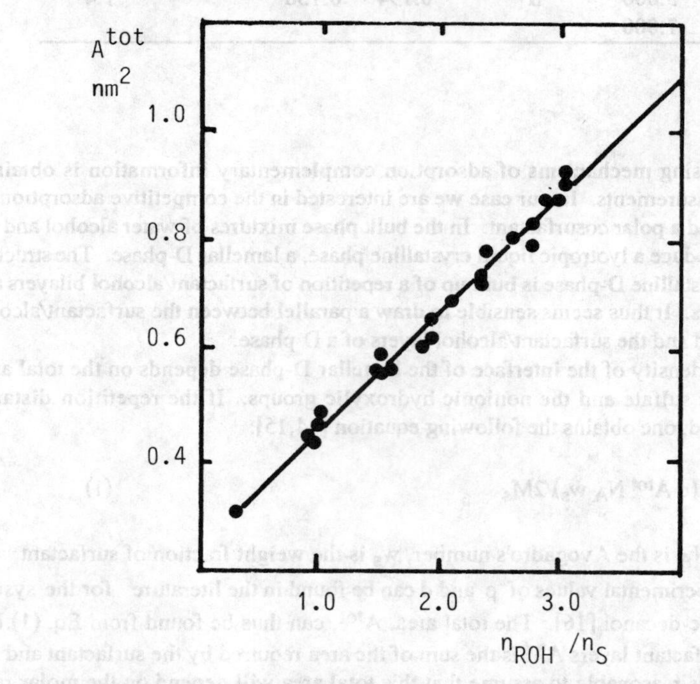

Figure 3. The total area per molecule on Al_2O_3 as a function of the molar ratio butanol to sodium dodecyl sulfate.

A straight line is observed and molecular areas can thus be calculated. For SDS it is 0.39 nm^2 and for butanol 0.30 nm^2. This comparison suggests that the area requirements are largest in the adsorbed state which is consistent with the hypothesis that adsorbed molecules will be slightly horizontally orientated at the solid surface, more so than for a D-phase [17].

Since we have not carried out a detailed analysis of the microemulsion phases, it is difficult to compare adsorption from microemulsion phases with the results from SDS-butanol solutions above. The II- phase is generally thought to consist of micelles, and one would expect SDS adsorption close to that observed without added oil, as indeed it does. It is somewhat surprising that SDS adsorption from the Winsor III middle phase is higher than from the II- phase. However, as shown above, one has to take into account the competitive adsorption of butanol, and this has not been analysed for the microemulsion systems yet. The upper phase II+ microemulsion presumably is closed structures, reversed micelles in an oil continuous medium. SDS still adsorbs on alumina, but as could be expected, the adsorption is least from this phase. It seems to decrease with alcohol content which is in accordance with findings for competitive adsorption.

Acknowledgement

This work has been supported by the National SPOR program.

4. References

1. Somasundaran, P. and Fuerstenau, D.W. (1966) J. Phys Chem. 70, 90.

2. Wakamatsu, T. and Fuerstenau, D.W. (1968) Adv. Chem. Ser. 79, 161.

3. Dick, S.G., Fuerstenau, D.W. and Healy, T.W. (1971) J. Colloid Interface Sci. 37, 595.

4. Rosen, M.J. and Nakamura, Y. (1977) J. Phys. Chem. 81, 873.

5. Scamehorn, J.F., Schecter, R.S., and Wade, W.H. (1982) J. Colloid Interface Sci. 85, 463.

6. Fuerstenua, D.W. (1971) 'The Chemistry of Biosurfaces' Vol 1 in M.L. Hair (ed.) Marcel Dekker, London, pp. 143.

7. Chander, S., Fuerstenau, D.W., and Stigter, D. (1983) 'Adsorption from Solution' in R.H. Ottewill, C.H. Rochester, and A.L. Smith (Eds.) Academic Press, New York, pp. 97.

8. Somasundaran, P. and Goddard, E.D. (1979) 'Modern Aspects of Electrochemistry' Vol. 13 in J. O'M. Bockris and B.E. Conway (Eds.) Plenum Press, New York pp. 207.

9. Trogus, F.J., Schecter, C.S., and Wade, W.H. (1979) J. Colloid Interface Sci. 70, 293.

10. Tamamushi, B. and Tamaki, K. (1957) Proc. 2nd. Int. Congr. Surface Activity, Butterworths, London, 3, 449.

11. Reid, V.W., Longman, G.F., and Heinert, E. (1968) Tenside 5, 90.

12. Choppin, G.R. and Rydberg, J. (1980) 'Nuclear Chemistry, Theory and Applications' Pegamon Press, New York, pp. 90.

13. Winsor, P.A. (1968) Chem. Rev. 68, 1.

14. Danielsson, I., Friman, R., and Sjøblom, J. (1982) J. Colloid Interface Sci. 85, 442.

15. Friman, R., Danielsson, I., and Stenius, P. (1982) J. Colloid Interface Sci. 86, 501.

16. Fontell, K., Mandell, L., Lehtinen, H., and Ekwall, P. (1968) Acta Polytech. Scand. Chem. Lett. Metall. Ser. III, 74.

17. Partyka, S., Rudzinski, W., Brun, B., and Clint, J.H. (1989) Langmuir 5, 297.

ASSOCIATION MODELS FOR PHENOTHIAZINE DRUGS IN AQUEOUS SOLUTION

D. ATTWOOD[1], E. BOITARD[2], J.P. DUBÈS[2], H. TACHOIRE[2], V. MOSQUERA[3] and V. PEREZ VILLAR[3]

[1]*Department of Pharmacy, University of Manchester, Manchester, M13 9PL, U.K.*
[2]*Laboratoire de Thermochimie, Université de Provence, F-13331 Marseille, Cedex 3, France.*
[3]*Departmento de Fisica de la Materia Condensada, Universidad de Santiago, Santiago de Compostela, Spain.*

ABSTRACT. The association characteristics of the phenothiazine drugs, promethazine and chlorpromazine hydrochloride in water and in solutions of added electrolyte have been examined by time-average light scattering, vapour pressure osmometry and heat conduction calorimetry. The concentration dependence of the osmotic coefficients in water could be quantitatively described by a mass action model of association but low or negative values for ion interaction coefficients suggested limited association before the first critical concentration, c_1. Discontinuities in light scattering data at a second critical concentration, c_2, indicated restructuring of the aggregates at high solution concentration (approximately 0.2 mol kg^{-1}). Calorimetric measurements on solutions of promethazine in the presence of high concentration of electrolyte (in excess of 0.1 mol dm^{-3} NaCl) have shown the formation of trimers or tetramers by a continuous association process below the first critical concentration. Subsequent growth of these primary aggregates with increase of concentration above c_1 in the presence of high concentration of electrolyte is by the stepwise addition of monomers according to a co-operative association scheme.

1. Introduction

The mode of association of amphiphilic molecules is thought to be strongly influenced by the nature of the hydrophobic group [1]. The flexibility of the hydrocarbon chains of typical surfactants is conducive to the formation of spheroidal micelles, whilst rigid planar molecules such as cationic dyes associate by face-to-face stacking in a continuous association pattern. Whereas in micellar association there is a predominance of aggregates of an energetically preferred size, in continuous association the products of association are generally multimers with a broad size distribution.

Many drugs are amphiphilic and self-associate in aqueous solution [2]. Their hydrophobic moieties are usually aromatic and of limited flexibility compared with the hydrocarbon chains of typical surfactants. On the other hand, they differ from the cationic dyes in that they generally

D. M. Bloor and E. Wyn-Jones (eds.),
The Structure, Dynamics and Equilibrium Properties of Colloidal Systems, 103–129.

possess charge-bearing side chains of appreciable length attached to their ring systems, which limits their ability to form stacks in solution. Drugs, therefore, form an interesting group of compounds, the study of which provides insights into the factors influencing the association properties of amphiphilic compounds.

Previous studies [2] have established a micellar mode of association for a large number of drugs possessing a diphenylmethane hydrophobic group. Rotation around the central carbon atom of diphenylmethane clearly hinders stacking of these molecules. Linking of the two phenyl rings of such molecules to form a rigid fluorene group, changed the association from micellar to a continuous association pattern [3]. On the basis of such studies it would be expected that drugs based on the rigid, almost planar, phenothiazine ring system would associate by the stacking of monomers giving rise to a continuous association pattern. Although the nmr studies of Florence and Parfitt [4] have provided evidence of stacking in solution, many workers have reported abrupt discontinuities in the concentration-dependence of the physicochemical properties of aqueous solutions of phenothiazine drugs which they have identified with the critical micelle concentrations (CMC) of typically micellar systems [2].

This apparent anomaly between the expected and observed properties of aqueous solutions of the phenothiazine drugs has led us to further investigate their association characteristics. This paper reports on studies of the thermodynamic properties of two phenothiazine drugs, promethazine (I) [10-(2-dimethylaminopropyl) phenothiazine] and chlorpromazine (II) [2-chloro-10-(3-dimethylaminopropyl) phenothiazine] in water and in the presence of added electrolyte.

R
N
X
S

I : R = $-CH_2CH(CH_3)N(CH_3)_2$; X = H

II : R = $-CH_2CH_2CH_2N(CH_3)_2$; X = Cl

2. Results and Discussion

2.1 ASSOCIATION IN AQUEOUS SOLUTION

2.1.1 *Vapour Pressure Measurements.* The concentration dependence of the rational osmotic coefficient, g, as determined from vapour pressure measurements [5] on aqueous solutions of the drugs at 303K using a Knauer osmometer (model 11.00), shows discontinuities at well defined critical concentrations (Fig 1). For both drugs, these critical concentrations are in good agreement with values from light scattering techniques (see Table 1).

TABLE 1. CMC values and parameters derived from simulation of the concentration dependence of osmotic coefficients

	Promethazine	Chlorpromazine
CMC[a] mol kg^{-1}	0.050 (0.058)[b]	0.021 (0.022)[c]
n[a]	7(12)[b]	8 (12)[c]
log K	15.4	22.5
$B_{1\gamma}$ kg mol^{-1}	-1.6	0.2
$B_{n\gamma}$ kg mol^{-1}	1.4	4.9
δ	0.47	0.40
σ^d	0.014	0.010

[a] Literature values given in parenthesis. [b] Reference 6. [c] Reference 7
[d] Root-mean-square deviation of fit.

Mean ion activity coefficients, $\gamma_\pm$, derived from eq.1 are plotted as a function of solution molality in Fig 2.

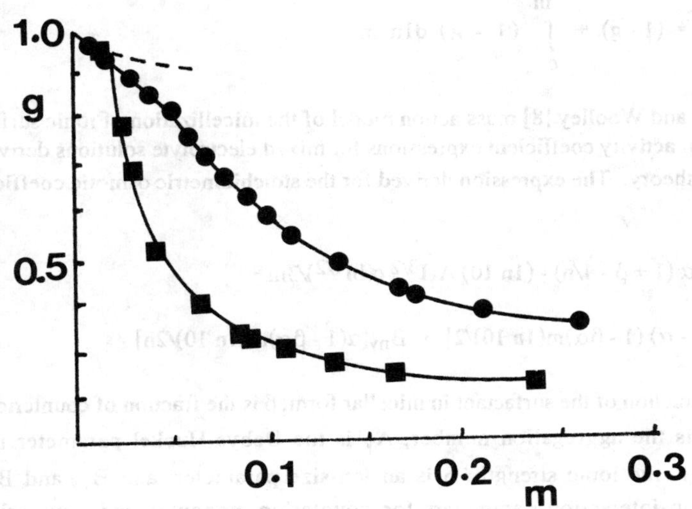

Figure 1 Variation of the rational osmotic coefficient g with molality m for (●) promethazine, and (■) chlorpromazine at 30°C: Continuous lines calculated from eq 2 and 5. (---) values for NaCl.

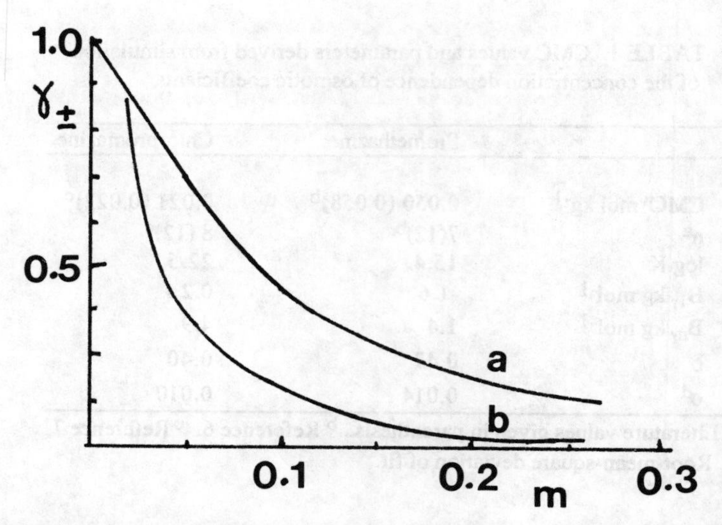

Figure 2 Molal activity coefficients, $\gamma_{\pm}$ as a function of molality, m, for (a) promethazine and (b) chlorpromazine at 30°C

$$-\ln \gamma_{\pm} = (1 - g) + \int_{0}^{m} (1 - g)\, d\ln m \tag{1}$$

The Burchfield and Woolley [8] mass action model of the micellization of ionic surfactants uses the Guggenheim activity coefficient expressions for mixed electrolyte solutions derived from the Debye-Huckel theory. The expression derived for the stoichiometric osmotic coefficient, ϕ, for the solution is

$$\phi = 1 - \alpha\,(1 + \beta - 1/n) - (\ln 10)\, A_{\gamma} I^{3/2}\, \sigma(bI^{1/2})/3m +$$

$$B_{1\gamma}\,[(1 - \alpha)\,(1 - \beta\alpha)m(\ln 10)/2] + B_{n\gamma}[\alpha(1 - \beta\alpha)m(\ln 10)/2n] \tag{2}$$

where α is the fraction of the surfactant in micellar form, β is the fraction of counterion bound to the micelle, n is the aggregation number, A_{γ} is the Debye-Huckel parameter for activity coefficients, I is the ionic strength, b is an ion-size parameter, and $B_{1\gamma}$ and $B_{n\gamma}$ are the Guggenheim ion interaction parameters for counterion-monomer and counterion-micelle interactions, respectively. ϕ is related to the rational osmotic coefficient, g, by the expression.

$$g \ln (x_1) = -(\upsilon_2 m_2 M_1/1000)\phi \tag{3}$$

where M_1 is the molar mass of the solvent, m_2 is the molality of the solute, x_1 is the mole fraction of solvent and υ_2 is the number of ions produced by one molecule of surfactant.

In the application of eq 2 to the osmotic coefficient data, a value of unity was assumed for b as suggested by Guggenheim [9] for simple electrolytes, β values were those determined previously from light scattering measurements, and a value of 0.5150 $kg^{1/2}mol^{-1/2}$ was used for A_γ. The effective ionic strength, I, of the micellar electrolyte was calculated from

$$I = \sum \frac{m_i z_i^2}{2} = [2(1-\alpha) + n\delta^2(1-\beta)^2\alpha + (1-\beta)\alpha]m/2 \tag{4}$$

where δ is a shielding factor introduced by Burchfield and Woolley to reduce the effect of micellar charge on the ionic strength of the solution.

The value of α used in eq 2 was determined by iterative solution of the equation

$$K = \left\{ \alpha/[n(1-\beta\alpha)^{n\beta}(1-\alpha)^n m^{(n\beta+n-1)}] \right\} (\gamma_B/\gamma_M{}^{n\beta}\gamma_A{}^n) \tag{5}$$

where K is the equilibrium constant for micelle formation, and γ_B, γ_M and γ_A are the activity coefficients for the species $[M_{n\beta}A_n{}^{-n(1-\beta)}]$, M^+, and A^-, respectively, as given by the Guggenheim equations

$$\log \gamma_B = -n^2(1-\beta)^2\delta^2 A_\gamma I^{1/2}/(1 + b_B I^{1/2}) + B_{n\gamma}(1-\beta\alpha)m \tag{6}$$

$$\log \gamma_M = -A_\gamma I^{1/2}/(1 + b_m I^{1/2}) + B_{1\gamma}(1-\alpha)m + B_{n\gamma}\alpha m/n \tag{7}$$

$$\log \gamma_A = -A_\gamma I^{1/2}/(1 + b_A I^{1/2}) + B_{1\gamma}(1-\beta\alpha)m \tag{8}$$

Equations 2 and 5 were iterated in combination to produce the best fit to the experimental osmotic coefficient data with K, n, δ, $B_{1\gamma}$ and $B_{n\gamma}$ as variables. The best fit values attained for the variables are summarised in Table 1. The predicted aggregation numbers and critical micelle concentrations are in reasonable agreement with literature values and the derived values of both K and $B_{n\gamma}$ are higher for chlorpromazine than promethazine, as might be expected from the greater hydrophobicity of this compound.

Although the mass action theory fits the data above the inflection point and generates reasonable values for the micellar parameters, there is evidence from the data below the CMC that limited association may be occurring. Figure 1 shows appreciable deviation of the osmotic coefficients below the CMC from values for a 1:1 electrolyte at similar concentrations. Although precise measurement at such low concentrations is difficult it is thought that this effect is real and in excess of experimental error. The ion interaction coefficient, $B_{1\gamma}$, is determined mainly by the data in the pre CMC region. The value for promethazine, which of the two compounds is probably the more reliable because of the higher CMC, is strongly negative suggesting the possibility of premicellar association. Full details of these measurements and of similar measurements on these drugs at higher temperatures are given in references 10 and 11.

2.1.2 *Light Scattering Measurements.* A recent investigation of the physicochemical properties of aqueous solutions of these phenothiazine drugs at high solution concentrations has shown that the association pattern is more complex than previously thought. Figure 3 shows time-average light scattering data determined using a Fica 42000 photogoniodiffusometer at 546 nm. The results are expressed as the ratio, S_{90}, of the intensity of light scattered from the drug solution to that from a calibrated standard, against solution molality, m. Two inflection points are clearly visible in each plot, the second of which occurs at concentrations c_2 which are outside the concentration range examined by previous workers. The first inflection point occurs at concentrations c_1 corresponding to the literature values of the CMC. Plots of $(m-c_1)/\Delta S_{90}$ against $m-c_1$ (where ΔS_{90} is the scattering ratio for solutions of concentration m in excess of that at the first critical concentration) show two linear segments with an intersection at c_2, suggesting that aggregates are of a stable size over each concentration range.

Additional inflection points such as those shown in Fig 3 have been noted in many surfactant systems and are frequently referred to as 'second CMCs'. Clearly there is a restructuring of the aggregates at c_2 and further work is in progress to define more clearly the nature of the structural changes occurring in solutions of these drugs.

2.2 ASSOCIATION IN CONCENTRATED ELECTROLYTE SOLUTION

2.2.1 *Light Scattering Measurements.* The concentration dependence of the light scattering ratio, S_{90}, for promethazine hydrochloride in aqueous solutions of sodium chloride shows abrupt discontinuities at well defined critical concentrations, c_1, at all electrolyte concentrations (Fig 4). Plots of $(m-c_1)/\Delta S_{90}$ against $(m-c_1)$ were linear with positive slopes, only in the presence of dilute electrolyte (Fig 5). When the concentration of added sodium chloride exceeded 0.1 mol dm^{-3}, such plots showed pronounced downward curvature indicative of increase of mean aggregate size with increase of solution concentration. Similar plots to those of Fig 4 and 5 were obtained for solutions of chlorpromazine hydrochloride under identical conditions.

A detailed analysis of the light scattering data for promethazine in the presence of electrolyte concentrations of 0.4 to 0.8 mol dm^{-3} has been presented [6]. It was assumed that the first step of the association occurred at c_1 by a single step process in which n monomers formed the primary micelle. Increase of solution concentration was assumed to lead to growth of the primary micelle by the stepwise addition of monomers. Both co-operative and antico-operative models were considered in which the magnitude of the stepwise association constants was related to the aggregation number by simple empirical relationships. For each model an expression was derived for the weight average aggregation number of the micellar species, n_w, as a function of the monomer concentration, m_1, and a global equilibrium constant, K. The following models were considered.

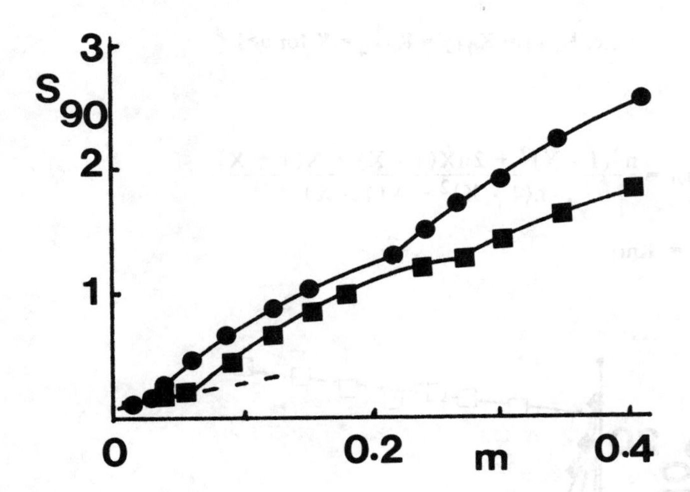

Figure 3 Variation of the scattering ratio, S_{90}, with molality, m, for • chlorpromazine hydrochloride and ■, promethazine hydrochloride in water at 30°C; (--------------) monomer line.

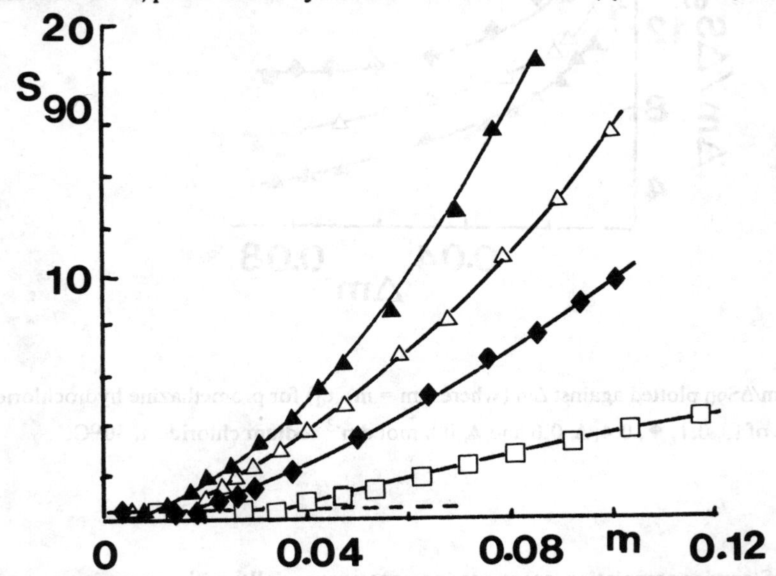

Figure 4 Variation of the scattering ratio, S_{90}, with molality for promethazine hydrochloride in the presence of □, 0.1; ♦, 0.4; Δ , 0.6 and ▲, 0.8 mol dm^{-3} sodium chloride at 30°C. (--------) monomer line.

Model 1 The equality of all stepwise association constants was assumed,

$$\text{i.e. } K_{n+1} = K_{n+2} = K_{n+q} = K \text{ for } q \geq 1$$

for which

$$n_w = \frac{n^2(1 - X)^2 + 2nX(1 - X) + X(1 + X)}{n(1 - X)^2 + X(1 - X)} \tag{9}$$

where $X = Km_1$

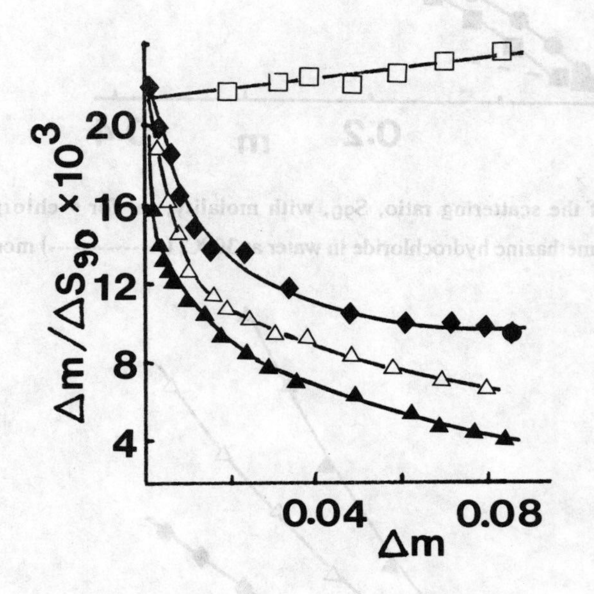

Figure 5 $\Delta m/\Delta S_{90}$ plotted against Δm (where $\Delta m = m - c_1$) for promethazine hydrochloride in the presence of □, 0.1; ♦, 0.4; Δ, 0.6 and ▲, 0.8 mol dm^{-3} sodium chloride at 30°C.

Model 2 Stepwise association constants increase sequentially with increasing aggregation number according to

$$K_{n+q} = K(n+q-1)/(n+q) \qquad \text{for } q \geq 1$$

leading to

$$n_w = n + \left(\frac{X}{1-X}\right) \tag{10}$$

Model 3 Stepwise association constants decrease sequentially with increasing aggregation number according to

$$K_{n+q} = K(n+q)/(n+q-1) \qquad \text{for } q \geq 1$$

leading to

$$n_w = \frac{n^3(1-X)^3 + 3n^2X(1-X)^2 + 3nX(1-X^2) + X(1+4X+X^2)}{n_2(1-X)^3 + 2nX(1-X)^2 + X(1-X^2)} \tag{11}$$

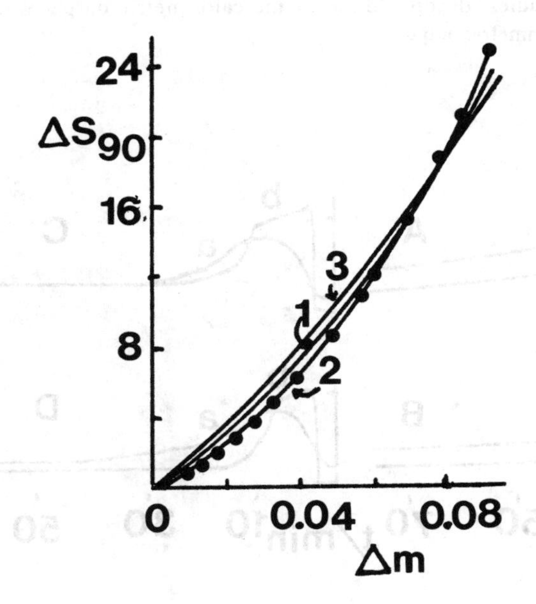

Figure 6 Curves representing the best fit of experimental light scattering data (●) for promethazine hydrochloride in 0.8 mol dm^{-3} NaCl, as derived from stepwise association models, 1,2 and 3.

For a particular model and for selected values of K, Equations 9-11 enable the concentration dependence of n_w and hence of S_{90} to be predicted and compared with experimental curves. The best fit to the experimental data for each model was determined by an iterative procedure. Of the three models considered, the closest fit to the data for promethazine in sodium chloride solutions ≥ 0.4 mol dm^{-3} was provided by the co-operative model, Model 2 (see Fig 6).

Although Fig 6 shows a good fit to the data for solution concentrations greater than c_1, no information is provided in the pre CMC region because of difficulties in obtaining precise light scattering data at low solution concentration. For this reason the same systems were studied using the more sensitive technique of heat conduction calorimetry.

2.2.2 Calorimetric measurements

Calorimetric measurements were performed at 303K on a modified Arion-Electronique conduction calorimeter, the design, operation and calibration of which have been described previously [12]. The transfer function of such equipment may lead to signal deformation and, in order to measure the instantaneous power, P, absorbed by the environment of the reaction during the course of an experiment, it is essential to deconvolve the response by compensation of the principal time constants [13]. In these studies, deconvolution of the calorimetric output was achieved by analogue filtering of the calorimetric output.

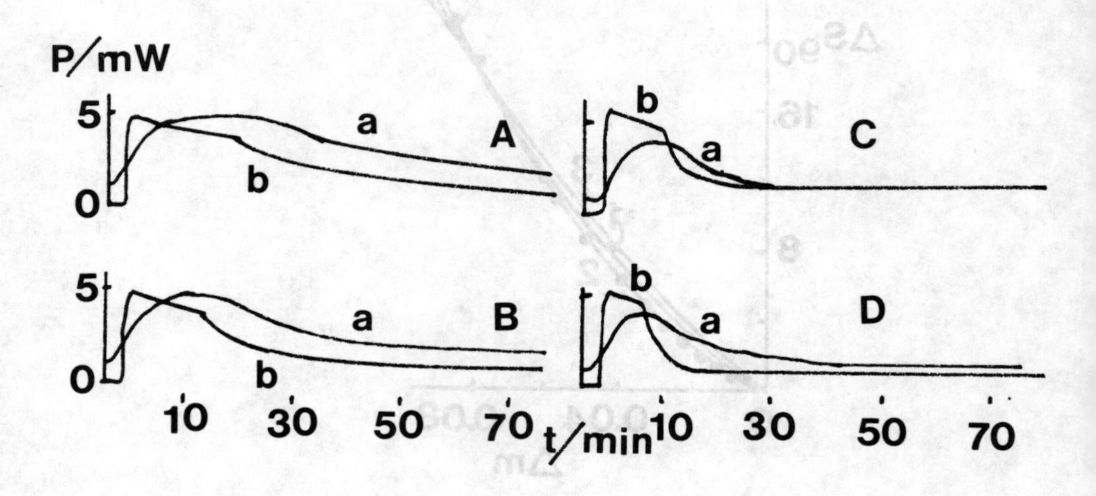

Figure 7 Thermograms at 30°C for the dilution of aqueous solutions of promethazine hydrochloride (0.25 mol kg^{-1}) in A, 0.1; B, 0.2; C, 0.4 and D, 0.6 mol dm^{-3} NaCl. Curves a represent uncorrected instrumental response, curves b are reconstituted thermograms after deconvolution.

Figure 7 presents calorimetric data for promethazine hydrochloride in a range of concentrations of added electrolyte, both as raw thermograms (curve a) and after deconvolution of the signal by inverse filtering (curve b). A comparison of curves a and b shows the importance of deconvolution of data for these systems. Inflection points noted in the deconvolved thermograms are in good agreement with the apparent CMCs as determined previously by light scattering techniques.

In the light of the evidence from osmotic data of possible premicellar association, theoretical treatments of the calorimetric data have been developed in which primary micelles are formed, not only by a single step association as in the light scattering treatment, but also by continuous association processes. Full details of the derivation of the equations for these schemes have been previously reported [14-16] and only a summary is given here.

2.2.3. *Single Step Formation of Primary Micelle.* The association scheme may be represented as follows. Assuming that the micellar aggregates have a negligible concentration below the CMC, formation of the primary micelle may be assumed to be the single step process

$$nA_1 \rightleftarrows A_n \qquad (12)$$

The equilibrium constant for the formation of the primary micelle, A_n from n monomeric units A_1 is K_n and the molar enthalpy of formation of this micelle is ΔH_n.

The growth of this micelle by increase of concentration above the CMC may be represented by

$$A_n + A_1 \rightleftarrows A_{n+1}$$

$$A_{n+1} + A_1 \rightleftarrows A_{n+2} \qquad (13)$$

$$A_{n+(q-1)} + A_1 \rightleftarrows A_{n+q}$$

with corresponding equilibrium constants, $K_{n+1}, K_{n+2} \ldots K_{n+q}$ and molar enthalpies of formation, $\Delta H_{n+1}, \Delta H_{n+2}, \ldots \Delta H_{n+q}$.

Addition of successive equilibria gives the global equilibrium for the formation of the secondary micelle A_{n+q} from the primary micelle thus,

$$A_n + qA_1 \rightleftarrows A_{n+q} \qquad (14)$$

The global equilibrium constant K^*_{n+q} and molar enthalpy of fomation for this process ΔH^*_{n+q} are given by

$$K^{*}_{n+q} = \prod_{i=1}^{q} K_{n+i} \qquad (15)$$

and

$$\Delta H^{*}_{n+q} = \sum_{i=1}^{q} \Delta H_{n+i} \qquad (16)$$

Addition of the equilibria of eq. 12 and 14 gives the global equilibrium for the formation of the secondary micelle A_{n+q} from monomers thus

$$(n + q)A_1 \rightleftharpoons A_{n+q} \qquad (17)$$

The global equilibrium constant for the above equilibrium is $K^*_{n,n+q}$ and the corresponding molar enthalpy of formation, $\Delta H^*_{n,n+q}$ where

$$K^{*}_{n,n+q} = K_n K^*_{n+q} = K_n \prod_{i=1}^{q} K_{n+i} \qquad (18)$$

$$\Delta^{*}_{n,n+q} = \Delta H_n + \Delta H^{*}_{n+q} = \Delta H_n + \sum_{i=1}^{q} \Delta H_{n+i} \qquad (19)$$

It can be shown [14] that the total power, $P_T(t)$, associated with the formation of all the associated species present in a solution of molality m(t) is

$$P_T(t) = \Delta H_n \left(M_T(t)\frac{dY}{dt} + Yd_1\right) + \Delta H \left(M_T(t)\frac{dZ}{dt} + Zd_1\right) \qquad (20)$$

where

$$Y = K_n [m_1(t)]^n \left(1 + \sum_{i=1}^{q} K^{*}_{n+i} [m_1(t)]^i\right) \qquad (21)$$

$$Z = K_n [m_1(t)]^n \sum_{i=1}^{q} i K^{*}_{n+i} [m_1(t)]^i \qquad (22)$$

and ΔH is the molar enthalpy change for monomer addition, which is the same for each step of the process.

i.e. $\Delta H_{n+1} = \Delta H_{n+2} = \Delta H_{n+q} = \Delta H$

$M_T(t)$ is the total mass of solvent at time t in the calorimeter as given by

$$M_T(t) = M_0 + d_1 t \tag{23}$$

where M_o is the initial mass of solvent in the cell and d_1 the rate of addition of solvent.

Similarly it can be shown that the total molality of all species in solution at time t is given by

$$m(t) = m_1(t) + nY + Z = N_T(t)/M_T(t) \tag{24}$$

where $N_T(t)$ is the total number of moles at time t.

The experimental data may be fitted to the preceding equations by assuming a fixed relationship between the stepwise equilibrium constants. Two association models were considered.

Model 1

The equality of all stepwise association models is assumed

i.e. $$K_{n+1} = K_{n+2} = K_{n+q} = K \tag{25}$$

Introducing this relationship into eq 21 and 22 gives

$$Y = \frac{K_n[m_1(t)]^n}{1-K\,m_1(t)} = \frac{aX^n}{1-X} \tag{26}$$

where $a = K_n/K^n$ and $X = Km_1(t)$

and $$Z = \frac{KK_n\,[m_1(t)]^{n+1}}{[1-Km_1(t)]^2} = \frac{aX^{n+1}}{(1-X)^2} \tag{27}$$

From eq 24 the molality of the solution may be written

$$m(t) = \frac{X}{K} + \frac{naX^n}{1-X} + \frac{aX^{n+1}}{(1-X)^2} \tag{28}$$

Substituting dY/dt and dZ/dt into eq 20 yields

$$P_T(t) = \Delta H_n \left[\frac{M_T(t) a X^n [n(1-X)+X]}{X(1-X)^2} \frac{dX}{dt} + d_1 \left(\frac{aX^n}{1-X} \right) \right]$$

$$+ \Delta H \left[\frac{M_T(t) a X^n [n(1-X)+X+1]}{(1-X)^3} \frac{dX}{dt} + d_1 \frac{aX^{n+1}}{(1-X)^2} \right] \tag{29}$$

with
$$\frac{dx}{dt} = \frac{[d_2 - d_1\, m(t)] K\, (1-X)^3}{M_T(t) \{ aKX^{n-1} \{ [n(1-X)+X]^2 + X \} + (1-X)^3 \}} \tag{30}$$

Model 2

Stepwise association constants are related to the aggregation number by the expression

$$K_{n+q} = K(n+q-1)/(n+q) \qquad \text{with } q \geq 1 \tag{31}$$

Introducing this relationship into eq 10 and 11 gives

$$Y = aX^n + an \sum_{i=1}^{q} \frac{X^{n+i}}{n+i} \tag{32}$$

$$Z = an \sum_{i=1}^{q} i \frac{X^{n+i}}{n+i} \tag{33}$$

The molality of the solution at time t is given by

$$m(t) = \frac{X(1-X) + K\, anX^n}{K\,(1-X)} \tag{34}$$

Substituting dY/dt and dZ/dt into eq 20 yields

$$P_T(t) = \Delta H_n \left\{ \frac{M_T(t)\, anX^n}{X(1-X)} \frac{dX}{dt} + d_1 \left[aX^n + an \sum_{i=1}^{q} \frac{X^{n+i}}{n+i} \right] \right\}$$

$$+ \Delta H \left\{ \frac{M_T(t)\, anX^n}{(1-X)^2} \frac{dX}{dt} + d_1\, an \sum_{i=1}^{q} \frac{iX^{n+i}}{n+i} \right\} \qquad (35)$$

$$\text{with} \quad \frac{dX}{dt} = \frac{(d_2 - d_1\, m(t))\, kX(1-X)}{[KN(t) - M_T(t)][n-(n-1)X] + M_T(t)X(1-X)} \qquad (36)$$

In the association schemes outlined above, expressions have been derived for the total power, P_T. This parameter is related to the experimentally determined instantaneous power P by the relationship

$$P_T = P - P^{\infty}$$

and hence

$$P/d_2 = P_T/d_2 - (P/d_2)^{\infty}$$

where P^{∞} is the power at infinite dilution and d_2 is the rate of addition of the drug. Simulation of the concentration dependence of P/d_2 has been used in this study as a means of determination of the thermodynamic parameters of association. For each association model, values of P/d_2 were computed as a function of molality using the relevant equations for m(t) and $P_T(t)$. Iteration of the parameters n, K, K_n, ΔH and ΔH_n gave best fit curves for each association model.

Figs 8 and 9 show the best fit curves to the experimental thermograms obtained using the two single step association models. Model 2 which is identical to the model used to fit the light

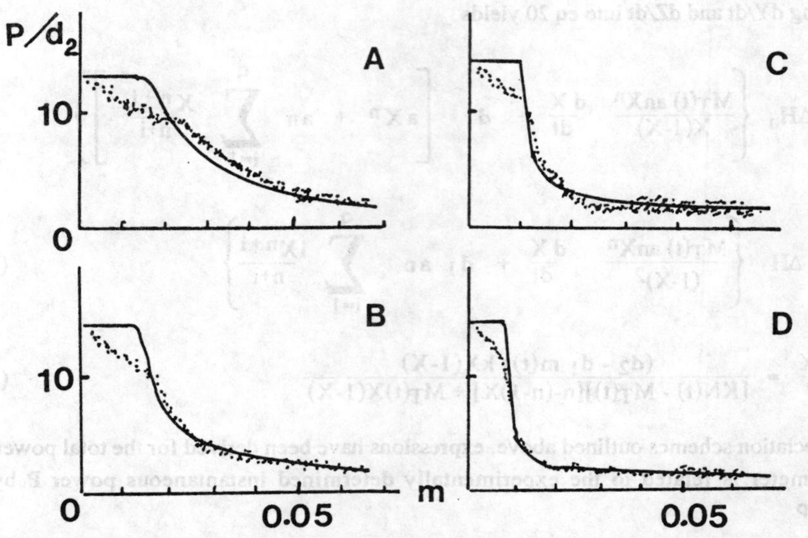

Figure 8 Simulation of the dependence of P/d_2 (kJ mol^{-1}) on molality, m, for promethazine hydrochloride in A, 0.1; B, 0.2; C, 0.4 and D, 0.6 mol dm^{-3} NaCl using Model 1.

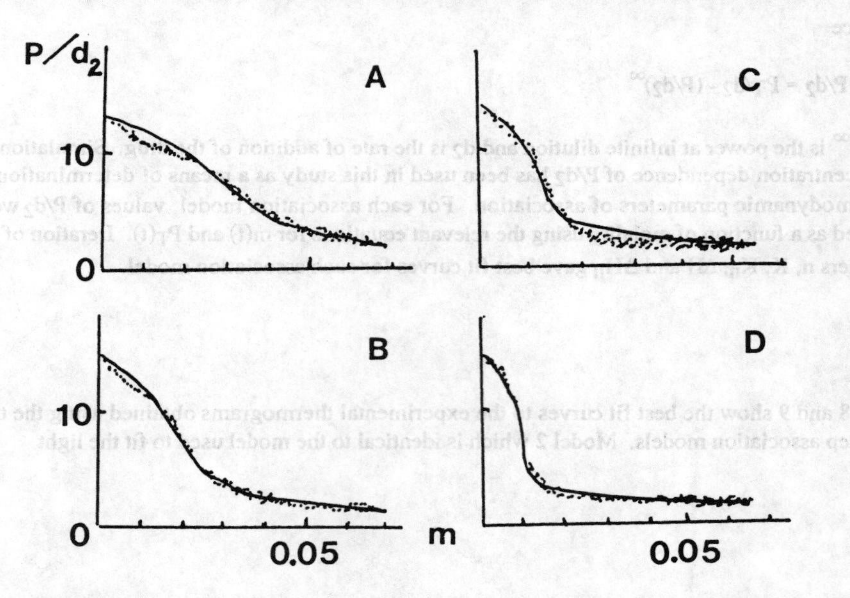

Figure 9 Simulation of the dependence of P/d_2 (kJ mol^{-1}) on molality, m, for promethazine hydrochloride in A, 0.1; B, 0.2; C, 0.4 and D, 0.6 mol dm^{-3} NaCl using Model 2.

scattering data clearly provided a better representation of the data than the association model in which K was independent of concentration. However, examination of the fit in the pre CMC region, particularly in systems of lower salt concentration shows the inability of this model to correctly predict the curvature of the plots in the premicellar region. The discrepancy between experimental and predicted values of P/d_2 in the low concentration region was in excess of the experimental uncertainty and the model was consequently rejected.

2.2.4 *Limited Premicellar Association.* In the following treatment the primary micelle is assumed to be formed by limited stepwise association according to

$$A_1 + A_1 \rightleftarrows A_2$$

$$A_2 + A_1 \rightleftarrows A_3 \qquad (37)$$

$$A_{n-1} + A_n \rightleftarrows A_n$$

with corresponding equilibrium constants, K_2, K_3 .....K_n and molar enthalpies of formation, ΔH_2, ΔH_3 ..... ΔH_n. Addition of successive equilibria gives the global equilibrium for the formation of the primary micelle, thus

$$nA_1 \rightleftarrows A_n \qquad (38)$$

The global equilibrium constant K^*_n for the formation of the aggregate A_n and the corresponding molar heat of formation ΔH^*_n are given by

$$K^*_n = \prod_{j=2}^{n} K_j \qquad (39)$$

$$\Delta H^*_n = \sum_{j=2}^{n} \Delta H_j \qquad (40)$$

Subsequent growth of the primary micelle A_n is assumed to occur according to eq 13 with global equilibrium constants and molar enthalpy of formation given by eq 15 and 16 respectively.

Addition of the equilibria represented by eq 14 and 38 gives the equilibrium for the formation of the aggregates A_{n+q} from monomeric species, thus

$$(n+q)\, A_1 \rightleftarrows A_{n+q} \qquad (41)$$

The global equilibrium constant $K^*_{n,n+q}$ and the molar enthalpy of formation $\Delta H^*_{n,n+q}$ are given by

$$K^*_{n,n+q} = K^*_n K^*_{n+q} \tag{42}$$

$$\Delta H^*_{n,n+q} = \Delta H^*_n + \Delta H^*_{n+q} \tag{43}$$

The total molality m(t) of all the species present in the solution at time t can be shown to be [15]

$$m(t) = m_1(t) + nY + Z + R_1 = \frac{N(t)}{M_T(t)} \tag{44}$$

with

$$Y = \sum_{i=1}^{q} K^*_{n,n+i}\,[m_1(t)]^{n+i} = K^*_n\,[m_1(t)]^n \sum_{i=1}^{q} K^*_{n+i}\,[m_1(t)]^i \tag{45}$$

$$Z = \sum_{i=1}^{q} iK^*_{n,n+i}\,[m_1(t)]^{n+i} = K^*_n\,[m_1(t)]^n \sum_{i=1}^{q} iK^*_{n+i}\,[m_1(t)]^i \tag{46}$$

$$R_1 = \sum_{j=2}^{n} jK^*_j\,[m_1(t)]^j \tag{47}$$

Similarly it can be shown that the total power, $P_T(t)$, associated with the formation of all the associated species present in a solution of molality m(t) is

$$P_T(t) = \Delta H_A \left\{ d_1\,[R_2 + (n-1)Y] + M_T(t)\,\frac{dR_2}{dT} + (n-1)\,\frac{dY}{dt} \right\} + \Delta H \left[d_1\,Z + M_T(t)\,\frac{dZ}{dt} \right] \tag{48}$$

where R_2 is defined as

$$R_2 = \sum_{j=2}^{n} (j-1)\ K_j^* \ [m_1(t)]^j \tag{49}$$

and ΔH_A and ΔH are respectively the molar enthalpy changes for monomer addition during the formation of the primary micelle and during its subsequent growth.

Equations 44 and 48 are general expressions for this type of association process and may be solved by assuming a relationship between aggregation number and equilibrium constant. Two

such relationships were considered.

Model 3

This model assumes the equality of all stepwise association constants in the premicellar region.

$$K_j = K_{j+1} = K_A \tag{50}$$

Above the cmc, the aggregate A_n is assumed to grow such that the equilibrium constants increase with aggregation number according to

$$K_{n+i} = K\,(n+i-1)/(n+i) \tag{51}$$

From these two hypotheses we may write

$$K_j^* = \prod_{j=2}^{j} k_j = K_A{}^{j-1} \quad \text{and} \quad K_n^* = K_A^{n-1} \tag{52}$$

with j = 2,3 ....n

$$K_{n+i}^* = \prod_{i=1}^{i} K_{n+i} = nK^i / (n+i) \tag{53}$$

with i = 1,2 ....i ....∞

The total molality m(t) may be shown to be

$$m(t) = m_1(t) \left\{ 1 + \sum_{j=1}^{n-1} (j+1) \; [K_A m_1(t)]^j + \frac{nK[K_A m_1(t)]^n}{K_A[1-Km_1(t)]} \right\} = \frac{N(t)}{M_T(t)} \qquad (54)$$

Solution of eq 54 for given values of the parameters n_1 K_A and K permits the evaluation of $m_1(t)$. The total power absorbed $P_T(t)$ at a molality m(t) may then be determined from eq 48 using the following expressions for R_2, Y and Z and the derivatives of these functions.

$$R_2 = \frac{1}{K_A} \sum_{j=2}^{n} (j-1)[K_A \; m_1(t)]^j \qquad (55)$$

$$Y = \frac{n[K_A m_1(t)]^n}{K_A} \sum_{i=1}^{\infty} \frac{[Km_1(t)]^i}{n+i} \qquad (56)$$

$$Z = \frac{n[K_A m_1(t)]^n}{K_A} \sum_{i=1}^{\infty} \frac{i[K \; m_1(t)]^i}{n+i} \qquad (57)$$

$$\frac{dR_2}{dt} = \frac{dm_1}{dt} \sum_{j=1}^{n-1} j(j+1) \; [K_A \; m_1(t)]^j \qquad (58)$$

$$\frac{dY}{dt} = \frac{n[K_A m_1(t)]^{n-1}}{1-Km_1(t)} \; \frac{dm_1}{dt} \qquad (59)$$

and $$\frac{dZ}{dt} = \frac{nK \; [K_A \; m_1(t)]^n}{K_A \; [1-Km_1(t)]^2} \; \frac{dm_1}{dt} \qquad (60)$$

with $$\frac{dm_1}{dt} = \frac{d_2 - d_1 m(t)}{A \; M_T(t)}$$

and $$A = 1 + \sum_{j=1}^{n-1} (j+1)^2 \, [K_A m_1(t)]^j + \frac{n\,[K_A m_1(t)]^{n-1}[n(1-Km_1(t)) + Km_1(t)]}{[1 - Km_1(t)]^2}$$

Model 4

In this model it is assumed that in the premicellar region the aggregates form according to a continuous association mode of n steps, in which the equilibrium constants are such that

$$K = K_A \, j/(j-1) \tag{61}$$

As in Model 3, the primary aggregate is assumed to grow in a co-operative manner according to eq 51.

The global constants arising from these two hypotheses are

$$K_j^* = \prod_{j=2}^{j} K_j = \frac{K_A^{j-1}}{j} \quad \text{and} \quad K_n^* = \frac{K_A^{n-1}}{n} \tag{62}$$

with j = 2,3 .... n

and K^*_{n+i} as defined by eq. 53

The total molality m(t) calculated from eq 44 is

$$m(t) = m_1(t)\left\{1 + \sum_{j=1}^{n-1} [K_A \, m_1(t)]^j + \frac{K[K_A \, m_1(t)]^n}{K_A[1-K\,m_1(t)]}\right\} = \frac{N(t)}{M_T(t)} \tag{63}$$

Eq 63 allows the determination of $m_1(t)$ for a given set of the parameters n, K_A and K.

The total power absorbed, $P_T(t)$ at a molality m(t) may be determined from eq 48 using the following expressions for R_2, Y and Z and their derivatives.

$$R_2 = \frac{1}{K_A} \sum_{j=1}^{n-1} \frac{j}{j+1} [K_A m_1(t)]^{j+1} \tag{64}$$

$$Y = \frac{[K_A\, m_1(t)]^n}{K_A} \sum_{i=1}^{\infty} \frac{[K m_1(t)]^i}{n+i} \tag{65}$$

$$Z = \frac{[K_A m_1(t)]^n}{K_A} \sum_{i=1}^{\infty} \frac{i[K m_1(t)]^i}{n+i} \tag{66}$$

$$\frac{dR_2}{dt} = \frac{dm_1}{dt} \sum_{j=1}^{n-1} j[K_A m_1(t)]^j \tag{67}$$

$$\frac{dY}{dt} = \frac{K[K_A m_1(t)]^n}{K_A[1-K\, m_1(t)]} \frac{dm_1}{dt} \tag{68}$$

$$\frac{dZ}{dt} = \frac{[K_A\, m_1(t)]^n}{K\, K_A\, [1-K\, m_1(t)]^2} \frac{dm_1}{dt} \tag{69}$$

with dm_1/dt as given in eq 60 and

$$A = 1 + \sum_{j=1}^{n-1} (j+1)\, [K_A\, m_1(t)]^j + \frac{K_A m_1(t)]^n}{K_A\, K\, [1-K m_1(t)]} [nK^2\, [1-K m_1(t)] + 1] \tag{70}$$

Iteration of the parameters n, K, K_A, ΔH and ΔH_A gave best fit curves for each of these two association schemes. Fig 10 and 11 show an excellent representation of the experimental data over the whole concentration range for each system. This fit was maintained over the whole range of electrolyte concentration studied with no dependence of the value of n on the concentration of

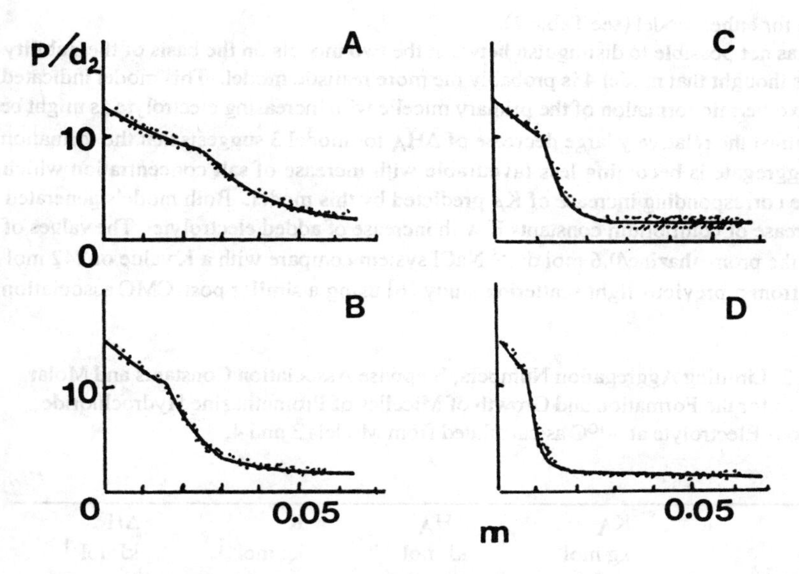

Figure 10 Simulation of the dependence of P/d_2 (kJ mol^{-1}) on molality, m, for promethazine hydrochloride in A, 0.1; B, 0.2; C, 0.4 and D, 0.6 mol dm^{-3} NaCl using Model 3.

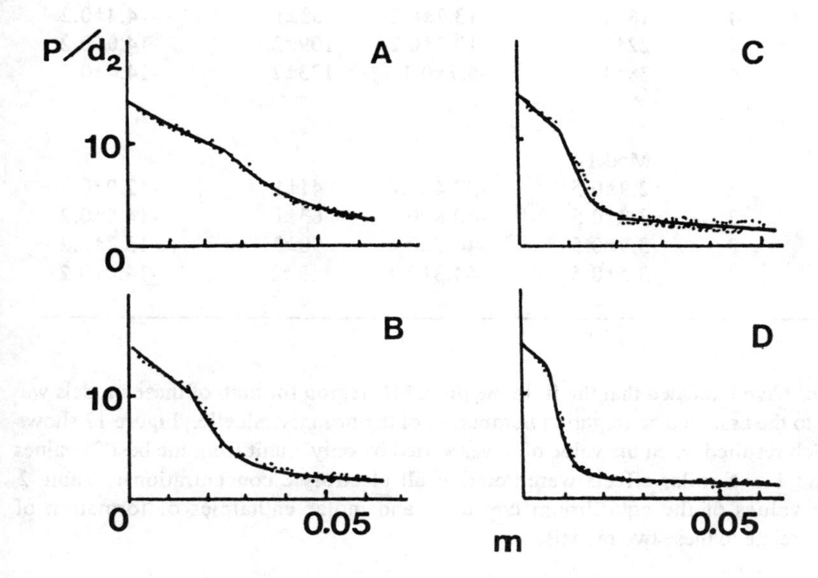

Figure 11 Simulation of the dependence of P/d_2 (kJ mol^{-1}) on molality, m, for promethazine hydrochloride in A, 0.1; B, 0.2; C, 0.4 and D, 0.6 mol dm^{-3} NaCl using Model 4.

added electrolyte for either model (see Table 2).

Although it was not possible to distinguish between the two models on the basis of their ability to fit the data it is thought that model 4 is probably the more realistic model. This model indicated an increasingly exothermic formation of the primary micelle with increasing electrolyte as might be expected. In contrast the relatively large decrease of ΔH_A for model 3 suggests that the formation of the primary aggregate is becoming less favourable with increase of salt concentration which conflicts with the corresponding increase of K_A predicted by this model. Both models generated the expected increase of equilibrium constants K with increase of added electrolyte. The values of K calculated for the promethazine/0.6 mol dm^{-3} NaCl system compare with a K value of 242 mol kg^{-1} as derived from a previous light scattering study [6] using a similar post-CMC association scheme.

TABLE 2 Limiting Aggregation Numbers, Stepwise Association Constants and Molar Enthalpies for the Formation and Growth of Micelles of Promethazine Hydrochloride in Aqueous Electrolyte at 30°C as calculated from Models 3 and 4.

NaCl concn mol dm^{-3}	n	K_A kg mol^{-1}	ΔH_A kJ mol^{-1}	K kg mol^{-1}	ΔH kJmol^{-1}
		Model 3			
0.1	4	12±1	-12.5±0.2	47±1	-13.9±0.2
0.2	4	16±1	-13.7±0.2	68±1	-14.4±0.2
0.4	4	22±1	-10.2±0.2	109±2	-14.0±0.2
0.6	4	38±1	-5.7±0.1	173±2	-14.4±0.2
		Model 4			
0.1	3	2.9±0.5	-37.4±0.4	41±1	-12.9±0.2
0.2	3	3.8±0.5	-40.8±0.4	63±1	-14.0±0.2
0.4	3	3.9±0.5	-40.7±0.4	90±2	-13.7±0.2
0.6	3	3.5±0.5	-44.3±0.4	119±2	-14.0±0.2

Our computations have indicated that the fit in the pre-CMC region for both of these models was highly sensitive to the assumed aggregation number, n, of the primary micelle. Figure 12 shows the poor fits which resulted when the value of n was varied by only 1 unit from the best fit values of figures 10 and 11. Similar effects were noted at all electrolyte concentrations. Table 2 summarises the values of the equilibrium constants and molar enthalpies of formation of aggregates which relate to these two models.

3. Summary

Our studies have provided evidence of limited premicellar association in aqueous solutions of the phenothiazine drugs, chlorpromazine and promethazine hydrochloride. In the presence of high concentrations of added electrolyte, this association has been shown to involve the formation of units of 3 or 4 monomers by a continuous association process. Evidence for premicellar association of the drugs in the absence of added electrolyte is from the significant departure of the osmotic coefficients from the values for typical 1:1 electrolytes and from the low or negative values of ion interaction coefficients, $B_{1\gamma}$. No detailed analysis of the premicellar association process in these systems has yet been attempted.

At the first critical concentration (referred to as the CMC) aggregates are formed from these small units. Analysis of the osmotic and light scattering data for phenothiazine drugs in water suggests that these aggregates are of a stable size over a concentration range to approximately 0.2 mol kg^{-1}, when restructuring of the aggregates occurs at a second critical concentration. In contrast, in the presence of high concentrations of electrolyte (>0.1 mol dm^{-3}) the aggregates formed at the first critical concentration increase in size with increase of solution concentration by a continuous association process, giving rise to a high polydispersity of aggregate sizes in solution.

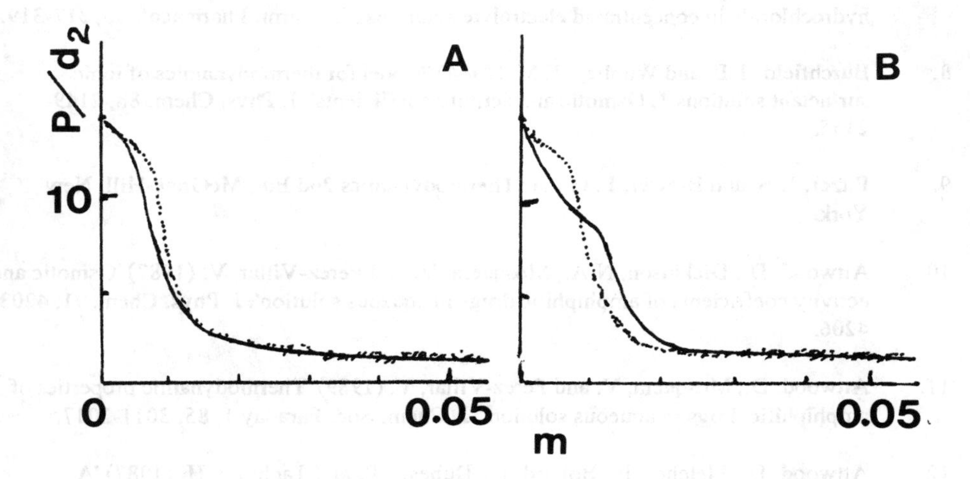

Figure 12 Influence of the value of n on the best fit of Model 4 to graphs of P/d_2 (kJ mol^{-1}) as a function of molality for promethazine hydrochloride in 0.6 mol dm^{-3} NaCl. A, n = 2; B, n=4.

4. References

1. Mukerjee, P. (1974) 'Micellar properties of drugs : micellar and nonmicellar patterns of self-association of hydrophobic solutes of different molecular structures - monomer fraction, availability and misuses of micellar hypothesis', J. Pharm. Sci., 63, 972-981.

2. Attwood, D. and Florence, A.T. (1983) Surfactant Systems, Chapman and Hall, London, Chapter 4.

3. Attwood, D., Agarwal, S.P., and Waigh, R.D., (1980) 'Effect of the nature of the hydrophobic group on the mode of association of amphiphilic molecules in aqueous solution', J. Chem. Soc. Faraday 1, 76, 2187-2193.

4. Florence, A.T., and Parfitt, R.T. (1971) 'Micelle formation by some phenothiazine derivatives. II Nuclear magnetic resonance studies in deuterium oxide', J. Phys. Chem. 75, 3554-3560.

5. Brady, A.P., Huff, H. and McBain, J.W. (1951) 'Measurement of vapour pressures by means of matched thermistors', J. Phys. Colloid Chem. 55, 304-311.

6. Attwood, D. (1983) 'Self-association of phenothiazine drugs in aqueous solutions of high ionic strength', J. Chem. Soc. Faraday 1, 79, 2669-2677.

7. Attwood, D. and Natarajan, R. (1983) 'Micellar properties of chlorpromazine hydrochloride in concentrated electrolyte solutions', J. Pharm. Pharmacol. 35, 317-319.

8. Burchfield, T.E. and Woolley, E.M. (1984) 'Model for thermodynamics of ionic surfactant solutions 1. Osmotic and activity coefficients', J. Phys. Chem. 88, 2149-2155.

9. Pitzer, K.S. and Brewer, L. (1961) Thermodynamics 2nd Ed., McGraw-Hill, New York.

10. Attwood, D., Dickinson, N.A., Mosquera, V. and Perez-Villar, V. (1987) 'Osmotic and activity coefficients of amphiphilic drugs in aqueous solution', J. Phys. Chem. 91, 4203-4206.

11. Attwood, D., Mosquera, V. and Perez-Villar, V. (1989) 'Thermodynamic properties of amphiphilic drugs in aqueous solution', J. Chem. Soc. Faraday 1, 85, 3011-3017.

12. Attwood, D., Fletcher, P., Boitard, E., Dubès, J.P. and Tachoire, H. (1987) 'A calorimetric study on the self-association of amphiphilic drugs in aqueous solution', J. Phys. Chem. 91, 2970-2975.

13. Tachoire, H., Macqueron, J.L. and Torra, V. (1984) 'Thermogenesis by heat conduction calorimetry', in M.A.V. Ribeiro da Silva (ed.), Thermochemistry and its Applications to Chemical and Biochemical Systems, D. Reidel Publishing Co., Dordrecht, pp 77-126.

14. Attwood, D., Fletcher, P., Boitard, E., Dubès, J.P. and Tachoire, H. (1989) 'Calorimetric study on the self association of promethazine hydrochloride in aqueous solutions of high ionic strength', in K. Mittal (ed.), Surfactants in Solution, Plenum

Press, New York, Vol 7 pp 262-276.

15. Attwood, D., Boitard, E., Dubès, J.P. and Tachoire, H. (1990) 'Association models for an amphiphilic drug in aqueous solutions of high ionic strength from calorimetric studies', Colloids and Surfaces. in press.

16. Attwood, D., Fletcher, P., Boitard, E., Dubès, J.P. and Tachoire, H. (1990) 'A calorimetric study on the self-association of an amphiphilic phenothiazine drug in aqueous electrolyte solutions', J. Phys. Chem. in press.

POLYMER-SURFACTANT SYSTEMS

B. LINDMAN AND G. KARLSTRÖM
Physical Chemistry 1 and Theoretical Chemistry
Chemical Centre
Lund University
P.O. Box 124
S-221 00 Lund
Sweden

ABSTRACT. The phase behaviour of aqueous polymer - ionic surfactant systems is discussed. Both ionic and nonionic polymers are considered. A two conformational description is found essential for the ethylene oxide containing nonionic polymers. Theoretical model calculations show that the observed phase behaviour can be understood in terms of simple interactions between the different parts of the polymer and surfactant molecules.

1. Introduction

Polymers and surfactants are extensively employed both in industry and in everyday life. Frequently they occur together, in particular in complex colloidal systems, to achieve colloid stability, emulsification or flocculation, structuring and suspending properties as well as rheology control. Examples of their uses are in cosmetic products, paints, detergent liquids, foods, polymer synthesis, drug formulations, formulations for crop disease control and high power solid state batteries. In addition to this, polymer-amphiphile mixtures are the fundamental units of almost all biological cells.

The behaviour of surfactants and polymers separately in aqueous solution has become better understood recently. While both can form a range of solutions and liquid crystalline phases, the surfactant/water systems are particularly rich in polymorphism [1]. As the practical applications require a knowledge and control of the phase properties, a lot of emphasis in research has been on phase diagram determination. For mixed surfactant or mixed polymer systems, the phase behaviour can become quite complex but is to a considerable extent understood.

The characterization of polymer-surfactant-solvent systems has also received considerable interest over several years [2-24], but our understanding is rather limited. A key feature appears to be the extent to which polymers and surfactants form separate or mixed complexes. Much emphasis in previous work has been on dilute systems, in particular in investigations of how physico-chemical properties of a polymer solution change as a function of the concentration of added surfactant. Binding isotherms, as determined inter alia by surfactant-selective electrodes, is one rather broadly studied aspect [3, 6, 10-12, 16] and rheology, because of its practical significance, is another [10, 11, 14]. Structural studies, with notable exceptions [24], are largely lacking and an aspect virtually untouched is that of phase equilibria. Investigations are generally limited to noting phase separation, e g precipitation, while systematic determinations of phase diagrams, in particular over wider domains of concentration of polymer and surfactant, have not been made. On the other hand, phase diagrams have for surfactant-solvent as well as for

D. M. Bloor and E. Wyn-Jones (eds.),
The Structure, Dynamics and Equilibrium Properties of Colloidal Systems, 131–147.

polymer-solvent systems been instrumental in advancing our understanding, both as regards fundamental aspects and as a basis for designing systems for technical applications.

The interaction in solution between a polymer and small added molecules, notably surfactants, is thus a research field of large, both technical and scientific importance. The technical importance
quite often comes from the fact that polymer solutions, which are used for some technical purpose are on the limit of phase separation. This means that addition of a cosolute may drastically increase or decrease the solubility of the polymer. Part of the scientific interest comes also just from this fact. Systems of this type can, therefore, be regarded as very sensitive probes for the intermolecular interactions present. When another kind of molecule is added to the system, this addition will affect the polymer solubility . From such changes, which are very easy to monitor experimentally, information about the effective intermolecular interactions present in the system can be obtained.

The theoretical modelling of the thermodynamic behaviour of polymers in non-aqueous solutions is normally based on the Flory-Huggins theory [28]. For aqueous solutions the situation is different. It is often claimed that the decrease in solubility observed for many polymers in water when the temperature is increased can not be explained within the Flory-Huggins model. Recently, it has, however, been shown that this type of behaviour can be obtained if a simple model is used, which assumes that a polymer segment can exist in either a polar conformation or a nonpolar one, and that the nonpolar conformations have a larger statistical weight (are entropically favoured) [25, 27]. In fact, a similar result can be obtained by invoking that the polymer-water interaction may be either hydrogen-bonded or nonhydrogen-bonded [29].

It is not the purpose of this paper to thoroughly describe and analyse thermodynamic models for a polymer-solvent-additive system, but one such model will be described below and a few applications of the model will be given. In these cases, the reader is kindly asked to look for the details of the theoretical modelling in the original papers. The main objective is instead to discuss the mechanisms behind the effective polymer - additive interaction, the different type of aggregates that may occur in these types of systems, and the effect changes in temperature and electrolyte concentration may induce in these quantities. We will deal with water as a solvent, but the mechanisms behind the behaviour are quite general although their relative importance is system dependent. The experimental results presented here are mainly taken from investigations made at the Department of Physical Chemistry 1 in Lund. The main reason for this is that for these systems both experimental and model data are easily accessible to us.

The outline of this article will be as follows. In the next section we will discuss the intermolecular interactions present in these systems, and in the third section we will describe a crude theoretical model capable of explaining the phase behaviour observed. In the fourth and fifth sections the pure polymer-solvent and the pure surfactant-solvent systems will be analyzed. Finally, in the last section we will discuss what is happening when these three components and eventually salt are mixed together.

2. Intermolecular interactions in a polymer-ionic surfactant-polar solvent-salt system

The purpose of this section is to analyse the interactions between the different molecules or parts of molecules occurring in a system built up from a polymer, which may be ionic or nonionic, an ionic surfactant with a counterion, a polar solvent (water) and salt. The discussion will focus on the attractive part of the interactions but also the repulsive part will be considered. The main concept in this discussion will be polarity, but it could be replaced by hydrophilicity. It will also be assumed that polar substances like to interact with polar substances and nonpolar substances like to interact with nonpolar ones, and the larger the difference in polarity the larger the effective repulsion or the smaller the effective attraction between the two substances.

The substances constituting this system can be classified accordingly. Ordinary salts as NaCl are the most polar type of species. Obviously, the salt is dissociated into ions so it is actually these ions that are the polar entities. The second most polar species in the system, which is considered here, is the solvent (water). For the purpose of this work it is sufficient to classify the segments of the nonionic polymer as less polar than the solvent (water). In order to understand the properties of an ionic polymer or an ionic surfactant, it is instructive to formally divide these species into an ionic part, which is more polar than the solvent, and a nonionic part which is less polar and, in the case of the surfactant, almost unpolar. From these types of considerations one may conclude that there exist minima in the potential of mean force curves between uncharged species or between one charged and one uncharged species, provided that the low polarity parts are large enough and the medium is polar enough. If both the solute particles are charged (with the same charge) there will also be a repulsive component in their interaction and it is impossible to a priori conclude if there exists a minimum in their potential of mean force. If both species are charged with charges of opposite sign then there will be a strong attraction between the two species. Finally, one may conclude by saying that the effect of salt in the system is to screen all coulombic interactions, attractive or repulsive, and to increase the effective attraction between the hydrophobic parts of the molecules, since it makes the effective solvent more polar.

3. Theoretical model

One of the standard ways to model a polymer solution, is to use Flory-Huggins (F-H) theory [28]. Within this model the mixing enthalpy can be written

$$U = N \Phi_1 \Phi_2 w_{12} \tag{1}$$

where w_{12} is the effective interaction parameter between a polymer segment and a neighbouring solvent molecule. Note that the w used in this work differs from the ordinary χ parameters used in F-H theory, in that the trivial temperature dependence has been removed. N is the total number of cells (solvent molecules + polymer segments) in the system and Φ_1 and Φ_2 the volume fractions of solvent and solute. The effective interaction parameter w_{12} is related to the direct interaction parameters $w'_{\alpha\beta}$ according to

$$w_{12} = w'_{12} - (w'_{11} + w'_{22})/2 \tag{2}$$

The corresponding entropy expression is

$$S = -kN\,(\Phi_1 \ln \Phi_1 + \Phi_2/M \ln \Phi_2) \tag{3}$$

k in equation 3 is Boltzmann's constant and M the degree of polymerization of the polymer. Equations 1 and 3 can easily be modified to include a two conformational description of the polymer segments. This means that each polymer segment can exist in either a polar conformation which interacts with one set of interaction parameters or in a less polar conformation which interacts with another set of interaction parameters. The probability of a polar conformation will be denoted P in the following. One then obtains.

$$U = N(\Phi_1 \Phi_2 (P w_{1p} + (1-P) w_{1u}) + \Phi_2 \Phi_2 (P(1-P) w_{pu} + (1-P)^2 w_{uu}/2)) \quad (1')$$

and

$$S = -k N (\Phi_1 \ln \Phi_1 + \Phi_2/M \ln \Phi_2 + \Phi_2 (P \ln P + (1-P) \ln ((1-P)/F)$$

Subscripts u and p in equation 1′ refer to the nonpolar and polar conformations of the polymer. The parameter F appearing in equation 3′ is a measure of the size of phase space corresponding to the nonpolar conformations relative to that of the polar conformations. In more physical terms this means that there may be several conformations of a nonpolar type and only one or a few of a polar type. F measures the ratio between these numbers. Equations 1′ and 3′ can be further generalized to include the effect of yet another polymer. One then obtains

$$U = U' + \Phi_1 \Phi_3 w_{13} + \Phi_2 \Phi_3 (P w_{p3} + (1-P) w_{u3}) \quad (1'')$$

$$S = S' - kN\Phi_3/M_3 \ln \Phi_3 \quad (3'')$$

U′ and S′ in these equations are defined as U and S according to equations 1′ and 3′ . These equations are used in our theoretical modeling in the following sections. Technically this is done by forming the free energy A=U-TS and minimizing this quantity with respect to the compositions and amounts of the three possible phases. If all the three solutions are the same the stable realization of the system will be a one phase system, if two are the same but the third different, the system prefers to separate into two phases. Finally, if the compositions of all three are different, the system is a three-phase system according to the model.

The model suggested above is extremely crude in its treatment of the interactions in the system. First of all we assume that the total interaction energy in a system can be written as a polynomial of second degree in system concentrations even in a system with charged polymers or micelles, despite the fact that the electrostatic energy is nonlocal. Technically this corresponds to an expansion of the electrostatic part of the interaction energy in a Taylor expansion of second degree around a suitable point in phase space. Another weak point in the modeling is that in some of the applications we will model a surfactant system where micelles are present as if the micelles were polymers. The reason why this can be done is that, since both polymers and micelles are large entities, the entropy of mixing is small compared to the interaction energy and is of importance only at very low concentrations. This part of the entropy is reasonably well described by any model. Thus it is clear that it is the interaction part which is important. Obviously, aggregation, which is not described in the model, will also occur in the real systems. But again, it may be argued that most likely it is possible to reasonably well describe the interaction part of the free energy in a Taylor expansion of second degree around a suitable point in phase space.

4. The Polymer - Solvent System

In this section we will describe the properties of the polymer-solvent system and the applications of the theoretical model to this type of system. Two types of polymers will be discussed in this work. The first type is of nonionic character and contains ethyleneoxide groups, and the other kind of polymer is ionic. In the former case, the solubility of the polymers (in water and formamide) decreases with temperature. This must mean that the effective interaction between the polymer segments gets more attractive at higher temperatures. In our model this is a consequence of the equilibrium between the polar and nonpolar conformations of the ethyleneoxide groups. When the temperature is increased, entropy populates the nonpolar conformations, since they can be realized in more ways than the polar ones, and thus the effective polarity of the polymer decreases with increasing temperature. In figure 1 we show the theoretical and experimental phase diagrams for the water-poly(ethylene oxide) system [26, 30].

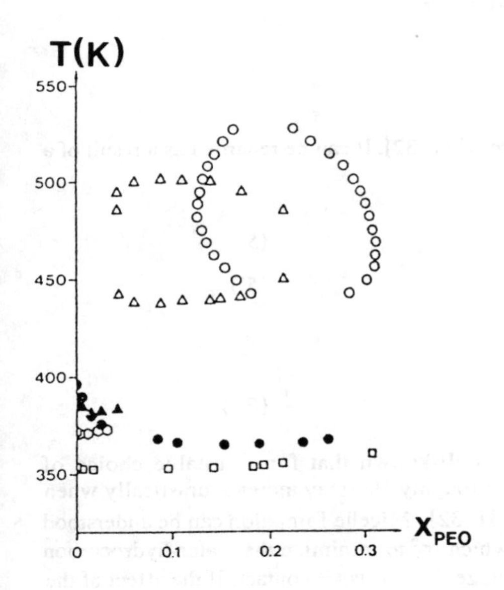

Figure 1. Phase diagram for the PEO-Water system. Theoretical results from [26] and experimental results from [30]. Used interaction parameters $w_{lt,lt}$ = 10172 J/mol, $w_{lt,ht}$ = 6353 J/mol, $w_{lt,H2O}$ = 650.8 J/mol, $w_{ht,H2O}$ = 10654 J/mol. The parameter F=8. Theoretical (□) and experimental (O) points for M = 23200, theoretical (●) and experimental (▲) points for M = 330, theoretical (o) and experimental (Δ) points for M = 51.2. Theoretical data from [26], experimental data from [30]. Reprinted with permission of the copyright owner from [26].

Parameter values are given in the figure captions. In an alternative picture one may regard the phase separation as a consequence of a series of equilibria similar to those responsible for micelle formation (see next section).

$$P + P \rightleftarrows P_2 \qquad (4)$$

$$P + P_2 \rightleftarrows P_3 \qquad (4')$$

$$\vdots \qquad \vdots$$

$$P + P_n \rightleftarrows P_{n+1} \qquad (4'')$$

Note that we have regarded the phase separation process occurring in a polymer-solvent solution as an aggregate formation. This is only correct for infinite n values. A problem with equation 4 is the meaning of the concept complexes - P_2, P_3 etc. - in the semidilute concentration region ,but once the definition of these concepts has been agreed on it is obvious that an increased effective attraction between the polymer segments will favour the right hand side of equation 4.

Similar effects as those observed for the nonionic polymers discussed above on temperature increase, can be observed for ionic polymers when salt is added to the system. Salt will screen the electrostatic repulsion between the polymer molecules and if the backbone of the polymer system is hydrophobic enough phase separation may occur.

5. The Surfactant - Solvent System

Micellization in water is a highly cooperative process [31, 32]. It can be regarded as a result of a set of equilibria

$$S + S \rightleftarrows S_2 \qquad (5)$$

$$S + S_2 \rightleftarrows S_3 \qquad (5')$$

$$\vdots \qquad \vdots$$

$$S + S_n \rightleftarrows S_{n+1} \qquad (5'')$$

In systems which are coupled in this way it is well-known that for a suitable choice of equilibrium constants, the concentration of S_n for n roughly 100 may increase drastically when the free S concentration is only slightly increased [31, 32]. Micelle formation can be understood from the interplay between hydrophobic forces, which try to minimize the water-hydrocarbon contact, and electrostatic forces, which try to maximize the ion-water contact. If the effect of the counterions on the micellization process is considered, an observation, which at first may seem somewhat confusing, can be made [33]. It is then possible to write the equilibrium conditions for micelle formation.

$$n\,S + \alpha\, n\, I \rightleftarrows S_n\, I_{\alpha n} \qquad (6)$$

$$[S_n I_{\alpha n}] / [S]^n [I]^{\alpha n} = K \quad (7)$$

α in equations 6 and 7 is the degree of counterion binding. However, such a treatment does not predict some striking features of ionic surfactant solutions. This is due to not considering the long-range electrostatic interactions. One important observation is that of a closely concentration invariant degree of counterion association, a phenomena also encountered for polyelectrolyte systems and usually denoted counterion condensation. Another is that of a decreased surfactant monomer concentration and activity at increasing surfactant concentrations above the critical micelle concentration (CMC). This can be understood from the unfavourable contribution to micellization, which arises from the negative entropy term arising from the inhomogeneous counterion distribution, and its partial elimination with increasing surfactant concentration due to the increased counterion concentration in the bulk [33].

6. Systems containing polymer and surfactant

In this section we will first consider the phase behaviour of solutions of nonionic polymers containing ethylene oxide groups when an ionic surfactant is added and later the phase behaviour of solutions of an ionic polymer when an ionic surfactant of opposite charge is added. The starting point for the discussion of the interactions between a nonionic polymer and an ionic surfactant is that both these species contain fairly hydrophobic parts, and that they may take part in cooperative processes, which are induced by the interaction between these hydrophobic domains. Since the origin of the cooperativity is the same (hydrophobicity) it seems reasonable to assume that there may be an interplay between the processes behind the cooperativity. Thus it seems plausible to write in accordance with equations 4 and 5

$$S + P \rightleftarrows SP \quad (8)$$

$$S + SP \rightleftarrows S_2P \quad (8')$$

$$\vdots$$

$$S + S_n \rightleftarrows S_{n+1}P \quad (8'')$$

The process illustrated in equation 8 is the formation of a micelle bound to a polymer. Normally it is assumed that the aggregation number for these micelles is somewhat smaller than that for free micelles. It is also standard to assume that the amount of aggregates of type S_2P, S_3P and so on, where no real micelles are formed, is so small that their contribution to the amount of bound surfactant can be neglected. On the other hand, we may also write in accordance with equations 4 and 5:

$$P + S \rightleftarrows PS \tag{9}$$

$$P + PS \rightleftarrows P_2S \tag{9'}$$

$$\vdots$$

$$P + P_nS \rightleftarrows P_{n+1}S \tag{9''}$$

Here we have considered the effect of the surfactant on the clouding phenomena for the polymer-solvent system. If the ionic part of the surfactant is ignored one would thus expect that addition of a nonpolar molecule like hexanol would lower the cloudpoint temperature. In figure 2, we show a cut in a three-component diagram for the EHEC-water-hexanol system [22]. (EHEC is a cellulose-based polymer, where the cellulose has been substituted with ethyl and ethyleneoxide groups.)

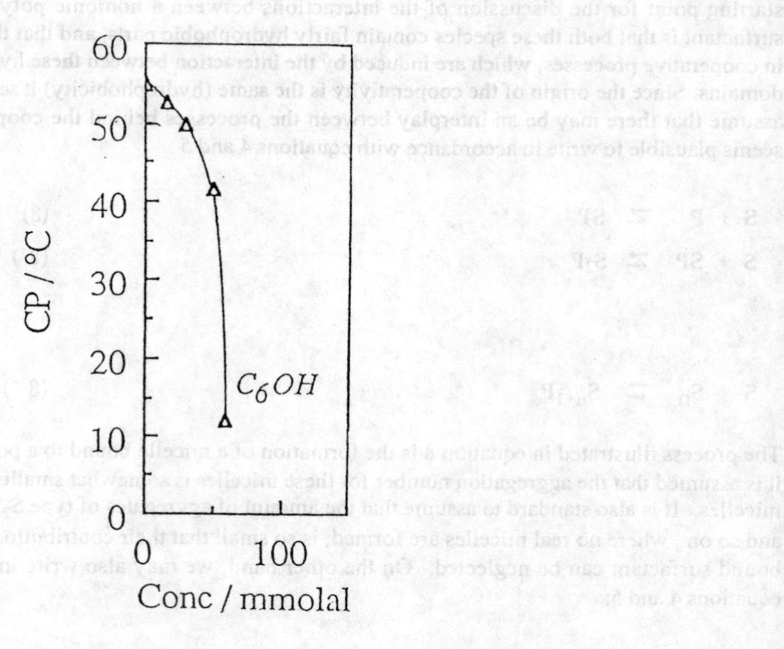

Figure 2. Cloud point of a salt-free 1.0% EHEC - water solution vs. the concentration of hexanol. Reprinted with permission of the copyright owner from [22].

From figure 2 and the above reasoning, it seems clear that the hydrophobic part of the surfactant favours phase separation and induced aggregation in polymers of the type discussed. If an ionic surfactant in very low concentrations is added, it seems reasonable to assume that it induces the same type of aggregation, but that the electrostatic repulsion between the aggregates prevents phase separation. Salt may, however, screen this long range repulsion. In figure 3, we show similar cuts as in figure 2, but for sodium dodecylsulfate (SDS) as an additive, for different salt concentrations [18].

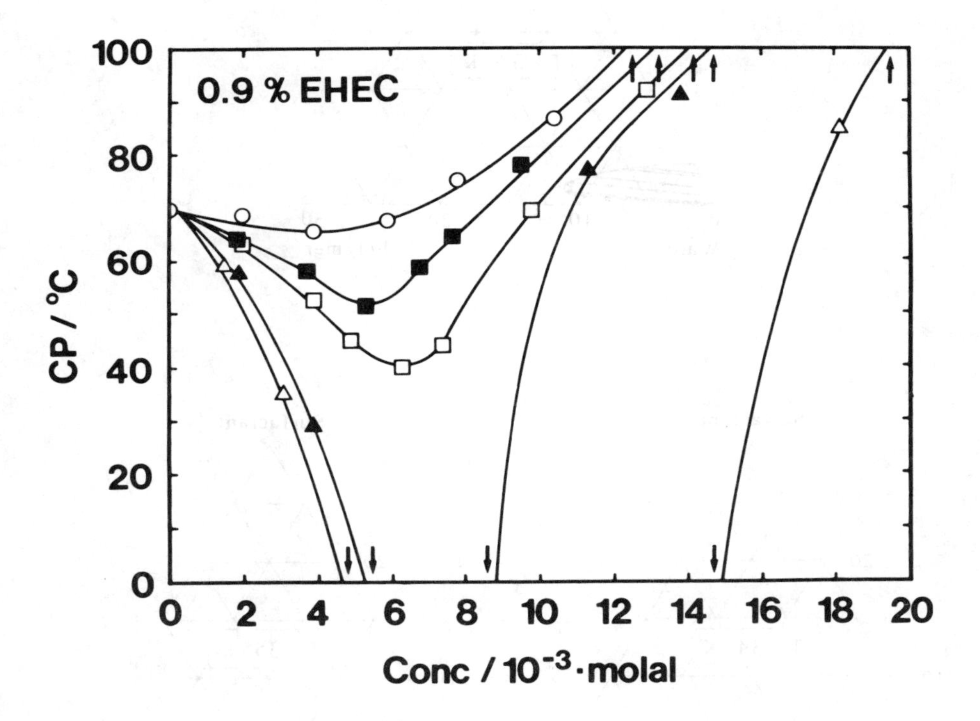

Figure 3. Cloud point of a 0.9% EHEC - water solution vs. the SDS concentration. Different amounts of NaCl is added: (o) 0, ( ) 0.009%, ( ) 0.019%, (Δ) 0.046%, (Δ) 0.11%. Arrows represent clouding occurring below 0 °C or above 100 °C. Reprinted with permission of the copyright owner from [18].

In figure 4, we show theoretically calculated phase diagrams for a solvent - polymer - surfactant system at a few different temperatures [22]. In figure 5a, we show a similar cut as in figure 3 corresponding to the system presented in figure 4 [22]. In figure 5b, we show the effect on the cut of making the attraction between the polymer and the surfactant 400 J/mole larger [22]. This is the way the effect of addition of salt can be incorporated into the model, since salt will decrease the repulsion between the different aggregates.

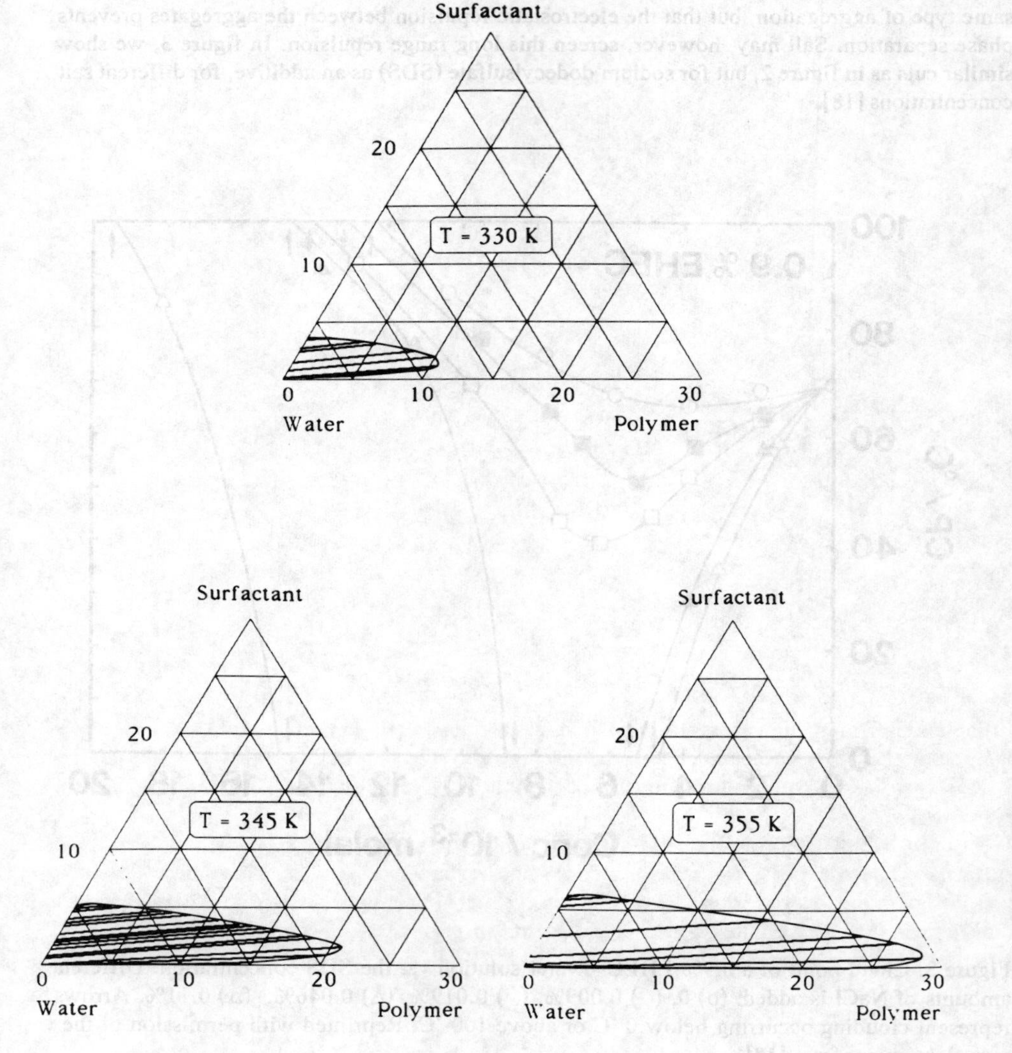

Figure 4. Full three component phase diagrams (at three different temperatures) calculated for a model system where the surfactant is treated as a polymer (degree of polymerization = 100). Interaction parameters describing the water polymer system are those used in figure 1 but they are scaled by 338/352 in order to give the right clouding temperature. Parameters used to describe the surfactant interactions $w_{surf,H2O}$ = -3750 J/mol, $w_{lt,surf}$ = -7100 J/mol, $w_{ht,surf}$ = 1100 J/mole. The degree of polymerization of the polymer is 100000. Reprinted with permission of the copyright owner from [22].

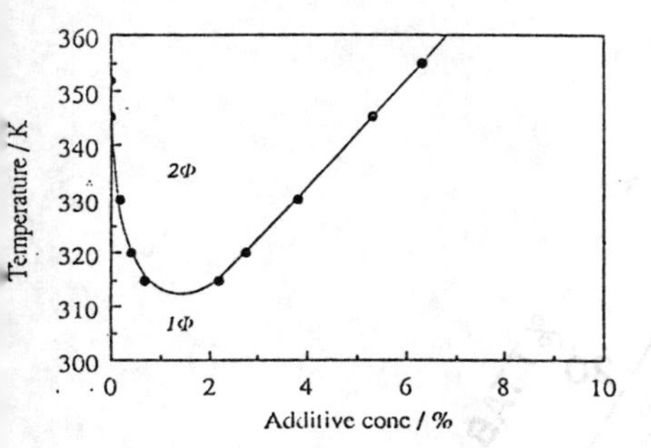

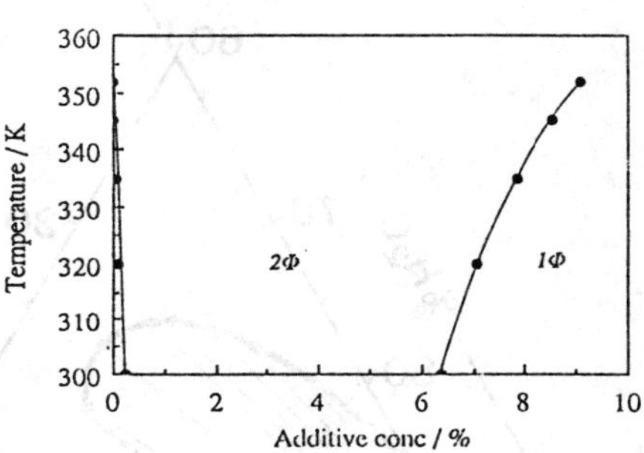

Figure 5a. A similar cut as obtained experimentally in figure 3 but obtained from the theoretical phase diagrams in figure 4. Reprinted with permission of the copyright owner from [22].

Figure 5b. A similar cut as in figure 5a but the polymer surfactant interaction parameter has been made 400 J/mol more attractive. Reprinted with permission of the copyright owner from [22].

From what has been said above it seems necessary to consider the cooperativities in both the polymer system and in the surfactant system and write for the general complex formed.

$$nS + mP \rightleftarrows S_n P_m \qquad (10)$$

Complexes with large m values and small n values are apparently important close to the clouding temperature and low surfactant concentrations. One may also expect that complexes with high n values and low m values are most important at low temperatures and high surfactant concentrations.

Finally, we will concentrate on systems built up from an ionic polymer and an ionic surfactant. A typical phase diagram for such systems is shown in figure 6a (experimental results) and in figure 6b (theoretical model calculation) [23]. The characteristic feature of such phase diagrams is that they contain a two-phase region in the water-rich corner with the tie-lines pointing towards the water corner.

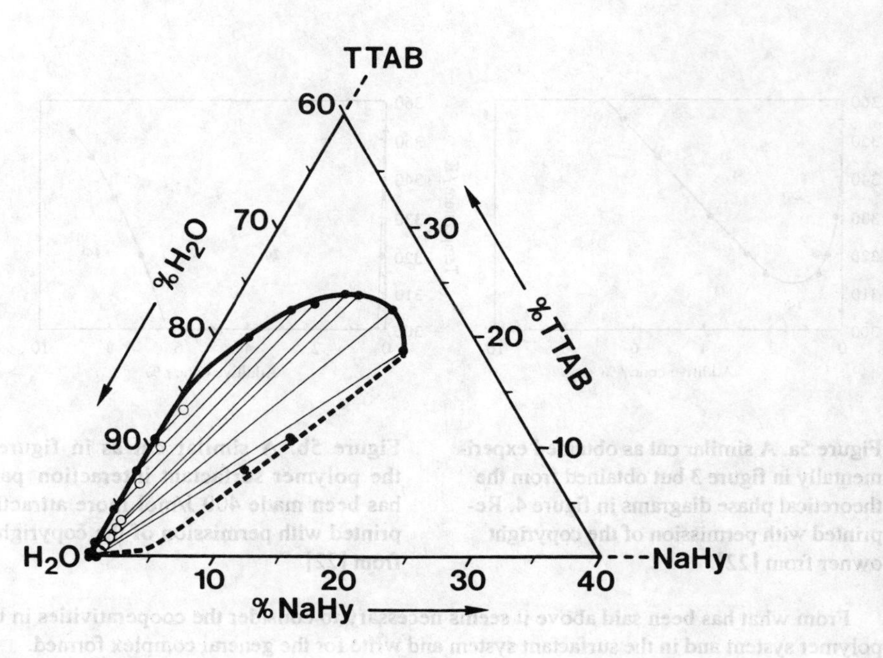

Figure 6a. Experimental 3-component phase diagram for the system Sodiumhyaluronan - Tetradecyltrimethylammonium bromide - water. The composition of some samples are indicated. Open circles refer to initial sample compositions and filled circles connected by tie-lines to the composition of the 2 phases in equilibrium. The dashed part of the phase boundary indicates larger uncertainty in this region. Reprinted with permission from the copyright owner from [22].

This type of phase diagram is typical for a situation where there is strong attraction between the two macromolecular species. The parameters used to calculate this phase diagram are shown in the figure caption. For comparison in figure 7 we show a phase diagram for a system made up from poly(ethyleneoxide), dextran and water [27].

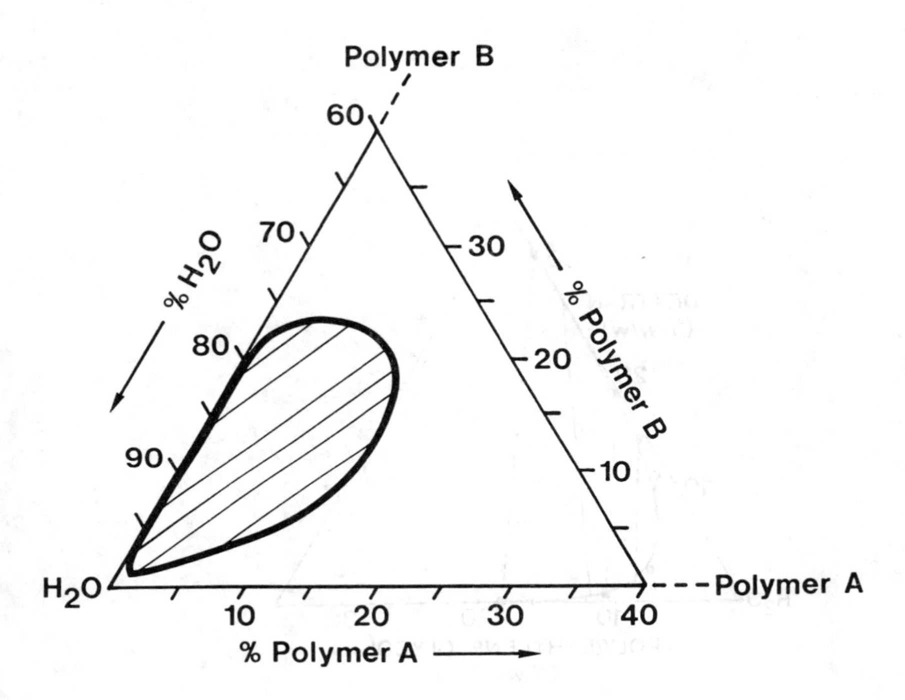

Figure 6b. Theoretically calculated phase diagram for a ternary system of 2 different polymers in a common solvent. $M_2 = 300$ and $M_3 = 25$. The following interaction parameters have been used $w_{12} = -200$ J/mol, $w_{13} = 1000$ J/mol, $w_{23} = -5200$ J/mol. Reprinted with permission from the copyright owner from [22].

From the model calculations it is clear that the two-phase region observed in figure 7 is a result of a repulsion between the two macromolecular species and that tie-lines here are almost perpendicular to those in figure 6. The easiest way to understand the phase behaviour observed in figure 6 is to adopt the ideas of Shinoda [34] and regard the polymer-surfactant complex as a solvent for water. The ability to incorporate solvent will obviously depend both on the effective charge density of the polymer-surfactant complex and on the amount of hydrophobic material present in the complex. On a molecular level, one may say that when surfactant is added to the polymer, micelles will start to form at the polymer, thus effectively reducing the charge density of the aggregates and at the same time starting an accumulation of hydrophobic material on the polymer. This will decrease the solubility of water in the polymer surfactant system and these two effects are the cause of the observed phase separation. When more surfactant is added the amount of polymer-surfactant complex will increase and finally it will be capable of dissolving all water molecules and the one phase region is reached again.

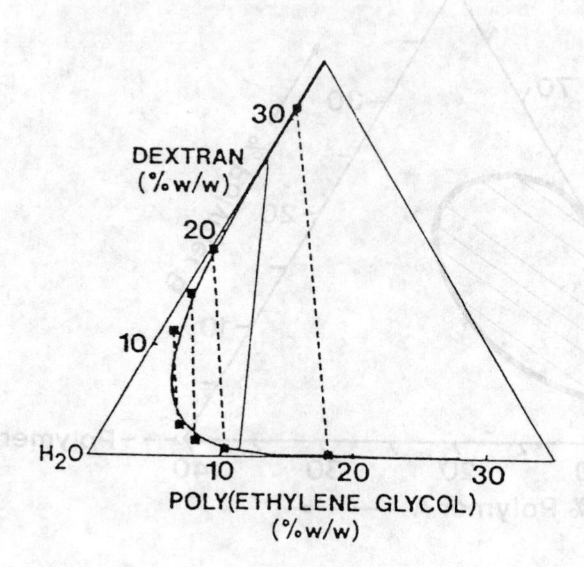

Figure 7. Calculated and experimental phase diagrams for the system Polyethyleneoxide - Dextrane - Water. Calculated curve (—), with tieline (—) and experimental points (■) with tielines (----) . Molecular weight for polyethyleneoxide = 20000, for dextrane 40000, temperature is 20 °C.
Parameters describing the polyethyleneoxide water system from figure 1. Other parameters w_{13} = 1096 J/mol, $w_{lt,3}$ = 757.5 J/mol, $w_{ht,3}$ = 6628 J/mol. Reprinted with permission of the copyright owner from [27].

Two alternative explanations are given for these phenomena in the literature. The first amounts to saying that when more surfactant is added, the charge of the polymer-surfactant complex is reversed. This will increase the effective polarity of the surfactant system. The other explanation is based on the fact that the free surfactant ion concentration decreases once free micelles start to form. Thus one may expect the binding of surfactant micelles to the polymer to decrease when the surfactant concentration is increased. Both these latter effects cannot simultaneously be true, and the directions of the tie-lines in the phase diagram suggest that it is the amount of polymer-surfactant complex which is the important effect. There may, however, be minor contributions from one of these latter phenomena. In this context it may be appropriate to issue a warning. As was mentioned in the section dealing with the polymer-solvent systems, the meaning of the concepts polymer complex is not unambiguous, and in a similar manner one must be careful with the meaning of the concept polymer-bound micelle in the semidilute region. If we use a lattice model and assume that 1% of the cells are occupied with polymer segments and that half the micelle surface is covered with ionic groups and half with nonpolar material, and we further assume that we have an aggregation number for the micelles of 50 and that each cell can contain a charged headgroup, then if we distribute the micelles randomly over the entire volume, 37% of the micelles will not be in direct contact with any polymer chain. If the polymer concentration is 2% the corresponding number is 13%. These numbers indicate that only a minor part of the volume is polymer-free even at low polymer concentrations.

Acknowledgements

The experimental results are due to our coworkers, notably Anders Carlsson(nonionic polymers), Kyrre Thalberg(ionic polymers) and Åke Sjöberg (the polyethylene oxide-dextrane system). We are grateful to Gordon Tiddy for part of the introduction.

References

1. Tiddy, G.J.T. (1980) ´Surfactant-Water Liquid Crystal Phases´, Physics Report, 57,1-46.
2. Hall, D.G. (1985) ´Thermodynamic of Ionic Surfactant Binding to Macromolecules in solution´, J.Chem. Soc., Faraday Trans. 1, 81, 885-911.
3. Hayakawa, K. and Kwak, J.C.T. ´Interactions between Polymers and Cationic Surfactants´ in D. Rubingh and P.M. Holland (eds), Cationic Surfactants: Physical Chemistry. Surfactant Series, in press.
4. Ruckenstein E., Huber G., and Hoffmann H. (1987) ´Surfactant Aggregation in the Presence of Polymers´, Langmuir 3, 382-387.
5. Hayakawa, K., Ohta, J., Maeda, T., Satake, I. and Kwak J.C.T. (1987) ´Spectroscopic Studies of Dye Solubilizates in Micellelike Complexes of Surfactant with Polyelectrolytes´, Langmuir 3, 377-382.
6. Shirahama, K., Masaki, T. and Takashima, K. (1985) ´Interaction between DNA and Dodecylpyridinium Cation´, in P. Dubin (ed), Microdomains in Polymer Solution, Plenum Publishing Corporation pp. 299-309.
7. Delville, A., Laszlo, P. and Schyns, R. (1986) ´Displacement of Sodium Ions by Surfactant Ions from DNA. A ^{23}Na NMR Investigation.´, Biophysical Chemistry 24, 121-133.
8. Tondre, Christian (1985) ´Interaction of Poly(ethyleneoxide) with Sodium Dodecyl Sulfate Micelles: A Fast Kinetic Study by Temperature Jump´, J. Phys. Chem. 89, 5101-5106.
9. Oakes, John (1974), ´Protein Surfactant Interactions´, J.Chem.Soc., Faraday Trans. 1, 70, 2200-2204.
10. Goddard, E.D. (1986) ´Polymer-Surfactant Interaction Part I. Uncharged Water-Soluble Polymers and Charged Surfactants´, Colloids and Surfaces 19, 255-300.
11. Goddard, E.D. (1986) ´Polymer-Surfactant Interaction Part II. Polymer and Surfactant of Opposite Charge. Colloids and Surfaces 19, 301-329.
12. Satake, I. and Yang, J.T. (1976) ´Interaction of Sodium Decyl Sulphate with Poly(L-ornithine) and Poly(L-lysine) in Aqueous Solution´, Biopolymers 15, 2263-2275.
13. Satake, I., Hayakawa, K., Komaki, M. and Maeda, T. (1984) ´The Cooperative Binding

Isotherms of Sodium Alkanesulfonates to Poly(1-methyl 1-4-vinylpyridinium chloride)', Bulletin of the Chemical Society of Japan 57, 2995-2996.
14. Saito, S. (1987) 'Polymer-Surfactant Interactions' in M.J. Schick (ed.) Nonionic Surfactants: Physical Chemistry. Surfactants Series, 15, 881-925.
15. Dubin, P. L., Thé, S. S., McQuigg, D. W. and Gan, C. H. (1988) 'Binding of Polyelectrolytes to Oppositely Charged Ionic Micelles at Critical Micelle Surface Charge Densities', Langmuir 5, 89-95.
16. Skerjanc, J., Kogej, K. and Vesnaver, G. (1988) 'Polyelectrolyte-Surfactant Interactions. Enthalpy of Binding of Dodecyl- and Cetylpyridinium Cations to Poly(styrenesulfonate) Anion', J. Phys. Chem. 92, 6382-6385.
17. Brackman, J. C., van Os, N. M. and Engberts, J.B.F.N. (1988) 'Polymer-Nonionic Micelle
Complexation. Formation of Poly(propylene oxide)-Complexed n-Octyl Thioglucoside Micelles', Langmuir 4, 1266-1269.
18. Carlsson, A., Karlström, G. and Lindman, B. (1986) 'Synergistic Surfactant-Electrolyte Effect in Polymer Solutions', Langmuir 2, 536-537.
19. Carlsson, A., Karlström, G., Lindman, B. and Stenberg, O. (1988) 'Interaction between Ethyl(hydroxyethyl)cellulose and Sodium Dodecyl Sulphate in Aqueous Solution', Colloid and Polym. Sci. 266, 1031-1036.
20. Carlsson, A., Karlström, G., Lindman, B. (1989) 'Characterization of the Interaction between a Nonionic Polymer and a Cationic Surfactant by the Fourier Transform NMR Self-Diffusion Technique', J. Phys. Chem. 93, 3673-3677.
21. Carlsson, A., Karlström, G. and Lindman, B. 'Thermal Gelation of Nonionic Cellulose Ethers and Ionic Surfactants in Water', J. Phys. Chem. (in press) .
22. Karlström, G., Carlsson, A., and Lindman, B. 'Phase Diagrams of Nonionic Polymer - Water Systems. Experimental and Theoretical Studies of the Effects of Surfactants and Other Cosolutes', J. Phys. Chem. (submitted).
23. Thalberg, K., Lindman, B., and Karlström, G. 'Phase Diagram of a System of Cationic Surfactant and Anionic Polyelectrolyte: Tetradecyltrimethylammoniumbromide-Hyaluronan-Water', J. Phys. Chem. (in press).
24. Cabane, B. (1977), ´Structure of Some Polymer-Detergent Aggregates in Water´, J. Phys. Chem. 81, 1639-1645.
25. Andersson, M. and Karlström, G. (1985) 'Conformational Structure of 1,2-Dimethoxyethene in Water and Other Dipolar Solvents Studied by Quantum Chemical, Reaction Field, and Statistical Mechanical Techniques', J. Phys. Chem. 89, 4957-4962.
26. Karlström, G. (1985) 'A New Model for Upper and Lower Critical Solution Temperatures in Poly(Ethylene Oxide) Solutions', J. Phys. Chem. 89, 4962-4964.
27. Sjöberg, Å. and Karlström, G. (1989) 'Temperature Dependence of the Phase Equilibria for the System Poly(Ethylene Glycol)/Dextran/Water. A Theoretical and Experimental Study', Macromolecules 22, 1325-1330.
28. Hill, T.L. (1960) ´Introduction to Statistical Thermodynamics´, Addison-Wesley MA , Chapter 21.
29. Goldstein, R.E. (1984) ´On the Theory of Lower Critical Solution Points in Hydrogenbonded Mixtures´, J. Chem. Phys. 80, 5340-5341.
30. Sakei, S., Kawahara, N., Nakata, M. and Kaneko M. (1976) ´Upper and Lower Critical Solution Temperatures in Poly(ethyleneglycol) solutions´, Polymer 17, 685-689.
31. Israelachvili, J.N. (1985), ´Intermolecular and Surface Forces´, Academic Press.
32. Lindman, B., and Wennerström, H. (1980) ´ Micelles. Amphiphile Aggregation in Aqueous

Solution.´, Topics in Current Chemistry, 87, 1-83.
33. Gunnarsson, G., Jönsson, B., andWennerström, H. (1980) ´Surfactant Association into Micelles. An Electrostatic Approach´, J. Phys. Chem. 84, 3114-3121.
34. Shinoda, K. (1978) ´Principles of Solution and Solubility´, Marcel Dekker Inc, p 135.

KINETIC AND EQUILIBRIUM STUDIES ASSOCIATED WITH POLYMER SURFACTANT INTERACTIONS

D.M. BLOOR, E. WYN-JONES
Department of Chemistry and Applied Chemisrtry,
University of Salford,
Salford
M5 4WT
England.

ABSTRACT Binding isotherms and kinetic measurements associated with the adsorption/desorption of the surfactants sodium hexadecylsulphate (SHS) and sodium decylsulphate(SDeS) onto the neutral polymer poly(N-vinylpyrrolidone) are considered. Surfactant ion selective electrodes were used to construct equilibrium binding isotherms from emf measurements. The kinetic data were measured with the pressure-jump relaxation technique (for SHS) and also by ultrasonic absorption (for SDeS). Application of the linear phenomenological theory to the binding and kinetic data shows clearly that the desorption rate is proportional to the amount of bound surfactant.

1. Introduction

It is well known that aqueous solutions containing surfactants and polymers have many pharmaceutical and industrial applications [1, 2]. Due to their importance many of these systems have been studied at fundamental levels using a variety of experimental techniques. The vast majority of studies have involved equilibrium measurements where the objective is to construct a binding isotherm which essentially relates to the distribution of surfactant between the monomer species in solution and those bound on to the polymer. The aim of the investigations in this laboratory has been to obtain information about the mechanistic aspects of binding and in these circumstances it is necessary to carry out kinetic measurements to complement the equilibrium data. It is well known that the dynamics associated with surfactant exchange processes are extremely fast and can not be measured by conventional kinetic methods. As a result chemical relaxation techniques have to be used [3, 4].

This report describes the use of pressure jump and ultrasonic relaxation techniques to investigate the binding of the anionic surfactants sodium hexadecyl sulphate (SHS) and sodium decyl sulphate (SDeS) to poly(N-vinyl pyrrolidone) (PVP). Information on binding isotherms have been obtained by the application of surfactant selective electrodes which we have extensively used to investigate the partitioning of the surfactant between free monomers in solution and those bound to various additives in different formulations [3, 4].

D. M. Bloor and E. Wyn-Jones (eds.),
The Structure, Dynamics and Equilibrium Properties of Colloidal Systems, 149–160.

2. Experimental

2.1 SURFACTANT SELECTIVE ELECTRODES

The surfactant selective electrodes used in this work are a much improved version to those described in previous publications [5-7]. The principle of this electrode is that the monomer surfactant activity in various solutions can be obtained from emf measurements from the following cell:

Surfactant electrode	Test solution containing 10^{-5} or 10^{-4} mol dm^{-3} sodium bromide	electrode reversible to sodium bromide

The surfactant and bromide ions carry the same charge, hence measurements using the above cell give a ratio (surfactant ion activity)/(bromide activity). The activity coefficient of the surfactant and its counterions are expected to be approximately equal, in which case the ratio (surfactant monomer concentration)/(bromide ion concentration) is measured.

The surfactant membrane electrode was a modified version of an EDT research electrode (ISE310). In principle the Calcium membrane of the electrode tip was drilled out and replaced by a specially conditioned polyvinylchloride (PVC) membrane which was attached to the electrode tip using tetrahydrofuran; the internal silver coil of the Calcium electrode was replaced by a silver rod used as a Ag/AgBr internal reference electrode. The key factor which has contributed to the success of this electrode is the specially conditioned polyvinylchloride membrane which for anionic surfactant is prepared according to the recipe described by Davidson [7].

For anionic surfactants the polyvinylchloride has positively charged trimethylammonium groups which are neutralised by the surfactant anions before use. The modified PVC is then mixed with a commercially available polymer plasticiser to form the membrane. The potentiometric measurements were made using digital pH millivolt meter (Corning Ion analyser 50). In all measurements the temperature was controlled to within ± 0.1% by circulating thermostatted water through a double walled glass cell containing the sample solution which was continually stirred using an air driven magnetic stirrer. During the emf measurements the concentration of the test sample solution was changed successively by adding a known amount of solution in to the initial sample (30 cm^3) using an Aglar microcylinder system.

The monomer surfactant concentrations in the presence of polymer were evaluated as follows: First, the e.m.f. of the surfactant selective electrode relative to the bromide reference electrode was measured for the pure surfactant at increasing surfactant concentrations. At surfactant concentrations above the critical micelle concentration (cmc) micellar aggregates are formed and the concentration of monomer surfactant is found to decrease. At concentrations below the cmc all the surfactant is in its monomeric form and in this range the emf data gave good Nernstian response for the electrode as shown in Fig. 1.

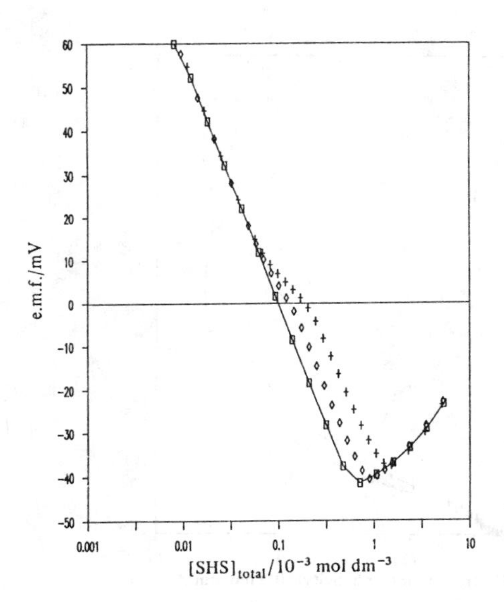

Fig. 1 Emf data for SHS and 0.0 (□), 0.01 (◇), 0.025 (+) (w/v%) PVP at 35°C in the presence of 10^{-5} mol dm^{-3} sodium bromide.

The experiment was then repeated by measuring the relative emf of the surfactant selective electrode in the presence of a constant amount of PVP (see Fig. 1.). These measurements on the polymer/surfactant solutions were carried out in such a way that the reversibility of the binding process was confirmed. When the emf measurements of the polymer/surfactant solutions were completed the emf of the electrode was then rechecked against monomer surfactant to ensure consistency. From these data it is possible to evaluate the monomer concentration m_1 at each total concentration of surfactant (C). The binding ratio $\Gamma = (C-m_1)/Cp$ can be derived from such data; where C_p is the concentration of polymer expressed in mol monomer dm^{-3}. Binding isotherms were constructed in terms of surfactant absorbed per mol monomer polymer plotted against m_1. For all measurements the solutions were doped with either 10^{-5} or 10^{-4} mol dm^{-3} sodium bromide for reference electrode purposes. Typical binding isotherms are shown in Figs. 2 and 3.

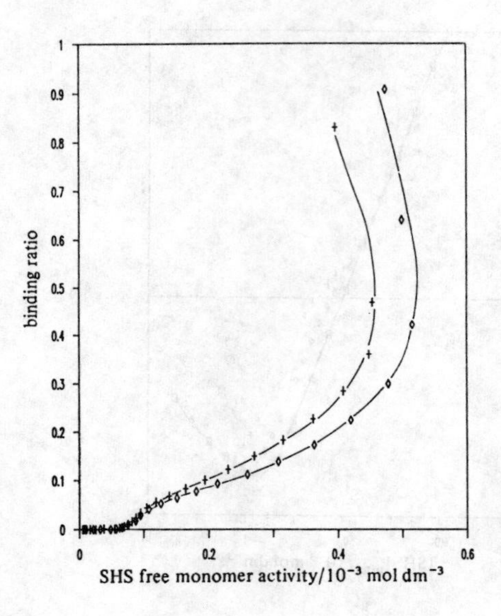

Fig 2. Binding isotherm for SHS/PVP system, (◇) 0.01, (+) 0.025, (w/v%) PVP at 35°C.

3. Relaxation Measurements

For the system SHS/PVP, kinetic measurements were undertaken using a Dialog pressure jump apparatus with conductivity detection incorporating a rapid data capture and analysis system [8, 9]. Relaxation times could be measured over the range 10^{-1}-10^{-4} seconds. At low concentrations of surfactant the amplitude of the relaxation signal was rather weak and in these circumstances several transients were measured for each solution and the stored data were averaged. All measurements were carried out at 35°C.

For the SDeS/PVP measurements two ultrasonic techniques were used. The Eggers Resonance method [10] for which the frequency range has recently been extended down to 0.2

MHz was used in the range 0.2-20 MHz and a standard pulse technique was used for frequencies up to [11] 95 MHz. Aqueous solutions of PVP are known to exhibit a weak ultrasonic relaxation which has been attributed to various conformational changes in the polymer [12]. However at a concentration of 1% (w/v) of the polymer the amplitude of this relaxation is negligible. When surfactant was added to 1% polymer solution a relaxation process associated with the binding process was observed and in all cases the data were consistent with a single relaxation equation.

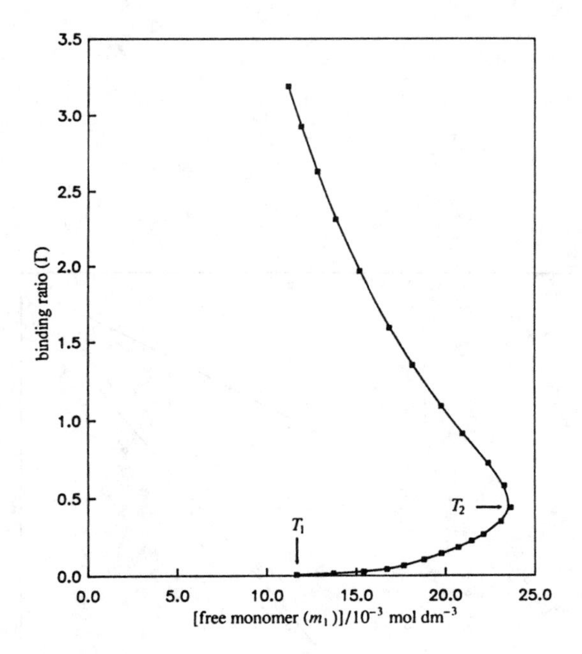

Fig 3. Binding isotherm for SDeS/PVP system, 1.0 (w/v%) PVP at 30°C.

4. Analysis of Data

4.1 ELECTRODE MEASUREMENTS

The binding isotherms for the systems investigated in the present work are shown in Figs. 2 and 3 and these data agree qualitatively with conclusions of other experiments involving the binding of alkyl sulphates with PVP [13-16]. In the present case as in previous studies the binding isotherms are characterised by two surfactant concentration regions [1] defined by T_1 and T_2 which are identified respectively as the total concentration of surfactant corresponding to the onset of binding, after which both the amount of surfactant bound to the polymer and monomer surfactant increases as total surfactant concentrated is increased, and T_2 which corresponds to the total surfactant concentration at which micelles are formed. The onset of T_2 is associated with a decrease in monomer concentration with increasing surfactant - a good indication of the formation of "proper" micelles .

5. Relaxation Data

For SHS and SDeS binding to PVP the range of surfactant concentrations from which it was possible to carry out relaxation measurements are indicated in Figs. 4 and 5 which show plots of the reciprocal relaxation time $\frac{1}{\tau}$ against total concentration.

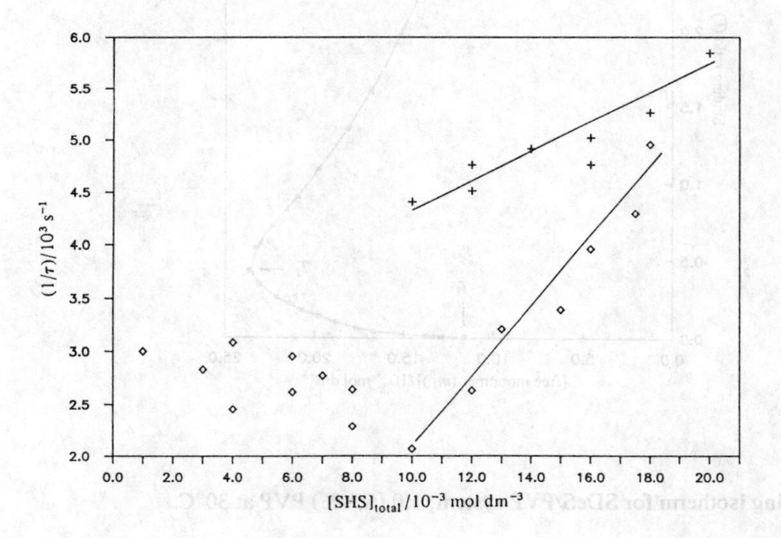

Fig. 4. The reciprocal relaxation time $(1/\tau)$ as a function of total SHS concentration for (◇) 0.01, (+) 0.025 (w/v%) PVP at 35°C.

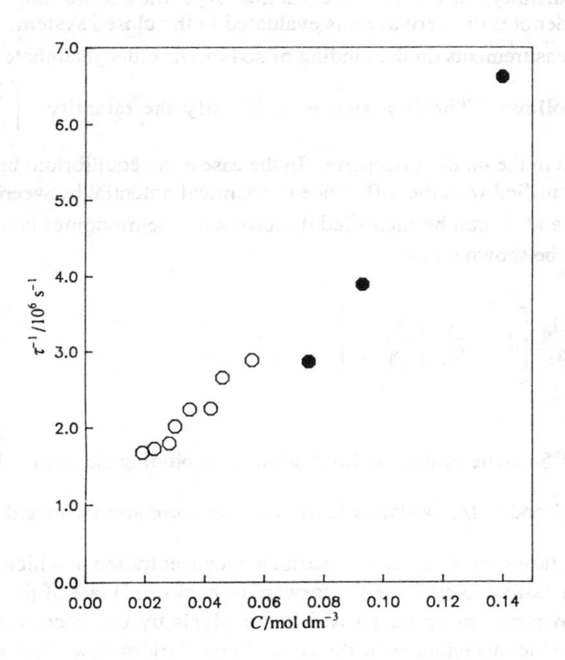

Fig. 5. The reciprocal relaxation time $\frac{1}{\tau}$ as a function of total SDeS concentration for 1.0(w/v%) PVP at 30°C. Open symbols, binding region; filled symbols, micellar region.

By referring to the binding isotherms it is clear that these measurements have been taken in the region T_1-T_2 and also in micellar solutions beyond T_2. The relaxation data in the present work has been analysed using a phenomenological treatment which has been developed in this laboratory to investigate relaxation data associated with different types of association phenomena [17]. The pre-requisite in the application of this treatment is that the association phenomenon in question can be defined by a general one step equilibrium and that the relaxation is a well defined

single process. In the present situation such an equilibrium can be defined as follows:

"Free" monomer $\rightleftarrows$ "Bound" Monomer

In order to apply the following treatment equilibrium data in the form of binding isotherms are also essential. In these circumstances the phenomenological equation of interest is:

$$\frac{1}{\tau} = \frac{R_f}{RT}\left(\frac{\partial A}{\partial \xi}\right)_c \qquad (1)$$

where A is the relaxation affinity, the extent of the reaction, R_f is the equilibrium forward rate (= backward rate R_b) and c denotes the derivation is evaluated in the closed system. In connection with the pressure jump measurements on the binding of sodium hexadecylsulphate to PVP we use the above equation as follows. The first step is to identify the quantity $\left(\frac{\partial A}{\partial \xi}\right)_c$ with the equilibrium data presented in the binding isotherm. In the case of an equilibrium between free and bound monomer, A is identified with the difference in chemical potential between the bound and free monomer surfactant and ξ can be identified directly with the monomer concentration. In these circumstances it can be shown that

$$\frac{\partial A}{\partial \xi} = \frac{RT}{m_1}\left[1 + \frac{1}{C_p}\left(\frac{\partial m_1}{\partial \Gamma}\right)_e\right] \qquad (2)$$

and the quantity on the RHS can be evaluated directly from the binding isotherm. This means that by reference to equations 1 and 2 $1/\tau$ is available from the pressure jump data and $\frac{\partial A}{\partial \xi}$ from the binding isotherm. This in turn means that at each surfactant concentration at which relaxation data have been measured it is possible to evaluate the forward (= backward) rate of the above process. Once these rates are known one can proceed with the analysis by using conventional kinetic considerations to examine the mechanisms of the general equilibrium described above. In these circumstances it is convenient to consider the backward process, that is, the step associated with the dissociation of the bound monomer surfactant to a free monomer in solution. In this situation the simplest kinetic consideration suggests that the backward rate should be proportional to the amount of bound surfactant $R_b = k^-$ (C-m_1) thus a plot of R_b against (C-m_1) should be expected to be a straight line passing through the origin. Such a plot is shown in Fig. 6 and clearly indicates that the backward process in the above equilibrium is a first order reaction.

In the SDeS/PVP system ultrasonic relaxation has been used. In these circumstances the application of the phenomenological treatment is even more straight forward than that described in the above discussion. For ultrasonic measurements a second phenomenological equation of interest is associated with the maximum absorption per wavelength μ_m. This equation is

$$\mu_m = \frac{\pi \Delta V^2}{2\kappa_s \left(\frac{\partial A}{\partial \xi}\right)} \tag{3}$$

where ΔV is the volume change and κ_s the adiabatic compressibility. If we divide equation (3) by equation (1) we have:

$$\frac{\mu_m}{\tau} = \left[\frac{\pi \Delta V^2}{2RT\kappa_s}\right] R_1 \tag{4}$$

in which the left hand side is known from ultrasonic relaxation experiment and the bracketed thermodynamic term on the right hand side can be evaluated through direct or indirect measurements. In the SDeS/PVP system data the backward rate $R_b(=R_1)$can now be evaluated from equation (4) and again $R_b$can be plotted against the amount of bound surfactant (C-m_1)

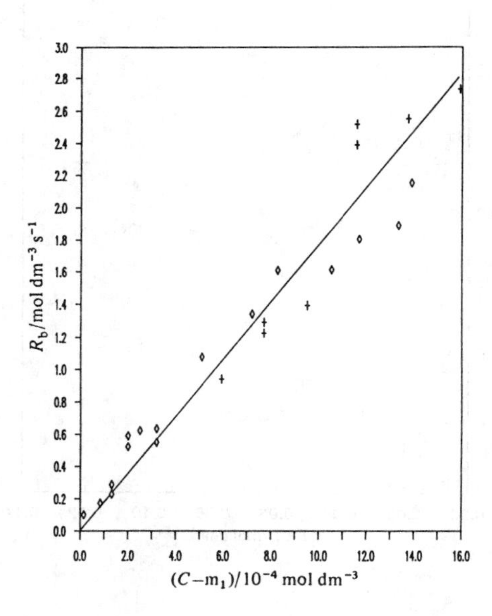

Fig. 6. The backward rate (R_b)as a function of the amount of bound surfactant for SHS in the presence of (◇) 0.01, (+) 0.025 (w/v%) PVP at 35°C

giving identical results for the SDeS/PVP system as shown in Fig. 7 again confirming that the backward rate is a simple first order process.

The most interesting feature concerning the data for both surfactants is that if we refer to the equilibrium binding measurements there is a distinct change in behaviour at the concentration corresponding to T_2. The relaxation data for both surfactants have been measured (over a concentration range spanning) the T_1-T_2 range and also at concentrations beyond T_2, i.e. in the binding region as well as when proper micelles are formed. Despite this the backward rate process is still a simple first order process only depending on the amount of bound surfactant. For the dissociation of bound surfactant the rate law appears to obey the same type of behaviour whether the surfactant is in the form of micelles or in T_1-T_2 region. This implies that in the T_1-T_2 region the surfactant exists as small aggregates. Further evidence for the occurrence of small aggregates can be obtained if we carry out additional electrode measurements in which one

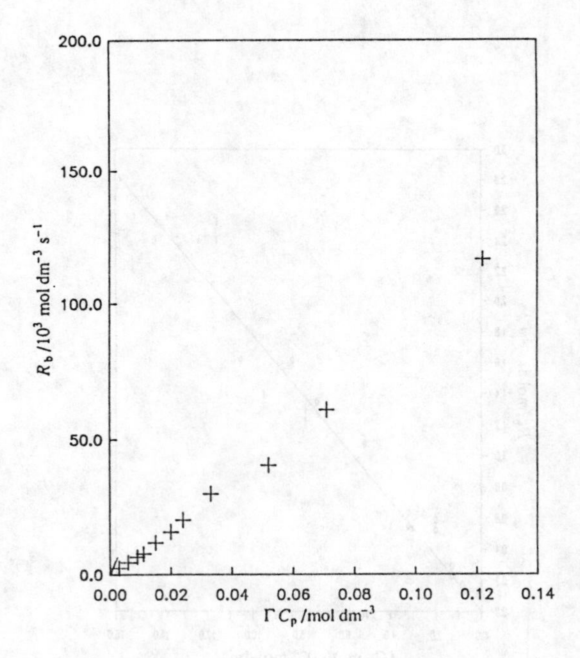

Fig. 7 The backward rate (R_b) as a function of the amount of bound surfactant for SDeS in the presence of 1.0 (w/v %) PVP at 30°C.

measures the emf of the surfactant electrode in the polymer/surfactant solutions against both the bromide and sodium electrode. These measurements show that once binding occurs at T_1 the sodium counter ion also starts to bind to the polymer with a binding constant slightly higher than expected for a micellar solution. The implication here again is that once T_1 is reached the bound surfactant is in the form of small aggregates.

6. Concluding Remarks

At this stage it is interesting to mention one of the noteworthy features in the application of chemical relaxation methods on systems such as polymer/surfactant solutions by using these relaxation methods over a wide time range - our experimental methods cover approximately 10 to 10^{-9} seconds. In principle the different interactions which are observed during the process of surfactant binding to polymer can occur at different time scales and therefore with the use of chemical relaxation methods it is possible to tune in and resolve the various individual processes. In the above discussion we have given examples of how monomer/aggregate equilibrium can be investigated. We have carried out preliminary measurements on other processes in these systems - we describe briefly two of these:

(i) It is well known that when the onset of binding occurs at T_1 there are distinct conformational changes in the polymer [1]. The main experimental evidence for this intramolecular process has been that associated with dramatic changes in viscosity. In principle conformational changes in a polymer can be monitored using ultrasonic relaxation methods. In the system sodium dodecylsulphate/PVP at surfactant concentrations slightly exceeding T_1 a relaxation process has been observed. This relaxation is not associated with the binding process since at the time scale of measurement the exchange rate for the sodium dodecylsulphate is frozen out. Our results suggest that the relaxation, which coincides with the dramatic changes in viscosity, is due to a conformational change in the polymer as a result of surfactant binding.

(ii) At surfactant concentrations exceeding T_2 micelles proper are formed. In relaxation experiments sodium dodecylsulphate micelles have a slow relaxation time denoted τ_2 in the milli-second range which is associated with dissolution/formation of complete micelles [18]. When this slow relaxation time is investigated in the presence of polymer the experimental data shows that initially the reciprocal relaxation time ($1/\tau_2$) decreases, reaches a minimum and then increases dramatically with addition of further polymer. We believe that this minimum is associated with a situation in which the polymer is fully saturated. The addition of more polymer introduces more binding sites causing the aggregation number of the bound micelles to decrease and consequently ($1/\tau_2$) to increase.

References

1. E.D. Goddard, Colloids Surf., 1986, 19 255 and references therein.
2. M.M. Brewer and I.D. Robb, Chem. Ind., 1972, 530.
3. N. Takisawa, P. Brown, D.M. Bloor, D.G. Hall, E. Wyn-Jones, J. Chem. Soc., Faraday Trans (1), 1989, 85 (8) 2099.
4. D.M. Painter, D.M. Bloor, N. Takisawa, D.G. Hall, E. Wyn-Jones, J. Chem. Soc., Faraday Trans (1), 1988, 84 (6), 2087.
5. S.G. Cutler, D.G. Hall, P. Mears, J. Electroanal. Chem., 1977, 85, 145.
6. S.G. Cutler, D.G. Hall, P. Mears, J. Chem. Soc. Faraday Trans. 1, 1978, 74, 1758.
7. C.J. Davidson, Ph.D Thesis (University of Aberdeen 1983).
8. W. Knoche, G. Weise, Chem. Instrum., 1973 5, 91.
9. M. Kriznam, H. Strehlow, Chem. Instrum., 1973 5, 99.
10. F. Eggers, Acoustica, 1968, 19, 323.
11. F. Eggers, Th. Funk, K.N. Richmann, Rev. Sci. Instrum., 1976, 47, 361.
12. M.C. Pererria, P.L. Jobling, E. Wyn-Jones, E.R. Morris and R.A. Pethrick, J. Chem. Soc. Faraday Trans. 2, 1983, 79, 977.
13. M.L. Fishman, F.R. Eirich, J. Phys. Chem., 1971 75, 3135.
14. H. Arai, M. Murata, K. Shinoda, J. Colloid Interface Sci., 1971, 37, 223.
15. G.C. Kresheck, I. Constantinides, Anal. Chem., 1984, 56, 152.
16. D.G. Hall, J. Chem. Soc. Faraday Trans. 1, 1985, 81, 885.
17. D.G. Hall, J. Gormally, E. Wyn-Jones, J. Chem. Soc., Faraday Trans 22, 1983, 79, 645.
18. E.A.G. Aniansson, S.N. Wall, N. Almgren, H. Hoffmann, I. Kiebmann, W. Ulbricht, R. Zana, J. Lang, C. Tondre, J. Phys. Chem. 1976, 80 905.

THE INTERACTION OF AMPHIPHILES WITH MOLECULAR ASSEMBLIES AND POLYMERS

K. SHIRAHAMA, T. WATANABE, and M. HARADA
Department of Chemistry
Faculty of Science and Engineering
Saga University
Saga 840, Japan

ABSTRACT. The interactions of amphiphiles (surfactant and drug) with polymers and molecular assemblies were studied by using the amphiphile-selective electrode to construct binding isotherms, from which the characteristics of the various interactions can be extracted.

1. Introduction

Amphiphiles may be characterised by their particular dual affinity toward water and hydrophobic compounds. Most biomaterials belong to this class of compound. It is therefore very important to study the solution equilibrium properties of amphiphiles. Surfactants are probably one of the best model compounds which typically mimic amphiphilic biomaterials. In this paper, we will describe some aspects of surfactant as well as drug interactions with polymers and molecular assemblies.

There are many methods available to follow these interactions. We prefer a binding isotherm, or an equation of state which is direct evidence of interaction and can be interpreted by such theoretical methods as thermodynamics and statistical mechanics.

Experimentally, we employed amphiphile-selective electrodes which allowed us to successfully investigate the interaction of surfactants and drugs with molecular assemblies i.e. micelles and vesicles, and with polymers.

2. Electrodes Selective to Amphiphiles

Electrodes selective to cationic surfactants are used throughout the work mentioned below. The electrode we used is based on a concentration cell as schematically shown in Fig. 1, where C_0 and C_1 are concentrations of a surfactant on both sides of an ion-selective membrane, **M**, and the electromotive force, E_m, is picked up by a pair of silver/silver chloride reference electrodes. The surfactant ion-selective membrane contains 75-80% dioctylphthalate and 25-20% poly(vinyl chloride) both of which have been dissolved into tetrahydrofuran and cast on a flat glass plate [1-3]. The membrane responds well to cationic surfactants although no ion-carrier is actively added. The membrane itself may have a property of taking up a surfactant ion.

D. M. Bloor and E. Wyn-Jones (eds.),
The Structure, Dynamics and Equilibrium Properties of Colloidal Systems, 161–172.

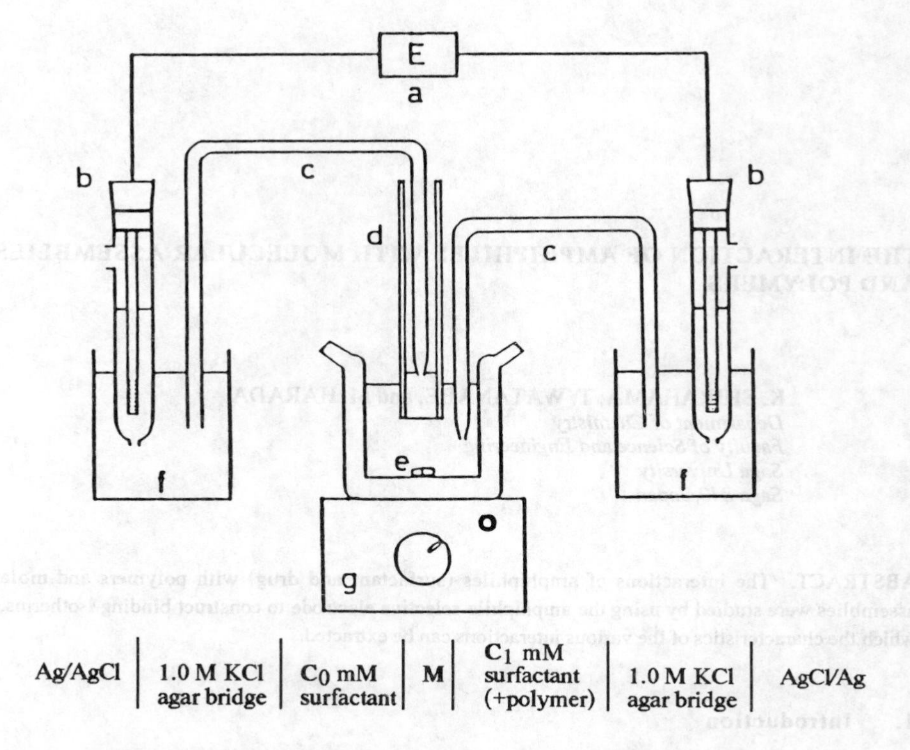

Figure 1. A schematic diagram of titration setup for amphiphile-selective electrode. a; potentiometer, b; Ag/AgCl electrode, c; salt bridge, d; membrane electrode, e; stirring bar, f; sat. KCl solution, g; stirrer.

It was rather difficult to prepare an electrode responsive to anionic surfactant [1], therefore only the electrode sensitive to cationics is used throughout the work.

3. Binding to Polymers

3.1. BINDING OF A CATIONIC SURFACTANT TO AN ANIONIC POLYELECTROLYTE WITH DIFFERENT MOLECULAR WEIGHT

There have been a lot of papers that describe the interactions between ionic surfactants and macroions [1, 2, 4-6]. In short, ionic surfactants interact with macroions with opposite charges very strongly and cooperatively, where hydrophobic chains are playing crucially important role. The strong affinity is , however, diminished by added electrolyte due to electrostatic shielding. The effect of temperature on the interaction is rather small as compared with that of added salt, and the temperature coefficient even changes sign. All these aspects are excellently reviewed [7, 8].

Among the many problems associated with these systems, there remains the effect of molecular weight of macroion on surfactant-macroion interaction. Most works mentioned above assume implicitly that the molecular weight is so large that the interaction does not depend on the size of macroion. It is very interesting to see how the size of macroion affects the interaction.

3.1.1. *Dextran Sulfate and Surfactant.* The dextran sulfate with various molecular weights were prepared by chopping the original dextran sulfate (MW = approx. 5×10^5) by air-oxidation, followed by fractionation by Sephadex gel chromatography. The molecular weight of each fraction was estimated on the chromatogram and shown in TABLE 1, where the numbers of sulfate groups per macroion molecule as measured by gravimetry combined with a colloid titration are seen.

The cationic surfactant, dodecylpyridinium bromide (DoPBr) was synthesized by treating 1-dodecane bromide with dried pyridine. The crude surfactant was decolored by active charcoal in methanol solution, and finally purified three times by recrystallization from acetone.

TABLE 1. Various parameters of dextran sulfate with different molecular weight

DS no.[a]	M W	K/dm^3mmol^{-1}	u
DS1300	5.0×10^5	0.23	100
DS450	1.7×10^5	0.23	100
DS60	2.2×10^4	0.13	100
DS22	8.5×10^3	0.08	100

(a) Figures after DS mean numbers of sulfate groups on a dextran sulfate molecules

3.1.2. *Binding Isotherms.* The electromotive force of the surfactant-selective electrode is plotted against added DoPBr concentration to obtain a straight line in accordance with the Nernst equation.

$$E_m = (RT/F)\ln(C_1/C_0) \tag{1}$$

In the presence of dextran sulfate however, there is a deviation from the linear response which is caused by a partial uptake of the added surfactant onto the polyion. By following the arrows in Fig. 2, two important quantities, the amount of bound surfactant, C_b, and the equilibrium surfactant concentration, C_f are obtained. Binding isotherms are constructed by plotting the amount of bound surfactant per sulfate group on the macroion (X) versus C_f for dextran sulfate with different molecular weights as shown in Fig. 3.

Let us first look at the isotherm for the dextran sulfate with molecular weight , 5×10^5, or the original polymer (DS1300). It is noted that binding sets up suddenly and grows fairly steeply with a shoulder around X = 1. This means that the binding is cooperative, and the primary binding sites are anionic sulfate groups. The binding at values of $X \geq 1$ may be ascribed to an interaction of the cationic surfactant monomer with the dextran sulfate already saturated with cationic surfactant. For dextran sulfate with the least binding sites (DS22), binding grows less steeply implying a weaker apparent cooperativity. The binding isotherm of DS60 lies in-between. DS450 nearly coincides with DS1300 although it deviates at higher C_f. The deviation is probably caused by precipitation which sometimes occurs near neutralization. Although the binding at values of $X \geq 1$ may be interesting, present analysis is focused on the primary binding, due to its increased reproducibility.

The binding is viewed as a transfer of cationic surfactant from the aqueous bulk phase onto a one dimensional array of binding sites, or the sulfate groups. The partition function of the system is expressed as [9],

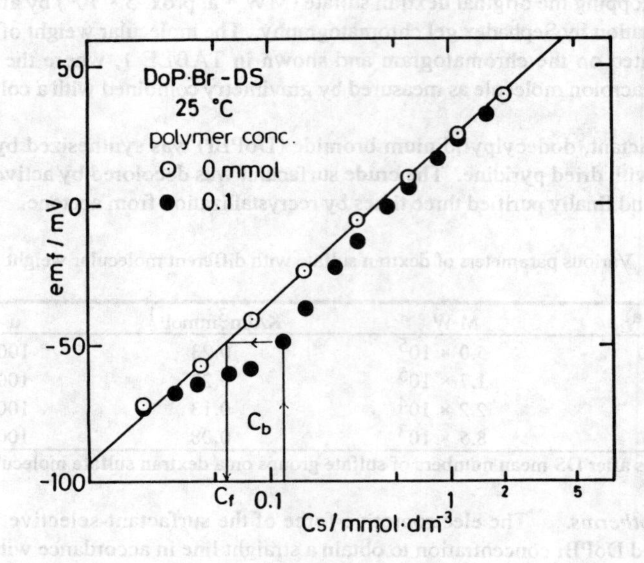

Figure 2. A potentiogram for dodecylpyridinium bromide in the absence and presence of dextran sulfate.

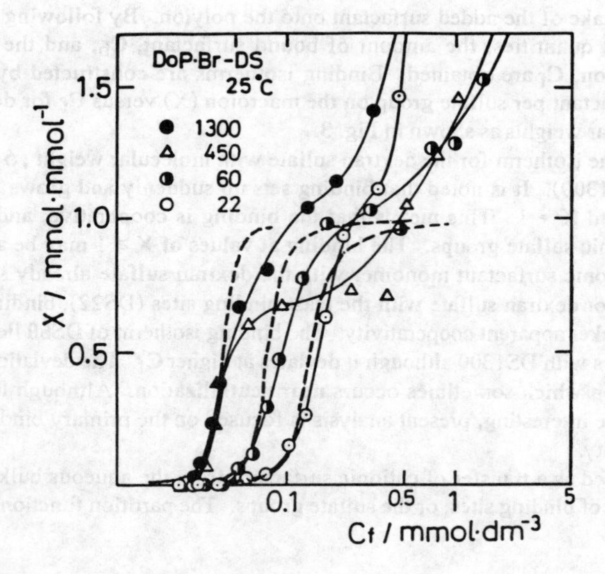

Figure 3. Binding isotherms of DoPBr to dextran sulfate with various molecular weights.

$$Z = aM^n a^* \tag{2}$$

where $M = \begin{pmatrix} 1 & 1 \\ s/u & s \end{pmatrix}$ $a = (1,1)$, and $a^* = (1,0)^*$ with $a = KuC_f$. Here K is a binding constant

for the process of surfactant being transferred from the aqueous bulk phase to an isolated binding site on a macroion, and u a cooperativity parameter, or a kind of equilibrium constant for an aggregation process of surfactant on macroion. A binding isotherm is calculated by using this partition function as,

$$X = [d\ln Z/d\ln s]/n \tag{3}$$

Equation (3) becomes an explicit form when n is sufficiently large.

$$X = [1-(1-s)/\sqrt{(1-s)^2+4s/u}]/2 \tag{4}$$

Equation (4) was fitted to experimental results of DS1300 and DS450, since both give nearly equal results at sufficiently large n values. Curve-fitting was carried out by choosing iteratively u and K parameters using lower half of the binding isotherms, where the electrostatic potential of macroion is supposed to be practically constant according to the Manning's ion condensation theory [10]. Once a set of u and K values are determined, the u value was used to analyze the experimental results of DS with fewer binding sites by directly applying equations (2) and (3), since the interaction force between surfactant molecules bound on macroion should be of short-range and thus may not depend on the size of polyelectrolyte. Table 1 also contains the set of binding constants, which reproduce binding isotherms as shown (dotted line) in Fig. 3. Fitting for the lower half is satisfactory, and the discrepancy in the upper half is expected since the electrostatic potential should decrease on surfactant binding, i.e., a negative cooperativity. It is shown that a single u parameter is enough to describe the decrease in apparent cooperativity as seen in Fig. 3. It is the end effect that causes a decrease in apparent cooperativity, while the interaction with neighbours remains the same because of the short-range nature of the interaction. The K value decreases as the size of polyelectrolyte becomes smaller. This may be due to a decreased superposition of electrostatic potential on reducing the size of polyelectrolyte.

3.2. BINDING OF A CATIONIC SURFACTANT TO AMYLOSE

It has been generally recognized that cationic surfactants have very little affinity to neutral water-soluble polymers such as poly(vinyl pyrrolidone) and poly(ethylene oxide); [11] which, surprisingly, interact very strongly with anionic surfactants [7]. It has also been shown that anionic surfactants bind to amylose [12]. It is very interesting to see if this asymmetry holds in a system containing amylose. The interaction between a cationic surfactant, tetradecylpyridinium bromide (TDPBr) and amylose was studied by using the surfactant-selective electrode.

3.2.1. *Preparation of amylose and TDPBr.* Amylose was precipitated from potato starch (Sigma) by the method of Schoch [13], and purified on a gel chromatography, and the molecular weight was estimated as 1.9×10^4. TDPBr was synthesized in a similar manner as DoPBr as described in the previous section.

3.2.2. *Binding Isotherms.* A potentiogram was obtained by the same method as described above, and binding isotherms at different temperatures are shown in Fig. 4. It is seen that the lower the temperature, the stronger the binding affinity is. An example of the Scatchard plot is shown in Fig. 5, where two straight lines appear suggesting at least two binding mechanisms.

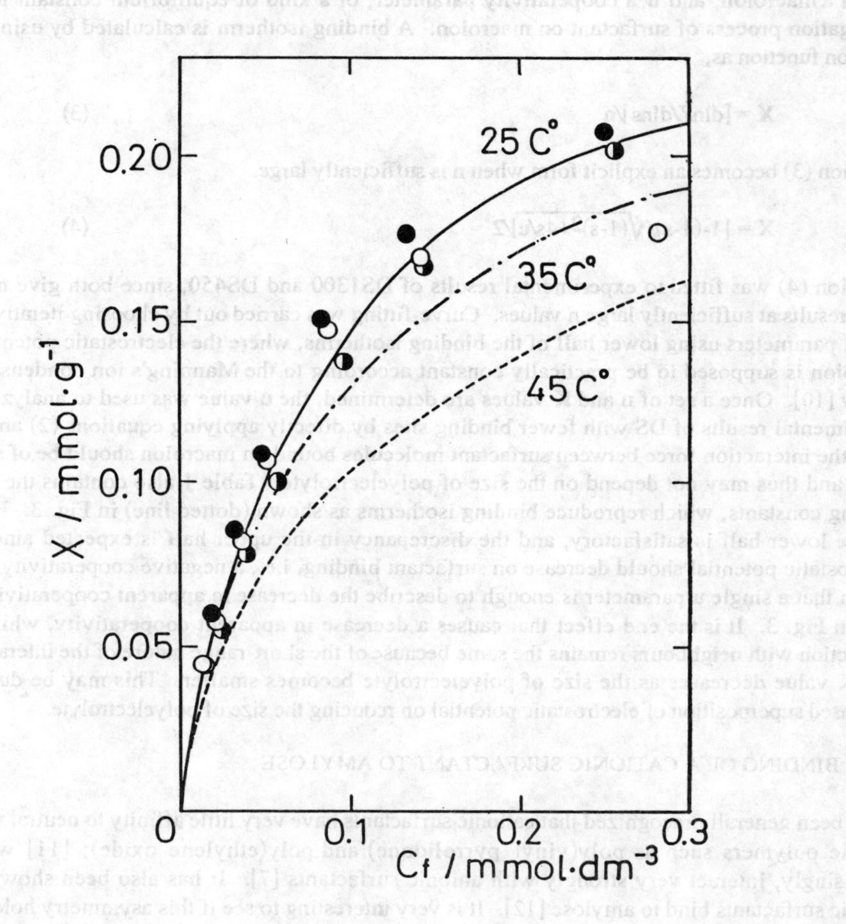

Figure 4. Binding isotherms of TDPBr-amylose at different temperatures.

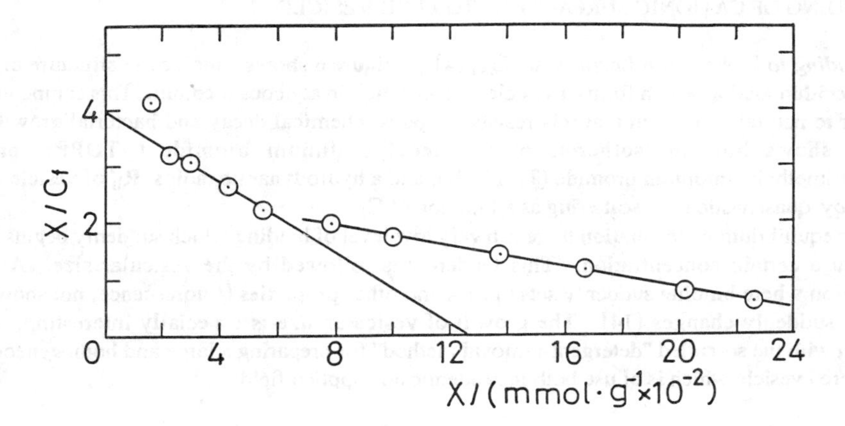

Figure 5. The Scatchard plot for a TDPBr-amylose system.

TABLE 2. Binding parameters of a TDPBr-amylose system

Temperature/ °C	K_1/dm^3 mmol^{-1}	m_1/mmol g^{-1}
25	36.2	0.13
35	30.7	0.13
45	20.7	0.13

The results of the Scatchard plot are summarized in TABLE 2.

It may be noted in the first step binding that the amount of saturation, m_1, does not depend on temperature, while the binding constant, K_1, decreases with increasing temperature. Binding parameters of the second binding were estimated by subtracting the amount of the first binding from the total binding. Although the results are subject to a large experimental errors, it may be concluded that m_2 is approximately same value as m_1, and K_2 only a fraction of K_1. Temperature dependence gives an enthalpy of binding = -20 kJmol^{-1}, which strongly suggests that the binding mechanism is inclusion of surfactant molecules into the helical structure of amylose. A large exothermicity is rather rare in the interaction of surfactant probably because hydrophobic effect is usually overwhelming. The value $m_1 = 0.13$ mmol/g amylose corresponds to a binding of about 2.4 TDPBr molecules per amylose molecule, or about 49 glucose units for one surfactant molecule taking the molecular weight of the amylose (1.9×10^4) into account. The total amount of binding is doubled even when the second binding is completed. The binding is so scarce and cannot be of cooperative nature in accordance with the Scatchard plot. It may be supposed that helices including surfactants are separated by unhelical glucose units.

4. Binding to Molecular Assemblies

Surfactants interact with molecular assemblies such as vesicles and micelles as well as self aggregate to form molecular assemblies. In this section, the interaction with vesicles will be briefly shown by binding isotherms as obtained by a surfactant-selective electrode, although experimental details are not described here because they have been published elsewhere [14].

For interaction with nonionic micelles, readers should refer to our previous paper [15].

4.1. BINDING OF CATIONIC SURFACTANT TO LIPID VESICLE

4.1.1. *Binding to Vesicle of a Lecithin-analog*[14]. Figure 6 shows a molecular structure of a synthetic lecithin-analog which forms a vesicle on sonication in aqueous medium. This compound is superior to natural lecithin in that it is resistant against chemical decay and bacterial growth. Figure 7 shows binding isotherms of tetradecylpyridinium bromide (TDPBr) and tetradecyltrimethylammonium bromide (TDTMBr), and a hydrodynamic radius, R_h of vesicle as measured by quasielastic light scattering as a function of C_f.

At lower equilibrium concentration there is a very low level of binding which suddenly begins to grow up at a certain concentration. This tendency is followed by the vesicular size. At a concentration where binding suddenly sets up, R_h and other properties (fluorescence, not shown here) also suddenly changes [14]. The growth of vesicular size is especially interesting, in connection with the so-called "detergent-removal method" for preparing a large and homogeneous single layered vesicle which is of use both in academic and applied fields.

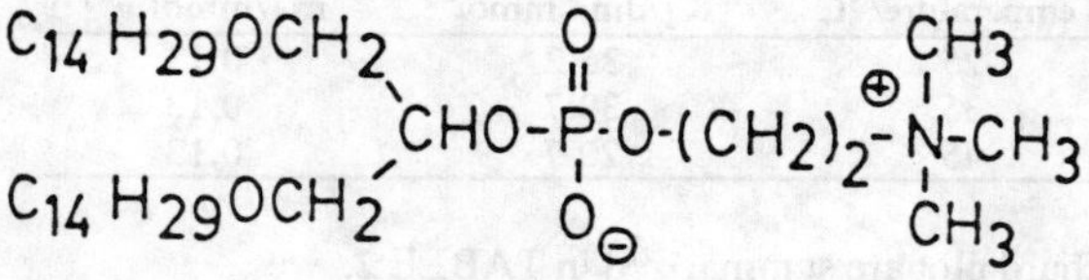

$$\begin{array}{l} C_{14}H_{29}OCH_2 \\ \qquad\qquad\qquad \diagdown \\ \qquad\qquad\qquad\ CHO\text{-}\overset{\overset{\displaystyle O}{\|}}{\underset{\underset{\displaystyle O^{\ominus}}{|}}{P}}\text{-}O\text{-}(CH_2)_2\text{-}\overset{\overset{\displaystyle CH_3}{|}}{\underset{\underset{\displaystyle CH_3}{|}}{\overset{\oplus}{N}}}\text{-}CH_3 \\ \qquad\qquad\qquad \diagup \\ C_{14}H_{29}OCH_2 \end{array}$$

Figure 6. The chemical structure of a lecithin analog, 1,1-ditetradecyl-rac-glycero-phosphatidyl choline.

4.1.2. *Binding to Synthetic and Natural Lecithin Vesicles.* Another example is the interaction of the cationic surfactant (TDPBr) with dimyristoylphosphatidylcholine (DMPC) and egg yolk lecithin (EYL). The interaction was confirmed by a binding isotherm which had been obtained by the surfactant-selective electrode. Figure 8 displays two examples of Scatchard plots implying at least a two step binding mechanism again. Figure 9 shows a van't Hoff plot in which a relationship between binding constant for the first binding step and temperature is plotted.

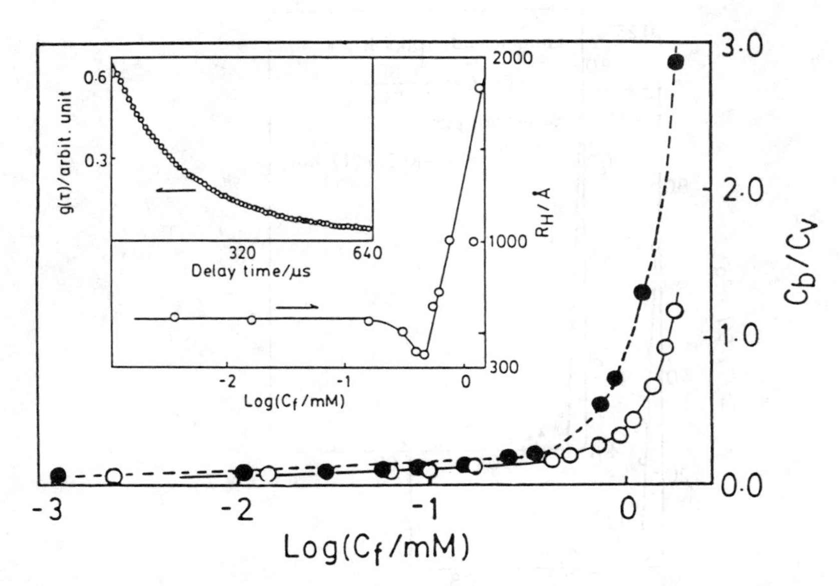

Figure 7. Binding isotherms of two cationic surfactants to the lecithin analog vesicle. Filled circles; TDPBr, and open circles; TDTMBr. Insert shows change of hydrodynamic radius as a function of equilibrium TDPBr concentration.

For EYL which has a gel-to-liquid crystal transition temperature nearly equal to 0 °C, there is a straight line giving an enthalpy on binding -25 kJ/mol, while two straight lines come out for DMPC whose transition temperature was 24 °C as determined by the I/III intensity ratio of pyrene vibronic spectrum which is also included in Fig. 9. The two straight lines are disconnected at the transition temperature suggesting a different binding mechanism above and below the temperature. Since the slopes of the straight lines are nearly equal, thus giving the same value of enthalpy of binding, it is an entropy change that causes a sudden jump of the binding affinity reflecting the difference in the two states of the vesicle.

5. Drugs as an Amphiphile

On examining the chemical structures of drugs, it was noted that most of them are hydrophobic cations, or amphiphiles. Actually, the electrode described in Section 2 senses as well drugs, such as hyamines (disinfectant, cationic surfactant itself in chemical structure), valethamate bromide (anticholigenic), chlorpromazine (antipsychotic), and local anesthetics.

The application of a drug-selective electrode will be shown. Figure 10 shows a binding isotherm of dibucaine (a local anaesthetic) to anionic dimyristoyl phosphatidylglycerol (DMPG) vesicle at 35 °C. From the Scatchard plot it was found that the total isotherm may be divided into two parts. The dashed line in Fig. 10 is the first binding step (Langmuir binding) separated from the total isotherm. Here it is interesting to note that a change of binding mechanism, probably a change in original vesicular structure, occurs on binding a certain amount of amphiphile irrespective of surfactant or drug. Dimyristoylphosphatidylcholine vesicle was also found to bind dibucaine but to a much lesser degree (not shown here) suggesting the importance of electrostatic effects in the interaction.

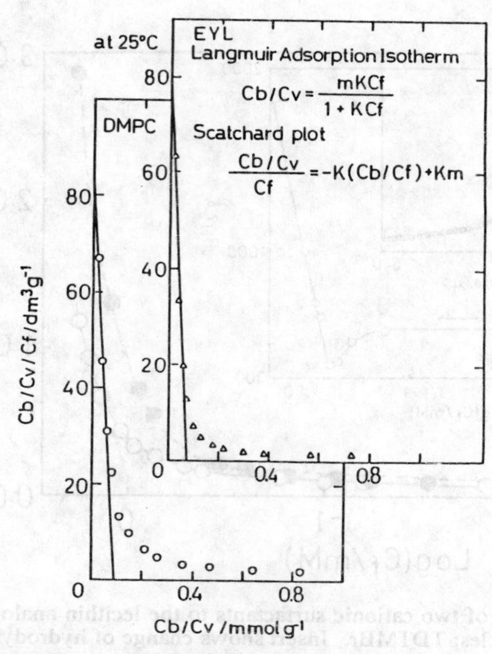

Figure 8. Scatchard plots for the TDPBr-lecithin systems.

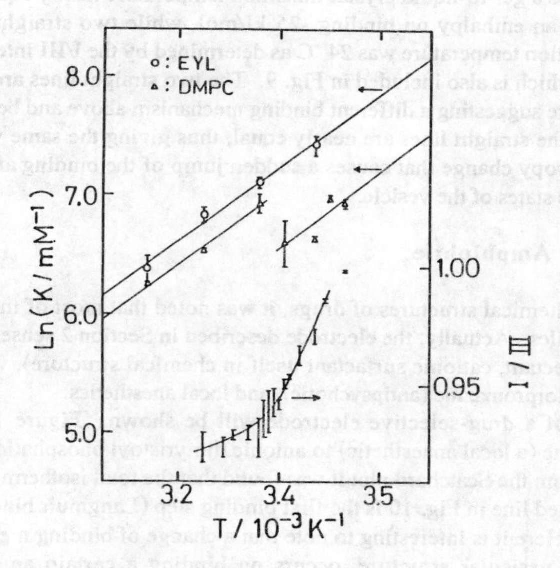

Figure 9. van't Hoff plots for the TDPBr-lecithin systems. Circles; egg yolk lecithin, triangles; dimyristoylphosphatidylcholine.

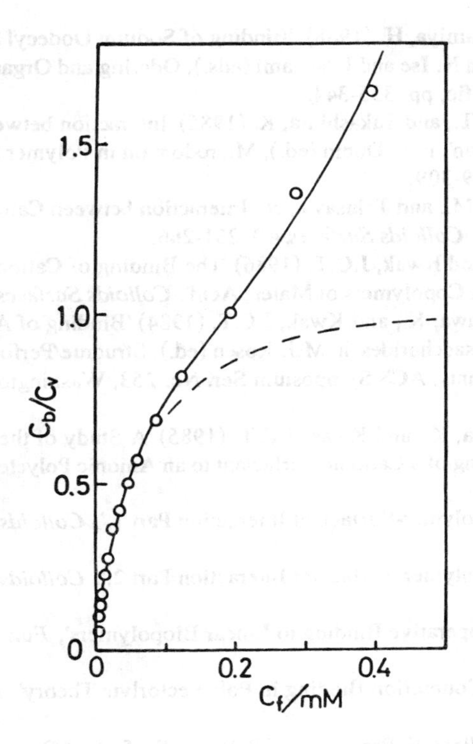

Figure 10. Binding isotherm of dibucaine-dimyristoyl phosphatidylglycerol system in 0.15 M NaCl.

6. Concluding Remarks

The amphiphile-sensitive electrode has been shown to be very effective in studying interactions of amphiphiles with polymers and molecular assemblies, although there may be a limitation in that the amphiphile must be ionic. Resulting binding isotherms are useful to analyze molecular interactions. It was observed that binding begins initially with a statistical independent mechanism (Langmuir type) which eventually triggers a second step binding which may be cooperative or noncooperative, and may bring about changes in the host (polymers and molecular assemblies) structures. In addition, it is advisable to apply such physical measurements as NMR, ultrasonic technique, and other spectroscopic methods to obtain a deeper insight into the phenomena.

7. Acknowledgement

We thank the Asahi Glass Industry Foundation for partially supporting this work.

8. References

[1] Shirahama, K., and Nakamiya, H. (1988) 'Binding of Sodium Dodecyl Sulfate to N-Methylglycolchitosan', in N. Ise and I. Sogami (eds.), Odering and Organisation in Ionic Solutions, World Scientific, pp. 335-344.

[2] Shirahama, K., Masaki, T., and Takashima, K. (1985) 'Interaction between DNA and Dodecylpyridinium Cation', in P. Dubin (ed.), Microdomain in Polymer Solutions, Plenum Pub., New York, pp. 299-309.

[3] Shirahama, K., Oh-ishi, M., and Takisawa, N. 'Interaction between Cationic Surfactants and Poly(vinyl alcohol)', *Colloids Surfaces*, **40**, 261-266.

[4] Shimizu, T., Seki, M., and Kwak, J.C.T. (1986) 'The Binding of Cationic Surfactants by Hydrophobic Alternating Copolymers of Maleic Acid', *Colloids Surfaces*, **20**, 289-301.

[5] Malovikova, A., Hayakawa, K., and Kwak, J.C.T. (1984) 'Binding of Alkylpyridinium Cations by Anionic Polysaccharides' in M.J. Rosen (ed.), Strucrue/Performance Relationships in Surfactants, ACS Symposium Ser. No. 253, Washington D.C., pp.225-239.

[6] Santerre, J.P., Hayakawa, K., and Kwak, J.C.T. (1985) 'A Study of the Temperature Dependence of the Binding of a Cationic Surfactant to an Anionic Polyelectrolyte', *Colloids Surfaces*, **13**, 35-45.

[7] Goddard, E.D. (1986) 'Polymer-Surfactant Interaction Part 1.', *Colloids Surfaces*, **19**, 255-300.

[8] Goddard, E.D. (1986) 'Polymer-Surfactant Interaction Part 2.', *Colloids Surfaces*, **19**, 301-329.

[9] Schwarz, G. (1970) 'Cooperative Binding to Linear Biopolymers', *Eur. J. Biochem*, **12**, 442-453.

[10] Manning, G.S. (1979) 'Counterion Binding in Polyelectorlyte Theory', *Acc. Chem. Res.*, **12**, 443-449.

[11] Saito, S. (1967) ' Solubilization Properties of Polymer-Surfactant Complexes', *J. Colloid Interface Sci*, **24**, 227-234.

[12] Yamamoto, M., Sano, T., Harada, S., and Yasunaga, T. (1983) 'Cooperativity in the Binding of Sodium Dodecyl Sulfate to Amylose', *Bull. Chem. Soc. Jpn.*, **56**, 2643-2646.

[13] Schoch, T.J. (1942) 'Fractionation of Starch by Selective Precipitation with Butanol', *J. Am. Chem. Soc.*, **64**, 2957-2961.

[14] Takasaki, M., Takisawa, N., Shirahama, K. (1987) 'Interaction between a Synthetic Phospholipid Vesicle and Cationic Surfactants', *Bull. Chem. Soc. Jpn.*, **60**, 3849-3853.

[15] Shirahama. K., Nishiyama, Y., and Takisawa, N. (1987) 'Binding of N-Alkylpyridinium Chlorides to Nonionic Micelles', *J. Phys. Chem*, **91**, 5928-5930.

STEP-WEIGHTED RANDOM WALK STATISTICS, AS APPLIED TO ASSOCIATION COLLOIDS

J. LYKLEMA,
Department of Physical and Colloid Chemistry,
Dreijenplein 6,
6703 HB Wageningen,
The Netherlands.

ABSTRACT. After treating some properties of completely random, or stochastic, processes, a discussion is given on how the insights into these phenomena can be modified to mimick properties of long chain molecules. Some emphasis is put on lattice theories, the incorporation of neighbour interactions and the presence of an adsorbate. The models obtained are suited to describe a variety of phenomena, including adsorption of polymers and the formation of association colloids, such as bilayer membranes, micelles and vesicles.

1. Introduction. Random Walks

In the present paper it will be discussed how random walk statistics can be used to describe conformations of free and adsorbed long chain molecules.

By way of introduction, consider figure 1. It is redrawn from old, but famous work by Perrin [1] and gives the positions of a mastic particle (the "precursor" of our present latices), undergoing Brownian motion. Straight lines have been drawn between the positions taken after successive intervals of 30 seconds. No physical implication is involved: the lines are there just as a visual aid. If one of these steps would have been broken down into, say 30 subdivisions, the resulting "subpicture" would look like the entire figure 1. In other other terms, Brownian motion is *self-similar.*

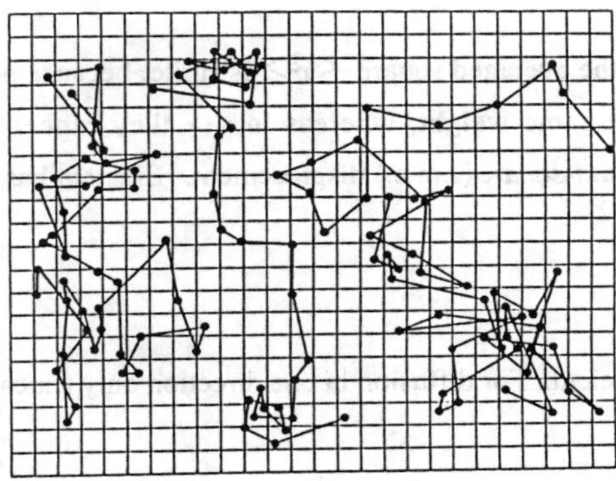

Figure 1. Brownian motion of mastic particles after Perrin (1909).

D. M. Bloor and E. Wyn-Jones (eds.),
The Structure, Dynamics and Equilibrium Properties of Colloidal Systems, 173–179.

At the time Perrin did his experiments, the very existence of molecules was at issue and even the title of his paper reflected this. Later, other issues have drawn due attention. One of them was, and still is, the question of how short the time interval between two observations may be made to lose the random nature of this movement. For our purpose this interesting problem is not relevant. We recognize that each trace in figure 1 may be considered as the two-dimensional projection of the configuration that a polymer coil in a solution or melt can assume. In other words, random walks of particles resemble polymer configurations and hence the mathematics of random walks may be used, with due adjustments, to mimick long chain molecules.

What are these "due adjustments"? A polymer coil is not usually completely random. Among other factors, polymer segments cannot rotate freely, they have a certain intrinsic volume denying part of the available space to others, and the segments will generally interact with each other. These problems become even more prominent when association colloids are considered: micelles, vesicles, adsorbates of surfactants and lipid bilayer membranes. For such systems, the walk cannot be fully random, but this does not preclude the use of modification where the basic pattern is a random walk, but where the extent of randomness is harnessed by imposing certain rules to account for the above (and any other) complications.

One of the typical features of a random walk is that the vectorial distance travelled (r) is zero, if averaged over the time t. (The vector r gives distance and direction of a position after t seconds minus the same at $t = 0$). This is in contrast to a particle moving at constant velocity v in a given direction, say under the influence of an applied force. Then $r = vt$, r and v have the same direction.

Although $<r> = 0$ and, for that matter, $<v> = 0$, we know that there is a certain non-zero displacement. For a process that is entirely devoid of sense of direction or order, (a *stochastic* process), the probability of finding the particle at a given distance from the position at $t = 0$, *disregarding the direction*, is proportional to $t^{1/2}$, rather than to t. Regarding our theme, this is an important observation because for isolated polymers dissolved in a θ-solvent the size of the molecule scales with the square root of the molecular weight, $M^{1/2}$ rather than with M, as should be expected for the length of a fully stretched chain. it is typical for θ-solvents that, due to compensation of excluded volume and segment attraction, polymers in it behave as if they are random, provided the chains are long enough. Hence, a clear analogy has been established.

Let us discuss the $t^{1/2}$ dependence quantitatively. We define the mean displacement Δ as

$$\Delta^2 = <r^2> \tag{1}$$

It is realized that, unlike $<r>$ the averaged square $<r^2>$ is finite, because positive and negative vectors are now assigned the same weight, whereas in $<r>$ they cancel. The quantity $\Delta = <r^2>^{1/2}$ is called the *root-mean-square* (r.m.s.) displacement. Einstein has derived that Δ is in the following way related to $t^{1/2}$

$$\Delta^2 = 6Dt \tag{2}$$

where D is the diffusion coefficient. For diffusion in one direction only (linear diffusion)

$$\Delta^2 = 2Dt \tag{3}$$

The relation between (2) and (3) is easy: for random motion, the x, y and z direction each contribute equally.

The equation of motion for a particle undergoing Brownian motion contains three contributions.

First there is the Newton term mass x acceleration, $m\,dv/dt$. it is equal to the sum of the forces acting on it. There are two of such forces, a friction force $-fv$ and a rapidly fluctuating random force $F_r(t)$, due to the "kicks" the particle receives from the collisions the solvent molecules. The constant f is a friction coefficient. Combining the three contributions leads to

$$m\frac{dv}{dt} = -fv + F_r(t) \tag{4}$$

which is known as the *Langevin equation.* If there were no friction and $F_r(t)$ would be replaced by a constant applied force F, (4) would reduce to Newton's familiar law $F = ma$. On the other hand, in stationary processes such as sedimentation or electrophoresis, there are no inertia effects, the rate is constant and proportional to the force.

Langevin's equation must lead to the $\Delta(t^{1/2})$ dependence. In fact this is not so difficult to prove and we shall give the derivation for the one-dimensional case where displacements take place in the x direction only. As v is now placed by dx/dt, (the vectorial nature of the displacement can be ignored), we have from (4)

$$m\frac{d^2x}{dt^2} = -f\frac{dx}{dt} + F_x(t) \tag{5}$$

We are interested in $\dfrac{d\langle x^2\rangle}{dt} = 2\left\langle \dfrac{x\,dx}{dt}\right\rangle$. To obtain this derivative, multiply both hands of (5) by x and take time averages

$$m\left\langle x\frac{d^2x}{dt^2}\right\rangle + f\left\langle x\frac{dx}{dt}\right\rangle = \langle xF_x(t)\rangle \tag{6}$$

In this expression,the second term on the l.h.s. equals $\dfrac{1}{2}f\dfrac{d\langle x^2\rangle}{dt}$; the r.h.s. is zero because the position x of the particle and the force active on it $F_x(t)$ are uncoupled: hence if their product is averaged over time nothing can result. With this taken into account, (6) reduces to

$$\frac{d\langle x^2\rangle}{dt} = \frac{2m}{f}\left\langle x\frac{d^2x}{dt^2}\right\rangle \tag{7}$$

The r.h.s. of (7) can be further simplified. Realizing that

$$mx\frac{\mathrm{d}^2x}{\mathrm{d}t^2} = m\frac{d}{\mathrm{d}t}\left(x\frac{\mathrm{d}x}{\mathrm{d}t}\right) - m\left(\frac{\mathrm{d}x}{\mathrm{d}t}\right)^2$$

we can take averages

$$m\left\langle x\frac{\mathrm{d}^2x}{\mathrm{d}t^2}\right\rangle = m\frac{\mathrm{d}}{\mathrm{d}t}\left\langle x\frac{\mathrm{d}x}{\mathrm{d}t}\right\rangle - m\left\langle\frac{\mathrm{d}x}{\mathrm{d}t}\right\rangle^2 \tag{8}$$

and recognize that the first term on the r.h.s. is again zero (position x and rate $v_x = \mathrm{d}x/\mathrm{d}t$ are uncoupled), whereas the second has a clear physical meaning: it equals $-m\langle v_x\rangle^2$, which, because of equipartition must be equal to $-kT$. (Recall that in the three dimensional analogue there are three degrees of freedom, then $\frac{1}{2}m\langle v^2\rangle = \frac{3}{2}kT$).

All told, we obtain

$$\frac{\mathrm{d}\langle x^2\rangle}{\mathrm{d}t} = \frac{2kT}{f} \tag{9}$$

and for the three dimensional case

$$\frac{\mathrm{d}\langle r^2\rangle}{\mathrm{d}t} = \frac{\mathrm{d}\Delta^2}{\mathrm{d}t} = \frac{6kT}{f} \tag{10}$$

which can be integrated to give

$$\Delta = \left(\frac{6kT}{f}\right)^{1/2} t^{1/2} \tag{11}$$

This result is identical to the Einstein expression (2) provided

$$D = \frac{kT}{f} \tag{12}$$

This is a well-known equation. For spheres, Stokes has derived that $f = 6\pi\eta a$

2. Application to Polymers and Association Colloids

In order to apply the statistics, described above, to long chain molecules some modifications have to be introduced. Let us for the moment consider homopolymers. In such systems all bonds have the same length. Hence, the complete randomness pictured in figure 1 is not adequate. First, some "taming" is necessary by imposing equality of all step lengths. In the second place, the segments are not freely-jointed. For a simple chain like that of alkanes, the allowed angles are about the tetrahedral angle. The two impositions of fixed bond length and restricted rotation has lead to the introduction of *lattice theories.* Figure 2 gives an example. It is a very simple cubic system. A hexagonal lattice, or rather a tetrahedral one would be at least as adequate. However, experience has shown that it does not matter very much what kind of lattice is selected for the final equations, provided the chain is not too short.

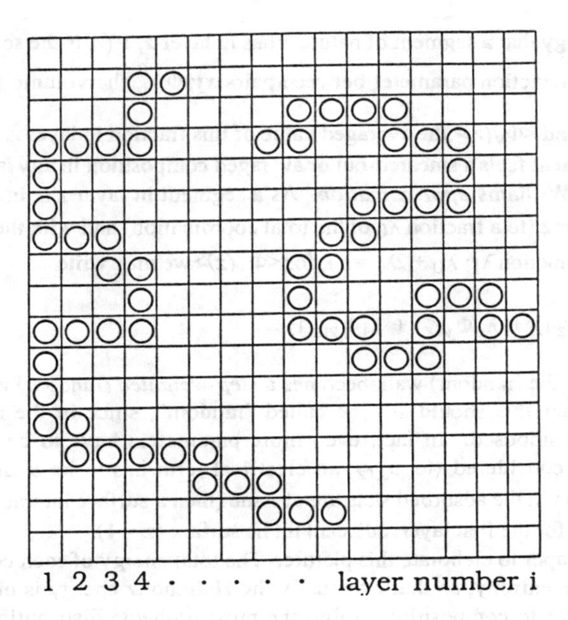

Figure 2 Two polymer molecules, mimicked as a random walk in a square lattice. The spheres represent chain segments, the other sites contain solvent molecules (not shown).

A walk on a lattice is still called "random" if the direction of each step is entirely uncorrelated to the preceding one. Another term is *first order Markov process*, and the result if a *first order Markov chain*, where "chain" is not only used to denote the generated polymer but also in the more figurative sense to indicate the nature of the process.

It is immediately appreciated that such a chain is not yet an adequate reflection of reality, a.o. because backfolding is allowed.

This last problem can be overcome by accounting for the difference between gauche and trans configurations, preferably on a tetrahedral lattice. In this *second order Markov process* the growing polymer, making the *n*th step "remembers" the direction of steps *n*-1 and *n*-2. In the relevant professional literature this refinement is known as *rotational isomeric state*, usually abbreviated R.I.S.

For an isolated long chain molecule, the second order Markov approach, described above, gives a satisfactory picture. However, if application to association colloids is required one can no longer ignore the interaction with neighbours. If a chain is generated by adding segments to it, the probability of finding the (n+1)st in layer (z+1), z or (z-1), if the nth segment is in layer z, depends not only on the type of the lattice and the occupancy of each of these layers, but also on the interaction energy that the (n+1)st segment has with the segments already present. The result is that a weighting factor has to be included, typically including Flory-Huggins types of pair interaction parameters χ and averaged volume fractions. Mathematically one could write these as

$$u_x(z) = u'(z) + kT\,\Sigma_y \chi_{xy}\left[<\Phi_y(z)> - \Phi_y^b \right] \tag{13}$$

where $u_x(z)$ is the energy that a segment of nature x has in layer z, $u'(z)$ is the so-called hard core interaction, χ_{xy} the interaction parameter between pairs xy, Φ_y^b the volume fraction of type y segments in the bulk and $<\Phi_y(z)>$ the averaged value of this fraction in layer z. The assumption is made that each segment feels a smeared-out or averaged composition in any layer z. This is the *mean field* or *Bragg-Williams approximation.* As a segment in layer z is in interaction with neighbours inside layer z (to a fraction λ_0 of the total coordination) and with those in layers z+1, (fraction λ_1) and z-1 (fraction λ_1; $\lambda_0 + 2\lambda_1 = 1$), for $<\Phi_y(z)>$ we may write

$$<\Phi_y(z)> = \lambda_1 \Phi_y(z{-}1) + \lambda_o \Phi_y(z) + \lambda_1 \Phi_y(z{-}1) \tag{14}$$

With this in mind, the (random) walk becomes a *step-weighted (random) walk* It remains a matter of taste whether this should still be called "random", since in the mean time many constraints have been imposed. In fact, even more parameters have to be introduced when charged systems are considered (to $u_x(z)$ an electrical term has to be added) and/or when surfactants or polymers in the adsorbed state are at issue (then a surface interaction parameter χ_s has to be incorporated for the first layer, adjacent tot he surface (z = 1).

It is beyond this paper to elaborate this picture. The total energy of each conformation must be computed, then the entropy, so that eventually the Helmholtz energy is obtained, which is minimized with respect to composition to find the most probable distribution. A number of mathematical difficulties have to be overcome, including dealing with the problem that in [14] the various Γ_y's are not a *priori* known, but follow from the computation. This last issue requires an iteration.

Regarding the application of this model to polymer adsorption, see the contribution of M. Cohen Stuart to this Study Institute [2]. The basic theory for the self assembly of bilayer membranes is in a paper by Leermakers *et al* [3]. This paper does not yet include RIS. Elaborations for other symmetries (spheres, cylinders [4,5] have for flat membranes been followed by more detailed theory including RIS [6] and an interpretation of the sol-gel transition [7]. RIS has now also been used for vesicles [8]. Adsorption of nonionic surfactants has been treated by this method by Böhmer and Koopal [9] and a paper considering the incorporation of foreign molecules ("hosts") into membranes is in course of publication [10]. These papers also contain a discussion on alternative theories and include a comparison with experimental data.

It is anticipated that the potentialities of this approach are by no means exhausted.

References

1. J. Perrin, Ann. Chim. Phys., (8) *18* (1909) 1.
2. M.A. Cohen Stuart, this ASI,
3. F.A.M. Leermakers, J.M.H.M. Scheutjens and J. Lyklema, Biophys. Chem. *18* (1983) 353.
4. J.M.H.M. Scheutjens, F.A.M. Leermakers, N.A.M. Besseling and J. Lyklema, Proc. Int. Symp. Surfactants, New Delhi (1986) in press.
5. F.A.M. Leermakers, J.M.H.M. Scheutjens, P.P.A.M. van der Schoot and J. Lyklema, Proc. Int. Symp. Surfactants, New Delhi (1986) in press.
6. F.A.M. Leermakers, J.M.H.M. Scheutjens, J. Chem. Phys. *89* (1988) 3264.
7. F.A.M. Leermakers, J.M.H.M. Scheutjens, J. Chem. Phys. *89* (1988) 6912.
8. F.A.M. Leermakers, J.M.H.M. Scheutjens, J. Chem. Phys. *93* (1989) 7417.
9. M. Böhmer, L.K. Koopal, Langmuir, accepted.
10. F.A.M. Leermakers, J.M.H.M. Scheutjens and J. Lyklema, Biophys. Biochim. Acta., submitted.

A POPULATION BALANCE MODEL FOR THE DETERMINATION OF SOLUBILIZATE EXCHANGE RATE CONSTANTS IN REVERSED MICELLAR SYSTEMS

A.S. BOMMARIUS[@], J. F. HOLZWARTH[#*], D. I.C WANG[@], and T. A HATTON[@*]

[@] *Dept. of Chemical Engineering, Massachusetts Institute of Technology, 25 Ames St., Cambridge/MA 02139, U.S.A.*
[#] *Fritz Haber-Institut der Max-Planck Gesellschaft, Faradayweg 4-6, 1000 Berlin 33, West Germany*

* to whom correspondence should be addressed

ABSTRACT. Upon coalescence and decoalescence of reversed micellar aggregates, molecules solubilized in the water cores are exchanged and redistributed between aggregates. Investigation of the solubilizate exchange phenomenon is of interest to understand the mechanism of coalescence and to assess the possibility of transport limitations for chemical and biochemical reactions. We have studied solubilizate exchange in the system dodecyltrimethylammoniumchloride (DTAC)/hexanol/n-heptane/water through electron transfer indicator reactions with the Continuous Flow Method with Integrating Observation (CFMIO). Solubilizate exchange rate constants k_{ex} of 10^6-10^7 $(Ms)^{-1}$ were obtained with a novel population balance model incorporating distribution effects of probe molecules across reversed micellar aggregates. Such rate constants are two to three orders of magnitude slower than those for molecular diffusion. The results are consistent with opening of the surfactant layer upon coalescence as the rate-determining step.

1. Introduction

Reversed micelles are nanometer-sized water droplets surrounded by a surfactant layer in a water-immiscible organic medium. Because of their size (in the 1 - 10 nm range), their movement is not controlled by the hydrodynamics in the solution but rather by Brownian motion which causes the aggregates to continually collide, temporarily coalesce, and decoalesce again. Upon coalescence and decoalescence, probe molecules solubilized in the water pool are redistributed between the two newly formed micellar cores on a rapid timescale with a second-order rate constant k_{ex}.

D. M. Bloor and E. Wyn-Jones (eds.),
The Structure, Dynamics and Equilibrium Properties of Colloidal Systems, 181–199.

Owing to the short timescale of coalescence and decoalescence and thus very low concentration of transient dimers, the reversed micellar system cannot be observed directly during collisions or in the state of the dimer. Instead, indicator reactions between probe molecules are used to obtain information about the coalescence process. To investigate solubilizate exchange between micellar cores, the probe molecules must be partitioning exclusively to the pool and reaction progress must be observable.

The collision and coalescence process of reversed micelles probed by an indicator reaction consists of five sequential elementary steps:

i) diffusional approach of the micellar aggregates,
ii) opening of the micellar walls,
iii) diffusion of indicator ions in the temporary dimeric aggregate,
iv) chemical reaction of the indicator ions, and
v) decoalescence of the temporary dimeric aggregate.

In a sequence of rate processes, the overall, observed rate constant k_{obs} relates to the rate constants of the individual elementary processes as written in equ. (1) [1]:

$$1/k_{obs} = 1/k_{diff,agg} + 1/k_{opening} + 1/k_{diff,ions} + 1/k_{reac} + 1/k_{decoal} \qquad (1)$$

The time constants of all elementary steps except (ii) and (v) are known to a reasonable degree of accuracy.

1.1 DIFFUSION

Collisions between micellar aggregates as well as probe molecules are governed by diffusion, the fastest process of encounter between two species in solution. The rate constant for diffusion, k_{diff}, the upper bound for second-order processes such as reversed micellar coalescence, can be calculated by the simple Smoluchowski equation [2] (equ. (2))

$$k_{diff} = 4(D_A+D_B)(r_A+r_B)N_A/1000 \qquad (2),$$

and has been determined by Holzwarth [3] and later by Sutin [4] with electron-transfer reactions to be 3.2 10^9 $(Ms)^{-1}$ for equally-sized diffusing species with no interaction in aqueous solution at 25°C. If every collision led to coalescence, the solubilizate exchange rate constant k_{ex} could also be calculated by the Smoluchowski equation. If the ratio k_{ex}/k_{diff} is smaller than one, it reflects the fraction of "successful" collisions resulting in coalescence.

1.2 REACTION

To investigate the coalescence process by solubilizate exchange of probe molecules, the timescale of the indicator reaction must at least match or preferentially be faster than the timescale of coalescence or associated solubilizate exchange. Diffusion-controlled indicator reactions are most suitable because the coalescence process is rate-limiting under these conditions if its time constant is slower than that of the diffusion process. In this work, we employ the predominantly diffusion-controlled, irreversible reaction between hexacyanoferrate(II), $Fe(CN)_6^{4-}$, and osmium(trisbipyridyl)(II), $Os(bipy)_3^{2+}$, [3]:

$$Fe(CN)_6^{4-} + Os(bipy)_3^{2+} \rightarrow Fe(CN)_6^{3-} + Os(bipy)_3^{3+} \qquad (3)$$

as well as the totally diffusion-controlled association of p- nitrophenolate with the hydrogen ion, H^+:

$$p\text{-}NO_2\text{-}Ph\text{-}O^- + H^+ \rightarrow p\text{-}NO_2\text{-}Ph\text{-}OH \qquad (4)$$

1.3 DECOALESCENCE

Since the rate constant of decoalescence k_{decoal} is not known, an assumption about its magnitude has to be incorporated in the model to interpret rate data (see below).

For an assessment of possible transport limitations from micellar coalescence on the intrinsic reaction rate, the timescales of all processes of interest have to be compared. For an n-th order reaction, the time constant τ is given by:

$$\tau_n = 1/(kc^{n-1}) \qquad (5)$$

while for an enzymatic reaction, assuming validity of the Michaelis-Menten rate law, we have:

$$\tau_{enz} = 1/k_{cat} \qquad (6)$$

Reversed micellar coalescence is a second-order process dependent on the concentration or number density N of reversed micelles, expressed in M (= mol/l). Thus, the rate of reversed micellar coalescence is $r_{coal} = k_{ex}N^2$, with k_{ex} representing the solubilizate rate constant. With typical values of $N = 10^{-3}$ M and $k_{ex} = 10^7$ (Ms)$^{-1}$ [5], the order of magnitude for the time constant of the transport process is:

$$\tau_{coal} = 1/(k_{ex}N) = 10^{-4}\ s^{+1} \qquad (7)$$

Comparison of time constants for reactions of between 10^{-3} s^{-1} for slow and about 10^7-10^9 s^{-1} for very fast chemical and enzymatic [6] reactions with expected time constants for reversed micellar coalescence demonstrates that reversed micellar transport phenomena could very well be important or even dominating in case of fast chemical or enzymatic reactions. If either reaction or micellar coalescence is dominating, the result is one of two limiting scenarios:

i) If the intrinsic rate is controlling, i.e. is much slower than the coalescence rate, all reversed micellar pools can be regarded as equilibrated with respect to solubilizate concentration on the timescale of the intrinsic rate.
ii) If the intrinsic rate process is considerably faster than coalescence, an unequal distribution is frozen in the pools between coalescence events, and only excess probe molecules remain after reaction: no accumulation occurs in any micellar pool.

2. Previous Micellar Coalescence Research and Goals of this Work

Much of the work on dynamics of both oil-in-water and water-in-oil microemulsions has been reviewed by Zana and Lang in 1988 [5]. Qualitative studies demonstrated that solubilizate exchange between reversed micelles precedes reaction if

i) water-soluble probe molecules are initially solubilized in different populations of aggregates [7], that
ii) this exchange occurs on a timescale comparable to the mixing process but
iii) does not occur if microemulsions containing different agents in the pools are separated by low molecular weight cut-off dialysis membranes [8].

In the system AOT/hydrocarbon/water, the magnitude of the exchange rate constant k_{ex} of about 10^7 $(Ms)^{-1}$ has been obtained several times [9-15]. A k_{ex} of 1.7 10^7 $(Ms)^{-1}$ was measured by Atik and Thomas in heptane with fluorescence quenching of $Ru(bipy)_3^{2+}$ by $Fe(CN)_6^{4-}$ [11], and a similar result was found by Pileni et al. in isooctane with quenching of hydrated electrons by nitrate ions [12]. Likewise, in a very extensive study in 1987, Fletcher et al. compared three different indicator reactions, electron transfer, proton transfer, and metal-ligand complexation, with stopped-flow- and continuous-flow techniques to obtain k_{ex}-values ranging from 1 10^7 $(Ms)^{-1}$ for pentane to 3 10^7 $(Ms)^{-1}$ for dodecane [13].

However, in another study Pileni et al. found a k_{ex} of 2 10^8 $(Ms)^{-1}$ in AOT/isooctane with fluorescence quenching reaction of magnesium porphyrin oxidized by substituted viologens [14], and Lang et al. used fluorescence quenching of $Ru(bipy)_3^{2+}$ by $Fe(CN)_6^{4-}$ to find values ranging from 10^8 $(Ms)^{-1}$ at low temperature with hexane and heptane to 6.9 10^9 $(Ms)^{-1}$ at high temperature in decane [15]. These numbers are so close to the diffusion limit that solubilizate exchange may no longer be rate-limiting (k_{diff} equals 3.2 10^9 $(Ms)^{-1}$ at ionic strength $I \to \infty$ but equals 1.2 10^{10} $(Ms)^{-1}$ at $I = 0$ [3,16] for a reaction with charge product of -6). Solubilizate exchange between water pools can only be investigated with probes solely located in the pools but molecules such as porphyrins [14] or the fluorescor $Ru(bipy)_3^{2+}$ [16,17] are known to be bound electrostatically or by hydrophobic forces to the interfacial layer. Transfer between reversed micelles upon collision even if no solubilizate exchange between the cores occurs could markedly increase k_{ex} values.

Far less work has been done on surfactant systems other than AOT. Atik and Thomas [18] found that with potassium oleate surfactant hexanol, but not pentanol, as cosurfactant led to the formation of discrete water droplets: with hexanol, k_{ex} was 1-3 10^8 $(Ms)^{-1}$ in hexadecane and 1 10^8 $(Ms)^{-1}$ in dodecane. The system CTAB/ hexanol/dodecane/water yielded very similar results [19]. Both systems were investigated using a luminescence quenching reaction of $Ru(bipy)_3^{2+}$ with $Fe(CN)_6^{4-}$, thus raising the problem of location of the Ru^{2+}-species.

The present work focuses on the question whether four-component reversed micellar systems exchange solubilizates faster, i.e. whether they are more fluid, than standard three-component systems such as those with AOT. Experimentally, this work differs from most others by employing a continuous fast-flow method with an indicator reaction with verified exclusive location of both probe molecules, $Fe(CN)_6^{4-}$ and $Os(bipy)_3^{2+}$, in the pool so that only solubilizate exchange between cores is measured.

Mechanistically, the distribution of solubilizates among reversed micellar cores is addressed by proposing a population balance model to explain the exchange of solubilizates between reversed micellar cores to improve understanding of rate processes on a colloidal length scale and a diffusion time scale.

3. Experimental

3.1 MATERIALS

Dodecyltrimethylammoniumchloride (DTAC) was obtained from Tokyo Kasei (Tokyo) in 99.6% purity. Hexanol and n-heptane (synthesis grade), $K_4Fe(CN)_6$ x 3 H_2O (p.a.), and KCl (Suprapur grade) were from Merck (Darmstadt, F.R.G.) and were used as received. $Os(bipy)_3(ClO_4)_3$ was synthesized according to common literature methods [20]. Water was deionized and doubly distiled. The reagents were dissolved in 0.1 M KCl aqueous solution of neutral pH, unless otherwise specified, for constant ionic strength. Aqueous solution was pipetted into clear organic solutions and dispersed by ultrasonication for 1 to 2 minutes.

3.2 SURFACTANT SYSTEM

2.64 g (10 mmol) DTAC surfactant was suspended per liter of n-heptane and hexanol added to bring the resulting mixture with n-heptane to 10 vol-% in alcohol (0.8 M). The surfactant was dissolved by shaking the mixture for 5 to 10 minutes. An aqueous $Os(bipy)_3(ClO_4)_3$ solution of 1 mM pool concentration was freshly prepared and added to the organic solution of surfactant. The other aqueous solution contained $K_4Fe(CN)_6$ x 3 H_2O of 1 mM to 6 mM pool concentration, depending on reactant ratio q. Reaction progress was monitored by observing the appearance of the $Os(bipy)_3^{2+}$-complex at $\lambda = 470$ nm.

3.3 M-Q PLOTS

Water-surfactant-ratios of 5, 10, 20, 25, 30, and 70 were tested at reactant ratios $q = \mu_{Fe}/\mu_{Os} = [Fe]/[Os]_{pool}$ ranging from 0.167 to unity. Lower q-values yielded results dominated by turbulent mixing rather than coalescence.

3.4 MEASUREMENT OF ARRHENIUS PARAMETERS

3.6 ml of a 1 mM $Os(bipy)_3(ClO_4)_3$ solution and 3.6 ml of a 1.25 mM $K_4Fe(CN)_6 \times 3\,H_2O$ solution, each in 0.1 M KCl, were added to 1 l each of a 10 mM DTAC solution in 9:1 n-heptane/ hexanol ($w_0 = 20$ and $q = 0.8$).

3.5 COALESCENCE EXPERIMENTS WITH p-NITROPHENOL

A 1 mM nitrophenolate solution of pH 9.45 was prepared in 10 mM DTAC in 9:1 n-heptane/ hexanol and brought into contact with the same reversed micellar solution containing a solution of pH 3 in the pool. The reaction was monitored by the disappearance of p-nitrophenolate at $\lambda = 400$ nm.

3.6 FAST KINETIC METHOD

Since the timescale of diffusion-controlled encounters ($\tau = 1/(k_{diff}N)$ is of the order of microseconds, we have employed the Continuous Flow Method with Integrating Observation (CFMIO) developed by Holzwarth [21] (Figure 1). The method has been described elsewhere [22], so only the basic features are mentioned here. The two streams containing reactants A and B are turbulently mixed at the entrance of the tube with a time constant of 4 μs [22,23] and react while passing through it. The measures of the tube were: diameter 2 mm, length 3 cm, fluid velocity 2.1 m/s. The change of concentration along the length of the tube was measured with a source of white light shining through the circular top and bottom of the tube onto a detection system. Behind the tube, the signals were passed through a wavelength filter, and collected on an SEM. The amplified signal was recorded by a photon counter.

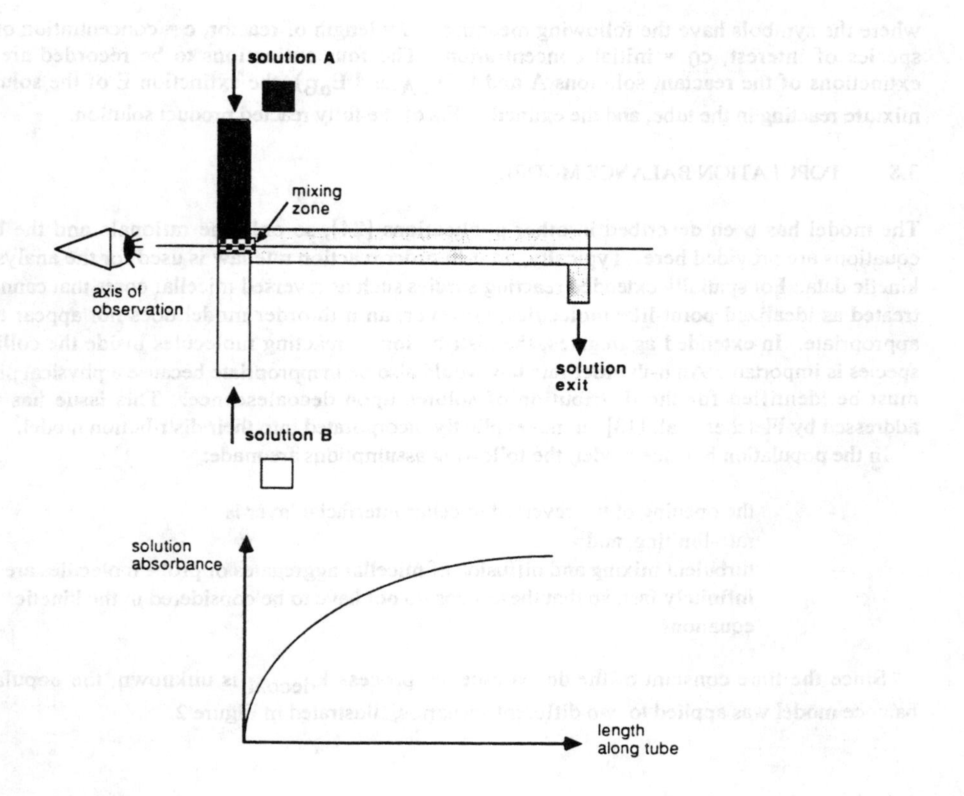

Figure 1: Schematic of the CFMIO-apparatus

3.7 DATA ANALYSIS

The quantity obtained from CFMIO experiments is an average concentration over the length of the tube scaled to the initial concentration. It is a dimensionless parameter called M-value scaled between 0 (all reactant is converted) and 1 (no reaction occurs) defined by:

$$M = \frac{E - E_\infty}{(E_{oA} + E_{oB})/2 - E_\infty} = \frac{1}{C_0 l} \int_0^1 c\,dx \tag{8}$$

where the symbols have the following meaning: l = length of reactor, c = concentration of the species of interest, c_0 = initial concentration. The four extinctions to be recorded are the extinctions of the reactant solutions A and B (E_{oA} and E_{oB}), the extinction E of the solution mixture reacting in the tube, and the extinction E_∞ of the fully reacted product solution.

3.8 POPULATION BALANCE MODEL

The model has been described in other publications [24], so only the rationale and the basic equations are provided here. Typically, an n-th order reaction rate law is used for the analysis of kinetic data. For spatially extended reacting species such as reversed micellar cores that cannot be treated as idealized point-like molecules, however, an n-th order model does not appear to be appropriate. In extended aggregates, the distribution of reacting molecules inside the colliding species is important. An n-th order rate law would also be inappropriate because a physical picture must be identified for the distribution of solutes upon decoalescence. This issue has been addressed by Fletcher et al. [13] but not explicitly incorporated into their distribution model.

In the population balance model, the following assumptions are made:

- the opening of the reversed micellar interfacial layer is rate-limiting, and
- turbulent mixing and diffusion of micellar aggregates or probe molecules are infinitely fast, so that these steps do not have to be considered in the kinetic equations.

Since the time constant of the decoalescence process k_{decoal} is unknown, the population balance model was applied to two different scenarios, illustrated in Figure 2:

Model Picture

Homonuclear collisions:

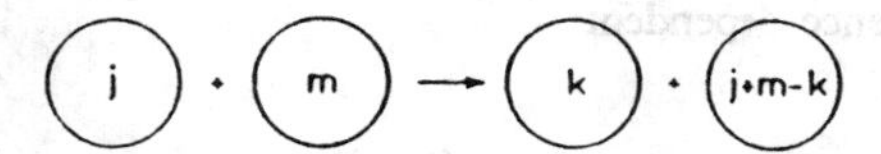

Heteronuclear collisions:

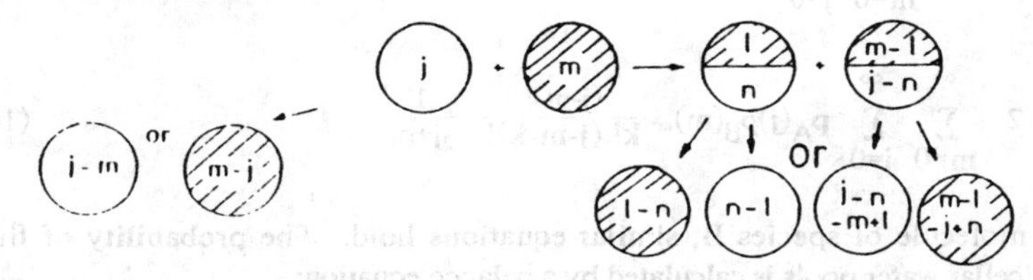

Decoalescence–dependence Reaction–dependence

Figure 2: **The slowest step in distribution of probe molecules is either decoalescence or reaction**

3.8.1 ***Distribution is dependent on reaction*: reseparation is fast compared to the reaction step. The reactants separate into either of the two newly formed micelles and react within those micellar pools.**

3.8.2 ***Distribution is dependent on decoalescence*: slow reseparation is assumed compared to diffusion-controlled reaction. Reaction occurs in the dimer; products separate into newly formed aggregates.**

For the two scenarios, the population balance equations are written as follows (for details, see ref. [24]):

Distribution is reaction-dependent:

$$\frac{dp_A(k)}{d\tau} = \sum_{m=0}^{\infty} \sum_{j=0}^{\infty} pA(j)pA(m) \frac{(j+m)!}{k!\,(j+m-k)!} \frac{1}{2^{j+m}} - p_A(k)$$

$$+ 2 \sum_{m=0}^{\infty} \sum_{j=0}^{\infty} p_A(j)p_B(m) \sum_{n=0}^{\infty} \frac{j!}{(k+n)!\,(j-k-n)!} \frac{j!}{n!\,(m-n)!} \frac{1}{2^{j+m}} \tag{9}$$

Distribution is decoalescence-dependent:

$$\frac{dp_A(k)}{d\tau} = \sum_{m=0}^{\infty} \sum_{j=0}^{\infty} p_A(j)p_A(m) \frac{(j+m)!}{k!\,(j+m-k)!} \frac{1}{2^{j+m}} - pA(k)$$

$$+ 2 \sum_{m=0}^{\infty} \sum_{j=0}^{\infty} p_A(j)p_B(m) \frac{(j-m)!}{k!\,(j-m-k)!} \frac{1}{2^{j+m}} \tag{10}$$

For probe molecule of species B, similar equations hold. The probability of finding empty reversed micellar water pools is calculated by a balance equation:

$$p_0 = 1 - \left\{ \sum_{k=1}^{\infty} pA(k) + \sum_{k=1}^{\infty} pB(k) \right\} \tag{11}$$

Solving the steady state equation with the heterogeneous collision terms (last terms in equs. (9) and (10)) suppressed yields the initial conditions on A which are subject to constraints of conservations of mass and additivity of probabilities to one. The initial conditions to be used in the solution of the differential equations (9) and (10) were found to be a Poisson distribution over p_A and p_B, respectively.

$$p_A(k;0) = \Phi_A \, \{\mu_A{}^k \exp(-\mu_A)\}/k! \tag{12}$$

$$p_B(k;0) = (1 - \Phi_A) \, \{\mu_B{}^k \exp(-\mu_B)\}/k! \tag{13}$$

$$p_0(0) = \Phi_A \exp(-\mu_A) + (1 - \Phi_A) \exp(-\mu_B) \tag{14}$$

Φ_A = fraction of reversed micelles initially containing molecule A

The rate constant k_{ex} was determined with this procedure by finding the dimensionless time, at which the limiting component had reacted to the extent that the sum of probabilities over k matched the measured M-value. After calculating an exchange rate constant k_{ex} for each experimental datum point, the best fit for k_{ex} at each water surfactant ratio w_0 was determined across all measured reactant ratios q.

4. Results

4.1 SOLUBILIZATE EXCHANGE RATE CONSTANTS FROM M-Q-PLOTS

The magnitude of the exchange rate constant k_{ex} as a function of water-surfactant ratio w_0 and reactant ratio q was the foremost issue of this investigation. At constant w_0 and therefore constant size and curvature of the reversed micelles [25], but different reactant ratios q, the rate constant k_{ex} is assumed to be the same. Table 1 lists the obtained exchange rate constants as a function of pool size; Figure 3 illustrates the results.

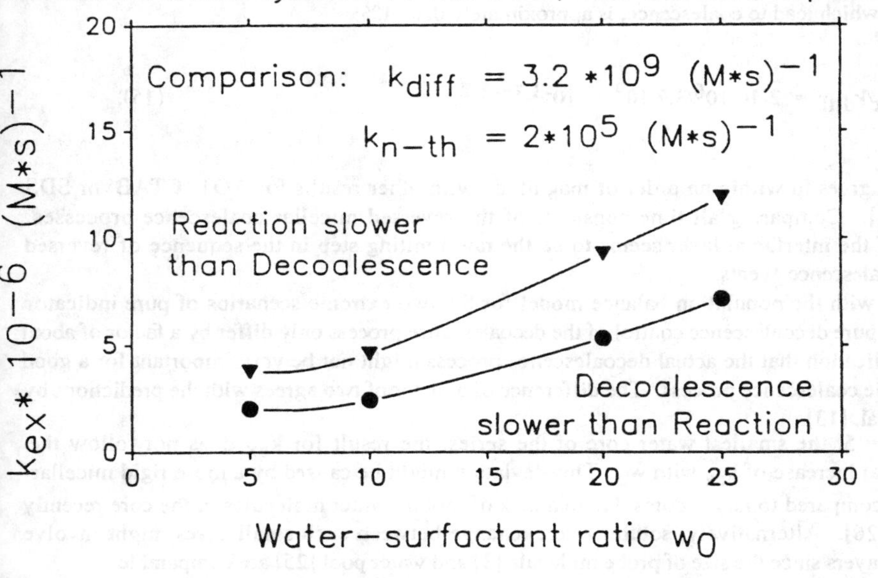

Figure 3: **Exchange rate constants k_{ex} at different pool sizes Reversed micellar system: 10 mM DTAC, 10% hexanol, n-heptane, 0.1 M KCl solution**

Table 1: Solubilizate exchange rate constants k_{ex} with both assumptions of distribution dependence on reaction and decoalescence

DTAC/hexanol/n-heptane/buffer		
w_0	reaction k_{ex} 10^{-6} $(Ms)^{-1}$	decoalescence & σ (k_{ex})
5	5.47 ± 1.37	2.70 ± 0.68
10	4.61 ± .93	2.44 ± 0.43
20	9.23 ± 1.37	5.32 ± 0.78
25	10.91 ± 2.11	7.16 ± 1.20

σ = standard deviation

Comparison of the magnitude of the exchange rate constant k_{ex}, of order 10^6 $(Ms)^{-1}$, with the rate constant for molecular diffusion, $k_{diff} = 3.2\ 10^9$ $(Ms)^{-1}$, reveals that the ratio of the two rate constants, k_{ex}/k_{diff}, representing the fraction of "successful" collisions between micellar aggregates which lead to coalescence, is approximately 0.1 - 1%:

$$k_{ex}/k_{diff} = 2\text{-}10\ 10^6/3.2\ 10^9 = 10^{-3}\text{-}10^{-2} \quad (15).$$

This ratio agrees to within an order of magnitude with other results for AOT, CTAB, or SDS systems [5]. Comparing all time constants of the reversed micellar coalescence processes, opening of the interfacial layer seems to be the rate-limiting step in the sequence of reversed micellar coalescence events.

Results with the population balance model for the two extreme scenarios of pure indicator reaction or pure decoalescence control of the decoalescence process only differ by a factor of about two, an indication that the actual decoalescence process might not be very important for a good model of the coalescence process. The difference of a factor of two agrees with the predictions by Fletcher et al. [13].

At w_0 = 5, the smallest water core of the series, the result for k_{ex} does not follow the proportional increase of k_{ex} with w_0. This deviation might be caused by a more rigid micellar aggregate compared to larger cores due to a lack of mobile water molecules in the core recently found [25,26]. Alternatively, solubilizate exchange between such small cores might involve surfactant layers since the size of probe molecule [3] and water pool [25] are comparable.

4.2 COMPARISON OF POPULATION BALANCE MODEL WITH AN N-TH ORDER MODEL

If solubilizate exchange was viewed as a coalescence process driven by the number of micellar aggregates, a second-order rate law would suffice. If solubilizate exchange was driven by the order of the indicator reaction, its order would vary from second order ($n = 2$) at equal concentrations of reactant (reactant ratio $q = 1$) to pseudo-first order ($n = 1$) at large excess of one reagent (reactant ratio $q \rightarrow 0$). Table 2 lists solubilizate exchange rate constants obtained with both the population balance and the n-th order model.

Table 2: Comparison of exchange rate constants by different kinetic models

w_0	pop. bal. model k_{ex} $(Ms)^{-1}$	n-th order model k_{ex} $(Ms)^{-1}$
5	2.70 10^6	2.09 10^5
10	2.44 10^6	3.06 10^5
20	5.32 10^6	
25	7.16 10^6	

Experiments at varying ratios q of probe molecules can distinguish between the proposed descriptions for solubilizate exchange. Both second-order and n-th order model predict that the average conversion M should not depend on the reactant ratio q: the number of micellar aggregates is constant in the first case and the indicator reaction is diffusion-controlled in the second case. Figure 4 shows the experimental results of M as a function of q and the good fit of the population balance model to these data. It is evident that the population balance model provides a more realistic description of the exchange processes than does the n-th order model.

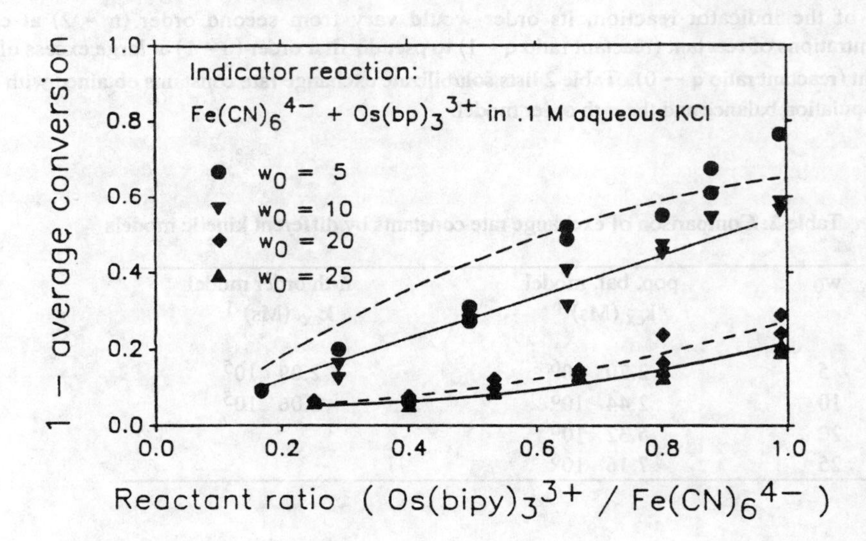

Figure 4: **Experimental data and fit of population balance equations in a conversion-reactant ratio plot. Reversed micellar system: 10 mM DTAC, 10% hexanol, n-heptane, 0.1 M KCl sol.**

The n-th order model underpredicts the exchange rate constant k_{ex} by about one order of magnitude, a significant difference because applying the n-th order model would have more reactions be apparently controlled by coalescence, with consequences for reactor design and control. Development of the population balance model seems justified from this comparison. Extrapolation to other reaction conditions is straightforward and unambiguous with the population balance model whereas the n-th order model cannot be used for this purpose.

4.3 SOLUBILIZATE EXCHANGE OF INTERFACIALLY BOUND PROBE MOLECULES

Coalescence experiments with association of p-nitrophenol with the hydrogen ion, H^+, as indicator reaction resulted in such low M-values that turbulent mixing instead of solubilizate exchange was rate limiting. However, if it is considered that p-nitrophenol is known to reside predominantly at the interfacial layer in other charged surfactant systems [7,27], and that strong interfacial association of p-nitrophenolate with DTAC aggregates has been measured in this work, the explanation is plausible that p-nitrophenolate can redistribute from one reversed micelle to the next without coalescence of the water pools. Therefore, much higher exchange rate constants can be found. As mentioned above, for several other interfacially bound probes exchange rate constants at or near the molecular diffusion limit have been measured [14,15]. However, for measuring coalescence of water pools, the probes have to be solubilized exclusively in the pool.

4.4 TEMPERATURE DEPENDENCE OF k_{ex}

It has been suggested that a simple Arrhenius model as applied to membrane pore opening could also be applied to reversed micelles [28,29]. To be applicable, one of the sequential steps in the reversed micellar coalescence process must be rate-limiting because the Arrhenius model describes elementary reactions only. The rate constant k_{ex} was determined at several temperatures from 18.5 to 41.5°C at a water-surfactant ratio $w_0 = 20$ and a reactant ratio $q = 0.8$ as shown in Figure 5. The Arrhenius parameters were determined to be $\Delta H^{\neq} = 63.5$ kJ/mol and $\Delta S^{\neq} = 352.4$ J/(K mol). Compared to an activation energy of the reaction $E_{a,rxn} = 7.9$ kJ/mol [30] and of the diffusion steps of $E_{a,diff} = 10\text{-}17$ kJ/mol [31], the values found point to opening of the interfacial layer as rate-determining step. In the AOT/isooctane system, a higher value for the activation enthalpy ΔH was found, but a much lower one for the activation entropy $\Delta S^{\neq}$ [13], which corroborates the idea that AOT forms a particularly ordered and stable structure compared to the four-component system of DTAC.

5. Discussion

In the range of $10 < w_0 < 25$, k_{ex} is found to increase proportionately with w_0. Two potential reasons for this functionality are a change in the composition or a change in stiffness of the interfacial layer. The composition of the interfacial layer, however, has been found to be the same at all w_0- values in the DTAC system: the ratio of hexanol to DTAC molecules is 2.85 [25]. The stiffness of the layer can be influenced by hydrophobic or electrostatic effects [32] but composition of the interfacial layer and ionic strength in the water pool are constant throughout all experiments. Change of water core structure is not a likely explanation either because the mobility of water resembles that of liquid bulk water already above $w_0 = 5$ [25].

Even at constant interfacial composition and thus hydrophobicity, constant ionic strength in the pool, and constant water structure, the rigidity of the micellar interface can be influenced by its degree of curvature. With increasing w_0-value, curvature decreases and could potentially result in a higher flexibility and thus permeability of the interfacial layer. There are several models of surfactant surfaces incorporating the effects of curvature, such as an equation by Auvray [33] on the internal curvature energy E_{curv} which is linearly dependent on radius:

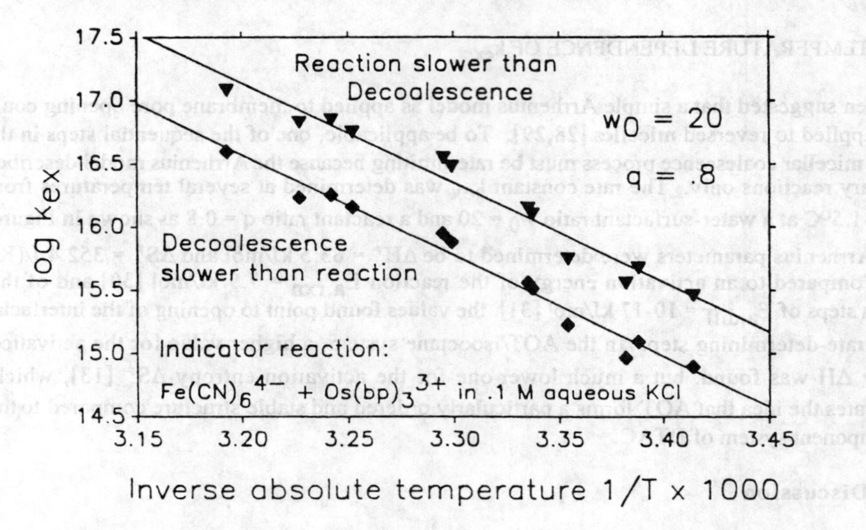

Figure 5: **Arrhenius plot of reversed micellar coalescence process Reversed micellar system: 10 mM DTAC, 10% hexanol, n-heptane, 0.1 M KCl solution**

$$E_{curv} = 8\pi K (1 - c_0 r) \qquad (16).$$

The spontaneous film curvature c_0 and the rigidity constant K can be assumed to be constant at constant interfacial composition.

The data presented in this work do not in themselves allow application of a model incorporating effects of the surfactant surface. However, they demonstrate that such effects are in all probability important to understand the magnitude and dependencies of the exchange rate constant. Further studies over a larger range of micellar radii and varying surface properties such as nature of the surfactant and ionic strength are strongly suggested as a result of this work.

The similarity between the exchange rate constants k_{ex} in both DTAC and other investigated systems is evident: all are on the order 10^6 to 10^7 $(Ms)^{-1}$, the fraction of collisions leading to coalescence k_{ex}/k_{diff} is of order 10^{-3}. Based on the literature on other systems, this ratio seems to be common to most reversed micellar systems [5]. It is nuclear, however, whether

interpretations of experimental data without taking into account distribution effects properly influence the results other groups have obtained. Studies of different surfactant systems under the same conditions as in this work including indicator reactions, ionic strength, and interpretations of data by the population balance model are strongly suggested.

6. Conclusions

Reversed micellar coalescence introduces an upper rate limit for pool-solubilized probe molecules: the solubilizate exchange rate constant between pools k_{ex} is two to three orders of magnitude lower than the rate constant for molecular diffusion. The rate constant depends linearly on reversed micellar radius.

Only chemical or biochemical reactions with a timescale faster than 10^{-3} to 10^{-4} s are affected by intermicellar transport. For reactions with slower timescales, the concentration of substrates in the pools can be regarded as equilibrated.

These results are obtained by applying a population balance model to experimental data of solubilizate exchange gathered with the Continuous Flow Method with Integrating Observation (CFMIO), which is much more sensitive than ordinary continuous flow methods.

The population balance model is a novel approach applied to reversed micellar systems, which is expected to find further applications for phenomena other than solubilizate exchange or interfacial transport and for other systems too small for continuum theory.

7. Acknowledgements

The authors wish to acknowledge the Biotechnology Process Engineering Center funded under the NSF-ERC Cooperative Agreement CDR 85 0003 for financial support. Additional funding was provided by Dow Chemical Co. and an NSF Presidential Young Investigator Award to T.A.H., Contract No. 8451593. The authors also wish to thank Mark G. Allen and Reiner F.H. Musier for assistance with the population balance program.

8. References

[1] Noyes, R.M., Progress in Reaction Kinetics, 1961, **1**, 129

[2] Smoluchowski, M. v., Z. physik. Chem., 1917, **92**, 129

[3] Holzwarth, J.F., and Jürgensen, H., Ber. Bunsenges. physik. Chem., 1974, **78**, 526

[4] a) Lin, C.-T., Böttcher, W., Chou, M., Creutz, C., and Sutin, N., J. Amer. Chem. Soc., 1976, **98**, 6536; b) Lin, C.-T., and Sutin, N., J. Phys. Chem., 1976, **80**, 97; c) Creutz C., and Sutin, N., J. Amer. Chem. Soc., 1977, **99**, 241

[5] Zana., R., and Lang, J., "Dynamics of Microemulsions", in: "Microemulsions: Structure and Dynamics", S. Friberg and P. Bothorel (eds.), CRC Press, Boca Raton,

Florida, 1988

[6] Fersht, A., "Enzyme Structure and Mechanism", 2nd edition, Freeman & Comp., 1985, p. 147

[7] Menger, F.M., Donohue, J.A., and Williams, R.F., J. Am. Chem. Soc., 1973, 95, 286

[8] Eicke, H.-F., Shepherd, J.C.W., and Steinemann, A., J. Colloid Interface Sci., 1976, 56, 168

[9] Fletcher, P.D.I., and Robinson, B.H., Ber. Bunsenges. Phys. Chem., 1981, 85, 863

[10] Atik, S.S., and Thomas, J.K., Chem. Phys. Lett., 1981, 79, 351

[11] Atik, S.S., and Thomas, J.K., J. Amer. Chem. Soc., 1981, 103, 3543

[12] Pileni, M.P., Brochette, P., Hickel, B., and Lerebours, B., J. Interface Colloid Sci., 1984, 98, 549

[13] Fletcher, P.D.I., Howe, A.M., and Robinson, B.H., J. Chem. Soc., Faraday Trans. 1, 1987, 83, 985

[14] Furois, J.M., Brochette, P., and Pileni, M.P., J. Colloid Interface Sci., 1984, 97, 552

[15] Lang, J., Jada, A., and Malliaris, A., J. Phys. Chem., 1988, 92, 1946

[16] Holzwarth, J.F., unpublished results

[17] Bruhn, H., and Holzwarth, J.F., Ber. Bunsenges. physik. Chem., 1978, 82, 1006

[18] a) Atik, S.S., and Thomas, J.K., J. Phys. Chem., 1981, 85, 3921; b) Atik, S.S., and Thomas, J.K., J. Am. Chem. Soc., 1981, 103, 7403; c) Almgren, M, Grieser, F., and Thomas, J.K., J. Amer. Chem. Soc., 1980, 102, 3188; d) Gregoritch, S.J., and Thomas, J.K., J. Phys. Chem., 1980, 84, 1491

[19] Lianos, P., Lang, J., Cazabat, A.-M., and Zana, R., "Luminescent probe study of W/O microemulsions", in: Surfactants in Solution, eds.: Mittal, K.L., and Bothorel, P., Plenum Press, New York, 1986

[20] Burstall, F.H., Dwyer, F.P., and Gyarfas, E.C., J. Chem. Soc., 1950, 953

[21] Gerischer, H., Holzwarth, J.F., Seifert, D., Strohmaier, L., Ber. Bunsenges. physik.Chem., 1969, 73, 952-5

[22] Holzwarth, J.F., "Fast-Continuous-Flow", and "Electron-transfer reactions", in: Techniques and Applications of Fast Reactions in Solution, eds.: Gettins, W.J., and Wyn-Jones, E., 1979, D. Reidel Publ. Comp., Dordrecht, pp. 13-24, and pp. 509-521

[23] a) Holzwarth, J.F., unpublished results; b) Gerischer, H., Holzwarth, J.F., Seifert, D., and Strohmaier, L., Ber. Bunsenges. physik. Chem., 1972, **76**, 11

[24] Bommarius, A.S., Holzwarth, J.F., Wang, D.I.C., and Hatton, T.A., submitted to J. Phys. Chem.

[25] Bommarius, A.S., Wang, D.I.C., Hatton, T.A., Petit, C., and Pileni, M.-P., to be submitted to Langmuir

[26] Verbeeck, A., Voortmans, G., Jackers, C., and de Schryver, F.C., Langmuir, 1989, 5, 766

[27] Magid, L.J., Kon-no, K., and Martin, C.A., J. Phys. Chem., 1981, **85**, 1434

[28] Siegel, D.P., J. Colloid Interface Sci., 1984, **99**, 201

[29] Siegel, D.P., Biophys. J., 1984, 45, 399

[30] Marcus, R.A., Ann. Rev. Phys. Chem., 1964, **15**, 155

[31] Satterfield, C.N., "Mass Transfer in Heterogeneous Catalysis", Robert E. Krieger Publ. Comp., Huntington, NY, 1981, p. 4-5

[32] Hou, M.J., Kim. M., and Shah, D.O., J. Colloid Interface Sci., 1988, **123**, 398

[33] Auvray, L., J. Physique Lett., 1985, **46**, L163

SELECTIVE SOLUBILISATION IN REVERSED MICELLES

E.B. LEODIDIS and T.A. HATTON
Department of Chemical Engineering
Massachusetts Institute of Technology
Cambridge, MA 02139
U.S.A.

ABSTRACT. The equilibrium distribution of polar solutes between a reversed micellar organic solution and an excess bulk water phase can provide detailed *microscopic* information on the solubilisation phenomenon. This information can be used to investigate specific ion effects in electrical double layers and to probe interfacial interactions in structured surfactant layers. A phenomenological electrostatic model indicates that charge, hydrated size, and polarisability are simultaneously responsible for the selective solubilisation of cations in anionic reversed micelles. Interfacial solubilisation of amino acids is driven by the hydrophobic effect and mediated by the presence of hydrogen-bonding groups on the amino acid side chains and by the possibility of solute-interface hydrogen-bonding. A significant electrolyte effect is attributed to a squeezing out of the amino acids from the interface as a result of increasing curvature as the micelles become smaller, and interfacial thermodynamic calculations indicate that the interfacial bending moment may play a significant role in this regard. An analysis of the solubilisation of charged amino acids indicates that the electrostatic potential close to the micelle wall is close to estimates provided by the classical Poisson-Boltzmann algorithm.

1. Introduction

Solubilisation is the increased solubility, in the presence of supramolecular surfactant aggregates, of a substance normally insoluble or only slightly soluble in a continuous solvent phase. A surprising number of ions, polar organic molecules, and even biopolymers are easily solubilised in organic solvents by being hosted within the polar cores of a class of supramolecular aggregates known as reversed micelles, surfactant-stabilised, nanometer-scale water droplets in an apolar solvent. These reversed micellar solutions can be considered as novel solvents possessing unique solubilising properties [1-3]. Their versatility can be ascribed to the fact that three distinctly different solubilisation environments are simultaneously present in reversed micellar solutions: (i) the organic continuum, (ii) the water pools, and (iii) the surfactant interface (Figure 1).

To date, very little is known about the thermodynamics or the kinetics of the solubilisation process in reversed micelles [2], despite the fact that this information would be of great value for the successful application of reversed micellar solutions as novel solvents with significant biotechnological potential [4]. A number of important questions must be addressed to improve our present understanding of solubilisation processes, and these can be stated as follows:

D. M. Bloor and E. Wyn-Jones (eds.),
The Structure, Dynamics and Equilibrium Properties of Colloidal Systems, 201–231.

(i) Why does solubilisation occur? What are the thermodynamic driving forces for solubilisation?

(ii) Where do the solubilised species reside in the reversed micellar phase? What is the nature of the average environment experienced by the individual solutes?

(iii) What happens when solute species are solubilised in reversed micelles? Do they induce any size or shape changes in their host aggregates?

(iv) How does solubilisation take place? What are the mechanisms and kinetics of the uptake of individual solutes by the surfactant assemblies?

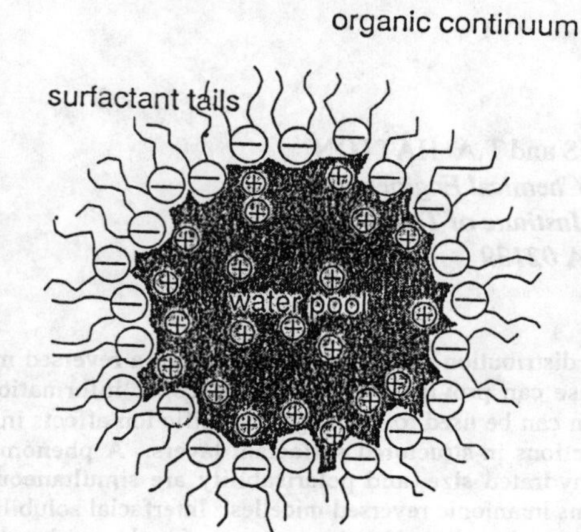

Figure 1. The three different solubilisation environments in a reversed micellar solution.

The present paper is concerned with the first three topics. In particular, it is shown that a surprisingly detailed picture of the solubilisation process can be constructed using the phase equilibrium experiment combined with appropriate mathematical models as an analytical tool [3]. With this approach, a significantly wider range of solutes can be investigated than is customarily possible using spectroscopic and other indirect methods. In particular, we have investigated the solubilisation of monovalent and divalent alkali and alkali-earth cations, and of a range of natural and synthetic amino acids. Interpretation of the results focuses on specific effects of different cations, through their contributions to electrical double layer structure, in the formation of reversed micelles, and on elucidation of the interplay between the hydrophobic effect, steric constraints and the effect of polar moieties on the interaction of amino acid residues with the surfactant interface. A more complete account of these results is presented elsewhere [5]. Earlier work by Fletcher [6] also recognised the value of the phase equilibrium method for exploring the interfacial association of solutes.

2. Experimental

The phase equilibrium experiment is presented schematically in Figure 2. An aqueous electrolyte solution was contacted with a surfactant solution (usually AOT) in isooctane solvent; typically 5 ml of each solution were used. Conditions were selected to ensure the formation of a Winsor II system, characterised by an excess aqueous phase in equilibrium with a conjugate reversed micellar solution. The two solutions were intimately mixed and placed in a temperature-controlled bath at 25°C to equilibrate and phase separate. The equilibrium distribution of the solute species between the phases was obtained from simple measurements of their concentrations in the final aqueous phase, and of the water uptake by the microemulsion phase of the Winsor II system.

2.1 MATERIALS

Aerosol-OT (bis(2-ethylhexyl) sodium sulfosuccinate) of 99% purity was obtained from Pfaltz and Bauer (Waterbury, CT), and from Sigma (St Louis, MO), and spectrophotometric grade isooctane from Mallinckrodt (St Louis, MO); both the surfactant and the organic solvent were used without further purification. The other surfactant used in this study, DTAC (dodecyltrimethylammonium chloride), was purchased from American Tokyo Kasei (Portland, OR), and was recrystallized from ether and methanol prior to use.

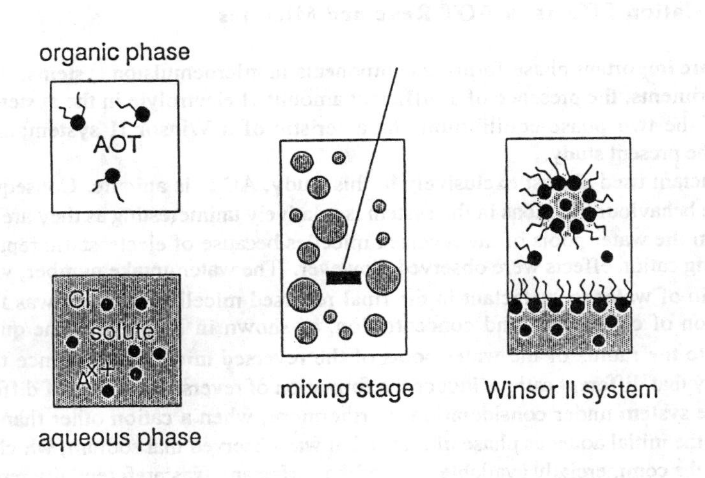

Figure 2. A schematic of the phase transfer experiment.

The electrolytes used in this work were obtained from Mallinckrodt and used as received. Almost all the amino acids were obtained from Sigma, exceptions being homophenylalanine and para-hydroxy-phenylglycine, obtained from Chemical Dynamics (South Plainfield, NJ); these were also used as received.

Atomic absorption standards were obtained from Aldrich (Milwaukee, WI). Karl Fischer solvent and titrant were obtained from Crescent Chemical Co. (Hauppauge, NY). Ninhydrin reagent solution was obtained from Sigma. The water used for the preparation of electrolyte solutions was doubly distiled and deionized.

2.2 ANALYSIS

The water content of the reversed micellar phase was measured by Karl Fischer titration, with a Mettler DL 18 automatic titrator. The cationic concentrations in the final aqueous phase ([Na^+] and [M^{z+}]) were determined on a Perkin-Elmer Plasma 40 atomic emission spectrometer. The chloride content of the final aqueous phase was occasionally measured. No attempt was made to measure the ionic contents of the organic phase.

The residual amino acid concentration in the final aqueous phase was measured using the standard ninhydrin assay [7]. The final pH in the excess aqueous phase was also measured to ensure that no significant pH shift occurred during the contacting experiment. This was found to be the case for most samples when phosphate buffers were used as the phase-forming electrolytes. The maximum observed pH shift was 0.2 to 0.3 pH units in this case, which created no concern for the purpose of the present work.

3. Specific Cation Effects in AOT Reversed Micelles

Electrolytes are important phase-forming components in microemulsion systems. In the phase-transfer experiments, the presence of a sufficient amount of electrolyte in the system ensures the formation of the two phase-equilibrium characteristic of a Winsor II system; this has been exploited in the present study.

The surfactant used almost exclusively in this study, AOT, is anionic. Consequently it was found that the behaviour of anions in the system is relatively uninteresting as they are significantly excluded from the water pools of the reversed micelles because of electrostatic repulsions. Two very interesting cation effects were observed, however. The water uptake number, w_o, defined as the molar ratio of water to surfactant in the final reversed micellar solution, was found to be a strong function of cation type and concentration, as shown in Figure 3. The quantity w_o is proportional to the radius of the water pools of the reversed micelles [6], hence the results of Figure 3 imply that different cations induce the formation of reversed micelles of different sizes in the two-phase system under consideration. Furthermore, when a cation other than sodium was introduced in the initial aqueous phase of Figure 2 it was observed that sodium, which is the initial counterion of the commercially available form of the surfactant, was preferentially replaced by this new cation inside the water pools (see Figure 4 for results obtained with potassium chloride as the excess phase electrolyte).

The specific cation effect on reversed micellar sizes and the selectivity of the reversed micelles for certain cations over others can be ascribed to electrostatic phenomena. A phenomenological model has been proposed by Leodidis and Hatton [8] to account for the observed effects. The key components of this model which set it apart from the classical Poisson-Boltzmann approximation are a recognition of the diffuse nature of the surfactant interface, wherein the surfactant heads are distributed over a spherical layer of width 5-7 Å [9], an accounting for the finite size of the hydrated ions using volume fraction statistics for the entropy of mixing in a way originally

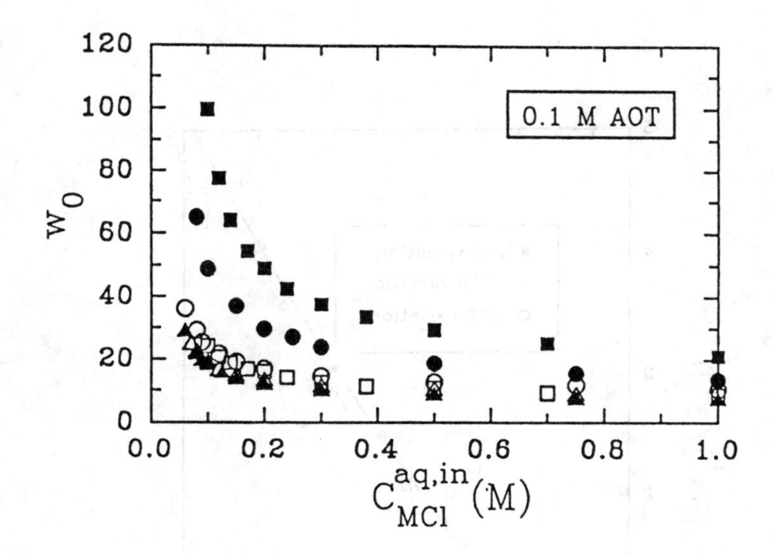

Figure 3. The water uptake number, w_o, as a function of the electrolyte concentration and type in the initial aqueous phase. The cations used were (■) ammonium, (●) sodium, (o) cesium, (□) silver, (Δ) rubidium, and (▲) potassium

proposed by Ruckenstein and Schiby [10], and an allowance for the variation of the dielectric constant of water inside the water pools with the local electric field according to Booth's equation, originally developed for pure water [11]. With these approximations, the concentration law to be used in Poisson's equation for a dielectric medium becomes

$$C_i^{mwp}(r) = C_i^{aq,f}\left(\frac{\phi_w^{mwp}(r)}{\phi_w^{aq,f}}\right)^{\tau_i} \exp\left[-\frac{z_i e\psi(r)}{k_B T} - A_i\left(\frac{1}{\varepsilon^{mwp}(r)} - \frac{1}{\varepsilon^{aq,f}}\right)\right] \quad (1)$$

where C_i is the concentration of ionic species i, ϕ_w is the volume fraction of space occupied by free water molecules (i.e., water molecules that do not participate in the formation of the hydration sheath of the ions or the surfactant heads), $\psi(r)$ is the local electrostatic potential inside the water pools, ε is the dielectric constant, A_i is the electrostatic part of the free energy of hydration of ionic species i, and the superscripts mwp and aq,f refer to the micellar water pools and the excess equilibrium aqueous phase respectively. The parameter τ_i plays an important role in this model; it is a measure of the hydrated size of the ionic species i.

This model contains only a single adjustable parameter, the hydrated-size parameter for sodium, which was obtained by reconciling the model predictions with a single experimental datum for the distribution ratio of sodium to potassium ions between the reversed micellar water

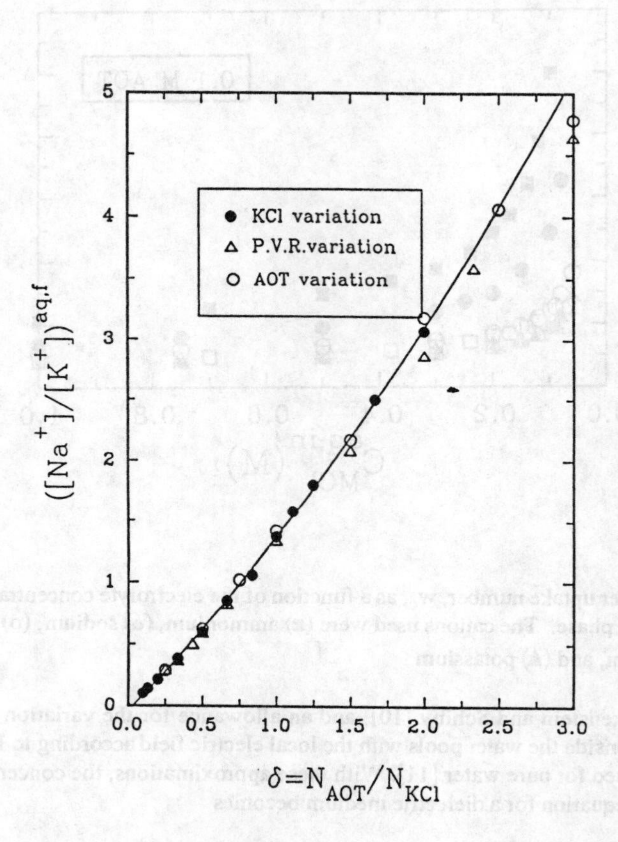

Figure 4. The ratio of the concentration of sodium to potassium in the final aqueous phase as a function of the total moles surfactant to moles potassium chloride in the initial aqueous phase. The results are for (o) variation of the initial KCl concentration, (Δ) variation of the volume ratio of the initial aqueous to the initial organic phase, and (o) variation of the AOT concentration in the initial organic phase.

pools and the excess aqueous phase; all the other τ_i were calculated from this parameter and the ratios of hydrated volumes reported in the literature [12]. The model performed extremely well in predicting cation distributions between the two phases for both monovalent and divalent cations, as is shown in Figure 5. The concentration profiles of cations inside the water pools for one particular case are presented in Figure 6. It is clear that smaller hydrated cations or cations with smaller electrostatic free energy of hydration screen the double layer more effectively, and are therefore able to promote the formation of smaller reversed micelles than are the larger hydrated ions.

A satisfying aspect of these model calculations is that the number of bound counterions, i.e.,

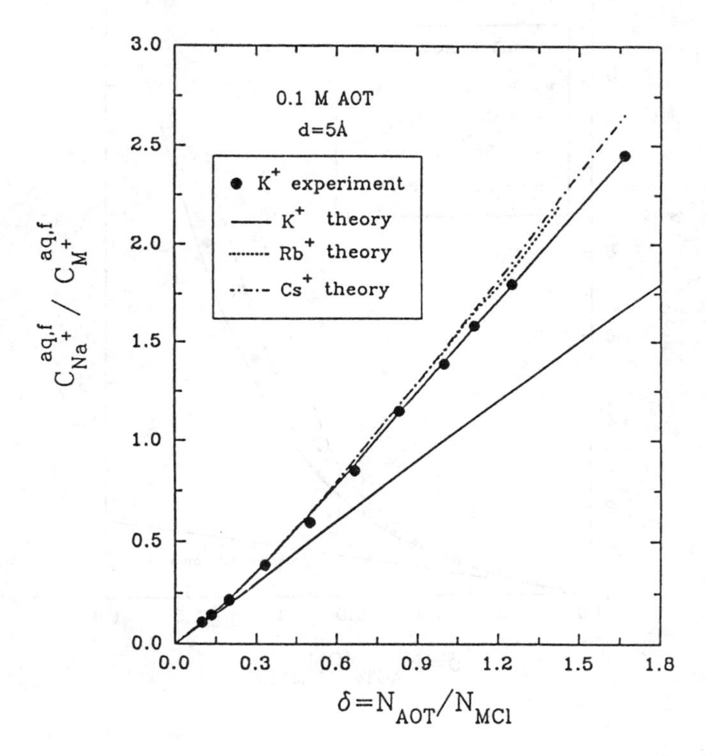

Figure 5. Cation solubilisation curves predicted by the phenomenological model. (a) monovalent ions, with experimental data shown for potassium.

the degree of association of the surfactants, can be identified as those ions in the headgroup region. This permits a calculation of the effective degree of surfactant dissociation without the need to invoke specific binding interactions [8]; the calculated degree of dissociation of between 20 and 30 percent agreed well with reported experimental results. Moreover, the ratio of the concentrations of the two ions in the interfacial region relative to their ratio in the final aqueous phase can be interpreted as the ion exchange constant for the surfactant interface. Again, the calculated values for the potassium-sodium case are close to experimental values reported for other micellar systems [5]. These observations give some confidence in the model formulation.

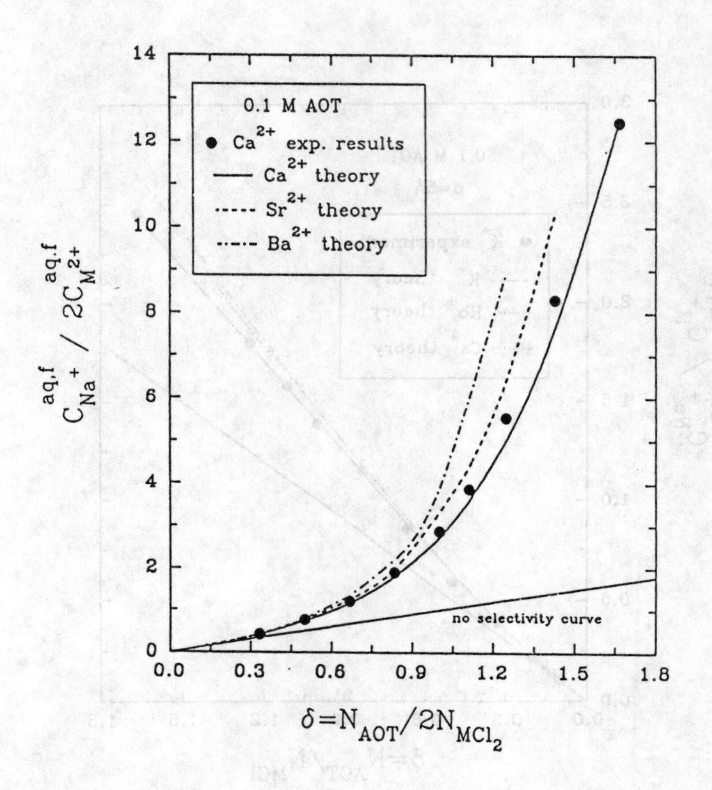

Figure 5 (contd.) Cation solubilisation curves predicted by the phenomenological model. (b) divalent ions, with experimental results shown for calcium.

4. Amino Acids in Reversed Micelles

Amino acids have a number of desirable properties that make them ideal model compounds for the study of interfacial solubilisation phenomena in reversed micellar systems. They are insoluble in organic solvents, but reasonably soluble in water, they have zero net charge under neutral conditions of pH and therefore do not interact strongly with the double layer in the water pools, and a large variety of amino acids is available commercially, with side-chains of widely-differing properties. Any difference in interfacial behaviour can be ascribed to the different side-chain characteristics. The analysis and interpretation of the solubilisation results relied on the acceptance of a simple solubilisation model in which the reversed micellar pools were considered

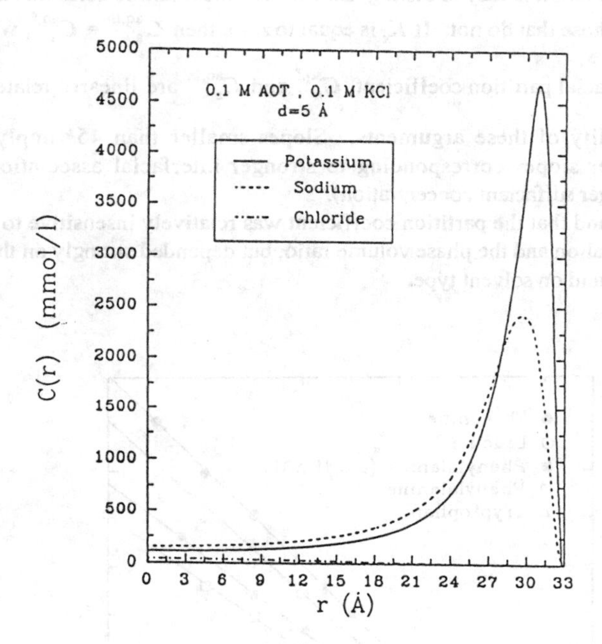

Figure 6. Ion concentration profiles inside the water pool of a reversed micelle for 0.1 M AOT in the initial organic phase and 0.1 M KCl in the initial aqueous phase.

as a uniform, average solubilisation environment, and the average concentration in the water pools was assumed to be equal to the excess aqueous phase amino acid concentration. This latter assumption was justified *a posteriori*. These assumptions, together with simple material balance constraints, enabled a mole-fraction-based interfacial partition coefficient to be calculated directly from the measured aqueous phase concentrations following the phase equilibration experiment, using the equation

$$K_x^{\infty} = \frac{C_w V^{aq,in}}{N_s^{tot}} \left(\frac{C_a^{aq,in} - C_a^{aq,f}}{C_a^{aq,f}} \right) \qquad (2)$$

where C_w is the average concentration of water (55.5 M), N_s^{tot} is total number of moles of surfactant in the system, $V^{aq,in}$ is the initial volume of the aqueous phase, and C_a is the concentration of amino acid in the initial (in) or final (f) excess aqueous phase.

This equation shows that it is easy to distinguish between those amino acids that associate with the interface, and those that do not. If K_x is equal to zero, then $C_a^{aq,in} = C_a^{aq,f}$, while for non-zero values of the interfacial partition coefficient, $C_a^{aq,f}$ and $C_a^{aq,in}$ are linearly related. Figure 7 demonstrates the validity of these arguments. Slopes smaller than 45° imply interfacial association, the smaller slopes corresponding to stronger interfacial association, or larger interfacial area (i.e., larger surfactant concentration).

It was generally found that the partition coefficient was relatively insensitive to variations in the amino acid concentration and the phase volume ratio, but depended strongly on the electrolyte type and concentration, and on solvent type.

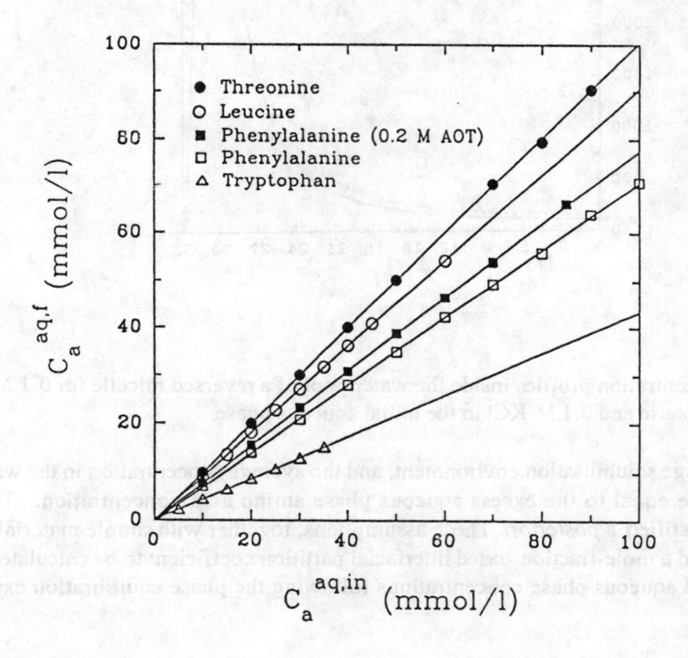

Figure 7. The relationship between the amino acid concentrations in the final and initial aqueous phases for a range of amino acids of varing interfacial activity.

4.1 THE HYDROPHOBIC EFFECT AS A DRIVING FORCE FOR INTERFACIAL SOLUBILISATION

The driving forces for amino acid interfacial solubilisation were consistent with the general guidelines provided by the excellent papers of Katz, Diamond and Wright [13,14]. The interfacial partition coefficients of 45 amino acids were found to correlate well with a variety of hydrophobicity scales, including solution scales, such as the octanol-water (π) scale, and molecular size-based scales such as the van der Waals molecular area or volume. The correlation with van der Waals volume is shown in Figure 8.

Solutes found above the hydrophobic line of Figure 8 can be classified into two groups: (i) solutes whose side chains contain one or more moieties capable of hydrogen bonding with water, and (ii) large solutes which may potentially aggregate in the aqueous phase. It can be concluded that the principle driving force for amino acid solubilisation is the hydrophobic effect, and that the interface provides an accommodating environment for hydrophobic amino acid side chains. No specific chemical effects indicating "binding" were detected for any of the solutes used. Hydrogen bonding with water increases the solute hydrophilicity and decreases K_x. Thus hydrogen-bonding is another important factor in interfacial solubilisation, acting in a direction opposite to the hydrophobic effect.

The existence of one additional important driving force for interfacial solubilisation was shown using a second, cationic W/O microemulsion system. The interface of dodecyltrimethylammonium chloride/hexanol reversed micelles contains a significant number of alcohol molecules, which render the interface more polar, leading to a reduced hydrophobic driving force to the interface. Alcohol molecules are also able to form hydrogen bonds with specific side-chain moieties on the more hydrophilic amino acids, counteracting the hydrogen-bonding effects in the aqueous pools, and increasing the interfacial partition coefficients of these molecules. These effects are evident from the results shown in Figure 9. The free energies of transfer for the purely non-polar solutes are well-correlated with those for the AOT system, although they are significantly higher because of the more polar nature of the DTAC/hexanol interface. Those solutes which exhibit a free energy below the correlation line are polar amino acids, and are therefore capable of forming hydrogen bonds with the DTAC interface, enhancing their drive to partition to this interface.

4.2 DEPENDENCE OF INTERFACIAL PARTITIONING ON ELECTROLYTE CONCENTRATION

The strong effect of salt concentration on K_x shown in Figure 10(a) cannot be attributed to direct interactions of the amino acid dipole with the local electric field [5]. It is evident though that when expressed in terms of the curvature of the surfactant interface, which is inversely proportional to w_0, the dependency on the specific salt type is significantly reduced, but that there is a very strong effect of interfacial curvature on the interfacial transfer free energy (Figure 10(b)). It was concluded that the electrolyte effect on K_x for amino acids is a *squeezing out* effect. According to this interpretation, it is hypothesised that a locus of pivotal points exists in the interface, over which the interfacial surfactant density remains essentially independent of curvature. The interfacial density increases with curvature between this locus and the water pools, while it decreases towards the tail region. Increased chain density and increased rigidity of the interface lead to a gradual expulsion of amino acid molecules from the curved interface, as shown schematically in Figure 11. The opposite effect is observed for solutes that are preferrentially solubilised further back in the surfactant interface, such as dichloroindophenol (DCIP), an interfacially active hydrophobic molecule. These effects are analogous to the results for mono- and bilayers reported in the recent literature [15,16].

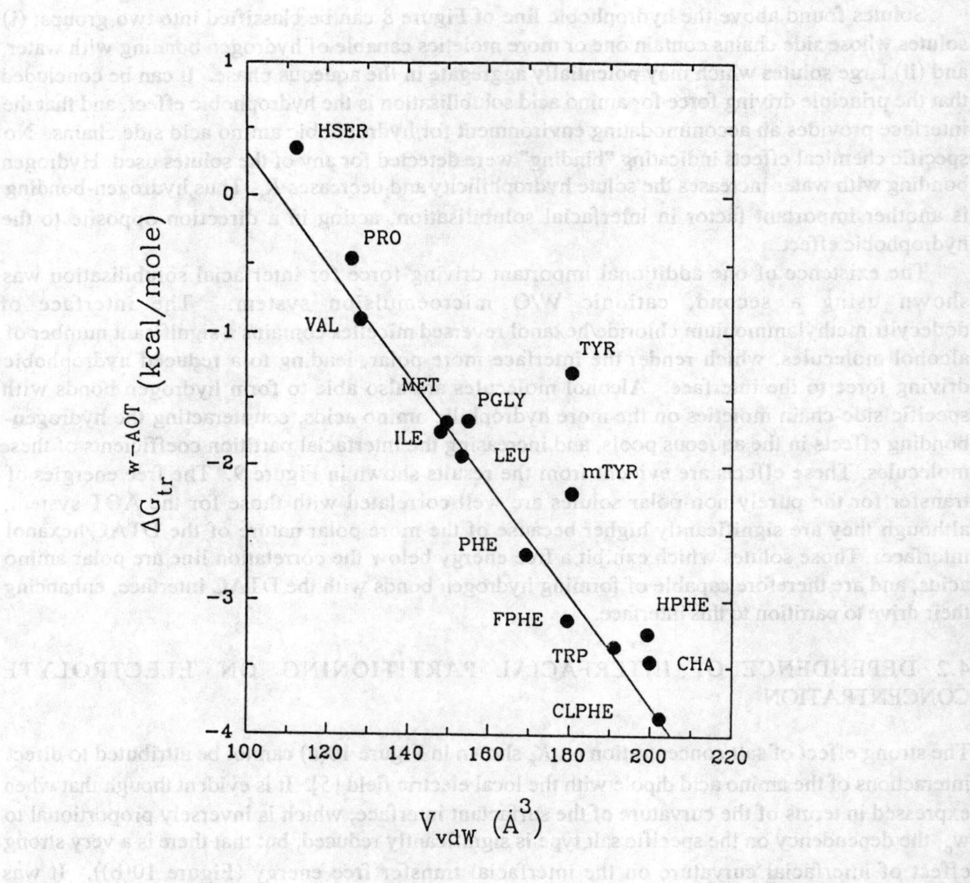

Figure 8. Correlation of the free energy of transfer of amino acids from water to the AOT microemulsion interface with the van der Waals molecular volume. The solid line represents the behaviour if only hydrophobic effects are operative.

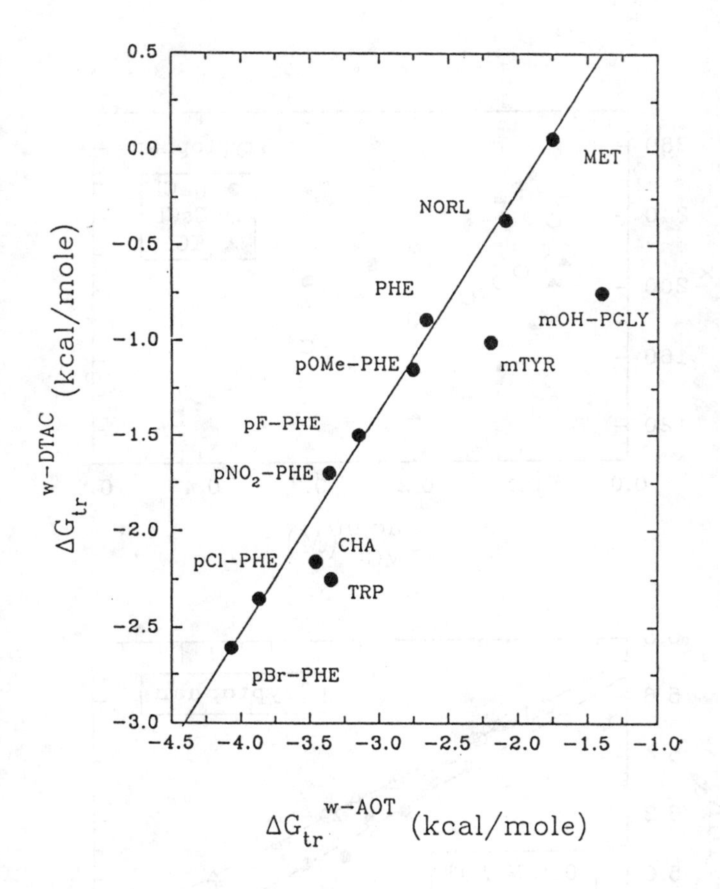

Figure 9. Correlation of the free energies of transfer from water to the DTAC and AOT microemulsion interfaces.

An extensive experimental study using a wide range of molecules yielded results entirely consistent with this hypothesis. Significant specific steric interactions between amino acids and the AOT interface have been observed, such that branched solutes or solutes containing bulky moieties are more strongly expelled from the interface as its curvature increases. A more

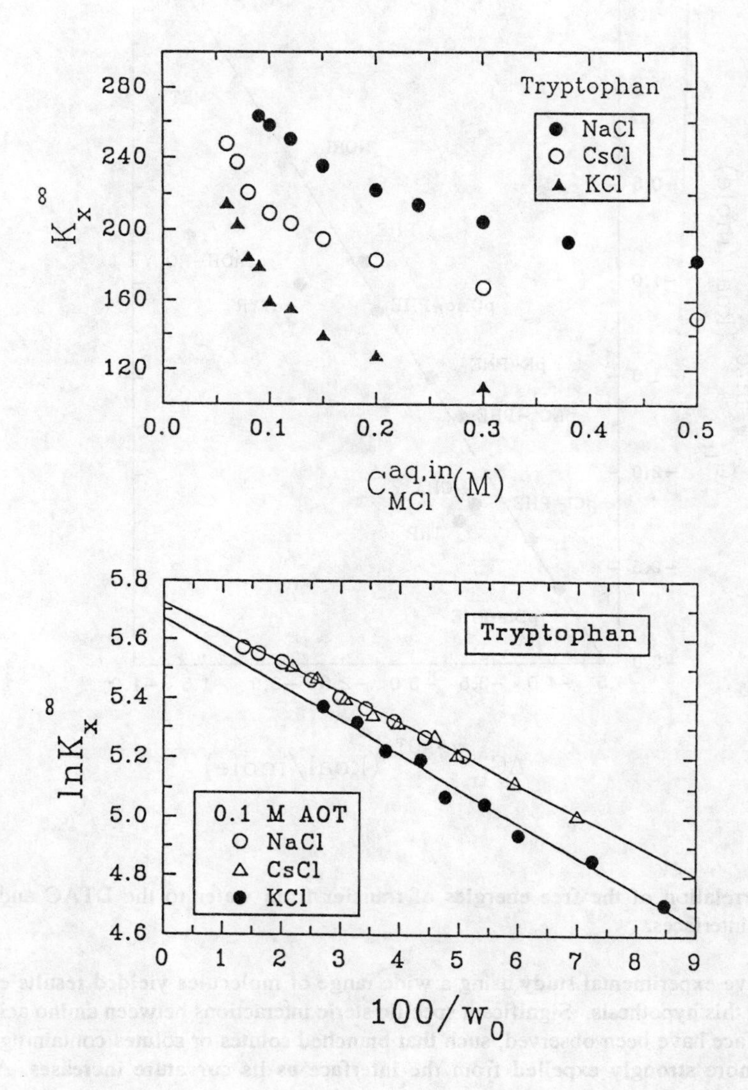

Figure 10. Dependence of the interfacial partition coefficient for tryptophan on (a) the electrolyte concentration and type in the initial aqueous phase, and (b) the inverse water uptake number, w_o.

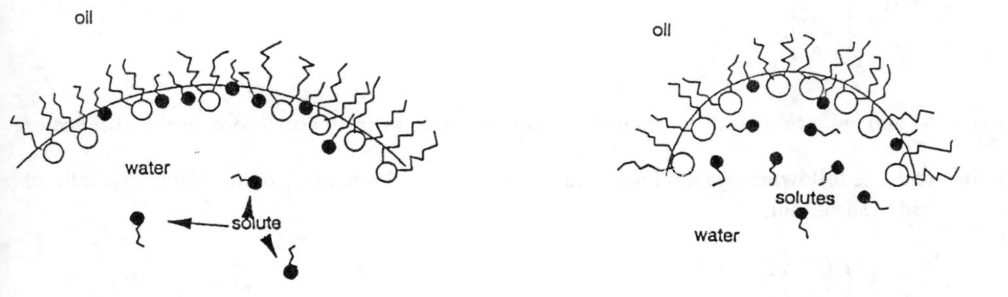

Figure 11. A molecular interpretation of the "squeezing-out" effect.

quantitative interpretation of the squeezing-out effect was attempted using the principles of interfacial thermodynamics, which yielded the following dependence of the chemical potential of an interfacially-solubilised solute on interfacial curvature:

$$\mu_a^{\infty}(\gamma_d, R_w) = \mu_a^{\infty}(\gamma_{\infty}, \infty) - \int_{\gamma_{\infty}}^{\gamma} f_a^{\infty}\, d\gamma - \int_{\infty}^{R_w} \frac{2}{r^2} \left(\frac{\partial(HA_d)}{\partial N_a^{\infty}} \right) dr \qquad (3)$$

where f_a^{∞} is the amino acid molecular area at a flat interface, γ is the surface tension, H the interfacial bending moment, and A_d is the droplet surface area. This equation leads to a negative correlation between $\ln K_x$ and $1/w_o$ whose slope is of order

$$\frac{2 f_a f_s H}{3 v_w k_B T} \qquad (4)$$

where f_s is the molecular area of the surfactant at the curved interface and v_w is the molecular volume of water. The predicted values for the slope are of the same order of magnitude as those observed experimentally. Moreover, this result indicates that amino acids with larger molecular area projections will be squeezed-out more strongly, in accord with our experimental observations.

4.3 AMINO ACIDS AS COSURFACTANTS

Interfacially solubilised amino acids increase the area per surfactant molecule at the reversed micellar interface, thus acting as cosurfactants. Simple geometric arguments based on a model by Oakenful [17] indicate that the equilibrium sizes of reversed micelles should increase at constant external salinity as the amino acid content of the system increases. Assuming a power law dependence of the water pool radius on w_o, i.e.,

$$R_w = R_w^{\infty} \left(\frac{w_o}{w_o^{\infty}} \right)^{x} \tag{5}$$

where R_w^{∞} and w_o^{∞} are quantities obtained under the same salinity conditions in the absence of amino acid, the following equation is obtained for the dependence of w_o on the interfacial ratio of amino acid to surfactant, :

$$\ln\left(\frac{w_o}{w_o^{\infty}} \right) = \left(\frac{\mathcal{L}_a/\mathcal{L}_s}{1-x} \right) \lambda \tag{6}$$

The exponent x is a measure of interfacial rigidity. It is expected to decrease with increasing salinity, and be approximately independent of surfactant concentration. These hypotheses are verified by the experimental results of Figure 12. A direct calculation of the exponent x using interfacial thermodynamics provided the following result

$$x = \frac{1}{1 + \frac{2}{m} \left(\frac{\mathcal{L}_a \gamma_d^o}{k_B T} \right)} \tag{7}$$

where γ_d^o is the droplet surface tension at the reference salinity conditions in the absence of amino acid, and m is a numerical coefficient related to the curvature dependent adsorption of co-ions at the reversed micellar interface, which can be calculated once a suitable model for the electrical double layer has been identified. It was concluded that x must be very close to unity. A preliminary small angle X-ray scattering analysis has provided qualitative backing for these arguments, but the results were of insufficient precision for quantitative analysis purposes.

4.4 EFFECT OF AVERAGE MOLECULAR CHARGE ON PARTITIONING

A modified form of the simple solubilisation theory for amino acids was applied to a different class of experiments, in which the composition of the system was kept constant and the pH of the initial aqueous phase was varied. A straightforward calculation coupled with some simplifying assumptions led to the result

$$K_{meas} \left(1 + \frac{[H^+]}{K_{a1}} \right) = K_{\pm} + K_{+} \exp\left(- \frac{e\psi}{k_B T} \right) \frac{[H^+]}{K_{a1}} \tag{8}$$

where K_{meas} is an effective partition coefficient using experimental data and equation (2), ψ is the electrostatic potential at the average solubilisation site for positively charged amino acids, K_{a1} is

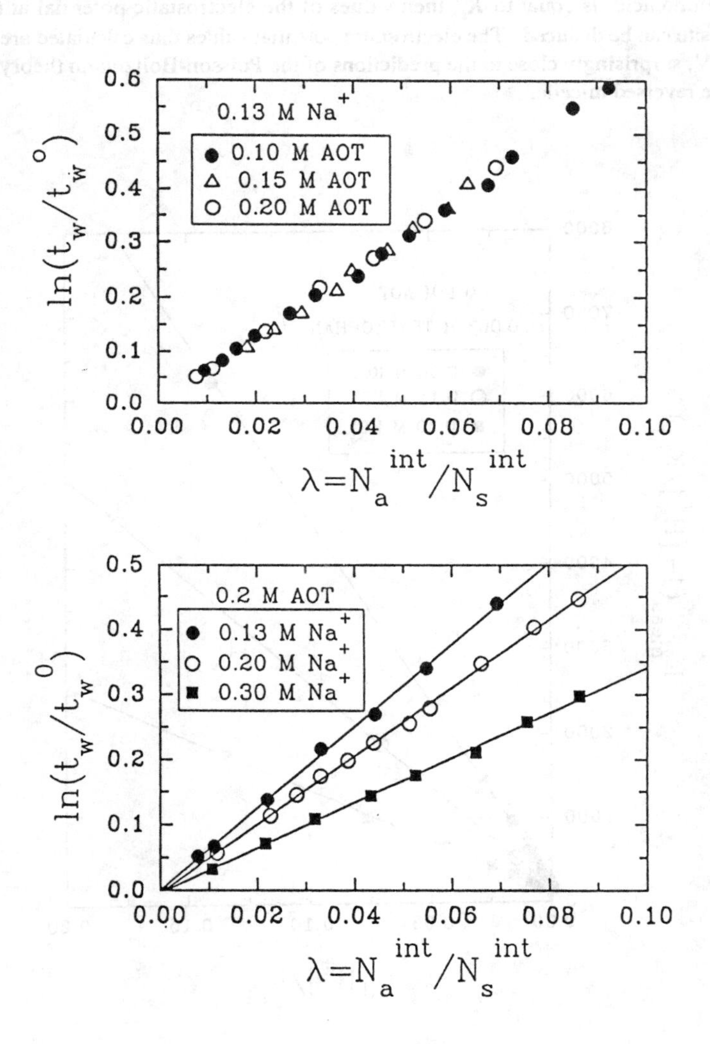

Figure 12. The effect of the interfacial amino acid to surfactant ratio on the water uptake number for varying (a) AOT concentration, and (b) excess aqueous phase salinity.

the acidic dissociation constant of the amino acid, and $K_{\pm}$ and K_{+} are the true interfacial partition coefficients of the zwitterionic and cationic forms of the amino acid molecule. The validity of this expression is evident in Figure 13, from which estimates of the individual partition coefficients can be made. K_{+} and cannot be calculated independently, however, because they appear combined in Equation (8). If we assume that K_{+}, the chemical part of the partition coefficient for the cationic form of the amino acid, is equal to $K_{\pm}$, then values of the electrostatic potential at the average solubilisation site can be deduced. The electrostatic potential values thus calculated are in the range of 120-150 mV, surprisingly close to the predictions of the Poisson-Boltzmann theory for the wall potential of the reversed micelle.

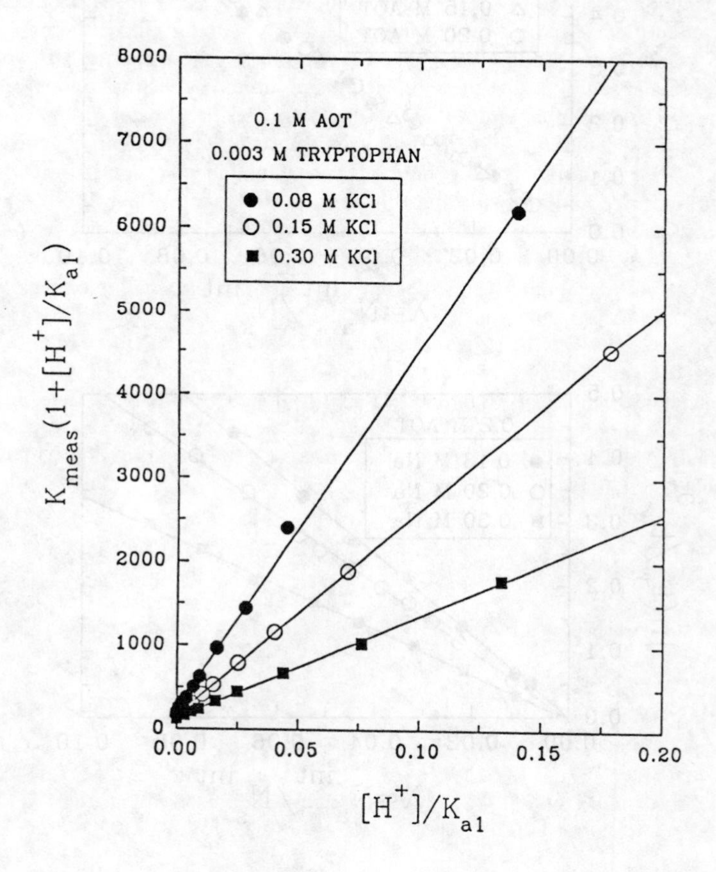

Figure 13. The effect of the average molecular charge on interfacial partitioning of tryptophan for varying initial aqueous phase salt concentrations.

5. Conclusions

The solubilisation of a range of hydrophilic solutes in AOT reversed micelles has provided a clearer understanding of the thermodynamic driving forces for solubilisation, and of the interfacial interactions and solute location in these surfactant aggregates. The phase transfer equilibrium experiments used here have provided a greater degree of flexibility in terms of experimental conditions and solutes investigated than is possible using the more standard spectroscopic techniques. The coupling of the experimental investigation with electrostatic and themodynamic arguments has shown that mean-field continuum theories can be applied with significant success to molecular-scale systems such as reversed micelles. These arguments have been crucial elements in the interpretation of the solubilisation phenomena, and in their quantification.

The most important driving forces for solubilisation in a reversed micellar aggregate have been identified and quantified. Charged species interact strongly with the electrical double layer, their concentration gradients within the water pools depending on their charge, size and electronic properties. Interfacial solubilisation of amino acids is promoted by the hydrophobic properties of the solute (e.g., its hydrophobic molecular area), and opposed by the possibility of solute-water hydrogen-bonding. The polarity of the interface itself affects these two significant driving forces.

Interfacially-active amino acids behave as cosurfactants in reversed micellar systems, expanding the interface and allowing the solubilisation of greater quantities of water. Their effectiveness as cosurfactants is related directly to their interfacial partitioning behaviour. It would appear that a detailed study of this cosurfactant character of amino acids could yield important information on the interfacial rigidity and other properties of the surfactant interface.

Electrolytes can have a pronounced effect on the interfacial partitioning of amino acids, an effect which has been shown to be curvature-mediated. With increasing interfacial curvature there is a squeezing-out of the amino acid from the interface, the effects being more pronounced for bulky amino acid side chains, or for amino acids with branched moieties. The solubilisation of charged amino acids can provide information directly on the potential of the average solubilisation site, and estimated values of this potential are consistent with those predicted by the Poisson-Boltzmann theory.

6. Acknowledgements

This work was supported in part by the NSF Biotechnology Process Engineering Research Centre at MIT.

7. References

1. Luisi, P.L. and Magid, L.J. (1986) "Solubilization of Enzymes and Nucleic Acids in Hydrocarbon Micellar Solutions", *CRC Crit. Rev. Biochem.* 20(4), 409.

2. Luisi, P.L., Giomini, M., Pileni, M.P. and Robinson, B.H. (1988) "Reverse Micelles as Hosts for Proteins and Small Molecules", *Biochim. Biophys. Acta* 947, 209

3. Leodidis, E.B. and Hatton, T.A. (1989) "Interphase Transfer for Selective Solubilization of Ions, Amino Acids, and Proteins in Reversed Micelles", in Pileni, M.P., (ed.), "Structure and Reactivity in Reversed Micelles", *Elsevier, Amsterdam, pp. 270-302.*

4. Hilhorst, R. (1989) "Applications of Enzyme Containing Reversed Micelles", in Pileni, M.P., (ed.), "Structure and Reactivity in Reversed Micelles", *Elsevier, Amsterdam, pp. 323-341.*

5. Leodidis, E.B. (1990) "Solubilisation of Ions and Amino Acids in AOT Reversed Micellar Solutions", *Ph.D. Thesis, Massachusetts Institute of Technology.*

6. Fletcher, P.D.I. (1986) "The Partitioning of Solutes Between Water-in-Oil Microemulsions and Conjugate Aqueous Phases", *J. Chem. Soc. Faraday Trans. 1 82, 2651.*

7. Rosenthal, G.A. (1985) in Barrett, G.C, (ed.), "Chemistry and Biochemistry of the Amino Acids", *Chapman and Hall, London.*

8. Leodidis , E.B. and Hatton, T.A. (1989) "Specific Ion Effects in Electrical Double Layers: Selective Solubilization of Cations in Aerosol-OT Reversed Micelles", *Langmuir 5, 741.*

9. Aniansson, G.E.A. (1978) "Dynamics and Structure of Micelles and Other Amphiphile Structures", *J. Phys. Chem. 82, 2805.*

10. Ruckenstein, E. and Schiby, D. (1985) "Effect of the Excluded Volume of the Hydrated Ions on Double-Layer Forces", *Langmuir 1, 612.*

11. Booth, F. (1951) "The Dielectric Constant of Water and the Saturation Effect", *J. Chem. Phys. 23, 391.*

12. Conway, B.E. (1981) "Ionic Hydration in Chemistry and Biophysics", *Elsevier, Amsterdam.*

13. Diamond, J.M. and Wright, E.M. (1969) "Molecular Forces Governing Nonelectrolyte Permeation through Cell Membranes", *Proc. Roy. Soc. B 172, 273.*

14. Katz, Y. and Diamond, J.M. (1974) "Thermodynamic Constants for Nonelectrolyte Partition between Dimiristoyl Lecithin and Water", *J. Membr. Biol. 17, 101.*

15. Marqusee, J.A. and Dill, K.A. (1986) "Solute Partitioning into Chain Molecule Interfaces: Monolayers, Bilayer Membranes, and Micelles", *J. Chem. Phys. 85, 434.*

16. De Young, L.R. and Dill, K.A. (1988) "Solute Partitioning into Lipid Bilayer Membranes", *Biochemistry 27, 5281.*

17. Oakenful, D. (1980) "Constraints of Molecular Packing on the Size and Stability of Microemulsion Droplets", *J. Chem. Soc. Faraday Trans. 1 76, 1875.*

WHEN IS A MICROEMULSION A MICROEMULSION?

STIG E. FRIBERG and KHOWLA QAMHEYE
Chemistry Department
Clarkson University
Potsdam, NY 13676

ABSTRACT. Microemulsions are transparent liquids, which contain large amounts of water and hydrocarbons. They are stabilized by a surfactant or a combination of an ionic surfactant and a nonionic cosurfactant. They are distinguished from solutions because they contain droplets or long lived aggregates of colloidal size.

This article is a brief overview of microemulsions which do not contain droplets.

1. Introduction

Microemulsions, transparent liquid systems of water and hydrocarbon, are by now well known and a great number of articles and biographies [1-5] are available. The treatment of the phenomenon has varied from regarding systems of spherical particles from a fundamental point of view [2] to empirical treatments based on normal and inverse micelles [1].

It is essential to notice that these treatments reflect the essence of the microemulsion concept, the fact that the microemulsions contain droplets. This distinction from molecular solutions which also are optically transparent is actually not only of fundamental significance but of decisive practical importance. The fact that the dispersed compound is in the form of droplets with a defined interface means that the chemical potential of the dispersed compound remains close to its value in free form. This in turn means that the maximum solubility of the dispersed compound is not as much influenced by the total composition as is the case in a solution.

Fig. 1 illustrates this fact. Dissolving water and hydrocarbon into each other using a solvent such as isopropanol leads to a solubility curve for one component that is drastically reduced by addition of the second compound. So is water and isopropanol completely soluble in each other, but a 1/1 solution of hydrocarbon and isopropanol will dissolve a small percentage of water. In the same manner hydrocarbon is but little soluble in a 1/1 solution of water and isopropanol in spite of the fact that hydrocarbon and isopropanol are mutually soluble. Isopropanol as the solubility agent is always present in greater amount than the dissolved compound.

D. M. Bloor and E. Wyn-Jones (eds.),
The Structure, Dynamics and Equilibrium Properties of Colloidal Systems, 221–231.

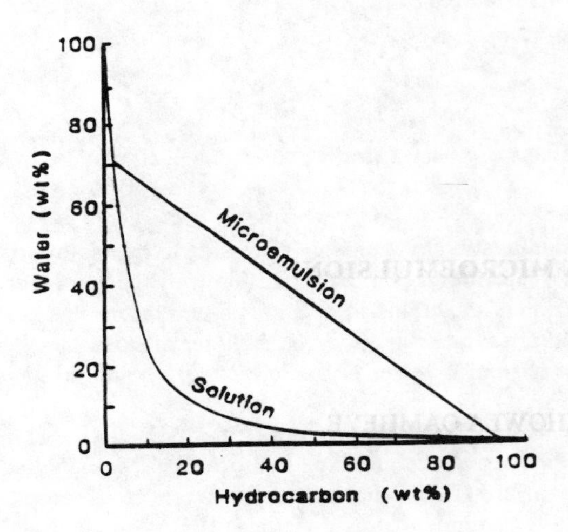

Figure 1. The mutual solubility of water and hydrocarbon are much greater in a microemulsion than in a solution

A microemulsion system, Fig. 1, is different. Although the water has only a limited solubility in the microemulsion base, a much higher solubility at 50% hydrocarbon is found because of the almost linear reduction of the water maximum solubility with hydrocarbon content. It is illustrative to compare the solubility of water in the two vehicles at 50% hydrocarbon. The solution contains a maximum of 5% water while the microemulsion contains 32%. In the first case the stabilizer/dispersed compound mass ratio is 9.0; in the microemulsion it is 0.56 . These numbers reveal the justification for microemulsion application. Microemulsions are less expensive than solutions and any solvent based formulation should be replaced by a microemulsion because of cost considerations.

There is one exception to this rule; also due to the fact that the chemical potential of the dispersed compound remains high when the fraction of dispersed phase is low. This fact means that vapour pressure remains high even at low content of the dispersed compound. The obvious consequence is that dispersing a flammable compound into water in a microemulsion does not make it less hazardous. On the other hand, a fragrance compound dispersed in a microemulsion retains its impact with small changes down to low concentrations.

With these features in mind, it is appropriate to emphasize those cases in which the "microemulsion" does not contain defined droplets. Such compositions exist both for aqueous and non-aqueous systems. The former ones are described first.

2. Aqueous Systems

Although microemulsions exist both as water continuous (oil-in-water, O/W) and oil continuous (water-in-oil, W/O) the latter ones are more relevant to this treatment, because the association in oil media is less sudden than is the case in aqueous environment. In aqueous systems association of surfactant to micelles takes place at a certain concentration, the critical micellization concentration, cmc, and compositions in excess of this concentration contain monomers at the critical concentration and micelles. In the oil medium the association is more complex and hence of more interest from the microemulsion aspect.

2.1 PREMICELLAR AGGREGATES IN W/O SYSTEMS

The simplest form of microemulsion systems use only a single stabilizer which may be of the ionic kind, such as Aerosol OT, or a nonionic one such as the polyethyleneglycol alkyl ethers.

2.1.1 *Single Surfactant Systems.* The stabilization by Aerosol OT was early treated by Eiche [6,7] who calculated the optimal size of the inverse micelle and showed the consecutive association from monomers to dimers to trimers a.s.o. This association pattern means that at low water concentrations there are no water droplets; the water bound to a dimer of surfactants does have a very low vapour pressure and the system is better characterized as a solution than an emulsion.

The association of inverse micelles from nonionic surfactant shows a similar pattern, but now the association is strongly influenced by the presence of aromatic hydrocarbons. As a matter of fact, the association process takes place in a different manner in the presence of aromatic compounds [8]. Fig. 2 shows an illustrative example of this difference [9]. With an aliphatic hydrocarbon the association and the accompanying solubilization of water is initiated at low water concentrations, while the association from the aromatic hydrocarbon requires significantly higher amounts of the surfactant.

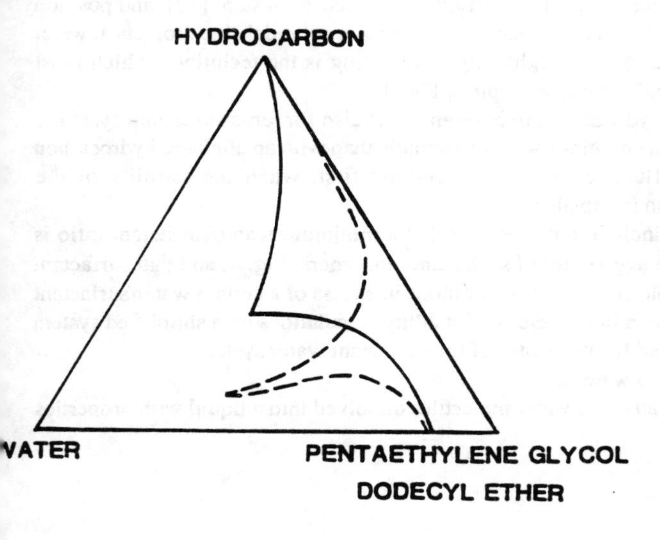

Figure 2. The onset of inverse micellization for a nonionic surfactant takes place at a lower concentration in an aliphatic hydrocarbon than in an aromatic one.

——— aliphatic hydrocarbon
– – – aromatic hydrocarbon

The explanation for this difference was found after investigations combining NMR spectroscopy [8] and calorimetry [9]. The NMR results showed the aromatic hydrocarbon to be preferentially located along the polar part of the surfactant and that self-association of the surfactant began in the system without water at surfactant amounts of the order of 70%. This is the concentration, Fig. 2, at which the solubilization of water began an increase to levels indicating the presence of micelles. Hence, in this case the self-association of the surfactant in the water free part of the system was a prerequisite for water solubilization through the formation of inverse micelles.

With the surfactant dissolved in an aliphatic hydrocarbon the situation was different. The association to inverse micelles now took place at low concentrations and from the monomers without significant interaction between the surfactants in the water free systems.

The difference in behaviour is based on the interaction between the benzene molecules and the oxyethylene groups. Calorimetry [10] revealed this interaction to be moderately strong, approximately 1 kcal/mol. This interaction is apparently of sufficient strength to modify the approach of water molecules to act as an association agent for the polar groups of the surfactant. In both these examples of surfactants monomeric species exist at low water concentrations, and the water concentrations may be rather large, 12% [8], before association to micellization occurs.

Similar conditions exist in the combinations of a water soluble ionic surfactant and a medium chain length alcohol or carboxylic acid. However, the structures are more complex in this case and a more detailed discussion is necessary.

2.2 COMBINATION SYSTEMS

The basis for these W/O microemulsions is the cosurfactant solution in the water-surfactant-cosurfactant system. Light scattering [11], determinations of dielectric constant [12], and position annihilation techniques [13] have shown the systems not to contain colloidal size droplets at water concentrations below a level of 25% by weight. Light scattering is the technique which most directly illustrates the absence of colloidal size droplets, Fig. 3.

The results show the aromatic hydrocarbon to have an effect also for ionic surfactant systems, association took place at lower concentrations with an aromatic than with an aliphatic hydrocarbon in the microemulsions. This influence may be understood first, when the stability of the aggregates in the sub-colloid region is explained.

Such an explanation needs to include both the fact that a minimum water/surfactant ratio is necessary to confer stability to the aggregates of surfactant monomers, Fig. 4, and that surfactant oligomer aggregates are more stable at water concentrations in excess of a certain water/surfactant ratio. Both the features were included in the results of stability calculations on a simplified system [14]. The free energy was compared for two states of the surfactant water system.

State 1. Crystalline soap + liquid water and

State 2. Monomeric soap with attached water molecules dissolved into a liquid with properties similar to those of pentanol.

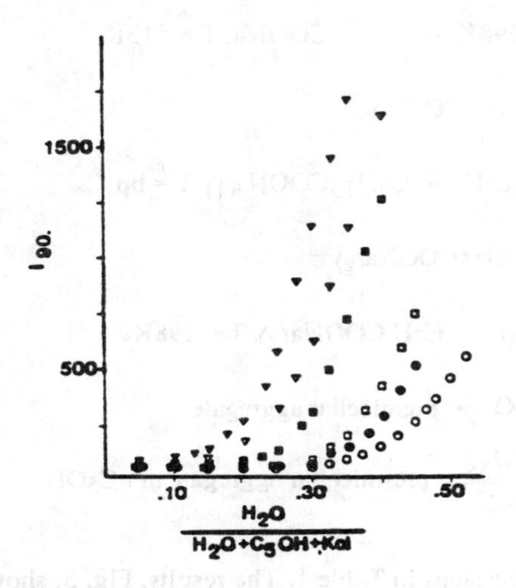

Figure 3. The minimum water concentration to form inverse micelles from an ionic surfactant/ alcohol cosurfactant is lower in an aromatic hydrocarbon than in an aliphatic one. (O) 0% hydrocarbon; (●) 25% decane; (□) 50% decane; (■) 25% benzene; (∇) 50% dodecylbenzene; (▼) 50% benzene.

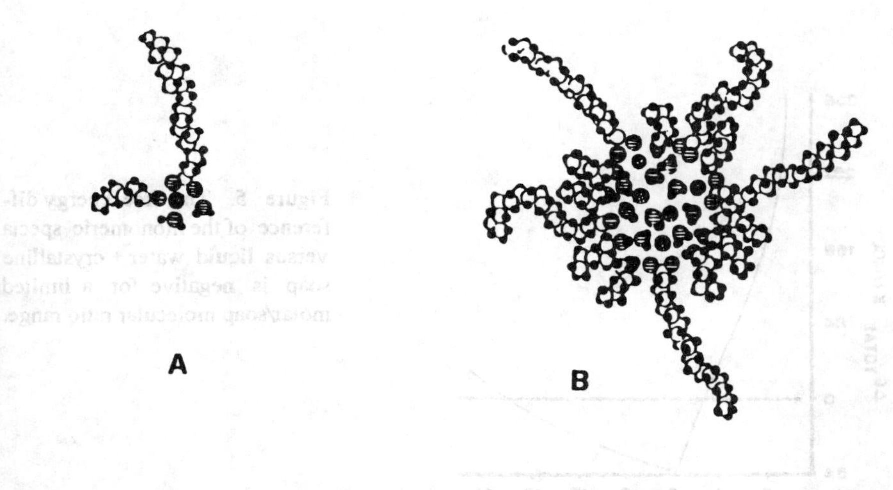

Figure 4. The premicellar aggregate (A) in a W/O microemulsion contains only one surfactant molecule and the water molecules are water of hydration. In the microemulsion droplet (B) the water molecules are in the form of a microdroplet.

TABLE 1. Thermocycle for Pre-micellar Aggregate Stability

1. $n\,H_2O_{(1)} \rightarrow n\,H_2O_{(g)}$

2. $C_7H_{15}COONa_{(S)}$, T = 298K $\rightarrow$ $C_7H_{15}COONa$, T = 513K

3. $C_7H_{15}COONa_{(s)} \rightarrow C_7H_{15}COONa$

4. $C_7H_{15}COONa_{(1)}$, T = 513K $\rightarrow$ $C_7H_{15}COONa_{(1)}$ T = bp

5. $C_7H_{15}COONa_{(1)} \rightarrow C_7H_{15}COONa_{(g)}$

6. $C_7H_{15}COONa_{(g)}$, T = bp $\rightarrow$ $C_7H\,COONa_{(g)}$, T = 298K

7. $C_7H_{15}COONa_{(g)} + nH_2O \rightarrow$ premicellar aggregate

8. $nH_2O + C_7H_{15}COONa \xrightarrow{T\Delta S}$ pre-micellar aggregate in nC_5OH

The free energy difference was obtained by the steps in Table 1. The results, Fig. 5, showed two interesting features. A minimum water content was necessary to obtain a negative free energy

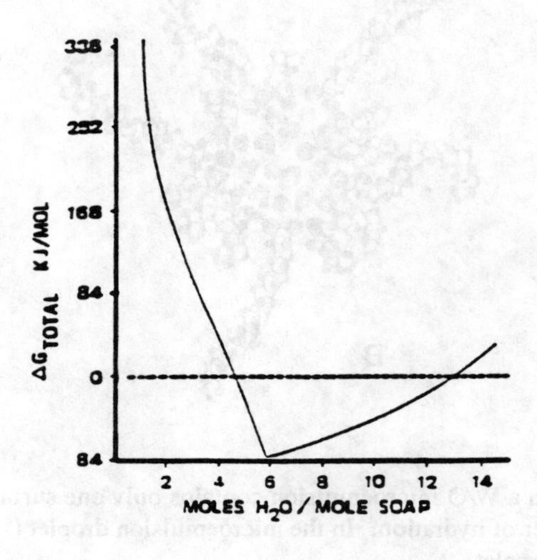

Figure 5. The free energy difference of the monomeric specia versus liquid water + crystalline soap is negative for a limited molar/soap molecular ratio range.

difference between State 1 and 2, and the negative values were found for a range of water/surfactant ratios of 4-11 only. Hence, the calculations indicated the surfactant monomer aggregates to be stable in only a small part of the total water/surfactant range for the W/O microemulsion. The latter typically reaches over a water/surfactant molecular ratio range of 6-50.

The limitation for maximum water surfactant ratio of 11 as well as the minimum value as explained by the experimental results, Fig. 3. One finds, Fig. 2, that a minimum water/surfactant ratio of six is needed to dissolve the surfactant. This value corresponds to the lowest ratio for a negative free energy, Fig. 5. The higher limit, Fig. 5, represents the transition from the monomeric surfactant species to dimers, Fig. 2. The excellent agreement of the experimental and calculated results are, of course, fortuitous, but it indicates the possibility of making exact calculations of this kind when more exact expressions for the interaction potentials are available.

These experimental results and the calculations demonstrate the fact that a large part of the W/O microemulsion region in reality does not contain droplets but actually is a molecular solution. This fact has significance for the capacity of the microemulsion to dissolve water soluble substances.

Medium chain length carboxylic acids may also be used as cosurfactants, but the difference between the alcohol and the carboxylic acid system, Fig. 6 deserves some comments. The minimum water/surfactant ratio to dissolve water characterizing the alcohol system is not a condition for the carboxylic acid/soap combination. The soap is directly soluble in a liquid carboxylic acid to a soap/acid molecular ratio of 1/2.

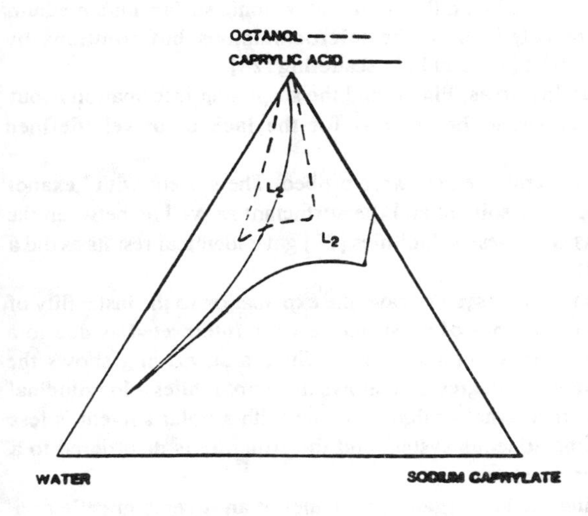

Figure 6. The W/O microemulsion base for a soap/carboxylic acid system requires no minimum water content for solubility contrary to the soap/alcohol system

The interesting feature is the kind of aggregate formed by the soap and the acid and the bonds that provide stability to the combination. IR, NMR and osmotic pressure measurements showed the aggregate to consist of a 4/2 acid/soap combination [15,16], Fig. 6. Its stability was analyzed by semi-empirical quantum mechanical calculations showing the strength of the hydrogen band at the level of 20 cal/mol resulting in excellent stability of the aggregate. Also in this case did larger aggregates form with greater amounts of water as shown by Lindman and collaborators [18]. Hence, the W/O microemulsions stabilized by a carboxylic acid/soap combination also show a large region in which the water molecules are not gathered into droplets, but, instead, are strongly bound in small surfactant/cosurfactant aggregates.

These examples have illustrated the fact that the W/O microemulsion systems show monomeric or oligomeric association structures at water contents to the magnitude of 30% by weight. Replacing the water with a polar organic solvent has led to vehicles with even more extended regions without droplet formation.

3. Non-aqueous Microemulsions

Microemulsions in which a polar solvent is dispersed in a hydrocarbon were introduced rather late [19-21] in spite of the fact that micellization in polar organic solvents had been investigated for some time [22-24].

Glycerol-in-hydrocarbon microemulsions were analyzed by Robinson and coworkers [21] and found to contain discrete microdroplets. Similar results have been reported by Lattes and Rico [19] who have made extensive investigations into chemical reactions in such microemulsions. Wärnheim has reported on microemulsions stabilized by nonionic surfactants [25-26].

On the contrary, "microemulsions" stabilized with the combination ionic surfactant/a medium chain alcohol as cosurfactant were revealed not to be microemulsions but solutions by investigations into their structure, using MNR [27] and light scattering [28].

The combinations showed large solubility areas, Fig. 7, and the surprising information about the structure prompted investigations about the reasons for the lack of a well defined hydrocarbon/polar solvent interface.

The explanation was obtained after several avenues were probed. The system with hexanol showed virtual insolubility between the polar solvent and the surfactant as well as between the alcohol and the surfactant. Other systems with these solubilities [29] gave identical results as did a system with a lamellar liquid crystal [30].

The properties of the liquid crystal in the latter system gave the explanation to the instability of the interface [31]. NMR and XRD investigations of its structure were interpreted as due to a structure with the amphiphiles significantly staggered, Fig. 8. Such a staggering shows the potential well at the interface to be shallow to a degree that allows the amphiphiles a longitudinal degree of freedom. Hence the interface in a lamellar liquid crystal with a polar solvent is less defined than the one in the corresponding aqueous system and the structure is disordered to a maximum degree allowed by the lamellar structure.

With this fact in mind and considering the less organized manner of an inverse micelle or a microemulsion droplet, the instability of the droplet interface appears a reasonable and, actually, expected consequence.

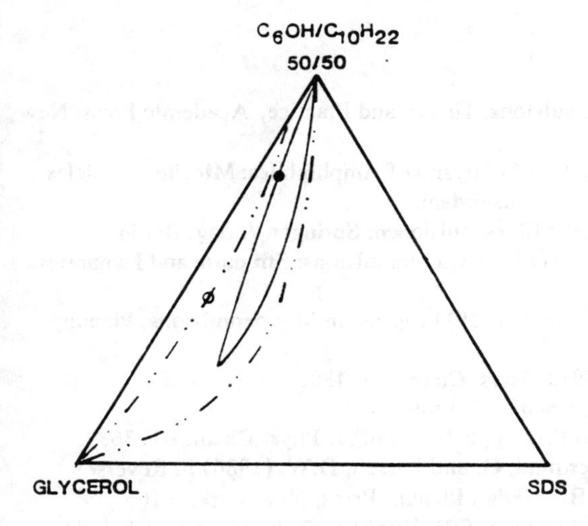

Figure 7. The "microemulsion" region in a glycerol-in-oil system is large.

$\underline{C_6OH/C_{10}H}$

— · · — 1/0

——— 1/1

— · · — ⊖——● plait points.

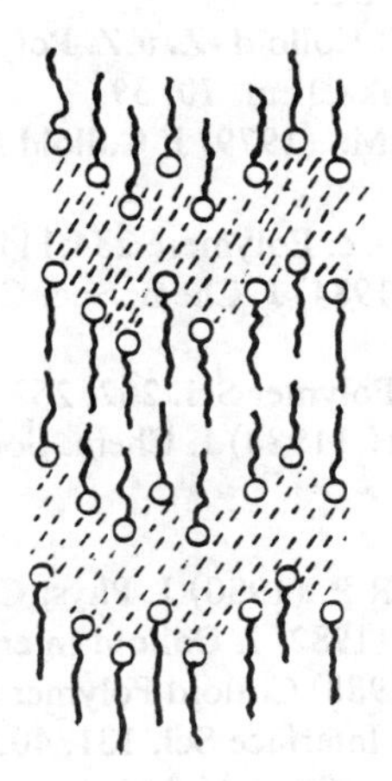

Figure 8. For a lamellar liquid crystal with a polar organic solvent the amphiphiles are staggered.

4. Acknowledgement

This research was supported by an NSF Grant #CBP-8819140.

5. References

1. Prince, L.M. (ed.) (1977) Microemulsions. Theory and Practice, Academic Press, New York.
2. Degiorgio, V. and Corti, M. (eds.) (1985) Physics of Amphiphiles: Micelles, Vesicles and Microemulsions., North Holland, Amsterdam.
3. Pfüller, U. (1986) Mizellen-Vesikel-Mikroemulsioncn, Springer Verlag, Berlin.
4. Friberg, S.E. and Bothorel, P. (eds.) (1987) Microemulsions: Structure and Dynamics. CRC Press, Boca Raton, FL.
5. Martelucci, S. and Chester, A.N. (eds.) (1989) Progress in Microemulsions, Plenum Press, New York (1989).
6. Zulauf, M. and Eiche, H.-F. (1979) J. Phys. Chem. 83, 480.
7. Eicke, H.-F. (1980) Pure & Appl. Chem. 52, 1349.
8. Christenson, H. Larsen, D.W. and Friberg, S.E. (1980) J. Phys. Chem. 84, 3633.
9. Friberg, S.E., Christenson, H., Bertrand, G. and Larsen, D.W. (1984) in Reverse Micelles, Luisi, P.L. and Straub, B.E. (eds.) Plenum Press, New York, p 105.
10. Nakamura, M., Bertrand, G.L. and Friberg, S.E. (1983) J. Colloid Interface Sci. 91, 516.
11. Sjöblom, E. and Friberg, S.E. (1978) J. Colloid Interface Sci. 67, 16.
12. Clausse, M. and Rayer, R. (1976) Colloid and Surface Science, Kerker, M. (ed.), Vol II, Academic Press, New York.
13. Jean, Y. and Ache, H.J. (1978) J. Am. Chem. Soc. 100, 6320.
14. Friberg, S.E. and Flaim, T.D. (1982) in Reactions in Micellar Solutions, Holt, S.L. (ed.) ACS Symposium Series #177, ACS Washington DC.
15. Friberg, S.E., Mandell, L. and Ekwall, P. (1969) Kolloid -Z. u.Z. Polymere, 233, 955.
16. Söderlund, G. and Friberg, S.E. (1970) Z. Physik Chem. 70, 39.
17. Bendiksen, B., Friberg, S.E. and Plummer, P.L.M., (1979) J. Colloid Interface Sci. 72, 495.
18. Lindman, B. and Ekwall, P. (1969) Kolloid -Z. u.Z. Polymere 234,1115.
19. Escoula, B. Hajjaji, N., Rico, I. and Lattes, A. (1984) J. Chem. Soc. Chem. Comm. 12,33.
20. Friberg, S.E. and Podzimek, M. (1984) Colloid Polymer Sci. 262, 252.
21. Fletcher, P.D.E., Galal, M.F. and Robinson, B.H. (1984) J. Chem. Soc. Faraday Trans. I, 80, 3307.
22. Reinsborough, V. (1970) Austr. J. Chem.
23 1471. 23. Sing, H.N., Salem, S.M. and Sing, R.P. (1980) J. Phys. Chem. 84, 2191.
24. Evans, D.F., Yamamuchi, A. and Casassa, R.E. (1982) J. Colloid Interface Sci. 88, 89.
25. Wärnheim, T., Bokström, J. and Williams. Y. (1988) Colloid Polymer Sci. 266, 562.
26. Wärnheim, T. and Sjöberg, M. (1989) J. Colloid Interface Sci. 131, 402.
27. Friberg, S.E. and Liang, Y.-C. (1987) Colloids Surfaces 24,325.

28. Das, K.P., Celglie, A., Lindman, B. and Friberg, S.E. (1987) J. Colloid Interface Sci. 116, 390.
29. Friberg, S.E. and Liang, Y.C. (1989) in Progress in Microemulsions, Martelucci, S. and Chester. A.N. (eds.) Plenum Press, New York, p. 73.
30. Friberg, S.E. and Rong, G. (1988) J. Phys. Chem. 92, 7247.
31. Friberg, S.E., Ward, A.J.I. and Larsen, D.W. (1987). Langmuir 3,735.

UNI- AND BICONTINUOUS MICROEMULSIONS

U. OLSSON and B. LINDMAN,
Physical Chemistry 1,
Chemical Center,
Lund University,
P.O. Box 124,
S-221 00 Lund,
Sweden.

ABSTRACT. The structure of liquid surfactant-water-oil mixtures, so called microemulsions, is discussed. These solutions are microstructured into separate polar and apolar domains (typical length scales are of the order of 10-400 Å) separated by a monolayer of oriented surfactant molecules. The structure may be either uni- or bicontinuous. It is stressed that multicomponent self-diffusion measurements, most easily performed with the pulsed gradient FT NMR technique, is presently the most suitable technique for studying microemulsion structure. Molecular self-diffusion properties of uni- and bicontinuous microemulsion structure are presented. Various surfactant-water-oil systems are reviewed with emphasis on their liquid microstructure.

1. Introduction

Microemulsions [1-4] are thermodynamically stable liquid phases of oil, water (or an alternative polar solvent) and surfactant [5]. In many cases some additional component, cosurfactant and/or electrolyte, is needed for the formation of a microemulsion. Stable microemulsions form in a large variety of surfactant systems [6-21]. Very intriguing from both practical and theoretical points of view are microemulsions in which quite small amounts of surfactant molecules can mix comparable amounts of water and oil [22].

Of main concern when first introduced to a liquid phase in a surfactant(S)-water(W)-oil(O) system is the question of microstructure. In solution the surfactant molecules form a monolayer of oriented molecules that acts as dividing surface between water and oil domains. Depending on the conditions, such surfaces may enclose a finite volume, as in micelles, or be continuous in one, two or three dimensions. The interestingly rich phase behaviour of S/W/O systems is related to the numerous ways in which space may be divided into polar and apolar regions for a given surface to volume ratio. It is these properties of the surfactant monolayer we refer to when concerned with the liquid microstructure of microemulsions.

D. M. Bloor and E. Wyn-Jones (eds.),
The Structure, Dynamics and Equilibrium Properties of Colloidal Systems, 233–242.

2. How To Study Microemulsion Structure

Several experimental techniques have been invoked in studies that address the structure of microemulsions. Scattering techniques [23-28] are well suited for determining structural dimensions in cases where the structure already has been determined, but has not yet been shown to discriminate between uni- and bicontinuous structures. Difficulties are associated with the simultaneous high volume fractions of water and oil, and the fact that, in contrast to many other colloidal solutions, dilution procedures are not appropriate for microemulsions: these structures are produced as a result of reversible self-assembly, which depends on the concentrations of all components in the system.

Electron micrographs of microemulsions are often plagued by artifacts, arising from the sample preparation procedure (for a critical review, see [29-31]). Another problem is obviously associated with the very long freezing time - of the order of several ms - compared to the estimated timescale for rearrangements of the surfactant films (≈μs). In principle it is impossible to prove or disprove bicontinuity from a section of a three dimensional structure; bicontinuity cannot, in fact, exist in a plane. Nevertheless, micrographs of microemulsions, showing reliable structures, have been presented in recent years [32].

Probably, the technique that offers the most insight to the structure of microemulsions is by measuring the self-diffusion constants of the various components of the system. This technique easily discriminates between droplet and bicontinuous structures. The closely related technique of measuring conductivity can discriminate between water continuous and discontinuous structures. However, the strength of the self-diffusion technique is that it allows for measuring, simultaneously, the transport properties on the two sides of the surfactant monolayer (as well as in the surfactant film itself).

3. Self-diffusion

In an NMR FTPGSE experiment [33,34] one measures the mean squared displacement in one dimension. For Gaussian diffusion this is related to the diffusion constant, D, and the observation time, τ, by

$$\langle x^2 \rangle = 2 D \tau \quad (1)$$

The observation time in the experiment is of the order of 100 ms and in typical microemulsions one measures a macroscopic root mean squared displacement of the order of 1 - 10 μm, which is usually much larger than any typical structural length scales in the solution. The validity of eq. (1), which may be verified experimentally, implies that nonrestricted diffusion [35] occurs on this time and length scale. We note, however, that restricted diffusion was observed in the very careful experiments of Bodet *et al*. [36] in a concentrated O/W microemulsion.

In the mid-1970s, the technique of NMR self-diffusion proved extremely valuable in establishing the bicontinuity of certain cubic l.c. phases, and the discreteness of certain other cubic phases [37]. In these self-diffusion studies, structural deductions are made from the restrictions in the long-range translational mobilities of the constituent molecules. The success of this method led us to apply the self-diffusion approach to microemulsion systems [38,39]. While in the first study a combination of radio-tracer and classical NMR spin-echo techniques was used [40], much of the progress can be referred to the development of the Fourier transform NMR technique [33] which permits the rapid and precise simultaneous determination of the components' self-diffusion in a complex mixture.

The principles of using molecular diffusivities for obtaining information on average equilibrium solution structure are simple and straightforward without any delicate assumptions. Self-diffusion coefficients depend on the size of the diffusing entity, friction, obstruction effects, etc., but for microemulsions, and especially those with small contents of surfactants, simple geometrical considerations generally take us a long way. Obstruction-effect considerations can be used to predict the oil and water self-diffusion behaviour; the diffusion equation is solved within a microstructure defined by a model for the surfactant-rich dividing surface. Such results can then be compared with experiment. The problem of solvent-molecule diffusion has been solved by Jönsson *et al.* [41] for the case of droplet structures, with different droplet shapes, and by Anderson and Wennerström [42] for periodic bicontinuous structures. Jönsson *et al.* [41] also investigated the rate of diffusion of the droplet-forming solvent.

3.1. UNICONTINUOUS MICROEMULSIONS

In W/O and O/W microemulsions one of the solvents forms a continuous medium, whereas the other is confined to discrete droplets. The continuous solvent has a diffusion constant near that of the neat liquid, D_0, only slightly reduced owing to obstruction by the dispersed droplets. In the case of the dispersed solvent, on the other hand, the process for the macroscopic transport of molecules is the thermal translation of the discrete droplet and, consequently, its diffusion constant is relatively low. If the solubility of surfactant and dispersed solvent in the continuous solvent is negligible, their diffusion constants are equal. According to the Stokes-Einstein equation for the self-diffusion of spheres in a viscous continuum, D is simply given by $D = kT/(6\eta R)$. With a viscosity of 1 cP and a diffusing particle having the size of a small molecule, i.e. ≈ 2 Å, we obtain $D = 10^{-9}\ m^2\ s^{-1}$ while a microemulsion droplet of R = 200 Å would have $D = 10^{-11}\ m^2\ s^{-1}$; this value would be reduced in the presence of interdroplet interactions.

3.2. BICONTINUOUS MICROEMULSIONS

In bicontinuous microemulsions, oil and water domains may be assumed to have the same properties as the neat liquids, with self-diffusion coefficients close to the values of the neat liquids. It has been explicitly shown, that the observed diffusion process of water [43] and oil [44] is a molecular diffusion in a medium similar to the neat liquid. With this in mind, any deviation of D/D_0 from unity may be ascribed to obstruction by the surfactant-rich dividing surface.

Calculations of diffusion properties in triply periodic bicontinuous structures, resembling bicontinuous cubic phases, have recently been performed [42]. In microemulsions the structure lacks long-range order and is undergoing continual thermal disruption. However, as regards transport properties, ordered structures seem to serve as good model systems for average liquid properties, as judged from the good agreement observed in the limiting case of a balanced zero average mean curvature dividing film. Both for balanced microemulsions, of equal volume fractions of water and oil, and in the related bilayer continuous L_3 phase, quantitative agreement with the ordered model is obtained [42].

3.3 MIXED SOLVENTS

Besides absolute and relative diffusion constants of water and oil, additional information can be achieved from studying systems,where the solvents are mixtures of two species. By comparing the diffusion constants of, say, two oils, information regarding the dominating diffusion process is obtained. If micelle diffusion is the dominating process, the ratio between the two diffusion constants should equal unity. If the dominating process is molecular diffusion in a continuous oil medium, the ratio should be the same as in the neat oil mixture. Studies of this kind have been performed in several different systems [43,44]. In some of these, an interesting transition region could be found where exchange of molecules between aggregates were important for the rate of macroscopic displacement.

4. Structure

4.1 UNICONTINUOUS MICROEMULSIONS

In unicontinuous microemulsions one of the solvents (usually the less abundant) is confined to micelles, dispersed in the other solvent. These solutions, which are also referred to as droplet microemulsions, may contain oil-swollen normal micelles in an aqueous continuum or alternatively water-swollen reversed micelles dispersed in oil. The notations oil-in-water (O/W) and water-in-oil (W/O) microemulsions, respectively, are also frequently used.

4.1.1 *O/W microemulsions.* The capacity of aqueous surfactant solutions to solubilize non-water-soluble compounds was known long ago [45]. However, studies of these phenomena were mostly restricted to the relatively weak solubilization power of aqueous solutions of ionic surfactant. Considerable swelling of normal micelles may, however, take place in systems with nonionic surfactant [22,44] or when cosurfactant and electrolyte is added to an ionic surfactant system [46].

4.1.2 *W/O microemulsions.* Much more studies have been performed on reversed micelles, in particular, for the L_2 phase in ternary systems with Aerosol OT (AOT) where relatively large amounts of water can be solubilized [47-51]. Self-diffusion and conductivity studies have shown that closed water droplets occur over a large concentration range [52,53]. However, as shown by kinetic studies, exchange of material between micelles can take place [54]. A large number of small angle scattering studies have been performed on these systems, in particular SANS studies with variable contrast [55]. The general picture of a reversed AOT micelle is a spherical water droplet coated with a monolayer of surfactant. Most scattering studies are also interpreted in terms of spherical micelles, where the absence of Bessel-function minima in the scattered intensity is ascribed to a size polydispersity. This picture was challenged in a recent NMR relaxation study [57]. Here it was suggested that the micelles rather are prolate shaped with an aspect ratio in the range 2-3. Most likely, the micellar size and shape are temperature dependent.

4.2 BICONTINUOUS MICROEMULSIONS

More intriguing are those solutions where the surfactant-rich film forms a three dimensionally continuous surface of highly connected topology (is continuous in three dimensions). Here, both water and oil are simultaneously continuous and the structure is referred to as bicontinuous. Most microemulsions that are bicontinuous contain comparable amounts of water and oil and most interesting are those systems that are stabilized by only small amounts of surfactant [22]. However, systems also exist where bicontinuity appears at low water content [17,18,57].

The first suggestions that microemulsions may be bicontinuous, and the first suggested structures, were based on analogues with liquid-crystal (l.c.) structures. Shinoda and Saito [58] suggested a structure in terms of a thermally roiled lamellar l.c. structure, while Scriven [59] argued that microemulsions may have a structure that corresponds to that of a bicontinuous cubic l.c. structure. The first experimental demonstration that microemulsions may indeed be bicontinuous came shortly after from NMR self-diffusion studies [40]. A further possible bicontinuous microstructure, considered recently in relation to systems with low water content, is one having analogies with a hexagonal l.c. structure [60]. In the proposed structure one of the solvents lies inside interconnected tubules.

In many microemulsions the surfactant film is at comparable volume fractions of water and oil characterized by a low area averaged mean curvature, <H>, that also may be zero. This is known from self-diffusion experiments that show that transport properties of both sides of the surfactant film are similar (or equal if <H>=0) [61]. It has also been demonstrated by SANS contrast variation experiments [23]. (The mean curvature, H, at a given point on a surface is one-half the sum of the signed principal curvatures.)

It has been argued that the structure of bicontinuous liquid phases (microemulsions and the related bilayer continuous L_3 phase) tends towards a homogeneous mean curvature over the dividing surface, which is the spontaneous mean curvature of the surfactant monolayer. On changing parameters, such as salinity in ionic surfactant cases or temperature in nonionic surfactant cases, the spontaneous mean curvature is changed and there is an accompanying change in the oil-water ratio in the microemulsion phase. Low mean curvature arises from cancelling principal curvatures, i.e. the surface is predominantly saddle-like and has a highly connected topology.

5. Systems

5.1 TERNARY SYSTEMS

5.1.1 *Single-chained ionic surfactants.* Single-chained ionic surfactants usually display a wide solution phase, L_1, on the binary surfactant-water axes. These solutions however, have very low tendency of dissolving oil [62]. Moreover, no oil-rich solution phase, L_2, appears in ternary system with oil. However, as first shown by Schulman, ionic surfactants may mix water and oil if alcohol (cosurfactant) also is added [63], *vide infra*.

5.1.2 *Double-chained ionic surfactants.* For double-chained surfactants the L_1 phase is usually very small. On the other hand, an extensive L_2 phase is often present in the ternary phase diagram. The two most studied surfactants are the anionic AOT [64-68] and the cationic didodecyldimethyl-ammonium bromide (DDAB) [17,18,59,69,70].

In ternary systems with AOT, reversed micelles are formed over a large composition range in the L_2 phase. The structure in these solutions is, however, slightly temperature dependent. Reversed micelles are formed at lower temperatures while tending towards bicontinuity at elevated temperatures [43]. The temperature is dependent on the hydrocarbon chain length of the oil.

DDAB forms extensive L_2 regions in ternary systems. In contrast to AOT systems the microstructure in these solutions depends rather strongly on the composition [59,69,70]. The structure is bicontinuous at low water content while tending towards closed reversed micelles at high water content.

5.1.3 *Nonionic surfactants.* The properties of nonionic surfactants are strongly temperature dependent. This is seen already in the binary phase diagrams with water that displays a miscibility gap at higher temperatures [71-73]. The change of phase behaviour in ternary systems with temperature is well known. In particular, it has been shown that at a constant (and for efficient surfactants, low) surfactant concentration an isotropic one phase solution channel exists, connecting the two binary solutions surfactant/water and surfactant/oil [4,22,44,74,75]. Water-rich solutions are stable at lower temperatures while solutions, rich in oil, are stable at higher temperatures. At intermediate temperatures the surfactant can stabilize equal volume fractions of water and oil. The microstructure has been shown to progress from O/W at high water content to W/O at low water content via a bicontinuous microemulsion around equal volume fractions of water and oil [44,74].

5.2 MULTICOMPONENT SYSTEMS

Essentially the same phase behaviour as in ternary systems with nonionic surfactant can be found in quatenary systems with single-chained ionic surfactant, cosurfactant (short chained alcohol) and electrolyte [46]. Here the electrolyte concentration or the surfactant to cosurfactant ratio play a similar role as temperature in the nonionic systems. Also molecular properties are similar in the two types of systems as shown by self-diffusion and NMR relaxation [46].

References

1. Friberg, Stig E. and Bothorel, Pierre (eds.) (1987) Microemulsions: Structure and Dynamics, CRC Press, Boca Raton.
2. Bourrel, Maurice and Schechter, Robert S. (1988) Microemulsions and related systems, Marcel Dekker, New York.
3. Martelucci, S. and Chester, A.N. (eds.) (1989) Progress in Microemulsions, Plenum Press, New York.
4. Shinoda, K and Friberg, S. (1986) Emulsions and Solubilization, Wiley-Interscience, New York.
5. Danielsson, I. and Lindman, B. (1981) 'The definition of microemulsion', Colloids and Surfaces 3, 391-392.
6. Shinoda, K. and Yutaka, S. (1986) 'Principles for the attainment of minimum oil-water interfacial tension by surfactants: The characteristics of organized surfactant phase', Colloids and Surfaces 19, 185-196.
7. Kunieda, H. and Shinoda, K. (1980) 'Solution Behavior and Hydrophile-Lipophile Balance Temperature in the Aerosol OT-Isooctane-Brine System: Correlation between Microemulsions and Ultralow Interfacial Tensions', J. Colloid Interface Sci. 75, 601-606.
8. Kunieda, H. and Shinoda, K. (1979) 'Solution Behavior of Aerosol OT/Water/Oil System', J. Colloid Interface Sci. 70, 577-583.
9. Shinoda, K., Kunieda, H., Arai, T. and Saijo, H. (1984) 'Principles of Attaining Very Large Solubilization (Microemulsion): Inclusive Understanding of the Solubilization of Oil and Water in Aqueous and Hydrocarbon Media', J. Phys. Chem. 88, 5126-5129.
10. Shinoda, K. (1985) 'The Significance and Characteristics of Organized Solutions', J. Phys. Chem. 89, 2429-2431.
11. Kahlweit, M., Strey, R. and Busse, G. (1990) 'Microemulsions - A Qualitative Thermodynamic Approach', submitted to J. Phys. Chem.

12. Kahlweit, M., Strey, R., Firman, P., Haase, D., Jen, J. and Shomäcker, R. (1988) 'General Patterns of the Phase Behavior of Mixtures of H_2O, Nonpolar Solvents, Amphiphiles, and Electrolytes. 1', Langmuir 4, 499-511.
13. Kahlweit, M., Strey, R., Shomäcker, R. and Haase, D. (1989) 'General Patterns of the Phase Behavior of Mixtures of H_2O, Nonpolar Solvents, Amphiphiles, and Electrolytes. 2', Langmuir 5, 305-315.
14. Kahlweit, M. and Strey, R (1985) 'Phase Behavior of Ternary Systems of the Type H_2O Oil-Nonionic Amphiphile (Microemulsions)', Angew. Chem. Int. Ed. Engl. 24, 654-668.
15. Kahlweit, M. (1988) 'Microemulsions', Science 240, 617-621.
16. Kilpatrick, P.K., Gorman, C.A., Davis, H.T., Scriven, L.E. and Miller, W.G. (1986) 'Patterns of Phase Behavior in Ternary Ethoxylated Alcohol-n-Alkane-Water Mixtures', J. Phys. Chem. 90, 5292-5299.
17. Evans, D.F., Mitchell, D.J. and Ninham, B.W. (1986) 'Oil, Water, and Surfactant: Properties and Conjectured Structure of Simple Microemulsions', J. Phys. Chem. 90, 2817 2825.
18. Chen, S.J., Evans, D.F. and Ninham, B.W. (1984) 'Properties and Structure of Three Component Ionic Microeemulsions', J. Phys. Chem. 88, 1631-1634.
19. Ninham, B.W., Chen, S.J. and Evans, D.F. (1984) 'Role of Oils and Other Factors in Microemulsion Design', J. Phys. Chem . 88, 5855-5857.
20. Mitchell, D.J. and Ninham, B.W. (1981) 'Micelles, Vesicles and Microemulsions', J. Chem. Soc., Faraday Trans. 2, 77, 601-629.
21. Shah, D.O. (1985) 'Macro- and Microemulsions: Theory and Applications', ACS symposium series, v. 272.
22. Shinoda, K. and Lindman, B. (1987) 'Organized Surfactant systems: Microemulsions', Langmuir 3, 135-149.
23. Auvray, L., Cotton, J., Ober, R. and Taupin, C. (1984) 'Evidence for Zero Mean Curvature Microemulsions', J. Phys. Chem. 88, 4586-4589.
Auvray, L., Cotton, J., Ober, R. and Taupin, C. (1984) 'Concentrated Winsor microemulsions: a small angle X-ray scattering study', J. Physique 45, 913-928.
24. Zemb, T.N., Hyde, S.T., Derian, P.-J., Barnes, I.S. and Ninham, B.W. (1987 'Microstructure from X-ray Scattering: The Disordered Open Connected Model of Microemulsions', J. Phys. Chem. 91, 3814-3820.
25. Milner, S.T., Safran, S.A., Andelman, D., Cates, M.E. and Roux, D. (1988) 'Correlations and structure factor of bicontinuous microemulsions', J. Phys. France 49, 1065-1076.
26. Vonk, C.G., Billman, J.F. and Kaler, E.W. (1988) 'Small angle scattering of bicontinuous structures in microemulsions', J. Chem. Phys. 88, 3970-3975.
27. Chang, N.J. and Kaler, E.W. (1986) 'Quasi-Elastic Light Scattering Study of Five Component Microemulsions', Langmuir 2, 184-190.
28. Teubner, M. and Strey, R. (1987) 'Origin of the scattering peak in microemulsions', J. Chem. Phys. 87, 3195-3200.
29. Talmon, Y. (1986) 'Electron microscopy in the research of surfactants in solution', in K.L. Mittal and P. Bothorel (eds.), Surfactants in Solution, Vol. 6, Plenum Press, New York, pp. 1581-1588.
30. Talmon, Y. (1986) 'Imaging surfactant dispersions by electron microscopy of vitrified specimens', Colloids and Surfaces 19, 237-248.
31. Talmon, Y. (1983) 'Staining and Drying-Induced Artifacts in Electron Microscopy of Surfactant Dispersions', J. Colloid Interface Sci. 93, 366-382.
32. Jahn, W. and Strey, R. (1988) 'Microstructure of Microemulsions by Freeze Fracture

Electron Microscopy', J. Phys. Chem. 92, 2294-2301.
33. Stilbs, P. (1987) 'Fourier Transform Pulsed-Gradient Spin-Echo Studies of Molecular Diffusion', Prog. Nucl. Magn. Reson. Spectrosc. 19, 1-45.
34. Callaghan, P.T. (1984) 'Pulsed Field Gradient Nuclear Magnetic Resonance as a Probe of Liquid State Molecular Organization', Aust. J. Phys. 37, 359-387.
35. Tanner, J.E. and Stejskal, E.O. (1968) 'Restricted Self-Diffusion of Protons in Colloidal Systems by the Pulsed-Gradient, Spin-Echo Method', J. Chem. Phys. 49, 1768-1777.
36. Bodet, J.-F., Bellare, J.R., Davis, H.T., Scriven, L.E. and Miller, W.G. (1988) 'Fluid Microstructure Transition from Globular to Bicontinuous in Midrange Microemulsion', J. Phys. Chem. 92, 1898-1902.
37. Bull, T. and Lindman, B. (1974) 'Amphiphile diffusion in cubic lyotropic mesophases', Mol. Cryst. Liq. Cryst. 28, 155-160.
38. Lindman, B., Shinoda, K., Olsson, U., Anderson, D., Karlström, G. and Wennerström, H. (1989) 'On the Demonstration of Bicontinuous Structures in Microemulsions', Colloids and Surfaces 38, 205-224.
39. Lindman, B. and Stilbs, P. (1987) 'Molecular diffusion in microemulsions', in S. Friberg and P. Bothorel (eds.), Microemulsions, CRC Press, Boca Raton, pp. 119-152.
40. Lindman, B., Kamenka, N., Kathopoulis, T.-M., Brun, B. and Nilsson, P.-G. (1980) 'Translational Diffusion and Solution Structure of Microemulsions', J. Phys. Chem. 84, 2485-2490.
41. Jönsson, B., Wennerström, H., Nilsson, P.-G. and Linse, P. (1986) 'Self-diffusion of small molecules in colloidal systems', Colloid Polym. Sci. 264, 77-88.
42. Anderson, D. and Wennerström, H. (1990) 'Self-Diffusion in Bicontinuous Cubic Phases, L_3 Phases and Microemulsions', J. Phys. Chem., in press.
43. Jonströmer, M., Parker, W.O. and Olsson, U. (1990) , in prep.
44. Olsson, U., Nagai, K. and Wennerström, H (1988) 'Microemulsions with Nonionic Surfactants. 1. Diffusion Process of Oil Molecules´, J. Phys. Chem. 92, 6675-6679.
45. Persoz (1846) 'Traité théoretique et practique de l'impression des tissus', Vol. 1, p. 354, quoted by F. Krafft in Ber. 27, 1755 (1894).
46. Olsson, U., Ström, P., Söderman, O. and Wennerström, H. (1989) 'Phase Behavior, Self-Diffusion, and ^2H NMR Relaxation Studies in an Ionic Surfactant System Containing Cosurfactant and Salt. A Comparison with Nonionic Surfactant Systems', J. Phys. Chem. 93, 4572-4580.
47. Ekwall, P., Mandell, L. and Fontell, K. (1970) 'Some Observations on Binary and Ternary Aerosol OT Systems', J. Colloid Interface Sci. 33, 215-235.
48. Fontell, K. (1973) 'The Structure of the Liquid Crystalline Optical Isotropic Viscous Phase Occurring in Some Aerosol OT Systems', J. Colloid Interface Sci. 43, 156-164.
49. Zulauf, M. and Eicke, H.-F. (1979) 'Inverted Micelles and Microemulsions in the Ternary System H_2O/Aerosol-OT/Isooctane as Studied by Photon Correlation Spectroscopy', J. Phys. Chem. 83, 480-486.
50. Eicke, H.-F. (1980) 'Aggregation in surfactant solutions: formation and properties of micelles and microemulsions', Pure & Appl. Chem. 52, 1349-1357.
51. Chen, S.H. (1986) 'Small angle neutron scattering studies of the structure and interaction in micellar and microemulsion systems', Ann. Rev. Phys. Chem. 37, 351-399.
52. Stilbs, P. and Lindman, B. (1984) 'Aerosol OT aggregation in water and hydrocarbon solution from NMR self-diffusion measurements´, J. Colloid Interface Sci. 99, 290-293.
53. Eicke, H.-F., Borkovec, M. and Das-Gupta, B. (1989) 'Conductivity of Water-in-Oil Microemulsions: A Quantitative Charge Fluctuation Model', J. Phys. Chem. 93, 314-317.

54. Fletcher, P.D.I., Howe, A.M. and Robinson, B.H. (1987) 'The Kinetics of Solubilisate Trans. 1, 83, 985-1006.
55. Kotlarchyk, M., Huang, J.S. and Chen, S.-H. (1985) 'Structure of AOT Reversed Micelles Determined by Small-Angle Neutron Scattering', J. Phys. Chem. 89, 4382-4386.
56. Carlström, G. and Halle, B. (1989) 'Shape Fluctuations and Water Diffusion in Microemulsion Droplets. A Nuclear Spin Relaxation Study', J. Phys. Chem. 93, 3287-3299.
57. Fontell, K., Ceglie, A., Lindman, B. and Ninham, B. (1986) 'Some observations on phase diagrams and structure in binary and ternary systems of didodecyldimethylammonium bromide', Acta Chem. Scand. A49, 247-256.
58. Saito, H. and Shinoda, K. (1970) 'The Stability of W/O Type Emulsions as a Function of Temperature and of the Hydrophilic Chani Length of the Emulsifier', J. Colloid Interface Sci. 32, 647-651.
59. Scriven, L.E. (1976) 'Equilibrium bicontinuous structure', Nature 263, 123-125.
60. Anderson, D.M. (1986) 'Studies in the Microstructure of Microemulsions', Ph. D. Thesis, University of Minnesota, Minneapolis.
61. Olsson, U. (1988) 'Surfactant Organization and Dynamics in Micellar Solutions and Microemulsions', Ph. D. Thesis, University of Lund, Lund.
62. Ekwall, P., Mandell, L. and Fontell, K. (1969) 'Solubilization in Micelles and Mesophases and the Transition from Normal to Reversed Structures', Mol. Cryst. Liquid Cryst. 8, 157-213.
63. Hoar, T.P. and Schulman, J.H. (1943) , Nature 152, 102.
64. v. Dijk, M.A., Joosten, J.G.H., Levine, Y.K. and Bedeaux, D. (1989) 'Dielectric Study of Temperature-Dependent Aerosol OT/Water/Isooctane Microemulsion Structure', J. Phys. Chem. 93, 2506-2512.
65. Kotlarchyk, M., Stephens, R.B. and Huang, J.S. (1988) 'Study of Schultz Distribution to Model Polydispersity of Microemulsion Droplets', J. Phys. Chem. 92, 1533-1538.
66. Robinson, B.H., Toprakcioglu, C. and Dore, J.C. (1984) 'Small-angle Neutron-scattering Study of Microemulsions Stabilized by Aerosol-OT Part 1.-Solvent and Concentration Variation', J. Chem. Soc., Faraday Trans. 1, 80, 13-27.
67. Toprakcioglu, C., Dore, J.C. and Robinson, B.H. (1984) 'Small-angle Neutron-scattering Studies of Microemulsions Stabilized by Aerosol-OT Part 2.-Critical Scattering and Phase Stability', J. Chem. Soc., Faraday Trans. 1, 80, 413-422.
68. Eicke, H.F. (1982) 'Self-Organization of Amphiphilic Molecules: Micelles and Micro-Phases', Chimia 36, 241-246.
69. Blum, F.D., Pickup, S., Ninham, B., Chen, S.J. and Evans, D.F. (1985) 'Structure and Dynamics in Three-Component Microemulsions', J. Phys. Chem. 89, 711-713.
70. Chen, S.J., Evans, D.F., Ninham, B.W., Mitchell, D.J., Blum, F.D. and Pickup, S. (1986) 'Curvature as a Determinant of Microstructure and Microemulsions', J. Phys. Chem. 90, 842-847.
71. Lang, J.C. and Morgan, R.D. (1980) 'Nonionic surfactant mixtures. I. Phase equilibria in $C_{10}E_4$-H_2O and closed-loop coexistence', J. Chem. Phys. 73, 5849-5861.
72. Mitchell, D.J., Tiddy, G.J.T., Waring, L., Bostock, T. and McDonald, M.P. (1983) 'Phase Behavior of Polyoxyethylene Surfactants with Water', J. Chem. Soc., Faraday Trans. 1, 79, 975-1000.
73. Strey, R., Schomäcker, R., Roux, D., Nallet, F. and Olsson, U. (1990) 'On the Dilute Lamellar and L_3 Phases in the Binary Water-$C_{12}E_5$ System', J. Chem. Soc., Faraday Trans. 1, in press.

74. Olsson, U., Shinoda, K. and Lindman, B. (1986) 'Change of the Structure of Microemulsions with the Hydrophile-Lipophile Balance of Nonionic Surfactant As Revealed by NMR Self-Diffusion Studies', J. Phys. Chem. 90, 4083-4088.
75. Kahlweit, M., Strey, R., Haase, D., Kunieda, H., Schmeling, T., Faulhaber, B., Borkovec, M., Eicke, H.-F., Busse, G., Eggers, F., Funck, TH., Richmann, H., Magid, L., Söderman, O., Stilbs, P., Winkler, J., Dittrich, A. and Jahn, W. (1987) 'How to Study Microemulsions', J. Colloid Interface Sci. 118, 436-453.

EXPERIMENTAL AND COMPUTATIONAL ASPECTS OF THE TIME-CORRELATED SINGLE PHOTON COUNTING TECHNIQUE

A. MALLIARIS
N.R.C. "Demokritos"
Athens 153 10
Greece

ABSTRACT. Two important aspects of the time-correlated single photon counting technique are presented. The first, concerning the extraction of the true decay curve from the experimental data, deals with the "reference convolution method". The second is related to the treatment of decay data by means of the "simultaneous analysis" of fluorescence decay curves. It allows for the best parameter recovery and very accurate model testing. Applications involving aqueous micellar systems, using simulated data, are briefly reviewed.

1. Introduction

Time-correlated single photon counting methods have been extensively used for determining decay rates of electronically excited molecular states [1-3]. The success of the technique, apart from its high sensitivity, relies on the fact that the error distribution of the experimental data is well-established. Indeed it is known that the error function follows Poisson statistics. Such knowledge allows correct statistical analysis of the curve fitting and therefore reliable and properly evaluated decay parameters. In general, the technique involves high repetition rate lasers as pulse sources [4], fast detecting devices [5], and powerful computing machines and software [6]. With present day equipment, fluorescence lifetimes of only few picoseconds can be established, and emission intensities spanning up to ca. five orders of magnitude can be easily measured.

The fundamental principle of the technique is the following. A sample S is repeatedly excited (several thousand times per second) by means of a narrow flash produced by a pulsed laser or a discharge lamp P. Each flash also starts a time to amplitude converter (TAC). When a fast and sensitive photodetector receives the very first photon emitted from the excited sample it sends a signal which stops the TAC. The time interval between the flash and the first photon detected, comes out of the TAC as a pulse of amplitude (voltage) directly proportional to this time interval. This pulse is then fed to a multichannel analyser (MCA), where it is registered, as a single "fluorescent count", in the appropriate channel according to its amplitude. If this process, i.e. flash-excitation-emission-detection-registration, is repeated until good statistics is obtained, and if the fluorescence count rate is kept very small (<ca. 2%>) compared to the frequency of the flash generator, the histogram of the number of counts per channel vs channel number of the MCA is

D. M. Bloor and E. Wyn-Jones (eds.),
The Structure, Dynamics and Equilibrium Properties of Colloidal Systems, 243–251.

directly analogous to the decay of the sample under examination. Note that in these measurements the axis of the channel number, which in fact corresponds to time, is assumed to be errorless. All experimental errors are assumed to occur along the fluorescence intensity axis.

Evidently, the fluorescence decay data, obtained in a real experiment, do not give the true fluorescence decay profile. The response of the particular experimental set up (including the shape of the exciting pulse, the response of the detecting system, etc, affect the detailed structure of the decay curve. It is therefore necessary to eliminate all outside influences on the true decay curve. This is accomplished by means of deconvolution methods which separate the true fluorescence decay from the experimental artifacts. The obvious requirement, in order to apply deconvolution of the experimental decay, is to have an independent knowledge of the flash shape and of the instrument response function. However, such knowledge is very difficult to be obtained experimentally. [7,8].

2. Deconvolution of the Decay Curve

The fluorescence intensity $F_S(\lambda_{ex},\lambda_{em},t)$ of an emitting sample S is the convolution of the true sample response function $F(\lambda_{ex}\lambda_{em},t)$ with the instrument response function $I(\lambda_{ex},,\lambda_{em},t)$. This fact is formulated in eq.1,where λ_{ex} and λ_{em} are the wavelengths of the exciting and the

$$F_S(\lambda_{ex}, \lambda_{em}, t) = I(\lambda_{ex}, \lambda_{em}, t)*F(\lambda_{ex}, \lambda_{em}, t) \qquad (1)$$

emitted photons respectively, t represents the time at which the functions have their particular values, and "*" denotes the convolution operator. Note that the instrument response function $I(\lambda_{ex}, \lambda_{em}, t)$ is also the result of the convolution of the function of the exciting pulse $P(\lambda_{ex},t)$ and the detection response function $D(\lambda_{em},t)$, i.e. the function which describes the detecting system (photomultiplier etc). This convolution is expressed in eq.2.

$$I(\lambda_{ex},\lambda_{em},t)=P(\lambda_{ex},t)*D(\lambda_{em},t) \qquad (2)$$

In the ideal experiment one expects to measure the instrument response function $I(\lambda_{ex},\lambda_{em},t)$ and the sample decay $F_s(\lambda_{ex}, \lambda_{em},t)$ at various times t, and then calculate the true sample response function $F(\lambda_{ex}, \lambda_{em},t)$ by deconvolution according to eq.1. In practice, however, this is not plausible, simply because the instrument response function $I(\lambda_{ex}, \lambda_{em},t)$ cannot be measured directly. Alternatively, the functions $I(\lambda_{ex}, \lambda_{ex},t)$ or $I(\lambda_{em},\lambda_{em},t)$, measured with a scatterer, have been employed instead of the correct function $I(\lambda_{ex},\lambda_{em},t)$. On the other hand, it is known that the detection response function $D(\lambda,t)$, depends on the wavelength i.e. $D(\lambda_{ex},t)=D(\lambda_{em},t)$, primarily because of the wavelength dependence of the photomultiplier response [9]. Consequently, it is quite wrong to employ either $I(\lambda_{ex},\lambda_{ex},t)$ or $I(\lambda_{em},\lambda_{em},t)$ in the place of $I(\lambda_{ex},\lambda_{em},t)$. Therefore, it is impossible to obtain accurate data for the true sample response function by means of eq.1, since $I(\lambda_{ex},\lambda_{em},t)$ is not directly available. It has been shown that the best way to overcome this serious problem is to work with the so-called "reference convolution method" [10,11].

2.1. THE REFERENCE CONVOLUTION METHOD

According to the reference convolution method, a reference fluorescent compound R is used, which absorbs and emits light at the same wavelengths λ_{ex} and λ_{em} as the sample compound. Moreover, the fluorescence decay of the reference compound $R(\lambda_{ex},\lambda_{em},t)$ must be single exponential as expressed by eq.3.

$$R(\lambda_{ex},\lambda_{em},t)=a_r\exp(-t/\tau_r) \quad (3)$$

In eq.3 a_r is a scaling factor and τ_r is the fluorescence lifetime of the reference compound R, which must be rather short. If the true fluorescence decay $F_r(\lambda_{ex},\lambda_{em},t)$, of the reference compound, is measured under the same experimental conditions as the sample compound, one can define a modified sample response function $F_m(\lambda_{ex},\lambda_{em},t)$ which satisfies eq.4.

$$F_s(\lambda_{ex},\lambda_{em},t) = F_r(\lambda_{ex},\lambda_{em},t) * F_m(\lambda_{ex},\lambda_{em},t) \quad (4)$$

This modified function can be obtained from $F_s(\lambda_{ex},\lambda_{em},t)$ and $F_r(\lambda_{ex},\lambda_{em},t)$ by standard deconvolution methods via eq.4. Furthermore, it has been shown that if the decay of the reference compound indeed satisfies eq.3 then the modified function $F_m(\lambda_{ex},\lambda_{em},t)$ (hereafter denoted simply as $F_m(t)$) is given by eq.5, where $\delta(t)$ is the Dirac delta function and $f'(t)$ is the

$$F_m(t) = a_r^{-1}\,[F(0)\delta(t) + F'(t) + F(t)/\tau_r] \quad (5)$$

first time derivative of F(t) [7,12].

For the particular case of decay rates distributed according to Poisson statistics, e.g. the case of immobile fluorophores and quenchers [Q] distributed among micelles [M], the true sample response function has the form of eq.6 [13], and the associated $F_m(t)$ function has

$$F(t)=a_1\exp(-A_2t)\exp\{-A_3[1-\exp(-A_4t]\} \quad (6)$$

the form of eq.7.

$$F_m(t) = A_1\{\delta(t) + [1/\tau_r-1/\tau_o-A_3A_4\exp(-A_4t)]\exp(-t/\tau_o) \quad (7)$$
$$\exp\{-A_3[1-\exp(-A_4t)]\}$$

In these equations $A_1=a_1/a_r$, $A_2=1/\tau_o$ (where τ_o is the fluorescence lifetime of the fluorophor), A_3 is the average number of quenchers per micelle, i.e. [Q]/[M] and A_4 is the intramicellar fluorescence quenching rate constant for only one quencher per micelle.

It is worth mentioning that at very low A_3 values ($A_3 \ll 1$), eq.6 simplifies to eq.8 (using the approximation exp(-A)=1-A for A<<1), which in fact describes a biexponential decay.

$$F(t)=a_1\{A_3\exp[-t(1/A_2+A_4)]+(1-A_3)\exp(-A_2t)\} \quad (8)$$

The fluorescence parameters of the two decays in the biexponential eq.8 are related to the micellar parameters of the original eq.7, through eqs.9.

scaling factor of decay 1 : $\alpha_1=a_1A_3$

lifetime of decay 1 : $\tau_1=1/(1/A_2+A_4)$ (9)

scaling factor of decay 2 : $\alpha_2=a_1(1-A_3)$

lifetime of decay 2 : $\tau_2=A_2$

and therefore:

$$A_2=\tau_2$$

$$A_3=\alpha_1/(\alpha_1+\alpha_2) \quad (10)$$

$$A_4=1/\tau_1-1/\tau_2$$

Through eqs.10 one can obtain the micellar parameters A_2, A_3 and A_4 from the fluorescence parameters of the biexponential decay.

Evidently, at low values of A_3 the distinction between the two decay modes, viz. micellar (eq.6) and biexponential (eq.8), becomes very difficult, practically impossible. Such serious problems involving model testing, are successfully dealt with by means of the so-called "simultaneous fluorescence analysis", also referred to as "global fluorescence analysis".

3. Simultaneous Fluorescence Analysis

The deduction of photophysical mechanisms usually involves recording several fluorescence decay curves under changing experimental conditions, e.g. temperature, solvent, wavelength, etc. The decay parameters thus obtained, are subsequently interrelated to extract the optimum physical model which describes the situation. This popular, and usually effective, method, has a rather serious handicap. Namely, it does not take into account relationships which may exist among the individual decay curves. A way to overcome this problem is to use "simultaneous (or global) analysis" of decay curves. According to this method one tries to combine different experimental decays, in the sense that in a fit not only constant and free-running parameters are used, but also there exist parameters which although not fixed, they are however held common, they are linked, among the related decay curves. In fact, it has been shown that additional information can be obtained if different experiments are combined together [14].

Fitting a theoretical curve to experimental decay data consists of consecutive iterative changes until an acceptable minimization of deviations between theory and experiment is achieved. In global fluorescence analysis the same procedure is followed, except that in this case, instead of a single curve a set of decay curves is involved in the minimization of deviations. This is obtained by means of global mapping vectors which link the local parameters of each single curve to the corresponding global parameters.

A simple example which illustrates the notion of global analysis is the case of a solution of two fluorescent molecules which decay monoexponentially. The expression for the overall fluorescence decay of the solution is given by eq.11.

$$F(t)=a_1 exp(-t/\tau_1)+a_2 exp(-t/\tau_2) \quad (11)$$

In a set of n experiments, where the composition of the solution varies, there will be 4n independent parameters, because there are four unknowns in each decay curve, i.e. a_1, a_2, τ_1 and τ_2. However, if the lifetimes τ_1 and τ_2 are assumed constant among the various decay curves, the 4n independent parameters become 2n+2, since only the pre-exponential factors a_1 and a_2 (equal to 2n) will change from one solution to another. These 2n+2 global parameters (indicated by the letter g) are related to the single curve parameters (a and τ)as shown below

decay curve 1: $g_1=a_1,\ g_2=\tau_1,\ g_3=a_2,\ g_4=\tau_2$
decay curve 2: $g_5=a_1,\ g_2=\tau_1,\ g_6=a_2,\ g_4=\tau_2$ (12)

.

.

decay curve n: $g_{2n+1}=a_1, g_2=\tau_1, g_{2n+2}=a_2, g_2=\tau_2$

Therefore the decay for the nth curve will be described by eq.13.

$$F_n(t)= g_{2n+1}exp(-t/g_2)+g_{2n+1}exp(-t/g_4) \quad (13)$$

To facilitate manipulations one can usually construct a mapping matrix A which links local parameters, i.e. parameters of a specific decay, to global ones. This matrix is constructed by copying the global parameter indices as they appear in eqs.13. For instance, in the above case the linking matrix A will be

$$A = \begin{vmatrix} 1 & 5 & . & . & . & 2n+1 \\ 2 & 2 & . & . & . & 2 \\ 3 & 6 & . & . & . & 2n+2 \\ 4 & 4 & . & . & . & 4 \end{vmatrix}$$

By replacing local by global parameters, with the help of the linking matrix A, one can apply any adequate algorithm, e.g. Marquardt's algorithm for non-linear least squares, to obtain simultaneous decay curve fitting. Having described the reference convolution method and the global fluorescence analysis, we now proceed with the presentation of some simple applications, using simulated decay curves, to demonstrate the power of this data analysis.

4. Single Curve vs Global Fluorescence Analysis

In this Section we will compare the performance between single curve and global fluorescence analysis. The data and calculations discused here are based on references 12 and 15. In both situations, of single and multiple decay curve analysis, the reference convolution method will be employed. The physical system under investigation is an aqueous micellar solution of micellar concentration [M], in which a fluorophor and a quencher, both immobile, have been solubilized. Such system of Poisson distributed decay rates is adequately described by eq.6. Therefore, according to eq.1, synthetic data were produced by convolution of F(t) (from eq.6) with a measured instrument response function I(t). In this way, a number of sample response functions $F_s(t)$ were produced. For their production a set of values for the micellar parameters A_2=200 ns, $[M]=10^{-3}$M (A_3=[Q]/[M]) and $A_4=10^7 s^{-1}$ was used, along with several quencher concentrations [Q]. On the other hand, simulated reference decays $F_r(t)$, were generated by convolution of a monoexponential decay (eq.3) with the same instrument response function I(t). The preexponential factors a_1 and a_r occurring in eqs.6 and 3 respectively, were chosen such as to obtain the desired number of counts at the peak channel. Finally, to every simulated $F_s(t)$ and $F_r(t)$ function, independent Poisson noise was added.

The thus obtained functions $F_s(t)$ and $F_r(t)$, along with the expression for the modified sample response function F(t), given by eq.9, were used to extract the micellar parameters (see eq.4). The expression for F(t) (eq.7) involves five parameters, i.e. A_1, A_2, [M], A_4 and τ_r. Note however that the parameter A_1 does not convey any important physical meaning since it is only a measure of the total number of counts in the peak channel. The micellar parameters were determined by both single curve and simultaneous analysis. In the single curve analysis all parameters were free running, except τ_r which was kept constant at its known value. In the case of global analysis the parameters A_2, [M] and A_4 were allowed to change but they were linked among the various decay curves, τ_r was again constant, while A_1 was treated as a free running parameter. In a set of decay curves according to eq.8, involving e.g. five experiments, the linking matrix will be the following:

experiment No:	1	2	3	4	5	Parameter
A =	1	6	7	8	9	A_1 (free-running)
	2	2	2	2	2	A_2 (linked)
	3	3	3	3	3	[M] (linked)
	4	4	4	4	4	A_4 (linked)
	5	5	5	5	5	T_r (constant)

4.1 PARAMETER RECOVERY

Here the objective is to compare micellar parameters A_2, [M] and A_4, extracted from the simulated curves $F_s(t)$, by means of both single curve and global fluorescence analysis. Characteristic values of one of these parameters (e.g. [M] vs [Q]) are plotted in Figs.3-5 of reference 12. They all correspond to statistically acceptable fits as indicated by the global reduced chi-square (≈1) and its normal deviate Z chi-square (<3) for the global analysis, and by the same statistical factors plus the Durbin-Watson test, for the single curve analysis.

In a simulated experiment with 10^3 counts at the peak channel and total number of channels = 256, the improvement of the recovered [M] values when global analysis is used is impressive [12]. Thus, with only three experiments combined the global analysis gives $[M]=9x10^{-4}$, whereas the single curve analysis gives [M] values between ca. $2x10^{-4}$ and $5x10^{-4}$ (true $[M]=1x10^{-3}$). When the number of data channels was 512 and the count number in the peak channel was kept equal to 10^3, the single curve analysis gave better [M] values but still the global analysis was far more accurate. Similarly, simulated decays with 256 channels and 10^4 counts in the peak channel improved single curve parameter recovery but the simultaneous analysis gave much more accurate [M] values. In all cases the superiority of the multiple over the single curve analysis becomes particularly evident for low quencher concentrations.

It should be noticed here that in real experiments the number of counts in the peak channel is approximately 10^3 while the number of data channels is usually 256 or 512. Only in exceptional cases curves with peak channel counts of the order of 10^4 and channel numbers more than 512 are recorded. On the other hand, it is desirable to keep the quencher concentration as low as possible in order to prevent disturbances of micellar structure by solubilized quencher molecules. It is therefore obvious that under usual experimental conditions, i.e. peak channel counts ca. 1000, channel number 256 and low [Q], the accuracy of the parameters recovered is much higher with global than single curve analysis. Note that in some extreme cases single curve data analysis did not even converge [12].

4.2 MODEL TESTING

Although parameter recovery is greatly improved by multiple curve analysis, the most important application of global fluorescence analysis is in model testing. Obviously, it is very important to be able to distinguish between different possible physical models which involve very similar values of the parameters. Unfortunately, very long data accumulation times involve introduction of random errors which mask the underlying physical model.

As an example we will discuss below the case of the micellar decay of eq.6 which degenerates to the biexponential decay of eq.8 for very small quencher concentrations. Indeed, it was shown above that when A_3 is very small eq.6 which describes the fluorescence decay in a system with Poisson distribution of decay rates, simplifies to the biexponential eq.8. Evidently, two different physical models, the one with Poisson distributed decay rates and the other with two distinct monoexponential decays, give the same micellar parameters. It is therefore very important to be able to distinguish between different photophysical mechanisms which, when fitted to experimental data, produce similar values for the parameters of the system.

Synthetic sample $F_s(t)$ and reference $F_r(t)$ decay curves, simulated with 10^4 counts in the peak channel, 512 channels and A_3 values typical of the ones used in real experiments, viz. 0.5-1, were used to fit eqs. 8 and 10. As judged by the statistical goodness-of-fit tests, single curve analysis cannot distinguish between the biexponential and the micellar decay model. Either one gives good fits. The same seems to be the case with global analysis, i.e. either one of the models gives equally good fits for A_3 values in the range 0.5 to 1. However, if in the multiple decay analysis, in addition to the decays with $A_3>0$ the decay with $A_3=0$ is included, the situation changes drastically. Thus, global analysis of decays including $A_3=0$ gives the acceptable Z chi-square value equal to -0.13 for fits according to eq.8, but Z chi-square equal to 7.6 for fits according to the biexponetial decay of eq.8, indicating unsatisfactory fit in the latter case.

Acknowledgement

The author is grateful to Prof. F.C. DeSchryver and to Drs N.Boens and M.van der Auweraer for the time he spent in their laboratory (Chemistry Dept., Catholic University, Leuven, Belgium).

5. References

1. Luo H.L., Boens N., Van der Aweraer M., DeSchryver F.C and Malliaris A. (1989), 'Simultaneous analysis of time-resolved fluorescence quenching data in aqueous micellar systems in the presence and absence of added alcohol', J.Phys.Chem, 93, 3244.

2. Thomas J.K. (1987), 'Characterization of surfaces by excited states', J.Phys.Chem., 91, 267-276.

3. Malliaris A. (1989), 'Fluorescence probing in aqueous micellar systems: an overview', Intern.Rev.Phys.Chem., 7, 95-121.

4. Koester V.J. (1978), 'Subnanosecond single photon counting fluorescence spectroscopy', Rev.Sci.Instrum., 49, 1186-1191.

5. Yamazaki I.N., Tamai K., Kume H., Tsuchiya H. and Oba K. (1985), 'Michrochannel-plate photomultiplier applicability to the time-correlated photonon-counting method', Rev.Sci. Instrum., 56, 1187-1194.

6. Van den Zegel M., Boens N., Daems D. and DeSchryver F.C. (1986) 'Possibilities and limitations of the time-correlated single photon counting technique: A comparative study of correction methods for the wavelength dependence of the instrument response function', Chem.Phys., 101, 311-335.

7. Zuker M., Szabo A.G., Bramall L., Krajcaski D.T. and Selinger B. (1985), 'Delta function convolution method (DFCM) for fluorescence decay experiments', Rev.Sci.Instrum. 56, 14-22.

8. Boens N., Ameloot M, Yamazaki I. and DeSchryver F.C. (1988) 'On the use and the performance of the delta function convolution method for the estimation of fluorescence decay parameters', Chem.Phys., 121, 73-86.

9. James D.R., Demmer D.R.M., Verrall R.E. and Steer R.P. (1983), 'Excitation pulse-shape mimic technique for improving picosecond-laser-excited time-correlated single-photon

counting deconvolutions', Rev.Sci.Instrum., 54, 1121-1130.

10. Boens N., Van den Zegel M. DeSchryver F.C. and Desie G. (1986) 'The time-correlated single photon counting technique as a tool in pho tobiology', Photobiochem. Photobiophys. Suppl., 93-108.

11. Arcioni A. and Zannoni C. (1984), 'Intensity deconvolution in fluorescence depolarization studies of liquids, liquid crystals and membranes', 113-128.

12. Boens N., Malliaris A., Van der Auweraer M., Luo H. and DeSchryver F.C. (1988), 'Simultaneous analysis of single-photon timing data with a reference method: Application to a Poisson distribution of decay rates', Chem.Phys. 121, 199-209.

13. Ameloot M., Beecham J.M. and Brand L. (1986), 'Simultaneous analysis of multiple fluorescence decay curves by Laplace transforms, Deconvolution with reference or excitation profiles', Biophys.Chem., 23, 155-171.

14. Knutson J.R., Beechem J.M. and Brand L. (1983), 'Simultaneous analysis of multiple fluorescence decay curves: A global approach' Chem. Phys.Lett. 102, 501-507.

15. Boens N., Luo H., Van der Auweraer M., Reekmans S., DeSchryver F.C. and Malliaris A. (1988), 'Simultaneous analysis of fluorescence decay curves for the one-step determination of the mean aggregation number of aqueous micelles', Chem.Phys.Lett. 146, 337-341.

DROPLET SIZE AND DYNAMICS IN WATER IN OIL MICROEMULSIONS. CORRELATIONS BETWEEN RESULTS FROM TIME-RESOLVED FLUORESCENCE QUENCHING, QUASIELASTIC LIGHT SCATTERING, ELECTRICAL CONDUCTIVITY AND WATER SOLUBILITY MEASUREMENTS

J. LANG, R. ZANA and N. LALEM
Institut Charles Sadron (CRM-EAHP), CNRS-ULP
6, rue Boussingault
67083 Strasbourg Cédex
France

ABSTRACT. The effect of the oil, surfactant and alcohol chain length, of the temperature, and of the alcohol concentration on the water droplet size and interdroplet interaction in water-in-oil (w/o) microemulsions has been investigated by time-resolved fluorescence quenching, quasielastic light scattering, electrical conductivity and water solubility (partial phase behavior). The results obtained with these different methods correlate perfectly well and indicate increases in droplet size and interdroplet attractive interactions as the oil(alkane) chain length and the temperature increase and as the surfactant and alcohol chain length and alcohol concentration decrease, in agreement with the predictions of recent theories of the stability of w/o microemulsions. The rate constant k_e associated with the exchange of material between droplets upon collisions with transient merging was found to be at least equal to or larger than $(1\text{-}2)\times10^9\ M^{-1}s^{-1}$ for all of the systems where electrical percolation occurred upon increase of the fraction of disperse phase or temperature. This requirement indicates that above the percolation threshold the high electrical conductivity is due to the motion of counterions through water channels and/or fusion between droplets in droplet clusters, rather than to hopping of surfactant ions between droplets upon droplet collisions. The results also show that from the variation of the electrical conductivity with any of the parameters which characterize the microemulsion one can predict qualitatively the resulting variations of droplet size, interdroplet attractive interactions, and rate of exchange of material between droplets via droplet collisions.

1. Introduction

In recent years water-in-oil (w/o) microemulsions have been investigated by means of several techniques in order to examine the influence of various parameters as for example the structure of the oil and surfactant molecules, the temperature, the water content and the nature of additives like alcohol molecules, on the size of the w/o microdroplets, on the interdroplet interactions and on two macroscopic properties of the microemulsions, namely the water solubility and the electrical conductivity. The rate of exchange of material between droplets occurring upon collision and transient merging of two droplets, has also been investigated.

General rules concerning the effect of the above parameters on the structure, dynamics, water

D. M. Bloor and E. Wyn-Jones (eds.),
The Structure, Dynamics and Equilibrium Properties of Colloidal Systems, 253–278.

solubility and electrical conductivity of w/o microemulsions have been recently inferred from studies in which one parameter at a time was systematically varied [1-5]. The aim of this paper is to summarize the results of these studies, which concern ternary as well as more complex (quaternary and even quinary) w/o microemulsions. In the case of ternary water/surfactant/oil microemulsions, the following parameters have been systematically varied: (i) the oil structure, (ii) the temperature and (iii) the surfactant structure. In the case of quaternary water/surfactant/oil/ alcohol microemulsions the structure and concentration of the alcohol molecule have been investigated.

With respect to previous studies the main interest of our investigations lies in the use of several techniques which have yielded results showing good correlations between properties differing very much in nature, some of macroscopic character, others of microscopic character. A deeper understanding of the effect of the above parameters on droplet size, interactions and dynamics was thus reached. This now offers the possibility of making predictions concerning the changes of droplet properties as one of the above parameters is varied by using a fairly inexpensive, readily available technique, namely the electrical conductivity technique.

The effects of the oil nature and temperature are presented first. These studies involved microemulsions based on sodium bis(2-ethylhexyl)sulfosuccinate (AOT) as surfactant. The effects of the surfactant structure and of the structure and concentration of the alcohol have been investigated for microemulsions based on cationic surfactants. These results are presented next. Finally it is shown that some of the conclusions drawn from the results obtained with these systems also hold for w/o microemulsions where the oil continuous phase is made of a mixture of organic molecules.

2. Material and Methods

The origin and purification of all of the chemicals (surfactants, solvents), probe (ruthenium (II) trisbipyridyl chloride ($Ru(bpy)_3^{2+}$)) and quenchers (methylviologen chloride, (MV^{2+}), and potassium ferricyanide, ($Fe(CN)_3^{2+}$))has been reported elsewhere [1-5].

All solutions used in the time-resolved fluorescence quenching measurements were thoroughly deoxygenated by at least four freeze-pump- thaw cycles.

2.2. METHOD

2.2.1. *Time-Resolved Fluorescence Quenching (TRFQ).* The TRFQ method has been described in this volume [6] and elsewhere [7]. $Ru(bpy)_3^{2+}$ was used as fluorescent probe. The quencher was $Fe(CN)_6^{3-}$ for the microemulsions containing the anionic surfactant AOT and MV^{2+} for the microemulsions based on cationic surfactants [8]. The probe and the quenchers, referred to as reactants in the following, are all solubilized in the water pools of the droplets. The probe was used at a $[Ru(bpy)_3^{2+}]/[Droplet]$ molar concentration ratio between 0.01 and 0.08. The quencher concentration, [Q], was such that the molar concentration ratio R = [Q]/[Droplet] was between 0.5 and 1.5. The fluorescence decay data were collected by means of a single-photon counting apparatus [9]. The excitation wavelength was 480 nm, and the emission was monitored above 530 nm with a high-pass cutoff filter.

The fluorescence decay curves obey the equation [10,11]:

$$I(t) = I(o)\exp\{-A_2t - A_3[1 - \exp(-A_4t)]\} \qquad (1)$$

where I(t) and I(o) are the fluorescence intensities at time t and t = 0, respectively, following the exciting pulse. A_2, A_3 and A_4 are time independent parameters that are obtained, together with I(o), by fitting eq.1 to the fluorescence decay data, using a nonlinear weighted least-squares procedure.

When there is no interdroplet exchange of reactants on the probe fluorescence time scale A_2, A_3 and A_4 write [12]:

$$A_2 = k_0 \quad ; \quad A_3 = [Q]/[M] \quad ; \quad A_4 = k_q \qquad (2)$$

where k_0 is the fluorescence decay rate constant of the probe in droplets without quencher, k_q the pseudo-first-order rate constant for intradroplet fluorescence quenching and [M] = [Droplet]. In this case the mean surfactant aggregation number N (average number of surfactants per droplet) is given by:

$$N = C/[M] = C\,A_3/[Q] \qquad (3)$$

where C is the total surfactant concentration. Equation 3 assumes that all the surfactant is used in making up droplets.

When interdroplet exchange of quencher takes place on the fluorescence time scale, by collisions between droplets, as shown in Figure 1, A_2, A_3 and A_4 are given by [8,13-15]:

$$A_2 = k_0 + k_e k_q [Q]/A_4 \qquad (4)$$

$$A_3 = (k_q/A_4)^2 [Q]/[M] \qquad (5)$$

$$A_4 = k_q + k_e [M] \qquad (6)$$

where k_e is the second-order rate constant associated with collisions giving rise to reactant migration between droplets.

From eqs.4-6, N, k_q and k_e are obtained from the fitting quantities A_2, A_3, A_4 and k_0 as:

$$N = \frac{C}{[M]} = \frac{C}{[Q]}\,\frac{(A_3A_4 + A_2 - k_0)^2}{A_3A_4^2} \qquad (7)$$

$$k_q = A_3A_4^2/[A_3A_4 + A_2 - k_0] \qquad (8)$$

$$k_e = A_4(A_2 - k_0)/(k_q[Q]) \quad (9)$$

The fluorescence decay rate constant k_0 was obtained, for each system investigated, from an independent fluorescence decay experiment in the absence of quencher in the solution.

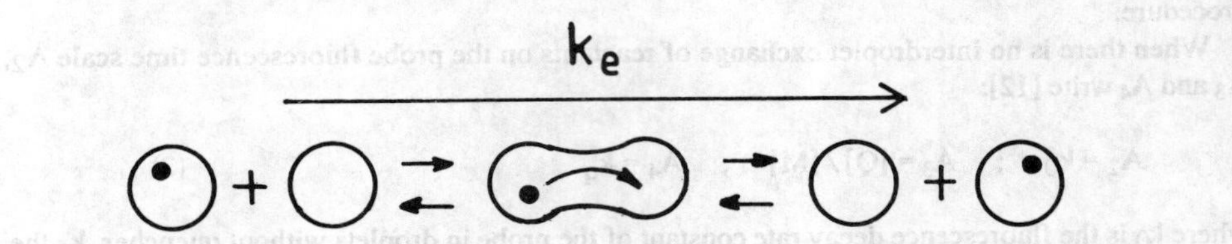

Figure 1. Exchange of material between droplets through collision with temporary merging. The exchange is illustrated by the transfer of the black dot from one droplet to the other and is characterized by the overall second-order rate constant k_e.

Recall that two other types of exchange, besides the one shown in Fig.1, have been observed in surfactant containing systems, namely exchange through the continuous phase [10,14-19] and exchange by fragmentation-coagulation [20-22]. In the w/o microemulsions studied here, exchange through the oil phase can be neglected. Indeed it is "frozen" on the fluorescence time scale since the reactants used are ions nearly insoluble in the organic solvents constituting the interdroplet solution. As for the exchange by fragmentation-coagulation it occurs chiefly in polydisperse systems [20,23]. It is known, and it has been shown as a part of this work, that the droplets in the investigated w/o microemulsions are relatively monodisperse. Therefore this exchange also has not been considered in the present studies.

Almgren et al.[24,25] have shown that eqs.4-6 are not valid if the exchange of excited probes by interdroplet collisions is taken into account. However eqs.7 and 8 remain unchanged and only eq.9, which gives k_e, is then modified. It has been found [1] in studies of w/o microemulsions based on AOT with $[Q]/[M] \approx 1$, that the k_e values calculated from eq.9 were slightly lower, by less than 20%, than those obtained from the data analysis proposed by Almgren et al. Therefore, since this analysis is not straightforward, eq.9 has been used throughout this work.

2.2.2. *Quasielastic Light Scattering (QELS).* Details concerning the QELS technique used are given elsewhere [4]. This method measures the translational diffusion coefficient, D, of the droplets. Recall that for interacting droplets, D can be written, in the low concentration range, as:

$$D = D_0 (1 + \alpha\Phi) \qquad (10)$$

where D_0 is the self-diffusion coefficient at infinite dilution and Φ the droplet volume fraction in the microemulsion, given by:

$$\Phi = (V_w + V_s)/V = 10^{-3}C(18\omega + v_s) \qquad (11)$$

where ω is the molar concentration ratio $[H_2O]/C$ in the solution, C being the surfactant concentration in mole/liter, V_w and V_s are the volumes of water and surfactant, respectively, in a volume V of microemulsion, and v_s is the apparent molar volume of the surfactant expressed in cm^3/mole. In the following, values of v_s equal to 369, 406 and 443 cm^3/mole have been used for the $C_mH_{2m+1}(C_6H_5CH_2)^+N(CH_3)_2\ Cl^-$ surfactants (alkylbenzyldimethylammonium chlorides referred to as Nm,1ø,1,1,Cl) with m = 12,14 and 16, respectively. In eq.11, α is the virial coefficient associated to the diffusion coefficient D. Recall that for hard-sphere-type interactions $\alpha = 1.5$ and that negative values of α correspond to attractive interactions.

2.2.3. *Electrical Conductivity.* Electrical conductivity measurements were performed using an automated autobalanced conductivity bridge (Wayne-Kerr type B905), operating at a frequency of 1 kHz.

2.2.4. *Structural Parameters of the Droplets.* Knowing N from the TRFQ measurements (with eq.3 or 7), some characteristics of the droplets can be derived by means of an appropriate working model. We have assumed that all the water is inside the droplets, that the droplets are spherical, monodisperse and separated from the continuous oil phase by a monolayer of N surfactant ions. In the case of ternary w/o microemulsions the radius of the water pool, R_w, and the spherical surface area, σ, occupied by each surfactant ion at the surface of the water pool are then given by:

$$R_w = [3N(\omega v_w + v_{ci})/4\pi]^{1/3} \qquad (12)$$

$$\sigma = 4\pi R_w^2/N \qquad (13)$$

where v_w in the molecular volume of a water molecule (29.9 Å^3 at 25°C)and v_{ci} the apparent molecular volume of the counterion.

In the case of quaternary microemulsions containing a cosurfactant which is distributed between the water pool, the interfacial film and the oil continuous phase, as in the case of alcohol molecules in oil/water/surfactant/alcohol microemulsions, the expression of R_w is:

$$R_w = [3N(\omega v_w + v_{ci} + fzv_a)/4\pi]^{1/3} \qquad (14)$$

and σ is still given by eq.13. In eq.14 z is the molar concentration ratio [alcohol]/C in the solution, v_a is the apparent molecular volume of the alcohol molecule in water and f the fraction of alcohol in the water pool. Since f is usually difficult to determine experimentally, eq.15 has been used instead of eq.14.

$$R_w = [3N(\omega v_w + v_{ci} + q\omega v_a)/4\pi]^{1/3} \quad (15)$$

with

$$q = 18\, C'/(1000 - C'M_a/d) \quad (16)$$

In eq.16 C' represents the solubility of the alcohol in water, in mole per liter, M_a is the alcohol molecular weight and d its density. Of course, the real value of R_w is between those calculated from eq.12 (no alcohol in the water pool) and eq.15 (alcohol-saturated water pool).

The values of D from the QELS measurements can be used for the calculation of the apparent hydrodynamic radius R_H, assuming spherical droplets, from the Stokes-Einstein equation:

$$R_H = kT/6\pi\eta D \quad (17)$$

where η is the solvent viscosity. Notice that R_H involves a contribution due to interdroplet interactions. As the droplet volume fraction Φ goes to zero, D becomes equal to D_0 and R_H becomes close to the true hydrodynamic radius of the droplet, R^o_H. This value can then be compared to the overall droplet radius, R_M, derived from the TRFQ measurements and given by:

$$R_M = R_w + l \quad (18)$$

where l represents the length of the fully extended surfactant ion. Values of l = 20.4, 22.9 and 25.5 Å have been used for the surfactant ions Nm,1∅,1,1 with m = 12,14 and 16, respectively.

3. Results and Discussion

3.1. TERNARY W/O MICROEMULSIONS

3.1.1. *AOT/Water/n-Alkane.* Figure 2 shows the effect of temperature and alkane chain length on the AOT aggregation number, N, and on the interdroplet exchange rate constant, k_e. These results show that N and k_e increase in parallel, as either the temperature or the oil chain length increases. Note also that, at any temperature, N and k_e decrease when the linear n-octane is replaced by the branched isooctane. This further emphasizes the importance of the length of the oil alkyl chain on the values of N and k_e. Similar results have been obtained for values of ω and C other than those in Fig.2 [1].

The interdroplet exchange rate constant k_e is a parameter which is sensitive to the attractive interdroplet interactions. Indeed the process characterized by k_e corresponds to the opening of the surfactant layers separating the water cores of colliding droplets. Therefore the increase of k_e with the oil chain length or temperature indicates that the collisions between droplets become more efficient for interdroplet exchange of reactants and, thus, that the attractive interactions between droplets increase as the above two parameters are increased. Therefore the increase of k_e with the alkane chain length and temperature can be used to infer an increase in attractive interdroplet interactions. This conclusion is in agreement with the results obtained by Hou et al. [26] from

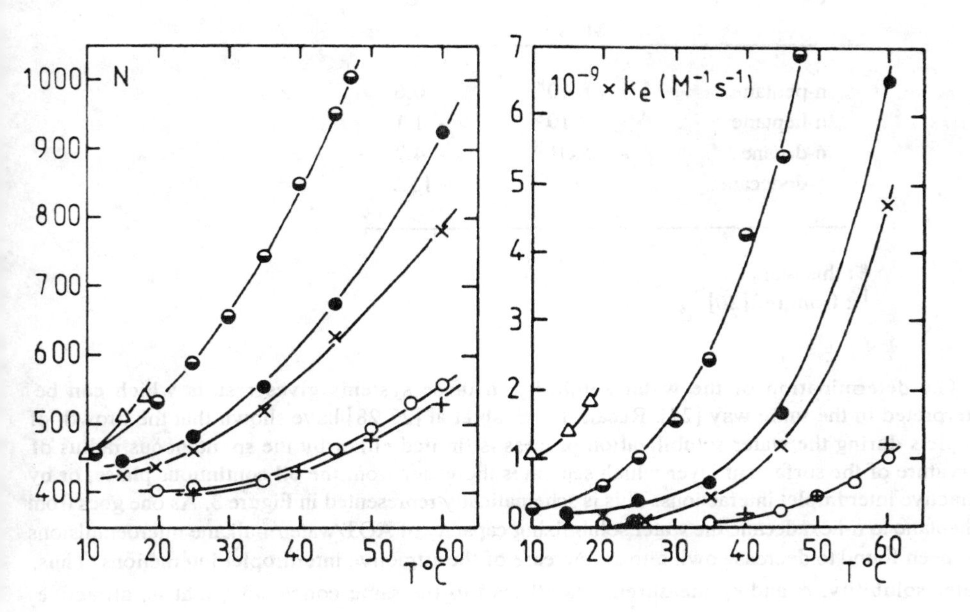

Figure 2. Variations of the surfactant aggregation number N and of the rate constant k_e for microemulsions with n-hexane (+), n-heptane (O),isooctane (X), n-octane (●), n-decane (⊖), and n-dodecane (Δ). C =0.182 M ; ω = 26.3.

QELS measurements. Thus Table 1 shows that the value of the virial coefficient α of the droplet diffusion coefficient D, from QELS, becomes more negative indicating an increase of attractive interdroplet interactions with the oil chain length, in agreement with the parallel increase of k_e, from TRFQ

TABLE 1. Variations of the interdroplet exchange rate constant (k_e) and the virial coefficient (α) of the droplet translational diffusion coefficient with the oil chain length for water/AOT/alkane microemulsions at $\omega = 20$.

Oil	k_e[a] $M^{-1}s^{-1}$	α[b]
n-pentane	$< 10^8$	- 0.6
n-heptane	$< 10^8$	- 1.1
n-decane	$5x10^8$	- 4.7
n-dodecane	$2x10^9$	- 13.9

[a] : this work
[b] : from ref.[26].

The determination of the water solubility in these systems gives results which can be interpreted in the same way [27]. Recall that Shah et al [27,28] have shown that the growth of droplets during the water solubilization process is limited either by the spontaneous radius of curvature of the surfactant layer which separates the water from the oil continuous phase, or by attractive interdroplet interactions. This is schematically represented in Figure 3. As one goes from n-heptane to n-hexadecane the water solubilizing capacity of AOT/water/n-alkane microemulsions has been found to decrease owing to an increase of the attractive interdroplet interactions. Thus, water solubility, α and k_e measurements all lead to the same conclusion, that is, attractive interdroplet interactions increase with the oil chain length.

The influence of the alkane chain length on the electrical conductivity, K, is shown in Figure 4. This figure represents the variations of K and also k_e as a function of ω for three water/AOT/n-alkane microemulsions differing by the nature of the n-alkane. The microemulsions with n-octane and n-decane present an electrical percolation phenomenon but there is not such an effect with n-heptane. Moreover, the onset of electrical percolation appears for an ω-value lower with n-decane than with n-octane. Remember that in microemulsions electrical percolation results from the formation of macroscopic (infinite) clusters of droplets where the droplets are in contact thereby allowing an easy motion of surfactant ions or counterions (see discussion below) over macroscopic distances. Percolation is favoured by attractive interactions between droplets, the larger these interactions, the lower the droplet concentration threshold for the onset of electrical percolation. In the experiments reported in Fig.4, where K varies as a function of ω it is via the increase of droplet size that the percolation phenomenon takes place. Thus, the results of electrical conductivity measurements are in agreement with the results from TRFQ, QELS and water solubility determination, namely the attractive interdroplet interactions increase as the oil chain length increases.

Percolation can be also induced by an increase of temperature as can be seen in Figure 5. An increase of k_e parallels the increase of K. Thus here again the TRFQ results agree with the conductivity results. Both indicate that the interdroplet interactions become more attractive as the temperature increases.

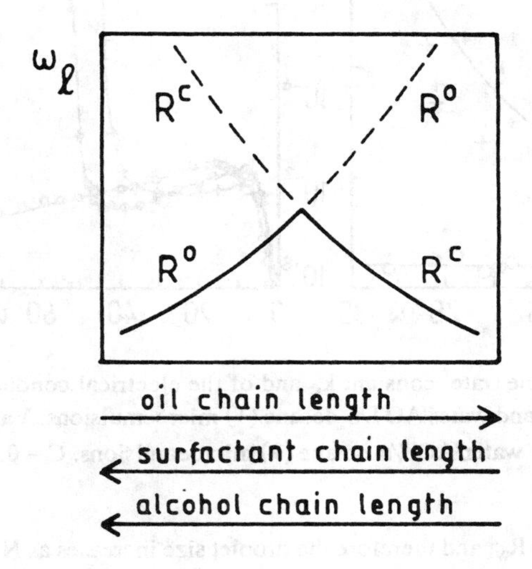

Figure 3. Schematic representation of the variation of the solubility of water (ω_l) in w/o microemulsions with various parameters, according to Shah et al [27,28]. The arrow indicates the direction of increase of the value of the parameter. For a given microemulsion, as ω increases, the growth of the droplet radius R is limited by the lowest of the two following radii which characterize the droplet : the radius of spontaneous curvature (R^o-branch) and the critical radius which originates from interdroplet attractive interactions (R^c-branch).

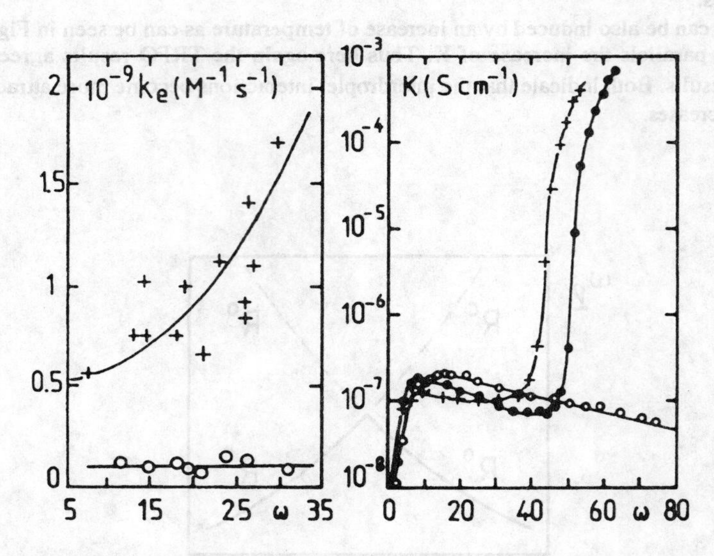

Figure 4. Variations of the rate constant k_e and of the electrical conductivity K with ω for water/AOT/n-heptane (O) and water/AOT/n-decane (+) microemulsions. Variation of the electrical conductivity K with ω for water/AOT/n-octane (•) microemulsions. C = 0.182 M. T = 25°C.

Equation 12 shows that R_w, and therefore the droplet size increases as N increases. The values of R_w corresponding to the N-values shown in Fig.2 have been given elsewhere together with the values of σ calculated from eq.13 [1].

A speculative explanation, based on geometrical considerations, has been given for the increase in droplet sizes and attractive interactions with the oil chain length or temperature [1].

Consider first the droplet size. As the length of the oil alkyl chain is increased, it becomes increasingly coiled and therefore its penetration in the surfactant layer which is made of closely-packed AOT alkyl chain becomes more difficult. The penetration of shorter oil molecules is easier and results in the formation of additional interfacial area. Since the experiments were performed at constant volume of water, this results in smaller but more numerous droplets. Similarly when the temperature is raised at constant oil chain length the increasing thermal motion reduces the oil penetration in the surfactant layer thereby resulting in a decrease of interfacial surface area and thus in an increase of droplet size. Note that considerations based on the packing ratio $v/a_M l$ (where v is the volume of the surfactant alkyl chain plus that of the oil per surfactant molecule in the interfacial layer, l is the length of the fully extended surfactant chain and a_M is the optimal surface area per surfactant head group) lead to the same conclusion. As more oil penetrates in the surfactant layer, v and thus the packing ratio increase and the droplet radius decreases.

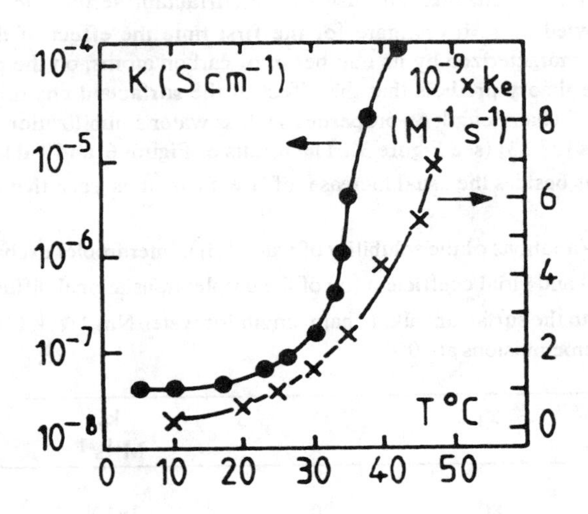

Figure 5. Variations of the rate constant k_e (X) and of the electrical conductivity K (•) with temperature for the water/AOT/n-decane microemulsions with C = 0.182 M and ω = 26.3.

We now turn to interdroplet interactions. The increases of k_e, $|\alpha|$ and K (percolation phenomenon) with the oil chain length or the temperature, all indicate an increase of interdroplet attractive interactions which can be explained in the following way. The extent of attractive interactions depends on the relative strength of the interactions between AOT tails of one droplet with AOT tails of another droplet, and the interactions between AOT tails and oil. The attractive interactions between droplets are understood to originate from the overlap of surfactant tails belonging to collided droplets [29-32]. As argued above, the longer the oil molecule or the higher the temperature, the more difficult for the oil molecules to orient parallel to the surfactant tails in order to maximize their interactions. On the contrary, the short AOT tails of two collided droplets are always more or less parallel to each other in the overlap region. Therefore, as the oil chain length or the temperature increases, the tail-tail interaction becomes stronger than the surfactant tail-oil molecule interaction. Consequently, attractive forces develop between colliding droplets, and k_e,$|\alpha|$ and K increase with the oil chain length or temperature. Notice that an increase

of droplet size produces an increase of the overlap region between two collided droplets which, in turn, results in an increase of attractive interactions. This effect is at the origin of the occurrence of the electrical percolation in Fig.4 which shows conductivity measurements carried out upon increasing water content at constant C, i.e., upon increasing droplet size.

3.1.2. *Cationic Surfactant/Water/Oil.* The microemulsions investigated were made of water/Nm,1∅,1,1,Cl/chlorobenzene. The use of the surfactant series Nm,1∅,1,1,Cl (see paragraph 2.2.2.) allowed us to investigate for the first time the effect of the length of the surfactant alkyl chain, characterized by its number m of carbon atoms, on the properties of the system. Recall that the theory predicts that the effect of the surfactant chain length should be opposite to that of the oil chain length on properties such as water solubilization, droplet size and interdroplet interactions [27,33] (see Figure 3). The results of Figure 6 and Table 2 fully confirm these expectations. Thus besides the usual increase of N with ω, it is seen that N decreases, the

TABLE 2. Variations of the solubility of water (ω_1), interdroplet exchange rate constant (k_e) and virial coefficient (α) of the droplet translational diffusion co-efficient with the surfactant alkyl chain length for water/Nm,1∅,1,1, Cl/chloro-benzene microemulsions at 20°C.

m	ω_1	ω	k_e $M^{-1}s^{-1}$	α
16	80	20	$2x10^8$	0
		40	$2x10^8$	0
14	65	10	$0.7x10^9$	- 3.9
		40	$3.6x10^9$	- 4.8
12	25	20	$5.2x10^9$	- 12.4

water solubility increases and the strength of the attractive interdroplet interactions (α) decreases as m is increased, all other parameters remaining unchanged. The theory also predicts that systems where the water solubility is determined by the strength of the attractive interdroplet interactions phase-separate into two w/o microemulsion phases of comparable volumes when the water solubility is exceeded. This prediction has been experimentally verified for the water/Nm,1∅,1,1,Cl/chlorobenzene systems [3]. Also the results show that at a given value of ω and for a given oil, the rate constant k_e increases very much as m is decreased (see Figure 6) which reflects an increase of interdroplet attractive interactions. Moreover the electrical conductivity data represented in Figure 6 show that percolation takes place for the systems with short chain surfactants(m=12 and 14) but not with the long chain surfactant, m=16. Also, the ω-value corresponding to the percolation threshold decreases with m. All of the results concerning k_e and the percolation characteristics show that the effect of the surfactant chain length is opposite to that of the oil chain length. However, when either of these two parameters is varied we find the

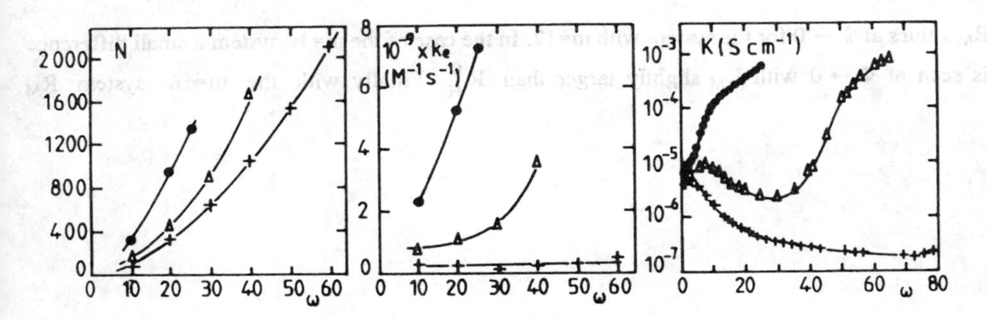

Figure 6. Variations of the surfactant aggregation number N, of the rate constant k_e, and of the electrical conductivity K with ω for water/ Nm,1Ø,1,1,Cl/chlorobenzene microemulsions with m = 12 (•), 14 (Δ) and 16 (+). C = 0.27 M. T = 20°C.

same correlation between size and k_e values from TRFQ, interdroplet interactions from QELS, water solubility, occurrence of percolation and percolation threshold from electrical conductivity. Thus an increase of oil chain length or a decrease of surfactant chain length, all other parameters being constant, brings about an increase of droplet size, interdroplet interactions, and rate of exchange of material between droplets, a decrease of water solubility and the occurrence of percolation, with a decreasing value of the percolation threshold.

As for the case of anionic surfactant microemulsions examined in paragraph 3.1.1. the changes of droplet size and interdroplet interactions can be explained at least qualitatively in terms of changes of packing ratio $v/a_M l$ and of penetration of a given oil in the interfacial surfactant layer, with the surfactant chain length. The reasoning is however somewhat more complex than in the case of AOT containing microemulsions. Indeed as m increases, the volume of the surfactant chain increases linearly with m (increase of v), whereas the penetration of the oil is increasingly hindered (decrease of v) owing to increasing coiling of the surfactant chain. These two effects are opposite and, therefore, the overall variation of v is probably small. The increasing coiling of the surfactant chain with m tends to decrease l, thereby resulting in an increase of packing ratio and thus in smaller droplets, since a_M is expected to change only little for a surfactant series with constant head group size. The coiling of the surfactant chains prevents their interpenetration upon collisions and an increase of m will therefore be accompanied by a decrease of interdroplet attractive interactions as is indeed observed [4]. The above interpretation rests on the assumption of a strong increase of surfactant chain coiling upon increasing m. The following QELS results strongly support this assumption. Figure 7 shows the variation of the hydrodynamic radius R_H calculated from eq.17 and of the droplet radius R_M calculated from eq.18 as a function of the droplet volume fraction Φ for the water/Nm,1ø,1,1,Cl/chlorobenzene microemulsions with m=12,14 and 16. It can be seen that the R_H values extrapolated to $\Phi = 0$ (R_H^o) are very close to the R_M values at $\Phi \rightarrow 0$ for the system with m=12. In the case of the m=14 system a small difference is seen at $\Phi \rightarrow 0$ with R_M slightly larger than R_H^o. Finally with the m=16 system R_M

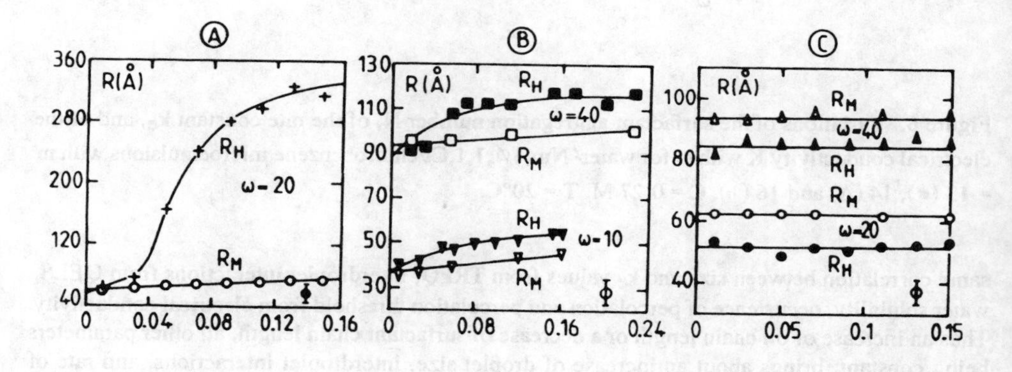

Figure 7. Variations of the hydrodynamic radius R_H and of the overall droplet radius R_M with the volume fraction Φ of the dispersed phase for water/Nm,1ø,1,1,Cl/chlorobenzene microemulsions with m = 12 (A), m = 14 (B), and m = 16 (C). T = 20.5±0.5°C.

is larger than R_H^o by 8-10 Å. Recall that R_H^o includes a solvation layer constituted by the solvent molecules moving with the droplets during their brownian jumps. Therefore R^{o}_{H} should be larger than R_M, if the surfactant chains are fully extended, as was assumed in the calculation of R_M (eq.18). The fact that R^{o}_{H} is found to be slightly smaller than R_M for m=14 and well below R_M for m=16 is clearly reflecting the rapidly increasing coiling of the surfactant chain with increasing m, responsible for the remarkable changes of properties described above.

The effect of the nature of the oil on N, k_e, α and K has also been investigated by means of TRFQ [4], QELS [34] and electrical conductivity [34] measurements for the system water/N16,1∅,1,1,Cl/oil where the oil was either chlorobenzene or benzene. The results clearly show that as benzene substitutes chlorobenzene, droplet sizes, attractive interdroplet interactions and k_e-values are increased much and that electrical percolation then occurs. However the parameter which determines the effect of the oil does not appear to be solely the oil molecular volume. The oil dielectric constant seems to play an equally important if not superior role [4].

The effect of the nature of the surfactant head-group on droplet sizes, interdroplet interactions and water solubility has also been studied. The results are reported elsewhere [3,4].

3.2. QUATERNARY W/O MICROEMULSIONS

3.2.1. *Effect of the Alcohol Chain Length.* The variations of the water solubility ω_l with the alcohol chain length, characterized by its number n_c of carbon atoms in the microemulsions water/chlorobenzene/alcohol/dodecyl, tetradecyl and hexadecyl trimethylammonium bromides(DTAB, TTAB and HTAB, respectively) are shown in Figure 8 at constant molar concentration ratio z = [alcohol]/C = 2.5. As the alcohol chain length increases ω_l goes through a maximum. Similar results have been found with other systems [27,35]. It has been shown [27,28] (see Fig.3) that in systems at the left hand side with respect to the maximum of ω_l, the droplet growth is limited by the spontaneous radius of curvature,R^o, of the surfactant layer, as ω is increased. In systems at the right hand side of the maximum, droplet growth is limited by attractive interactions between droplets. For these systems the limiting droplet radius R^c, corresponding to ω_l, is lower than R^o. Thus the effect of alcohol chain length is qualitatively similar to that of the surfactant chain length.

Figures 3 and 8 indicate that at constant ω the droplet size should increase as the alcohol chain length decreases. Indeed, it is reasonable to assume that the spontaneous radius of curvature of the interfacial layer (surfactant+alcohol) determines the size taken by the droplets for any value of ω below ω_l, even for systems where the droplet growth is limited by interdroplet interactions (R^c-branch of the plots in Figs.3 and 8). Experimentally an increase of N and thus of R_w, at constant

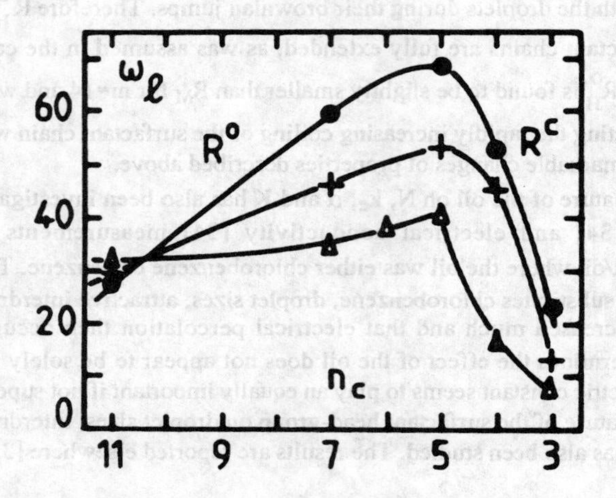

Figure 8. Variations of the water solubility (ω_1) with the alcohol chain length (n_c = number of carbon atoms in the alcohol alkyl chain) for water/DTAB/chlorobenzene (Δ), water/TTAB/chlorobenzene (+) and water/HTAB/chlorobenzene (•) microemulsions. C = 0.27 M ; z = 2.5; T = 20°C.

ω is indeed found as the alcohol length decreases as shown in Figs.9 and 10 which give the results for quaternary microemulsions based on DTAB and HTAB, respectively. Note that for all of the systems investigated, we observed the expected increase of N, and therefore of R_w, with ω. The value of R_w and σ calculated for these systems via eqs.15 and 13 are given elsewhere [36].

The variations of the electrical conductivity with ω in Figs.9 and 10 can be correlated to the water solubility results in Fig.8. Indeed the variations of K with ω show that the electrical percolation occurs only in the microemulsions containing the shortest alcohol molecules, that is precisely those belonging to the R^c-branch of the water solubility curves in Fig.8. Moreover as the alcohol chain length decreases ω_l decreases and the onset of electrical percolation appears for decreasing values of ω. Both results indicate an increase of attractive interactions between droplets as the alcohol chain length decreases. Note also that for all of the alcohols investigated one observes a decrease of K, at constant ω, as the alcohol chain length increases (see Figs.9 and 10). This is easily understood if one recalls that the increase in R^c in Figs.3 and 8 reflects the decrease in interdroplet attractive interactions even for systems for which ω_l falls on the R^o-branch, i.e. systems made with alcohols having in the present study, more than five carbon atoms. Therefore the decrease of electrical conductivity with increasing alcohol chain length is again due to a decrease in interdroplet attractive interactions.

The increase of interdroplet attractive interactions as the alcohol chain length decreases is also evidenced from the variation of k_e shown in Figs.9 and 10. It is seen that at constant ω, k_e

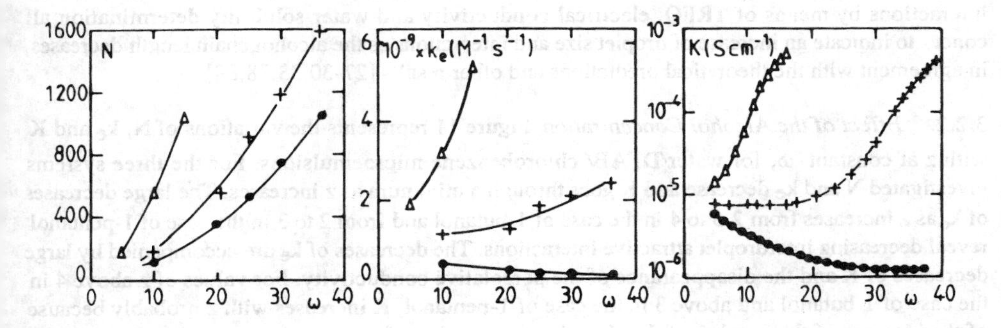

Figure 9. Variations of the surfactant aggregation number N, of the rate constant k_e, and of the electrical conductivity K with ω for water/ DTAB/chlorobenzene/alcohol microemulsions with 1-butanol (Δ) ; 1- pentanol (+) and 1-hexanol (•). C = 0.27 M ; z = 2.5 ; T = 20°C.

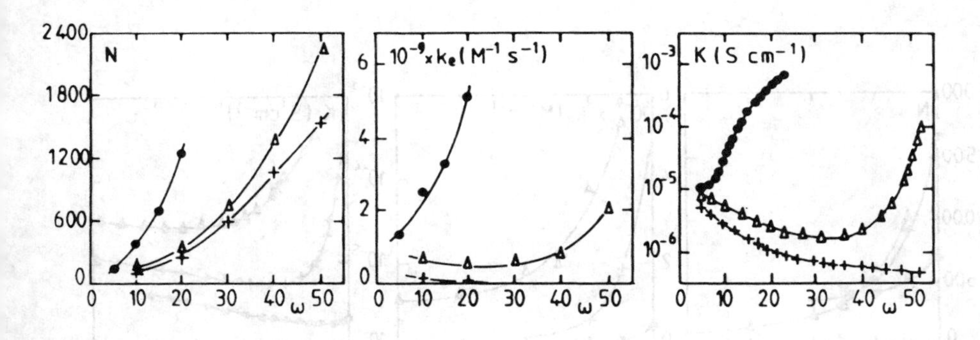

Figure 10. Variations of the surfactant aggregation number N, of the rate constant k_e, and of the electrical conductivity K with ω for water/ HTAB/chlorobenzene/alcohol microemulsions with 1-propanol (•) ; 1- butanol (Δ) and 1-pentanol (+). C = 0.27 M ; z = 2.5 ; T = 20°C.

increases as the alcohol chain length decreases. An increase of the rate of exchange of material between colliding droplets as the alcohol chain length decreases has also been found by Atik and Thomas [37] for potassium oleate based quaternary microemulsions.

The results of the studies of the effect of alcohol chain length on droplet size and interdroplet interactions by means of TRFQ, electrical conductivity and water solubility determination all concur to indicate an increase of droplet size and interactions as the alcohol chain length decreases, in agreement with the theoretical predictions and other results [27-30,33,38,39].

3.2.2. *Effect of the Alcohol Concentration.* Figure 11 represents the variations of N, k_e and K with z at constant ω, for water/DTAB/ chlorobenzene microemulsions. For the three systems investigated N and k_e decrease and K goes through a minimum as z increases. The large decreases of k_e as z increases from 2.5 to 4 in the case of 1-butanol and from 2 to 3 in the case of 1-pentanol reveal decreasing interdroplet attractive interactions. The decreases of k_e are accompanied by large decreases of K and the disappearance of the percolative conductivity. For values of z above 4 in the case of 1-butanol and above 3 in the case of 1-pentanol, K increases with z probably because of the increase of the number of droplets, i.e. the number of charge carriers, with z. Indeed, the decrease of N in Fig.11 corresponds to an increase of the droplet concentration since ω is constant. Thus the minimum observed in the variation of K with z seems due to two antagonistic effects : the large decrease of the percolative electrical conductivity which occurs at low z and the increase of droplet concentration which affects slightly K and is apparent only in the z-range where no percolative conductivity is present.

The decrease of the percolative electrical conductivity and therefore of the interdroplet attractive interactions upon increasing z can be attributed to the decrease in droplet size (see Fig.11). Indeed

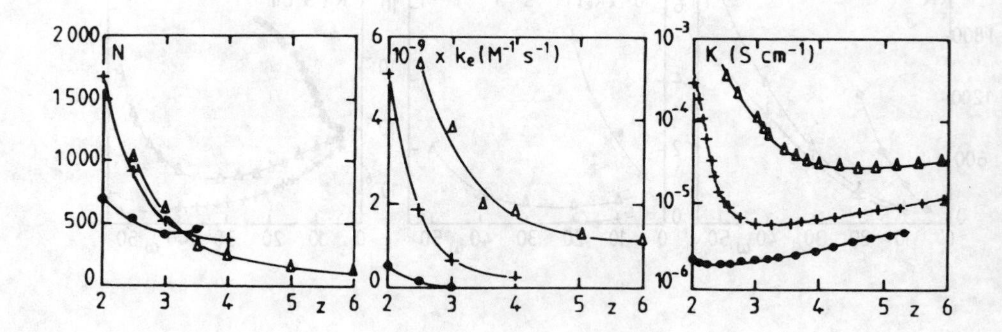

Figure 11. Variations of the surfactant aggregation number N, of the rate constant k_e, and of the electrical conductivity K with z for water/ DTAB/chlorobenzene/alcohol microemulsions with 1-butanol (Δ) at ω= 15 ; 1-pentanol (+) and 1-hexanol (•) at ω = 25. C = 0.27 M ; T = 20°C

collided droplets then have a smaller overlap region and therefore interact more weakly. The decrease of droplet size upon increasing z arises from the increase in the surface area available to form the droplet interfacial layer since part of the added alcohol molecules go to the interface. Thus, here again a correlation appears between variation of droplet size, value of k_e and electrical percolation. As z increases the droplets become smaller which produces a decrease of the interdroplet attractive interactions and, in turn, of k_e and of the percolative conductivity. However for a given droplet size (see for instance the N- values for the systems with 1-butanol and 1-pentanol at z = 2.5) the percolative conductivity is larger for the system with shorter alcohol chain length. This is probably the result of a larger overlap area of the interfacial layers of collided droplets with the alcohol having shorter chain length.

3.3. ELECTRICAL CONDUCTIVITY

Two aspects of the electrical conductivity results must be emphasized.

The first one concerns the correlation between the interdroplet exchange rate constant, k_e, and the onset of electrical percolation. In all of the systems investigated we have seen that the occurrence of electrical percolation is accompanied by an increase of k_e-value. A close examination of the results (see Figs.4,5,6,9,10 and 11) reveals that percolation takes place upon increasing ω and temperature or decreasing z only when k_e has reached a sufficiently high value, of about $(1\text{-}2)\times10^9\ M^{-1}s^{-1}$. The systems where the k_e-value remains below this limit show no percolation, even at very high values of ω or low values of z.

The requirement that k_e be large enough for the occurrence of electrical percolation appears to apply to a wide variety of w/o microemulsions. In fact, all of the systems where k_e and K measurements were available have been found to fulfil this condition . We have previously mentioned that the substitution of benzene for chlorobenzene as oil in the water/N16,1∅,1,1,Cl/oil system increases the attractive interactions between droplets. We have found for the water/N16,1∅,1,1,Cl/benzene microemulsions k_e values larger than $1.5\times10^9\ M^{-1}s^{-1}$ for $\omega > 10$ [4]. On the other hand Chatenay et al.[34] have reported the occurrence of electrical percolation in this same system. Notice that the results in Fig.6 show low values of k_e, below $(1\text{-}2)\times10^9\ M^{-1}s^{-1}$ and no electrical percolation for the water/N16,1∅,1,1,Cl/chlorobenzene microemulsions. Geiger and Eicke [40] have reported the occurrence of percolation as the temperature of water/AOT/isooctane microemulsions is raised. The extrapolation of their data yields a threshold temperature of 47°C at $\omega = 26$. The corresponding k_e- value is $(1\text{-}2)\times10^9\ M^{-1}s^{-1}$ (see Fig.2). For the system water/AOT/n-decane with $\omega = 25$, Dutkievicz and Robinson [41] reported the onset of electrical percolation upon increasing temperature at 17°C. Our results in Fig.2 for the same system with $\omega = 26.3$ show that k_e is equal to $(1\text{-}2)\times10^9\ M^{-1}s^{-1}$ at T = 17°C.

The above results pertaining to ternary and quaternary w/o microemulsions clearly show that electrical conductivity percolation occurs whenever the rate constant k_e for interdroplet collisions with exchange of material becomes larger than $(1\text{-}2)\times10^9\ M^{-1}s^{-1}$, irrespective of the nature of the surfactant (anionic, cationic), oil (alkanes, arenes), temperature, size of the droplet (N, ω) and volume fraction of the dispersed phase (C, ω). The results represented in Fig. 12 reveal that the condition $k_e > (1\text{-}2)\times10^9\ M^{-1}s^{-1}$ also holds for the occurrence of electrical percolation in w/o microemulsions made of five components :water/HTAB/chloroform/isooctane/cetylbromide [5]. Notice that the system without cetylbromide shows no electrical percolation and its k_e-values are below $10^8\ M^{-1}s^{-1}$. It seems therefore that the condition $k_e > (1\text{-}2)\times10^9\ M^{-1}s^{-1}$ for the occurrence of electrical percolation as a parameter like the volume fraction of the dispersed phase or the temperature is increased will remain valid for a great variety of w/o microemulsions of various compositions.

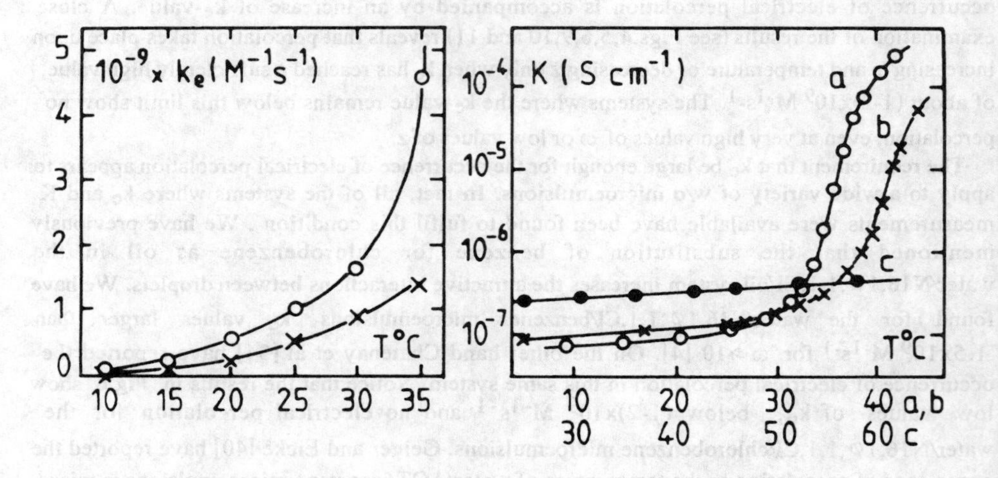

Figure 12. Variations of the rate constant k_e and of the electrical conductivity K with temperature for water/HTAB/cetylbromide (CB)/chloroform ($CHCl_3$)/isooctane (iC_8)($CHCl_3/iC_8$,2/1,v/v) microemulsions with [HTAB] = 0.15 M, [CB] = 0.77 M and ω = 20 (X) and 25 (O). Variation of the electrical conductivity K with temperature for the water/HTAB/ chloroform/isooctane ($CHCl_3/iC_8$,2/1,v/v) microemulsions (•) with [HTAB] = 0.15 M and ω = 25.

The correlation found between the value of k_e and the occurrence of electrical conductivity percolation has some implications on the mechanism of electrical conductivity in w/o microemulsions. Various mechanisms have been proposed to explain the large conductivity of these systems above percolation threshold [41-44]. One of the proposed mechanisms involves the motion of surfactant ions on the surface of the water droplets and surfactant ion hopping from droplet to droplet in the droplet clusters present in the system or upon droplet collisions[42,43]. Such a mechanism, however, does not explain why percolation is observed only when $k_e > (1\text{-}2)\times10^9\ M^{-1}s^{-1}$. Indeed the rate constant k_e is associated with the interdroplet transfer of reactants which are relatively large ions ($Ru(bpy)_3^{2+}$, $Fe(CN)_6^{3+}$, MV^{2+}). This transfer certainly involves at least a partial merging of the collided droplets. Therefore the above condition on k_e is better understood if one assumes that above the experimental threshold the percolative conductivity is due to the motion of counterions within transient water tubes formed in droplet clusters upon opening of the surfactant layers separating adjacent water cores. Note that the opening of the surfactant layers between two collided droplets should not be visualized as resulting in droplet coalescence. Rather, it yields a droplet dimer with connected aqueous cores [41,44,45]. The interpenetration of the surfactant layers which takes place prior to their opening is directly related to interdroplet attractive interactions which, as shown above, increase as k_e increases.

The second aspect of electrical conductivity measurements which must be emphasized is that the conductivity is directly related to the values of α and k_e. This fact can be used to make a number of qualitative predictions about the variations of α k_e and also ω_l with a given parameter, such as the surfactant, oil or alcohol chain length or the temperature, from the observed change of conductivity with this parameter. This is illustrated by the results in Fig.9, for example : the conductivity curves obtained with various alcohols in water/DTAB/chlorobenzene/alcohol microemulsions show that K increases as the alcohol chain length decreases at constant ω. From this result it can be predicted that, at constant ω : (i) k_e increases and a becomes more negative as the alcohol chain length decreases and (ii) the value of k_e is around $(1\text{-}2)\times10^9\ M^{-1}s^{-1}$ at the onset of electrical percolation. Notice that in Fig.9, the electrical percolation occurs only for the microemulsions with 1-butanol and 1-pentanol and that the onset of electrical percolation corresponds to ω values increasing with the alcohol chain length. This result suggests that the qualitative variation of ω_l with the alcohol chain length can be also predicted. The results for the shorter alcohols will fall on the R^c-branch with a decrease of ω_l as the alcohol chain length decreases. The results for longer alcohols will fall on the R^o-branch with a decrease of ω_l as the alcohol chain length increases (see Fig.3).

Another prediction which may be made from conductivity measurements is the relative variation of droplet size with a given parameter. Indeed in all of the examples shown above the surfactant aggregation number N and, therefore, the droplet size (R_w) varied qualitatively in the same manner as the rate constant k_e. Thus, since the variations of k_e can be predicted from electrical conductivity measurements, the relative variations of N and R_w can be predicted as well.

Finally it must be mentioned that other parameters, besides those examined here, can affect the droplet sizes and dynamics, as for instance the ionic strength inside the water pool, the dielectric constant of the oil, the nature of the cosurfactant. From what has been put forward here concerning the predictions which can be made from simple conductivity measurements we suggest to first perform such measurements before undertaking measurements with more sophisticated equipment. The conductivity measurements will probably reveal whether the variation of the parameter investigated induces or not large modifications in droplet size and interactions.

4. Conclusions

Time-resolved fluorescence quenching, quasielastic light scattering, electrical conductivity and water solubility measurements have been used to show that droplet size and interdroplet attractive interactions increase as the oil chain length (alkane) and the temperature increase or as the surfactant or alcohol chain length and concentration decrease. It has been shown that an electrical percolation phenomenon occurs upon increasing ω and temperature or decreasing z only when the interdroplet exchange rate constant k_e has reached a sufficiently high value of about $(1\text{-}2)\times10^9$ $M^{-1}s^{-1}$. This requirement has been found to be valid irrespective of the nature of the oil (alkanes, arenes), surfactant (cationic, anionic), alcohol, temperature, droplet size and volume fraction of the dispersed phase. Good correlations have been obtained between the results from the different methods used here. This has allowed us to show how the results from simple electrical conductivity measurements as a function of one of the above parameters which characterize the microemulsion can be used to predict the qualitative variations of droplet size, interdroplet attractive interactions and exchange rate constant k_e, with this parameter.

5. References

[1] Lang, J., Jada, A. and Malliaris, A. (1988) "Structure and dynamics of water-in-oil droplets stabilized by sodium bis(2-ethylhexyl)sulfosuccinate", J. Phys. Chem. 92, 1946-1953.

[2] Jada, A., Lang, J. and Zana, R. (1989) "Relation between electrical percolation and rate constant for exchange of material between droplets in water-in-oil microemulsions", J. Phys. Chem.93, 10-12.

[3] Jada, A., Lang, J. and Zana, R. (1990) "Ternary water-in-oil microemulsions made of cationic surfactants, water, and aromatic solvents. 1. Water solubility studies", J. Phys. Chem. 94, 381- 387.

[4] Jada, A., Lang, J., Zana, R., Makhloufi, R., Hirsch, E. and Candau, S.J. (1990) "Ternary water-in-oil microemulsions made of cationic surfactants, water, and aromatic solvents. 2. Droplet sizes and interactions and exchange of material between droplets",J. Phys. Chem. 94, 387-395.

[5] Lang, J., Mascolo, G., Zana, R., Luisi, P.L. (1990) "Structure and dynamics of cetyltrimethylammonium bromide water-in-oil microemulsions", J. Phys. Chem. 94, 3069-3074.

[6] Lang, J. (1990) "The time-resolved fluorescence quenching method for the study of micellar systems and microemulsions : Principle and limitations of the method", this volume.

[7] Zana, R. (1987) "Luminescence probing methods", in R. Zana (ed.),Surfactant Solutions: New Methods of Investigation, Marcel Dekker,New-York, pp 241-294.

[8] Atik, S.S. and Thomas, J.K. (1981) "Transport of photoproduced ions in water in oil microemulsions: movement of ions from one water pool to another", J. Am. Chem. Soc. 103, 3543-3550.

[9] Pfeffer, G., Lami, H., Laustriat, G. and Coche, A. (1963)"Détermination des constantes de temps de scintillateurs", C.R.Hebd. Séances Acad. Sci. 257, 434-437.

[10] Infelta, P.P., Grätzel, M. and Thomas, J.K. (1974) "Luminescence decay of hydrophobic molecules solubilized in aqueous micellar systems. A kinetic model", J. Phys. Chem. 78, 190-195.

[11] Tachiya, M. (1975) "Application of a generating function to reaction kinetics in micelles. Kinetics of quenching of luminescent probes in micelles", Chem. Phys. Lett. 33, 289-292.

[12] Atik, S.S., Nam, M. and Singer, L. (1979) "Transient studies on intramicellar excimer formation. A useful probe of the host micelle", Chem. Phys. Lett. 67, 75-80.

[13] Atik, S.S. and Thomas, J.K. (1981) "Transport of ions between water pools in alkanes", Chem. Phys. Lett. 79, 351-354.

[14] Dederen, J.C. and Van der Auweraer, M. and De Schryver, F.C.(1979) "Quenching of 1-methylpyrene by Cu^{2+} in sodium dodecylsulfate. A more general kinetic model", Chem. Phys. Lett.68,451-454.

[15] Grieser, F. and Tausch-Treml, R. (1980) "Quenching of pyrene fluorescence by single and multivalent metal ions in micellar solutions", J. Am. Chem. Soc. 102, 7258-7264.

[16] Dederen, J.C., Van der Auweraer, M. and De Schryver, F.C. (1981)"Fluorescence quenching of solubilized pyrene and pyrene derivatives by metal ions in SDS micelles", J. Phys. Chem. 85,1198-1202.

[17] Grieser, F. (1981) "The dynamic behaviour of I^- in aqueous dodecyltrimethylammonium chloride solutions. A model for counter-ion movement in ionic micellar systems", Chem. Phys. Lett. 83, 59-64.

[18] Löfroth, J.-E. and Almgren, M. (1982) "Quenching of pyrene fluorescence by alkyl

iodides in sodium dodecyl sulfate micelles",J. Phys. Chem. 86, 1636-1641.

[19] Croonen, Y., Geladé, E., Van der Zegel, M., Van der Auweraer, M.,Vandendriessche, H., De Schryver, F.C. and Almgren, M. (1983)"Influence of salt, detergent concentration and temperature on the fluorescence quenching of 1-methylpyrene in sodium dodecyl sulfate with m-dicyanobenzene", J. Phys. Chem. 87, 1426-1431.

[20] Malliaris, A., Lang, J., Sturm, J. and Zana, R. (1987)"Intermicellar migration of reactants : effect of additions of alcohols, oils and electrolytes", J. Phys. Chem. 91, 1475-1481.

[21] Fletcher, P.D.I. (1988) "Time-resolved fluorescence study of the structure and dynamics of the cubic I1 lyotropic mesophase of dodecyltrimethylammonium chloride", Mol. Cryst. Liq. Cryst. 154,323-333.

[22] Luo, H., Boens, N., Van der Auweraer, M., De Schryver, F.C. and Malliaris, A. (1989) "Simulations analysis of time-resolved fluorescence quenching data in aqueous micellar systems in the presence and absence of added alcohol", J. Phys. Chem. 93, 3244-3250.

[23] Lang, J., Zana, R. and Candau, S. (1987) "Study of intermicellar migration through reactions of fragmentation-coagulation. Dynamics of micellar systems", Ann. Chim. (Rome), 77, 103-115.

[24] Almgren, M., Löfroth, J.-E. and Van Stam, J. (1986) "Fluorescence decay kinetics in monodisperse confinements with exchange of probes and quenchers", J. Phys. Chem. 90, 4431-4437.

[25] Almgren, M., Van Stam, J., Swarup, S. and Löfroth, J.-E. (1986) "Structure and transport in the microemulsion phase of the system Triton X-100-toluene-water", Langmuir, 2, 432-438.

[26] Hou, M.J., Kim, M. and Shah, D.O. (1988) "A light scattering study on the droplet size and interdroplet interaction in microemulsions of AOT-oil-water system", J. Colloid Interface Sci. 123, 398-412.

[27] Hou, M.J. and Shah, D.O. (1987) "Effects of the molecular structure of the interface and continuous phase on solubilization of water in water/oil microemulsions", Langmuir, 3, 1086-1096.

[28] Leung, R. and Shah, D.O. (1987) "Solubilization and phase equilibria of water-in-oil microemulsions. 1. Effects of spontaneous curvature and elasticity of interfacial films", J. Colloid Interface Sci. 120, 320-329.

[29] Lemaire, B., Bothorel, P. and Roux, D. (1983) "Micellar interactions in water-in-oil microemulsions. 1. Calculated interaction potential", J. Phys. Chem. 87, 1023-1028.

[30] Brunetti, S., Roux, D., Bellocq, A.M., Fourche, G. and Bothorel,P.(1983) "Micellar interactions in water-in-oil microemulsions",J. Phys. Chem. 87, 1028-1034.

[31] Huang, J.S., Safran, S.A., Kim, M.W., Grest, G.S., Kotlarchyk, M.and Quinke, N. (1984) "Attractive interactions in micelles and microemulsions", Phys. Rev. Lett. 53, 592-595.

[32] Huang, J.S. (1985) "Surfactant interactions in oil continuous microemulsions", J. Chem. Phys. 82, 480-484.

[33] Mukherjee, S., Miller, C.A. and Fort, Jr., M. (1983) "Theory of drop size and phase continuity in microemulsions. 1. Bending effects with uncharged surfactants", J. Colloid Interface Sci. 91,223-243.

[34] Chatenay, D., Urbach, W., Cazabat, A.M. et Langevin, D. (1985) "Onset of droplet aggregation from self-diffusion measurements in microemulsions", Phys. Rev. Lett. 54, 2253-2256.

[35] Bansal, V.K., Shah, D.O. and O'Connell, J.P. (1980) "Influence of alkyl chain length compatibility on microemulsion structure and solubilization", J. Colloid Interface Sci. 75, 462-475.

[36] Lalem, N., Lang, J. and Zana, R. (1990) "Quaternary water-in-oil microemulsions. 1. Effect of alcohol chain length and concentration on droplet size and exchange of material between droplets" in preparation.

[37] Atik, S.S. and Thomas, J.K. (1981) "Abnormally high ion exchange in pentanol microemulsions compared to hexanol microemulsions", J. Phys. Chem. 85, 3921-3924.

[38] Leung, R. and Shah, D.O. (1987) "Solubilization and phase equilibria of water-in-oil microemulsions. 2. Effects of alcohols, oils, and salinity on single-chain surfactant systems", J. Colloid Interface Sci. 120, 330-344.

[39] Roux, D., Bellocq, A.M. and Bothorel, P. (1984) "Effect of the molecular structure of components on micellar interactions in microemulsions", in K.L. Mittal and B. Lindman (eds.), Surfactants in Solution, Plenum Press, New York, pp.1843-1865.

[40] Geiger, S. and Eicke, H.F. (1986) "The macrofluid concept versus the molecular mixture: A spin-echo-NMR study of the water/Aerosol OT/oil system", J. Colloid Interface Sci. 110, 181-187.

[41] Dutkiewicz, E. and Robinson, B.H. (1988) "The electrical conductivity of a water-in-oil microemulsion system containing an ionic surfactant. Part I. Temperature effect", J. Electroanal. Chem. 251, 11-20.

[42] Hilfiker, R., Eicke, H.F., Geiger, S. and Furler, G. (1985) "Optical studies of critical phenomena in macrofluid like three component microemulsions", J. Colloid Interface Sci. 105, 378-387.

[43] Bhattacharya, S., Stockes, J.P., Kim, M.W. and Huang, J.S. (1985) "Percolation in an oil-continuous microemulsion", Phys. Rev. Lett. 55,1884-1887.

[44] Matthew, C., Patanjali, P.K., Nabi, A. and Maitra, A. (1988) "On the concept of percolative conduction in water-in-oil microemulsions", Colloids Surf. 30, 253-263.

[45] Fletcher, P.D.I. and Robinson, B.H. (1981) "Dynamic processes in water-in-oil microemulsions", Ber. Bunsen-Ges. Phys. Chem. 85, 863-867.

THE INFLUENCE OF THE DISTRIBUTION OF SALT ON THE PHASE BEHAVIOUR OF MICROEMULSIONS WITH IONIC SURFACTANTS

G.A. VAN AKEN,
Van 't Hoff Laboratory,
University of Utrecht,
Padualaan 8,
3584 CH Utrecht,
The Netherlands.

ABSTRACT. Partitioning of salt between the microemulsion phase and the excess aqueous phase is studied for Winsor II type microemulsion systems. Double layer theory is used to explain the observed effect that salt is expelled from the interior of the water-rich regions in the microemulsion phase. The resulting increase of the salt concentration in the excess water phase strongly influences the phase behaviour of the microemulsion system.

1. Introduction

Many authors [1-6] observed that the concentration of salt in the excess water phase of Winsor II or III microemulsion systems is significantly higher than the salt concentration in the initial water phase before mixing. Meanwhile the concentration of salt inside the water rich regions dispersed in the microemulsion phase is lower than in the excess water phase. This difference is caused by the extremely large ratio of the interfacial area of the water pools and their volumes, so that interfacial effects become important. Salt is negatively adsorbed in the electrical double layer of the interface, and therefore is "expelled" from the dispersed water pools into the excess phase. Because the phase behaviour of microemulsion systems depends on the concentration of salt in the aqueous phase, the initial solution of salt in water may not be regarded as a pseudo-component of the system. In fact the salt expulsion appears to be of major importance to the ultimate interpretation of the phase behaviour.

2. Theory

The presence of the charged heads of the surfactant ions in the interface of the water droplets inside the microemulsion phase results in the establishment of an electrical double layer at this interface. The counter-ions (sodium in this study), are attracted to the charged interface. The co-ions (chloride in this study), are expelled from the charged interface, and are said to be negatively adsorbed. An accurate analytical expression for $\Gamma_{co\text{-}ion}$ for 1:1-electrolytes, based on the Poisson-Boltzmann equation inside a sphere of radius a, was derived by van Aken *et al.* [7] . Defining the scaled adsorption as follows $\mathcal{A}$

D. M. Bloor and E. Wyn-Jones (eds.),
The Structure, Dynamics and Equilibrium Properties of Colloidal Systems, 279–294.

$$\Gamma_{\text{co-ion}} = -2\,\kappa^{-1}\,c_{el}\mathcal{A}, \tag{1}$$

$\mathcal{A}$ is given by

$$\mathcal{A} = |p| + 1 - q + \frac{1}{\kappa a}\left(-2\,\frac{q-1}{q}\right) + \left(\frac{1}{\kappa a}\right)^2 \left[-1 - \frac{2}{q(q+1)} + \frac{2}{q^3} + \frac{2}{q}\ln\left(\frac{1+q}{2}\right) + \int_{z=\frac{2}{q+1}}^{z=1} \frac{\ln z}{z-1}\,dz\right], \tag{2}$$

where κ is the inverse of the Debye screening length,

$$\kappa = \left(\frac{2F^2\,c_{el}}{\varepsilon_r\,\varepsilon_o\,RT}\right)^{1/2} \tag{3}$$

The parameters p and q are related to the electrolyte concentration c_{el} and to the surface charge density σ of the charged surface, by

$$p = \frac{\sigma}{\sqrt{8\,\varepsilon_r\,\varepsilon_o\,RT\,c_{el}}}, \tag{4}$$

and

$$q = \sqrt{p^2 + 1}\;. \tag{5}$$

We assume that the dielectric permittivity $\varepsilon_r\varepsilon_o$ is a constant and equals the value for pure water. It was shown [7] that eq.(2) yields results that lie within 2% of exact numerical results for radii with $\kappa a \geq 4$. For the present discussion we will from now on assume negligible curvature of the charged interface, by setting $1/\kappa a = 0$ in eq.(2). For the present discussion this approximation is justified, because it does not introduce large errors, and leads to the same qualitative results. Elsewhere [8] we applied eq.(2) in its untruncated form to interpret the distribution of salt in Winsor II phase equilibria at several salt concentrations and for pentanol and heptanol as the cosurfactant.

On assuming complete ionization of the adsorbed anionic surfactant molecules, the surface charge density σ equals

$$\sigma = -F\Gamma_{sa}, \tag{6}$$

where Γ_{sa} is the surface excess concentration or adsorption density of the surfactant anion. Substitution of eq.(6) into eqs.(1)-(5) yields an expression for the co-ion adsorption in terms of experimentally accessible variables

$$\Gamma_{co\text{-}ion} = -2\kappa^{-1} c_{el} - \frac{1}{2}\Gamma_{sa} + \sqrt{\left(2\kappa^{-1} c_{el}\right)^2 + \left(\frac{1}{2}\Gamma_{sa}\right)^2} \,. \tag{7}$$

The principal feature of the co-ion adsorption is easily found by setting $\Gamma_{sa} >> 4\kappa^{-1} c_{el}$ in eq.(7), yielding the limiting expression

$$\Gamma_{co\text{-}ion} = -2c_{el}\,\kappa^{-1}. \tag{8}$$

The negative adsorption is then proportional to the square root of the salt concentration, and is seen to equal the amount of salt enclosed in a film of unit cross-section and of thickness twice the Debye length extending from the planar surface.

In the Winsor II microemulsion system, the co-ions that are negatively adsorbed at the internal interface of the microemulsion phase are expelled to the center of the water droplets and to the excess water phase. As this aqueous phase must remain electroneutral, an equal amount of counter-ions accompanies the transport of the co-ions. As a result the salt concentration in the excess water phase will exceed the mean salt concentration in the water pools of the microemulsion phase. This partitioning of salt between the two phase of the Winsor II system is quantified by writing the mass balance

$$c^{i}_{el} V^{i}_{w} = c_{el} V^{ex}_{w} + n^{M}_{el} \quad , \tag{9}$$

where c^{i}_{el} is the initial concentration of salt in the brine of volume V_w^{i} used to prepare the two phase system. After equilibration this volume separates into two volumes, V_w^{ex} the volume of the excess water phase, and V_w^{M} the volume of brine dispersed in the microemulsion phase. The dispersed phase contains n_{el}^{M} moles of salt. In writing eq.(9) we have not taken into account the small solubility of cosurfactant (pentanol) in the excess water phase nor the uptake of salt-free water in the continuous oil phase. The former will increase and the latter will decrease V_w^{ex}. Both effects are however small and easily corrected for if necessary.

The equilibrium amount of salt in the dispersed phase is related to the adsorption density of the co-ions

$$n^{M}_{el} = V^{M}_{w}\, c_{el} + \Gamma_{co\text{-}ion} \cdot A \,, \tag{10}$$

where A is the total area of the internal interface of the microemulsion phase. To find A we divide the total amount of surfactant by the adsorption density of the surfactant

$$A = \frac{n_{sa}}{\Gamma_{sa}} \quad . \tag{11}$$

This expression for A is justified here, because at the high surfactant concentrations used in these experiments the surfactant is almost completely adsorbed.

Combination of eqs.(9), (10) and (11) yields the expression for the salt concentration in the excess water phase

$$c_{el} = c_{el}^{i} - \frac{\Gamma_{co\text{-}ion}}{\Gamma_{sa}} \cdot c_{sa} \quad , \tag{12}$$

where by definition $c_{sa} = n_{sa}/V_w^i$. Since $\Gamma_{co\text{-}ion}$ is negative, it follows from eq.(12) that c_{el} is always larger than c_{el}^i. At low surfactant concentration, the increase of the salt concentration in the excess aqueous phase will be small. We may then substitute c_{el}^i for c_{el} in eq.(7), so that $\Gamma_{co\text{-}ion}$ becomes constant, to find a linear relationship between the salt concentration and the surfactant concentration. On the other hand, at high surfactant concentration, c_{el} is considerably larger than c_{el}^i and then this substitution is not allowed. Solutions are then easily found by iteration. Since $\Gamma_{co\text{-}ion}$ becomes more negative with increasing salt concentration (eq.(7)) we expect a plot of c_{el} against c_{sa} to curve upwards.

It is instructive to substitute the limiting expression eq.(8) for $\Gamma_{co\text{-}ion}$ in eq.(12). After some rearrangements we then find for low surfactant concentrations

$$c_{el} \cong c_{el}^{i} + B\sqrt{c_{el}^{i}} \cdot c_{sa} \quad , \tag{13}$$

where B is a constant,

$$B = \frac{\sqrt{\varepsilon_r \varepsilon_o RT}}{\sqrt{2}\, F\Gamma_{sa}} \tag{14}$$

From eq.(13) it is clear that the increase of the salt concentration in the excess phase is approximately proportional to the surfactant concentration, and to the square root of the salt concentration. For microemulsion systems we are mainly interested in the relative increase of the salt concentration. From eq.(13) we see that the relative increase is given

$$\frac{c_{el} - c_{el}^{i}}{c_{el}^{i}} \cong \frac{B\, c_{sa}}{\sqrt{c_{el}^{i}}} , \tag{15}$$

by showing that the relative increase of the salt concentration is larger for lower salt concentrations.

3. Materials and Methods

3.1. MATERIALS

The surfactant was sodium dodecyl sulfate (SDS) of "specially pure" grade purchased from BDH, and was used as received.

The oil, cosurfactant and salt were pentanol, cyclohexane and NaCl respectively. These chemicals were of "analyzed" grade and purchased from Baker. Pentanol and cyclohexane were used without any further purification, NaCl was heated for one hour at 150°C before use.

Water was deionized water that was distilled three times before use.

3.2. METHODS

3.2.1 *Sample preparation.* We prepared samples by mixing 10 grams of initial water phase (NaCl and water), 8 grams of oil phase (pentanol and cyclohexane), and the desired amount of surfactant (SDS). We used a procedure by which we obtained equilibrated microemulsion systems within a few days. This procedure is described in detail elsewhere [8,9].

3.2.2 *Salt concentrations.* We measured the salt concentration in the excess water phase by potentiometric titration of the chloride ion with silver nitrate. The indicator electrode consisted of a small silver plate, submerged in the solution. The reference electrode was a calomel electrode, which was connected to the solution with a Vycor tip (EG+G/Parc). The titrations were performed in an excess electrolyte concentration of 3% KNO_3 to maintain a constant ionic strength and to keep a constant diffusion potential at the salt bridge. We achieved an accuracy of approximately 0.1 percent.

3.3.3 *Dilution experiments.* W/O microemulsions can be diluted with their oil-continuous phase. The underlying assumption is that the composition of the droplets, their interface, and the continuous medium are independent of the volume fraction of the microemulsion droplets. Assuming a monodisperse system of spherical droplets, the radius a of the water cores of these droplets is given by

$$a = \frac{3\, V_w^M}{A} = \frac{3\, V_w^M\, \Gamma_{sa}}{n_{sa}} , \tag{16}$$

and is constant during the dilution experiment.

An initial coarse emulsion is prepared by mixing respectively surfactant, cosurfactant, oil and desired amount of brine, and stirring well. Then the coarse emulsion is titrated with cosurfactant until the L_2-region is entered. The L_2-region is the single phase region of W/O microemulsions. At this point we can either add more oil or more cosurfactant to leave the single phase microemulsion region. In the procedure that is usually reported in literature [10-19], more oil is added. The procedures are illustrated in the sketch of Fig.1. After an addition of a certain amount of oil a sharp increase of the turbidity indicates that the microemulsion system is no longer stable. By repeated addition of, in turn, cosurfactant and oil, the microemulsion is diluted along the "lower" phase boundary of the L_2-region, determined by a deficiency of cosurfactant.

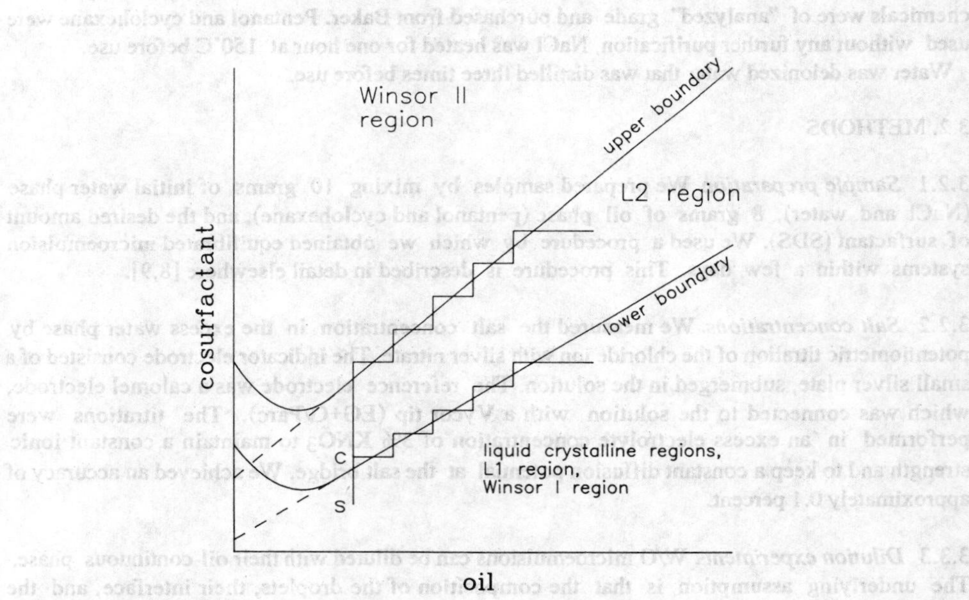

Fig.1. Sketch of the dilution procedure for microemulsions close to the phase boundaries of the L_2-region. Drawn lines represent the lower and upper boundaries of the L_2-region, dotted lines represent the extrapolation to zero oil content. At starting point (s) a coarse emulsion is present. After addition of a certain amount of cosurfactant, one reaches point (c), where one can choose to determine either the lower or the upper phase boundary.

In this titration procedure the L_2-region is always entered by the addition of pentanol. For microemulsions along this lower phase boundary, the ratio of cosurfactant and oil of the oil continuous medium can be found from the slope of a plot of the total amount of cosurfactant against the total amount of oil. The intercept, found by extrapolation of the straight line to zero oil content, gives the composition of the droplets plus the interface. Coarse emulsions that are obtained by the addition of a slight excess of oil to a microemulsion at this phase boundary separate to give either a Winsor I system, or an equilibrium system composed of an L_2 region and a birefringent liquid crystalline phase. Therefore, the microemulsions at this phase boundary differ from the microemulsions in Winsor II systems. Winsor II systems are formed by addition of an excess amount of cosurfactant, and therefore correspond to the "upper" phase boundary of the L_2- region (see Fig.1.). This upper phase boundary can be titrated in a similar way as the lower phase boundary, the difference being that now the L_2-region is entered by the addition of oil, and is left by the addition of cosurfactant [20,21]. Addition of a small excess of cosurfactant to microemulsions at the upper phase boundary always produced Winsor II equilibria. The interpretation of the slope and intercept of the straight line that represents the upper phase boundary of the L_2-region are the same as that for the lower phase boundary.

4. Results and Discussion

4.1. MEASUREMENTS OF THE INCREASE OF THE SALT CONCENTRATION

Figure 2. is a plot of the measured salt concentration in the excess aqueous phase as a function of the surfactant concentration, for samples with an initial salt concentration of 0.200 M, and a pentanol fraction in the continuous oil phase of approximately 0.20 (w/w). We observe an almost linear increase of the salt concentration as a function of the SDS concentration. By application of eq.(12) we obtain from the slope of this line $-\Gamma_{co\text{-}ion}/\Gamma_{sa} = 0.127 \pm 0.002$. Using the surfactant adsorption as the only adjustable parameter in eq.(7) for the co-ion adsorption, we find $\Gamma_{sa} = 1.82$ mmol/m^2. This value is equal to the value reported by Verhoeckx *et al.*[18], who obtained $\Gamma_{sa} = 1.82 \pm 0.10$ mmol/m^2 from interfacial tension measurements, for systems with $c_{el} = 0.30$ M NaCl and a pentanol fraction in the oil of 0.20 (w/w). Although the accuracy of the agreement should be considered more or less a coincidence, this shows that the salt partitioning can be attributed to the electrical double layer.

In Figure 3 we plotted the relative increase of the salt concentration at $c_{sa} = 0.30$ M as a function of the initial salt concentration. The drawn line is a best fit of the inverse square root of the initial salt concentration, according to the approximate expression eq.(15). Also here we find qualitative agreement between theory and experiment, showing that the relative increase of the salt concentration is largest at low salt concentrations.

More extensive work, including measurements at several salt concentrations and measurements with heptanol as cosurfactant, is reported elsewhere [8]. There we also used the more accurate expression for the co-ion adsorption inside a spherical cavity to interpret the experimental values.

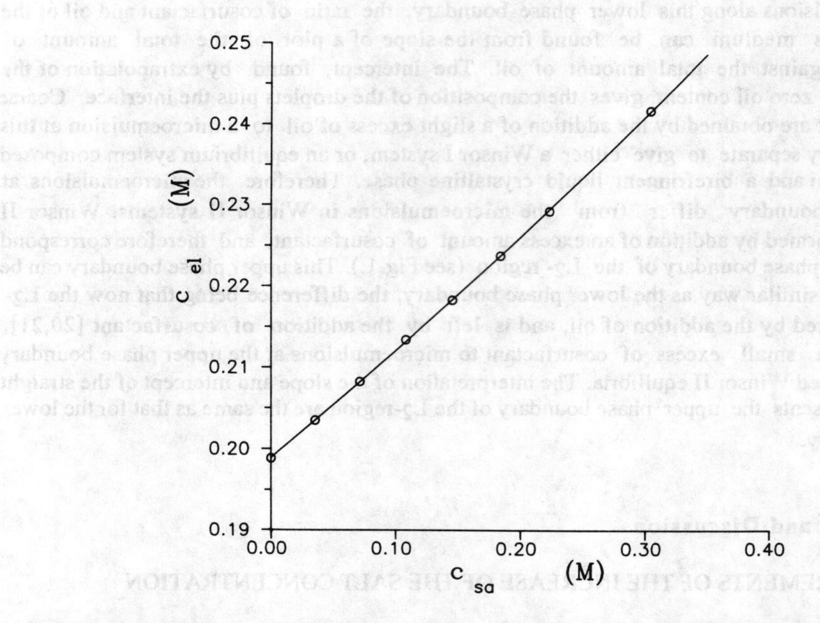

Fig.2. Concentration of salt in the excess aqueous phase after equilibration, as a function of the SDS concentration. The initial electrolyte concentration was 0.200 M. The measured electrolyte concentration at zero SDS concentration is slightly lower than 0.200 M due to the solubility of pentanol in water.

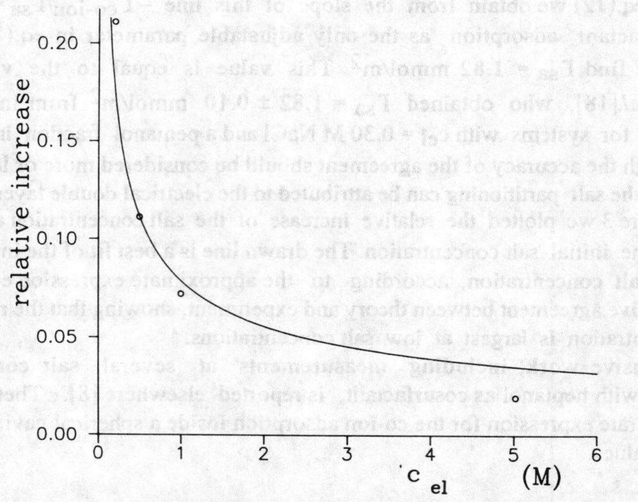

Fig.3. The relative increase $(c_{el} - c_{el}^i)/c_{el}^i$ plotted as a function of the initial salt concentration. SDS concentration is constant at 0.30 M. The plotted line is a best fit according to $(c_{el} - c_{el}^i)/c_{el}^i \sim 1/\sqrt{c_{el}^i}$.

4.2. THE INFLUENCE OF THE DISTRIBUTION OF SALT ON THE PHASE BEHAVIOUR OF MICROEMULSIONS SYSTEMS

It is a well established fact that both the salt and the cosurfactant concentrations strongly influence the equilibrium droplet radii. The droplet radius of a W/O microemulsion is related to the w/sa-ratio in the microemulsion phase by eq.(16). Fig.4 shows the w/sa-ratio as a function of the salt concentration for Winsor II microemulsion systems. The w/sa-ratios are limiting values, as found by extrapolation to $c_{sa} = 0$, so that the increase of the salt concentration is absent here. The influence of the salt partitioning on the phase behaviour of microemulsions will be illustrated by the following three examples.

4.2.1 *Example I.* In Fig.5 we show the dependence of the w/sa-ratio, related to the droplet radius through eq.(16), with the SDS concentration. In preparing these samples, we corrected for

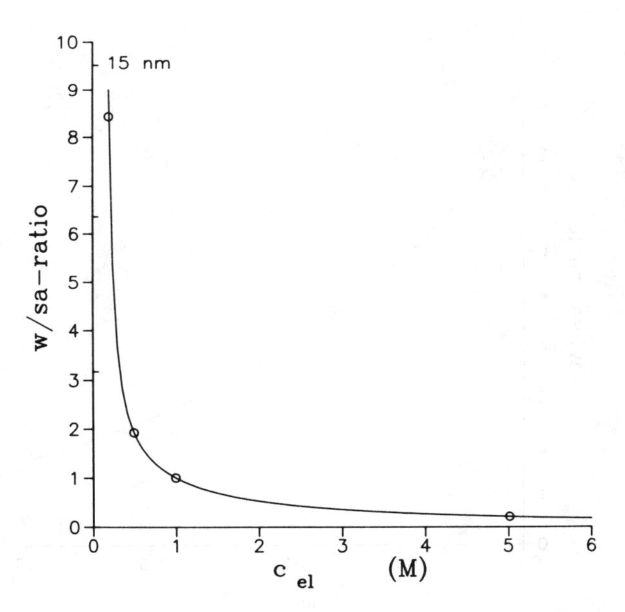

Fig.4. Water/surfactant ratio and droplet radius of the microemulsion phase of Winsor II systems as a function of the salt concentration in the excess phase. Data points are found by extrapolation to zero surfactant concentration of Winsor II equilibria, with an equilibrium pentanol concentration of 20% (w/w) in the continuous oil phase. The droplet radius is calculated with eq.(16), assuming $\Gamma_{sa} = 1.82\ \mu mol/m^2$.

the uptake of pentanol in the droplet interface by the addition of 0.7 gram of extra pentanol per gram SDS. This value was obtained from dilution experiments, but is unfortunately somewhat too low, because the dilution experiments refer to the lower boundary of the L_2-region. The w/sa-ratio is seen to decrease on increasing the SDS contents, due to the increase of the concentration of salt in the brine. If we would have added the correct (larger) amount of pentanol, the decrease of the w/sa-ratio would have been even more prominent. At the highest surfactant concentration of Fig.5 (at 1.98 moles SDS per dm^3 brine), the brine is almost completely taken up by the microemulsion phase. For this sample the increase of the salt concentration is extremely large, and as a result the w/sa-ratio, 1.75 (w/w), is now only a fraction of the value measured in the sample with the lowest SDS concentration, 8.42 (w/w).

In this sample we added the correct amount of extra pentanol to correct for the uptake of pentanol in the droplet interface, namely 2.0 gram pentanol per gram SDS, valid for $x_p^o = 0.20$ and $c_{el} = 0.76$ M, as obtained from dilution experiments at the upper boundary of the L_2- region [21]. The assumption that the observed strong decrease of the mean w/sa-ratio is mainly due to the increase of the salt concentration in the excess brine, is supported by comparison with w/sa-ratios measured in Winsor II microemulsion systems at low SDS concentrations. From interpolation of the measurements of Fig.4 we obtained an equilibrium value of the w/sa-ratio of 1.28 (w/w) at $x_p^o = 0.20$ (w/w) and 0.76 M NaCl, which is reasonably close to the previously reported value of 1.75 (w/w) at $x_p^o = 0.20$ (w/w), $c_{el}^i = 0.20$ M and $c_{SDS} = 1.98$ M.

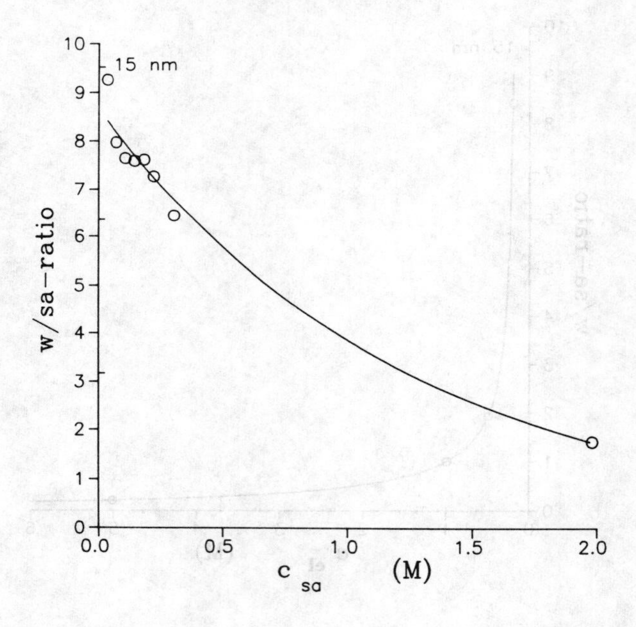

Fig.5. Water/surfactant ratio and droplet radius of the microemulsion phase of Winsor II systems as a function of the surfactant concentration. The initial salt concentration was 0.200 M, and the equilibrium pentanol concentration in the continuous oil phase was held at a constant value of approximately 20% (w/w). The droplet radius is calculated with eq.(16), assuming Γ_{sa} = 1.82 $\mu mol/m^2$.

4.2.2 *Example II.* We will now consider the experimental results of dilution experiments. In Fig.6 values for the pentanol fraction in the continuous oil phase of microemulsions at the upper boundary of the L_2-region are plotted as a function of the initial salt concentration in the brine. This is done for four different mean droplet radii, respectively 2.5, 5.0, 7.5 and 10 nm, calculated from the w/sa-ratio by using eq.(16) and setting Γ_{sa} = 1.82 μmol/m^2. We see that, for constant mean droplet radius, the pentanol fraction in the continuous oil phase always decreases on increasing the salt concentration. This is in accordance with the behaviour previously described for the Winsor II systems, since an increase of the salt concentration has the same effect as an increase of the cosurfactant concentration. The mean droplet radius can therefore be kept constant by increasing the pentanol concentration and decreasing the salt concentration, and vice versa. However, three of the four curves for different mean droplet radii intersect. For salt concentrations below the point of intersection, the mean droplet radius is seen to increase on increasing the pentanol concentration. At first sight this seems to be in contradiction with the influence of the pentanol concentration on the Winsor II equilibria, where we observed the opposite effect. The explanation of this apparent contradiction lies in the distribution of salt.

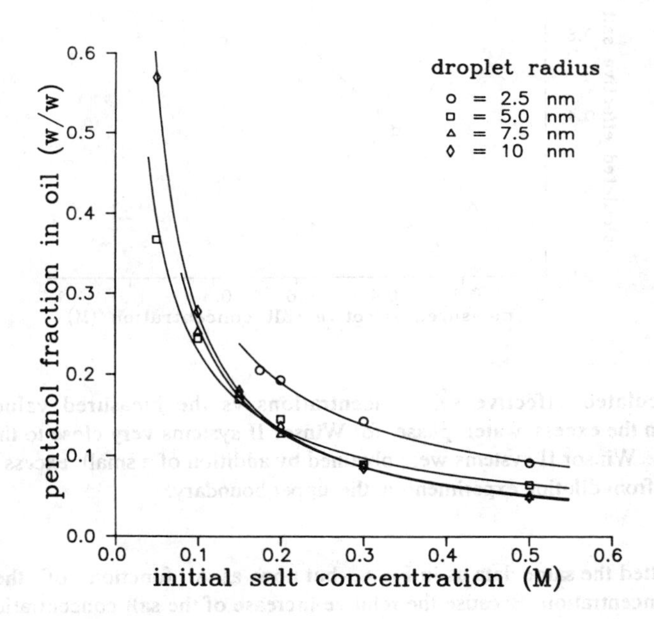

Fig.6. Pentanol concentration in the continuous oil phase as a function of the initial salt concentration. Data obtained from slopc of the upper boundary of the L_2-region and were measured by dilution experiments.

We added a very small excess of pentanol to microemulsions with a composition at the upper phase boundary, thus achieving the separation of a very small amount of excess brine.

Measurements of the concentration of salt in this excess brine revealed that the salt concentration is considerably increased with respect to the concentration in the initial brine. Fig.7 shows that the salt concentration in this excess brine can still be calculated accurately using the theoretical equations for the distribution of salt, with the exception for the droplet radius of 2.5 nm, where the expression for the co-ion adsorption in a flat electrical double layer slightly underestimates the effective salt concentration. We conclude therefore that for microemulsions with a composition inside the L_2-region, the effect of the distribution of salt is still present and leads to a considerably higher "effective" salt concentration than in the initial brine.

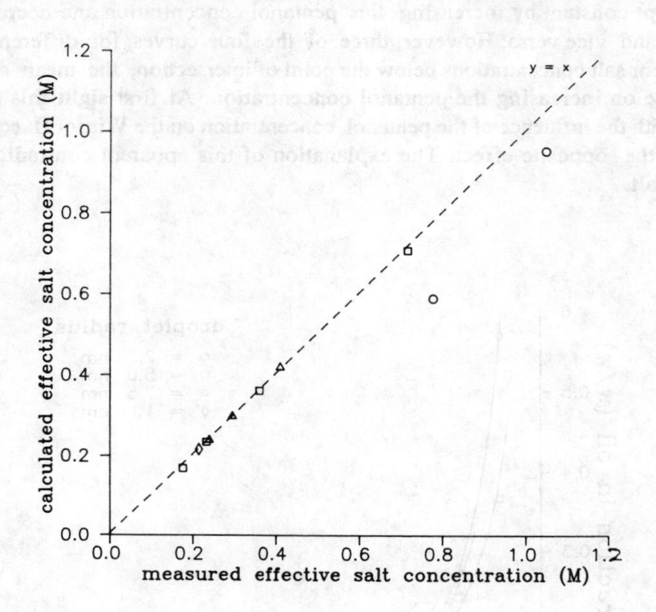

Fig.7. Calculated effective salt concentrations vs the measured values of the salt concentrations in the excess water phase for Winsor II systems very close to the upper phase boundary. These Winsor II systems were obtained by addition of a small excess of pentanol to microemulsions from dilution experiments at the upper boundary.

In Fig.8 we plotted the same data as in Fig.6, but now as a function of the (calculated) effective salt concentration. Because the relative increase of the salt concentration is larger for lower initial salt concentrations, and smaller droplet radii (according to eq.(16) this corresponds to higher SDS concentrations), the data points move up in such a way that now regular dependence of the droplet radii on the pentanol and salt concentration is obtained.

We further note that at low surfactant concentrations Winsor II systems cannot be formed at salt concentrations below approximately 0.10 M [22,23]. However, as a result of the increase of the effective salt concentration, we were able to measure the upper boundary of the L_2-region (thereby producing Winsor II microemulsion systems) at initial salt concentrations as low as 0.050 M. The effective salt concentrations for these microemulsion systems were always higher than the lower limit of 0.10 M.

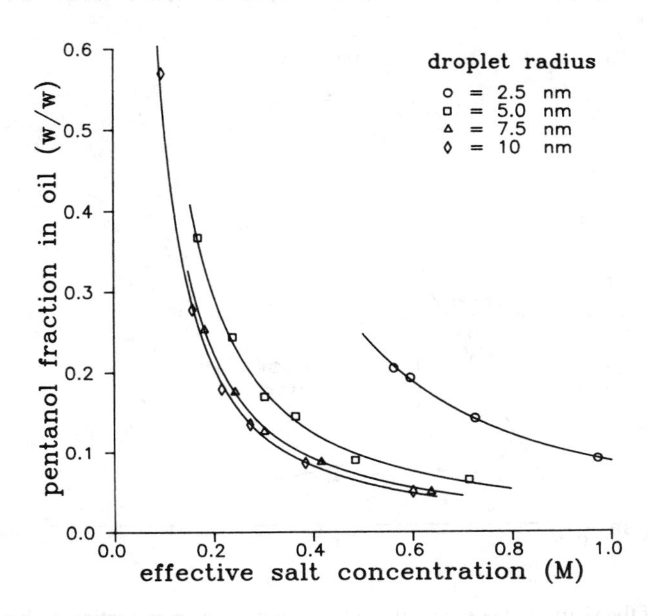

Fig.8. Same as Fig.6, but now as a function of the calculated effective salt concentration.

4.2.3 *Example III.* In this last example we show the consequence of the increase of the effective salt concentration for cross-sections of constant oil/surfactant-ratio through the L_2-region. Fig.9 shows the general shape of such a cross section. The upper and lower boundaries of the L_2-region in this picture correspond to the upper and lower boundaries in the dilution experiment, and are respectively due to a deficiency and an excess of pentanol in the system.

Notice that the lower phase boundary can be determined by a titration procedure similar to the procedure that was used to obtain the dilution line, but now with brine and pentanol as titrants. Repetitive titration is possible since the tangent of this lower phase boundary intersects the brine/pentanol axis of this composition diagram.

Since in such a cross-section of the composition diagram the w/sa-ratio is variable, the effective salt concentration is not a constant, and therefore the brine cannot be considered a pseudo-component. This is illustrated in Fig.10, where we compare the upper and lower phase boundaries at a constant initial salt concentration of 0.20 M to the corrected phase boundaries for a constant effective salt concentration of 0.20 M. The data points of this figure are obtained from dilution experiments at the upper and lower boundaries of the L_2-region. Because these dilution experiments were performed at several initial salt concentrations, we were able to determine by interpolation the positions of the phase boundaries at constant effective salt concentration. From Fig.10 we see that the phase boundaries move up to higher pentanol fractions. In addition the phase boundaries are also seen to rotate, because increase of the effective salt concentration is larger at smaller w/sa-ratio. This has the interesting consequence that, whereas in practice repeated titration with brine and pentanol are possible for the lower boundary, at constant effective salt concentration repeated titration would not be possible.

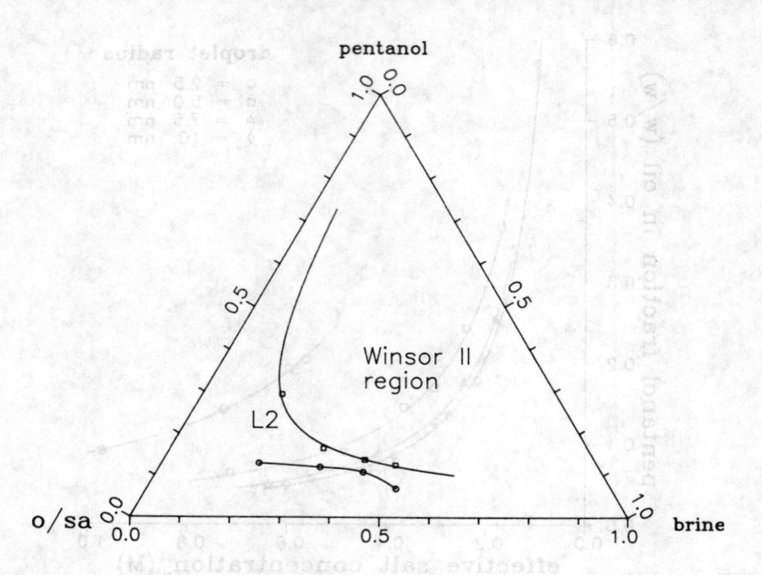

Fig.9. Sketch of the shape and position of the L_2-region in a cross-section with an oil/surfactant ratio of 5 (w/w), and for brine with an initial salt concentration of 0.20 M. Data points are obtained from dilution titrations, and correspond to the lower boundary (circles) and the upper boundary (squares) respectively.

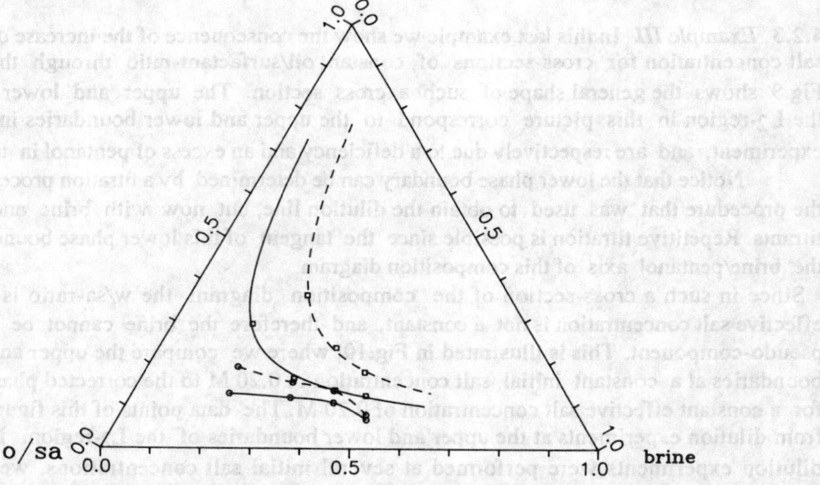

Fig.10. Same as Fig.9, but now we compare the shape and position for constant initial salt concentration of 0.20 M (solid line) to the shape and position for constant effective salt concentration of 0.20 M (dotted line).

Repeated titration of the lower boundary in a cross-section with constant oil/surfactant-ratio is therefore possible by the virtue of the variation of the effective salt concentration.

5. Conclusions

The salt partitioning between the excess water phase and the water-rich regions inside the microemulsions phase of Winsor II microemulsion systems is due to the large internal interfacial area of the microemulsion phase. We were able to derive a fairly accurate equation for this partitioning based on the negative adsorption of co-ions in the internal interface, by use of the Gouy-Chapman theory for the electrical double layer. The relative increase of the salt concentration in the excess aqueous phase is at largest for low electrolyte concentration, and small water/surfactant ratios of the system. The effect of this salt partitioning on Winsor II systems is that the uptake of water is reduced at high surfactant content. Although for single phase microemulsion systems co-ion adsorption is not directly measurable in the form of increase of the salt concentration, the effect on the position of the single phase L_2-region is evident.

The obvious consequence of this study is that the brine should not be regarded as a pseudo-component of microemulsion systems. It is however possible to correct the experimental results with relatively simple equations for the partitioning of salt.

ACKNOWLEDGEMENT This investigation was supported by the Netherlands Foundation of Chemical Research (SON) with financial aid from the Netherlands Organization for the Advancement of Pure Research (NWO). The author wishes to thank Mr. J.J. van 't Veld for performing the upper boundary dilution titrations, and is indebted to Professor P.L. de Bruyn, Professor H.N.W. Lekkerkerker and Professor J.Th.G. Overbeek for kind and helpful advise.

REFERENCES

1. Adamson, A.W. (1968) "A model for micellar emulsions", J. Colloid Interface Sci. 29, 261-267.
2. Tosh, W.C., Jones, S.C. and Adamson, A.W. (1969) "Distribution equilibria in a micellar solution system", J. Colloid Interface Sci. 31, 297-306.
3. van Nieuwkoop, J. and Snoei, G. (1985) "Phase diagrams and composition analysis in the system sodium dodecyl sulphate/butanol/water/sodium chloride/heptane", J. Colloid Interface Sci. 103, 400-416.
4. Fletcher, P.D.I. (1986) "The partitioning of solutes between water-in-oil microemulsions and conjugate aqueous phases", J. Chem. Soc. Faraday Trans. 1, 82, 2651-2664.
5. Biais, J., Barthe, M., Bourrel, M., Clin, B. and Lalanne, P. (1986) "Salt partitioning in Winsor type II systems", J. Colloid Interface Sci. 109, 576-585.
6. de Bruyn, P.L., Overbeek, J.Th.G. and Verhoeckx, G.J. (1989) "On understanding microemulsions III", J. Colloid Interface Sci. 127, 244-255.
7. van Aken, G.A., Lekkerkerker, H.N.W., Overbeek, J.Th.G. and de Bruyn, P.L. "Adsorption of monovalent ions in thin spherical and cylindrical diffuse electrical double layers", submitted for publication in J. Phys. Chem.
8. van Aken, G.A., de Bruyn, P.L., Overbeek, J.Th.G. and Lekkerkerker, H.N.W. "Partitioning of salt in Winsor II microemulsion systems with an ionic surfactant", submitted

for publication in J. Phys. Chem.
9. van Aken, G.A. (1990) Thesis, Utrecht.
10. Bowcott, J.E. and Schulman, J.H. (1955) "Emulsions", Z. Elektrochem. 59, 283-290.
11. Caljé, A.A., Agterof, W.G.M. and Vrij, A. (1977) "Light scattering of a concentrated w/o microemulsion, application of modern fluid theories", in K.L. Mittal (ed.), Micellization, Solubilization and Microemulsions 2, Plenum Press, New York, 779-790.
12. Vrij, A., Nieuwenhuis, E.A., Fijnaut, H.M. and Agterof, W.G.M. (1978) "Application of modern concepts in liquid state theory to concentrated particle dispersions", Faraday Disc. Chem. Soc. 65, 101-113.
13. Bansal, V.K., Shah, D.O. and O'Connell, J.P. (1980) "Influence of alkyl chain length compatibility on microemulsion structure and solubilization", J. Colloid Interface Sci. 75, 462-475.
14. Pouchelon, A., Chatenay, D., Meunier, J. and Langevin D. (1981) "Origin of low interfacial tensions in systems involving microemulsion phases", J. Colloid Interface Sci. 82, 418-422.
15. Cazabat, A.M. (1983) "Water in oil microemulsions: criteria for dilution at constant droplet size", J. Phys. Lett. 44, 593-599.
16. Graciaa, A., Lachaise, J., Bourrel, M., Schechter, R.S. and Wade, W.H. (1986) "Determination of microemulsion composition using dialysis", J. Colloid Interface Sci. 113, 583-584.
17. Overbeek, J.Th.G., de Bruyn, P.L. and Verhoeckx, G.J. (1984) "Microemulsions", in Th.F. Tadros (ed.), Surfactants, Academic Press, London, 111-132.
18. Verhoeckx, G.J., de Bruyn, P.L. and Overbeek, J.Th.G. (1987) "On understanding microemulsions.I", J. Colloid Interface Sci. 119, 409-421.
19. Graciaa, A., Lachaise, J. Bourrel, M., Schechter, R.S. and Wade, W.H. (1988) "Differentiation of continuous and discontinuous phases of water-in-oil microemulsions using dialysis", J. Colloid Interface Sci. 122, 83-91.
20. Dominguez, J.G., Willhite, G.P. and Green, D.W. (1979) "Phase behaviour of microemulsion systems with emphasis on effects of paraffinic hydrocarbons and alcohols", in K.L. Mittal (ed.), Solution chemistry of surfactants 2, Plenum Press, New York, 673-697.
21. van Aken, G.A., Lekkerkerker, H.N.W., Overbeek, J.Th.G. and de Bruyn, P.L., to be published.
22. Overbeek, J.Th.G., Verhoeckx, G.J. de Bruyn, P.L. and Lekkerkerker, H.N.W. (1987) "On understanding microemulsions. II", J. Colloid Interface Sci. 119, 422-441.
23. Verhoeckx, G.J. (1986) "On understanding microemulsions", Thesis, Utrecht.

A STUDY OF MICROEMULSION STABILITY

J. EASTOE and B.H. ROBINSON,
School of Chemical Sciences,
University of East Anglia
NORWICH NR4 7TJ, UK.

D.C. STEYTLER,
Institute of Food Research,
Colney Lane,
NORWICH NR4 7TJ, UK

ABSTRACT. We report P/T stability studies of the sodium bis(2-ethylhexyl) sulphosuccinate (AOT)/water/n-alkane microemulsion droplet system using the straight chain alkanes from the homologous series propane to n-decane as oil. On progressively decreasing the carbon number of the oil the P/T phase boundaries of these systems exhibit a reversal of behaviour. By plotting microemulsion stability on the oil density-temperature surface we have normalised the phase behaviour observed in P/T space for AOT/water/n-alkane systems. Further, the oil density-temperature plots suggest a minimum oil density of ~0.5 x 10^3 kg m^{-3} is required for AOT-stabilised microemulsion droplets to exist as discrete entities.

1. Introduction

A microemulsion is a thermodynamically stable, optically clear dispersion of two immiscible liquids stabilised by a monolayer of surfactant(s). The preferred surfactant configuration at the interface influences the microstructure, which may be at one extreme bi-continuous[1] or at the other discrete, spherical droplets (Fig. 1a)[1-3]. When the weight fraction of oil is ≥80% microemulsions stabilised by the surfactant sodium bis(2-ethylhexyl) sulphosuccinate (AOT) (Fig. 1a) are known to be a dispersion of discrete water droplets[2,3]. The droplet size is directly controlled by w_O, where

$$w_O = [H_2O]/[AOT].$$

At fixed [AOT] the 3D. ternary composition-temperature 'prism' may be reduced to a 2D. w_O-temperature diagram (Fig. 1b). For a given composition (w_O) two phase transition temperatures (PTT's) are evident; an upper PTT (T_U) and a lower PTT (T_L). Here we present a study in pressure-temperature (P-T) space of the phase behaviour of AOT-stabilised water droplets dispersed in the series of n-alkanes, ethane (C2)→ n-decane (n-C10). Our study indicates that two different droplet association mechanisms are responsible for the phase separations at each PTT.

D. M. Bloor and E. Wyn-Jones (eds.),
The Structure, Dynamics and Equilibrium Properties of Colloidal Systems, 295–301.

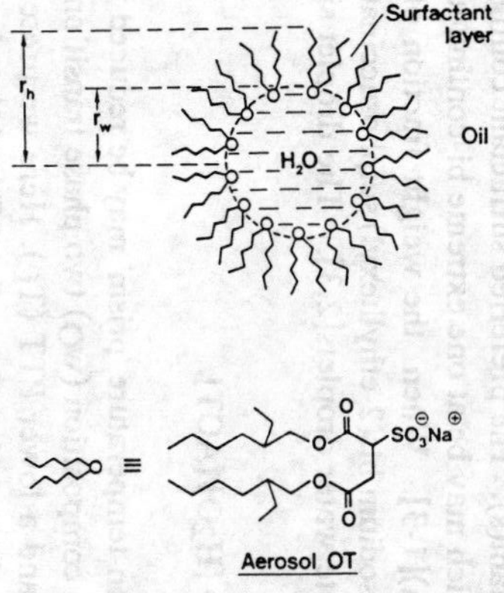

Fig 1a. Schematic AOT Stabilised Microemulsion Droplet

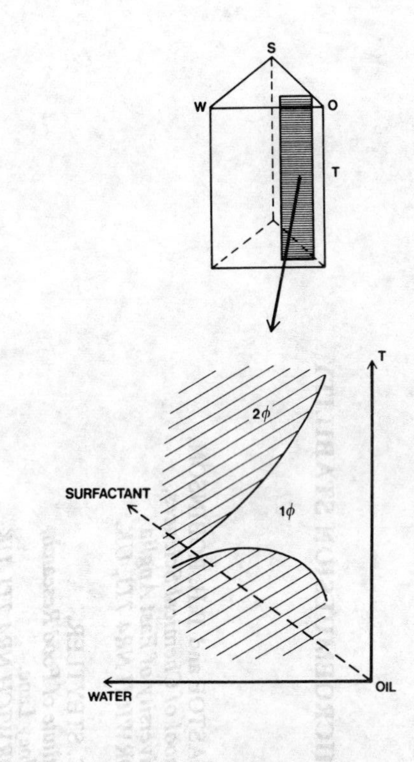

Fig 1b. Constant [Surfactant] gives rise to a 2D W_O-temperature diagram
1Ø indicates stable microemulsion
2Ø denotes phase separated system.

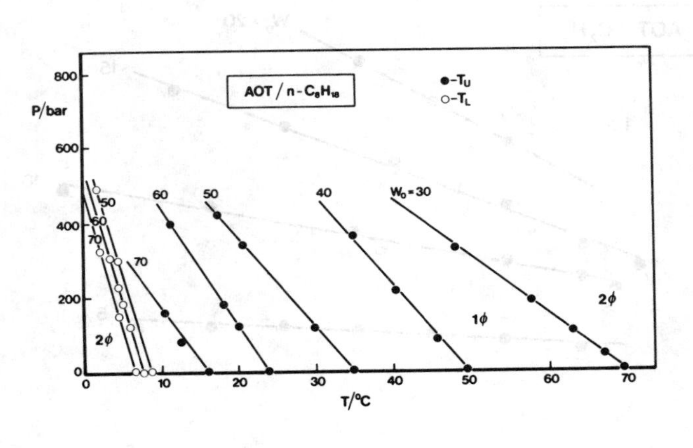

Fig. 2a. The P.T. Behaviour of AOT Microemulsions Dispersed in n-octane [AOT] = 0.1 mol dm^{-3}.

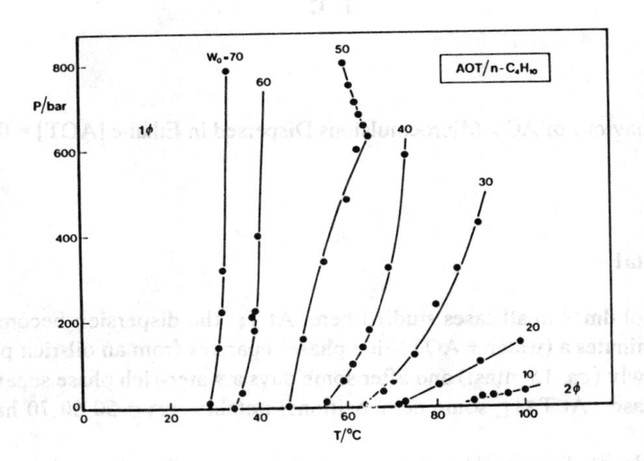

Fig. 2b. The P.T. Behaviour of AOT Microemulsions in n-butane [AOT] = 0.1 mol dm^{-3}

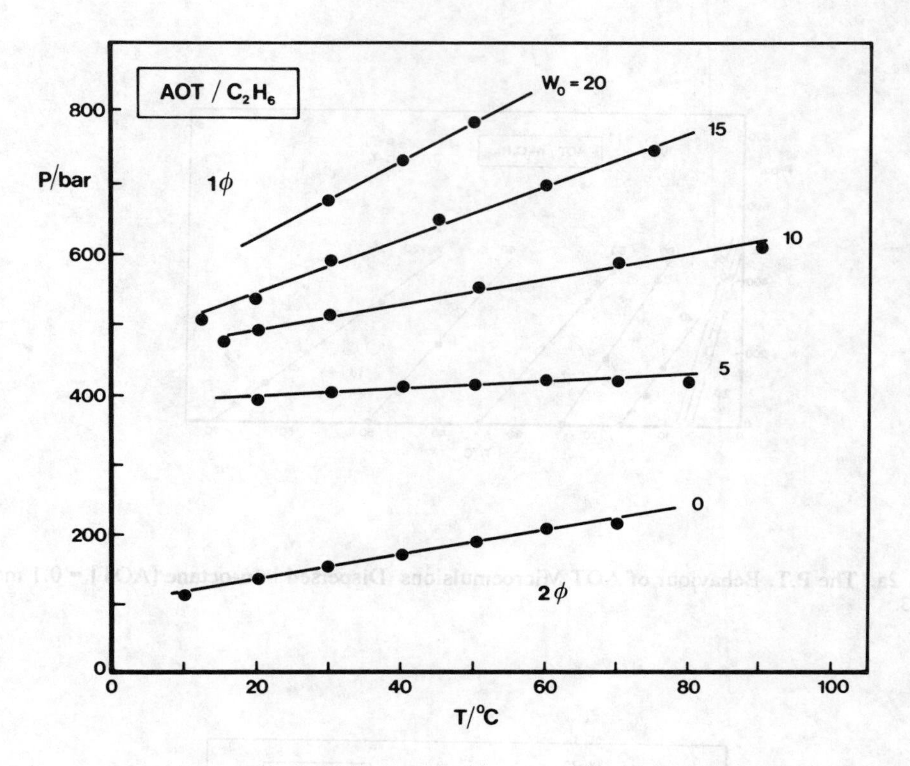

Fig. 2c. The PT Behaviour of AOT Microemulsions Dispersed in Ethane [AOT] = 0.1 mol dm^{-3}

2. Experimental

The [AOT] is 0.1 mol dm^{-3} in all cases studied here. At T_U the dispersion becomes instantly opaque, after *ca.* 5 minutes a (water + AOT)-rich phase separates from an oil-rich phase. At T_L opacity develops slowly (*ca.* 15 mins.) and after some days a water-rich phase separates from an (AOT + oil)-rich phase. At $T \le T_L$ some compositions, notably w_O = 50,60,70 have a strong orange colouration.

Measurements at elevated pressures and temperatures were made using a thermostatted high pressure cell fitted with sapphire windows allowing visual detection of the phase boundaries. A composite detector directly monitors pressure and temperature. At elevated pressures the T_U boundary is easily observed (± 5 bar), whereas there is some difficulty in establishing the T_L boundary, so data are approximate.

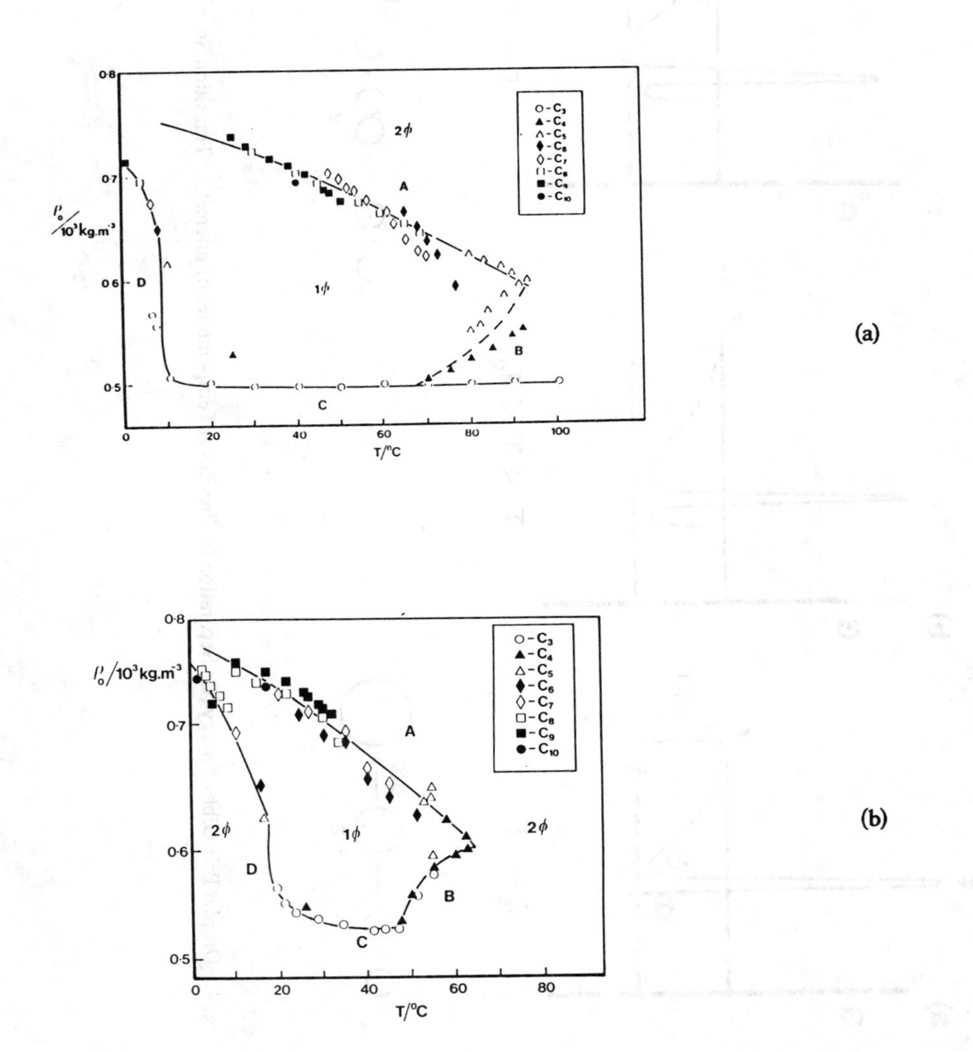

Fig. 3. Bulk Oil Density (ρ_o) - Temperature Stability Map for AOT/Water/n-alkane Microemulsions [AOT] = 0.1 mol dm^{-3} (a) w_O = 30, (b) w_O = 50.

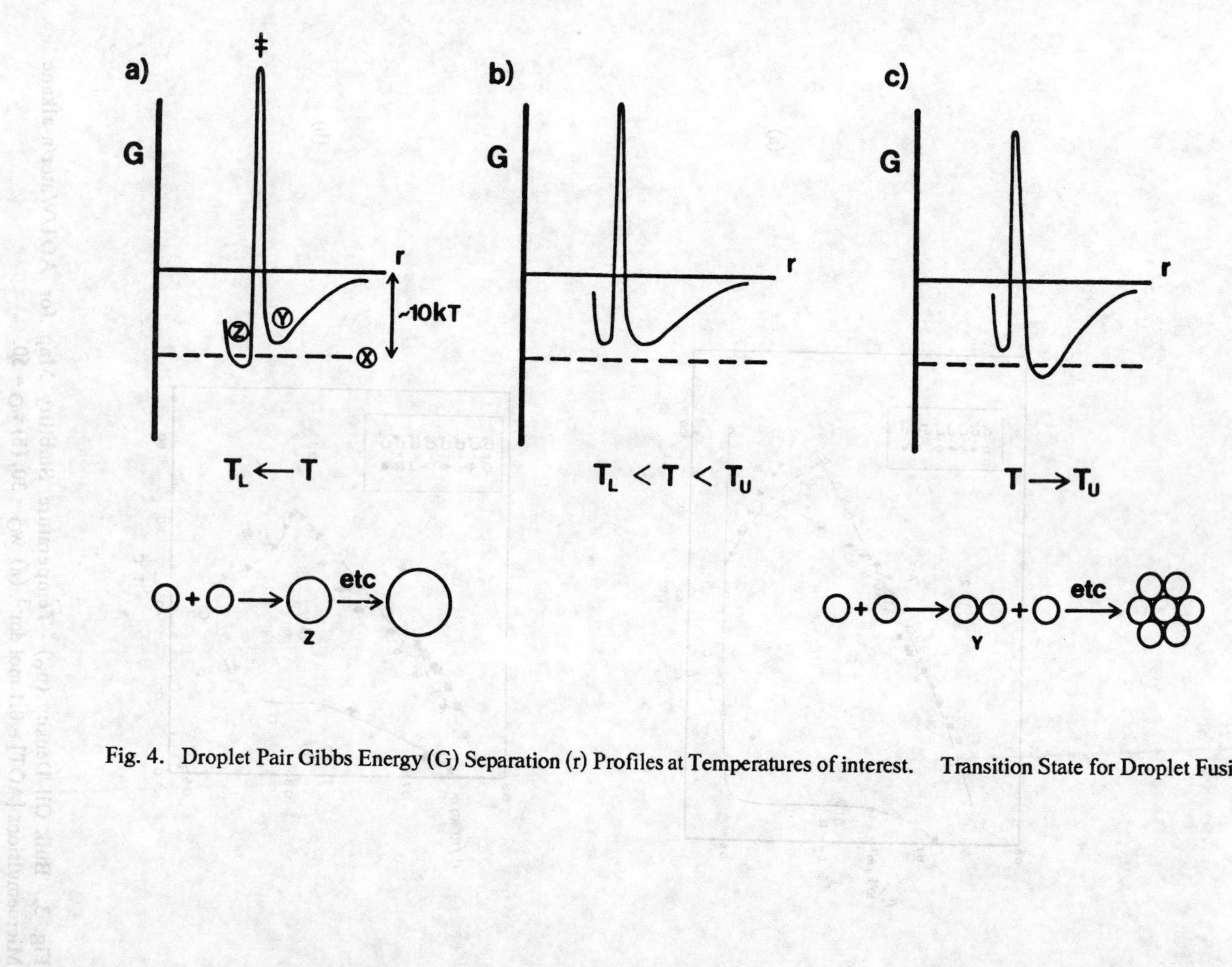

Fig. 4. Droplet Pair Gibbs Energy (G) Separation (r) Profiles at Temperatures of interest. Transition State for Droplet Fusion.

3. Discussion

Figure 2a shows the phase behaviour of the AOT/water/n-octane system; 1ϕ denotes a one phase microemulsion whilst 2ϕ indicates a two phase system. Slopes of the boundaries are negative *i.e.* increasing pressure decreases the PTT. Figure 2b shows a similar diagram for the AOT/water/n-butane system. The T_U boundaries are curves of positive slope *i.e.* increasing pressure increases T_U. The curve for $w_O = 50$ is noteworthy in that it has a negative slope at pressures greater than 600 bar. Figure 2c shows the one phase boundaries for the AOT/water/ethane system which are lines of positive slope.

On traversing the homologous series C2 → n-C10 the chemical nature of the oil is not changed but there is a regular increase in physical properties *e.g.* density and viscosity. Furthermore an increase in pressure can be interpreted as an increase in oil density. We have normalised the three different types of P-T phase behaviour observed in Figure 2 by treating bulk oil density (ρ_o) as a parameter. ρ_o for the n-alkanes used was calculated at pressures and temperatures corresponding to phase boundaries (*e.g.* in Fig. 2) using the appropriate method[5,6]. In Figure 3 data are plotted on the ρ_o-temperature surface for two different wO values. Four distinct boundaries are apparent.

𝔸 corresponds to phase boundaries of negative slope (*e.g* Fig. 2a) and suggests that at a given temperature a maximum oil density exists for droplet stability.
𝔹 corresponds to phase boundaries which are curves of positive slope *e.g.* Figure 2b. Note data from the 'kinked' wO = 50 boundary of Fig. 2b fall on the high-temperature 'nose' of Fig. 3b.
ℂ suggests that a minimum oil density exists for droplet stability. Oil densities that correspond to T_L values fall on line 𝔻.

We may interpret these results using a model based on droplet-droplet interactions.

Figure 4 shows schematic profiles of the droplet pair Gibbs energy (G) as a function of separation (r) at temperatures of interest. The association mechanism and nature of phase separation is determined by the relative depths of minima Y and Z; the rates of phase separation are determined by the free energy barriers (≈100 kJ mol^{-1}). X is a system dependent threshold of the order 10kT. At T_L (Fig. 4a) the rate of phase separation is slow as the droplet pair must overcome the free energy barrier to fusion.

Fused aggregates with dimensions of the order of 100 nm will scatter blue light, thus explaining the orange colouration seen at T_L. Figure 4b accounts for the one phase microemulsion region, neither droplet association process is favoured. Figure 4c represents the profile for diffusion-controlled droplet aggregation at T_U. The rate of phase separation is rapid as there is no activation energy associated with a contact pair (state Y). It may be that pressure results in a lowering of minimum Y destabilising droplets dispersed in n-alkanes n-C6 → n-C10. Whilst pressure raises minimum Y stabilising droplets dispersed in oils n-C5 → C2. For more information see[4].

4. References

1. Barnes, Hyde, Ninham, Denian, Drifford and Zemb, (1988) J. Phys. Chem. 92, 2286-2293.
2. Jahn and Strey, (1988) J. Phys. Chem , 92, 2294-2301.
3. Kotlarchyk, Chen, Huang, Kim, (1984) Phys. Rev. A. 29, 2054-2069.
4. J. Eastoe, B.H. Robinson and D.C. Steytler, J. Chem. Soc. Faraday Trans. 1990, 96, 511-19.
5. Scaife, Lyons, (1980), Proc. Roy. Soc. Lon. A, 370, 193-211.
6. Kumar, Starling, (1982) Ind. Eng. Chem. Fundam. , 21, 255.

STRUCTURE OF MICROEMULSION-BASED ORGANO-GELS

P.J. ATKINSON, S.J. HOLLAND, and B.H. ROBINSON,
School of Chemical Sciences,
University of East Anglia,
Norwich NR4 7TJ, UK.

D.C. CLARK,
AFRC Institute of Food Research,
Colney Lane, Norwich,
NR4 7UA, UK.

R.K. HEENAN
Rutherford-Appleton Lab.
Chilton, Didcot,
Oxon, OX11 OQX, UK.

A.M. HOWE,
Surface Science Group,
Research Division,
Kodak Ltd., Harrow,
Middlesex, HA1 4TY.

1. Introduction

Water-in-oil microemulsions containing 10-20% v/v water dispersed in heptane by the anionic surfactant Aerosol-OT (AOT) may be completely gelled on the addition of the biopolymer gelatin[1,2]. This gelation process, which is thermoreversible, is effected by dissolving the gelatin (a few wt%) in the microemulsion at 50°C, followed by shaking and cooling to below 30°C. The organo-gels thus formed are optically clear, stable in contact with a coexisting oil phase and have reproducible physical properties. Enzymes such as lipases retain their activity when solubilised in these systems allowing them to be successfully employed in reverse-enzyme synthesis[3].

In this paper a small-angle neutron scattering (SANS) study of the structure of microemulsion-based organo-gels is reported, along with a preliminary investigation of the diffusion processes occurring within the gel studied using the technique of Fluorescence Recovery After Photobleaching (FRAP). Both approaches suggest that there are two distinct structural components in the organo-gel.

2. Materials and Methods

The gels were prepared as reported previously[4]: a weighed amount of gelatin powder (Sigma acid-hydrolysed porcine skin, Bloom 300) is dissolved in triply-distilled water at 50°C. An AOT (Sigma 99% sodium bis(2-ethylhexyl)sulphosuccinate) solution in n-heptane at the same temperature is added and the mixture is shaken vigorously to produce a homogeneous dispersion. On cooling to below 30°C, with occasional shaking to maintain a single phase, the gel is formed. For SANS, deuterated water (Aldrich 99.8% D) was used to

D. M. Bloor and E. Wyn-Jones (eds.),
The Structure, Dynamics and Equilibrium Properties of Colloidal Systems, 303–308.

provide contrast; aqueous fluorescein (Aldrich, disodium salt) was used as the fluorophore for FRAP.

SANS spectra were recorded on the LOQ instrument of ISIS, the spallation neutron source at the Rutherford-Appleton Laboratory. For SANS work a typical gel composition would be 0.1mol.dm^{-3}AOT, 10% v/v D_2O, 3.5% w/v gelatin in n-heptane(H). Samples were held in 1mm quartz cells, thermostatted at 25±0.1°C.

FRAP experiments were carried out at the Institute of Food Research, Norwich. The FRAP technique involves subjecting a small area of the sample (3.6μm radius spot) to a powerful laser pulse (0.1-0.3 Watts at 488nm from an Argon ion laser) which acts to permanently bleach the fluorescein in that area. The same laser, attenuated by a factor of 10^4, is then used to monitor the recovery of the fluorescence intensity as unbleached fluorescein diffuses into the bleached zone. Mathematical analysis of the fluorescence recovery curve yields a characteristic recovery time, from which a translational diffusion coefficient, D_T, can be calculated. The organo-gels used in the FRAP measurements reported here had composition 0.15 mol.dm^{-3} AOT, 15% v/v 5x10^{-7} mol.dm^{-3} aqueous fluorescein, 6% w/v gelatin in n-heptane. Gels were melted then poured into the cavity of a microscope slide and covered with a glass cover-slip, the sample thickness being about 1mm.

3. Results and Discussion

SANS spectra are plotted in the form of absolute scattered neutron intensity, I(Q)/cm^{-1}, vs. scattering vector, Q/Å^{-1} where Q is related to the neutron wavelength, λ, and scattering angle θ by $Q = (4\pi/\lambda)\sin(\theta/2)$.

The parent water-in-oil microemulsion (based on the water/surfactant mole ratio and before addition of gelatin) has a SANS spectrum characteristic of spherical droplets with radius around 100Å. On addition of gelatin, such that the system becomes gelled, the scattering pattern changes significantly at small angles, Fig.(1). Below 30°C in water, the preferred conformation of gelatin is a triple helix. The gelatin helices are rigid rods with a wide range of lengths, the average being ≈1000Å. The dramatic enhancement of the scattered neutron intensity at low angles indicates the formation of larger structures in the system. When plotted in double logarithmic form, one can see that there are two distinct gradients in the scattering profile, Fig.(2). At high Q, in the Porod region, the slope of the plot is -4, which is the same as that observed for the parent microemulsion, and is due to scattering from the relatively smooth AOT interface between the water and oil domains. At low Q the slope is -1 which is characteristic of one-dimensional structures ie. long, rigid rods. Initially the data were fitted using a rod form factor alone, but the agreement of the fit to the data was poor in the mid-Q region. It therefore appears that the structure is more complicated. Other considerations indicate the existence of droplets in the presence of the gelatin rod network. Adding a contribution to the scatter from a droplet form factor significantly improves the fit. Thus we propose that the organo-gel structure is made up of a network of rigid gelatin/water rods coexisting with microemulsion-like droplets, all stabilised by a monolayer of AOT. A detailed fitting of the data gives rod dimensions of 100Å radius and >700Å length with ≈70% of the scatter coming from polydisperse droplets of number-mean radius 56Å[4].

Of particular interest in the FRAP experiments has been the extent of recovery of fluorescence after photobleaching. For fluorescein solubilised in the droplets of the parent microemulsion and for fluorescein in a 20% w/v aqueous gelatin gel there is 100% recovery

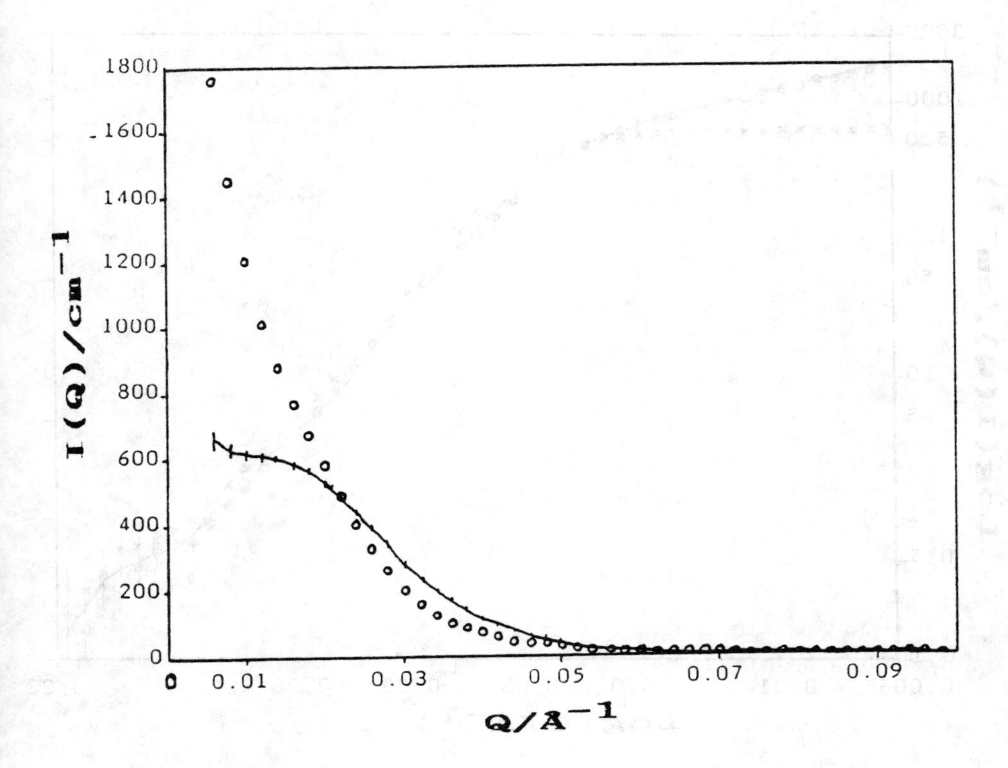

Figure 1. $I(Q)/cm^{-1}$ vs. $Q/Å^{-1}$ SANS spectrum of a 0.1 $mol.dm^{-3}$ AOT, 10% v/v D_20, n-Heptane(H) microemulsion gelled with 3.5% w/v gelatin (circles) and gelatin-free (solid line).

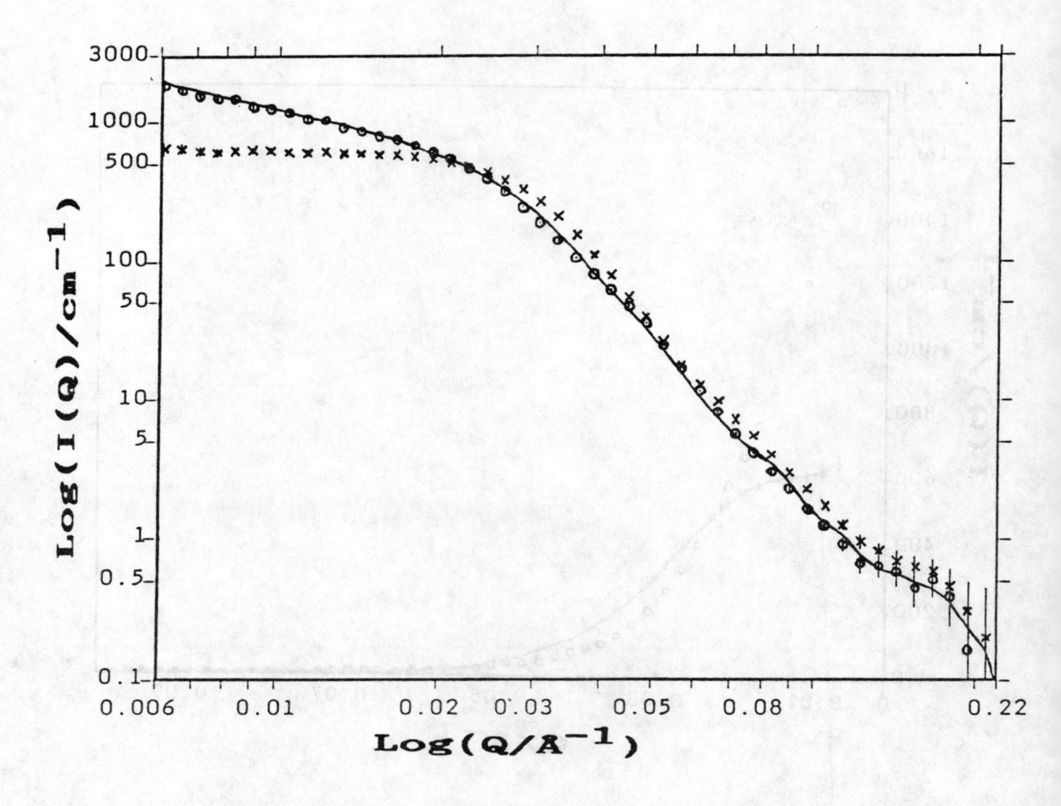

Figure 2. Log-Log plot of spectra in Fig.(1) with fit to gel curve (solid line through circles) as described in text; crosses denote parent microemulsion scattering.

of fluorescence after a few seconds, with diffusion coefficients of $5.5x10^{-7} cm^2 s^{-1}$ and $1.8x10^{-6} cm^2 s^{-1}$ respectively as determined by FRAP. These values are rather low, which suggests that there is significant droplet-droplet attraction, as expected at this water/surfactant mole ratio and volume fraction of dispersed phase [5]. For the aqueous gel, there is a gelatin-fluorescein interaction. FRAP experiments on strongly gelled microemulsions labelled with fluorescein show that there is only partial recovery of the initial fluorescence intensity after photobleaching, Fig.(3). This result suggests that there are two populations of fluorophore

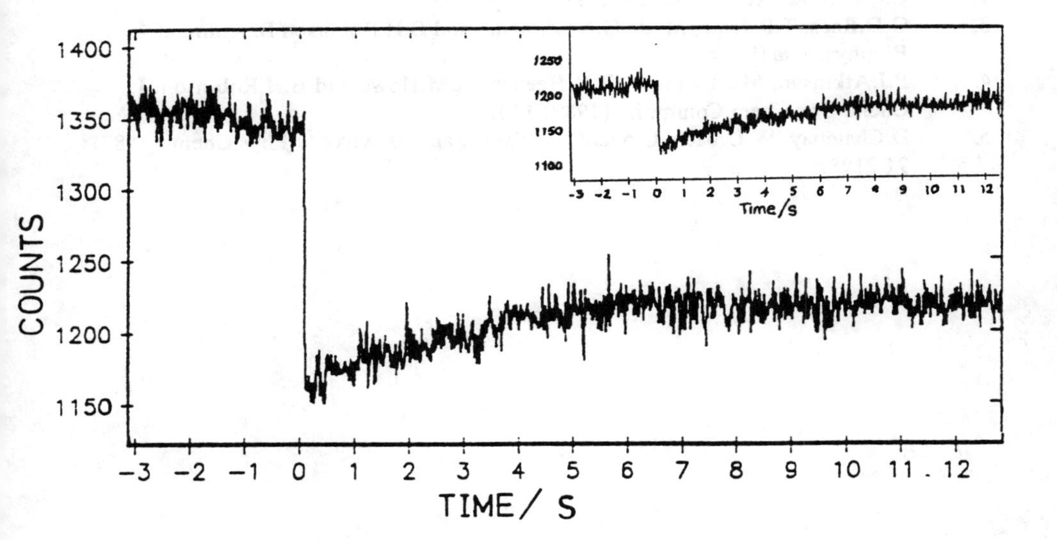

Figure 3. FRAP experimental data plot of the partial fluorescence recovery from an organo-gel of composition 0.15 mol.dm^{-3} AOT, 15% v/v $5x10^{-7}$ mol. dm^{-3} aqueous fluorescein, 6% w/v gelatin, n-Heptane. Inset: repeat bleach of same spot on same sample.

present: one essentially immobile and the other moving. If the same spot is bleached subsequently, recovery is complete and gives a similar diffusion time to the initial recovery (inset on Fig.(3)). The diffusion coefficient calculated from this partial recovery is 6.0×10^{-7} $cm^2 sec^{-1}$. This value of D_T is close to that of the parent microemulsion, but also consistent with diffusion of the fluorescein down the gelatin/water rod networks when the finite rod width and high aqueous gelatin concentration (40%) in the organo-gel are taken into account. The fitted SANS data suggest that 70% of the scattered intensity arises from microemulsion droplets, so if we assume that the fluorescein is evenly distributed throughout the aqueous domains, then the relative amounts of permanently bleached and recovered fluorescence intensity should provide a clue as to the location of the mobile part of the fluorescein. The fraction of immobile fluorescein obtained from Fig.(3) is 0.7, which when compared to the fitted SANS data suggests that it is the droplet component of the gel structure which is immobilised. It is expected that, as for the parent microemulsion, there will be interactions between the surfactant associated with the droplets and the surfactant associated with the gelatin/water/AOT mesh in which the droplets are located. Thus there appear to be severe constraints imposed on the droplet motion by the gel network structure.

4. References

1. G.Haering and P.L.Luisi, J.Phys.Chem. (1986) 90 5892.
2. C.Quellet and H-F.Eicke, Chimia (1986) 40 233.
3. G.D.Rees, T.R.Jenta, M. da G.Nascimento and B.H.Robinson Biochim. et Biophys. acta (in press).
4. P.J.Atkinson, M.J.Grimson, R.K.Heenan, A.M.Howe and B.H.Robinson, J. Chem. Soc.Chem.Commun., (1989) 1307.
5. D.Chatenay, W.Urbach, C.Nicot, M.Wacher and M.Waks, J.Phys.Chem. (1987) 91 2198.

LUMINESCENCE PROBE STUDY OF ORGANIZED ASSEMBLIES TREATED AS FRACTAL OBJECTS

P. LIANOS
University of Patras
School of Engineering
26000 Patras
GREECE

ABSTRACT. Diffusion and transport in microemulsions and lipid vesicles can be successfully described by utilizing ideas from the theory of random walks and the properties of fractal objects. A new parameter, the non-integer spectral dimension, governs the rate of diffusion and transport. Results are presented concerning three subjects: (1) relation between electric conductivity in percolating water-in-oil microemulsions and their fractal behaviour; (2) effect of a water-soluble polymer on water-in-oil droplets; and (3) structural variations in mixtures of phospholipid vesicles along a line of progressively varying microviscosity.

1. Introduction and Theory

Transport and interaction in non-homogeneous media can be well described by utilising ideas from the theory of fractal objects. The main condition which must be satisfied in order to represent a system by fractals is the existence of self-similarity, i.e. symmetry by dilation. Micelles, microemulsions, lipid vesicles and generally organized assemblies are of such nature that they favor self-similarity, at least in a statistical manner.

A quantitative expression [1] of self similarity is that the number of the available active sites increases as N^{d_f}, when the system is magnified by a factor of N. The exponent d_f is a static, geometric quantity and in homogeneous euclidean space is an integer number equal to the dimension of the system. Indeed, as seen in fig. 1a, when a linear (one dimensional) distribution of sites is magnified by a factor of two, the number of sites is doubled, i.e. it is multiplied by 2^1, as seen in fig. 1b. When the two-dimensional distribution of sites in fig. 1c is magnified again by a factor of two, the total number of sites is multiplied by 4, that is 2^2, as seen in fig. 1d. Likewise with a three dimensional distribution where magnification by two multiplies the number of sites by 8, i.e. 2^3. We notice then that in homogeneous systems where we always have self-similarity the number of sites increases as N^1 or N^2 or N^3 where 1, 2, 3 is the dimension and N is the magnification. Suppose now that the field of sites is no more homogeneous and isotropic but it follows the distribution of fig. 1e (dark squares). This field is embedded in a two-dimensional euclidean space but it is not two-dimensional itself, in the sense that it does not involve the total number of sites of the euclidean space.

D. M. Bloor and E. Wyn-Jones (eds.),
The Structure, Dynamics and Equilibrium Properties of Colloidal Systems, 309–321.

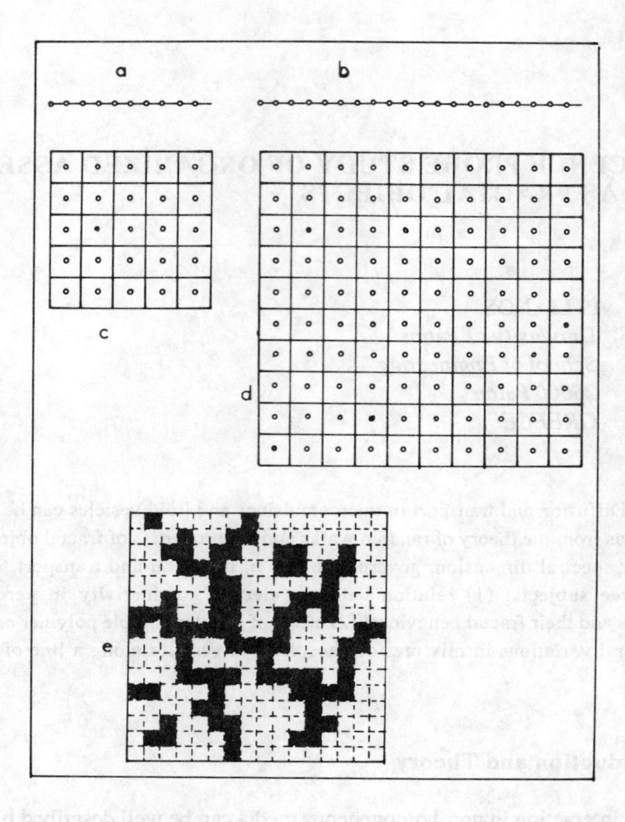

Figure 1. (a) One-dimensional field of ten sites. (b) Same field magnified by a factor of two. (c) Two dimensional field of twenty five sites. (d) Same field magnified by a factor of two. (e) A domain of non-integer dimensions embedded in a two dimensional euclidian space.

Magnification of this field under the condition of keeping self-similarity will increase its sites by a factor

$$S(N) \sim N^{d_f} \qquad (1)$$

where d_f is no more an integer number but it lies between one and two. Similarly, in three dimensions d_f will be a non-integer number smaller than three. Eq. 1 offers then the definition of an important new geometric quantity, the fractal dimension d_f. However, this quantity is static, that is, it cannot describe by itself a dynamic phenomenon such as diffusion and chemical reaction. What matters to interacting molecules is not only the number of the available sites but also the number of times each molecule visits a distinct site during its diffusive motion. But, of course, most of chemical reactions are diffusion limited, so the path of each reactant is of prime importance. In fractal modeling the reactant path is considered to be a random walk and in principle each reactant can visit a site an unlimited number of times. It is, therefore, necessary that this last number comes into consideration.

According to the classical diffusion equation, the mean square of the path followed by a diffusing molecule is $<r^2(t)> \sim t$. In a non-homogeneous environment, however, diffusion is anomalous and

$$<r^2(t)> \sim t^{2/d_w} \tag{2}$$

where $d_w>2$, since more time will be now needed for the same mean path [2]. Suppose that time t corresponds to N steps of the random walker. Obviously, $<r^2(t)> \sim N^{2/d_w}$ and

$$N \sim (\sqrt{<r^2(t)>})^{d_w} \tag{3}$$

Eq. 3 indicates that since N is the total number of sites visited and $\sqrt{<r^2(t)>}$ is the mean distance travelled, or the mean size of the space where the random walk is included, then according to the definition of Eq. 1, d_w is the fractal dimension of the random walk [2]. d_w is not a static parameter.

The total volume available within the diffusion distance is $V(t) \sim (\sqrt{<r^2(t)>})^{d_f}$, since $\sqrt{<r^2(t)>}$ is the mean displacement and d_f is the dimension of the fractal domain. The number of distinct sites visited by the walker after performing N random steps should be proportional to the volume V(t) which includes the random walk. Now, in a so-called "compact" exploration the random walker returns to its origin. The probability then of a compact exploration after N steps, P(N), will be higher when V(t) is smaller. Then we expect $P(N) \sim 1/V(t) \sim 1/(\sqrt{<r^2(t)>})^{d_f}$. Substituting $\sqrt{<r^2(t)>}$ from Eq. 3, we get

$$P(N) \sim N^{-d_f/d_w} \tag{4}$$

However, the probability of a compact exploration is related to the number of states $n(\varepsilon)$ by the equation [2] $P(N) = \int_0^\infty n(\varepsilon).\exp(-\varepsilon N)d\varepsilon$ which is a Laplace transform for $n(\varepsilon)$. It is then obvious that $n(\varepsilon) \sim \varepsilon^{d_f/d_w-1}$ which is similar to the corresponding relation [2] for homogeneous euclidean spaces $n(\varepsilon) \sim \varepsilon^{d/2-1}$, where d is the euclidean dimension. These expressions lead us to the definition of a new physical quantity

$$d_s = 2d_f/d_w$$

which is given the name spectral dimension. The density of states, which is proportional to the total number of times a random walker visits a distinct site scales in fractals according to the spectral and not the eucledian or the fractal dimension and according to the relation [2]

$$n(\varepsilon) \sim \varepsilon^{d_s/2-1} \tag{5}$$

The fractal dimension d_f is responsible for the distribution of sites. The spectral dimension d_s is responsible for the distribution of states, i.e. the number of times each site is visited by the random walker. As we have already said, the number of distinct sites visited will finally be given as $S(N) \sim V(t) \sim 1/P(N)$, i.e.

$$\text{(a)}\quad S(N) \sim N^{d_s/2} \qquad \text{for } d_s < 2$$
$$\text{(b)}\quad S(N) \sim N \qquad \text{for } d_s > 2 \tag{6}$$

The limit of 2 is obvious, since an exponent > 1 would have no physical meaning. The first of the Eq. 6 is valid for compact exploration in restricted spaces of fractal dimensions and the second is valid for homogeneous spaces.

Eq. 6 can be rewritten as $S(t) \sim t^f$ where $f = d_s/2$. In fact this is an asymptotic expression [3] obtained for $t \to \infty$, but it is accepted that the asymptotic limit is reached very fast and to a good approximation [4]. Let us now consider a reaction A+B → products, where the concentration of A is much smaller than the concentration of B (the so called target problem [5]). A can be an excited lumophore, L^*, and B a quencher Q, in a usual luminescence probing (quenching) procedure, i.e. $L^* + Q \to$ products. The time-dependent luminescence intensity following pulse excitation will be described by

$$I(t) = I_o \exp(-k_o t) \exp(-kQt) \tag{7}$$

where k_o is the decay constant in the absence of quencher, Q is the quencher concentration and k is the reaction rate. According to the above discussion and the fact that dynamic quenching is always diffusion limited k should be found as the first derivative with respect to time of S(t), the number of distinct sites visited by the random walker, Apparently, if $S(t) \sim t$, as in homogeneous spaces, k is a constant, the second-order reaction rate constant known so far. When, however, $S(t) \sim t^f$ then, as expected, k is a function of time, i.e. $k(t) = (dS(t)/dt) \sim t^{f-1}$, and Eq. 7 is modified into [6-11]

$$I(t) = I_o \exp(-k_o t) \exp(-k'Qt^f),\ f = d_s/2 < 1, \tag{8}$$

where k' is a constant. The exact expression for k(t), consistent with Eqs. (7) and (8) is actually [11]

$$k(t) = \frac{fk'}{t^{1-f}} \tag{9}$$

The decay then of a lumophore in the presence of quencher in a non-homogeneous environment treated as fractal is given by Eq. 8 while the reaction rate is time-dependent and it is given by Eq. 9. The exponent f is equal to one-half of the spectral dimension while it is proportional to the fractal dimension of the reaction domain. Inherent to the above analysis is the fact that no restriction is applied to the reacting molecules except the non-homogeneity of the reaction domain itself. In fact we assume that the domain is continuous, i.e. it is in the form of a percolation cluster. For this reason, we conclude that Eq. 8 is valid only for percolation clusters and it should, therefore, be a limit of a more general expression. Such an expression is derived in the following paragraph.

When the effective microviscosity of the reaction medium is very high or when the reactants are compartmentalized, quenching is ruled mainly by short-distance diffusion, even in cases of relatively long life times of the excited state. Then the reaction rates w_i are expected to depend explicitly on the distance r_i between the lumophore and the ith quencher [9]. In long-range interactions quenching is still possible even with immobile quenchers. Thus for singlet energy transfer $w_i = a/r_i^6$. Quenching by excited-state complex formation, however, as it is usually the case, is ruled by multipole interactions where $w_i = a/r_i^s$, $s > 6$, so that w_i decreases fast with distance. Immobile quenchers can then give no quenching at all, unless they are "in contact" with the lumophore. Nevertheless, for slowly moving quenchers a quenching model can still be formulated in the following manner. The moment a luminescent molecule is excited, there is a certain distribution of possible quenchers at distances r_{oi}. Since quenching is a matter of diffusion, we expect w_i to be proportional to the diffusion coefficient D and to decrease with increasing r_{oi}. Now, the experimentally observable quantity will be an average over all possible paths, therefore $r_{oi} = \sqrt{<r^2(t)>}$ and since $r^2(t) \sim Dt$ then we expect

$$w_i \sim \frac{D}{r_{oi}^2} \tag{10}$$

This expression is, of course, true only for homogeneous spaces while in the non homogeneous case the exponent of r_{oi} should be different from two [9]. Comparison with the discussion leading to Eqs. 2 and 3 suggests that 2 should in fact be substituted by d_w, the fractal dimension of the random walk. Having thus obtained an explicit expression for w_i we may proceed to calculate the survival probability of the excited lumophore and its luminescence intensity following pulsed excitation according to a procedure found by Allinger and Blumen [9, 12]:

$$I = I_o \exp(-k_o t) \exp(-kDQt^{d_f/d_w} + Q \sum_{m=1}^{\infty} D^m C_m t^{m-2m/d_w + d_f/d_w}) \tag{11}$$

An important remark should be made in view of this equation. d_f is the average dimension of the reaction domain and it is equal to the fractal dimension if the domain is fractal. d_w is the fractal dimension of the random walk and it contains the assumption that the random walk is a fractal. Apparently, Eq. 11 is still valid in a non-fractal restricted domain. In the latter case d_f and d_w should be properly redefined. Notice that $d_f/d_w = f = d_s/2$, also found in Eq. 8. Apparently, the case of Eq. 8 is the limit to which Eq. 11 tends when the terms contained in the infinite series tend to zero. The most general expression then for the reaction in restricted spaces is given by the complicated Eq. 11. If the domain possesses symmetry by the dilation then d_f is the fractal dimension. Eq. 11 can be presented in a more elegant form in another expression found by Blumen, Klafter and Zumofen [13-15]

$$I = I_o \exp(-k_o t) \exp(-\lambda \alpha t^f + \lambda^2 b t^{2f} + ...) \tag{12}$$

where $\lambda = \ln(1-p)$ and p is the occupation probability of the available sites by the quenchers. The terms inside the second exponential are an approximation of an infinite series. We notice then that kDQ is substituted by $\lambda.\alpha$ where α is a constant, while $\lambda^2 b t^{2f}$ is a first order approximation of the series, b is a constant, and $f = d_s/2$ as above. Obviously, Eq. 8 is an equivalent zero order approximation and Eq. 12 first order approximation of Eq. 11. The choice of the order of approximation can be made with the help of the values of λ and p. When the occupation probability p is very high, as in cases where the reaction space is limited, then ln(1-p) takes relatively large values and λ^2 is different from zero. Then Eq. 12 applies. When the restrictions are relaxed and the reactants are allowed to move greater distances during the excited state, then p is small, λ^2 tends to zero and Eq. 8 applies. Typical physical examples for the application of Eq. 12 is luminescence quenching in non-communicating or hardly communicating micelles [9,16], or lipid vesicles with very viscous lipid core [8,11]. For the application of Eq. 8, typical examples are found in percolating microemulsion [6,9,17] droplets. Such examples will be presented in detail below. In the present work, Eq. 12 is actually employed in the form [11]

$$I = I_o \exp(-k_o t)\exp(-k'Qt^f + k''Qt^{2f})$$
$$\text{where } k'Q = \lambda\alpha,\ k''Q = \lambda^2 b \tag{13}$$

2. Results and Discussion

2.1 RELATION BETWEEN THE FRACTAL BEHAVIOUR AND THE ELECTRIC CONDUCTIVITY OF ELECTRICALLY PERCOLATING WATER-IN-OIL MICROEMULSIONS

The application of fractal models to describe luminescence quenching in water-in-oil (w/o) microemulsions has been verified in our previous works [6,7,9,10]. Eq. 8 has been successfully applied to microemulsions above the percolation threshold, that is in microemulsions where the communication between droplets is sufficient enough to allow the interacting molecular species to move from droplet to droplet within the lifetime of the excited state. In this work we apply Eq. 8 to various quaternary w/o microemulsions and we deduce useful information by calculating the parameters k' and f and by relating them to data obtained with conductivity measurements.

The existence of electric percolation in w/o microemulsions is also an exchange phenomenon between droplets [18] no matter what is the specific mechanism of droplet communication [19-21]. In general, we find Eq. 8 applicable to w/o microemulsions where the interacting species exchange between droplets and at the same time the measured electric conductivity has an appreciable value.

In the present work, w/o microemulsions have been studied with luminescence probing by employing $Ru(bpy)_3^{2+}$ as lumophore and $Fe(CN)_6^{3-}$ as quencher. The microemulsions consisted of aliphatic oils, SDS, PeOH and water. Two series have been studied one with dodecane and other with heptane. The compositions were chosen for dodecane along the demixion line in the pseudoternary phase diagram and for heptane in analogy to dodecane. In both cases there exists an electric percolation when water is at least 5% by weight. However, the conductivity values are much higher in dodecane than they are in heptane, as it can be deduced by the semilogarithmic plot of Fig. 2a. Since conductivity values and exchange rates k_e between droplets should increase together, it is expected that k_e-values should be larger in the case of dodecane than in the case of heptane. k_e can be measured by analyzing the gradient of the luminescence decay profile at long times [22], and in that case it actually represents the quenching due to quencher transfer between droplets, while the quenching behaviour at short times is mainly due to short distance quenching mechanisms. However, the values coming out from long-time analyses are not accurate due to errors originating from electronic noise. Thus in Fig. 2b, no detectable difference exists between heptane and dodecane [23].

On the contrary, application of Eq. 8 which is simple and utilises a minimum number of parameters did give a distinction between the two types of microemulsions while the decay profile could then exploited at both short and long decay times without restrictions. The parameters f and k' were calculated by fitting Eq. 8 to the experimental decay profiles. The criteria for the quality of the fit were the same as in previous publications [6-10]. Figure 2c and Table 1 show all calculated f values. Notice that f are always larger in the case of dodecane than in the case of heptane. The differences in the f values between the two series are consistent with conductivity values. f is proportional to the spectral dimension which is proportional to the fractal dimension of the reaction domain, as said above. The latter obviously extends beyond a single droplet [6,7,9,10]. As f approaches unity the domain becomes more homogeneous which is equivalent to saying that the number of communication channels between droplets increases. Of course, in that case the conductivity attains higher values. Furthermore, this conclusion is consistent with the effect of temperature on the values of f. f is expected to increase with increasing temperatures since the communication between droplets is even easier at higher temperatures. This fact is demonstrated with the values of f in Table 2, given for dodecane. Similar results are obtained with heptane.

Table 1 contains also the calculated values of the constant k'. There is a problem with defining appropriate units for k' since time is raised to a non-integer exponent which is not the same for all cases analysed. In fact the second order reaction (quenching) rate constant k in restricted spaces is time-dependent and it is given by Eq. 9. k(t) is always measured as $M^{-1}s^{-1}$, no matter what the value of f is. k(t) decreases very fast at short times and it becomes almost constant at long times. This distinction is more pronounced when f is smaller Because of this variation with time, we have also calculated an average value for k(t) given in Table 1 as $\bar{k}$. The average has been calculated over 500 channels of time (2,16 ns/channel).

By comparing the values of $\bar{k}$ of Table 1 and the values of k_e of Fig. 2b we find that they are of about the same magnitude. This is consistent with the original assumption that the reaction domain extends beyond a single droplet therefore the reaction rate constant is mainly defined by the ability to transfer molecules from droplet to droplet. The fast increase of $\bar{k}$ at low water content is consistent with the fast increase of conductivity there.

The $\bar{k}$ values which are shown in Table 2 as a function of temperature increase with temperature. This again is a consistent result as reaction is expected to be facilitated at higher temperatures.

In conclusion, we have found an association between the fractal nature of the reaction domain in w/o microemulsions and the values of conductivity in these microemulsions. Conductivity values increase as the above domain becomes less fractal.

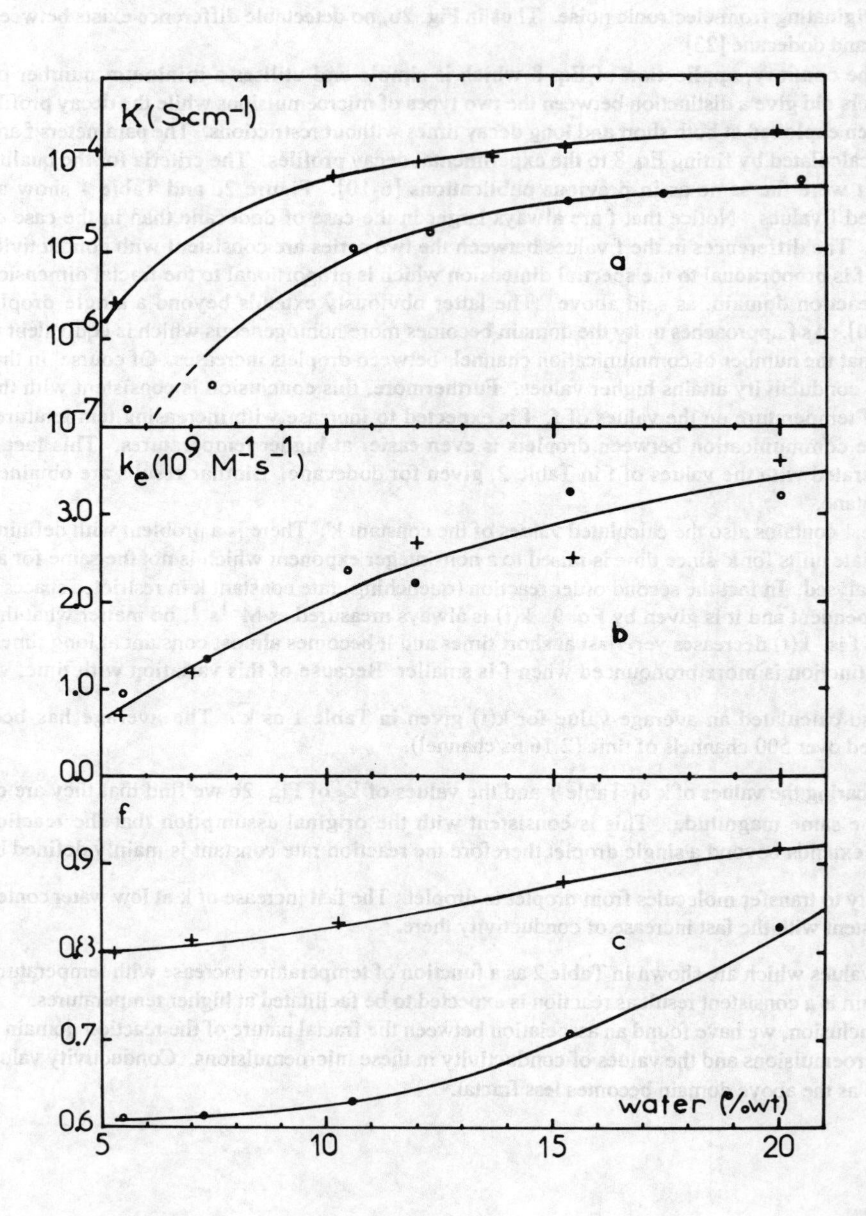

Figure 2. **Values of conductivity K (a), exchange rate constant k_e (b) and fractal exponent f (c) for heptane (o-o-o-o) and dodecane (+-+-+-+) w/o microemulsions.**

Table 1. Values of the fractal exponent f and the average second order quenching rate constant $\bar{k}$ for heptane and dodecane w/o microemulsions at various water contents.

Water content (% wt)	f	$\bar{k}$ ($10^9 M^{-1}s^{-1}$)
Heptane		
5.6	0.61	0.7
7.2	0.61	0.8
10.6	0.63	2.5
15.4	0.71	2.7
20.0	0.83	2.2
Dodecane		
5.7	0.80	1.0
7.0	0.81	1.2
10.1	0.83	3.1
15.3	0.88	3.0
20.0	0.92	2.9

Table 2. Values of the fractal exponent f and the average second order quenching rate constant $\bar{k}$ for a dodecane w/o microemulsion at 5.4% wt of water at various temperatures.

Temperature (°C)	f	$\bar{k}$ ($10^9 M^{-1}s^{-1}$)
20	0.80	1.0
25	0.80	1.0
29	0.84	1.1
37	0.89	1.3
45	0.91	1.6

2.2 EFFECT OF A WATER SOLUBLE POLYMER, POLYOXYETHYLENE GLYCOL, ON THE STRUCTURE OF W/O MICROEMULSIONS

A series of w/o microemulsions have been studied with the use of Eq. 13 by employing $Ru(bpy)_3^{2+}$ as lumophore and $Fe(CN)_6^{3-}$ as quencher. The scope of the present study was to investigate the effect of a water-soluble polymer solubilised in the dispersed phase on the structure of the microemulsion droplets. Polyoxyethylene glycol-6000 has been used. The microemulsions contained sodium dodecylsulfate, water, pentanol and one of the three oils: dodecane, toluene, cyclohexane. The ratio of water to surfactant (by weight) was always around 2,55. The compositions were chosen along the demixion line my mixing all products except alcohol and finally adding alcohol until clear solution. The polymer content was always 0,72% wt. Table 3 gives the results of the analysis by fitting Eq. 13 to the luminescence decay profiles. The average quenching constant $\bar{k}$ has been calculated as in the previous paragraph by employing Eq. 9. No data are shown for dodecane, which is an intermediate case between toluene and cyclohexane. Notice first the values of f obtained with cyclohexane in the absence of polymer. As the water content increases f also increases. This result is expected, since with all w/o microemulsions studied so far increase of water makes the reaction domain less fractal, as it has been found also in the previous paragraph. On the contrary, the values of $\bar{k}$ were rather unexpected since they show a decrease with increasing water content. In the previous paragraph, where w/o microemulsions in heptane or dodecane were studied, the ratio of water/surfactant increased with water while in the present case it remains constant. Apparently, the number of droplets increases, which leads to decreasing the ratio of interacting species per droplet, since the global concentration of the latter remains constant. This greater compartmentalization of the reactants decreases the reaction rates. In combination with the increasing f values we may conclude that the reaction domain increases with higher water content while the local reaction microconcentration, i.e. the ratio of reactant per available site decreases.

Table 3. Values of the fractal exponent f and the average second order quenching rate constant $\bar{k}$ for various w/o microemulsions in the absence and presence of polyoxyethylene glycol*.

Oil	Water content (% wt)	Polymer	f	$\bar{k}$ ($10^9 M^{-1} s^{-1}$)
Toluene	5.80	no	0.76	3.6
Toluene	5.80	yes	0.85	2.7
Cyclohexane	6.10	no	0.52	5.5
Cyclohexane	6.10	yes	0.92	4.8
Cyclohexane	12.19	no	0.93	4.0
Cyclohexane	12.27	yes	0.99	3.4
Cyclohexane	17.85	no	0.99	3.1
Cyclohexane	17.85	yes	1.00	2.6

* The microemulsions contain toluene or cyclohexane, pentanol, water and sodium dodecylsulfate (water/surfactant = 2.55).

In the presence of polymer we have detected a systematic increase of f and decrease of $\bar{k}$ both in toluene and cyclohexane and for all water contents. In view of the comments just given we may conclude that the polymer does just what the increase of surfactant and water do. It increases the number of droplets and extends the reaction domain. This problem is actually studied further in our laboratory.

2.3 STRUCTURAL STUDY OF VARIOUS PHOSPHOLIPID VESICLES TREATED AS FRACTAL OBJECTS. USE OF PYRENE AS FLUORESCENT PROBE

The fluorescence decay profiles of pyrene monomer in the presence of excimer, solubilised in various phospholipid mixtures have been analysed with Eq. 13. Deoxygenated pyrene samples at the concentration of 2 x 10^{-5} M have been used for recording pyrene fluorescence decay profiles. Steady-state fluorescence anisotropy measurements were made by using diphenylhexatriene-labelled vesicles. For details on preparation see ref. 8 and 11.

Table 4 and Fig. 3 show the results of the analysis of the fluorescence decay profiles of pyrene monomer in the presence of excimers for multilamellar vesicles obtained with varying proportions of dipalmitoylphosphatidylcholine and L-α-phoshatidylglycerol, henceforth appearing as DPPC-LαPG vesicles. The values of f are, as we said above, equal to $d_s/2$, the one half of the spectral dimension. Given also in Table 4 and Fig. 3 are the values of r, the corresponding fluorescence anisotropy. Fig. 3 shows clearly that by increasing the LαPG content in the vesicles, the fluorescence anisotropy undergoes a monotonous decrease. Since r is an index of the equivalent microviscosity in the vesicle lipidic core, the present results show that the variation of the composition results in a rather smooth variation of the fluidity of the vesicles. This is not the case with the values of the estimated spectral dimension of the (hydrophobic) reaction medium. These values increase rapidly with a rather small percentage of the "fluid" LαPG lipids and they obtain a value which varies little with further LαPG addition. As we have already said, d_s is a combination of the geometry of the medium and the dynamic aspect of pyrene-molecule diffusion, therefore the obtained f's and (d_s's) indicate an important restructuring of the vesicles in the presence of a rather small LαPG content. This fact cannot be detected by fluorescence anisotropy measurements.

Notice, that the constant k' and k", appearing in Eq. 13, cannot be given usual units because of the non-integer power of time that multiplies them. Even more, this power does not remain constant thus preventing comparison of the data obtained with mixtures of various compositions. For this reason and in order to get some representative reaction rates we have employed k(t) as given by Eq. 9.

To get an idea of how k(t) varies with time in the above experiments we have plotted k(t) for different values of k' and f against time, which varies between 1 to 750 ns (equivalent to 400 channels of time). Fig. 4 shows the results of five different k'-f pairs. Notice that k(t) varies extensively within a few first channels and it varies more slowly for the rest of the time. This trend is less marked in the cases where f is close to 1. At short times, k(t) decreases rapidly with increasing LαPG content. This result can be explained by assuming that in the presence of LαPG the reactants are distributed in a larger reaction domain thus decreasing the local quencher concentration, even though the global concentration remains constant. The local quencher concentration is not a measurable quantity but it affects the (overall) reaction rate. This result is in the same line as the variation of f's, since the latter are expected to approach unity when the reaction domain becomes larger and more homogeneous (less fractal).

Table 4. Values of the fluorescence anisotropy r, the fractal exponent f, the average quenching rates $\bar{k}'$ and $\bar{k}''$, the constant k" and the excimer-monomer intensity ratio I_E/I_M for multilamellar vesicles obtained with mixtures of dipalmitoylphosphatidyl-choline and L-α-phosphatidylglycerol (DPPC-LαPG)

composition DPPC/LαPG (w%)	r	f	$\bar{k}'(10^{11}M^{-1}s^{-1})$	k"	$\bar{k}''(10^{11}M^{-1}s^{-1})$	I_E/I_M
100/0	0.335	0.64	2.77	24.2	1.62	0.57
90/10	0.324	0.74	3.96	9.1	2.22	
80/20	0.306	0.83	3.05	2.3	1.68	0.96
70/30	0.280	0.89	3.28	1.0	1.71	0.96
60/40	0.255	0.90	2.59	0.5	0.98	0.85
50/50	0.218	0.94	2.80	0.26	0.86	0.85
40/60	0.178	0.96	2.80	0.17	0.76	0.91
30/70	0.142	0.99	3.05	0.12	0.82	
20/80	0.121	0.99	2.57	0.10	0.67	
0/100	0.106	0.99	3.29	0.10	0.90	1.19

At longer times we cannot observe a single trend in the variations of k(t) with composition. The curves of Fig. 4 intercross and the picture becomes very confusing - especially, if more curves were included. For this reason we have calculated the average values of k(t), appearing as $\bar{k}'$, and we show them in Table 4. Notice, then that $\bar{k}'$ remains of the same order of magnitude throughout the whole range of DPPC-LαPG vesicles studied, while a maximum is observed for 10% LαPG-90% DPPC. This behaviour can be rationalised in terms of the opposing effect of the decrease of the local reactant concentration and, at the same time, the decrease of the local microviscosity with increasing LαPG content. The reaction is hindered by the decrease of concentration and facilitated by the decrease of microviscosity. The calculated values of k(t) and $\bar{k}'$ may help rationalize some other rather unusual data obtained by measuring the ratio I_E/I_M of the excimer (480 nm) and the monomer (373 nm) intensity of pyrene fluorescence, and shown as last column in Table 4. Notice that there is a steep increase of excimer in going from 100 to 80% DPPC, then the ratio remains almost constant to increase a little bit further at 0% DPPC. It is, of course, expected that excimer is facilitated by decreasing microviscosity but obviously there is no close connection between the r values of Fig. 3 and Table 4 and the I_E/I_M values of Table 4. Nevertheless, the $\bar{k}'$ values vary in the same way as I_E/I_M, that is the two increase together, as expected.

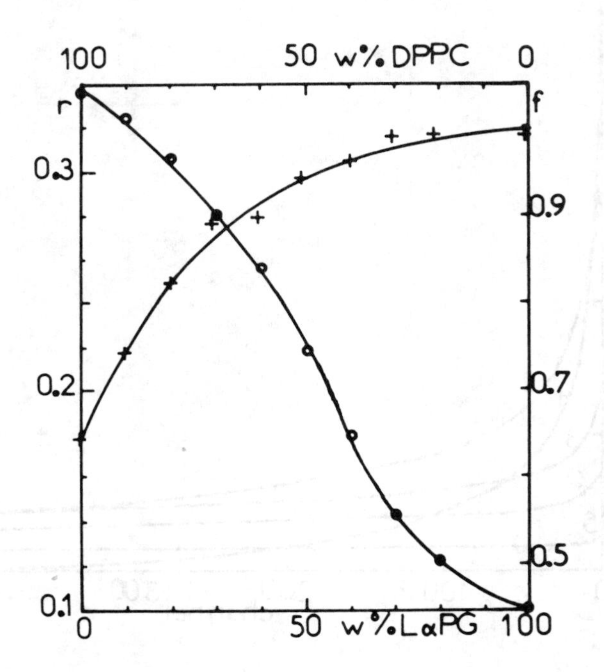

Figure 3. Plot of the variation of the fluorescence anisotropy r (oooo) and the fractal exponent f (+++++) with respect to the constitution of dipalmitoylphosphatidylcholine- L-α-phosphatidylglycerol (DPPC-LαPG) mixtures.

The relative values of k" given in Table 4 decrease rapidly at relatively small LαPG content, while with further variation of vesicle composition the decrease of k"slows down. Thus the k"= 0 domain is never reached and the data are described only by Eqs. 12 and 13 and never by Eq. 8.

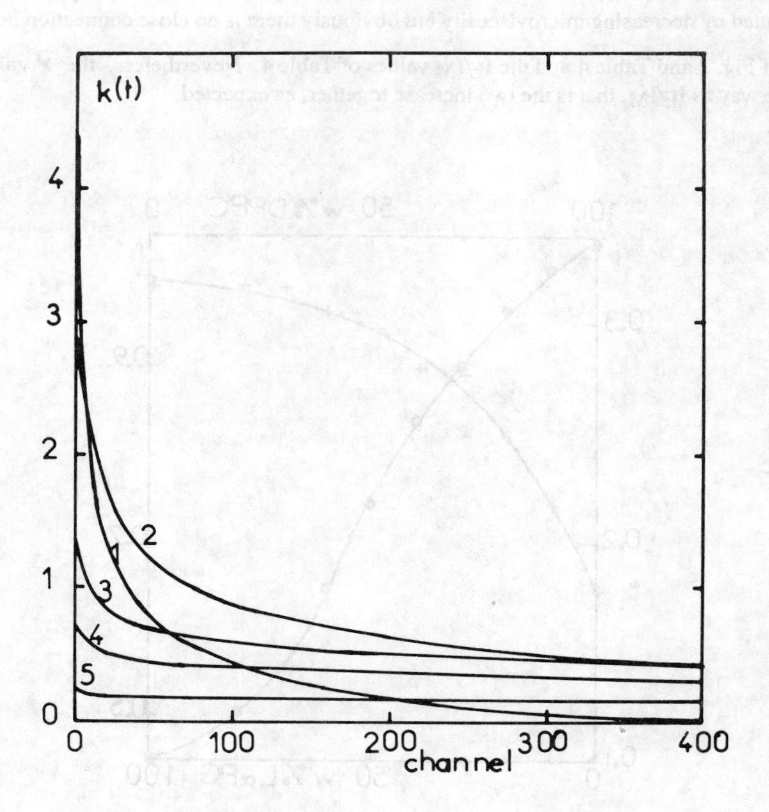

Figure 4. Plot of the variation of the quenching rate with time for various DPPC-LαPG mixtures (%wt): (1) 100/0, (2) 90/10, (3) 70/30, (4) 40/60 and (5) 20/80. Arbitrary units on the vertical axis.

This result shows that the reaction domain cannot be presented by a percolation cluster which, of course, is expected for isolated and non-communicating vesicles. However, Eq. 8 is also applicable to large vesicles with very low microviscosity in the lipid core [8], where the high "fluidity" allows averaging out the local microheterogeneity. Apparently, this is not the case with pure LαPG.

An average $\bar{k}''$ can be estimated by following a similar procedure as in the case of $\bar{k}'$. The $\bar{k}''$ values are also shown in Table 4. Notice that there exists a substantial decrease of $\bar{k}''$ with large amounts of LαPG. This result comes in support of the original conclusion that as LαPG increases the pyrene molecules are distributed in a larger domain thus decreasing local concentration. This is equivalent to saying that the occupation probability p decreases giving smaller λ, and, finally smaller $\bar{k}$" (see introduction).

In conclusion, we have successfully analysed the decay profile of pyrene monomer in the presence of excimers with the fractal model of eqs. 12 and 13. Pyrene was solubilised in a series of vesicles formed with binary mixtures of two lipids along a line of progressively decreasing microviscosity. We have detected an important restructuring of the vesicles in the presence of small percentages of LαPG, contrary to what is suggested by a smoothly varying microfluidity. The average reaction rates did not change much with decreasing microviscosity and this can be explained only by accepting existence of opposing trends of molecular redistribution and increased facility of molecular motion.

3. Acknowledgements

Financial aid from the Greek General Secretariat of Research and Technology and from NATO CR Grant No 0405/88 is gratefully acknowledged.

4. References

1. Even, U., Rademann, K., Jortner, J., Manor, N. and Reisfeld, R. (1984) 'Electronic energy transfer on fractals', Physical Review Letters 52, 2164-2167.

2. Havlin, S. and Ben-Avraham, D. (1987) 'Diffusion in disordered media', Advances in Physics 36, 695-798.

3. Keramiotis, A., Argyrakis, P. and Kopelman, R. (1985) 'Scaling and short-time corrections for random walks on two-dimensional exactly percolating clusters', Physical Review B 31, 4617-4621.

4. Anacker, L.W., Parson, R.P. and Kopelman, R. (1985) 'Diffusion-controlled reaction kinetics on fractal and euclidean lattices, transient and steady-state anhilation', J. Physical Chemistry 89, 4758-4761.

5. Klafter, J., Blumen, A., Zumofen, G. and Drake, J.M. (1987) 'Relaxation in restricted geometries', J. Luminescence 38, 113-115.

6. Lianos, P. and Modes, S. (1987) 'Fractal modelling of luminescence quenching in microemulsions', J. Physical Chemistry 91, 6088-6089.

7. Lianos, P. (1988) 'Microemulsions and other organised assemblies as media of fractal dimensions. A luminescence probe study', Progress in Colloid and Polymer Science 76, 140-143.

8. Duportail, G., and Lianos, P. (1988) 'Fractal modelling of pyrene excimer quenching in phospholipid vesicles', Chemical Physics Letters 149, 73-78.

9. Lianos, P. (1988) 'Luminescence quenching in organised assemblies treated as media of non integer dimensions', J. Chemical Physics 89, 5237-5241.

10. Modes, S. and Lianos, P. (1989) 'Luminescence probe study the conditions affecting colloidal semiconductor growth in reverse micelles and water-in-oil microemulsions', J. Physical Chemistry 93, 5854-5859.

11. Duportail, G. and Lianos, P. (1990) 'Phospholipid vesicles treated as fractal objects. A fluorescence probe study', Chemical Physics Letters, 165, 35-40.

12. Allinger, K. and Blumen, A. (1980) 'On the direct energy transfer to moving acceptors', J. Chemical Physics 72, 4608-4619.

13. Klafter, J. and Blumen, A. (1984) 'Fractal behaviour in trapping and reaction' J. Chemical Physics' 80, 875-877.

14. Blumen, A., Klafter J. and Zumofen, G. (1986) 'Influence of restricted geometries on the direct energy transfer', J. Chemical Physics 84, 1397-1401.

15. Klafter, J., Blumen, A. and Zumofen, G. (1984) 'Energy transfer in fractal structures', J. Luminescence 31 & 32, 627-633.

16. Lianos, P. and Argyrakis, P. (1989) 'Chemical reactions in restricted spaces: Decay in t he presence of quenchers', Physical Review A 39, 4170-4175.

17. Modes, S., Lianos, P. and Xenakis, A. 'Relation of fractal behaviour of luminescence quenching with electric percolation in water-in-oil microemulsions', paper submitted for publication.

18. Jada, A., Lang, J. and Zana, R. (1989) 'Relation between electrical percolation and rate constant for exchange of material between droplets in water in oil microemulsions', J. Physical Chemistry 1989, 10-12.

19. Kim Won, M. and Huang, S.J. (1986) 'Percolation-like phenomena in oil-continuous microemulsions', Physical Review A 34, 719-722.

20. Eicke, H.-F., Borkovec, M. and Das-Gupta, B. (1989) 'Conductivity of water-in-oil microemulsions: A quantitative charge fluctuation model', J. Physical Chemistry 93, 314-317.

21. Grest, G.S., Webman, I., Safran, S.A. and Bug, A.L.R. (1986) 'Dynamic percolation in microemulsions', Physical Review A 33, 2842-2845.

22. Lang, J., Jada, A. and Malliaris, A. (1988) 'Structure and dynamics of water-in-oil droplets stabilised by sodium bis(2-ethylhexyl)sulfosuccinate', J. Physical Chemistry 92, 1946-1953.

23. We have measured k_e by using a model applicable to quenching in micelles [10, 22].

ORGANIZED ASSEMBLIES IN CHEMICAL SEPARATIONS

E.PELIZZETTI
Dipartimento di Chimica-Fisica Applicata
Università di Parma
43100 Parma
Italy

V.MAURINO, C.MINERO and E.PRAMAURO
Dipartimento di Chimica Analitica
Università di Torino
10125 Torino
Italy

ABSTRACT. The use and application of surfactants and their organized assemblies in separation science is surveyed. The unique properties of the amphiphilic aggregates have been utilized to improve existing methods or to overcome problems associated with the development of new procedures. Specific areas include chromatography, membrane based separations, extractions, foam based separations and other fields for possible future development.

1. Introduction.

Chemical separations play a central role in various branches of science and technology. Within the last decade several surfactant-based separation processes were emerging, some of which have already found current applications whereas others are ready to be considered and commercially exploited.

Surfactant-based separations, in addition to some peculiarities, have a number of potential advantages over the traditional methods. In fact surfactants exhibit in general low toxicity and the separation process requires little energy.

The areas of application spread from the use in analytical techniques as well as in the solution of industrial problems, such as in biotechnology and in pollution control.

The aim of this overview is to summarize and update some previous review articles and books on this topic [1-9]. The essential of each technique will be presented and references for more details will be given.

D. M. Bloor and E. Wyn-Jones (eds.),
The Structure, Dynamics and Equilibrium Properties of Colloidal Systems, 325–353.

2. Surfactant-mediated chromatographic separations

2.1. MICELLES AS MOBILE PHASES IN LIQUID CHROMATOGRAPHY.

Surfactants have been successfully employed as additives in mobile phases in the so-called ion-pair or surfactant chromatography. In these applications, the experimental conditions (concentrations, presence of organic solvents) are such that micellar aggregates do not form. Information on the use of amphiphilic molecules as ion-pairing reagents in chromatography can be found in extensive and detailed reviews [10-12].

The use of micellar mobile phase originated with gel permeation chromatography (GPC) and thin layer chromatography (TLC), then further extended to the "high performance liquid chromatography" (HPLC).

The effect of solute partitioning between micellar and aqueous pseudophases during GPC was first recognized by Herries [13] and used to calculate partition coefficients. More recently some other applications have been reported [14,15], but the development of a high performance micellar GPC has to solve some problems such as packing stability, adsorption effects and lack of sufficient interstitial volume [3].

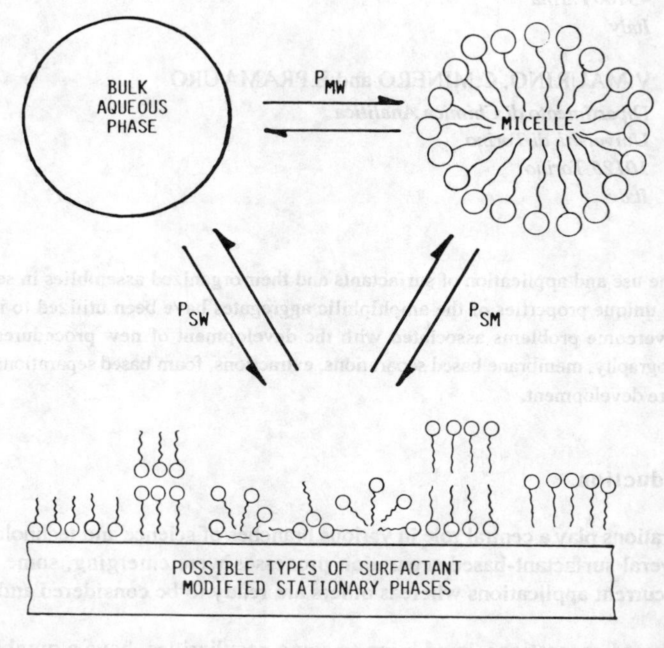

Figure 1. Schematic representation of the three "phase" model for MLC. The elution behaviour of a solute would be dependent on the combined effects of three partition coefficients (P_{MW} = partition coefficient between micelle and water, P_{SW} = partition coefficient between stationary phase and water, P_{SM} = partition coefficient between stationary phase and micelle).

The most popular utilization of surfactant aggregates in chromatographic separations concerns their use as mobile phase in TLC and HPLC. The presence of the micellar phase which is able to selectively solubilize and interact with solutes does not have the simple aim of introducing secondary equilibria to control the separation (see Figure 1). In fact the different types of interaction (hydrophobic, electrostatic, hydrogen bonding and their combination) cannot be realized by any one or a mixture of conventional organic solvents, used as the mobile phase.

A schematic view of the equilibria involved in micellar liquid chromatography is shown in Figure 1. A solute distributes between the three phases; the corresponding partition coefficients are related each other and consequently one can manipulate two partition coefficients (e.g. P_{MW} and P_{SW}) in order to achieve the desired separation.

The basic formulas relating these two partition coefficients and chromatographic parameters to micellar concentration are given in the following equations, for TLC and HPLC, respectively:

$$\frac{R_f}{1-R_f} = \frac{1}{\phi}\frac{K_{MW}C_M}{P_{SW}} + \frac{1}{P_{SW}} \tag{1}$$

$$\frac{1}{k'} = \frac{1}{\phi}\frac{K_{MW}C_M}{P_{SW}} + \frac{1}{P_{SW}} \tag{2}$$

where R_f and k' are the retardation and capacity factor, respectively; ϕ is the phase ratio (equal to V_s/V_o, where V_s and V_o are the stationary-phase and void volumes, respectively); C_M is the micellized surfactant concentration; K_{MW} is the micelle-solute binding constant (equal to $(P_{MW}-1)v$); P_{MW} and P_{SW} are defined in Figure 1.

Eq.(1) and (2) have been rearranged and presented in several forms. This is not only a simple mathematical exercise but relates to the convenience of obtaining different and more accurate data. The argument is exhaustively discussed in ref.3.

Eq.(1) and (2) hold for neutral solutes or for ionizable compounds, provided that only one form is present in the investigated conditions. For intermediate situations where both the acid and its conjugate base (or base and its conjugate acid) are present, the following equation predicts the dependence of k' upon pH and surfactant concentration (for weak acid HA)

$$k' = \frac{k_o'(1+K_{MW}^{HA}C_M) + k_1'(1 + K_{MW}^{A^-}C_M)K_{am}/[H^+]}{1 + K_{MW}^{HA}C_M + (1 + K_{MW}^{A^-}C_M)K_{am}/[H^+]} \tag{3}$$

where k_o' and k_1' are the limiting capacity factors of the weak acid (HA) and its conjugate base (A^-); K_{MW}^{HA} and $k_{MW}^{A^-}$ are the binding constant of the micelle with HA and A^- respectively; K_{am} is the apparent ionization constant for the weak acid. A similar equation can be derived for weak bases and their conjugate acids [16].

Examination of eq.(1)-(3) suggests immediately that retention can be controlled by surfactant concentration and, for ionizable species, by pH.

TABLE 1. Comparison of the general effect of variables on retention in Micellar Liquid Chromatography (MLC)[a].

Factor Varied	Effect upon retention
Nature of the polar head group	Retention of ionogenic compounds can change with nature and concentration of ionic surfactant [17].
Nature of the stationary phase	Different degree of coverage, modification of stationary phase properties affect retention and efficiency [18].
Concentration of surfactant mobile phase	Increasing concentration decreases retention (down to a limiting value) [1,6].
Presence of an organic modifier (added alcohol or acetonitrile)	Retention decreases with increasing concentration or hydrophobicity of the organic additive [19,20].
pH	Depends upon the nature (i.e. charge-type and concentration) of the surfactant micelle and ionizable solute; eq.(3) predicts a sigmoidal-type dependence between retention and pH (at constant surfactant concentration [17].
Temperature	Retention decreases slightly as temperature increases.
Ionic strength	Retention decreases as ionic strength increases; linear dependence between k'^{-1} vs. $\mu^{-1/2}$.

a. Adapted from ref. 1, 6.

Other crucial parameters are the partition coefficients which can be in turn manipulated by additives and other experimental conditions. Table 1 summarizes some of the factors which influence the retention in MLC.

TABLE 2. Comparison of MLC with classical hydro-organic mobile phase reverse chromatography.

Advantages	
Simultaneous separation of	- hydrophobic/hydrophilic solutes. - acidic, basic and neutral compounds [21].
Reduction in analysis time	- rapid gradient capability (of micelle concentration and pH without column regeneration). - direct sample injection (biological samples, wastewater) [22-24].
Better retention reproducibility	
Detection enhancement	- luminescence [25], amperometric [26].
Low cost	
Low toxicity	- ease of mobile phase disposal
Disadvantages	
Loss of efficiency compared to hydro-organic mobile phases	- slow mass transfer (improvement with small organic solvents and higher temperatures). - poor wetting of stationary phase [6,27-29].
Higher column backpressure	- more viscous
Separation of analyte from surfactant solution	- in preparative chromatography.

In the last few years MLC has received increasing attention and several applications have been reported. Particular attention is devoted in metal/anion speciation and biological/protein separations. Exhaustive reviews can be found [3,5,12,22,30,31].

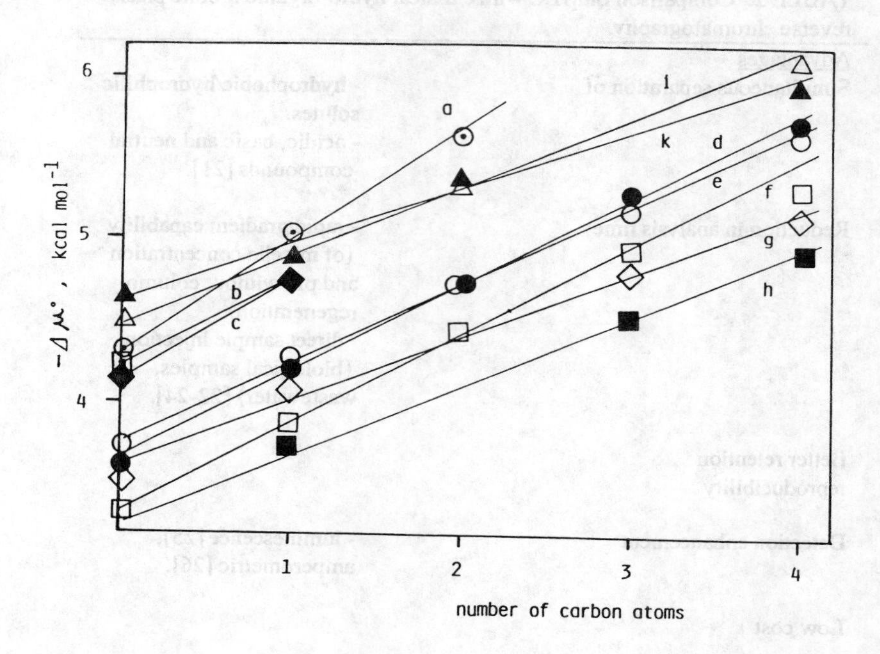

Figure 2. Free energies of transfer, defined as $\mu° = -RT\ln(55.5\ K_{MW})$, of aromatic solutes from water to SDS micelles as a function of the number of carbon atoms of the alkyl chain. Arenes: (a) solubility; (b) isopiestic method; (c) chromatography. Phenols: (d) dissociation constant; (e) chromatography; (f) ultrafiltration. Catechols: (g) chromatography; (h) kinetic. Benzoic acids: (i) dissociation constant; (k) ultrafiltration. Reproduced with permission from ref.32.

As a consequence of the growing number of applications, the advantages and limitations of this technique are becoming more definite. Although the basic mechanism is now fairly well understood some deviations can be observed. Concerning the comparison with "classical" chromatography, Table 2 attempts to report the major differences between the two techniques.

Finally MLC allows a convenient means of studying some properties of surfactant solutions (such as CMC) and, more important, the solute-micelle interactions. The prediction of hydrophobicity is of relevant importance in several field, from environmental (soil sorption, bioaccumulation) to biological areas (drug design, toxicology).

Excellent correlations between partition coefficients in micellar systems, determined by MLC and other techniques, and the hydrophobic contribution of substituents have been found as shown in Figure 2 and 3 and in Table 3 [32,33].

In the past the n-octanol/water partition coefficients (P_{OW}) have been extensively used for the same purpose. However, this parameter has been criticized since a two phase system is chosen to represent a biological structure and it is often difficult and tedious to measure.

TABLE 3. Standard free energy change for different substituent groups in various two-phase media[a].

Substituent	μ°_{t} (kJ mol^{-1})		
	SDS	1-Octanol/ water	Hydrocarbon/ water
Aliphatic -C-			
on phenol	-2.13	-2.63	-3.16[b]
on alkanes	-3.22		-3.68[c]
on arenes	-2.64	-2.82	-3.35[c]
4-Hydroxy	2.60	4.97	
Hydroxy (on arenes)	1.72	3.81	17.99[c]
4-Cyano	-1.01	-0.80	5.25[b],7.19[d]
4-Nitro	-1.17	-2.56	
4-Phenyl	-8.92		-9.53[b]
4-Fluoro	-1.38	-1.77	-0.95/-1.37
4-Chloro	-2.68	-5.30	-3.59[b]
4-Bromo	-4.31	-6.45	-4.51[b]
4-Iodo	-6.28	-8.27	-5.02[b]/-6.22[b]
4-Thio-(R)	-1.72	-2.05	
3-Trifluoromethyl	-6.36	-8.50	
2-Carboxamido	1.26	1.03	
2-N-Phenylcarboxamido	-7.98	-10.33	
4-Formyl	-0.55	-1.37/-1.99	
4-O-R	0.67	3.34	5.59[b]
4-CO-R	-0.42	3.25	
4-COO-R	-1.26	0.29	

(a)Adapted from ref.33. (b)Cyclohexane. (c)n-Heptane. (d)n-Hexane.

It is noteworthy that a correlation between P_{OW} and K_{SM} has been reported for a comprehensive series of compounds (see Figure 4) [33].

2.2. SURFACTANTS AS STATIONARY PHASES.

Surfactants have been immobilized or coated on stationary phases. Since in LC the surfactant will usually slowly elute from the support, most application are in GC and GLC [22]. Also immobilized surfactant vesicle bilayer has been reported to serve as stationary phase in GC separations [34]. Further work in these directions will lead to new GC supports and separation capabilities.

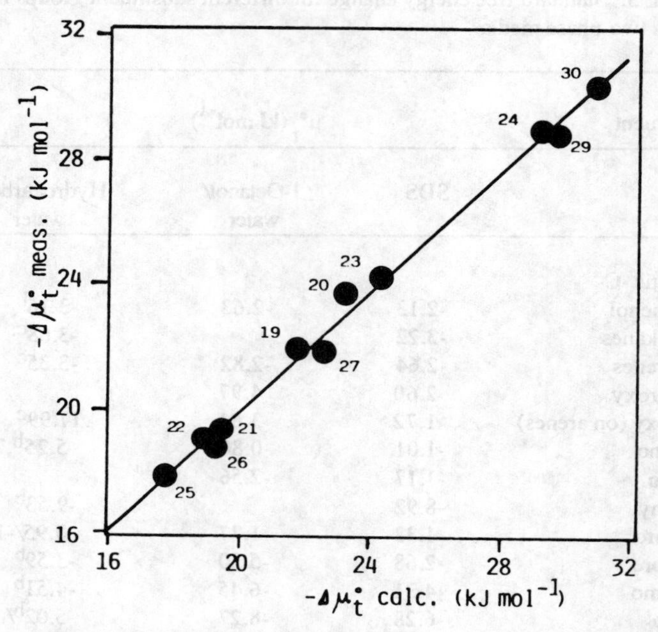

Figure 3. Comparison between calculated and measured free energies of transfer of polysubstituted phenols (correlation coefficient 0.996). (19) 3-trifluoromethylphenol; (20) 2-N-phenylcarboxamidophenol; (21) 2-nitro-4-chlorophenol; (22) 3-nitro-4-methylphenol; (23) 2-chloro-4-bromo-6-methylphenol; (24) 4-nitro-2,6-diiodophenol; (25) 3,5-dimethoxyphenol; (26) 2,6-dimethoxy-4-acetylphenol; (27) 2-acetamido-4,6-dichloro-5-methyl-phenol; (29) 2-N-butylcarboxamido-1-naphthol; (30) 2-N-phenylcarboxamido-1-naphthol. Reproduced with permission from ref.33 .

2.3. ADMICELLAR CHROMATOGRAPHY [35,36].

In the Admicellar-Enhanced Chromatography (AEC) an adsorbed layer or surfactant is formed reversibly on the surface of packing material.

Solute molecules form a stream partition into the surfactant layer in a phenomenon called adsolubilization, which is a surface analog of solubilization.

When the surfactant layer becomes saturated with the solute, the surfactant layer containing the solute can be stripped from the packing by changing a low-energy chemical/physical property (e.g. pH). Then the concentration of the solute in the stream from the bed can be much higher than the original concentration.

The degree of concentration of the adsolubilizate is not particularly remarkable, being typical of any fixed bed adsorption process, such as adsorption on activated carbon.

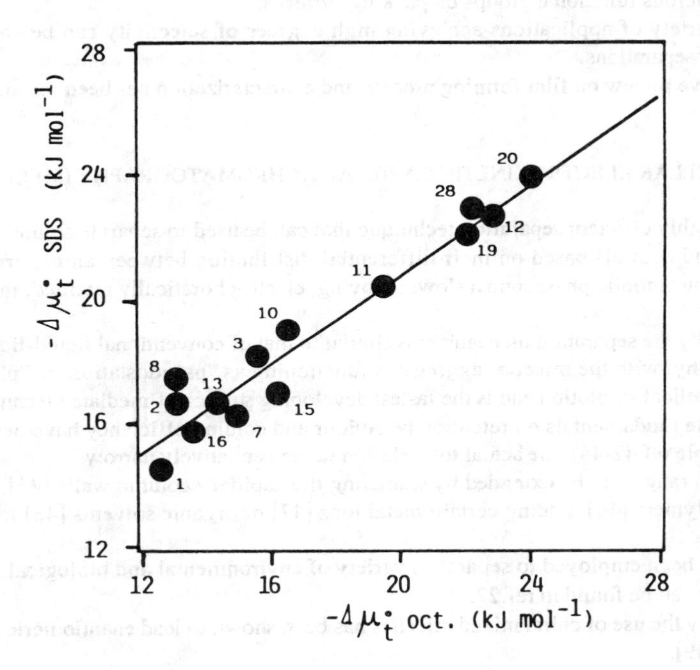

Figure 4. Correlation between the free energies of transfer of various substituted phenols in SDS micellar solution and in octanol/water (correlation coefficient 0.978). Reproduced with permission from ref.33 . (1) phenol; (2) 4-methoxyphenol; (3) 4-ethoxyphenol; (7) 4-formyl-phenol; (8) 4-acetylphenol; (10) 4-carboxymethylphenol; (11) 4-carboxyethylphenol; (12) 4-carboxypropylphenol; (13) 4-cyanophenol; (15) 4-nitrophenol; (16) 2-carboxamidophenol; (19) 3-trifluoromethylphenol; (20) 2-N-phenylcarboxamidophenol; (28) 1-naphthol;

The interest of AEC arises from the ease of removal of the saturated bilayer and the selectivity for one solubilizate over another.

Possible interesting applications have been reported in the concentration of phenolic compounds from a stream, isomers and optical isomers separation of estrogens and other steroidal hormones.

2.4. POLYMERIZED THIN-FILM SEPARATIONS.

Adsolubilization of monomer in adsorbed bilayers, followed by in situ polimerization results in formation of an insoluble, ultra-thin organic layer on the surface of an inorganic support [37-39]. The process is a surface analog of emulsion polymerization.

This surface process can modify the original support and can improve the feasibility of attacking numerous functional groups on packing surfaces.

A large variety of applications achieving high degrees of selectivity can be envisaged, for example in bioseparations.

An extensive review on film forming process and characterization has been recently published [40].

2.5. MICELLAR ELECTROKINETIC CAPILLARY CHROMATOGRAPHY (MECC).

MECC is a highly efficient separation technique that can be used to separate solutes in a mixture (both ionic and neutral) based on their differential distribution between an electroosmotically pumped aqueous mobile phase, and a slower moving, electrophoretically retarded, micellar phase [41].

Conceptually the separation mechanism is similar to that of conventional liquid-liquid partition chromatography, with the micellar aggregates functioning as "pseudostationary" phase. MECC resulted in excellent resolution and is the fastest developing surfactant-mediated technique.

Whereas the fundamentals on retention behaviour and column efficiency have been described (see for example ref.42-44), the actual total elution range is relatively narrow.

The elution range can be extended by silanating the capillary column walls [45], coating the walls with polymers [46], adding certain metal ions [47] or organic solvents [48] to the mobile phase.

MECC has been employed to separate a variety of environmental and biological mixtures; an exhaustive list can be found in ref.22.

Interestingly the use of chiral mixed micelles has been shown to lead enantiomeric resolution of amino acids [49].

3. Membrane Based Separations.

3.1. MICELLAR-ENHANCED ULTRAFILTRATION (MEUF).

MEUF has been recently proposed as an effective technique for preconcentration and removal of organic and inorganic pollutants from aqueous wastes [50-52].

A suitable surfactant at a concentration above its cmc is added to the solution containing the undesired components or the substrates to be concentrated. Provided a favourable partition coefficients of these compounds, their complete association to the host aggregates can be achieved.

The feed solution is then forced to pass through a membrane having a pore-size small enough to retain the micelles, so that pollutants-free permeate solutions can be obtained. The permeate contains very low amounts of surfactant (usually in the cmc range) and negligible amounts of unbound solute (the process is depicted in Figure 5).

For undissociated organic solutes, hydrophobic interactions are at the basis of MEUF performances, whereas for charged hydrophilic species (e.g. metal ions) the electrostatic interactions with ionic aggregates play a fundamental role. Both effects can be present when ionized solutes are partitioned between water and charged micelles, and in these cases the separation efficiency depends on pH and on the charge of the aggregates.

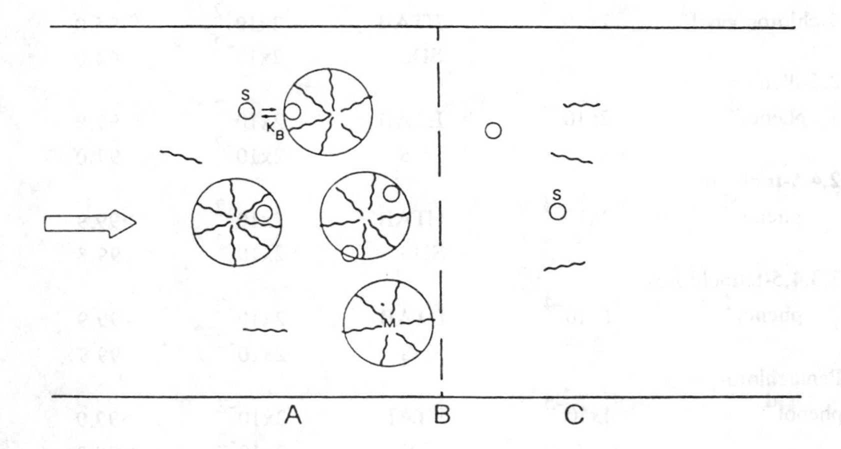

Figure 5. Scheme of micellar-enhanced ultrafiltration. (A) Retentate, (B) UF membrane, (C) permeate. O = solute (S) to be removed; = surfactant monomer; M = host micelle. Reproduced with permission from ref.53.

Table 4 summarizes the data concerning the separation of several organic substrates of environmental interest from aqueous media.

Irrespective of the micellar system used, quantitative separations of these solutes is achieved when their binding constants to the aggregates (K_B) are higher than ca.1000 M^{-1}. Then, the preliminar estimation of this parameter allows the prediction of the process efficiency.

TABLE 4. Removal of different pollutants from aqueous solutions using MEUF.

COMPOUND	CONCENTRATION (M)	SURFACTANT	SYSTEM (M)	%R
1-hexanol[a]	$7x10^{-3}$	HPC	$7x10^{-2}$	78
1-heptanol[a]	$7x10^{-3}$	HPC	$7x10^{-2}$	95
1-octanol[a]	$7x10^{-3}$	HPC	$7x10^{-2}$	98
Benzene[b]	$1x10^{-3}$	HPC	$5x10^{-2}$	89.9
		SDS	$5x10^{-2}$	80.4
Phenol	$1x10^{-2}$ [c]	HPC	$5x10^{-2}$	73.0
	$2x10^{-4}$ [d]	HTAB	$2x10^{-2}$	55.0
		SDS	$2x10^{-2}$	31.0
t-butylphenol[e]	$3.2x10^{-2}$	HPC	$3.2x10^{-2}$	99.7
4-chlorophenol[d]	$2x10^{-4}$	HTAB	$2x10^{-2}$	95.0
		SDS	$2x10^{-2}$	64.0
3,5-dichloro-phenol[d]	$2x10^{-4}$	HTAB	$2x10^{-2}$	99.9
		SDS	$2x10^{-2}$	97.0
2,4,5-trichloro-phenol[d]	$2x10^{-4}$	HTAB	$2x10^{-2}$	>99.9
		SDS	$2x10^{-2}$	99.8
2,3,4,5-tetrachloro-phenol[d]	$1x10^{-4}$	HTAB	$2x10^{-2}$	>99.9
		SDS	$2x10^{-2}$	>99.9
Pentachloro-phenol[d]	$4x10^{-5}$	HTAB	$2x10^{-2}$	>99.9
		SDS	$2x10^{-2}$	>99.9
2,4-D[d]	$2x10^{-4}$	HTAB	$2x10^{-2}$	99.7
		SDS	$2x10^{-2}$	97.5
2,4,5-T[d]	$2x10^{-4}$	HTAB	$2x10^{-2}$	>99.9
		SDS	$2x10^{-2}$	99.8
DDT[d]	$2x10^{-5}$	HTAB	$2x10^{-2}$	>99.9
		SDS	$2x10^{-2}$	>99.9

(a)From ref.51; (b)from ref.54; (c)from ref.55; (d)from ref.56; (e)from ref.50; %R: percent of rejection under the reported experimental conditions.

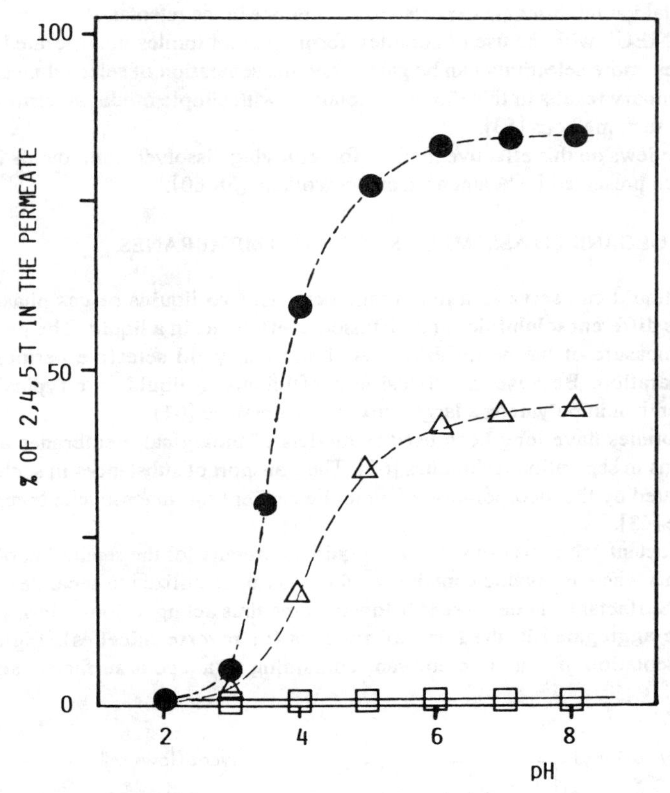

Figure 6. Variation of MEUF performances with pH for the pollutant 2,4,5-trichlorophenoxyacetic acid (2,4,5-T). ● : SDS 2x10^{-2} M; □ : HTAB 2x10^{-2} M; Δ : Brij 35 1x10^{-2} M. Adapted from ref.54.

Figure 6 shows the effect of pH on the separation of the carboxylic herbicide 2,4,5-T using MEUF. In this case, the use of cationic aggregates is essential for the effective removal of the pollutant in the pH range 2-8. The loss of hydrophobicity due to the ionization of the carboxylic group is largely compensated by the presence of the electrostatic attraction between the anion and the positively charged aggregate.

Metal ions present in aqueous streams have been also effectively using MEUF by working with anionic micelles [52] and polyelectrolyte systems [55]. In the presence of SDS micelles, the concentration of multivalent cations takes place in the Stern layer due to the electrostatic attraction and to the favourable competition with Na^+ counterions for the charged sites.

Rejection factors > 0.99 have been reported for Cu^{2+} and Zn^{2+} in the presence of SDS, starting from solutions containing these ions at the ppm level [57].

Although the efficiency of the above demetallation process is satisfactory, the method is not selective and metal ion mixtures are expected to accumulate in the retentate.

By coupling MEUF with the use of complex-forming amphiphiles incorporated into suitable carrier aggregates, more selectivity can be gained and the separation of selected metal ions can be achieved. Preliminary results in this direction obtained with simple model systems demonstrates the feasibility of such methods [53].

Exhaustive reviews on this effective method for removing dissolved compounds from aqueous streams have been presented by Scamehorn and coworkers [58-60].

3.2. USE OF ORGANIZED ASSEMBLIES AS LIQUID MEMBRANES.

An immiscible liquid can serve as a membrane between two liquids or gas phases. Different solutes will have different solubilities and diffusion coefficients in a liquid. The product of these two terms is a measure of the permeability. A liquid can yield selective permeabilities and, therefore, a separation. Because the diffusion coefficients in liquids are typically orders of magnitude higher than in polymers, a larger flux can be obtained [61].

Liquid membranes have long been used as models of biological membranes and have also found applications in separation techniques [62]. The transport of substances in such experiments is usually facilitated by the incorporation of a mobile carrier (e.g. macrocyclic compound) in the liquid membrane [63].

Aqueous surfactant solutions can serve as a liquid membrane for the separation of components in organic solvents whereas organic containing solutions can be utilized to separate components in aqueous media. Surfactant can be present below the cmc thus acting as ion-pair or phase transfer agent; or can be aggregated in the form of micelles (or reverse micelles). Figure 7 gives a schematic representation of a liquid membrane containing an aqueous surfactant solution as the active component.

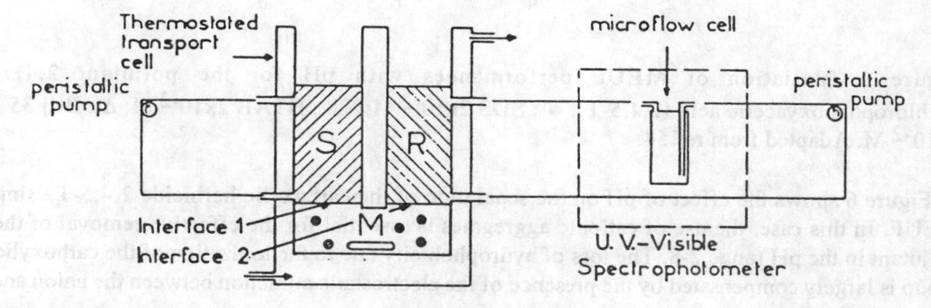

Figure 7. Schematic representation of the set-up suitable for transport experiments. S, R and M designate the source, receiver and membrane compartments. The hatched areas represent the oil (oil-rich continuous phase in S and R, dispersed phase in M). Reproduced with permission from ref.64.

The possibility of using microemulsion droplets as a carrier in liquid membranes has recently been explored. Contrary to the case of specific carriers, microemulsion droplets are not selective,

but are faster. From a practical point of view, the coupling between selective extractants and microemulsion carriers may lead to taking the advantage of selectivity and speed. Considering the transport of lipophilic solutes by an o/w microemulsion droplet [64], Scheme 1 can be proposed, which is analogous to that described for the interpretation of facilitated transport of ions by a carrier [65]. Figure 6 shows the set-up and defines the cell compartments.

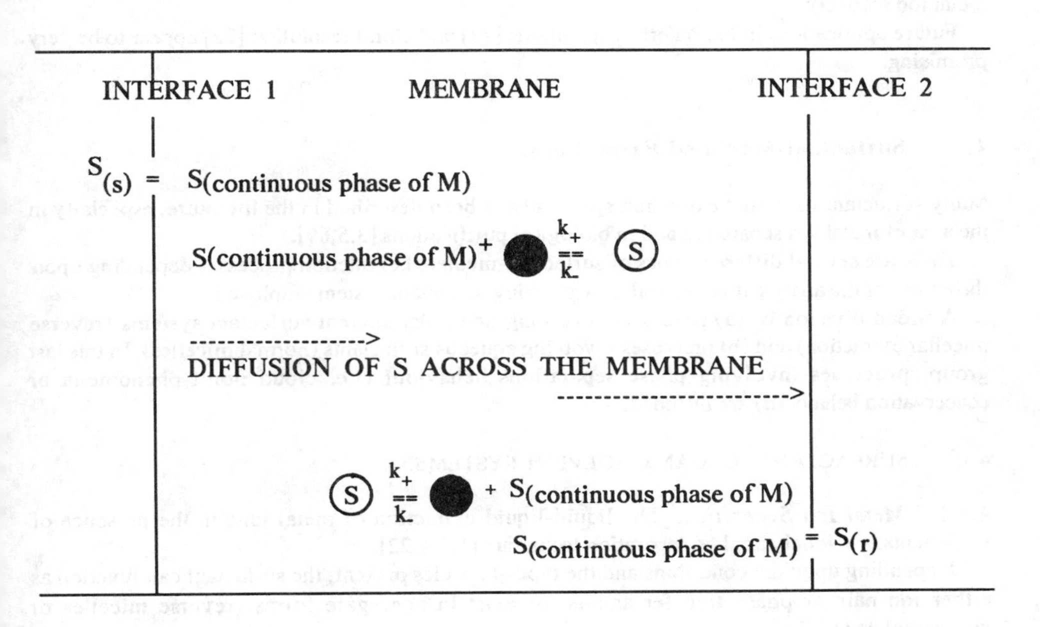

Scheme 1. Transport of lipophilic solutes by an oil-in-water microemulsion droplet. The subscript s, r, M have the meaning reported in the legend of Figure 6; S represent the solute and the circle the microemulsion droplet.

Following the previous reported considerations [65,66], the flux F of transported solute can be expressed as:

$$F = \frac{D_m K P C_D}{L} \frac{C_s^o}{1+KPC_s^o} \qquad (4)$$

where D_m is the diffusion coefficient of the microemulsion droplet, $K=k_+/k_-$, P is the partition coefficient of the solute between the oil phase and the continuous aqueous phase of the microemulsion, C_D the total droplet concentration, L the diffusion layer thickness and C_s^o the concentration outside the membrane in the source phase. Plots of 1/F versus $1/C_s^o$, according to eq.4, were found linear, thereby supporting the proposed scheme. The results are consistent with

a model in which the diffusion of the droplets is coupled with a fast solubilization-desolubilization process.

A similar model can also describe the transport of hydrophilic (ionic) species by using w/o microemulsion [67,68]. Then the separation to be accomplished dictates the experimental conditions and the form in which the surfactant acts. Since their introduction, liquid surfactant membranes have been utilized for many separations [22, 61, 62] both of organics as well as for metal ion recovery.

Future applications in biomaterial separations [22] and chiral resolution [22] appear to be very promising.

4. Surfactant-Mediated Extractions.

Many surfactant-mediated extraction systems have been described in the literature, especially in the area of metal ion separation and in biological purifications [3,5,69].

There are several different types of surfactant-mediated extraction processes depending upon the nature of the analyte mixture and the extracting surfactant system employed.

A broad division is: (a) processes involving non polar solvent-surfactant systems (reverse micellar extraction) and (b) processes involving aqueous surfactants (normal micelles). In this last group, processes involving phase separations behaviour (i.e. cloud point phenomena or coacervation behaviour) are included.

4.1. SURFACTANT - ORGANIC SOLVENT SYSTEMS.

4.1.1. *Metal Ion Separations.* The liquid-liquid extraction of metal ions in the presence of surfactants is a widely applied separation technique [1,3,4,22].

Depending upon the conditions and the type of species present, the surfactant can function as either ion-pair or phase transfer agents, or exist in aggregate forms (reverse micelles or microemulsions).

Recent accumulated evidence strongly supports that reverse micelles or microemulsions play a crucial role in the extraction process [22]. The nature and properties of the aggregates as well as the binding and the catalytic effect upon the ions to be extracted can be largely affected by changes in the solvent and experimental conditions (pH, ionic strength, temperature, etc) [70,71].

The presence of surfactants (either as phase transfer agents, reverse micelles or microemulsions) can favourably alter interfacial properties and facilitate the interphasic transfer of the species to be extracted.

In fact the surfactants can influence the rate of extraction which depends upon the kinetics of transfer of the species from the bulk water phase across a water-oil interface to the bulk organic phase.

An example can show the advantages of the presence of the aggregates. The extraction of gallium from alkaline solutions into benzene by means of the hydrophobic ligand 7-(1-ethenyl-3,3,5,5-tetramethylhexyl)-8-quinolinol can be improved if a carboxylate surfactant and a long chain alcohol are added [72]. The resulting water-in-oil microemulsion has a water content which depends on the surfactant concentration and on the co-surfactant chain length. The observed increase in the extraction rate (about 20 times) can be explained in term of the increased interfacial area of the microemulsion, faster mass transfer and concentration of the reagents in the droplets.

Most extraction schemes reported in the literature involve use of either cationic or anionic surfactants. Only few recent applications employed zwitterionic or non ionic surfactants. The interested reader is referred to several recent extensive compilations of extraction systems involving surfactants [22,73,74].

It is also remarkable that organic species that are capable to be ionized can also be extracted from aqueous media or solid matrices [75-77].

It should be noted that the extracted species in the organic reverse micellar phase can be back-extracted into the aqueous phase, usually by dilute acid (for metal ions) [78]. This stripping step is usually done by contacting the loaded organic surfactant medium with an aqueous acid solution at a relatively high organic/aqueous volume rate (e.g., 15:1). Typically, very good recovery of analytes is obtained along with a good concentration factor.

Finally mention should be made that many processes can not only be carried out at analytical or preparative scales but also at the process level.

The synthesis and characterization of new micelle-forming or specific hydrophobic ligands [79] together with detailed analysis of the extraction mechanism and kinetics could offer other perspectives in this field.

4.1.2. *Extraction of Proteins.* It has been demonstrated that a number of proteins can be solubilized in a polar solvents, then recovered with retention of their activity [71,80]. This phenomenon can form the basis in the separation, recovery and purification of biotechnological products.

The solubilization of a given protein and consequently its degree of extraction depends upon its structure, in particular the ionizable surface groups and the size. Other crucial parameters are the surfactant structure and concentration, nature of the organic phase, and the pH and electrolyte concentration of the aqueous phase. In general the transfer of the protein into the reverse micelle is favoured by a pH at which the surface groups are either protonated or deprotonated, by choosing the surfactant with an opposite charge to the protein. This choice of proper pH can result in selective solubilization of one protein from a mixture of proteins.

The proteins and bioproducts can be easily back-extracted from the reverse micellar organic phase to an aqueous phase. This can be realized simply by contacting the organic phase with an aqueous phase at high salt content. These observations suggest that minimization of the electrostatic contributions to the free energy of the system is the driving force for this process.

It is noteworthy that this general approach could be employed to extract proteins directly from solid matrices [81] and has the potential for operating in a continuous flow [82]. More information on the mechanistic features and applications are given in reviews by Hatton [82,83], also in a chapter of this book [84].

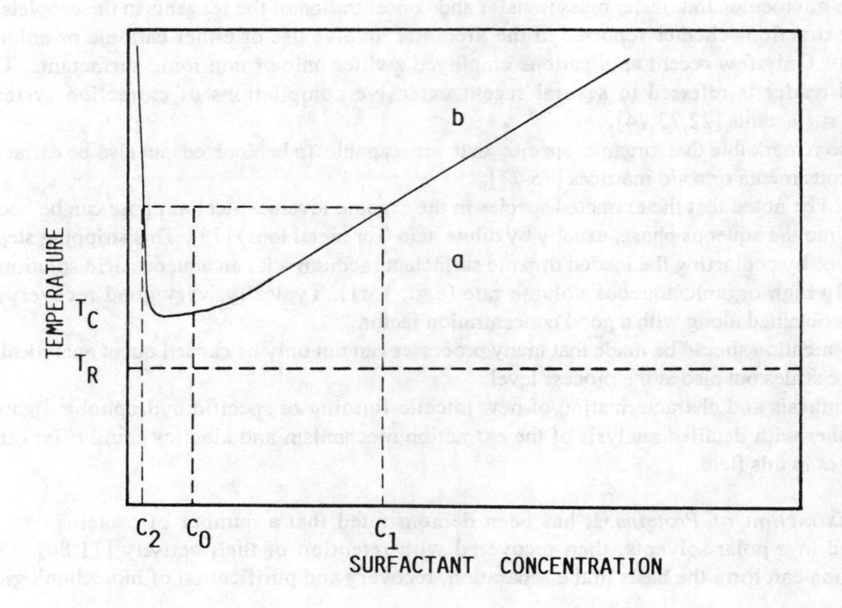

Figure 8. Typical phase diagram for a dilute aqueous solution of a nonionic surfactant. At temperature, below room temperature and at higher amphiphile concentrations, separation of solid phases can occur. (a) Monophasic region, (b) biphasic region.

4.2. AQUEOUS-SURFACTANT SOLUTION SYSTEMS.

4.2.1. ***Extractions with aqueous normal micelles.*** **Normal aqueous micelles can be employed to extract and purify membrane components, compounds from solid matrices or organic components from hydrocarbon matrices [22].**

In the solubilization of membrane components, the selectivity can be controlled through parameters as pH, ionic strength and the nature and concentration of surfactant.

Extraction of organic components from hydrocarbon matrices is based on the ability of aqueous micelles to exhibit differential solubilizing capability and rates with respect to different compounds. It has also to be noted that in most instances, the micellar extraction process is merely a preliminary step to further fractionation and purification. This area of separation science appears to hold great potential for large scale extraction schemes.

4.3. EXTRACTION BASED ON PHASE SEPARATION BEHAVIOR.

The extraction of components present in an aqueous mixture can be achieved by the use of surfactant systems capable of undergoing a phase separation as a result of altered conditions (i.e.

temperature or pressure changes, added salt, etc).

TABLE 5. Concentration and removal of hydrophobic pollutants using cloud point separation.

SOLUTES REMOVED	SURFACTANT	EFFICIENCY
Pesticides[a]:		
Endrin, Chlorodene, Endosultan, Lindane, Aldrin, BHC, DDT, Methoxychlor, Chloropyrifos,	PONPE-7.5	QR
2,4-D[b], 2,4,5-T[b].	BL-8SY/BL4.2	PR
Polycyclic hydrocarbons[a]		
Benzo[a]pyrene, Fluorene, Fluoranthene, Pyrene, Benzo[e]pyrene.	IGEPAL CO-630	QR
Chlorophenols[b]		
4-chlorophenol, 3,5-dichlorophenol, 2,4,5-trichlorophenol, 2,3,4,5-tetrachlorophenol, pentachlorophenol.	BL-8SY/BL-4.2	PR QR
PCB's[b]		
3-chlorobiphenyl, 3,3'-dichlorobiphenyl.	BL-8SY/BL-4.2	QR

a: from ref.88; b: from ref.56; QR: quantitative removal; PR: partial removal.

4.3.1. *Cloud Point Separations.* An important property of aqueous solutions of nonionic surfactants is that, when the temperature is raised, the solution becomes turbid due to the diminished solubility of the amphiphile in water [85]. This critical temperature is called "cloud point" (T_c) and depends on the amphiphile nature and concentration.

The phase diagram depicted in Figure 8 shows a consolution curve which separates the monophasic from the biphasic regions. Moreover, a minimum in the curve is present at low amphiphile concentrations (critical point). By operating at a given concentration (C_o) at room temperature (T_R) and heating above the cloud point T_c, separation into two liquid phases occurs. The composition of these phases is very different, one of them being rich in surfactant (C_1) and the other containing a dilute aqueous solution, with practically no micelles present (C_2).

Using suitable non ionic surfactants (or surfactant mixtures) and adjusting the experimental conditions (added solutes, buffers, ionic strength) phase separation can be obtained just above room temperature. If hydrophobic analytes are present in the solution, they will bind to the aggregates and concentrate in the low-volume micellar-rich layer. This procedure has been used for the effective extraction of metal ions as hydrophobic chelate complexes [86,87].

More recently, micellar extractions of organic pollutants, including pesticides, aromatic hydrocarbons, PCB's and chlorophenols, have been described [56,88,89] (see Table 5). Also in this extraction technique, the separation performances can be correlated to the partition coefficients (or the binding constants) of the extracted solutes.

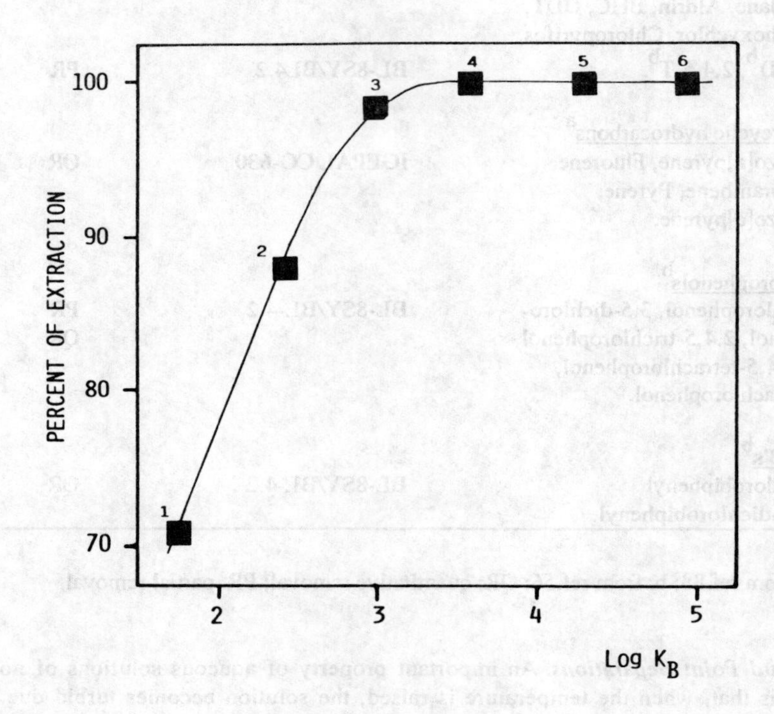

Figure 9. Extraction efficiency of chlorophenols as a function of their binding constant to nonionic micelles. (1): phenol; (2): 4-chloro-; (3): 3,5-dichloro-; (4): 2,4,5-trichloro-; (5): 2,3,4,5-tetrachloro-; (6): pentachloro-. Data taken from ref.56.

Figure 9 shows the variation of the extraction efficiency of some chlorophenols as a function of their hydrophobicity. Concentration factors comparable to those obtained using classic extraction methods with organic solvents can be achieved working with this method.

In addition to metal ions and environmentally important organic compounds also biological materials can be extracted using this approach.

Future work in this area should focus on further development of novel extraction schemes that exploit one or more of the advantages of the non-ionic cloud-point method [22].

4.3.2. *Separation Based on Coacervation.* Some ionic surfactants are also capable of separating into two liquid layers under proper conditions and this phenomenon has been termed coacervation [90].

In addition to cationic and anionic surfactant, aqueous solutions of proteins, synthetic polymers and microemulsions have been reported to exhibit this behaviour [91]. Coacervation can occur when a species of opposite charge is added to an ionic aqueous surfactant solution. Micelles aggregate to form "clusters" which can coalesce to give rise to microscopic droplets. Further coalescence forms a continuous surfactant-rich phase. Further addition of electrolyte can precipitate or flocculate the surfactant-rich phase.

Consequently, in principle, separations analogous to those described in the previous section should be possible. However few examples are reported in the literature [91-93]. Although some disadvantages of coacervate systems (compared to cloud point technique, e.g. in some cases solid phase must be redissolved), this area is open to further applications in separation science.

5. Separations Based on Emulsions.

5.1. MICROEMULSION-BASED SEPARATIONS.

The applicability of microemulsion systems in extraction processes has found a tremendous interest, related to tertiary oil recovery process [94-98].

Microemulsions are related to micellar solutions and some concepts and applications are similar to those referred to in section 4. The interested reader can find exhaustive information in the quoted overviews [94-98].

The mechanism of uptake of organic material (oil) from emulsion droplets into a microemulsion under rapid mixing was in fact found to be diffusion controlled by the collision between emulsion and microemulsion droplets [99]. This is a similar mechanism to that proposed for micellar solubilization [100,101]. Problems encountered in non-stirred extraction processes have also recently received attention [102].

5.2. ABSORPTION INTO EMULSIONS.

The use of emulsions for capturing volatile organic compounds from the exhaust gases of an automotive spray painting process has been proved to be a viable, cost-effective method of abatement of these fugitive compounds [103].

The process utilizes emulsion chemistry and the selection of surfactants allows one to optimize the operative conditions. The oil-soluble solvents are dissolved in the oil droplets while the water soluble components are dissolved in the continuous phase of the oil-in-water emulsion. A small pH variation breaks the emulsion, forming an oil phase, an aqueous phase and a sludge. The solvent removed from the gases are concentrated in the sludge and partially in the oil phase. Once the solvents are removed as sludge or from the liquid phases, the emulsion can be reformed by readjusting the pH.

This process shows great potential for the control of exhaust emission pollutants.

6. Separations Based on Foams.

6.1. FROTH FLOTATION.

Froth flotation is a separation process in which bubbles of air sparged into a column of water adhere to small particles of mineral. The particles then float to the top of the column where they collect in the froth. The froth can be simply skimmed off to complete the removal of the particulate matter.

Consequently, two broad types of reagents are required: reagents to be adsorbed on the particle to control the wettability and reagents to control foaming or frothing. Selectivity is accomplished by preferential adsorption of surfactants at the solid-liquid interface.

Ore flotation is actually the most important surfactant-based separation. As a consequence an extensive amount of excellent literature is available and the interested reader should refer to standard references [104-106].

6.2. BUBBLE SEPARATION PROCESSES.

In the previous section, flotation processes involved the removal of particulate by frothing. The same basic idea can be applied to the removal of dissolved material from a solution through adhesion to the surface of the bubbles. This process is termed foam fractionation.

Other separation techniques based on bubble adsorption are nonfoaming. These include bubble fractionation(use of a gas for adsorption and deposition of a concentrated solution at the top of the column) and solvent sublation (use of another liquid phase to adsorb and concentrate the solute at the top of the column).

In most of these processes, the dispersed phase consists of air bubbles and the continuous phase, generally water, contains the species being adsorbed [107-111].
Several factors have to be considered for an efficient separation in addition to the inherent surface activity or complexation of the species to be separated with a surface active compound: concentration of the species, flow rates of the phases, pH [111].

Examples of separation of biological materials (enzymes [112], albumin [113], penicillin [114], viruses [115], bacteria [116]), of cations and anions [110, 111, 117-119], organic compounds (e.g. phenols and chlorophenols) [120,121] have been reported. An extensive list of examples of foam separations appears in the quoted overviews [110,111,122]. Future work should be directed towards the use of stable foams and optimization of column design, coupled with more exhaustive theoretical treatments [123].

The use of a minute micro gas dispersion (in the order of 10 µm) or so-called gas aphrons shows promising applications since they have extremely large amounts of surface area and long residence times in liquid pools. The result is an almost complete separation of the adsorbed solute

[124]. Similar results have been obtained with minute liquid droplets or colloidal liquid aphrons.
A recent overview outlines the fundamentals and possible applications of this promising technique [125].

7. Other Surfactant-Based Separations.

7.1. SURFACTANT-ENHANCED CARBON REGENERATION.

Although this technique is not used for direct separation or concentration of pollutants from their environmental sources, it has been recently proposed as an effective regenerant procedure for spent sorbents used in water treatments. The solubilization of the adsorbed solutes by micelles is exploited in this method.

Although quite concentrated surfactant solutions are required to efficiently regenerate the exhausted carbon beds, the treatment is still cheap if compared with other reactivation procedures, such as thermal or solvent regeneration, and faster than most biological treatments. When SECR is used no heating is required, and the successive washing cycles with water are sufficient to complete the removal of the adsorbed surfactant.

The efficient recovery of 4-tert-butylphenol from loaded activated carbon has been reported using aqueous solutions of hexadecylpyridinium chloride (HPC) 0.4 M and operating under mild conditions [126].

A very important feature of SECR is that little volume of the regenerant amphiphilic solutions are able to dissolve a relatively large amount of adsorbate, thus allowing one to obtain high concentration factors (10^3-10^4) [127].

The combination of specific adsorption steps with functionalized adsorbent, followed by the use of SECR, may also find potential applications in selective recovery processes.

7.2. ENANTIOMERIC ENRICHMENT.

Micellar aggregates are able to influence the rates and mechanism of a large variety of reactions [70]. Evidence has been found that micellar catalysis can be different for reactants that are geometric, positional or optical isomers each other.

Although data indicate that micelles can stereoselectively catalyze some reactions [5], thus allowing partial resolution, the effect is in general relatively small to render the approach useful.

The use of chiral-micellar forming surfactants [70] or of coaggregates [128] (of a normal surfactant and a vesicle forming amphiphile) may remarkably enhance the selectivity.

7.3. ISOTOPIC ENRICHMENT.

Almost all present methods of isotope separation or enrichment are based upon differences in mass or related properties. However, in 1971, based on radical pair theory for magnetic polarization, Lawler and Evans predicted that the separation of isotopes based upon differences in their magnetic moments rather than masses should be possible [129]. The predicted effect was experimentally observed five years later [130]. However, the enrichment factors obtained in bulk solvent where very low. An exciting more recent development has been the utilization of micelles to enhance the enrichment factors for isotopes by Turro and co-workers [131-136].

The technique is based on the radical pair model of chemically induced dynamic nuclear polarization, CIDNP [134,137]. Theory predicts that the reactivity of chemical processes involving diradicaloids depends on the hyperfine interactions of the orbitally uncoupled electron of the radical pair with nuclear spin (magnetic isotope effect) or laboratory magnets (magnetic field effect). Magnetic isotopes can be separated from nonmagnetic ones in such photochemical reaction systems. The purpose of the surfactant micellar media is to provide "supercages" in which there is restricted dimensionality for diffusional excursions of the radical pair and a relatively high local viscosity (hindered fluidity) as well as to solubilize substrates.

An example is represented by the isotopic enhancement using the photolysis of dibenzylketone (DBK) in CTAB micelles. Initially the photolysis results in the formation of a triplet radical pair within the micellar "cage". The triplet radical pair containing the ^{13}C undergo nuclear hyperfine coupling intersystem crossing to a singlet radical pair that regenerates the starting material, DBK, by recombination or forms as a minor product 1-phenyl-4-methylacetophenone. On the other hand, the ^{12}C nucleus is non-magnetic and thus cannot undergo nuclear hyperfine coupling induced triplet to singlet radical pair conversion. Thus, a DBK molecule that goes through a recombination cycle becomes enriched in ^{13}C because ^{13}C-containing radicals undergo hyperfine induced intersystem crossing faster than radicals that do not contain ^{13}C (i.e. $13\text{-}k_{TS} > 12\text{-}k_{TS}$). The presence of micelles enhances the re-encounters of the radical pairs after the triplet-singlet conversion has occurred compared to that possible in solvents alone [132].

Further work is however required in order to asses the practical employment and the economic advantages of this technique for separation or enrichment of isotopes [5,132].

8. General Considerations in the Use of Surfactants in Separation Procedures.

Although in general for separation techniques the surfactants commercially available may be used as received, various procedures for their purifications are reported [138-140]. The most significant problems concern the recovery of the analyte from the surfactant media and the subsequent recovery of the surfactant for re-use.

Several methods are available for this purpose, including column chromatography, extraction-precipitation, use of "destructible" surfactants and distillation. Applicability and limitations have been extensively referred and discussed in ref.9 and 22.

9. Future Directions.

In addition to the areas already presented, some of which have already commercial interest and others with high potential for future practical utilization, other exciting possibilities can be envisaged about the use of surfactants in separation science.

Applications of surfactants in supercritical extraction systems or in supercritical fluid chromatography are exciting, as well as in field-flow fractionation and countercurrent chromatography.

Novel surfactant systems, such as fluoroamphiphiles or with a "functional" polar head (complexing, redox, chiral) could aid in the development of more selective procedures based on the specific possible interactions.

Finally, it has to be underlined that whereas there have been many practical applications using surfactants in chemical separations, mechanistic studies are often lacking. This is due to the complex nature of physicochemical processes involved and to the poor knowledge of the aggregates under the experimental conditions.

The rational design of separation systems will require an understanding of the processes involved in the separation procedures.

10. Acknowledgements.

Financial support from CNR, MPI and Eniricerche is gratefully acknowledged. We thank the publishers for permission to reproduces the figures.

11. References

1. W.L.Hinze, in "Solution Chemistry of Surfactants", K.L.Mittal (ed.), Plenum Press, New York, 1979, Vol.I, p.78.
2. L.J.Cline-Love, J.G.Habarta and J.G.Dorsey, Anal.Chem., 1984, 56, 1132A.
3. D.W.Armstrong, Sep.Purif.Methods, 1985, 14, 212.
4. E.Pelizzetti and E.Pramauro, Anal.Chim.Acta, 1985, 169, 1.
5. W.L.Hinze, Ann.Chim.(Rome), 1987, 77, 167.
6. W.L.Hinze and D.W.Armstrong (eds.), "Ordered Media in Chemical Separations", ACS Symp.Ser. 1987, n°342.
7. M.J.Rosen (ed.), "Surfactants in Emerging Technologies", M.Dekker, New York, 1987.
8. D.T.Wasan, M.E.Ginn and D.O.Shah (eds.), "Surfactants in Chemical/Process Engineering", M.Dekker, New York, 1988.
9. J.F.Scamehorn and J.H.Harwell (eds.), "Surfactant-based Separation Processes", M.Dekker, New York, 1989.
10. E.Tomlinson, T.M.Jefferies and C.M.Riley, J.Chromatogr., 1978, 159, 315.
11. R.Gloor and E.L. Johnson, J.Chromatogr.Sci., 1977, 15, 413.
12. F.G.P.Mullins, in ref.6, p.115.
13. D.G.Herries, W.Bishop and R.M.Richards, J.Phys.Chem., 1964, 64, 1842.
14. D.W.Armstrong and J.H.Fendler, Biochim.Biophys.Acta, 1977, 418, 75.
15. F.Maley and D.U.Guarino, Biochem.Biophys.Res.Comm., 1977, 77, 1425.
16. M.Arunyanart and L.J.Cline-Love, Anal.Chem., 1985, 57, 2837.
17. P.Yarmchuck, R.Weiberger, R.F.Hirsch and L.J.Cline-Love, Anal.Chem., 1982, 54, 2233.
18. A.Berthod, I.Girard and C.Gonnet, in ref.6, p.130.
19. M.G.Khaledi, E.Peuler and J.Ngeh-Ngwainbi, Anal.Chem., 1987, 59, 2738.
20. M.G.Khaledi, Anal.Chem., 1988, 60, 876.
21. Y.N.Kim and P.R.Brown, J.Chromatogr., 1987, 384, 209.
22. W.L.Hinze, in ref.6, p.2.
23. F.J.DeLuccia, M.Arunyanart and L.J.Cline-Love, Anal.Chem., 1985, 57, 1564.
24. L.P.Stratton, J.B.Hynes, D.G.Priest, M.T.Doig, D.A.Barron and G.L.Asleson, J.Chromatogr., 1986, 357, 183.

25. W.L.Hinze, H.N.Singh, Y.Baba and N.G.Harvey, Trends Anal.Chem., 1984, 3, 193.
26. M.G.Khaledi and J.D.Dorsey, Anal.Chem., 1985, 57, 2190.
27. J.G.Dorsey, M.T.DeEchegaray and J.S.Landy, Anal.Chem., 1983, 55, 924.
28. P.Yarmchuck, R.Weinberger, R.F.Hirsch and L.J.Cline-Love, Anal.Chem., 1985, 57, 1564.
29. J.G.Dorsey, in ref.6, p.105.
30. R.S.Matson and S.C.Goheen, LC-GC Mag., 1986, 4, 624.
31. M.G.Khaledi, Trends Anal.Chem., 1988, 7, 293.
32. E.Pelizzetti and E.Pramauro, J.Phys.Chem., 1984, 88, 990.
33. E.Pramauro, C.Minero, G.Saini, R.Graglia and E.Pelizzetti, Anal.Chim.Acta, 1988, 212, 171.
34. K.Taguchi, K.Hiratani and Y.Okahata, J.Chem.Soc. Chem.Comm, 1986, 364.
35. J.F.Scamehorn and J.H.Harwell, in ref.8, p.77.
36. J.H.Harwell and E.A.O'Rear, in ref.9, p.155.
37. J.Wu, J.H.Harwell and E.A.O'Rear, J.Phys.Chem., 1987, 91, 623.
38. J.Wu, J.H.Harwell and E.A.O'Rear, Colloids Surf., 1987, 26, 155.
39. J.Wu, J.H.Harwell, E.A.O'Rear and S.D.Christian, AIChE J., 1988, 34, 1511.
40. J.Wu, C.Lee, J.H.Harwell and E.A.O'Rear, in ref.9, p.173.
41. S.Terabe, K.Otsuka, K.Ichikawa, A.Tsuchiya and T.Ando, Anal.Chem., 1984, 56, 113.
42. S.Terabe, K.Otsuka and T.Ando, Anal.Chem., 1985, 57, 834.
43. M.J.Sepaniak and R.Cole, Anal.Chem., 1987, 59, 472.
44. D.E.Burton, M.J.Sepaniak and M.P.Maskarinec, Chromatographia, 1986, 21, 583.
45. A.T.Balchunas and M.J.Sepaniak, Anal.Chem., 1987, 59, 1466.
46. S.Terabe, H.Utsuni, K.Otsuka, T.Ando, T.Inomata, S.Kuze and Y.Hanaoka, HRC CC,J.High Resolut. Chromatogr. Chromatogr.Commun., 1986, 9, 666.
47. A.S.Cohen, S.Terabe, J.A.Smith and B.L.Karger, Anal.Chem., 1987, 59, 1021.
48. A.T.Balchunas and M.J.Sepaniak, Anal.Chem., 1988, 60, 617.
49. A.Dobashi, T.Ono and S.Hara, Anal.Chem., 1989, 61, 1984.
50. R.O.Dunn, J.F.Scamehorn and S.D.Christian, Sep.Sci.Technol., 1985, 20, 257.
51. L.Lane Gibbs, J.F.Scamehorn and S.D.Christian, J.Membrane Sci., 1987, 30, 67.
52. J.F.Scamehorn, R.T.Ellington, S.D.Christian, B.W.Penney, R.O.Dunn and S.N.Bhat, AIChE Symp.Ser., 1986, 82, 48.
53. E.Pramauro and E.Pelizzetti, Trends Anal. Chem., 1988, 7, 270.
54. G.A.Smith, S.D.Christian, E.E.Tucker and J.F.Scamehorn, from ref.6, p.184.
55. S.D.Christian, G.A.Smith, E.E.Tucker and J.F.Scamehorn, Langmuir, 1985, 1, 564.
56. E.Pramauro, Ann.Chim.(Rome), 1990, 80, 101.
57. K.J.Sasaki, S.L.Burnett, S.D.Christian, E.E.Tucker and J.F.Scamehorn, Langmuir, 1989, 5, 363.
58. J.F.Scamehorn and J.H.Harwell, in ref.8, p.77.
59. S.D.Christian and J.F.Scamehorn, in ref.9, p.3.
60. J.F.Scamehorn, S.D.Christian and R.T.Ellington, in ref.9, p.29.
61. R.D.Noble and J.D.Way (eds.), "Liquid Membranes. Theory and Applications", ACS Symp.Ser. 347, 1987.
62. N.N.Li and A.L.Shrier, in "Recent Developments in Separation Science", N.N.Li (ed.),

CRC Press, Cleveland, 1972.
63. M.Kirch and J.M.Lehn, Angew.Chem., Int.Ed.Engl., 1975, 14, 555.
64. A.Xenakis and C.Tondre, J.Phys.Chem., 1983, 87, 4737.
65. J.D.Lamb, J.J.Christensen, S.R.Izatt, K.Bedke, M.S.Astin andR.M.Izatt, .Am.Chem.Soc., 1980, 102, 3399.
66. W.J.Ward, AIChEJ, 1970, 16, 405.
67. A.Xenakis and C.Tondre, Faraday Discuss.Chem.Soc., 1984, 77, 115.
68. A.Xenakis, C.Selve and C.Tondre, Talanta, 1987, 34, 509.
69. C.H.Suelter, "A Practical Guide to Enzimology", Wiley, New York, 1985
70. J.H.Fendler, "Membrane Mimetic Chemistry", Wiley, New York, 1982.
71. P.L.Luisi and L.J.Magid, CRC Crit.Rev.Biochem., 1986, 20, 409.
72. P.Fourrè, B.Bauer and J.Lemerie, Anal.Chem., 1983, 55, 662.
73. N.N.Li (ed.), "Recent Developments in Separation Science", CRC Press, Florida, 1986, Vol.VIII.
74. J.Wisniak and A.Tamir (eds), "Liquid-Liquid Equilibrium and Extraction", Elsevier, New York, 1980 and 1987.
75. M.Puttermans, L.Dryon and D.L.Massart, Anal.Chim.Acta, 1985, 15, 189.
76. M.Puttermans, L.Dryon and D.L.Massart, J.Pharm.Biomed. Anal., 1985, 3, 503.
77. V.Graef, T.Banker, E.Zuruya and O.Nishikaze, Fresenius' Z.Anal.Chem., 1986, 324, 289
78. R.Swarup S.K.Patil, J.Inorg.Nucl.Chem., 1976, 38, 1203.
79. E.Pramauro, E.Pelizzetti, C.Minero, E.Barni, P.Savarino and G.Viscardi, Ann.Chim.(Rome), 1987, 77, 209.
80. P.L.Luisi and B.E.Straub (eds.), "Reversed Micelles", Plenum, New York, 1984.
81. M.E.Leser, G.Wei and P.L.Luisi, Biochem.Biophys.Res. Comm., 1986, 135, 629.
82. T.A.Hatton, in ref.9, p.55.
83. T.A.Hatton, in ref.6, p.170.
84. T.A.Hatton, this volume.
85. M.Corti, C.Minero and V.Degiorgio, J.Phys.Chem., 1984, 88, 309.
86. H.Watanabe, in "Solution Behavior of Surfactants", K.L.Mittal and E.J.Fendler (eds.), Plenum, New York, 1982, p.1305.
87. E.Pramauro, C.Minero and E.Pelizzetti, in ref.6, p.172.
88. W.L.Hinze, H.N.Singh, Z.S.Fu, R.W.Williams, D.J.Kippenberger, M.D.Morris and F.S.Sadek, in "Analysis of Polycyclic Aromatic Hydrocarbons", T.Vo-Dinh (ed.), Wiley, 1990, Ch. 5.
89. N.D.Gullickson, J.F.Scamehorn and J.H.Harwell, in ref.9, p.139.
90. A.E.Vassiliades, in "Cationic Surfactants", E.Jungermann (ed.), Dekker, New York, 1970, ch.12.
91. P.L.Dubin, T.D.Ross, I.Sharma and B.E.Yegerlehener, in ref.6, p.162.
92. A.M.De Trobriand, INIS Atom Index, 1979, 11, Report CEA-R-5009 (CA 1980, 521209).
93. W.A.Charewicz and J.Strzelbicki, J.Chem.Technol. Biotechnol., 1979, 29, 149.
94. D.O.Shah and R.S.Schechter (eds.), "Improved Oil Recovery by Surfactant and Polymer Flooding", Academic Press, New York, 1977.
95. D.O.Shah (ed.), "Surface Phenomena in Enhanced Oil Recovery", Plenum Press, New

York, 1981.
96. C.A.Miller and S.Qutubuddin, in "Interfacial Phenomena in Apolar Media", H.S.Eicke and G.D.Parfitt (eds.), Marcel Dekker, New York, 1987, p.117.
97. P.Neogy, in "Microemulsion: Structure and Dynamics", S.E.Friberg and P.Bothorel (eds.), CRC Press, Boca Raton, La., 1987, p.197.
98. E.L.Neustadter, in "Surfactants", Th.F.Tadros (ed.), Academic Press, London, 1984, p.277.
99. C.Tondre and R.Zana, J.Dispersion Sci.Technol., 1980, 1, 179.
100. B.J.Carroll, J.Colloid Interf.Sci., 1981, 79, 126.
101. B.J.Carroll, B.G.C.O'Rourke and A.J.I.Ward, J.Pharm.Pharmacol., 1982, 34, 287.
102. S.E.Friberg and P.Neogi, in ref.9, p.119.
103. W.H.Lindenberger, in ref.7, p.187.
104. P.Somasundaran and B.M.Moudgil (eds.), "Reagents in Mineral Technology", Dekker, New York, 1988.
105. J.Leja, "Surface Chemistry of Froth Flotation", Plenum Press, New York, 1982.
106. D.W.Fuerstenau and R.Herrera-Urbina, in ref.9, p.259.
107. B.L.Karger, R.B.Grieves and R.Lemlich, Sep.Sci., 1968, 3, 393.
108. F.Sebba, "Ion Flotation", Elsevier, Amsterdam, 1962.
109. R.Lemlich, Ind.Eng.Chem., 1968, 60, 16
110. P.Somasundaran, in "Separation and Purification Methods", E.S.Perry and C.J.van Oss (eds.), Dekker, New York, 1972, Vol.I, p.117.
111. T.E.Carleson, in ref.9, p.233.
112. S.E.Charm, J.Morningstar, C.C.Matteo and B.Paltiel, Anal.Biochem., 1966, 15, 498.
113. Z.Lalchev and D.Exerowa, Biotechnol.Bioeng., 1981, 23, 669.
114. R.D.Gehle and K.Schugerl, Appl.Microbiol.Biotechnol, 1984, 19, 373.
115. M.D.Guy, J.D.McIver and M.J.Lewis, Water Res., 1976, 10, 737.
116. P.R.Fields, P.J.Fryer, N.K.H.Slater and G.P.Woods, Chem.Eng.J., 1983, 27, B3.
117. R.B.Grieves, AICHE Symp.Ser., 1975, 71, 143.
118. P.Somasundaran, Sep.Sci., 1975, 10, 93.
119. K.Jurkewicz, Sep.Sci.Technol., 1985, 20, 1979.
120. R.B.Grieves, W.Charewicz and S.M.Brien, Anal.Chim.Acta, 1974, 73, 293.
121. G.A.Nyssen, G.S.Lovell, A.A.Simon, J.G.Smith and B.K.Tolar, Sep.Sci.Technol., 1987, 22, 2127.
122. R.B.Grieves, Chem.Eng.J., 1975, 9, 93.
123. A.N.Clarke and D.J.Wilson, "Foam Flotation: Theory and Applications", Dekker, New York, 1983.
124. F.Sebba, Sep.Purif.Methods, 1985, 14, 127.
125. F.Sebba, in ref.9, p.91.
126. J.F.Scamehorn, in ref.8, p.77.
127. D.L.Blakeburn and J.F.Scamehorn, in ref.9, p.205.
128. R.Ueoka, R.A.Moss, S.Swarup, Y.Amtsumoto, G.Strauss and Y.Murakami, J.Am.Chem.Soc., 1985, 107, 2185.
129. R.G.Lawler and G.T.Evans, Ind.Chem.Belg., 1971, 36, 1087.
130. A.L.Buchachenko, E.M.Galimov, V.V.Ershowv, G.A.Nikiforov and A.D.Pershin, Dokl.Akad.Nauk.SSSR., 1976, 228, 379.

131. N.J.Turro and B.Kraeutler, J.Am.Chem.Soc., 1978, 106, 7432.
132. N.J.Turro, D.R.Anderson and B.Kraeutler, Tetrahedron Lett., 1980, 21, 3.
133. N.J.Turro, B.Kraeutler and D.R.Anderson, J.Am.Chem.Soc., 1979, 101, 7435.
134. N.J.Turro and B.Kraeutler, Acc.Chem.Res., 1980, 13, 369.
135. M.B.Zimmt, C.Doubleday and N.J.Turro, J.Am.Chem.Soc., 1984, 106, 3363.
136. I.R.Gould, N.J.Turro and M.B.Zimmt, Adv.Phys.Org.Chem., 1984, 20, 1.
137. A.L.Buchachenko, V.F.Tarasov and V.I.Mal'tsev, J.Phys. Chem.(Russ), 1981, 55, 936.
138. D.Attwood and A.T.Florence, "Surfactant Systems", Chapman and Hall, London, 1983.
139. J.Cross (ed.), "Anionic Surfactants", Dekker, New York, 1977.
140. J.Cross (ed.), "Nonionic Surfactants", Dekker, New York, 1986.

STRUCTURAL CHANGES OF AOT REVERSE MICELLES BY THE PRESENCE OF PROTEINS: PERCOLATION PROCESS INDUCED BY CYTOCHROME C

M.P.PILENI, J.P. HURUGUEN, C. PETIT
-1-Université P et M Curie,
Laboratoire S.R.I.,
batiment de Chimie-Physique,
11 rue P et M Curie 75005 Paris,
France,
-2- C.E.N. Saclay, D.L.P.C., S.C.M., 91191 Gif sur Yvette,
France.

ABSTRACT. The formation of reverse micelles in non polar solvents have been investigated for many years [1,5]. The presence of water is necessary to form a large surfactant aggregate: close packing of the surfactant polar heads leaves an empty volume in the center of the micellar core which can only be filled with hydration (or bound) water. The reverse micelles considered here are then also microemulsions. We will not make differences between the terms swollen micelles and microemulsions, although certain authors use the first term for systems containing only bound water. For AOT, the maximum amount of bound water in the micelles correspond to a water-surfactant molar ratio w = $[H_2O]/[AOT]$ of about 10. Above this amount, part of the water is "free". Because AOT is soluble in many non polar solvents over a wide range of concentrations, a large number of studies of the structure of reverse micelles have been made with this surfactant.

Several models have been proposed to describe the solubilization of protein in reverse micelles. The models proposed take into account the size of the protein. In the case in which the protein size is smaller than the water pool two differents models are proposed: some groups claim that the presence of the protein gives rise to larger surfactant aggreggates than those formed in the absence of protein. Other groups suggest that the protein has no effect whatsoever on the size of the aggregate . In the case in which the protein size is larger than the water pool, very few models have been proposed.

In the present paper we will try to show that at relatively low water content, the solubilisation of protein [5] such as cytochrome c induces a very small perturbation whereas by increasing the water content a percolation process occurs with the appearance of a two phase transition.

1. Variation of the water pool size with the water content

The major geometrical constraint on reverse micelles is that the whole water content has to be shared by the dispersed water-pools. Each water-pool is surrounded by an interface of polar heads groups of AOT surfactant. The total volume of the dispersed aqueous phase (V) and the total interfacial area (S) are determined by the chemical composition of the sample. The interfacial area (s) per head group of the AOT molecule is almost independent of any other physico-chemical parameter. Thus, the micellar mass depends only on the total area S and on the volume V of the water-pool. Assuming that micelles are monodispersed, the water-pool volume, is directly related to the volume of water molecules (V_e): $V = V_e . W. N$, where N is the aggregation number and $W = [H_2O]/[AOT]$. If S is the total interfacial

D. M. Bloor and E. Wyn-Jones (eds.),
The Structure, Dynamics and Equilibrium Properties of Colloidal Systems, 355–371.

area, then $V/S = V_e$. W. N/s.N;s is about 60 $Å^2$ and V_e is easily calculated from the water density (V_e = 30 $Å^3$). Hence V/S = 0.5 W. For spherical aggregates, the volume-to-surface ratio is directly related to the radius of the water-pool, Rw by the following:

$$R = 3\ V/S \quad (1)$$

The characteristic radius of the inverse micelle is proportional to W, the molar ratio of water to surfactant molecules:

$$Rw(Å) = 1.5\ W \quad (2)$$

at high water content good agreement is obtained between eq. (2) and the various water-pool radii determined by different techniques (figure 1).

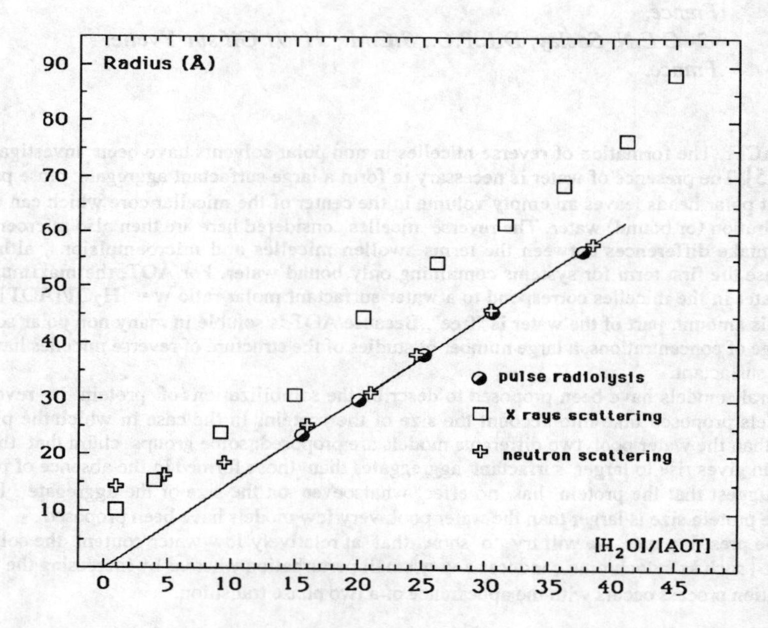

Figure 1. Variation of the water pool radius with the water content tested by small angle x ray scattering

At low water content (W< 10), the experimental radii are larger than those determined from eq. (1). This is due to the fact that in eq. (1), the volume of the polar head group is neglected. This latter approximation is not valid at low water content, where a more precise relation can be derived [4]. Since we dealt with a differential perturbation of the micellar mass by solutes at high W, we used only the linear relation.

2. Possible sites of solubilization in reverse micelles

2.1 GEOMETRICAL MODEL [6,7]

From eq. (1), addition of a solute in reverse micelles, could produce a small change in the volume V, noted dV, or in the interfacial area S, noted dS.
A priori, six different cases can occur for the localization of the added solute:

(A and B) If the solute is located in the hydrocarbon phase or in the external interface, the ratio V/S in unchanged, which implies that there is no perturbation of the characteristic radius of the micelles (figure 2A and 2B).

(C and D) The addition of solute inside the water-pool or in the internal interface(figure 2C and D), induces an increase of the volume, dV, which is due to the molecular volume of the solute added, whereas the interfacial area remains constant. Then the radius of the water-pool is : $Rw = (V+dV)/S$. In this case, the variation of the micellar mass concomitant to the solubilization of the hydrophilic component is equivalent to the perturbation observed upon addition of the same volume of water. If dV is not negligible with respect to V, the addition of a solute to the water-pool induces an increase of the characteristic radius.

(E) The addition of a solute anchored at the interface of the water-pool (figure 2E) induces an increase of the interfacial area, dS, whereas the volume of the water-pool remains constant. Then the water-pool radius is given by $Rw = V/(S+dS)$. Hence addition of a solute located at the interface of the droplet induces a decrease of the water-pool radius. In this case, the addition of a solute is formally equivalent to an increase of the surfactant concentration at constant water content: more interface is available to trap the same quantity of polar compound.

(F) Addition of solutes could induce the formation of small aggregates surrounding the solute. Hence two micellar populations could be in equilibrium (figure 2F). If the solute is surrounded by a coat of surfactant, the volume of the water pool remains constant, whereas the interfacial area of the empty micelles decreases: $Rw = V/(S-dS)$. Hence the water-pool radius of an empty micelle is expected to increase.
However, if the solute needs water molecules to form small aggregates no assumption on the evolution of the water-pool radius can be made. This model, implying two micellar populations (called " segregation" model), is very difficult to confirm. The only way to do this is to proceed by comparison between diffusion measurements, which average macroscopically over the whole sample, and indirect chemical techniques [5-9], which counts only the micelles where some reactives are present (or absent). Disagreement between physical and chemical techniques supports the segregation model.

Such simple models cannot take into account the case where the solute is partitioned between the water-pool (model C and D) and the interface (model E and F): if both S and V increase, our structural method cannot give any quantitative result, since the evolution of the water-pool radius depend on the partition coefficient of the solute between the interface and the water pool. We can only distinguish between a preponderant increase of V or S.

This geometrical model has been tested by SAXS, at w up to 20: On adding ether, mainly solubilized in the bulk non polar phase, no changes in the size of the water pool are observed (case A). On adding either methylviologen or chymotrypsin or ribonuclease at pH 4 and at pH 11, an increase in the water pool radius is obtained. From this result it can be deduced that the viologen and chymotrypsin and ribonuclease are located in the water pool. However the average location could be either at the internal interface or inside the water pool.

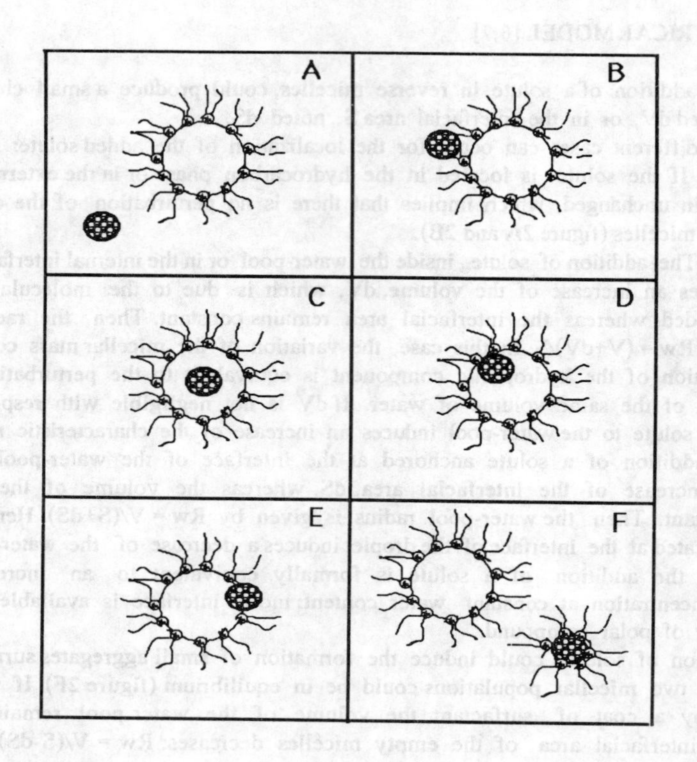

Figure 2. Geometric Models

2.2 KINETIC MODEL USING HYDRATED ELECTRON AS A PROBE [8]

The spectroscopic data for hydrated electrons are described in reference [1]. It has been previously shown that the distribution of probes in reverse micelles is a Poisson distribution [1]. This has been extended to the solubilization of proteins. The quenching rate constants of hydrated electrons for various quenchers have been determined kinetically. The choice of the quenchers used is directly related to their locations.

Using nitrate ion which is water soluble and negatively charged or using a water soluble zwitterionic propylviologen sulfonate, it seems reasonable to assume that these quenchers are located inside the water pool. On the other hand, copper lauryl sulfate, $Cu(LS)_2$, is expected to be located at the interface because of the electrostatic interactions between the surfactant head polar groups and copper ions and because of the hydrophobic character of the two alkyl chains [LS]. In the same way, copper ions and methylviologen which are positively charged are attracted to the micellar surface.The locations of these probes are confirmed by a geometrical model described previously and tested by small-angle X ray scattering. In all cases, the quenching rate constant of the hydrated electron is determined at various water contents and then at various water pool radii

(R_w = 1.5 w). By plotting, Ln kq over Ln w for a given quencher,the observed slope is close to 3 for nitrate ions and propylviologen, chymotrypsine and ribonuclease at pH 8 and close to 2 for copper ions, copper lauryl sulfate and methylviologen and ribonuclease at pH 4. Such differences in the slope can be explained in terms of the location of the quencher in the reverse micelles :It is assumed that in reverse micelles, the average distance between two species, r, is approximately equal to the water pool radius, R_w :

(i) When two species (hydrated electron and quencher) are located inside the droplet,it is experimentally found that kq varies with w^{-3}, and for a diffusion-controlled reaction, kq is expected to vary as r^{-3}, where r is the average distance between the two species. Because of the linear relationship between R_w and w, it is concluded that for probes located inside the water pool, the quenching rate constant depends on the volume of the droplet as is observed for nitrate ions and propylviologen sulfonate, chymotrypsine and ribonuclease at pH 8 .

(ii) For two species located in the water pool (hydrated electron) and at the interface (quencher), kq varies with w^{-2} and is expected to vary with r^{-2} as is observed using copper ions, methylviologen, copper lauryl sulfate and ribonuclease at pH 4. Then for such probes, located at the interface, the quenching rate constant depends on the surface of the droplet.

Theoretical studies compare the kinetic rate constants of the two reactants. One of the reactants freely diffuses inside the sphere and the other is either fixed at the centre or at the surface of the sphere. In the first case it is found that the kinetic rate constant is directly dependent on r^{-3} whereas in the latter case it depends on $r^{-2.5}$. The present results on the change of the variation of hydrated electron quenching rate constant with the water pool radius are in good agreement with those of theoretical calculations.

The geometrical model gives macroscopic information on the structural changes brought about by adding probes due to the location of the probes and the present model gives microscopic information from which the location is deduced. Using the two models, the average location of proteins which do not change drastically the micellar structure and have a hydrodynamic radius smaller than that of the water pool, can be deduced.

We have not discussed the cases in which the probes are located in the bulk hydrocarbon phase or form there own aggregates as the following. As has been described previously, the hydrated electron is formed exclusively in aqueous phase. To be observable, a certain number of water molecules in the water pool are required (w > 10). In the case in which the probe is located in the bulk hydrocarbon phase, the hydrated electron is never in contact with its quencher and the probe is not able to react with it. No quenching of hydrated electrons is observed.

The limitation of this kinetic model is that the structural changes of probes or proteins are negligible and the droplet can be considered as unchanged. This has been observed for chymotrypsin and ribonuclease. For cytochrome c, at high water content, the quenching rate constant does not change with the water pool size. This can be attributed to drastic structural changes.

2.3 LOCATION OF CYTOCHROME C

2.3.1 *From geometrical model*: The analysis of the diffusion spectra of reverse micelles containing various cytochrome c concentration shows a decrease in the radii as the cytochrome c concentration increases. The same behaviour, figure 3, observed for the gyration radius and for the characteristic radius indicates that this is due due to an effect of droplet interactions.

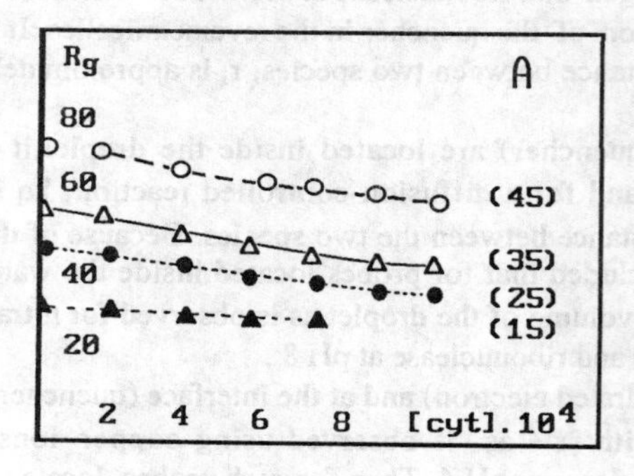

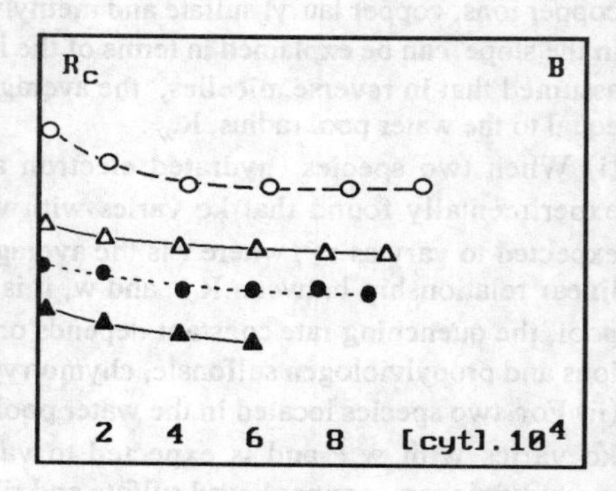

Figure 3: variation of the gyration radius (A) and the characteristic radius (B) at various w values with cytochrome c concentration.

The Porod plots with the increase of cytochrome c concentration clearly show a shift in the minimum indicating a decrease in the droplet size, and, an increase in the Porod limit indicating an increase in the total interface in the system (figure 4):

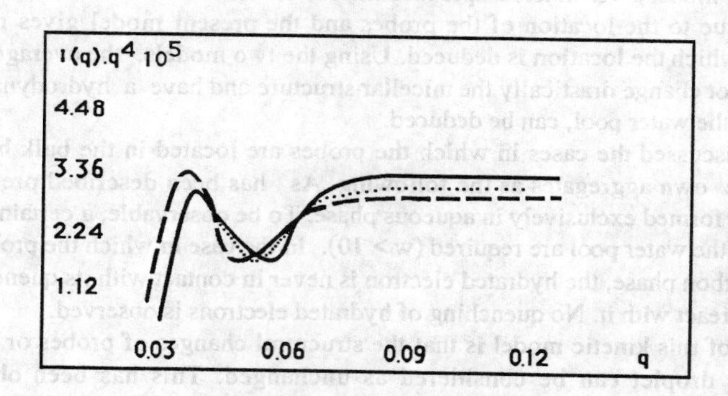

Figure 4: Porod plots of scattering spectra at various cytochrome c concentration.

According to the shape of the scattering patterns, fairly monodisperse spherical droplets can be assumed and the geometrical model previously proposed can be used. The radius is given by equation 1. V is the total polar volume entrapped by the total surface is partially due to the surfactant, S_{AOT}, and partially to the protein, S_{cyt} with a constant polar volume. The relationship of the radius of the droplet and cytochrome c concentration is the following:

$$\frac{1}{R} = \frac{S_{cyt}}{3\,(V_{AOT} + W.V_w).[AOT]} \cdot [cyt] + \frac{S_{AOT}}{V_{AOT} + W.V_w}$$

V_{AOT} is the polar volume of AOT and equal to 173 A^3, W is the water content equal to

$[H_2O]/[AOT]$ and V_w is the volume of water molecules.

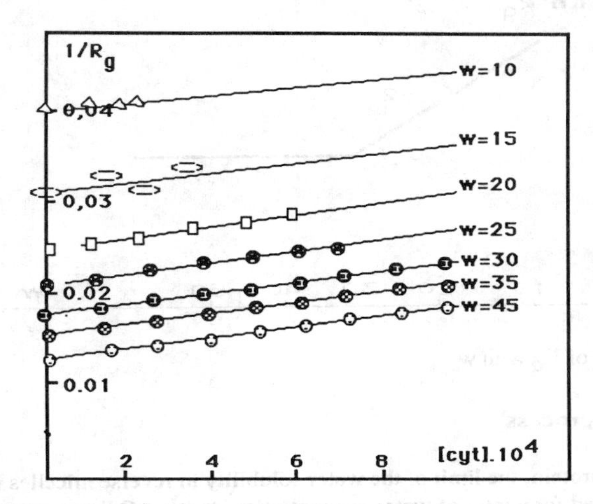

Figure 5: Variation of 1/R with cytochrome c concentration

From figure 5 the contribution to the interfacial area per cytochrome molecule is deduced and found to increase with the water content [7]. This phenomena is probably due to the increase in the fluidity of the interface by increasing the water content which favours the penetration of cytochrome c at the interface. This is consistent with data obtained on the interaction between monolayer surfactant and cytochrome [9].

2.3.2 *From kinetic model:* The reaction rate constant hydrated electron with cytochrome c were determined at various water content.

Figure 6 shows that , below w = 30, the slope of ln k_q over ln w is close to 2 and above W = 30 the k_q value does not change by increasing the water content . This indicates that at relatively high water content, the kinetic model cannot be used. This could be due either to the fact that the kinetic treatment is unavailable in this w range. The data presented below could explain such phenomena.

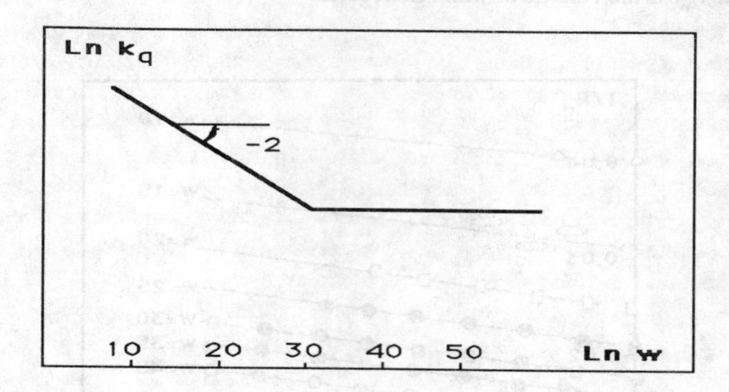

Figure 6: variation of K_q with w

3. Percolation process

In the absence of protein, the limit of the water solubility in reverse micelles in isooctane solution is generally obtained for a ratio of water concentration to the AOT concentration, w, equal to 60. By increasing w up to this value, a turbid solution is obtained corresponding to a mixture of the L_1 and L_2 phases [1]. At low cytochrome c concentrations (below 10^{-4}M), the limit of the solubility of water in reverse micelles remains unchanged and equivalent to w = 60. By increasing the cytochrome c concentration to 3 x 10^{-4}M, (T = 20°C), the limit in the solubility limit increases to reach w = 80 and the system goes towards a diphasic solution at higher water content, w.This increase in the solubility limit has been previously observed and explained in terms of the location of cytochrome c at the interface.

The conductivity of 0.1M AOT in isooctane reverse micelles is very low and remains unchanged on increasing the water content. In the presence of cytochrome c, by increasing the water content, a very low conductivity at low water content is followed by drastic increase in the conductivity.

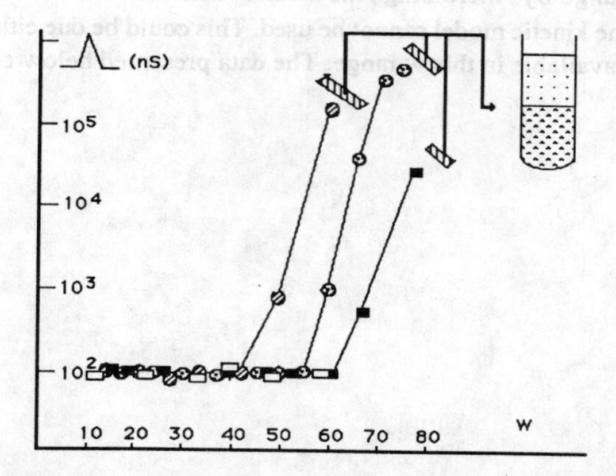

Figure 7: variation of the conductivity with the water content at various cytochrome c concentrations

Similar behaviour is observed under various experimental conditions by increasing increasing factors, such as, the water content, cytochrome c concentration, temperature or volume fraction and keeping the other factors constant. The increases on changing the water content at various cytochrome c concentrations are shown in figure 7. A sharp increase in the conductivity, by up to three orders of magnitude, is induced by raising the temperature above a critical value Tp(f_W = 8,7%, w = 40) (figure 8) The critical temperature value depends on the cytochrome c concentrations: at high concentration the critical value is obtained at relatively low temperatures whereas in the absence of cytochrome c such a value is obtained at high temperatures. From these experiments it can be seen that the conductivity onset obtained by increasing w or the temperature is lower than that obtained by increasing the cytochrome c concentration.

The permittivity can be defined through its complex expression by:

$$\varepsilon(\omega) = \varepsilon'(\omega) - i[\varepsilon''(\omega) + k/\omega\,\varepsilon_v].$$

In order to determine the static value of epsilon , the treatment of the experimental results has been done by Cole-Cole plots and the curves were fitted with a circle program (Rosenbrock). At or just after the onset of percolation, because of the conductivity increases that occurs, a deformation of the experimental Cole-Cole plots called Maxwell-Wagner effect is observed. The conductivity, $\sigma(\omega)$, is defined by its complex expression,

$$\sigma(\omega) = \sigma'(\omega) + i\sigma''(\omega)$$

with $\sigma(\omega) = i\omega\varepsilon_v\varepsilon(\omega)$.

The following can be derived:

$$\sigma'(\omega) = \omega\varepsilon_v.[\varepsilon''(\omega) + \kappa/\omega\varepsilon_v]$$

$$\sigma''(\omega) = \omega\varepsilon_v.\varepsilon'(\omega)$$

The real conductivity, $\sigma'(\omega)$, is plotted as a function of the imaginary form, $\sigma''(\omega)$. The same circle program has been used to fit and to obtain the low frequency limit of the conductivity, subtracted from the imaginary part of the permittivity.

Figure 8A shows an increase in the static dielectric constant with temperature. In the presence of cytochrome c ([cyt] = 4[M]), the static dielectric constant reaches a maximum at 35°C and then decreases. The maximum temperature is associated with the onset of the conductivity. Similar phenomena haves been previously observed in the absence of any solute at high temperatures [10-15]. The comparison of the same permittivity divergence with those of 0.1M AOT reverse micelles without protein is shown in figure 8A a critical temperature value is lower (35°C) in the presence than in the absence of cytochrome (50°C). There is a difference about 15 degrees between the two percolation thresholds. This indicates that cytochrome c has the ability to change the critical percolation temperature .

Figure 8B shows the variations of the permittivity versus the volume fraction for different protein concentrations. In the presence of protein, the divergence of the permittivity takes place at lower volume fraction as the protein concentration increases. Such a divergence of the permittivity was explained, as it has been observed in the absence of protein, as a capacity effect due to the presence of droplets in contact which keep their structure owing to sterical repulsion. This is attributed to the inability of the chains of AOT to penetrate between droplets.

These results show that the critical percolation thereshold is markedly changed by the presence of a solubilized protein at relatively low concentrations. It decreases on increasing water content or cytochrome c concentration or, for a given value of w and cytochrome c concentration, by increasing the temperature. These changes of the percolation threshold by the presence of protein could be attributed to the increase in the attractive interactions which reduces the distance between the droplets. However it is well known that addition of protein in biological membranes induces an increase in the permeability of the membrane. Such phenomena, would also, favour a percolation process and cannot be excluded.

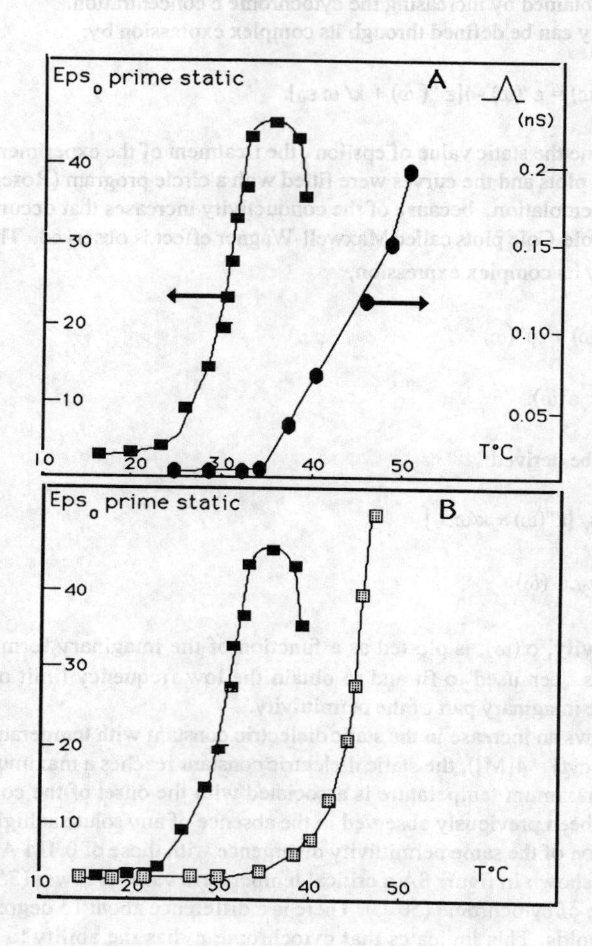

Figure 8: variation of the permitivity with the temperature in the presence and in the absence of cytochrome (A) and with the volume fraction (B)

By low angle X rays scattering measurements the water pool radius determined at various volume fraction is unchanged and found equal to 60Å in the absence of protein and to 48Å, using the Guinier plot, and 52Å, using the Porod plot, in the presence of cytochrome c ([cyt] = 4.[M]), at various volume fraction, (Figure 9).

The good agreement between these values confirms the presence of monodispersed droplets [6-7]. The decrease in the water pool radius observed by adding cytochrome c was previously observed in dilute solution (01.M AOT in isooctane). Such a decrease in the size of the droplet by protein addition was explained in terms of increase in the total surface area due to the contribution of the cytochrome at the interface.

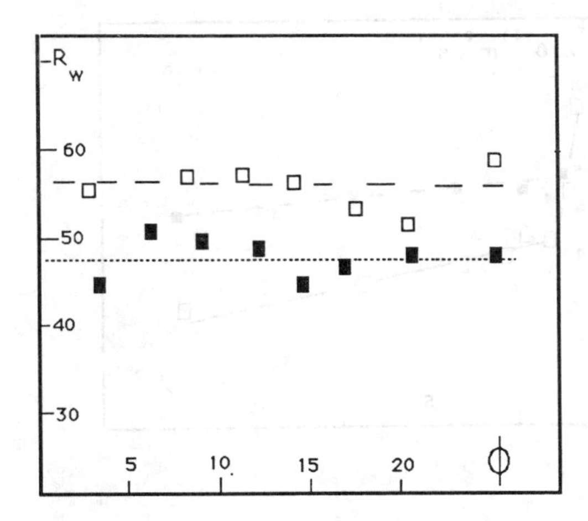

Figure 9: water pool radius in the absence and in the presence of cytochrome c determined from the Guinier plot.

The diffusion coefficient of AOT droplet is determined, by light scattering, in the absence and in the presence of cytochrome c , at w = 40, (Figure 10).

In the absence of cytochrome c, as has been observed previously, the hydrodynamic radius is found equal to 100A, at w = 40. Using a cytochrome c concentration equal to the micellar concentration the diffusion coefficient decreases markedly with the volume fraction. The value of the diffusion coefficient extrapolated zero volume fraction is greater than that obtained in the absence of protein. According to the Stokes-Einstein equation, the hydrodynamic radius decreases on protein addition. This is consistent with the data obtained by SAXS presented in this paper and with those previously published [6,7]. The diffusion coefficient is related to the volume fraction, of the droplet by: $D_c = D_o(1+ æf)$,where æ is the virial coefficient and is directly related to the interaction potential and D_o is the diffusion coefficient at infinite dilution obtained by extrapolation. From the initial slope of the curve in Figure 10 the virial coefficient is found equal to about -25. The negative value indicates attractive interactions between droplets [16-17]. These data are consistent with those previously published indicating an increase in the attractive interactions by solubilizing protein in reverse micelles. This is confirmed by the increase in the scattering intensity observed at low q by SAXS. Similar data were obtained in dilute solution in the presence of cytochrome c and were attributed to the existence of a negligible structure factor S(q), related to the presence of a strong interactions between micellar particles.

The viscosities were measured at various volume fractions and water contents in the presence and in the absence of cytochrome c. In the absence of protein the viscosity of AOT micellar solutions increases slowly with the water content (Figure 11A). In the presence 9.3×10^{-4}M of cytochrome c at w = 40, Figure 11A shows that the increase in the viscosity with the water content, w is greater than that obtained in the absence of protein. Similar data are obtained by measuring the viscosity in various volume fractions in the absence and in the presence of cytochrome c (w = 40; [cyt] =4[M]) (Figure 11B).

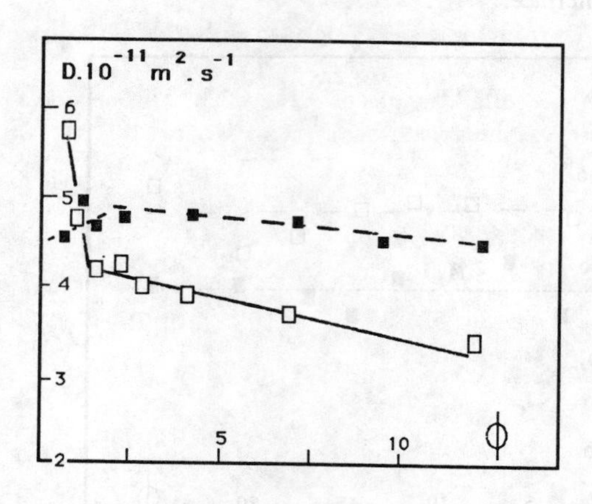

Figure 10: variation of the diffusion coefficient in the absence and in the presence of cytochrome c

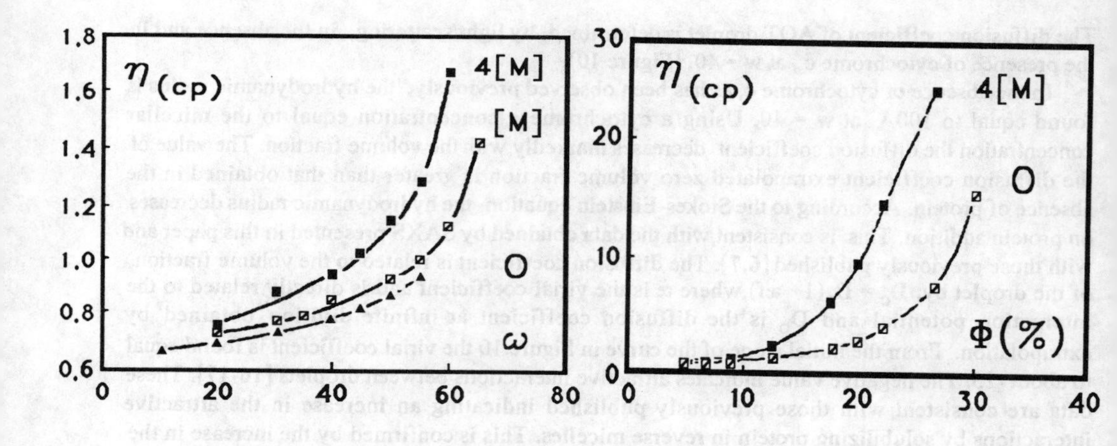

Figure 11: variation of the viscosity with the water content (A) and with the volume fraction (B) at various cytochrome c concentration

The strong increase in the viscosity with the volume fraction (Figure 11B) can be related to the increase in the static dielectric constant with volume fraction. Using these two techniques, the same critical volume fraction is deduced (around 17 or 18%). Such an effect cannot be related to the collision process between droplets since the increase in the water content induces a decrease in the droplet concentration which decreases the probability of collisions between droplets. This increase in the viscosity of micellar solution containing protein can be explained in term on clusters formation.

4. Phase diagram of "cytochrome-AOT-isooctane-water" solution at w = 80

A phase diagram gives the structural changes of 0.1M AOT in isooctane solution at w = 80 as a function of the temperature and the cytochrome c concentration. Figure 12 shows four different regions:

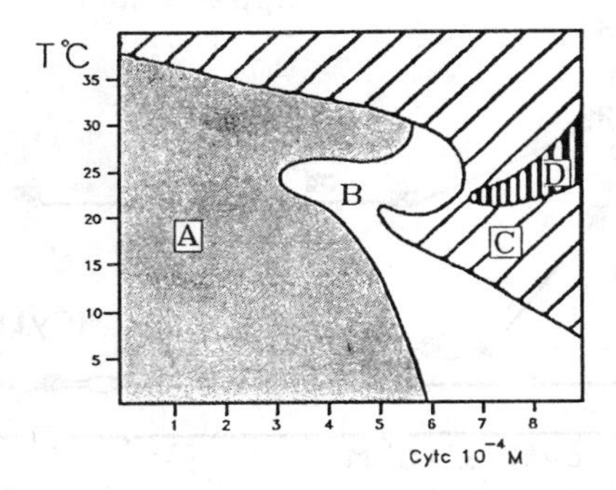

Figure 12: phase diagram

-1- region A is turbid as it is usually observed in the 0.1M AOT-isooctane-water system by solubilizing too many water molecules. It is usually attributed to a lamellar solution [1].

-2-region B is an optically transparent solution similar to that obtained at low water content .

-3- in region C two optically transparent phases appear. Qualitatively it can be seen that the most of the cytochrome c is located in the lower phase whereas the upper phase contains little or no cytochrome c.

-4- in region D two optically transparent phases are observed.

5. Two phase transition

The data presented in this section are those obtained in the C region, at 20°C. The composition of the two phases (Figure 13) obtained after demixion is the following: the lower phase contains the majority of the surfactant, the cytochrome c and the water molecules whereas the upper phase is very poor in these three components. It can be observed that, by increasing the cytochrome c concentrations, AOT and water molecules migrate from the upper phase to the lower phase and at cytochrome c concentration up to 6.10^{-4}M only isooctane remains in the upper phase. The residual concentration of water in the upper phase corresponds to the limit of the solubility of water in isooctane solution.

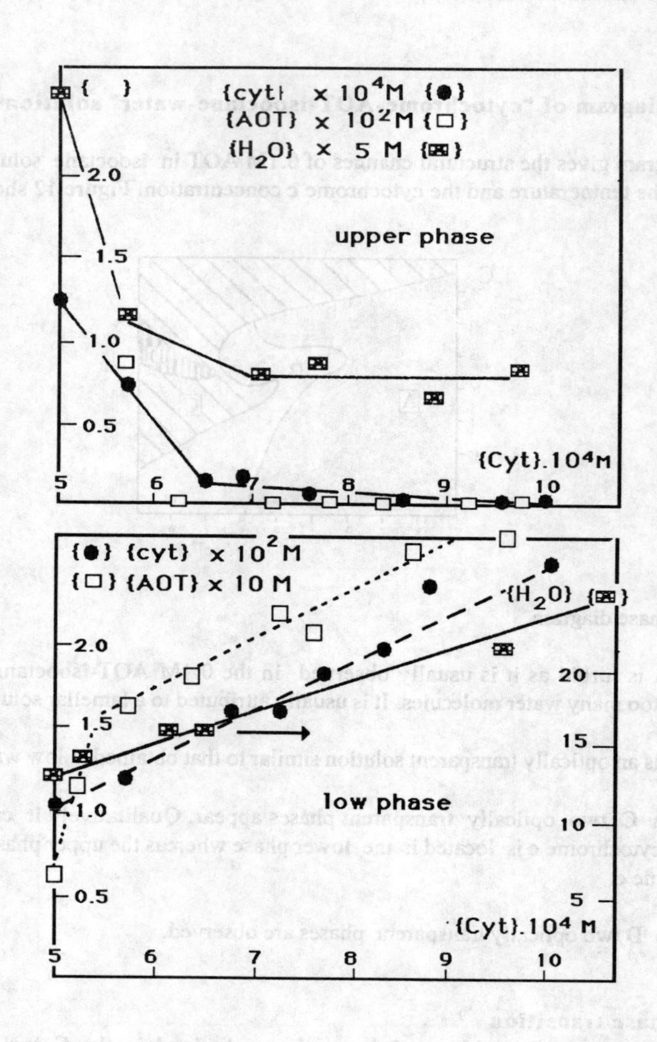

Figure 13: composition of the upper and lower phase

The conductivity measurements, (Figure 14) made at various cytochrome c concentrations in the two phases, shows:

i)In the upper phase a marked decrease in the conductivity to reach a value similar to that obtained in isooctane solution or in empty reverse micelles.

ii) In the lower phase, a very high conductivity is obtained and increases linearly with the cytochrome c concentration.

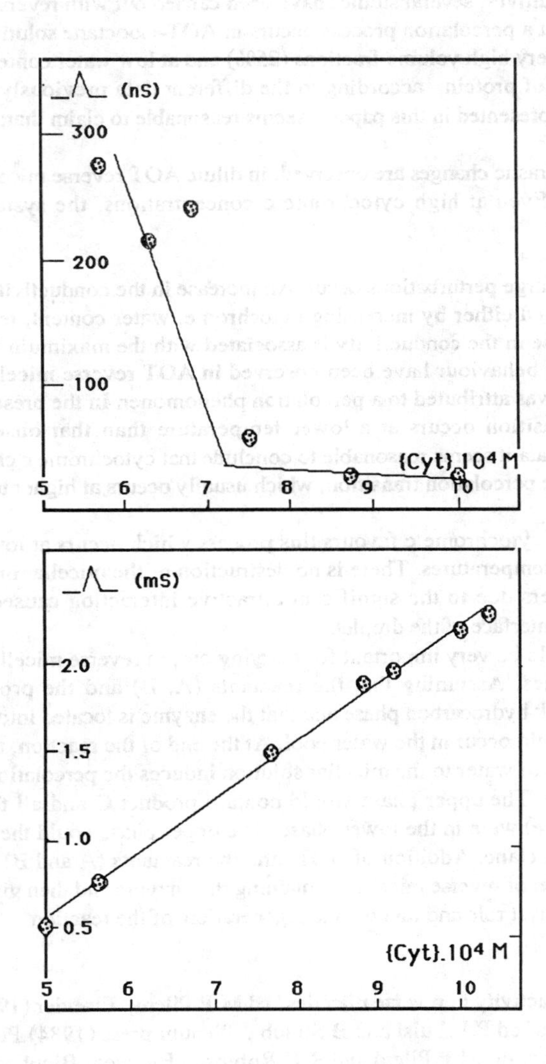

Figure 14: Variation of the conductivity of the upper and lower phase

The viscosity measured in the lower phase increases linearly with cytochrome c concentration . It can be seen that there is a large increase in the viscosity in comparison to that obtained in one phase.

6. Conclusions

In the absence of additives, several studies have been carried out with reverse micelles where it has be demonstrated that a percolation process occurs in AOT-isooctane solution. However, most of these were done at very high volume fractions (25%) and at low water contents.

In the presence of protein, according to the different data previously published by several groups and to those presented in this paper it seems reasonable to claim that:

i) At $10<w<30$ no drastic changes are observed. in dilute AOT reverse micellar solution ([AOT] = 0.1M, isooctane). Even at high cytochrome c concentrations, the system remains a reverse micellar solution

ii) At up to w = 30, large perturbations occur. An increase in the conductivity by several orders of magnitude is observed either by increasing cytochrome, water content, temperature or volume fraction. The increase in the conductivity is associated with the maximum in the static dielectric permitivity. Similar behaviour have been observed in AOT reverse micelles at a high volume fraction (25%) and was attributed to a percolation phenomonen In the presence of cytochrome c, the percolation transition occurs at a lower temperature than that observed in its absence. According to these data it seems reasonable to conclude that cytochrome c changes the percolation onset and favours the percolation transition, which usually occurs at higher temperatures and AOT concentrations.

The presence of cytochrome c favours this process which occurs at lower volume fractions, water contents, and temperatures. There is no destruction of the micellar interface when reverse micelles form clusters due to the significant attractive interaction caused by the presence of cytochrome c at the interface of the droplet.

This process could be very important for carrying out, in reverse micelles, chemical reaction catalyzed by enzymes. Assuming that the reactants (A, B) and the product (C) are mainly solubilized in the bulk hydrocarbon phase and that the enzyme is located into the micellar core, the chemical reaction could occur in the water pool. At the end of the reaction, the product C is in the bulk phase. Addition of water to the micellar solution induces the percolation and the appearance of a diphasic system. The upper phase would contain product C and all the other components (water, AOT, enzyme) were in the lower phase. The upper phase could then be removed and be replaced by pure isooctane. Addition of AOT and the reactants (A and B) to this system would favour the reformation of reverse micelles containing the enzyme and then gives the enzyme of the ability to play its catalyst role and favours the regeneration of the reaction.

7. References:

1. "Structure and reactivity in reverse micelles" ed M.P.Pileni, Elsevier (1989)
2. "Reverse micelles" ed P.L.Luisi and B.Straub , Plenum press (1984).Publisher
3. P.L.Luisi, M.Giomini, M.P.Pileni and R.H.Robinson Biochem.Biophysica Acta 947, 209, (1988).
4. P.D.I. Fletcher and B.H.Robinson Ber. Bunsenges.Phys.Chem., 85, 863, (1981).
5. M.P.Pileni Chem.Phys.Letters 81, 603, (1981).
6. M.P.Pileni, T.Zemb and C.Petit Chem.Phys.Letters 118, 414, (1985)
7. P.Brochette, C.Petit and M.P.Pileni J.Phys.Chem., 92, 3505, (1988).

8. C.Petit, P.Brochette and M.P.Pileni J.Phys.Chem., 90, 6517, (1986).
9. M.Saint-Pierre Chazalet, F.Billoudet and M.P.Pilenu Progress in Colloid and Polymer Sciences 5, 75, (1989)
10. M.A.Van Dijk, C.C.Boog, G.Casteleijn and Y.K.Levine Chem. Phys. Letters 111, 571, (1984).
11. M.A.Van Dijk, Phys Rev Letters 55, 1003, (1985).
12. M.A.Van Dijk, G.Casteleijn, J.G.Joosten and Y.K.Levine J.Chem.Phys, 85, 626, (1986).
13. M.A.Van Dijk, E.Broekman, J.G.H.Joosten and D.Bedeaux J.Physique 47, 727, (1986).
14. S.Bhattacharya, J.P.Stokes, M.W.Kim andJ.S.Huang Physical Review Letters 55, 1884, (1985).
15. S.A.Safran, I.Webman and G.S.Grest Phys.Review 32, 506, (1985).
16. D.Chatenay, W.Urbach, A.M.Cazabat and D.Langevin, Physical Review Letters 54, 2253, (1985).
17. A.M.Cazabat, D.Chatenay, D.Langevin and J.Meunier Faraday Discuss. Chem.Soc., 76, 291,(1982).

MICROPARTICLE FORMATION IN REVERSE MICELLES

A. KHAN-LODHI*, B.H. ROBINSON and T. TOWEY,
School of Chemical Sciences,
University of East Anglia,
Norwich NR4 7TJ, United Kingdom

C. HERRMANN, W. KNOCHE and U. THESING,
Fakultät für Chemie,
Universität Bielefeld,
Bielefeld, F.R.G.

***Present address: Unilever Research, Port Sunlight, Wirral, Cheshire.**

ABSTRACT. **In this article we desribe the preparation of microparticles of platinum and cadmium sulphide in reverse micellar media. The equilibrium particle size is rationalised in terms of the kinetics of the rate determining step. In the case of platinum this is the reduction step $Pt^{IV} \rightarrow Pt^{0}$. For CdS semiconductor microparticles the kinetics of microparticle formation has been followed as a function of wavelength using stopped-flow kinetics. The important step is shown to be the inter-droplet reagent transfer.**

1. Introduction

Micro-water droplets (reverse micelles or w/o microemulsions) in an oil solvent provide a convenient medium in which to prepare microparticles in the nanometre size range. In many situations w/o microemulsions are known to consist of essentially monodisperse water droplets, the size of which can be readily varied by changing the composition of the system. In fact, for the water/Aerosol-OT/alkane system, $r_W = 0.18R$ where r_W is the radius of the water droplet (encapsulated by a curved fluid-like monolayer) and R is the mole ratio of water to surfactant. Microparticles (*e.g.* metal catalysts such as Pt, Pd, [1,2] and semiconductors such as CdS, CdSe,[3-6]) are readily prepared in this compartmentalised liquid dispersion. It might be supposed that the equilibrium size of the microparticles which are generated in such a medium could be effectively limited by the water droplet dimensions if the protective surfactant layer around the droplets is sufficient to contain the growth process of the particles. To date, from experimental data, this correlation has been found to be at best only approximate for metal and semiconductor particle formation.

In this paper, we describe the preparation of microparticles of platinum in reverse micelles stabilised by Aerosol-OT. The correlation between equilibrium particle size and polydispersity with the size of the reverse-micelle droplets, which provide the 'microreactor' environment for particle growth, is examined. The weak correlation which is observed is discussed in terms of the

D. M. Bloor and E. Wyn-Jones (eds.),
The Structure, Dynamics and Equilibrium Properties of Colloidal Systems, 373–383.

mechanism of particle formation, giving particular emphasis to the significance of the rate-determining nucleation step.

In addition, we have carried out a systematic study of the mechanism of formation of semiconductor microparticles (CdS). This system is of particular interest since there is a direct relationship between particle size and the UV/VIS spectral properties of CdS. In particular, the small size of the particles (radius < 1 nm) results in spectral quantum size effects being directly observed [7,8]. For CdS, one observes a much better correlation between reverse micelle size and particle size. We have interpreted the kinetics of CdS particle formation in terms of a mechanism which gives particular emphasis to the dynamics of fast solubilizate exchange between droplets in the nucleation and growth regimes. In this way it can be shown that the rate of formation of particles is determined by the dynamic properties of the host reverse micelles (*i.e.* the process is essentially one of communication- control between droplets).

2. Preparation of Platinum and CdS Microparticles

Microparticles are easily prepared by mixing equal volumes of reverse-micelle solutions containing 0.1 mol dm^{-3} Aerosol-OT in n-heptane and different water contents (R values). In the case of the metal particles, the metal salt used was hexachloroplatinic acid ($H_2Pt^{IV}Cl_6$) together with a reducing agent (in our case hydrazine). For the preparation of CdS microparticles sodium sulphide and cadmium nitrate were solubilised in two reversed micelle systems and mixed together.

3. Experimental Methods

For platinum microparticles, particle sizes were determined by transmission electron microscopy.

The kinetics of cadmium sulphide formation were followed using a 'Hi-Tech Scientific' rapid mixing accessory (SFA11) in association with a Hewlett Packard diode array spectrophotometer (HP8452A). Using this arrangement complete spectra (as shown in Fig. 4) could be obtained in ~0.3 s for a 200 nm span. The dead time of the arrangement is ~ 50 ms. In our case the growth process was typically monitored for about 1 minute after mixing.

To follow the early stages of particle formation (*i.e.* nucleation and growth of particles to a radius < 1 nm) conventional single wavelength stopped flow measurements were made at 280 nm using a 'Hi-Tech Scientific' (SF-51) stopped flow unit. The dead time of the arrangement is < 5 ms.

4. Platinum Microparticles - Effects of Reactants Stoichiometry on Particle Size

From Fig. 1, it can be seen that for the case of Pt particles the final particle size is not directly related to the core droplet size. For a reactant stoichiometry $X = [N_2H_4] / [Pt^{IV}]$ of 10 the particles formed are slightly larger than the core size for low R values and smaller than the core size for the largest R value studied. However the particle size is found to be rather sensitive to the variation of X (see Fig. 2).

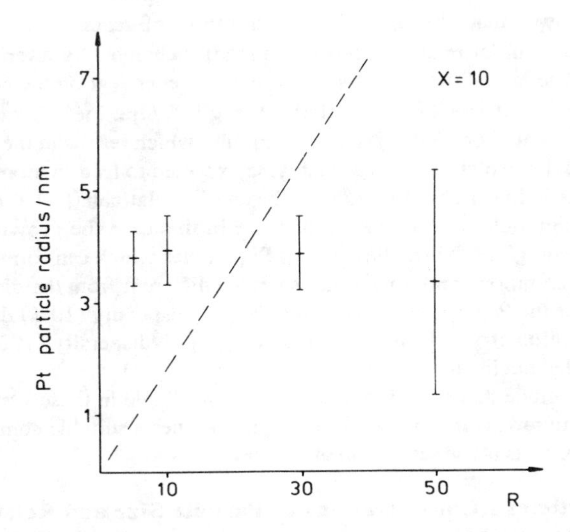

Figure 1. Correlation between water droplet size and particle size. Preparation conditions $[N_2H_4]$ = 4.5 mmol dm^{-3}, $[H_2PtCl_6]$ = 0.45 mmol dm^{-3}. The dotted line represents the microreactor (droplet) core radius. (All concentrations are overall concentrations and the bars represent the size spread: the polydispersity is expressed as a standard deviation).

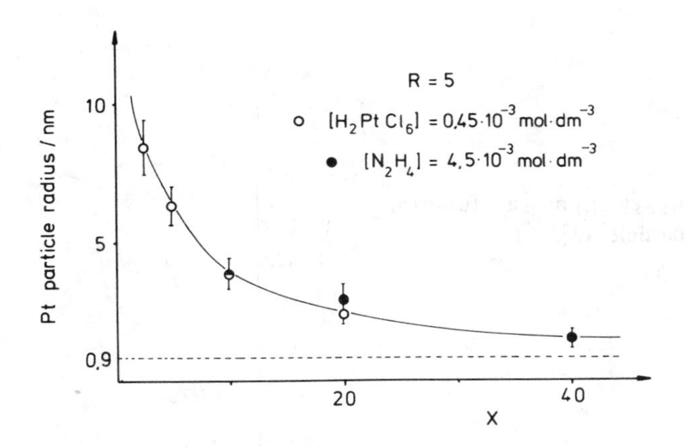

Figure 2. Effect of reactant stoichiometry X = $[N_2H_4]/[H_2PtCl_6]$ on particle size for platinum microparticles prepared in an R = 5 water/AOT/n-heptane system. (Dotted line represents microreactor core radius and the bars represent size spread).

An attempt to rationalise the data shown in Figs 1 and 2 is the following: for the case of Pt particle formation, the overall rate of particle formation is very dependent on the initial reduction step $Pt^{IV} \rightarrow Pt^0$ which is much slower than the rate of communication of reactants between droplets.

However the rate of this initial reduction process is readily changed by altering the reactant stoichiometry term X. The rate of the growth process is also dependent on the concentration of $Pt^{IV}Cl_6^{2-}$. From Fig. 2 it can be seen that at high hydrazine concentrations the hexachloroplatinate(IV) ion will be reduced relatively rapidly which results in the generation of a larger number of seed nuclei, which might reasonably be expected to lead to more, and therefore smaller particles overall. At lower concentrations of hexachloroplatinate(IV) at constant excess hydrazine again smaller particles are produced, because in this case the growth stage will be slowed down. It is interesting that the smallest size of Pt particles which can form under the most favourable reactant concentration conditions is not so much different from the dimensions of the containing droplet. For the R = 5 system the extent of polydispersity (10%) does not change much with reactant stoichiometry. However a much larger polydispersity (100%) is obtained when the reaction is carried out in larger (R = 50) droplets (see Fig. 1).

Therefore, we can conclude that since particle size is controllable in these systems, the main function of the droplet microreactor is achieved in that it provides a suitable compartmentalised medium which clearly prevents phase separation of the particles.

5. CdS Microparticles - Considerations of Particle Size and Relation to UV/VIS Spectrophotometry

For bulk CdS, the UV/VIS absorption threshold, as discussed by Henglein [8], is at 520 nm; as the size of the solid is decreased the absorption threshold shifts towards the UV and 'peaks' appear in the spectrum as a result of quantum size effects associated with the small size of the particles and subsequent confinement of the charge carriers [7,8]. It has been shown that there is a good correlation between the threshold absorption wavelength and particle size (Fig. 3). The general appearance of the spectra is shown later in Fig. 5.

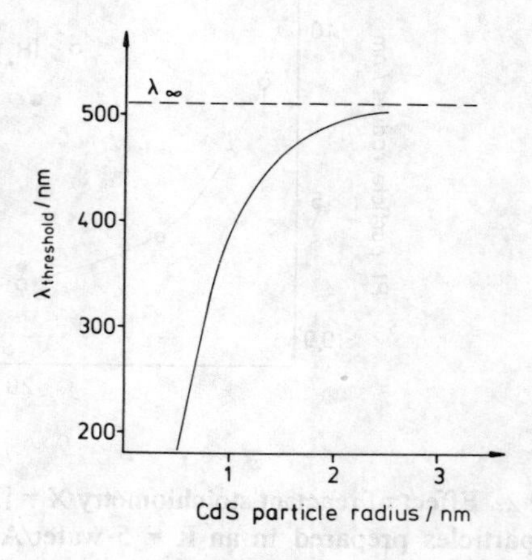

Figure 3. $\lambda_{threshold}$ as a function of radius of CdS particles [8].

6. Considerations of Microparticle Size in Relationto the Mechanism of CdS Particle Formation

For the case of CdS particle fomation the rate and mechanistic considerations are rather different. The necessary prerequisite for reaction in both cases is that the reactants communicate with each other. The communication process in the CdS case involves an encounter between ionic species Cd^{2+} and S^{2-}. This has been shown previously to occur on a very rapid time scale by transient fusion of droplets [9,10], with a second-order rate constant of 10^6 - 10^7 dm^3 mol^{-1} s^{-1} with some dependence on droplet size. Droplet concentrations are expressed as overall concentrations. The nucleation step involves an association process to form an inner-sphere complex of CdS by release of water from Cd^{2+}. This is also a very rapid process with a release rate constant of ~10^8 s^{-1} [11]. In addition no energetic restriction is envisaged to the growth process, at least for growth of small particles. Hence the kinetic and mechanistic considerations are different for Pt and CdS particle fomation.

In. Fig. 4 data are shown for CdS particle sizes (determined from the absorption spectrum) as a function of R value for $X' = [S^{2-}]/[Cd^{2+}] = 1$. It is clear that a better correlation between particle and droplet size is obtained, as compared with that for Pt particles.

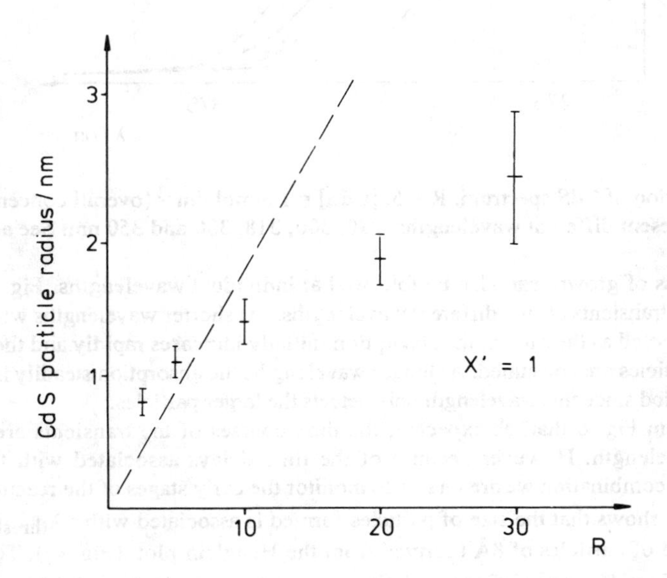

Figure 4. Effect of change of water droplet size on CdS microparticle size. Preparation conditions [CdS] = 0.1 mmol dm^{-3}, [AOT] = 0.1 mol dm^{-3}, T = 298 K. (Dotted line represents the microreactor core radius).

7. Detailed Considerations of the Mechanism of CdS Particle Formation

The growth of the particles can be readily monitored by the stopped-flow method. In Fig. 5 the spectrum of CdS evolving with time is shown, with the spectrum determined using the diode array spectrophotometer. Particle growth can be readily seen as the $\lambda_{threshold}$ shifts from ~ 320 nm after 2s to ~375 nm after 60s.

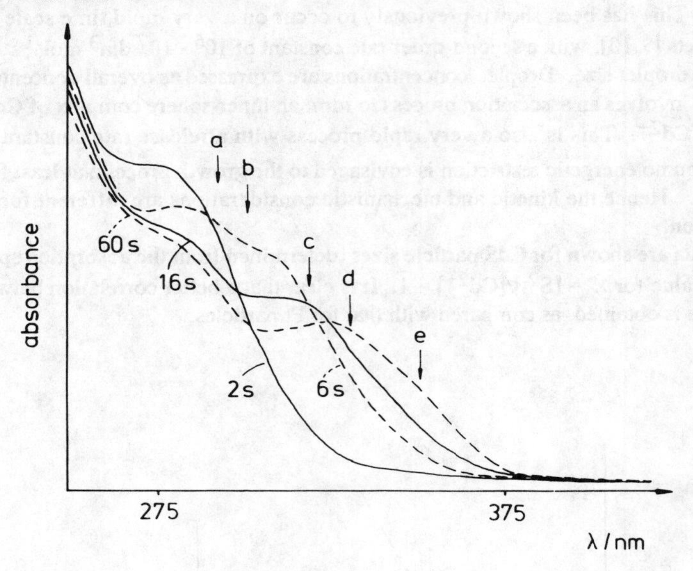

Figure 5. Evolution of CdS spectrum. R = 5, [CdS] = 1 mmol dm^{-3} (overall concentration). T = 288 K, a - e represent different wavelengths, 290, 300, 318, 330 and 350 nm. See also Fig.6.

The same process of growth can also be followed at individual wavelengths. Fig. 6 shows the absorbance/time transients at five different wavelengths. At shorter wavelengths where the small particles are observed as they form, the absorption initially increases rapidly and then tails off as these smaller particles are consumed. At longer wavelengths the absorption steadily increases over a longer time period since this wavelength only detects the larger particles.

It is clear from Fig. 6 that, as expected, the time courses of the transients are a sensitive function of wavelength. However because of the time delays associated with the stopped-flow/diode array combination we are unable to monitor the early stages of the reaction. In Fig. 5 the first transient shows that the size of particles formed is associated with a $\lambda_{threshold}$ of ~320 nm, giving a size of particles of 8Å (derived from the Henglein plot (Fig. 4)). The growth of particles in the first 10s or so of reaction (based on $\lambda_{threshold}$ values) is shown in Fig. 7. Agglomeration numbers are also shown; this number is associated with the number of CdS molecules in the particle.

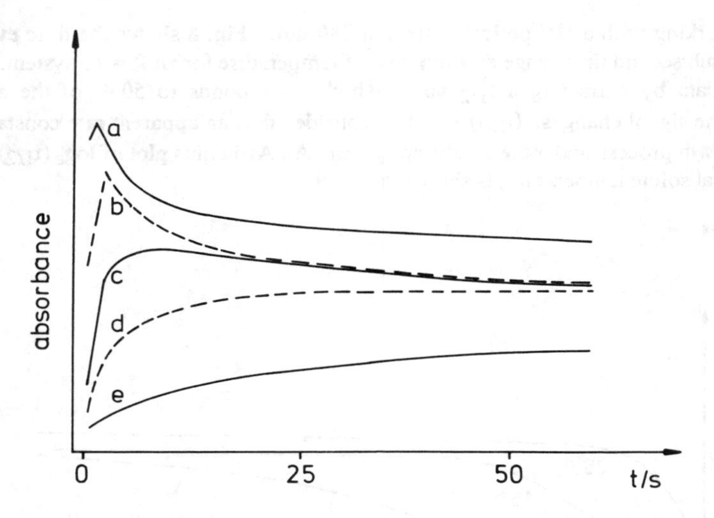

Figure 6. Absorbance/time transients for CdS microparticle formation at five different wavelengths as in Fig. 5.

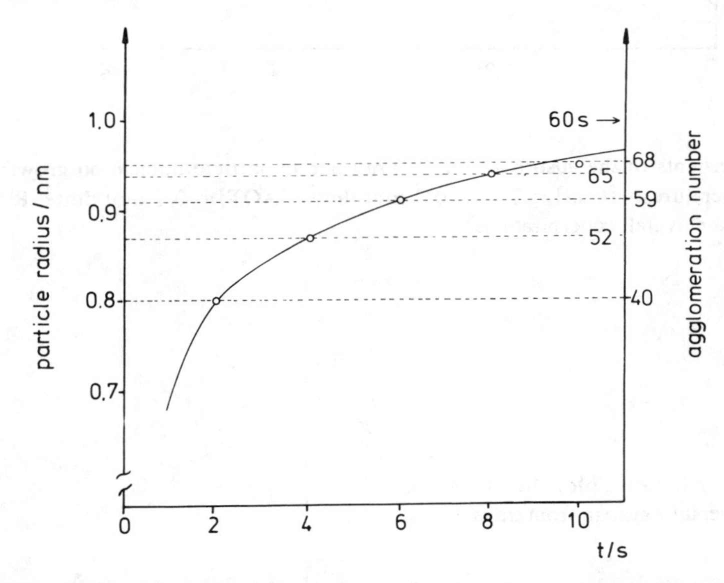

Figure 7. Plot of radius (determined from $\lambda_{threshold}$) and agglomeration number against time.

The data clearly indicate that a limiting value for the radius of approximately 1 nm is being achieved and hence a termination mechanism must be involved. The details of this are unclear but it is likely that the AOT plays a similar role to that of hexametaphosphate as an agent which limits growth of CdS in aqueous solution [12], so an interaction of AOT with the surface of the particle is to be expected as the particle size approaches the droplet size.

The early stages of particle growth have been monitored using a conventional stopped-flow

instrument, working with a UV optical system at 280 nm. Fig. 8 shows the time evolution of growth in the sub second time range as a function of temperature for an R = 10 system. We have analysed the data by extracting a $t_{1/2}$ value which corresponds to 50% of the absorbance amplitude of the signal changes. $(t_{1/2})^{-1}$ is then considered as an apparent rate constant for the nucleation/growth process under the conditions given. An Arrhenius plot of $\log_e (t_{1/2})^{-1}$ vs T^{-1}, where T is the absolute temperature, is shown in Fig. 9.

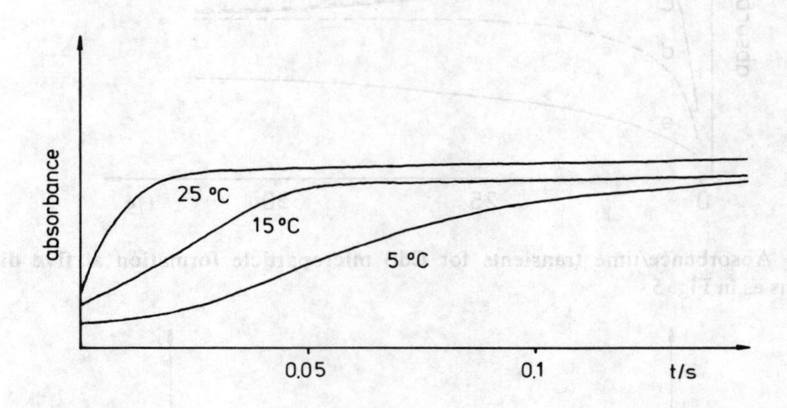

Figure 8. Transients of the absorbance at 280 nm due to particle nucleation/growth at three different temperatures. [CdS] = 2.5 x 10^{-5} mol dm^{-3}, [AOT] = 0.1 mol dm^{-3}, R = 10 (all concentrations are overall concentrations).

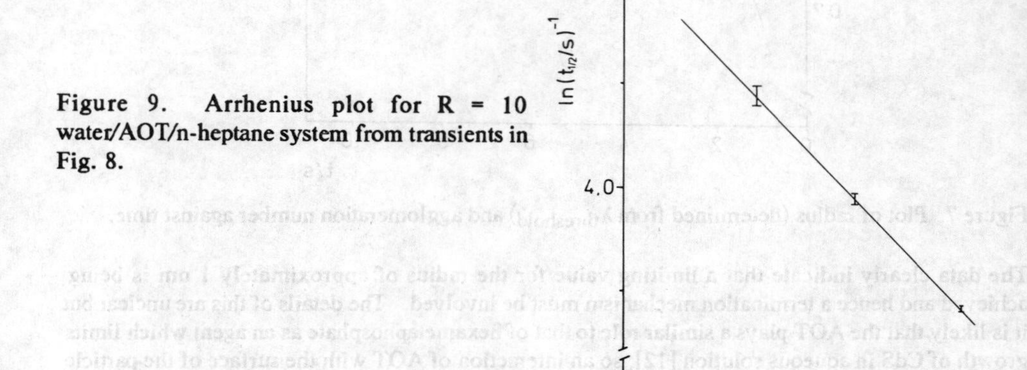

Figure 9. Arrhenius plot for R = 10 water/AOT/n-heptane system from transients in Fig. 8.

From the slope, the activation energy associated with the early stages of particle growth is determined to be 69±5 kJ mol^{-1}. This is in excellent agreement with the activation energy measured previously for small ion solubilizate exchange, where values of 70±10 kJ mol^{-1} and 67±7 kJ mol^{-1} were obtained for two types of solute transfer [10]. This result suggests that for the early stages of CdS particle formation, the rate determining step is exchange of $Cd^{2+}/S^{2-}/(CdS)_n$ species between droplets.

8. Mechanisms of Particle Growth

Different processes are expected to contribute to the formation of stable CdS microparticles.

1. Formation of the monomer CdS species:

$$Cd^{2+} + S^{2-} \xrightarrow{k_1} (CdS)_1$$

2. Formation of aggregates with higher agglomeration numbers from reactions of small particles.

$$(CdS)_m + (CdS)_n \xrightarrow{k_2} (CdS)_{n+m} \quad (m,n \geq 1)$$

3. Addition of Cd^{2+} and S^{2-} to growing particles. This process is expected to be particularly important when the reaction stoichiometry X' ≠ 1.

$$(CdS)_n + S^{2-} \xrightarrow{k_3} (Cd_nS_{n+1})^{2-}$$

$$(CdS)_n + Cd^{2+} \xrightarrow{k_3} (Cd_{n+1}S_n)^{2+}$$

Charged aggregates are also expected to react in the same way as 1:1 stoichiometric species indicated in 2.

"Ostwald ripening" may also occur but this process involves dissolution of the solid and is not on the time scale of the 'part' reactions we have discussed so far. Experiments are underway to determine the free Cd^{2+} during the particle formation process. Such data will then enable us to comment on this point in a more detailed way.

4. Termination. Simplistically we would predict that termination will occur when the particle size becomes comparable to the droplet size since then water is excluded from the droplet and an interaction between the surfactant and the particle surface must occur.

Although the mechanism appears complex in this compartmentalised system, in the CdS case the situation is greatly simplified since we have already established.that the rate of nucleation/growth reaction in the microemulsion medium is controlled by the rate of inter-droplet exchange. Therefore the rate constants for all the above-mentioned reactions are assumed to be the same *i.e.* k_{ex} . This statement of course does not apply to the termination step where $k \rightarrow 0$ for large values of m and n (>80).

For an R=5 reverse micelle system containing Cd^{2+}/S^{2-} ions , the concentration of droplets is about 10^{-3} mol dm^{-3} and the time to form a nucleus is ~ 1ms. These experimental data are consistent with the results of numerical calculations for the time required to form a cluster.

Acknowledgements

We thank the SERC for a studentship (to TT). We also thank NATO for a Travel Grant to WK and BHR.

References

1. Boutonnet, M., Kizling, J.,Stenius P. and Maire, G. (1982) 'The preparation of monodisperse colloidal metal particles from microemulsions', Colloid Surf. 5, 209-225.

2. Khan-Lodhi, A., Robinson, B.H. and Towey, T. (1989) 'Microparticle synthesis and characterisation in reverse micelles', in M. P.Pileni (ed.) Structure and Reactivity in Reverse Micelles, Elsevier Science Publishers, pp. 198-220.

3. Meyer, M., Wallberg, C., Kurihara K and Fendler, J.H. (1984) 'Photosensitized charge separation and hydrogen production in reversed micelle entrapped platinized colloidal cadmium sulphide', J. Chem. Soc. Chem. Comm., 90-91.

4. Lianos, P. and Thomas, J.K. (1986) 'Cadmium sulfide of small dimensions in inverted micelles', Chem. Phys. Lett., 125, 299-302.

5. Petit C. and Pileni M.P. (1988) 'Synthesis of cadmium sulfide in situ in reverse micelles and in hydrocarbon gels', J. Phys. Chem., 92, 2282-2286.

6. Steigerwald, M.L., Alivisatos, A.P, Gibson, J.M., Harris, T.D., Kortan, R., Muller, A.J., Thayer, A.M., Duncan, T.M., Douglas, D.C. and Brus, L.E. (1988) 'Surface derivatisation and isolation of semiconductor cluster molecules'. J. Am. Chem. Soc., 110, 3046-3050.

7. Rosetti, R., Nakahara, S. and Brus, L.E. (1983) 'Quantum size effects in the redox potentials, resonance Raman spectra and electronic spectra of CdS crystallites in aqueous solution', J. Chem. Phys., 79(2), 1085-1087.

8. Fojtik, A., Weller, H., Koch, U. and Henglein, A. (1984) 'Photochemistry of colloidal metal sulfides. Photophysics of extremely small CdS particles: Q-state CdS and magic agglomeration numbers', Ber. Bunsenges, Phys. Chem., 1988, 969-977.

9. Atik, S.S. and Thomas, J.K. (1981) 'Transport of photoproduced ions in water in oil microemulsions: movement of ions from one water pool to another', J.Am. Chem.Soc., 103, 3543-3550.

10. Fletcher, P.D.I., Howe, A.M. and Robinson B.H. (1987) 'The kinetics of solubilizate exchange between water droplets of a water-in-oil microemulsion', J.Chem.Soc.,Faraday Trans.1,83, 985-1006.

11. Eigen, M. (1963) 'Fast elementary steps in chemical reaction mechanisms', Pure Appl. Chem., 6, 97-115.

12. Fischer, Ch.-H., Weller, H., Fojtik, A., Lume-Pereira, C., Janata, E and Henglein A. (1986), 'Photochemistry of colloidal semiconductors. Exclusion chromatography and stop flow experiments on formation of extremely small CdS particles', Ber. Bunsenges, Phys. Chem., 90, 46-49.

ELECTRIC BIREFRINGENCE MEASUREMENTS IN MICELLAR AND COLLOIDAL SOLUTIONS

H. HOFFMANN, U. KRÄMER
Lehrstuhl für Physikalische Chemie I
der Universität Bayreuth
Postfach 101251, D-8580 Bayreuth

ABSTRACT It is shown that anisometric structures in solutions of rodlike micelles, clay particles or polyelectrolytes can be oriented by electric fields. Both AC and DC fields can be used for the measurements. Results on four different systems are reported and compared. The four different systems are micellar solutions of Tetradecyldimethylaminoxide and Hexadecyloctyldimethylammoniumbromide, and polyelectrolyte solutions of Poly(sodium p-styrenesulfonate) and Saponit. On the basis of the results different situations in the behaviour of the systems can be distinguished. In the dilute concentration range for all systems, information about the size, the size distribution, the induced dipole mechanism, and the optical and electrical anisotropy of the particles can be obtained. At the overlap concentration of the particles the behaviour of the charged systems differs dramatically from the uncharged systems. A second effect with opposite sign from the first one appears. This effect is referred to as the anomaly of the electric birefringence. Hence many differences between the uncharged Tetradecylaminoxide and the other three systems become apparent. It is concluded that particles orient in the anomaly region perpendicular to the electric field. At higher concentrations the behaviour is normal again and a network structure is formed, which can also be aligned in the electric field. The frequency and time dependent measurements reveal different concentration regions, which are characterized by critical concentrations, which depend on the axial ratio of the particles.

1. Introduction

In 1875 J. Kerr [1] observed that optical isotropic particles become birefringent in an electric field. Later Peterlin and Stuart [2] established a theory, which described this so called electric birefringence by the orientation of anisometric particles in the electric field. Most of the following investigations on colloidal systems were based on this theory and it was possible to obtain much information with electric birefringence measurements. The dependence of the birefringence as a function of the field strengths allows one to calculate the electric and the optical anisotropy of the particles. Broesma [3] established equations to calculate particle sizes from the time retarded decay of the birefringence. It is also possible to obtain information of the size distribution of the particles. O'Konski [4] explained for ionic particles without a permanent dipole moment the orientation in the electric field by the displacement of counterions along the main particle axis which leads to an induced dipole moment.

In micellar solutions different aggregation forms exist, but only solutions containing

D. M. Bloor and E. Wyn-Jones (eds.),
The Structure, Dynamics and Equilibrium Properties of Colloidal Systems, 385–396.

anisometric micelles can be aligned in the electric field. The electric birefringence is therefore a very convenient experimental method to distinguish between globular and anisometric micelles. Hence it is possible to investigate transitions from globular micelles to rodlike micelles which can occur by addition of cosurfactant or salt [5] . Otherwise one can solubilize hydrocarbons into rodlike micelles and a transition from rods to globules occurs [6] . Unfortunately most of the theoretical framework is only valid in dilute solutions and results at high concentrations are not well understood until now [7] . At higher concentrations interactions between the particles take place and the interpretation of the data becomes difficult. In this article we shall try to give a survey about the different behaviour of micellar solutions up to high concentrations and will classify the surfactants into charged ionic systems and uncharged zwitterionic or nonionic systems. We will observe many differences in the behaviour between micellar and colloidal systems. In micellar systems the interactions between micelles can easily be varied by mixing ionic and nonionic surfactants. Also the addition of salt changes drastically the interactions and generally gives rise to an increase of the size of the micelles. Therefore many investigations can be done on micellar systems and the results can be compared with the results on polyelectrolytes and other colloidal systems, which cannot change their size with the concentration.

Polyelectrolytes are macromolecules with a large number of charges distributed along the particle. Especially biological polyelectrolytes are often characterised by a great stiffness, which is due to secondary structures as can be seen for example on DNA or on tobacco mosaic virus. In contrast, synthetic macromolecules are usually flexible and their conformation depends on different factors like the ionic strength and the concentration. In dilute solutions these polyelectrolytes are very extended and electrostatic interactions increase dramatically with increasing concentration [8]. Inorganic colloidal particles like clays on the other hand are very stiff and their structure cannot change with the concentration. This is the case for some silicates like Saponit, which possess the shape of stiff disks.

In previous papers we have already explained the behaviour of ionic micellar systems and of polyelectrolytes in the electric field [9-11] . Hence in this paper we will try to give a general survey about the influence of electrostatic interactions on the orientation of anisometric particles in the electric field. We shall also compare these results with results of an uncharged system. We have chosen a well characterized zwitterionic surfactant [12] , which also forms rodlike micelles. It is the aim of the present study to explain the differences between charged and uncharged particles over the whole concentration region from dilute to concentrated solutions.

2. Experimental

2.1 MATERIALS

The surfactants Tetradecyldimethylaminoxide C_{14}DMAO and Hexadecyloctyldimethylammoniumbromide $C_{16}C_8$DMABr were a gift of Hoechst A.G., Gendorf. Both substances were recrystallised twice from acetone and contained no impurities, as expressed by the lack of a minimum in the surface tension curves at the critical micelle concentration (cmc). Saponit was donated from the Hoechst A.G., Frankfurt and was used without further purification. The Poly(sodium p-styrenesulfonate) (=NaPSS) is a product of Chemical Standard Service, Mainz and was received as a gift from BASF A.G., Ludwigshafen. BASF A.G. generously sent us the sample with a molecular weight of 400,000 and a molecular weight distribution of 1.06. Hence the problems arising from polydispersity can be neglected.

2.2 ELECTRIC BIREFRINGENCE INSTRUMENTS

The orientation of the particles was achieved by applying an rectangular electric pulse to the system (transient electric birefringence = TEB) or by use of an alternating electric field (dynamic electric birefringence = DEB). In the TEB measurements the high voltage pulses of short rise and decay time (~25ns) were long enough to reach the steady state value of the birefringence. The details of the used instruments have been well described [13]. In the TEB experiments the time constants of the rise and the decay of the birefringence and the stationary birefringence Δn_{st} were recorded.

For high viscous solutions the pulse length of the TEB was not long enough to reach Δn_{st}. Hence it was necessary to use sinusoidal pulses with varying frequency in the DEB apparatus. The frequency ω of the AC-field covered the range from 50 Hz to 700,000 Hz. Stationary amplitude, alternating amplitude and time constants of the decay were measured. For making measurements at very low field strength a Lock-in amplifier was used to detect the corresponding small birefringence values. For the determination of the sign of the birefringence a quarter wave device was placed between the analyzer and the cell. All measurements were carried out at a laser wavelength of 632.8 nm and a temperature of 25°C.

3. Results

The concentration range of ionic micellar systems, which form rodlike aggregates, can be subdivided into three different concentration ranges. The dilute concentration range extends from the cmc_2 (the concentration at which rodlike micelles are formed) to concentrations at which interactions between the particles first occur. The second semi-dilute concentration range is characterized by strong electrostatic and hard core steric interactions. Finally in the even more concentrated region a dynamic network is formed, which exists up to the liquid crystalline phase boundary. For nonionic or zwitterionic surfactants this subdivision between the semi-dilute and the concentrated region is not so clear, because the electrostatic interactions are comparatively small. Polyelectrolytes on the other hand behave quite differently, because they possess an inherent flexibility, which depends very strongly on the concentration. By means of electric birefringence measurements it is possible to obtain information about the different behaviour of these particles in the different concentration ranges.

3.1 DILUTE SOLUTIONS

It is well established that micellar solutions of $C_{14}DMAO$ and $C_{16}C_8DMABr$ form rodlike micelles above a certain concentration. These rods can easily be aligned in an electric field. In fig. 1a a typical TEB-signal for a 80 mM solution of $C_{14}DMAO$ is represented. The characteristic feature for nonionic and zwitterionic surfactants is the fact, that the increase of the birefringence follows a much slower time constant than the decrease of the stationary value. To compare this with an ionic surfactant in fig. 1b the TEB-signal of a 15 mM $C_{16}C_8DMABr$ solution is plotted.

It is already visible without making a fit, that the time constants for the build-up and the decay of the birefringence are quite similar. The same is true for the chosen NaPSS polyelectrolyte (c= 0.01 g/l) in fig. 1c and the Saponit (c=0.1 g/l) in fig. 1d.

In fig. 2 results for the same concentrations with the DEB-experiment are shown. The stationary value of the birefringence Δn_{st} is plotted as a function of the frequency. The $C_{16}C_8DMABr$ micelles and the NaPSS and Saponit polyelectrolytes behave similar and no frequency dependence is visible. However the $C_{14}DMAO$ shows a constant value of Δn_{st} up to a certain frequency, but then Δn_{st} decreases.

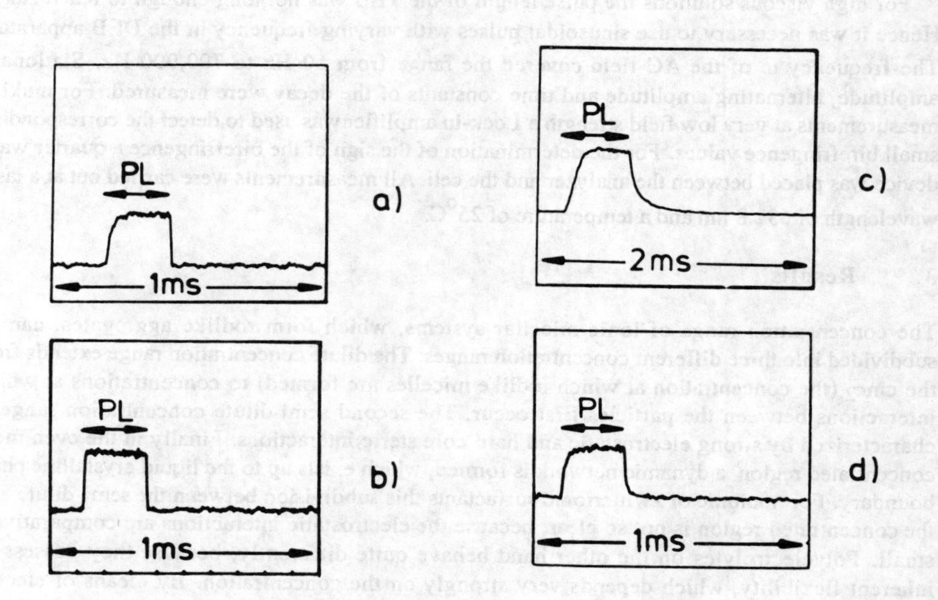

Fig. 1: TEB-signals of the dilute concentration regions. T = 25°C PL = Pulse Length

1a) 80 mM $C_{14}DMAO$ Δn_{st} = - 1.15 10^{-8} E = 7.5 10^4 V/m

1b) 15 mM $C_{16}C_8DMABr$ Δn_{st} = + 1.02 10^{-8} E = 14.3 10^4 V/m

1c) 0.01 g/l NaPSS Δn_{st} = - 2.44 10^{-8} E = 7.5 10^4 V/m

1d) 0.1 g/l Saponit Δn_{st} = + 1.15 10^{-8} E = 10^5 V/m

3.2. SEMI-DILUTE SOLUTIONS

From small angle neutron scattering (SANS) experiments it is well known that a strong correlation peak exists in ionic surfactant systems above a certain concentration, which is due to strong intermicellar interactions [14]. These electrostatic interactions cause a unique behaviour in the electric field called the anomaly of the electric birefringence. In fig. 3 the characteristic features of this anomalous behaviour are demonstrated in TEB-experiments. Up to a concentration of 20 mM $C_{16}C_8DMABr$ the behaviour is normal. That means that the birefringence increases at the beginning of the pulse up to a stationary value Δn_{st} and then remains constant until the pulse is

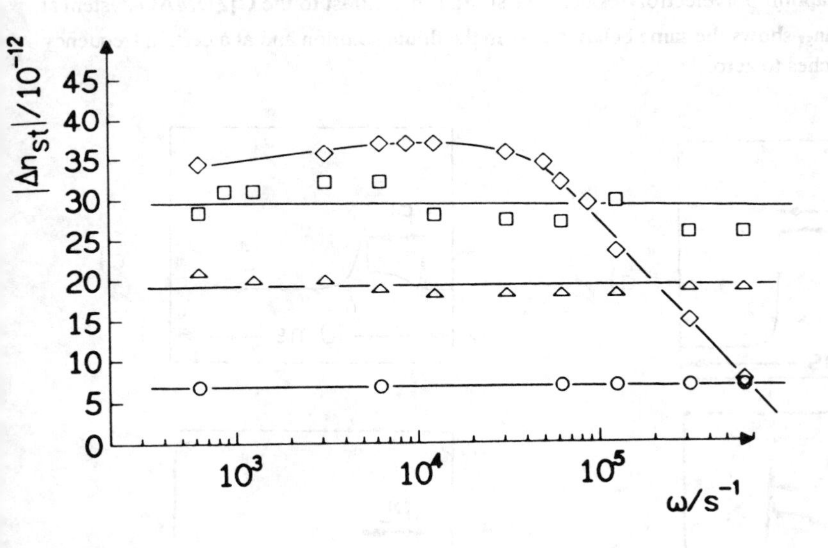

Fig. 2: Stationary birefringence Δn_{st} in DEB-measurements as a function of the frequency ω. T = 25°C

◇ : 80 mM $C_{14}DMAO$	E = 5.2 10^3 V/m
O : 15 mM $C_{16}C_8DMABr$	E = 2.9 10^3 V/m
□ : 0.01 g/l NaPSS	E = 2.9 10^3 V/m
Δ : 0.1 g/l Saponit	E = 2.9 10^3 V/m

switched off, as was already demonstrated in fig. 1a. At a concentration of 27.5 mM (fig. 3b) the birefringence reaches a maximum value and then becomes smaller again until a stationary value is reached. At the end of the pulse the birefringence decreases, passes through zero, increases again until it finally decreases to zero. The signal can be explained by a superposition of two effects: the normal first one and a second one, which is characterized by a much slower time constant and an opposite sign of birefringence. We have explained this behaviour in a previous paper [10] . In comparison to this the C_{14}DMAO system, also at concentrations above the overlapping of the rods show only a normal behaviour (fig. 3a). The NaPSS and Saponit polyelectrolytes resemble the behaviour of the $C_{16}C_8$DMABr micelles and an anomaly appears at concentrations greater than 0.05 g/l (fig. 3c) for NaPSS or greater than 0.5 g/l (fig. 3d) for Saponit.

In DEB-experiments one can also see the differences between the four solutions. In fig. 4 the Δn_{st} value of 27.5 mM $C_{16}C_8$DMABr is plotted as a function of the frequency. At high frequency Δn_{st} reaches a constant value, but at lower frequencies Δn_{st} passes through zero. The NaPSS and the Saponit polyelectrolytes behaves similar in contrast to the C_{14}DMAO system at 300 mM. There Δn_{st} shows the same behaviour as in the dilute solution and at a certain frequency the signal approaches to zero.

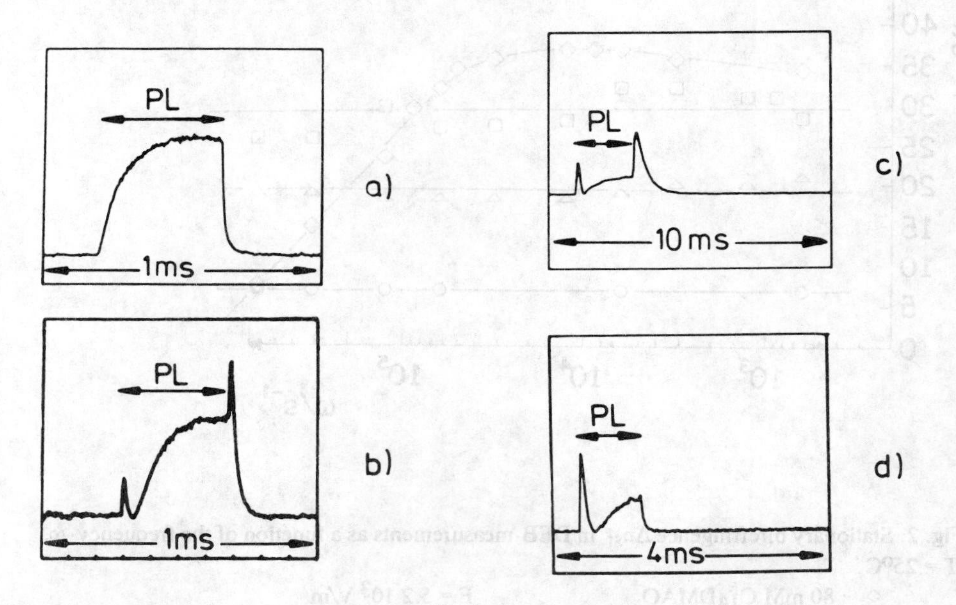

Fig. 3: TEB-signals of the semi-dilute concentration range.

3a) 300 mM C_{14}DMAO	$\Delta n_{st} = - 1.57\ 10^{-7}$	$E = 7.5\ 10^4$ V/m
3b) 27.5 mM $C_{16}C_8$DMABr	$\Delta n_{st} = - 1.86\ 10^{-8}$	$E = 4.8\ 10^4$ V/m
3c) 0.1 g/l NaPSS	$\Delta n_{st} = + 2.25\ 10^{-8}$	$E = 5.0\ 10^4$ V/m
3d) 2.0 g/l Saponit	$\Delta n_{st} = - 1.23\ 10^{-8}$	$E = 2.5\ 10^4$ V/m

4. Concentrated solutions

As already mentioned it is not possible for the $C_{14}DMAO$ system to distinguish between semi-dilute and concentrated solutions. On the other hand ionic micelles always show a very clear transition between the two concentration regions. The $C_{16}C_8DMABr$ micelles make this transition at 38 mM. Above this concentration the TEB- birefringence signal is normal again as shown in fig. 5a for a 40 mM solution. The same is true for the NaPSS polyelectrolyte in fig. 5b. Also in DEB-measurements a zero transition can no longer be observed as is demonstrated for the $C_{16}C_8DMABr$ and the NaPSS in fig. 6. The situation for Saponit is a little bit different from the other two charged systems, because above 17 g/l the birefringence vanishes completely and the birefringence does not change sign again.

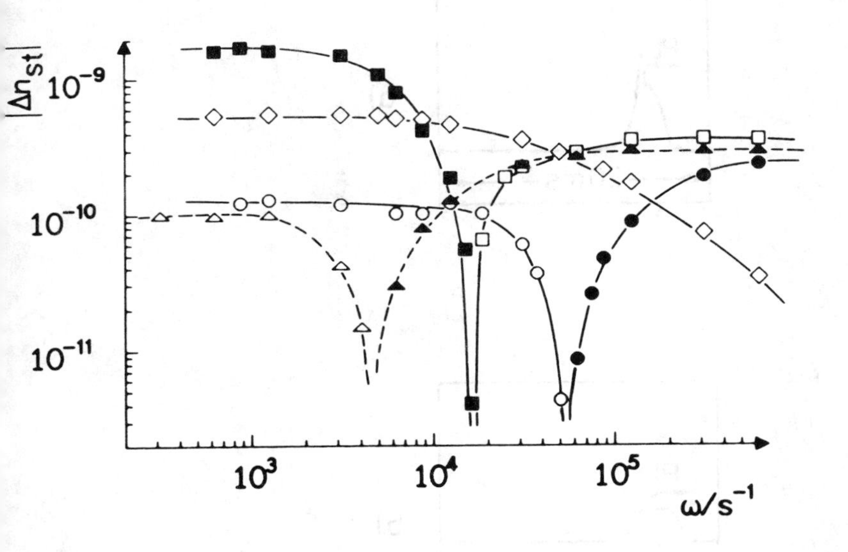

Fig. 4: Stationary birefringence Δn_{st} in DEB-measurements as a function of the frequency ω. The filled symbols represent a positive birefringence, while the empty symbols stand for negative birefringence. T = 25°C

◇	: 300 mM $C_{14}DMAO$	$E = 5.2\ 10^3$ V/m
O	: 27.5 mM $C_{16}C_8DMABr$	$E = 2.9\ 10^3$ V/m
□	: 0.1 g/l NaPSS	$E = 2.9\ 10^3$ V/m
Δ	: 2.0 g/l Saponit	$E = 2.9\ 10^3$ V/m

5. Discussion

A large difference in the electro-optical behaviour between ionic particles and nonionic or zwitterionic particles is already apparent in the dilute concentration range. The main reason for the different behaviour is due to the different ways in which a dipole moment is induced in the aggregates. Micellar systems do not possess a permanent dipole moment because the micelles have completely rotational symmetry. Polyelectrolytes often have a permanent dipole moment, but this is normally very small in comparison to an induced dipole moment and can therefore be neglected [15] . For ionic systems O'Konski has explained the build-up of a dipole by the displacement of the counterions along the main rod axis [4]. This mechanism is very quick and does not influence the time constant of the birefringence build-up. As a consequence the build-up and decay curve are completely symmetric, as it was theoretically explained by Benoit [16] .

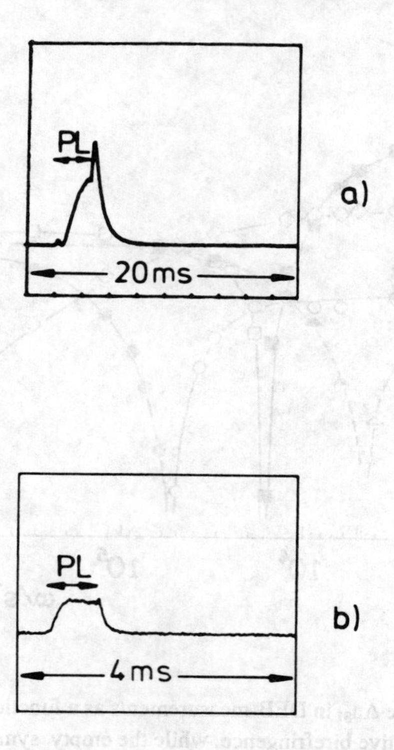

Fig. 5) TEB-signals of the concentrated region. $T = 25^{o}$ C
5a) 40 mM $C_{16}C_8DMABr$ $\Delta n_{st} = + 3.5\ 10^{-8}$ $E = 2.4\ 10^4$ V / m
5b) 5 g/l NaPSS $\Delta n_{st} = - 1.23\ 10^{-8}$ $E = 2.5\ 10^4$ V / m

The build-up is expressed by the simple equation:

$$\Delta n = \Delta n_{st} \; (1 - e^{-t/\tau})$$

For nonionic or zwitterionic systems a dipole can not be induced in such a manner. However the rotation of the single molecules in the micelles can cause also a dipole moment for the rods. But this rotation should be fast and it is until now not yet clear, why the build-up needs much more time than the decay. The behaviour seems however to be typical for all nonionic and zwitterionic systems. Perhaps the individual polarized micelles form larger aggregates of many micelles in the electric field. This seems only be possible for uncharged systems, when the electrostatic forces can be neglected. Hence first an orientation of the micelles occurs and then a translational motion between the micelles is responsible for the slow build-up process of the birefringence. At the end of the pulse the fast rotational motion causes the relaxation of the birefringence.

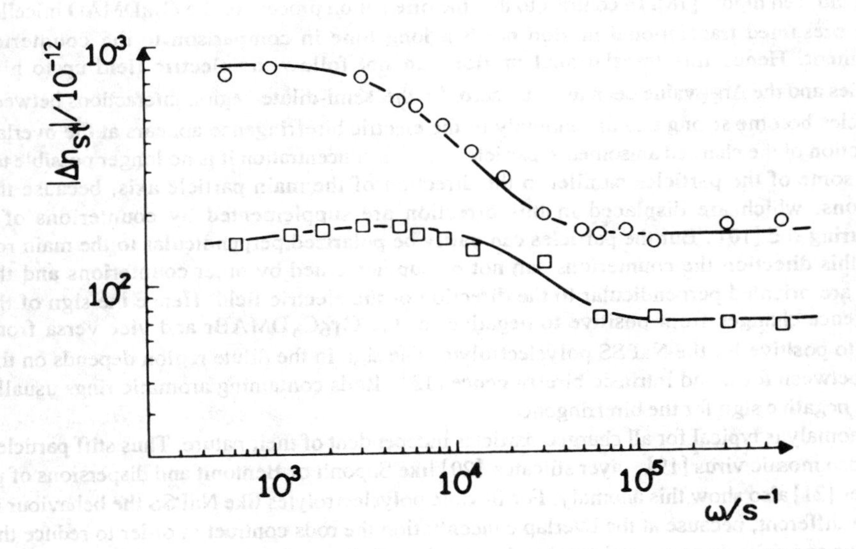

Fig. 6) Stationary birefringence Δn_{st} in DEB-measurements as a function of the frequency ω.
T = 25°C

O : 40 mM $C_{16}C_8DMABr$ E = 2.9 10^3 V/m

□: 5 g/l NaPSS E = 2.9 10^3 V/m

A further difference between the charged ionic micellar systems and the nonionic or zwitterionic systems is due to the electrostatic forces between the micelles, which are responsible for the monodispersity of ionic micelles. Therefore the decay of the birefringence is given by:

$$\Delta n = \Delta n_{st}\ e^{-t/\tau}$$

Nonionic and zwitterionic micellar systems are always polydisperse and the decay can only be fitted with a stretched exponential function [17] or with a sum of exponential functions. For polyelectrolytes the situation is a little bit different. Provided that the molecules are monodisperse the birefringence should decay also as a monoexponential function. But the NaPSS macromolecules possess a contour length l of about 4850 Å and the charge density of the polyelectrolyte is not so high as in ionic micelles. Therefore the repulsion between the charges is lower and the molecule more flexible in comparison to an ionic micelle. Consequently the molecules can exist in different conformations and the decay cannot be fitted with a monoexponential function. In contrast to that Saponit looks like a stiff disk with a broad molecular weight distribution. Hence the relaxation has also to be regarded as a sum of relaxation processes.

The frequency dependence in fig. 2 confirms the above explained different polarization mechanism. The displacements of counterions is very quick, so that the $C_{16}C_8$DMABr system and the NaPSS and Saponit polyelectrolytes show no frequency dependence, as it is characteristic for a fast induced dipole [18]. In contrast to that the orientation process of the C_{14}DMAO micelles with the presumed translational motion needs a long time in comparison to the counterion displacement. Hence this translational motion can not follow the electric field up to high frequencies and the Δn_{st} value decreases to zero. In the semi-dilute region interactions between the particles become strong and the anomaly of the electric birefringence appears at the overlap concentration of the charged anisometric particles. At this concentration it is no longer possible to polarize some of the particles parallel to the direction of the main particle axis, because the counterions, which are displaced in this direction are supplemented by counterions of a neighbouring rod [10] . But the particles can partly be polarized perpendicular to the main rod axis. In this direction the counterions can not be supplemented by other counterions and the particles are oriented perpendicular to the direction of the electric field. Hence the sign of the birefringence changes from positive to negative for the $C_{16}C_8$DMABr and vice versa from negative to positive for the NaPSS polyelectrolyte. The sign in the dilute region depends on the relation between form and intrinsic birefringence [13] . Rods containing aromatic rings usually possess a negative sign for the birefringence.

The anomaly is typical for all charged particles independent of their nature. Thus stiff particles like tobacco mosaic virus [19] , layer silicates [20] like Saponit or Bentonit and dispersions of p-TFE fibres [21] also show this anomaly. For flexible polyelectrolytes like NaPSS the behaviour is a little bit different, because at the overlap concentration the rods contract in order to reduce the interactions and only above a certain molecular weight and charge density an anomalous behaviour can be expected [11] . The increase of the ionic strength by addition of salt reduces the electrostatic interactions and the anomaly disappears [10] . Polyelectrolytes are coiled at low concentrations if the ionic strength is high enough [8,22] .

The frequency dependent measurements in fig. 4 underline the results of the TEB-measurements in fig. 3. At high frequencies the behaviour of all four systems corresponds to the dilute region. Thus the induced dipole in the C_{14}DMAO micelles can not follow the frequency of the electric field, while Δn_{st} for both other systems achieves a constant value at high frequencies. With decreasing frequency the second effect becomes apparent and at a certain frequency Δn_{st}

passes through zero and changes sign. At low frequencies both effects can follow the field and a stationary value is achieved.

In micellar solutions of rods a network with elastic properties is formed at a second critical concentration. At this point a third effect in solutions of charged particles becomes visible, which overcomes the anomalous behaviour. Hence the concentrated region is reached, which again shows normal behaviour in TEB-experiments (fig. 5). For the $C_{14}DMAO$ system this third region can not be distinguished from the semi-dilute region, because the electrostatic interactions, which control these characteristic concentrations can be neglected. The sign of this third effect is again opposite to the second effect because the network can only be oriented parallel to the electric field. The DEB-measurements of 40 mM $C_{16}C_8DMABr$ and 5 g/l NaPSS (fig. 6) support these results. At high frequencies only the single particles can be oriented, while at a certain frequency also the network is oriented and Δn_{st} increases to a constant value at low frequencies. As already mentioned Saponit does not again change the sign of the birefringence, but above 17 g/l the solutions possess a yield stress value and a orientation of the particles, which are connected to a network, is no longer possible. The network structure in the solution is probably electrostatically stabilized.

The behaviour of charged micelles and NaPSS polyelectrolytes was described to be very similar, but there exists a main difference concerning the flexibility of the NaPSS macromolecules. With increasing concentration the macromolecules become more flexible and above a certain concentration the macromolecules are completely coiled and a birefringence can no longer be measured [11]. The greater the contour length of the macromolecules the later the rod to coil transition occurs.

6 Conclusions

The electric birefringence method is a very useful tool for studying aqueous solutions of micellar or colloidal systems. The method can distinguish in dilute solutions between the behaviour of different systems. For nonionic or zwitterionic systems the increase of the birefringence is slower than the decay. This behaviour is typical for such systems. The multiexponential decay of the birefringence shows that the micelles are at all concentrations polydisperse and probably flexible. In contrast to that, charged micellar systems are characterized by monodispersity and stiffness. Rise and decay of the birefringence are symmetrically. For polyelectrolytes the latter is also true, but a further distinction is necessary. The NaPSS molecules are stretched in dilute solutions, but become coiled with increasing concentrations. In contrast to that Saponit possess at all concentrations a disk-like shape, which does not vary with the concentration.

At the overlap concentration further differences between charged and uncharged particles can be clearly recognized. All charged anisometric particles independent of their chemical nature show the anomaly of the electric birefringence in the semi-dilute concentration range, while uncharged micelles behaves similar as in the dilute concentration range. At higher concentrations usually a network with elastic properties is formed, which also can be oriented in the electric field and the behaviour of the charged systems is normal again and corresponds to the dilute concentration range. Hence at least two critical concentrations can be defined, which are typical for the size and flexibility of the investigated particles. Saponit differs in the concentrated region from both other charged systems in as much as a birefringence can above a certain concentration no longer be detected. A permanent network is formed, which is probably electrostatically stabilized which can not be oriented.

References

[1] J. Kerr, Phil. Mag. /4/ 50, 337 (1875)

[2] 2A. Peterlin, H. A. Stuart, Hand- und Jahrbuch der chemischen Physik, Bd. 8, p. 44, Leipzig 1943 A. Peterlin, H. A. Stuart, Z. Phys., 112, 129 (1939)

[3] S. Broesma, J. Chem. Phys., 32, 1626 (1960)

[4] C. T. O'Konski, J. Phys. Chem., 64, 605 (1960) M. Mandel in: S. Krause, ed, Molecular Electro-Optics, Plenum Publishing Corp., New York 1983

[5] K. G. Götz, K. Heckmann, Z. Phys. Chem., NF 20, 42 (1959) G. Porte, J. Appell, J. Phys. Chem., 85, 2511 (1981)

[6] H. Hoffmann, W. Ulbricht, Tenside Surfactants Detergents, 24, 1 (1987)

[7] M. E. Nash, B. R. Jennings, G. J. T. Tiddy, J. Coll. Interf. Sci., 120, 542 (1987)

[8] M. Mandel in: Encyclopedia of Polymer Science and Engineering, J. Wiley & Sons, Vol. 11, p. 739, New York 1986 V. N. Tsvetkov, L. N. Andreeva, Adv. Poly. Sci., 39, 95 (1981)

[9] M. Angel, Thesis, Bayreuth 1985 M. Angel, H. Hoffmann, U. Krämer, H. Thurn, Ber. Bunsenges. Phys. Chem., 93, 184 (1989)

[10] H. Hoffmann, U. Krämer, H. Thurn, submitted to J. Phys. Chem.

[11] U. Krämer, H. Hoffmann, submitted to Macromolecules

[12] H. Hoffmann, G. Oetter, B. Schwandner, Progr. Coll. Poly. Sci., 73, 95 (1987)

[13] W. Schorr, H. Hoffmann, J. Phys. Chem., 85, 3160 (1981) E. Fredericq, C. Houssier, Electric Dichroism and Electric Birefringence, Claredon Press, Oxford 1973

[14] J. Kalus, H. Hoffmann, J. Chem. Phys., 87, 714 (1987) H. Hoffmann, J. Kalus, H. Thurn, K. Ibel, Ber. Bunsenges. Phys. Chem., 87, 1120 (1983)

[15] K. Yamaoka, K. Ueda, J. Phys. Chem., 84, 1422 (1980)

[16] H. Benoit, J. Chim. Phys., 47, 719 (1950) H. Benoit, J. Chim. Phys., 48, 612 (1951)

[17] G. B. Thurston, D. I. Bowling, J. Coll. Interf. Sci., 30, 34 (1969)

[18] T. Bellini, F. Mantegazza, R. Piazza, V. Degiorgio, Europhys. Lett. 10, (5), 499, (1989).

[19] M. A. Laufer, J. Amer. Chem. Soc., 61, 2412 (1939)

[20] H. Mueller, B. W. Sackmann, Phys. Rev., 56, 615 (1939) H. Mueller, Phys. Rev., 55, 792 (1939) H. Mueller, Phys. Rev., 55, 508 (1939) F. J. Norton, Phys. Rev., 55, 668 (1939) N. J. Shah, D. C. Tompson, C. N. Hart, J. Phys. Chem., 67, 1170 (1963)

[21] M. Angel, H. Hoffmann, G. Huber, H. Rehage, Ber. Bunsenges. Phys. Chem., 92, 10 (1988)

[22] T. Odijk, Macromolecules, 12, 688 (1979) T. Odijk, J. Polm. Sci., Poly. Phys. Ed., 15, 477 (1979) J. Skolnick, M. Fixman, Macromolecules, 10, 944 (1977)

SURFACTANT LIQUID CRYSTALS

T.A. BLEASDALE and G.J.T. TIDDY
Department of Chemistry and Applied Chemistry
University of Salford
SALFORD M5 4WT

ABSTRACT. A review of the various lyotropic liquid crystalline phases formed by ionic and non-ionic surfactants/water is presented. The most common-types found are lamellar, hexagonal and cubic, whilst gel and netamic phases are more rare. The number and types of intermediate phases remains to be fully established and is the subject of much current research.

1. Introduction

Surfactants (or amphiphiles) are molecules containing two distinct moieties; a polar 'head group' which is water soluble, and a non-polar chain ('tail') which is oil soluble but only sparingly soluble in water. These substances adsorb strongly at air/water or oil/water interfaces where the polar head group orients itself in the water and the hydrocarbon tail resides at the air surface or in the oil. There is a large variety of surfactant types, both natural and synthetic, which vary in the type of headgroup and alkyl chain. The types of headgroup are ionic (cationic and anionic) and non-ionic, the latter also including dipolar zwitterionic compounds The minimum polarity required for a polar moiety to behave as a headgroup appears to be just larger than that of a single CH_2OH group [1].

The chemical structure of the surfactant tail is usually limited to hydrocarbon and (rarely) fluorocarbon chains, but recently attention has started to focus on polydimethylsiloxane derivatives [2, 3]. Whilst there is normally only one headgroup per surfactant molecule, there are frequently several non-polar tails. These can be linear or branched, the most common being single and linear, particularly for synthetic surfactants. Naturally occuring amphiphiles are usually of the di-alkyl type.

When surfactants are dissolved in water they form aggregates called micelles above a critical concentration, termed the critical micelle concentration (CMC). Typically a micelle comprises 50-120 monomers. Micelles only form if the surfactant is sufficiently soluble - i.e. above a specific temperature called the Krafft temperature. This is defined as the temperature at which the surfactant is just soluble enough to form micelles [4].

The formation of micelles can be attributed to the 'hydrophobic effect' [5]. Water-hydrocarbon attractions are considerably less than water-water attractions (free energy of attraction being -40 erg/cm^2 and -144 erg/cm^2 respectively [5]), thus accounting for their poor solubility in water. With amphiphiles this leads to weak Van der Waals bonding of the water molecules to the alkyl chain whilst maintaining a strong interaction between neighbouring water molecules. Consequently, there is a mono-layer of 'structured water' surrounding the tail. On surfactant aggregation the water/hydrocarbon interaction is reduced and the water mono-layer released, thus,

D. M. Bloor and E. Wyn-Jones (eds.),
The Structure, Dynamics and Equilibrium Properties of Colloidal Systems, 397–414.

increasing the entropy of the system. It is this increase in entropy that is the driving force for micellisation.

Above the CMC, added surfactant leads to an increase in the number of micelles. At even higher micelle concentration there is a disorder/order transition with the formation of lyotropic liquid crystals (or mesophases). The structure of the most common mesophases are described below, along with a qualitative theoretical description to account for their occurrence.

2. Micelles

There are two distinct categories of micelle; normal and reversed. Normal micelles are formed when surfactants are dissolved in water (polar solvent). The hydrophobic tails orient themselves towards the micelle interior whilst the head-groups remain hydrated. A schematic representation of a normal micelle is given in Fig. 1.

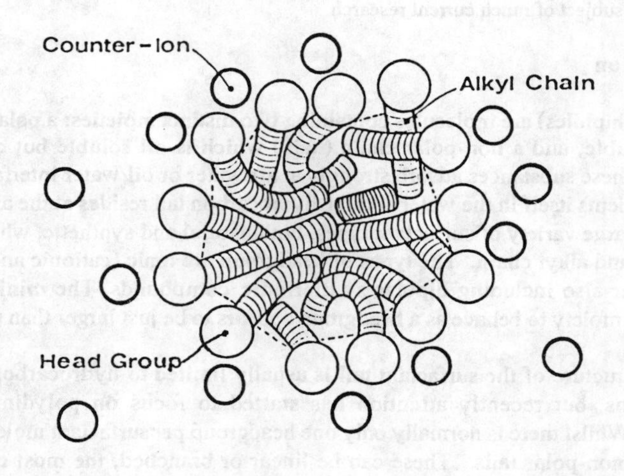

Figure 1. Schematic representation of a normal micelle. The dotted line indicates the oil/water interface.

Reversed micelles are formed when surfactants are dissolved in non-polar solvents. Here the polar headgroups reside in the micelle interior whilst the hydrocarbon chain is solvated. These will not be discussed further.

Since the hydrocarbon/water contact is minimised by micelle formation it is to be expected that the longer the chain length, the lower the CMC will be. Some typical values can be seen in Table 1.

Table 1. CMC values in water at 25°C

	SURFACTANTS	CMC (mol dm^{-3})
	IONIC	
$C_{12}OSO_3Na$	Sodium dodecylsulphate	8×10^{-3}
$C_{14}OSO_3Na$	Sodium tetradecylsulphate	2.1×10^{-3}
$C_{12}NMe_3Cl$	Dodecyltrimethylammonium chloride	2×10^{-2}
$C_{16}NMe_3Cl$	Hexadecyltrimethylammonium chloride	1.3×10^{-3}
$C_{14}PyBr$	Tetradecylpyridinium bromide	2.5×10^{-3}
$C_{16}PyBr$	Hexadecylpyridinium bromide	6.8×10^{-4}
	NON IONIC	
C_8EO_6	Hexaethyleneglycol mono n-octyl ether	1×10^{-2}
$C_{12}EO_6$	Hexaethyleneglycol mono n-dodecyl ether	7×10^{-5}
$C_{16}EO_6$	Hexaethyleneglycol mono n-hexadecyl ether	1×10^{-6}
$C_{12}EO_{12}$	Dodecaethyleneglycol mono n-dodecyl ether	12×10^{-5}

The higher CMC's for ionic surfactants can be attributed to the unfavourable electrostatic repulsion between adjacent headgroups at the micelle surface.

The dependence of chain length (C_n) on CMC for single alkyl chain surfactants is usually given by:

$$\log CMC = A - B(C_n)$$

where A and B are constants.

'A' depends on the nature of the head group. Typical values for B are ~ 0.5 for non-ionic and zwitterionic surfactants, while the value is 0.29 - 0.3 for monovalent ionic surfactants. For a given homologous series of surfactants, an increase in chain length of two CH_2 groups decreases the CMC by factors of ten and four for non-ionic and ionic headgroups respectively. If the ionic headgroup is multivalent, increasing chain length decreases the CMC to an even lesser extent. This is due to the greater degree of counterion dissociation since there are two (or more) counterions per alkyl chain.

The aggregation number of a micelle largely depends on the shape it adopts. Just above the CMC most surfactants form small globular (close to spherical) micelles with aggregation numbers in the range 40-200 [6, 7]. Long rod micelles with a circular cross-section occur for some longer chain surfactants such as N-hexadecyltrimethylammonium salicylate and hexadecylpyridinium salicylate [8]. These have aggregation numbers greater than 10^3. Aggregated, large disc micelles consisting of bilayers (separated lamellar phase -see below) are also quite common but conclusive evidence of small circular-disc micelles is lacking. Increasing the chain length leads to larger aggregation numbers, especially if a change in shape occurs, i.e. globular to rod. For all surfactants with a H_1 phase (see below), long rod micelles occur if C_n is sufficiently large. For mixed surfactants decreasing the average surface area per molecule (size of headgroup) often leads to rods, particularly if C_n is also increased [9].

Experimental methods for investigating aggregation numbers include light scattering, NMR/radio tracer self-diffusion measurements and fluorescent probes [6,7].

Qualitatively, the factors responsible for micelle shape are intermolecular interactions at the micelle surface and alkyl chain packing within the micelle interior. At the micelle surface, headgroup repulsions arising from electrostatic, solvation and steric effects act to increase the surface area per molecule (a). This is counterbalanced by the hydrocarbon/water repulsion which acts to reduce a.

The effect of alkyl chain packing on micelle shape is based on the assumption that the micelle radius cannot be longer than the all-trans hydrocarbon chain length [10]. This leads to a minimum area per molecule a_{min} for a particular shape of micelle. Assuming spheres, long circular rods and infinite bilayers with a smooth interface between alkyl chain and water regions, then it is easy to calculate a_{min} values for the various shapes. The values are given by [10]:

$$a_{min} = \frac{3V}{L_t} \quad \text{(sphere)}$$

$$a_{min} = \frac{2V}{L_t} \quad \text{(rod)}$$

$$a_{min} = \frac{V}{L_t} \quad \text{(bilayer)}$$

where V = volume of hydrocarbon chain
L_t = length of all-trans hydrocarbon chain.

For a C_{12} hydrocarbon chain, assuming L_t = 15 Å and with $V(CH_2)$ = 27 Å^3, $V(CH_3)$ = 54 Å^3,

a_{min} = 70 Å^2 (sphere)
a_{min} = 46.7 Å^2 (rod)
a_{min} = 23.3 Å^2 (bilayer)

Note that the a_{min} value decreases in the series sphere → rod → disc, and that the value does not alter significantly with chain length (because both V and L_t are roughly linearly dependent on the carbon number). In practice entropy favours the smallest possible micelle size, hence rods only occur with a values below a_{min} sphere, and disc (bilayers) only with a below a_{min} rod.

3. Lyotropic Liquid Crystals

There are four well established class of mesophase, these being lamellar, hexagonal, cubic (of which there are two types) and nematic [11-15]. The lamellar phase (L_α) is by far the most common lyotropic liquid crystal, consisting of surfactant bilayers separated by water layers (Fig.

2). The surfactant bilayer thickness is in the region 0.8 - 1.6 times the all-trans surfactant chain length (L_t), whilst the variation in water layer thickness is considerably greater, in the range 2 - 200 Å, and increasing with increased water content. The bilayers extend over large distances, typically a micron or more and thus have very high aggregation numbers. Over recent years it has become apparent that the L_α phase may also contain defects filled with water ('holes') within the bilayers [16].

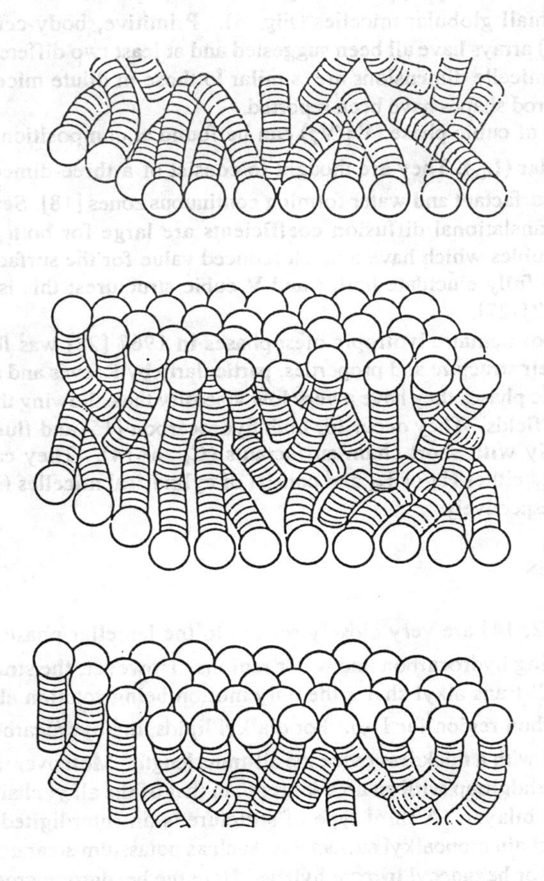

Figure 2. Schematic representation of a lamellar phase.

The hexagonal phase is the next most common structure with two distinct types being possible termed normal (H_1) and reversed (H_2). The H_1 phase consists of very long rod shaped micelles of circular cross section, having a diameter of 1.5 - 2.0 L_t (Fig. 3). The headgroups reside at the micelle surface with a continuous water region separating adjacent micelles. The intermicellar separation lies in the range 8 - 50 Å, again increasing with water concentration. The H_2 phase is composed of water cylinders packed in an hexagonal array surrounded by a continuous hydrocarbon environment. The diameter of the water cylinders is typically ~ 10 - 20 Å. Both L_α and H_1/H_2 are uniaxial, possessing a single axis of symmetry, and are optically anisotropic.

Phases with cubic symmetry are much less common, usually occuring over narrow composition regions. One class of cubic mesophases (I_1/I_2) can be found at compositions between micellar solution (L_1/L_2) and hexagonal phase (H_1/H_2). These consist of cubic close packed arrays of small globular micelles (Fig. 4). Primitive, body-centred (BCC) and face centred cubic (FCC) arrays have all been suggested and at least two different structures are found to occur [17]. The micelle dimensions are similar to those in dilute micellar solution and both spherical and small rod shapes have been reported.

The second type of cubic phase (V_1/V_2) can be found at compositions between hexagonal (H_1/H_2) and lamellar (L_α). They are thought to consist of a three-dimensional bicontinuous network with both surfactant and water forming continuous zones [18]. Several structures appear to be possible. Translational diffusion coefficients are large for both water and surfactant compared with I_1 cubics which have a much reduced value for the surfactant [19, 20]. Further work is required to fully elucidate both I and V cubic structures; this is a very active area of research at present [21-27].

The first report of nematic lyotropic mesophases in 1967 [28] was followed by extensive investigations on their structure and properties, particularly by Reeves and co-workers [15]. Like thermotropic nematic phases they have a very low viscosity thus allowing them to become aligned by strong magnetic fields. They occur for both hydrocarbon [15] and fluorocarbon derivatives [29] and particularly with short chain surfactants (C_n, $n<14$). They can be found between micellar solution and either H_1 or L_α and consist of cylindrical micelles (N_c) or ordered small disc micelles (N_d) respectively.

3.1 GEL PHASES

Gel phases (L_β) [12, 14] are very closely related to the lamellar phase, both having a layer structure of alternating hydrocarbon and water regions. However, the structure of the gel phase differs in its rigid all-trans alkyl chain, the only motion being rotation about its long axis (cf. liquid-like hydrocarbon region for L_α). For dialkyl lipids, the hydrocarbon chains are usually arranged in bi-layers with a thickness twice the all-trans length. Moreover the layers may be tilted to accommodate a headgroup with an area larger than that of the alkyl chains, thus also reducing the thickness of the bilayer. A third type of structure is an interdigited monolayer [30-32], occuring for longer chain monoalkyl surfactants, such as potassium stearate, hexadecyltrimethyl-ammonium chloride or hexadecyl trioxyethylene. Here the headgroup cross section is twice that of the alkyl chain.

Gel phases occur at temperatures below the other mesophases. On heating they usually melt to give a L_α, accompanied by a large melting enthalpy (typically about half that of a crystal → liquid transition). Gel phase stability clearly arises from short range attractions between the alkyl chains.

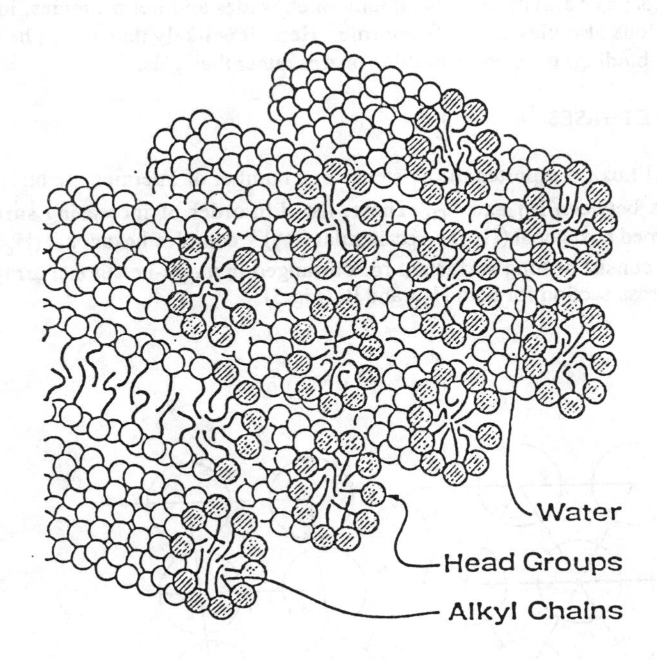

Figure 3. Schematic representation of a hexagonal phase

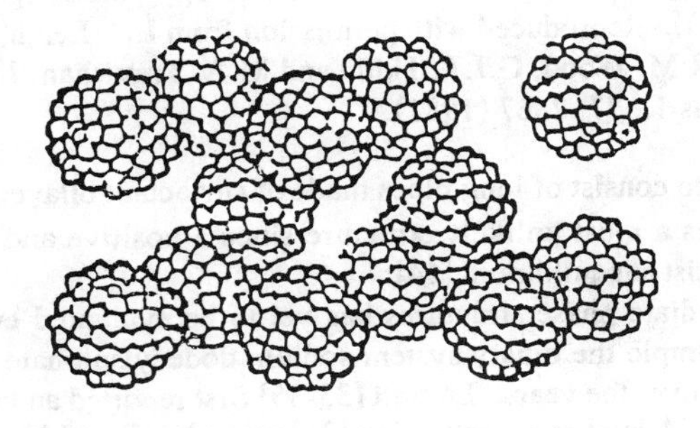

Figure 4. Schematic representation of an I_1 body centred cubic phase.

However, the occurrence of gels for surfactants with specific counterions such as potassium

soaps and not sodium soaps or alkyltrimethylammonium chlorides and not bromides, indicates that headgroup interactions also play a significant role. Here, it is likely that strong headgroup interactions (counterion binding) promote crystalline phases rather than gels.

3.2 INTERMEDIATE PHASES

As early as 1960 [33] Luzzati reported the existence of a number of intermediate birefringent phases at compositions between H_1 and L_α. These, listed in order of increasing surfactant concentration are deformed hexagonal (H_{1d}) < rectangular (R_1) < complex hexagonal (H_c). Both the H_{1d} and R_1 phases consist of long surfactant rods arranged in a non-hexagonal array. The rods have an eliptical cross-section for both H_{1d} and R_1(Fig.5).

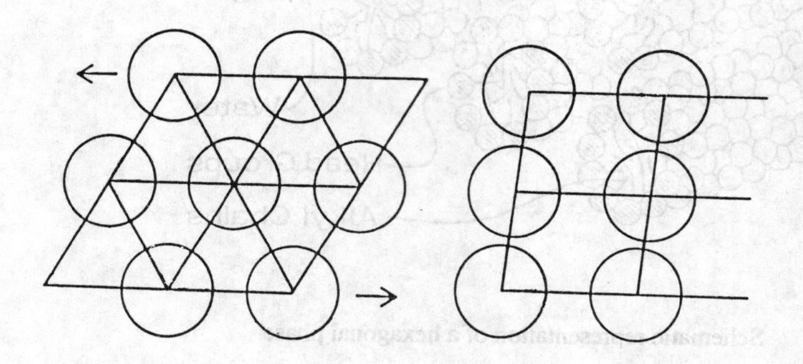

Figure 5. Schematic diagram of the deformed hexagonal (H_{1d}) phase (right) compared to H_1 phase (left). Reproduced with permission from I.D. Leigh, M.P. McDonald, R.M. Wood, G.J.T. Tiddy and M.A. Trevethan, J. Chem. Soc., Faraday Trans 1., 77, 2867 (1981).

The H_c phase was suggested to consist of long pipes made of surfactant bilayers filled with, and surrounded by water. This is a most unlikely structure since a positive and negative surface curvature would have to co-exist at equilibrium [14].

The full range of intermediate phase structures has yet to be evaluated even for the most common surfactants. For example the binary system sodium dodecylsulphate (SDS)/water has been extensively investigated over the years. Luzzati [33-35] first reported an intermediate phase and assigned it a H_c structure. A later re-investigation [36] using NMR and X-ray diffraction led to the assignment of H_{1d} for the intermediate structure. More recently a series of papers by Kékicheff and co-workers [37-40] reports the occurrence of four separate intermediate phases. In increasing surfactant concentration these are two dimensional monoclinic (M_α-2_d) < three dimensional monoclinic (M_α-3_d) < cubic phase (Q_α) < tetragonal phase (T_α). This follows an almost continuous decrease in surface curvature from cylindrical to planar geometry. The Q_α

phase is a cubic V_1 structure whilst the tetragonal phase is a 'lamellar phase with holes'. These structures were elucidated from X-ray diffraction data obtained using *inter alia* a sample cell with a concentration gradient across it.

Although intermediate phases have been frequently observed for anionic and less often for cationic surfactants [30], they have only recently been observed for non-ionic and zwitterionic surfactants [41, 42]. Moreover they only occur for longer chain surfactants with shorter chain lengths exhibiting a V_1 cubic phase. The change in behaviour occurs at *ca.*C_8-C_{12} for anionic surfactants [43] whereas it is C_{16} for many cationic, nonionic and zwitterionic derivatives. Surfactants having a transitional chain length form intermediate phases at lower temperatures which give way to V_1 cubic phases at higher temperatures. Intermediate phases are not restricted to hydrocarbons but also occur for short chain fluorocarbon surfactants [44]. At present it looks as if there are three distinct classes of intermediate phases, based on rod structures (similar to H_1), layer structures (L_α types) and three dimensional networks (V types).

4. Techniques

Evidence for the structures described above can be obtained from optical and electron microscopy, X-ray diffraction and NMR spectroscopy. Very often the information gained on a new system by any one technique is insufficiently unambiguous to allow full confidence in structure assignment. At least two techniques are usually required.

4.1. OPTICAL MICROSCOPY

The penetration technique of Lawrence [45] is most useful as an initial survey of the liquid crystals formed by a surfactant/solvent system. Pure surfactant is sandwiched between a microscope slide and cover slip and contacted by the solvent. At the contact region between pure surfactant and pure solvent a concentration gradient is present which becomes wider and less steep with time. Within this region all the mesophases develop as concentric bands with characteristic optical 'textures'. Anisotropic mesophases are optically birefringent when viewed between crossed polars, giving specific patterns ('textures') according to phase structure. The lamellar phase can be identified by the 'mosaic' or 'oily streaks' textures whilst hexagonal phases exhibit 'fan-like' or 'non-geometric' textures (Fig. 6) [46]. Additionally, the viscosities of the various phases are very different. This can be qualitatively assessed by gentle pressing on the cover-slip with a suitable implement. The hexagonal phase is very viscous whereas the lamellar phase is much less viscous and flows readily. Some high water content lamellar phases are able to flow under gravity. Cubic phases differ from the other mesophases by being isotropic (non-birefringent). They can be easily distinguished from isotropic solutions by the very high viscosities they exhibit. A general sequence of viscosities is $L_\alpha < H_2 < H_1 < V_1 \sim V_2 < I_1$.

4.2 X-RAY DIFFRACTION

Low angle X-ray diffraction (or neutron scattering) is by far the best technique for determining the phase structure. The lamellar phase exhibits diffraction lines in the ratio d : d/2 : d/3 : d/4 where d = d_o [12], the unit cell dimension. The unit cell comprises of one water layer and one surfactant

Fig. 6.1 Typical 'fan-like' textures exhibited by hexagonal phases. $C_{16}EO_8$/water (48 wt% surfactant)

Fig. 6.2 Typical 'mosaic' structure of the lamellar phase. $C_{16}EO_8$/water (76 wt% surfactant)

bilayer. The diffraction lines of the hexagonal phase are in the ratio $d : d/\sqrt{3} : d/\sqrt{4} : d/\sqrt{7}$ [12], where d is the separation between adjacent rows of rod micelles. Cubic phases display several patterns depending on the lattice type present (i.e. primitive, BCC, FCC). An example is $1/\sqrt{4}$: $1/\sqrt{5}$: $1/\sqrt{6}$: $1/\sqrt{8}$: $1/\sqrt{10}$ which corresponds to a primitive lattice [21, 22]. In practice however, only the first few reflections are usually seen due to the low intensity of the diffraction peaks and the high degree of scatter.

4.3 NMR

Information on order, rotation and diffusion of molecules within a liquid crystal can be obtained using NMR. A wide variety of nuclei can be used (^1H, ^2H, ^{13}C, ^{14}N, ^{19}F, ^{23}Na etc) enabling surfactant, solvent (particularly $^2H_20$) and counterions to be investigated.

For anisotropic samples (usually hexagonal, lamellar and gels), an order parameter (S) can be obtained from the splitting of the spectral lines (Δ) due to nuclear dipole ($I = \frac{1}{2}$) or quadrapole ($I \geq 1$) interactions. For a uniaxial mesophase the order parameter is related to the quadrapole splitting by [47]:

$$\Delta = \frac{3}{4} \chi S$$

where χ is the nuclear coupling constant. A similar expression holds for dipole-dipole interactions. In an isotropic liquid the tumbling motion of the molecules averages out any interaction, giving a single band with no splitting. Thus cubic and solution phases give a single peak, while hexagonal and lamellar phases give more complex spectra. Very often the Δ values for lamellar and hexagonal phases are related by:

$$\Delta(\text{Lam}) = 2\Delta(\text{Hex})$$

Spin lattice relaxation rates ($1/T_1$) allow an estimation of the correlation times (τ_c) related to molecular rotation. For single chain surfactants, rotation of the alkyl chain about the long axis is rapid ($\tau_c \sim 10^{-10}$ s), whilst dialkyl lipids have a τ_c value somewhat higher. A plot of $1/T_1$ versus carbon number shows little variation along the alkyl chain except for the terminal groups where τ_c values are considerably shorter. Since conformational changes are associated with rotation, this observation is to be expected as the trans-gauche barrier will reduce with increasing distance from the headgroup. Detailed analysis of T_1 values and self diffusion measurements allows micelle size and shape to be evaluated [48].

5. Theoretical Considerations

The two factors responsible for the formation and type of lyotropic liquid crystals are the intra- and intermicellar interactions. The intramicellar interactions are those dealt with previously for micelles and include alkyl chain packing constraints.

The contributions to the intermicellar interactions come from electrostatic, steric and solvation repulsions. For ionic surfactants the micelle surface is highly charged despite a considerable degree of counterion binding. This, plus counterion-counterion (solvent) interactions forms the electrostatic repulsion. Evidence for solvent interactions can be inferred from the similar mesophase behaviour of non-ionic and ionic surfactants. For non-ionic surfactants headgroup hydration results in intermicellar repulsions similar to the electrostatic effects for ionic systems [49]. The origin of these hydration forces is thought to arise from the polarization of a single layer of 'bound' water molecules at the interface. It should be noted that these hydration forces are also present in ionic systems. Whilst it is currently fashionable to explain much unusual behaviour of surfactant and colloid systems in terms of extensive 'water structure', hard experimental evidence to support such hypotheses is difficult to obtain. It is perhaps worthwhile to note the comments of G.S. Hartley [50], one of the giants of early surfactant research.

"There is a wide spread tendency to use 'hydration' in colloid chemistry as a sort of universal explanation of puzzling phenomena. Its inaccessibility to direct experimental determination fortifies this tendency..... Explanations based on great hydration should be regarded with extreme caution" [50].

Steric effects are only important for molecules with very large headgroups, such as non-ionic surfactants (eg polyoxyethylene surfactants).

With zwitterionic and non-ionic surfactants medium range attractive interactions can occur, which result in phase separation of surfactant-rich phases from dilute aqueous phases ('clouding'). For non-ionic surfactants the mechanism is probably related to changes in EO solubility with temperature (EO polymers have a lower consolute temperature). This, in turn, is probably related to changes in the EO conformation with temperature [51, 52]. With zwitterionic surfactants the attraction may arise from collective fluctuations in zwitterion conformation and orientation on opposing bilayers.

Intermicellar forces become important as the concentration of micelles is increased (there is virtually no interaction in very dilute micellar solution). The repulsive forces may cause a decrease in a which occasionally leads to a change in micelle shape (sphere to rod) but more often to micellar growth (rods). At high concentration, the strong repulsive forces eventually lead to a disorder/order transition, often with no change in micelle shape. Therefore the micelle shape just prior to mesophase formation determines the liquid crystal structure formed.

Spheres → Cubic (I_1)
Rods → Hexagonal (H_1)
Discs → Lamellar (L_α)

(note that 'disc' micelles may not be circular discs, but other less regular shapes such as 'rubber' or 'ruler' types).

Further increases in surfactant concentration within the mesophase region leads to a continuing reduction in the surface area per molecule at the micelle surface. When a falls below a_{min} for that particular shape, a phase change occurs to one of lower a_{min}. Therefore the expected mesophase transitions are

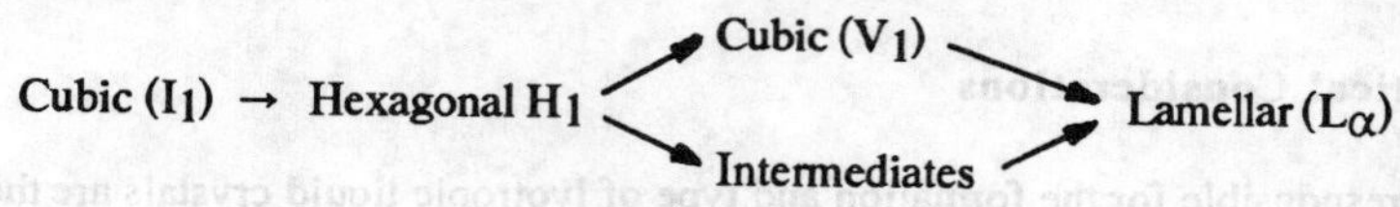

This is accompanied by a decrease in surface curvature from cubic (I_1) through to lamellar. Although the formation of fewer, larger micelles is entropically unfavourable, the intermicellar repulsion (be it hydration or electrostatic) is the dominant term.

6.1 NON-IONIC SURFACTANTS

A much studied series of non-ionic surfactants is the alkyl polyoxyethylene type ($C_nH_{2n+1}(OCH_2CH_2)_mOH$; C_nEO_m) [53]. They are a very useful model system since both headgroup and alkyl chain can be changed systematically. A characteristic feature of these surfactants is a decrease in headgroup hydration with increasing temperature, thus leading to the phenomena of clouding already mentioned. The effect of headgroup size is illustrated by the series $C_{12}EO_8$, $C_{12}EO_6$ and $C_{12}EO_4$ (Fig. 7). At low temperatures (<16°C) the most dilute mesophase formed is I_1, H_1 and L_α for $C_{12}EO_8$, $C_{12}EO_6$ and $C_{12}EO_4$ respectively. This is to be expected since with large headgroup size ($C_{12}EO_8$) the increased steric and hydration repulsions at the micelle surface will favour spherical micelles. Smaller headgroups ($C_{12}EO_4$) have much reduced steric and hydration repulsions thereby favouring bilayers. Moreover the phase diagrams show no H_1 phase for $C_{12}EO_4$ and only a very narrow L_α existence region for $C_{12}EO_8$. As expected $C_{12}EO_6$ shows mesophase behaviour midway between $C_{12}EO_8$ and $C_{12}EO_4$ with large existence regions for both H_1 and L_α.

The effect of increasing the temperature can also be seen with this series of surfactants. The most dilute mesophase for $C_{12}EO_8$ changes to H_1 above 16°C whilst at 37°C a L_α phase occurs

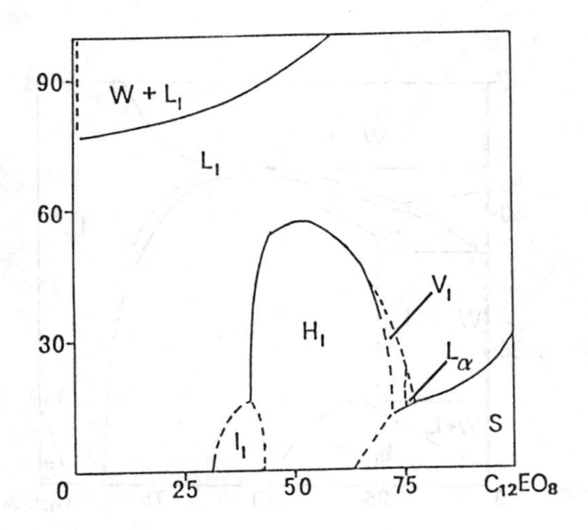

Figure 7.1 Phase diagram of the $C_{12}EO_8$/water system. Reproduced with permission from D.J. Mitchell, G.J.T. Tiddy, L. Waring, T. Bostock and M.P. McDonald. J. Chem. Soc., Faraday Trans 1., 79, 975 (1983).

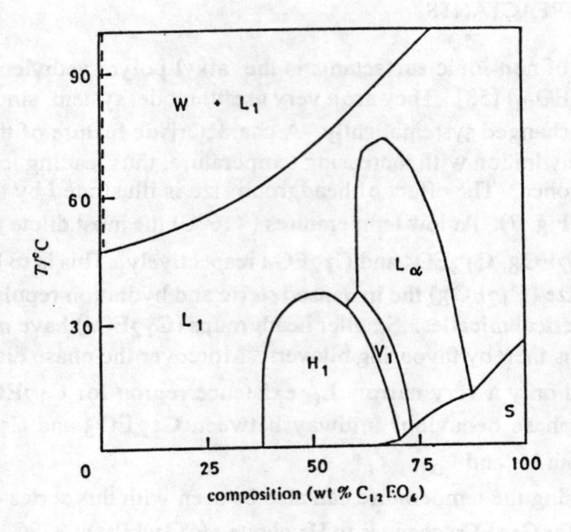

Figure 7.2. Phase diagram of the $C_{12}EO_6$/water system. Reproduced with permission from D.J. Mitchell, G.J.T. Tiddy, L. Waring, T. Bostock and M.P. McDonald. J. Chem. Soc., Faraday Trans 1., 79, 975 (1983).

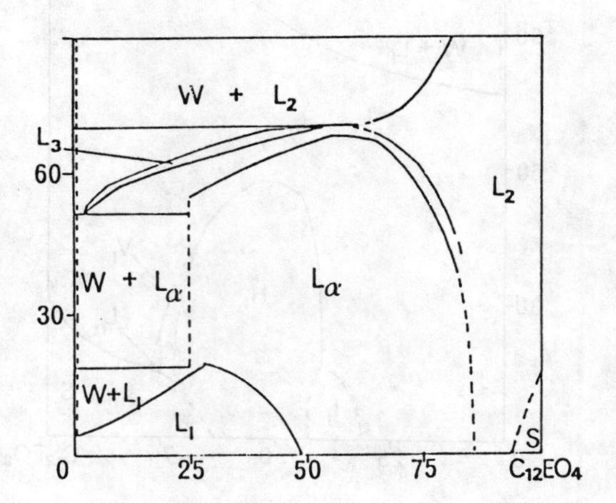

Figure 7.3 Phase diagram of the $C_{12}EO_4$/water system. Reproduced with permission from D.J. Mitchell, G.J.T. Tiddy, L. Waring, T. Bostock and M.P. McDonald. J. Chem. Soc., Faraday Trans 1., 79, 975 (1983).

for $C_{12}EO_6$. This is brought about by the reduced steric and hydration repulsion as the EO group becomes less hydrated.

C_{10} surfactants show a similar mesophase behaviour to C_{12} whilst no mesophase formation occurs above ~ 15°C for C_8 derivatives. Longer alkyl chain lengths tend to be similar to the C_{12} derivative but the mesophases persist to higher temperatures. In addition to the normal phases already mentioned, reversed cubic phases (V_2) occur for long alkyl chains (C_n, $n \geq 14$) and small headgroup ($\leq EO_4$). Additionally, gel phases (L_β) also occur for small headgroups (EO_3/EO_4) with alkyl chains of at least C_{16}.

As indicated in figure 7 polyethylene surfactant solutions exhibit a lower consolute curve where dilute and concentrated solutions co-exist. The lowest temperature at which phase seperation occurs is a critical point termed the cloud temperature. Because the curve is very flat in this region the cloud point (temperature) is often taken as the temperature where a 1% or 0.1% solution becomes turbid. Although the exact mechanism of clouding is not fully understood there appears to be two distinct groups of surfactants [54]. Firstly there are those which have a L_α phase at the cloud temperature but at higher surfactant concentrations ($C_{12}EO_4$, $C_{16}EO_6$). Here the phase separation is thought to arise from intermicellar EO-EO attractions of disc (ruler) micelles (which become more important as headgroup hydration decreases). At short range, the residual hydration repulsions are sufficiently strong to give a L_α phase. The second group form a $W + L_2$ region at temperatures higher than the mesophase melting temperatures (e.g. C_8EO_4, $C_{12}EO_8$). Here, the attractive EO-EO interaction between even small micelles is sufficient for phase separation.

6.2 IONIC SURFACTANTS

The phase behaviour of anionic surfactants has been extensively investigated over the years. Soaps (alkylcarboxylate salts) were amongst the first to be systematically studied, particularly by McBain and co-workers [see references in 11, 14]. The sodium and potassium derivatives form H_1 and L_α phases for chain lengths > C_8. A V_1 cubic phase is found for short chain lengths and at least one intermediate phase is formed for longer chain lengths. The changeover occurs for Na and K salts at chain lengths C_{10} and C_{12} respectively [43]. Sodium alkyl sulphates show a similar phase behaviour, with the V_1/intermediate transition occuring at C_8 chain length [43].

Mesophases formed by cationic surfactants are very similar to those of anionic surfactants. One notable difference however is the formation of I_1 cubic phases for alkyltrimethylammonium chlorides and alkylpyridinium chlorides with a chain length ($n \leq 14$) [30]. The absence of an I_1 phase for chain lengths longer than C_{14} is probably due to the decreased effective alkyl chain length. This arises from the increased number of alkyl chain conformations, thus reducing the probability of an all-trans chain length. The absence of an I_1 cubic phase for the bromide and iodide derivatives can be rationalised in terms of the degree of counterion binding ($Cl^- < Br^- < I^-$). Thus the low degree of chloride counterion binding leads to an increased value for a, thereby favouring spherical (globular) micelles.

7. Future Directions

Without a doubt one of the key areas for future research will be the substitution of water by other polar solvents. In the late 1960's Reinsborough and co-workers [55] investigated aggregation in pyridinium chloride, a molten electrolyte. Since then there have been numerous publications on micellization [56-65] and phase behaviour [66-73] in solvents such as ethylene glycol, glycerol, formamide and ethylammonium nitrate. Whilst it is clear that the surfactants investigated aggregate in these solvents, the CMC values are considerably higher (often one or two orders of magnitude) than the corresponding aqueous system. Moreover it would appear that a stepwise, less co-operative process is present [61, 65] giving much smaller micelles at the CMC. Indeed it has even been suggested that the term micelle is inappropriate for these systems in view of the results so far [68]. However, the widespread occurrence of mesophases for long chain surfactants (C_n, $n > 12$) [74] suggests that large surfactant aggregates are common in non-aqueous polar solvents. Thus despite the fact that the thermodynamics of aggregation is very different from that in water [75] we will probably continue to call them micelles for at least the next few years.

8. References

1. R.G. Laughlin in "Advances in Liquid Crystals", Ed. G.H. Brown, Academic Press, Vol. 3, p.42 (1978). Also p.99.
2. Cockett, S., Dodgeson, K., Simmonds, D., Tiddy, G.J.T., in preparation 1990.
3. Hall, D.G., Tiddy, G.J.T., in "Surfactant Science Series Vol. II, Ed. E.H. Lucassen-Reynders (Marcel Dekker, Inc. New York and Basel) 1981. p.55.
4. Murray, R.C., Hartley, G.S., Trans. Faraday Soc., 31, 183 (1935).
5. C. Tanford, "The Hydrophobic Effect", Wiley-Interscience, 1980.
6. Wennerström, H., Lindman, B., Physics Reports, 52, 1 (1979).
7. Lindman, B., Wennerström, H., "Topics in Current Chemistry" (Springer-Verlag, Berlin, Heidelberg, 1980) Vol. 87, p.1.
8. Hoffman, H., Platz, G., Rehage, H., Schorr, W., Ulbricht, W., Ber. Bunsenges. Phys. Chem., 85, 255 (1981).
9. Tiddy, G.J.T. Unpublished results.
10. Israelachvili, J.N., Mitchell, D.J., Ninham, B.W., J. Chem. Soc. Faraday Trans. 2, 72, 1525 (1976).
11. Ekwall, P. in "Advances in Liquid Crystals", Ed. G.H. Brown (Academic Press, New York, London), Vol. 1, Ch.1, p.1. (1971).
12. Luzatti, V. in "Biological Membranes" Ed. D. Chapman (Academic Press London and New York 1968) Ch, 3, p.71.
13. Winsor, P.A., Chem. Rev., 68, 1 (1968).
14. Tiddy, G.J.T., Physics Reports, 57, 1 (1980).
15. Forrest, B.L., Reeves, L.W., Chem. Rev., 81, 1 (1981).
16. Holmes, M.C., Charvolin, J., J. Phys. Chem., 88, 810 (1984).
17. Tiddy, G.J.T., Walsh, M.F., in "Aggregation Processes in Solution", Ed. E. Wyn-Jones and J. Gormally, Elsevier Scientific Publishing Co. (1983) p.151.
18. Scriven, L.E., Nature, 263, 123 (1976).
19. Bull, T., Lindman, B., Mol. Crystals Liquid Crystals, 28, 155 (1974).
20. Lidblom, G., Wennerström, H., Biophys. Chem., 6, 167 (1977).
21. Fontell, K., Colloid Polym. Sci., 268, 264 (1990).
22. Larsson, K., J. Phys. Chem., 93, 7304 (1989).
23. Gruner, M., J. Phys Chem., 93 7562 (1989).
24. Charvolin, J., Sadoc, J.F., Colloid Polym. Sci., 268, 190 (1990).
25. Seddon, J.M., Biochimica et Biophysica Acta, 1031, 1 (1990).
26. Lindblom, G., Rilfors, L., Biochimica et Biophysica Acta, 988, 221 (1989).

27. Andersson, S., Hyde, S.T., Larsson, K., Lidin, S., Chem. Rev., 88, 221 (1988).
28. Lawson, K.D., Flautt, T.J., J. Am. Chem. Soc., 89, 5489 (1967).
29. Boden, N., Jackson, P.H., McMullen, K., Holmes, M.C., Chem. Phys. Lett., 65, 476 (1979).
30. Blackmore, E.S., Tiddy, G.J.T., J. Chem. Soc., Faraday 2, 84, 1115 (1988).
31. Vincent, J.M., Skoulois, A.E., Acta Cryst., 20, 432, 437 (1966).
32. Adam, C.D., Durrant, J.A., Lowry, M.R., Tiddy, G.J.T., J. Chem. Soc. Faraday Trans. 1, 80, 789 (1984).
33. Luzzati, V., Mustacchi, H., Skoulios, A., Husson, F., Acta Cryst., 13, 660 (1960).
34. Husson, F., Mustacchi, H., Luzzati, V., Acta Crystallogr., 13, 668(1960).
35. Luzzati, V., Husson, F., J. Cell Biol., 12, 207 (1962).
36. Leigh, I.D., McDonald, M.P., Wood, R.M., Tiddy, G.J.T., Trevethan, M.A., J. Chem. Soc., Faraday Trans. 1, 77, 2867 (1981).
37. Kekicheff, P., Cabane, B., J. Phys., 48, 1571, (1987).
38. Kekicheff, P., Cabane, B., Acta Cryst., B 44, 395 (1988).
39. Kekicheff, P., Grabielle-Mendelmont, C., Ollivon, M., J. Colloid Int. Sci., 131, 112 (1989).
40. Kekicheff, P., J. Colloid Interface Sci., 131, 133 (1989).
41. Hall, C., Tiddy, G.J.T., Paper presented to the 6th International Symposium on Surfactants in Solution, New Delhi, Aug. 1986 in "Surfactants in Solution" Ed. K.L. Mittal, Plenum Press, Vol. 8, P9 (1989).
42. Tiddy, G.J.T., Unpublished Results.
43. Rendall, K., Tiddy, G.J.T., Treventhan, M.A., J. Chem. Soc. Faraday Trans. 1., 79, 637, (1983).
44. Kekicheff, P., Tiddy, G.J.T., J. Phys. Chem., 93, 2520 (1989).
45. Lawrence, A.S.C., in Liquid Crystals 2, Ed. G.H. Brown (Gordon and Breach, London, 1969) Part 1, p.1.
46. Rosevear, F.B., J. Am. Oil Chem. Soc., 31, 628 (1954).
47. Nuclear Magnetic Resonance of Liquid Crystals, Ed. J.W. Emsley (D. Reidel Publishing Co., Dordrecht, Boston and Lancaster, 1985).
48. Lindman, B., Söderman, O., Wennerström, H., Surfactant Solutions: New Methods of Investigation, Ed. R. Zana, (Dekker, New York) 1987.
49. Rand, R.P., Parsegian, V.A., Biochimica et Biophysica Acta, 988, 351 (1989).
50. G.S. Hartley, 'Aqueous Solutions of Parafin Chain Salts' (Published by Herman Cie, Paris 1936).
51. Andersson, M., Karlstrom, G., J. Phys. Chem., 87, 4957 (1985).
52. Karlstrom, G., J. Phys. Chem., 87, 4762 (1985).
53. Bostock, T., McDonald, M.P., Mitchell, D.J., Tiddy, G.J.T., Waring, L., J. Chem. Soc. Faraday Trans. 1, 79, 975 (1983).
54. Conroy, J.P., Hall, C., Lang, G.A., Rendall, K., Tiddy, G.J.T., Walsh, J., Lindblom, G., Progress in Colloid Polym. Sci. 1990 in press.
55. Reinsborough, V.C., Aust. J. Chem., 23, 1473 (1970).
56. Ray, A., J. Am. Chem. Soc., 91:23, 6511 (1969).
57. Evans, D.F., Yamauchi, A., Roman, R., Casassa, E., J. Colloid Interface Sci., 88, 89 (1982).
58. Evans, D.F., Yamauchi, A., Wei, G., Bloomfield, A., J. Phys. Chem., 87, 3537

(1983).
59. Rico, I., Lattes, A., J. Phys. Chem., 90, 5870 (1986).
60. Belmajdoub, A., ElBayed, K., Brondeau, J., Cannet, D., Rico, I., Lattes, A., J. Phys. Chem., 92, 3569 (1988).
61. Binana-Limbele, W., Zana, R., Colloid Polm. Sci., 267, 440 (1989).
62. Fletcher, P.D.I., Gilbert, P.J., J. Chem. Soc., Faraday Trans 1, 85, 147 (1989).
63. Backlund, S., Bergenstahl, B., Molander, O., Wärnheim, T., J. Colloid Int. Sci., 131, 393 (1989).
64. Sjoberg, M., Henriksson, U., Wärnheim, T., Langmuir, in press.
65. Gharibi, H., Palepu, R., Tiddy, G.J.T., Hall, D.J., Wyn-Jones, E., J. Chem. Soc., Chem. Commun., 115 (1990).
66. Wärnheim, T., Jönsson, A., J. Colloid Int. Sci., 125, 627 (1988).
67. Wärnheim, T., Bokström, J., Williams, Y., Colloid and Polymer Sci., 266, 562 (1988).
68. Jonströmer, M., Sjöberg, M., Wärnheim, T., J. Phys. Chem., in press.
69. Wärnheim, T., Sjöberg, M., J. Colloid Int. Sci., 131, 402 (1989).
70. Rong, G., Friberg, S.E., J. Disp. Sci. Tech., 9(4), 401 (1988).
71. Friberg, S.E., Rong, G., Langmuir, 4, 796 (1988).
72. Friberg, S.E., Blute, I., Stenius, P., J. Colloid Int. Sci., 127, 573 (1989).
73. Ward, A.J.I., Rong, G., Friberg, S.E., Colloid Polym. Sci., 267, 730 (1989).
74. Bleasdale, T.A., Tiddy, G.J.T., Unpublished observations.
75. Evans, D.F., Langmuir, 4, 3(1988).

THE PHASE BEHAVIOUR OF METAL(II) SOAPS IN ONE, TWO AND THREE COMPONENT SYSTEMS

H.D. BURROWS,
Departamento de Quimica,
Universidade de Coimbra,
3049 Coimbra,
Portugal.

ABSTRACT. A brief review is presented of the thermal behaviour of the long-chain carboxylates (soaps) of divalent metal ions, and some of the important factors influencing the phase behaviour are highlighted. A more detailed discussion is given of the behaviour of the lead(II) soaps, where a variety of physical techniques have been employed to characterize the phase structures. For the straight chain soaps, two liquid crystalline phases are observed between solid and isotropic liquid, whereas for longer chain homologues only one mesophase is observed. The effects of unsaturation and chain branching are discussed, and indications given of the major energetic contributions to the phase transitions. In some cases polymorphism is observed in the solid phases, and metastable phases produced on cooling may be stable for long times. The phase behaviour of divalent metal soaps in two and three component systems is examined, both in terms of the effect of additional components on phase structures, and of the solubility of soaps in a variety of polar and non-polar solvents.

1. Introduction

Although colloidal systems are normally considered to involve dispersion of one or more components in a bulk phase, certain single component systems, such as molten polymers, or fused phases of amphiphiles, share many of the characteristics of colloids, and a study of their structures and phase behaviour can be informative. One particular class will be considered, the divalent metal salts of long-chain carboxylic acids. These are commonly referred to as metal soaps, and have the general formula

$$(CH_3(CH_2)_nCO_2)_2M$$

where M is a divalent metal ion, and n is at least 2. The overall carbon chain length is (n + 2), and the most frequently observed metal soaps have even chain lengths, with between eight (octanoate) and eighteen (octadecanoate) carbon atoms.

These systems find extensive applications in lubricating greases, paint driers, polymer stabilisers, etc. (1), such that there is considerable technical interest in their phase behaviour. In addition, the use of studies on their phase structures in understanding behaviour in lipid-water systems has been highlighted (2). Some general aspects of their phase behaviour will now be considered.

D. M. Bloor and E. Wyn-Jones (eds.),
The Structure, Dynamics and Equilibrium Properties of Colloidal Systems, 415–426.

2. General Phase Behaviour

At room temperature, the pure divalent metal soaps are normally solid. Where certain commercial samples of soaps of straight chain acids are obtained as liquids, this is usually the result of impurities, such as excess free acid. In common with the corresponding hydrocarbons, and other lipid systems (3,4), the solid phases normally have the hydrocarbon chain in a fully extended all-trans conformation, which may be packed in monoclinic, triclinic, orthorhombic or hexagonal unit cells or subshells. Whilst it is generally difficult to get good enough single crystals for diffraction studies, this has been achieved for certain copper(II) carboxylates, such as the decanoate (5), and X-ray diffraction studies reveal packing of the hydrocarbon chains of copper(II) carboxylate dimers with a common triclinic subshell.

Table 1. Phase transition temperatures of some metal(II) octadecanoates [a]

Metal ion	Transition temperature/K	ref.
Mg(II)	solid (382) H_1 (468) H_2 (483) liquid	(1)
Ca(II)	solid (373) C_2 (396) D(T) (425) ... (452) H (>623) liquid	(1)
Sr(II)	solid (403) (443) Δ (R) (470) Q (519) H (>670) liquid	(1)
Ba(II)	solid (423) (493) Q (>673) liquid	(1)
Zn(II)	solid (403) liquid	(12)
Cd(II)	solid (372) (484) Q (>673) liquid	(1)[b]
Hg(II)	solid (372) (384) (389) liquid	(13)
Mn(II)	solid (361) (379) (392) liquid	(14)
Cu(II)	solid (387) (389) D (503) liquid	(15-17)
Pb(II)	solid (381) KM (387) liquid	(11,18)

[a] For a description of the phase nomenclature see ref. (19); [b] An alternative description is given in ref. (12).

On heating the metal soaps, one starts to observe one or more first-order phase transitions at temperatures around 100°C. These correspond to break-up of the intermolecular packing and onset of conformational disordering in the hydrocarbon chain. The solid may pass through various intermediate mesophases before forming an isotropic liquid, and eventually decomposing. Early knowledge of the phase behaviour of these systems came largely from studies by Lawrence (6), the Volds (7), Matsuura (8), Spegt and Skoulios (9), Luzzati (10), Adeosun and Sime (11). The actual phase sequence depends upon both the nature, or in particular the coordination behaviour, of the metal ion, and the structure of the organic moiety. With the group II metal ions, where the metal-carboxylate interactions are largely ionic, liquid crystalline phases are observed over a wide temperature range, and proposed structures include lamellae, discs, rods and ribbons in various one, two and three dimensional arrangements. The soaps of the other divalent metal ions tend to show a rather more limited range of intermediate phases. Examples of phases and transition temperatures for various octadecanoates (stearates) are shown in Table 1. A striking example of the effect of coordination on the phase behaviour is seen on going from the zinc(II) soaps, which normally melt directly to the isotropic liquid (12), to those of copper(II) (15) where discotic mesophases are observed over a wide temperature interval.

The effect of changes in the hydrocarbon chain length on phase behaviour shows up nicely with lead(II) carboxylates, where the short chain members exhibit two intermediate phases, whilst the longer chain ones show only one between solid and isotropic liquid (11,18,20,21). These are shown in Figure 1, using phase assignments from ref. (18). It is of interest that some odd-even variations are observed with the transition temperatures of the second transition. In common with other lipid systems, transition temperatures are expected to be lowered by chain branching or by the introduction of double bonds. Again this shows up well in studies on some lead(II)

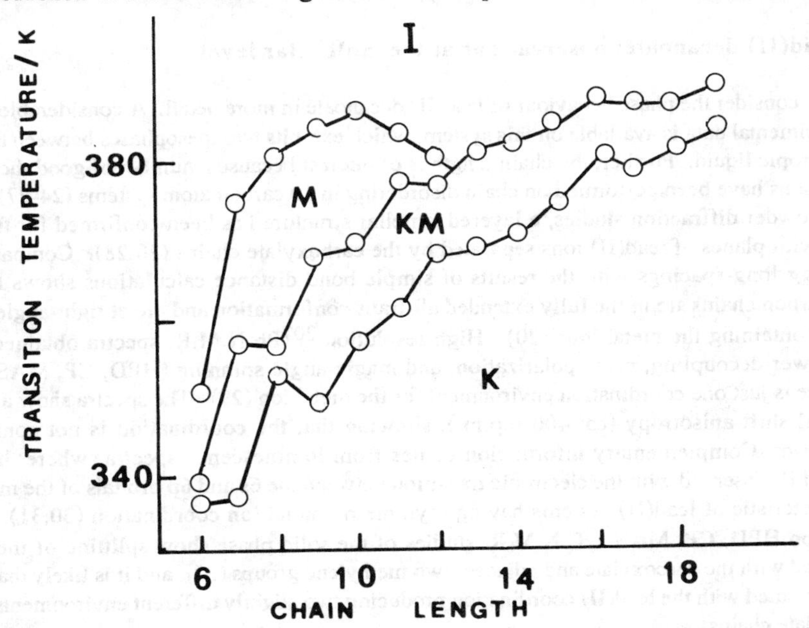

Figure 1. Phase transitions in lead(II) carboxylates as a function of chain length. Data from refs. 18 and 21. Phase description as in ref. 18.

carboxylates (Table 2). Also given in this Table are results for lead(II) dihydroxyoctadecanoate (22). The higher temperatures compared with the stearate may well reflect the importance of hydrogen bonding.

Table 2. Phase transition temperatures of some lead(II) carboxylates

Compound	Transition temperatures/K			ref.
lead(II) decanoate	357	369	387	(18)
lead(II) cyclohexyl-butyrate	336	360	378	unpublished data
lead(II) octadecanoate	381	387		(11)
lead(II) oleate	373			(23)
racemic lead(II) 9,10-dihydroxyoctadecanoate	405	412		(22)

3. Lead(II) decanoate: observations at the molecular level

We now consider the phase behaviour of lead(II) decanoate in more detail. A considerable amount of experimental data is available on this system, which exhibits two mesophases between the solid and isotropic liquid. Further, the chain length is of interest because a number of good theoretical calculations have been performed on chain disordering in 10 carbon atom systems (24-27). From X-ray powder diffraction studies, a layered lamellar structure has been confirmed for the solid phase, with planes of lead(II) ions separated by the carboxylate chains (20,28). Comparison of the Bragg long-spacings with the results of simple bond distance calculations shows that the hydrocarbon chains are in the fully extended all-trans conformation and are at right-angles to the planes containing the metal ions (20). High-resolution ^{207}Pb N.M.R. spectra obtained using high-power-decoupling, cross-polarization, and magic-angle-spinning (HPD, CP, MAS) show that there is just one coordination environment for the metal ion (29). The spectra show a modest chemical shift anisotropy (ca. 400 p.p.m.), showing that the coordination is not completely symmetric. Complementary information comes from luminescence spectra, where the large Stokes shift observed with the electronic transition between the 6s and 6p orbitals of the metal ion is characteristic of lead(II) systems having asymmetric metal ion coordination (30,31). High-resolution HPD, CP, MAS ^{13}C N.M.R. studies of the solid phase show splitting of the peaks associated with the carboxylate and adjacent two methylene groups (32), and it is likely that this is also associated with the lead(II) coordination producing two slightly different environments for the carboxylate chains.

From the ^{1}H N.M.R. spectra and the transverse and longitudinal relaxations of the proton signal, there are indications of the existence of a relatively amorphous region in the solid phase (32). The percentage of this relatively mobile fraction increases with temperature. It is worthy of note that the amount of this component at room temperature as seen by proton N.M.R. (ca. 20%) is virtually identical to the percentage of mobile methyl groups observed by infrared spectroscopy in studies on fatty acids (33). It is quite likely that the presence of some mobility in the hydrocarbon chains is a common feature in the solid phases of many such lipid systems.

At around 358 K, a new phase is formed which is optically anisotropic when observed upon a polarising microscope, and which can be sheared, although with some difficulty. From X-ray diffraction and dilatometry studies this appears to have a structure intermediate between a solid and a disordered smectic liquid crystal (18). Both these results and proton N.M.R. measurements (32) suggest that this new phase has an overall three dimensional lamellar structure based on ionic layers, but that within the hydrocarbon chain region there is partial disorder, and certain mobility of the chains. From ^{207}Pb N.M.R. (28) and a.c. impedance measurements (34) the metal ions are seen to have fairly restricted movement. Further information on this phase transition comes from Raman spectral studies (28). Studies in the C-C stretch region, where introduction of gauche bonds leads to changes in band intensities, indicate that there is relatively little conformational disordering accompanying this phase transition. In contrast, large changes in band intensities are observed in the C-H stretch region, where effects due to intermolecular packing are dominant. Thus the transition appears to involve increased lateral disorder, but only minor conformational disordering. This phase has many of the characteristics of the gel phases observed in phospholipid and other bilayers. Differential scanning calorimetry studies show differences in the phase transition with recrystallised and premelted samples, which will be discussed in Section 4.

A second optically anisotropic phase is observed above ~ 369 K. From the textures observed using a polarising microscope this has been assigned a lamellar L_α (smectic A) structure, in which the chains are disordered and there is relatively high lateral mobility (28). Although alternative assignments have been given to the phase structure (11,20), results from application of a variety of techniques (18,28,32) all support our assignment. From Raman spectra in the C-C stretch region (28), X-ray diffraction, dilatometry (18) and ^{13}C N.M.R. spectra (32) there is clear evidence for conformational disordering. Results from the spectral studies suggest an average of about 2 gauche bonds per chain (28,32). This is slightly higher than, but consistent with the results of molecular dynamics (25) and other (26) calculations. Evidence for relatively high mobility of the metal ion in this phase comes from ^{207}Pb N.M.R. spectra (28) and a.c. impedance measurements (34). Further, ^{207}Pb chemical shifts in the n.m.r. spectra are very sensitive to coordination environment, and the fact that there are only very small changes on going from the solid through the first mesophase to the L_α phase (28) shows that changes in coordination number do not seem to be involved. The minor discontinuities observed at the phase transitions are most probably associated with changes in paramagnetic shielding. From extrapolation of enthalpy data for phase transitions of a series of lead(II) soaps to the case for zero methylene groups there are indications that the lead(II)-carboxylate interactions may be rather stronger in the L_α phase than in the solid or first mesophase (18). It is likely that the increased fluidity of the chain region permits the ionic parts to maximize their interactions.

Lead(II) decanoate forms an isotropic liquid at 386-7 K (11,20,28). This phase has a relatively high viscosity (35), and there is evidence for lead(II) decanoate being present as small aggregates of about 4 molecules (11,35). Although a spherical micellar structure has been proposed for these (11,35), this seems unlikely both on steric grounds, and on the basis of the small entropy and enthalpy changes accompanying the $L_\alpha \rightarrow$ liquid transitions. A possible alternative structure (28) involves small discs, such as those suggested for the L_2 phase in monoglyceride-water systems (36). This transition shows up in small changes in chemical shift in the ^{207}Pb N.M.R. spectrum (28,29). However, there are only very minor modifications in the Raman (28) and ^{13}C N.M.R. spectra (32), suggesting that the extent of conformational disordering is similar in the liquid and L_α phases. Further, no significant discontinuities are observed in transverse or global proton relaxation in N.M.R. in this temperature region (32), suggesting that the local chain dynamics as seen by this technique are fairly similar. One final point on lead(II) decanoate, which is generally true for other lipid compounds over a similar temperature range, is that the dynamics of the terminal methyl group are very different from those of the other carbon atoms in the chain, and studies using ^{13}C N.M.R. show that this is involved in very rapid motion (probably rotation about the C-C bond) even in the solid phase (32).

4. Some thermodynamic and kinetic considerations

The phase transitions reported in Table 1 are accompanied by energy changes, which on heating the soaps are normally endothermic. These may conveniently be studied by differential scanning calorimetry (D.S.C.) or differential thermal analysis (D.T.A.). Whilst comparable results are obtained with the two techniques, the former is generally preferred as it provides direct measures of enthalpy changes. For the same chain length, the overall enthalpy changes for different soaps on going from solid to liquid often have similar magnitudes, and are comparable to those of the corresponding hydrocarbon (considered as that which has the same number of methyl or methylene groups as two carboxylate chains). As an example consider the overall enthalpies of fusion of the octadecanoates of Pb(II) (119.5 kJ mol^{-1}) (11)), Zn(II) (103 kJ mol^{-1} (12)) and the straight chain hydrocarbon having 34 carbon atoms (128 kJ mol^{-1} for the overall triclinic → liquid transition (3)). Where the reported values are considerably lower than these, as in the cadmium(II) carboxylates (12) it is possible that the overall transition observed is not to the isotropic liquid, but to some intermediate phase. For a series of soaps of the same metal ion and parity (i.e. even or odd number of carbon atoms) the overall enthalpy of fusion increases linearly with chain length. Taken together, these results agree with the observations at the molecular level that the dominant process in the overall solid → liquid phase transition in the metal(II) soaps involves melting of the aliphatic chains. Reasonable theoretical estimates can be made of the energetics of chain melting by considering the effects of changes in van der Waals interactions, chain conformation, etc (37,38). For the lead(II) carboxylates, we have applied the method developed by Nagle (38) and separated the enthalpy as

$$\Delta H\,(\text{solid} \rightarrow \text{liquid}) = \Delta U_{conf} + \Delta U_{vdW} + \Delta U_o + P\,(\Delta V)$$

where ΔU_{conf} and ΔU_{vdW} are the changes in energy associated with conformational disordering and van der Waals interactions, and ΔU_o incorporates all other energy terms. The P (ΔV) term can be ignored as it is very small compared with the other contributions. Using reasonable estimates for ΔU_{conf} and ΔU_{vdW} we have calculated the sum of these two terms for thecarboxylates having between eight and eighteen carbon atoms (18). Comparison of these calculations with experimental data for the overall solid → liquid phase transition in the even chain lead(II) carboxylates (Fig. 2) confirms the dominant role of chain melting. Changes in electrostatic interactions are much less important energetically.

Similar methods can be used for breaking down the entropy change for the overall solid → liquid transformation, although this decomposition is rather more suspect as the various contributions are interdependent (39). Using the treatment of Nagle (39) the change in entropy on fusion of lead(II) decanoate by considering conformational, volume expansion, excluded volume and other effects has been calculated as 177.8 J K^{-1} mol^{-1}, which is in reasonable agreement with the experimental value for the total change (solid → liquid) of 171.2 J K^{-1} mol^{-1} (18). A possibly more rigours method of treating the entropy change theoretically involves the use of molecular dynamics. Berendsen and co-workers (24,25) have treated the case of the gel → L_α transition in model 10-carbon atom bilayers with this technique. This transition is very similar to the mesophase → mesophase transition in the short chain lead(II) carboxylates. Combining the results of the molecular dynamics calculations with the increase in gauche population given by

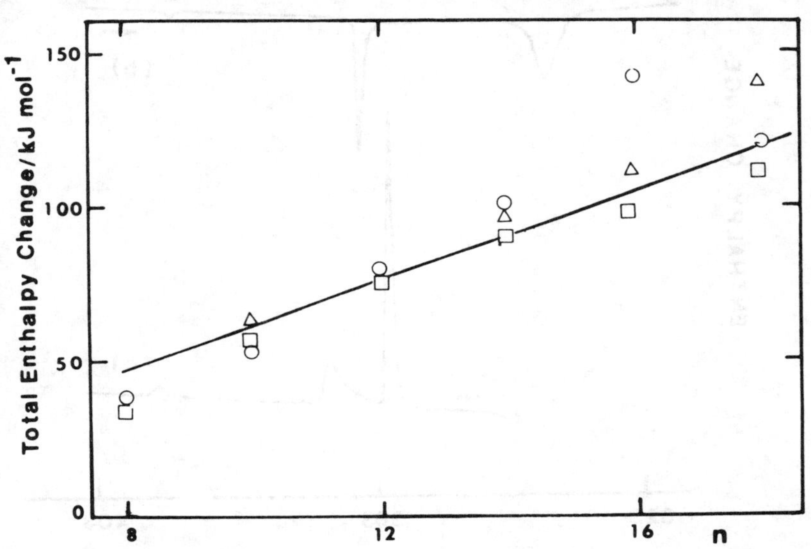

Figure 2. Plot of experimental (circles, squares and triangles) and theoretical (line) overall enthalpy change for solid → liquid phase transition for even chain lead(II) carboxylates. Reproduced from Liquid Crystals, 3, 1655, (1988) with the permission of Taylor and Francis Ltd.

Raman spectra gives an entropy change for the mesophase I → II transition in the decanoate of 54.6 J K^{-1} mol^{-1}, which is virtually identical to that observed by D.S.C. (54.4 J K^{-1} mol^{-1} (28)).

As noted earlier, differences in phase behaviour are sometimes observed between recrystallised and premelted samples. This shows up clearly with the lead(II) decanoate. With a freshly recrystallised sample, two very close transitions are observed by D.S.C. at 357-361 K (Figure 3). In our earlier study on this system (28) we were only able to see a fairly broad transition here. Further transitions are observed at 369 and 387 K corresponding to formation of the L_α and liquid phases. On cooling, only three of these phase transitions are observed, with thelowest temperature one showing considerable supercooling. On reheating, again only three transitions are observed, with the first having apparently collapsed into the second. However the enthalpy change for this transition on reheating is about 10-15% less than the sum of the enthalpy changes in this temperature region with the freshly prepared sample (28). This suggests that the solid phase produced on cooling is slightly different from that obtained by recrystallisaton.

Vibrational spectral studies confirm this (18), and suggests that the two solid phases both have the carboxylate chains in their fully extended all-trans conformation, but differ in their crystal packing. Over a period of several months, the polymorphic form produced on cooling does appear to convert back to the original form (18).

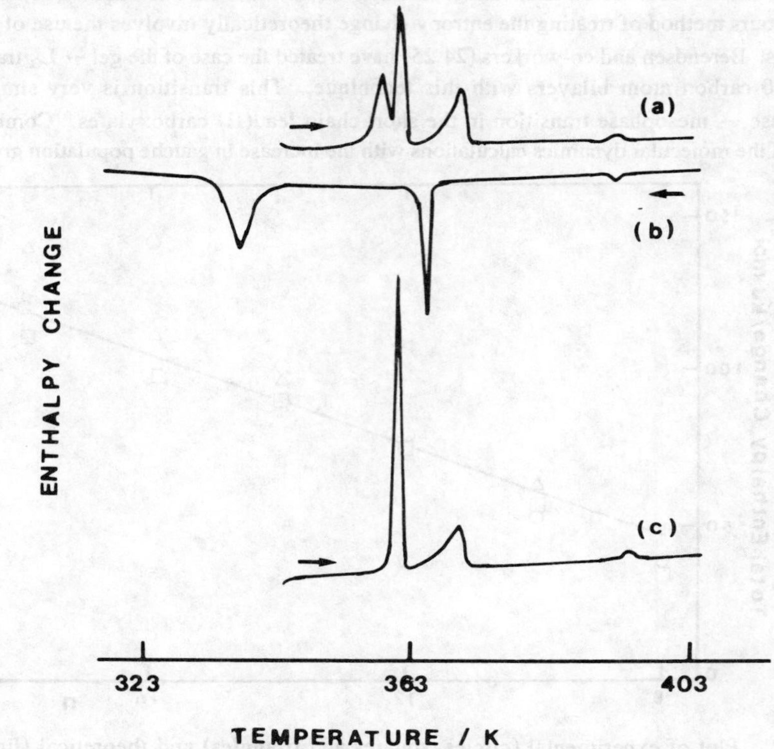

Figure 3. D.S.C. curves for lead(II) decanoate: (a) initial heating; (b) cooling; (c) reheating of same sample. From T.J.T. Pinheiro, M.Sc. Thesis, Universidade de Coimbra (1989).

Preliminary kinetic studies have been started on the other phase transitions in the lead(II) carboxylates (40). For these, isothermal scanning calorimetry has been used. One starts in this case on one side of a phase change, applies a heating or cooling ramp of a few degrees, and then measures the heat flow into or out of the system. Whilst the time resolution available is less than that observed in, for example, temperature jump studies, the information obtained is complementary, and does allow determination of the time scales involved. Results for the liquid → L_α transition in lead(II) decanoate show that it is complete in a few seconds, and that the kinetic curve has a sigmoidal shape. This is similar to what is observed for the corresponding transition in certain lyotropic liquid crystal systems (41), and implies a similar nucleation and growth mechanism.

5. Phase behaviour in two and three component systems

A number of studies have been reported on the effect of addition of a second component on the phase behaviour of metal(II) soaps. Representative examples are given in refs. 42-48. With mixtures of carboxylate of the same chain length, but with different metal ions, homogeneous mixing over the complete composition range without the formation of any new phase is reported for the octadecanoates of Pb(II)/Mn(II), Cd(II)/Zn(II), Cd(II)/Mn(II), Cd(II)/Hg(II) (42), and Ca(II)/Sr(II) (43). In other cases, either new phases were observed, or the metal carboxylates did not appear to be completely miscible (42,44). With mixtures of lead(II) decanoate and octadecanoate (45) small amounts of octadecanoate can be incorporated into the shorter chain soap without affecting the phase sequence, but above about 0.1 mole fraction one (or more) new phases start to be observed. One problem with studies on soaps of different chain lengths is that chain scrambling occurs, such that as well as dealing with mixtures of the two chain lengths, considerable amounts of metal(II) species having one chain of each length are also present (49).

The effects of PbO (46), n-undecane (46), decanoic acid (47), and lead(II) acetate (48) on the phase transitions of lead(II) carboxylates have been reported and attempts made to rationalise the observed behaviour using Winsor's R theory (19). All of these compounds show some solubility or miscibility with the liquid phase of the soap. Lead(II) oxide, undecane and decanoic acid decrease the stability of the L_α phase of the short chain carboxylates, which disappears at between 0.005 and 0.02 mole fraction of the second component. In some cases, new, and as yet uncharacterized phases are observed. The R theory considers the balance between the intermolecular forces making lamellar amphiphile structures becoming convex relative to their polar or hydrocarbon chain regions (19). The qualitative ideas appear to apply fairly well to these examples. The data reported for lead(II) dodecanoate/lead(II) acetate mixtures indicate rather different behaviour, with the lead(II) acetate apparently stabilizing the L_α phase (48).

We now consider the behaviour of low concentrations of the metal(II) soaps (less than ~ 0.01 mole fraction). At room temperature, the long chain carboxylates of divalent metal ions are virtually insoluble in all common solvents. Above a certain critical temperature, however, the solubility increases dramatically. This behaviour, which has been reviewed extensively elsewhere (50), is reminiscent of what is observed with surfactants in aqueous solution, where a big increase in solubility is observed at the Krafft point (51). The explanations in the two cases are, however, probably rather different. The Krafft point is normally associated with the formation of micelles, such that when the monomer solubility is equal to the critical micelle concentration, a dramatic increase is observed as further surfactant molecules dissolve to form micelles. With the solutions of metal(II) soaps in non-aqueous solvents, although there is good evidence for the existence of aggregates (50), these tend to be small and rather ill-defined. Different models have been presented for the solution formation in these systems (50,52,53) but more data is needed on the structure of aggregates present before any valid conclusions can be drawn. The observed solution temperatures depend upon both the metal ion and the chain length of the soap (54). For carboxylates of lead(II) (53,54), cadmium(II) and mercury(II) (53) in a series of alcohols, the solubility temperature also increases linearly with the alcohol chain length. With branched chain alcohols, higher solution temperatures are observed than with the straight chain ones. Such effects suggest significant contributions from hydrophobic interactions to the solubility.

Unsaturation or chain branching decreases the solubility temperatures of the divalent metal soaps in organic solvents, as may be anticipated from their lower melting points. In some cases these compounds may be reasonably soluble at room temperature. Indeed, differences in the solubility of their lead(II) salts have been used to separate saturated and unsaturated fatty acids (55).

The phase behaviour of a few three component systems involving metal(II) soaps has been studied. The most important cases, because of their practical applications, involve the system soap/organic solvent/water. Solutions of the soaps in organic solvents will frequently dissolve small quantities of water before phase separation occurs. For example, solutions of calcium(II), magnesium(II) or barium(II) phenylstearate in benzene will dissolve water up to a water/soap mole ratio of 1.5-2 before precipitation occurs (56). However, attempts to identify the nature of the separated phase in these cases have not been successful. One complication which may occur on addition of water to metal soap solutions is that hydrolysis of the carboxylate may lead to formation of basic soaps, having the general formula M(OH) (O_2CR). As with the well known case of the aluminium(III) soaps (57), these hydrolysed species may form large polymeric aggregates in solution. Studies have been reported, for example, on the aggregates formed by basic cobalt(II) oleate in nonpolar solvents, which may have weight average molecular weights up to 3.4×10^6 (58).

Another group of three component systems which have been studied involves metal soaps in organic solvents in the presence of organic bases, such as pyridine. The base tends to increase the solubility of the soap, and at the same time decrease the size of any aggregates present (59). Coordination between the base and metal ion is likely to be an important factor.

6. Conclusion

Studies on the phase behaviour of long chain carboxylates of divalent metal ions can provide information which is relevant to the understanding of aggregation phenomena in other amphiphile systems. One or more phase transitions are observed at relatively low temperatures. These involve predominantly the fusion of the hydrocarbon chain region. Such chain fusion involves both increased conformational disorder and decreased van der Waals interactions. The balance between interactions in the lipid and carboxylate regions leads to various structures for intermediate phases. Although the solid phases are characterized by three dimensional ordering, and a rigid metal-carboxylate region, the methyl groups still have considerable mobility. This may permit sliding of hydrocarbon planes, and provide an explanation for the excellent lubricant properties often observed with these compounds (60,61). This may have implications in other amphiphile systems, such as phospholipid bilayers.

7. Acknowledgements

The excellent experimental collaboration and stimulating discussions with my colleagues and co-workers in Coimbra, Braga, Lisbon, Strasbourg, Port Sunlight/Salford, Durham, Southampton, and Ile-Ife are gratefully acknowledged. Thanks are also due to I.N.I.C. for financial support.

References

1. F.J. Buono and M.L. Feldman, Kirk-Othmer Encyclopeadia of Chemical Technology, vol. 8, H.F. Mark, D.F. Othmer, C.G. Overberger and G.T. Seaborg, eds., Wiley, New York, 3rd edn., 1979, p. 34.
2. V. Luzzati in Biological Membranes, vol. 1, D. Chapman, ed., Academic, London, 1968, chapter 3.

3. M.G. Broadhurst, *J. Res. Natn. Bur. Stand. A*, **66**, 241 (1962).
4. R.W.G. Wycoff, Crystal Structures, vol. 5, *Interscience*, New York, 2nd edn., 1966.
5. T.R. Lomer and K. Perera, *Acta Cryst*, **B30**, 2912 (1974).
6. A.S.C. Lawrence, *Trans. Faraday Soc.*, **34**, 660 (1938).
7. G.S. Hattiangdi, M.J. Vold and R.D. Vold, *Ind. Eng. Chem.*, **41**, 2320 (1949).
8. Y. Koga and R. Matsuura, *Mem. Fac. Sci.*, Kyushu Univ., Ser. C4, 1 (1961).
9. P.A. Spegt and A. Skoulios, *Acta Cryst*, **17**, 198 (1964).
10. V. Luzzati, A. Tardieu and T. Gulik-Krzywicki, *Nature*, Lond., **217**, 1028 (1968).
11. S.O. Adeosun and S.J. Sime, *Thermochim Acta*, **17**, 351 (1976).
12. I. Konkoly-Thege, I. Ruff, S.O. Adeosun and S.J. Sime, *Thermochim. Acta*, **24**, 89 (1978).
13. H.A. Ellis, *Mol. Cryst. Liq. Cryst*, **138**, 321 (1986).
14. S.O. Adeosun, *Can. J. Chem.*, **57**, 151 (1979).
15. H. Abied, D. Guillon, A. Skoulios, P. Weber, A.M. Giroud-Godquin and J.C. Marchon, Liquid Crystals, **2**, 269 (1987).
16. M. Takekoshi, N. Watanabe and B. Tamamushi, *Colloid Polym. Sci.*, **256**, 588 (1978).
17. H.D. Burrows and H.A. Ellis, *Thermochim. Acta*, **52**, 121 (1982).
18. C.G. Bazuin, D. Guillon, A. Skoulios, A.M. Amorim da Costa, H.D. Burrows, C.F.G.C. Geraldes, J.J.C. Teixeira-Dias, E. Blackmore and G.J.T. Tiddy, Liquid Crystals, **3**, 1655 (1988).
19. P.A. Winsor, Liquid Crystals and Plastic Crystals, vol. 1, G.W. Gray and P.A. Winsor, eds., Ellis Horwood, Chichester, 1974, chapter 5.
20. H.A. Ellis, *Mol. Cryst. Liq. Cryst*, **139**, 281 (1986).
21. S.O. Adeosun and S.J. Sime, *Thermochim. Acta*, **27**, 319 (1978).
22. M.S. Akanni, *Thermochim. Acta*, **122**, 355 (1987).
23. S.O. Adeosun, A.O. Kehinde and G.A. Adesola, *Thermochim. Acta*, **28**, 133 (1979).
24. O. Edholm, H.J.C. Berendsen and P. van der Ploeg, *Mol. Phys.*, **48**, 379 (1983).
25. P. van der Pleog and H.J.C. Berendsen, *Mol. Phys.*, **49**, 233 (1983).
26. D.W.R. Gruen, *J. Phys. Chem.*, **89**, 146 (1985).
27. S. Marcelja, *J. Chem. Phys.*, **60**, 3599 (1974); *Biochem. Biophys. Acta*, **367**, 165 (1974).
28. A.M. Amorim da Costa, H.D. Burrows, C.F.G.C. Geraldes, C.G. Bazuin, D. Guillon, A. Skoulios, E. Blackmore, G.J.T. Tiddy and D.L. Turner, *Liquid Crystals*, **1**, 215 (1986).
29. H.D. Burrows, C.F.G.C. Geraldes, T.J.T. Pinheiro, R.K. Harris and A. Sebald, *Liquid Crystals*, **3**, 853 (1988).
30. H.D. Burrows, *Materials Lett.*, **6**, 191 (1988).
31. G. Blasse, *Chemistry of Materials*, **1**, 294 (1989).
32. G. Feio, H.D. Burrows, C.F.G.C. Geraldes and T.J.T. Pinheiro, manuscript in preparation.
33. G. Zerbi, G. Conti, G. Minoni, S. Pison and A. Bigotto, *J. Phys. Chem.*, **91**, 2386 (1987).
34. S.O. Adeosun and S.J. Sime, *J. Chem. Soc., Faraday Trans. 1*, **75**, 953 (1979).
35. U.J. Ekpe and S.J. Sime, *J. Chem. Soc., Faraday Trans. 1*, **72**, 1144 (1976).
36. K. Larsson, *J. Colloid Interface Sci.*, **72**, 152 (1979).
37. P.J. Flory, Statistical Mechanics of Chain Molecules, Carl Hanser Verlag, 1989.
38. J.F. Nagle, *Ann. Rev. Phys. Chem*, **31**, 157 (1980).
39. J.F. Nagle and M. Goldstein, *Macromolecules*, **18**, 2643 (1985).
40. H.D. Burrows, J.J.C. Cruz-Pinto, C.F.G.C. Geraldes, M.G.M. Miguel and T.J.T. Pinheiro, to be published.
41. P. Knight, E. Wyn-Jones and G.J.T. Tiddy, *J. Phys. Chem.*, **89**, 3447 (1985).
42. S.O. Adeosun and H.A. Ellis, *Thermochim. Acta*, **28**, 313 (1979).

43. P.A. Spegt and A. Skoulios, *J. Chim. Phys.*, **62**, 377 (1965).
44. P.A. Spegt and A. Skoulios, *Acta Cryst.*, **21**, 892 (1966).
45. H.A. Ellis, *Thermochim. Acta*, 130, 281 (1988).
46. S.O. Adeosun, W.J. Sime and S.J. Sime, *Thermochim. Acta*, **19**, 275 (1977).
47. S.O. Adeosun and M.S. Akanni, *Thermochim. Acta*, **27**, 133 (1978).
48. S.O. Adeosun, *Thermochim. Acta*, **25**, 333 (1978).
49. M.S. Akanni, E. Blackmore, H.D. Burrows and G.J.T. Tiddy, unpublished observations.
50. See, for example, N. Pilpel, *Chem. Rev.*, **63**, 221 (1963).
51. Definitions, terminology and symbols in colloid and surface chemistry, D.H. Everett, ed., *Pure Appl. Chem.*, 31, 613 (1972).
52. E.P. Martin and R.C. Pink, *J. Chem. Soc.*, 1750 (1948).
53. M.S. Akanni and N.A. Abass, *Liquid Crystals*, 6, 597 (1989).
54. M.S. Akanni, H.D. Burrows, H.A. Ellis, D.N. Asongwed, H.B. Babalola and P.O. Ojo, *J. Chem. Tech. Biotechnol., 34A, 127* (1984).
55. F. El Said Mohammed and M.M. Amer, Oils, Fats, Waxes and Surfactants, Anglo Egyptian, Cairo, 1965.
56. J.G. Honig and C.R. Singleterry, *J. Phys. Chem.*, **60**, 1108, 1114 (1956).
57. See, for example, A.E. Leger, R.L. Haines, C.E. Hubley, J.C. Hyde and H. Sheffer, *Can. J. Chem.*, 35, 799 (1957), and references therein.
58. Z. Zhou, Y. Georgalis, W. Liang, J. Li, R. Xu and B. Chu, *J. Colloid Interface Sci*, **116**, 473 (1987).
59. V.D. Tughan and R.C. Pink, *J. Chem. Soc.*, 1804 (1951).
60. N. Pilpel, *Adv. Colloid Interface Sci.*, 2, 261 (1969).
61. V. Novotny, J.D. Swalen and J.P. Rabe, Langmuir 5, 485 (1989).

RINGING GELS: THEIR STRUCTURE AND MACROSCOPIC PROPERTIES

M. GRADZIELSKI and H. HOFFMANN
Lehrstuhl für Physikalische Chemie I der Universität Bayreuth
Postfach 101251, D-8580 Bayreuth, Federal Republic of Germany

ABSTRACT : **Dimethyltetradecylaminoxide is a zwitterionic surfactant with a behavior in between that of nonionic and cationic surfactants. Solubilisation of hydrocarbon into the surfactant system leads to a rod-sphere transition. At surfactant concentrations of 35% and hydrocarbon concentrations of 5-10% a solid, transparent, optically isotropic single phase is formed. The position of this phase in the phase diagram is related to the interfacial tension. SANS, rheological, electric birefringence and light scattering experiments both in the isotropic solution and the gel phase were carried out to determine the structure of these phases. We conclude that globular micelles with solubilized hydrocarbon are present. For a given surfactant to hydrocarbon ratio these micelles retain a constant size independent of concentration. Scattering curves can be explained by a hard sphere model and this model also explains mechanical moduli via the osmotic compressibility.**

1. Introduction

The occurrence of single phase regions with gel-like properties in the phase diagram of binary and ternary surfactant systems is a quite common phenomenon [1] . Usually they occur in the range of high surfactant concentrations, but there are also well-known examples for gels with only very little surfactant content [2,3]. In the following we would like to concentrate on gels in ternary systems. They consist of three components: surfactant, hydrocarbon, water, and sometimes a cosurfactant can also be present. This composition is basically identical to that of microemulsions and for that reason these gels are also sometimes referred to as microemulsion gels [4,5] . Other names that were used in the literature are: 'viscous isotropic phase' [6,7] , 'transparent emulsions' [8], 'viscoelastic gel stage' [9] , and 'optically isotropic mesophase' [10] . These gels are thermodynamically stable, optically isotropic, transparent, highly viscous and of high elasticity. Some examples for systems where such an isotropic gel phase has been reported are:

- **Oleylpolyglycolether/Paraffin/Water [11]**
- **Dimethyltetradecylaminoxide/Decane/Water [12]**
- **AOT/Octanol/Water [13]**
- **Didodecyldimethylammonium Bromide/Octane/Water [14]**

D. M. Bloor and E. Wyn-Jones (eds.),
The Structure, Dynamics and Equilibrium Properties of Colloidal Systems, 427–448.

The gels that we are concerned with have common rheological features, are of a stiff nature and usually exhibit a relatively high yield stress. This yield stress is normally high enough to prevent a sample from flowing under its own weight. Furthermore they have elastic properties and this elasticity is also responsible for the name ringing gels. If samples of such gels are tapped with a soft object they will become mechanically excited and give off a ringing, metallic sound [6]. The mechanical damping of elastic waves in such systems must be fairly low and permits thousands of oscillations before the elastic energy becomes dissipated into frictional energy.

Here at this point we would also like to introduce the term cubic phase, which is more instructive in that sense that it already relates to the structure of the systems in question. One has to notice that the terms cubic phase and ringing gel cannot necessarily be used synonymous, because not all of the cubic phases show the typical resonance effect of the ringing gels and at the same time there are examples known for ringing gels that do not belong structurally to the class of cubic phases.

The question of the detailed structure of such isotropic gels is still a matter of current research. Originally they were proposed to be made up by a concentrated, random suspension of individual cylindrical micelles [6]. But for similar systems it was shown by SAXS studies that they are liquid crystalline cubic phases and probably composed of spherical aggregates [15,16]. Newer studies have lead to doubts again about the actual structure of the cubic phases [17,18,19,20] and currently one can distinguish two principally different models. In the first model the cubic phase is composed of individual particles and in the second one assumes a bicontinuous structure. The first model can again become subdivided into two classes, one where these particles are globular and form a densely packed array of spheres [21,22,23], similar to the well-known structures of metals, and a second where these individual particles are anisometric and the cubic phase is formed by a certain array of these anisometric aggregates [18]. At this time there is evidence for the existence of all of the above mentioned types of structures and there may well be the possibility that different systems with different locations in the phase diagram may accordingly possess fundamentally different structures.

2. Experimental

In our own investigations we concentrated on ternary systems, consisting of surfactant, hydrocarbon and water. The gel phase in these ternary systems always borders on the isotropic micellar solution L_1. As surfactant we used dimethyltetradecylaminoxide (C_{14}DMAO) or mixtures of it with dimethyltetradecylphosphinoxide (C_{14}DMPO). As hydrocarbons cyclohexane, octane, nonane, decane, dodecane and tetradecane were used.

2.1 PHASE BEHAVIOUR

C_{14}DMAO is a zwitterionic surfactant that exhibits a behaviour that is in between that of a typical nonionic and that of a typical cationic surfactant. Contrary to normal nonionic surfactants no cloud point is observed in this surfactant system. Its binary phase diagram has been reported before [24]. At room temperature it forms small rodlike micelles in the isotropic solution. Solubilisation of hydrocarbon into the micellar solution leads to a rod-sphere transition [22]. On the other hand the C_{14}DMPO, where the phase diagram has been reported by Pospischil [25], behaves like a typical nonionic surfactant and shows a cloud point at 15°C.

If one solubilizes hydrocarbon into an isotropic or liquid crystalline surfactant system which contains about 30-40% of surfactant one at first observes that the highly viscous solutions turn to low viscous solutions (because of the rod-sphere transition) and with a further increase of the hydrocarbon content these systems solidify and form a transparent, optically isotropic single phase. The ternary phase diagram for the system $C_{14}DMAO$/decane/water is given in fig. 1. Of special interest to us is the isotropic gel phase G and the isotropic L_1-phase, which is bordering on the gel phase in the phase diagram. In fig. 1 one finds that by extending the phase boundary between isotropic L_1-phase and two-phase

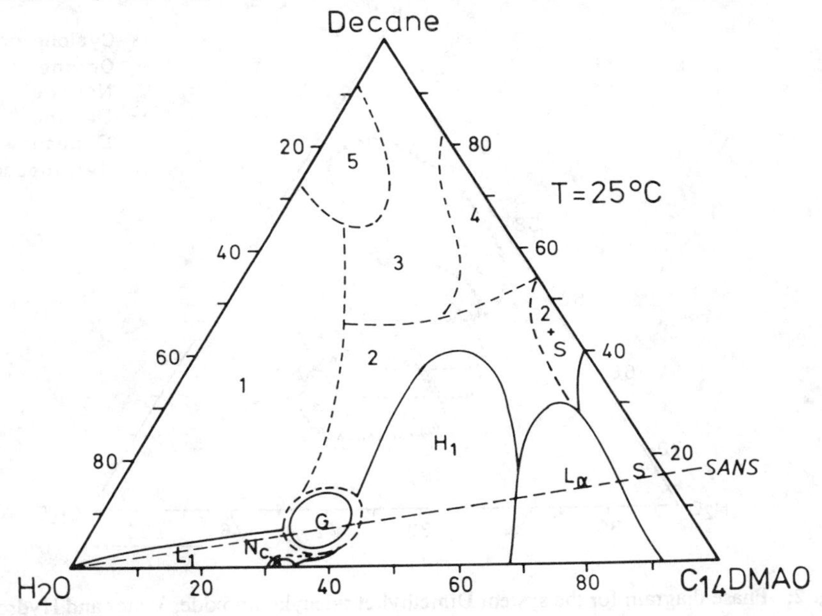

Fig. 1: Phase diagram for the system Dimethyltetradecylaminoxide, Decane, Water. One-phase regions : L_1: isotropic, water continuous phase; N_C: nematic phase; G: transparent, isotropic gel phase ("ringing gel"); H_1: hexagonal phase; L_α: lamellar phase. Multi-phase regions : S: crystalline; 1: liquid emulsion; 2: solid emulsion; 3: gel phase, saturated with Decane; 4: solid emulsion, saturated with Decane; 5: gel phase, rich in hydrocarbon.

region, that means the solubilisation capacity of the system, one arrives right in the center of the phase region of the ringing gel. This is a very general criterion for finding the gel phase in the phase diagram of such systems. In that way the line of the solubilisation capacity defines a search line for the ringing gel.

We have also investigated the phase behavior of such systems for the different hydrocarbons:

cyclohexane, octane, nonane, decane, dodecane and tetradecane. The location of the isotropic gel phase in the phase diagram is indicated in fig. 2 [12]. One finds that the longer the hydrocarbon chain the lower the hydrocarbon content of the gel phase region. This is consistent with other findings [26] that the solubilisation capacity increases for shorter hydrocarbon chains and as just stated one usually finds the isotropic gel phase in the extension of the line of the solubilisation capacity. The total amount of surfactant plus hydrocarbon in the gel phase remains always in the range of 38-48%.

After having investigated the phase behaviour as a function of the solubilised hydrocarbon we

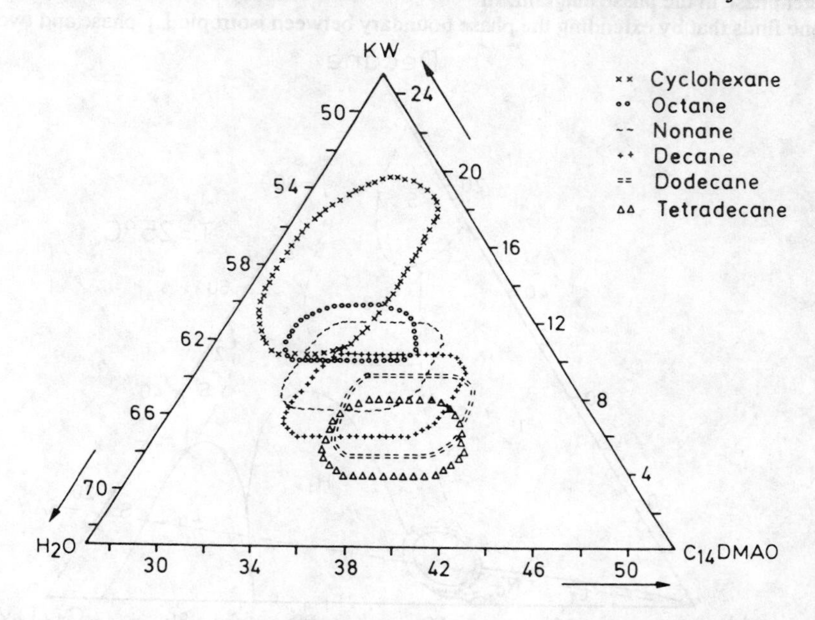

Fig. 2: Phase diagram for the system Dimethyltetradecylaminoxide, Water and Hydrocarbon, for the Hydrocarbons: Cyclohexane, Octane, Nonane, Decane, Dodecane and Tetradecane. Only the isotropic ringing gel phase is indicated.

also looked at variations in the surfactant. As a variation of the properties of the surfactant we investigated mixtures of C_{14}DMAO and C_{14}DMPO. The addition of the more hydrophobic phosphinoxide gives rise to a decrease in the interfacial tension, as can be seen in fig. 3, where we have plotted the interfacial tension γ as a function of the surfactant composition, as measured against decane [12]. All these values were determined with the spinning drop method [27]. As is known from studies on microemulsions such a decrease of the interfacial tension should correspond to an increase of the solubilisation capacity of the system, which again should lead to the formation of larger spherical aggregates.

In fig. 4 we present light scattering data [22] and have plotted the Raleigh factor $R(\Theta)$ for different mixtures of C14DMAO and C14DMPO as a capacity increases with increasing phosphin-

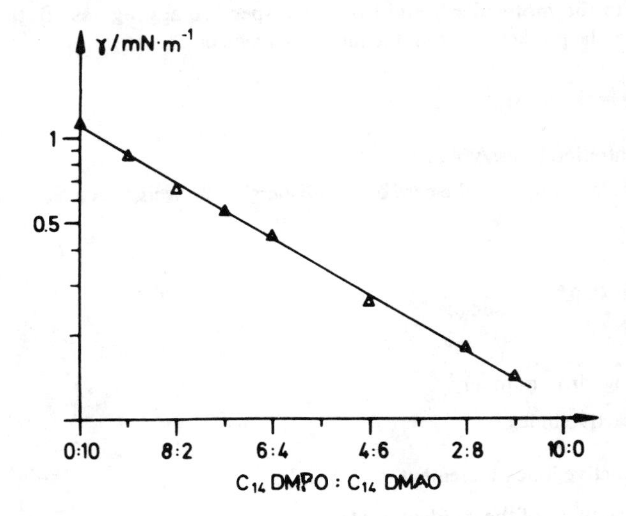

Fig. 3: Interfacial tension for mixtures of Dimethyltetradecylaminoxide and Dimethyltetradecylphosphinoxide. The overall surfactant concentration was always kept 10mM.

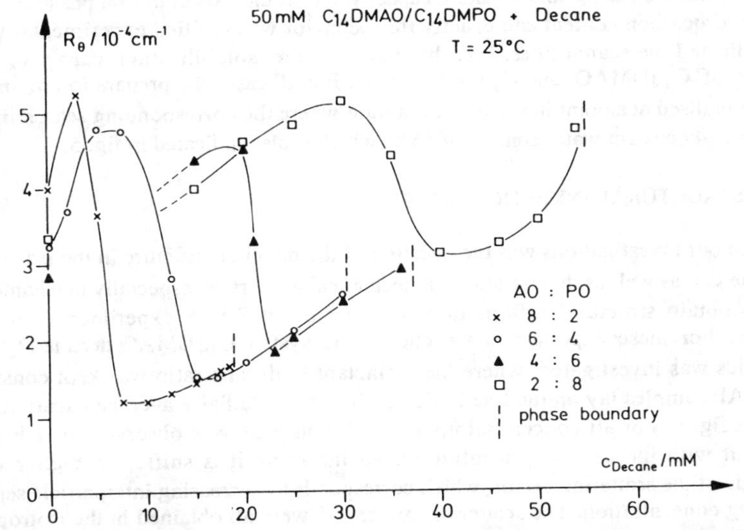

Fig. 4 : Rayleigh factor for surfactant solutions of mixtures of C14DMAO and C14DMPO as a function of the amount of solubilized Decane. The total surfactant concentration in all cases was 200 mM.

oxide content and at the same time we find an increase in the scattering intensity which is proportional to the molecular weight of the respective aggregates. Both these results are in agreement with the predictions from the interfacial tension data.

$$R(\Theta) = K \cdot c_w \cdot M_w \tag{1}$$

c_w : concentration (mass/volume)

M_w : molecular weight (in the case of polydispersity the weight average of the molecular weight)

$$K = \frac{4 \cdot \pi^2 \cdot n^2}{N_A \cdot \lambda^4} \cdot (dn/dc_w)^2$$

N_A : Avogadro's number

n : refractive index

dn/dc_w : refractive index increment

λ : wavelength of the incident light

Starting from these results one would expect to find changes in the location of the ringing gel in the phase diagram that correspond to the changes in the interfacial tension and the solubilisation capacity. With increasing solubilisation capacity one would expect the gel phase region to move to higher hydrocarbon content and exactly this behavior was verified experimentally. In fig. 5 we have indicated the search lines, i. e. the lines of the solubilisation capacity, for different mixtures of $C_{14}DMAO$ and $C_{14}DMPO$ only. For all cases the preparation of an isotropic gel could be realised at a point in the phase diagram where the corresponding search line intersected with a line of constant water content of 55%, which is also indicated in fig. 5.

2.2 STRUCTURAL INVESTIGATIONS

Central to our investigations was the question of the micellar structure in the isotropic L_1-phase and in the gel, as well their relation to mechanical properties, especially mechanical moduli. In order to obtain structural information we carried out SANS experiments at the I.L.L., Grenoble. For these experiments we choose the system $C_{14}DMAO$/decane/D_2O and a series of samples was investigated, where the surfactant to decane ratio was kept constant at 5.4 by weight. All samples lay on the line indicated in fig. 1. Radially averaged scattering curves are shown in fig. 6. For all concentrations a correlation peak was observed, which became more prominent with increasing concentration. At the same it is shifted to higher values of the magnitude of the scattering vector, which corresponds to decreasing interparticle separations with increasing concentration. The scattering curves 1-5 were all obtained in the isotropic L_1-phase, whereas the sample from which we obtained curve 6 was already located in the ringing gel phase.

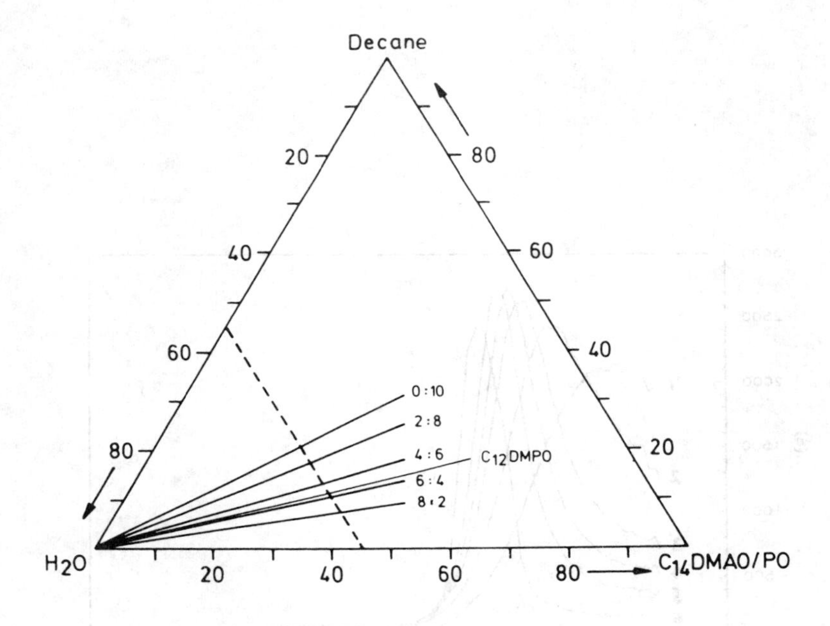

Fig. 5: Ternary phase diagram for different mixtures of C_{14}DMAO and C_{14}DMPO as surfactant. Only the search lines (solubilisation capacity) for the ringing gel phase and a line of constant water content of 55% are drawn.

We find that in fig. 6 the scattering function for isotropic solution and gel look very similar and no principal change is observed by going through the phase transition.

This is true for the series of samples described above. However for other samples from the gel phase region we observed scattering patterns that didn't show the radial symmetry.

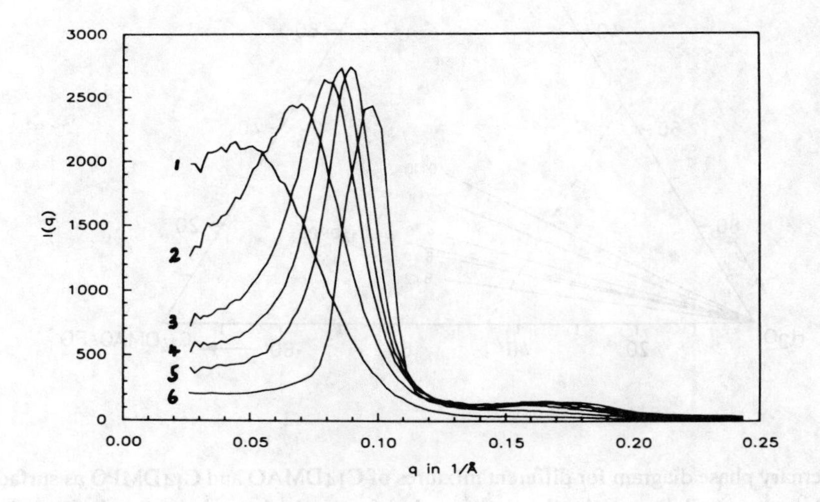

Fig.6: SANS intensity distributions for various concentrations of a system consisting of C_{14}DMAO/Decane = 5.4:1 (system 1). The sum concentration (by weight percent) of C_{14}DMAO plus Decane was : 1:8%; 2: 16%; 3: 24%; 4: 28%; 5: 32%; 6: 37.1%. The curves 1-5 are from samples in the L_1-phase, whereas curve 6 is from a sample in the isotropic gel phase region.

In fig. 7 we present a typical example for such a behaviour, where the scattering pattern was detected two-dimensionally and the scattering intensity is plotted vertically. We find more or less pronounced spikes, randomly distributed over the symmetrical correlation ring. However the position of these spikes is somewhat displaced from the maximum intensity of the isotropic scattering ring and they are randomly distributed around this scattering ring. This can best be seen in fig. 8, where a two-dimensional contour plot of equal scattering intensities is shown for the same sample. Similar results have been reported for systems made up from Brij96, paraffin and water [21]. There these results were explained by the presence of some large "crystallites".

Around twice the scattering vector of the correlation peak a weak second maximum is observed, which again becomes more prominent with increasing concentration. The occurrence of a second scattering maximum in our SANS data is quite unique in surfactant systems. Usually it is argued that polydispersity wipes out such a second peak. Therefore the presence of this second maximum is a first indication that the microemulsion droplets have to be fairly monodisperse under the investigated conditions. To confirm this result we also investigated a 1% sample, where the concentration refers to the sum concentration of surfactant and hydrocarbon and their ratio was again 5.4 by weight. The

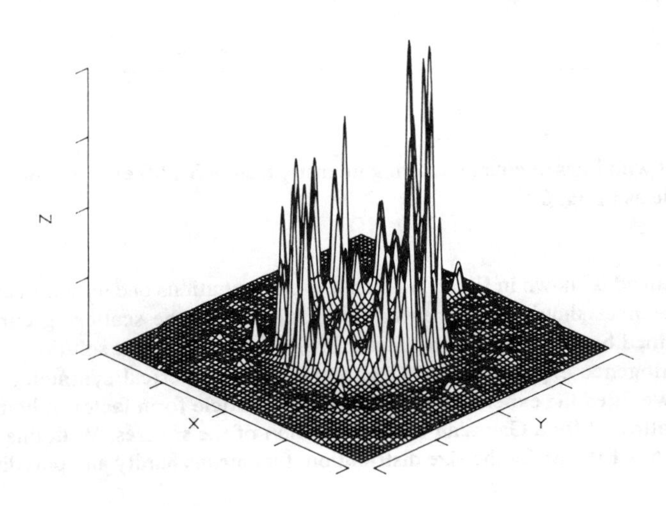

Fig. 7 : 3-dimensional plot of the scattered intensity in a SANS experiment for a sample in the isotropic gel phase consisting of 32.2% C_{14}DMAO, 6.0% Decane and 61.8% D_2O.

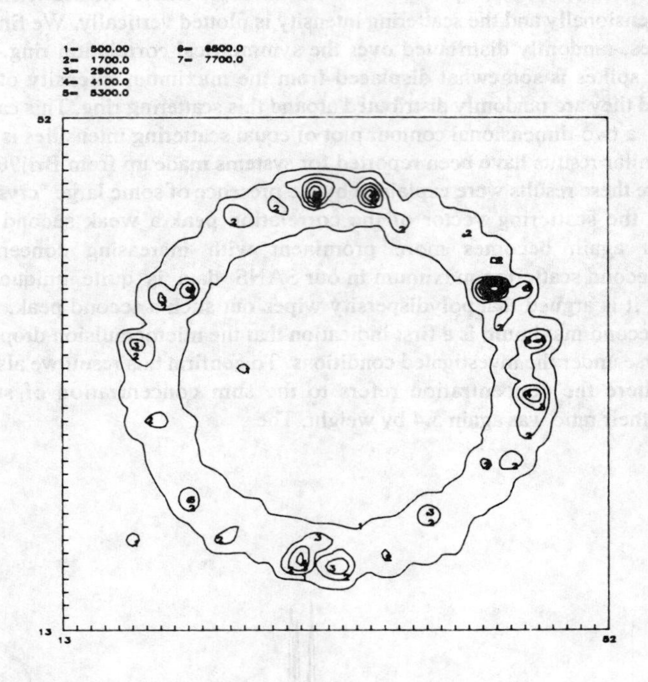

Fig. 8 : Contour plot with lines of equal scattering intensity from a SANS experiment for the same sample as in fig. 6.

scattering curve obtained is shown in fig. 9. At such low concentrations one may well assume that interparticle interferences should be of negligible influence and the scattering curve should basically be determined by the particle form factor P(q). For the isotropic solution it is evident from electric birefringence experiments that aggregates with spherical symmetry should be present. Therefore we fitted the experimental data with the particle form factor of homogeneous spheres, where we allowed for a Gaussian size distribution of the spheres. With this model we detected less than 1.5 Å FWHM for the size distribution, that means hardly any polydispersity at all.

$$I(q) \sim {}^{1}N.P(q).S(q) \qquad (2)$$

^{1}N : number density
P(q): particle form factor (accounts for intraparticle interferences)
S(q): structure factor (accounts for interparticle interferences)

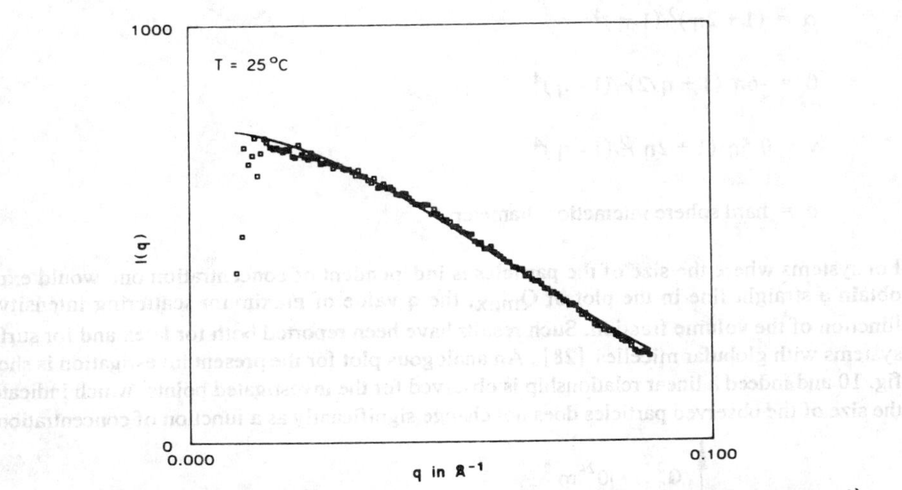

Fig. 9 : Plot of I(q) vs. q for a 1 wt% sample (C_{14}DMAO:Decane = 5.4:1, system 1). Comparison of the experimental points (squares) and the fitted curve for monodisperse spheres (solid line).

$$q = \frac{4 \cdot \pi}{\lambda} \cdot \sin(\Theta/2) \qquad \Theta\text{: scattering angle} \tag{3}$$

particle form factor for homogeneous spheres:

$$P(q) = \left(\frac{3 \cdot (\sin(q \cdot r) - q \cdot r \cos(q \cdot r))}{(q \cdot r)^3} \right)^2 \tag{4}$$

$$S(q) = \frac{1}{1 - {}^1N \cdot c(q)} \tag{5}$$

^1N: number density

$$\text{with } c(q) = -4\pi\,\sigma^3 \int_0^1 ds\; s^2 \,\frac{\sin(s \cdot q \cdot \sigma)}{s \cdot q \cdot \sigma}\,(\alpha + \beta \cdot s + \gamma \cdot s^3)$$

$$\eta = (\pi/6) \cdot {}^1N \cdot \sigma^3 \qquad \eta\text{ : volume fraction}$$

$$\alpha = (1 + 2\eta)^2/(1-\eta)^4$$

$$\beta = -6\eta \cdot (1 + \eta/2)^2/(1 - \eta)^4$$

$$\gamma = 0.5\eta\ (1 + 2\eta)^2/(1 - \eta)^4$$

σ = hard sphere interaction diameter

For systems where the size of the particles is independent of concentration one would expect to obtain a straight line in the plot of Q_{max}, the q-value of maximum scattering intensity, as a function of the volume fraction. Such results have been reported both for latex and for surfactant systems with globular micelles [28]. An analogous plot for the present investigation is shown in fig. 10 and indeed a linear relationship is observed for the investigated points, which indicates that the size of the observed particles does not change significantly as a function of concentration.

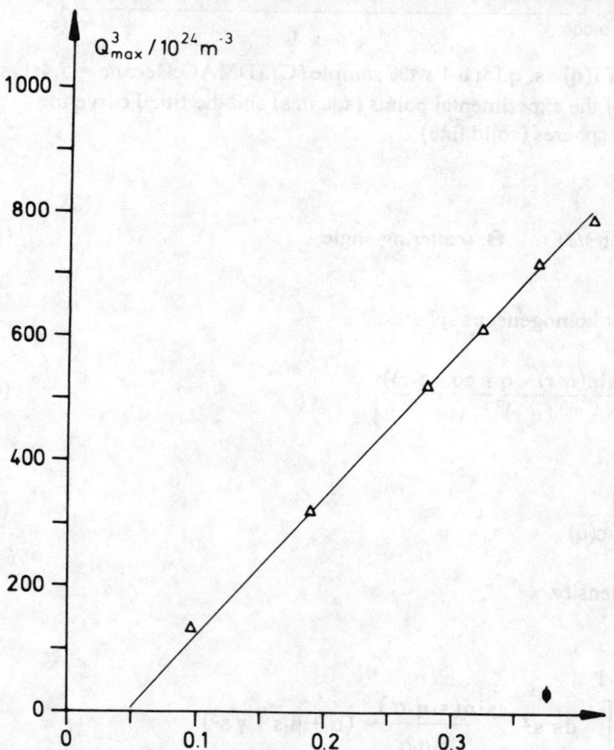

Fig. 10: $Q^3{}_{max}$ as a function of the volume fraction for systems made up from C_{14}DMAO:Decane = 5.4:1, data taken from the scattering curves as depicted in fig. 6.

In a next step we tried to explain our experimental scattering data with the help of a theoretical model. As mentioned before at least for the L_1-phase it is evident from electric birefringence and light scattering experiments, that spherical aggragates should be present. If one takes further into consideration that our surfactant is uncharged, one may well assume that the primary interaction in the system can be described by interactions of a hard sphere type. Using the Percus-Yevick [29] approximation for monodisperse hard spheres one obtains an analytic expression for the structure factor (eq. 5) [30,31]. One finds that the structure factor S(q) is only a function of number density N and effective hard sphere radius σ of the particles. In our approach we used the particle form factor for spheres (eq. 4) and the structure factor for hard spheres as given by eq. 5 to describe the angular dependence of the scattering intensity. Now the individual experimental scattering curves were fitted with a least square method to the above described model. There are three fit parameters: the radius for the particle form factor, R_p, the interaction radius for the structure factor, R_s, and the particle number density ^1N, which together with particle density gives another radius of the particle, R_ρ.

A typical example for such a fit is shown in fig. 11, where the squares represent the experimental data and the solid line is the fitted theoretical curve.

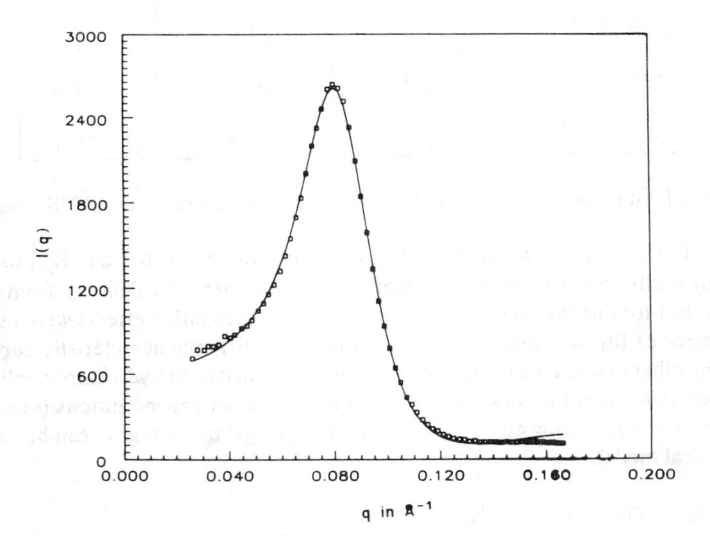

Fig. 11 : Plot of I(q) vs. q for a 24 wt% sample (C_{14}DMAO :Decane = 5.4:1, system 1). Comparison of the experimental points (squares) and the fitted curve for a hard sphere model (solid line).

Both are in good agreement. The fit parameters obtained, i. e. the three different radii, are summarized in table 1. These values are consistent with the proposed model, because the radii obtained from number density and particle form factor, R_p and R_ρ, are about equal (the radius for the particle form factor may well be somewhat smaller because hydration of the head groups will reduce the contrast of the outer part of the spheres and therefore reduce R_p), whereas the radius R_S is significantly larger. This is to be expected because the radius for the structure factor should be an effective interaction radius and should be larger than the 'real' radius because the spheres will already interact as soon as their shells of hydration approach each other. The data in table 1 is in full agreement with that description.

The most important point is that the obtained radii for all

weight percent Decane + C_{14}DMAO	R_p in Å	R_ρ in Å	R_S in Å
8	32.8	38.2	39.8
16	33.1	35.5	37.9
24	33.7	34.7	37.4
28	33.6	34.6	37.4
32	33.3	34.6	37.3
37.1	33.1	34.2	37.1

Table 1: fit parameter for a hard sphere model as obtained from SANS data

samples remain fairly constant. Even the somewhat larger values for R_p and R_ρ , for the most dilute sample are easily explained by the fact that both these values were deduced from the structure factor. But for this lowest concentration only small correlation effects were observed. Therefore the error of fitting a structure factor onto such a fairly uncharacteristic curve is much larger than in the other cases. That means that for a given surfactant to hydrocarbon ratio the radius of the particles present is constant over the whole range of concentrations investigated. At the same time even the scattering curve for the sample in the gel phase region can be explained by the same theoretical model.

2.3 MIXTURES OF C_{14}DMAO AND C_{14}DMPO

As we have seen before one can lower the interfacial tension between surfactant solution and a given hydrocarbon by gradually substituting C_{14}DMAO by the corresponding phosphinoxide. (fig. 3) By lowering the interfacial tension one increases the solubilisation capacity and should obtain larger aggregates correspondingly, as was indicated in fig. 4. Now we investigated three different surfactant systems (the pure aminoxide, and mixtures of it with the phosphinoxide in molar ratios of 6:4 and 2:8). Again we kept the surfactant to hydrocarbon ratio constant and our working line was always close to the solubilisation capacity. The compositions of the individual systems are given in table 2.

System	1	2	3
wt% C_{14} DMAO	84.37	46.45	12.81
wt% C_{14}DMPO	--	33.01	54.62
wt% Decane	15.63	20.54	32.57

Table 2: Composition of the investigated systems with surfactant mixtures from C_{14}DMAO and C_{14}DMPO. These ratios were kept constant while various amounts of water were added. The molar ratios of C_{14}DMAO to C_{14}DMPO were 10:0 in system 1, 6:4 in system 2 and 2:8 in system 3.

The results of light scattering experiments are presented in fig. 12, where the Rayleigh factor is plotted as a function of the total concentration of surfactant plus hydrocarbon.

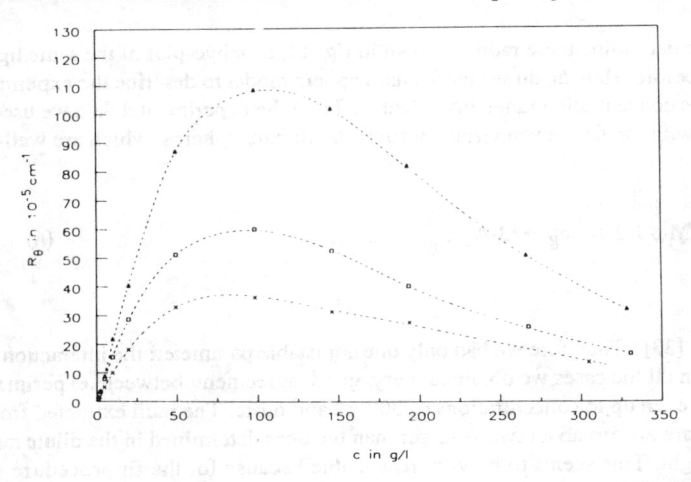

Fig. 12: Rayleigh factor as a function of concentration for various mixtures of C_{14}DMAO and C_{14}DMPO, with solubilized amounts of decane (according to table 2). The ratio of surfactant to hydrocarbon was kept constant. Molar ratios: Δ : C_{14}DMAO/C_{14}DMPO = 2:8 (system 3); □ : C_{14}DMAO/C_{14}DMPO = 6:4 (system 2); x : C_{14}DMAO/C_{14}DMPO = 10:0 (system 1).

Again one finds that the larger the phosphinoxide content the larger the particles. For all three systems the scattering intensity first increases linearly with the concentration, then reaches a maximum at around 10 wt% and finally decreases again because at even higher concentrations repulsive interactions make itself felt, which lead to the decrease in scattering intensity. The slope of the linear region is the higher the more hydrocarbon is solubilized [32] . From this linear region we determined the molecular weight and therefore also the radius of the respective particles. These values are given in table 3.

System	1	2	3
R(MW) in Å	33.5	38.7	45.9
R(Vir.) in Å	35.6	40.8	46.8
MW in 10^4 g/mol	8.2	12.3	20.1

Table 3: Radii and molecular weight of the aggregates in systems consisting of C_{14}DMAO/C_{14}DMPO as surfactant, decane and water, as determined from static light scattering.

Another method to determine these radii is shown in fig. 13, a Debye-plot of the same light scattering data as before. Here again we used a hard sphere model to describe the experimental data over the whole concentration range investigated. To fit the experimental data we used a virial expansion (eq. 6) with the first seven virial coefficients for hard spheres, which are well-known

$$\frac{K \cdot c_g}{R(\Theta)} = 1/M_w + 2 \cdot A_2 \cdot c_g + 3 \cdot A_3 \cdot c_g^2 + \ldots\ldots \qquad (6)$$

from the literature [33] . Therefore we had only one adjustable parameter: the interaction radius of the hard spheres. In all the cases we obtained very good agreement between experimental and theoretical values, even up to concentrations of 300 g/l and more. The radii extracted from this fit procedure (tab. 3) are always about two Å larger than the ones determined in the dilute range from the molecular weight. This seems to be very reasonable because for the fit procedure we again really have to use an effective hard sphere diameter, which should be a bit larger than the real diameter because of hydration effects. This result agrees well with the one from the SANS investigation. Moreover the good agreement between experimental and theoretical data proves that the aggregates can be treated as hard sphere particles. Furthermore it is shown that they do not change their size with increasing concentration. The reason for this unique behaviour is probably

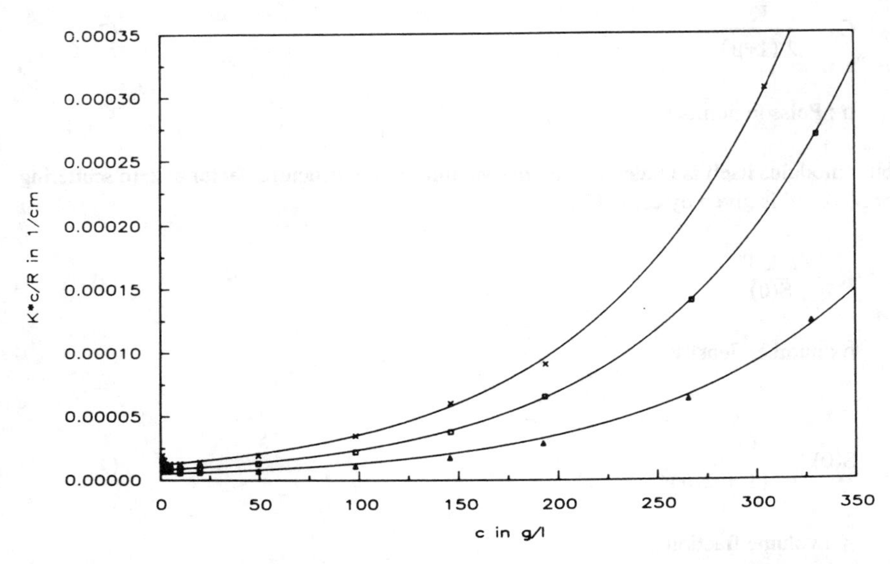

Fig. 13: Debye-plot of the static light scattering data for various mixtures of C_{14}DMAO and C_{14}DMPO, with solubilized amounts of oil (according to table 1). The ratio of surfactant to hydrocarbon was kept constant. Molar ratios: Δ : C_{14}DMAO/C_{14}DMPO = 2:8 (system 3); □: C_{14}DMAO/C_{14}DMPO = 6:4 (system 2); x: C_{14}DMAO/C_{14}DMPO = 10:0 (system 1). The solid lines are fits with a virial expansion for hard spheres.

the fact that the particles are uncharged, and therefore the ionic strength is independent of concentration. As a consequence, the interfacial tension of the globules also remains constant and hence also their size [32].

2.4 RELATION BETWEEN MECHANICAL PROPERTIES AND STRUCTURE

One main intention of our studies was to establish a connection between the structure of the gel and its mechanical properties, especially its mechanical moduli. Therefore shear moduli G were measured for gels with such a composition of surfactant and hydrocarbon, as were already characterized by light scattering or SANS in the more diluted region, i. e. systems with a composition as described in table 2. The total concentration of surfactant plus hydrocarbon was 43% for all gel samples. The shear modulus was determined by a rheological frequency sweep method.

The theory of solids gives us now a relation between the magnitude of the shear modulus G and the bulk modulus K (eq. 7), where μ is the Poisson number, which for our samples always seemed to be close to 0.5 (the limiting case of incompressibility).

$$G = \frac{K}{2 \cdot (1+\mu)} \tag{7}$$

μ : Poisson number

The bulk modulus itself is related by thermodynamics to the structure factor at zero scattering vector S(0) and is given by eq. 8 [34].

$$K = \frac{\rho \cdot k \cdot T}{S(0)} \tag{8}$$

ρ : number density

$$S(0) = \frac{(1 - \eta)^4}{(1 + 2.\eta)^2} \tag{9}$$

η : volume fraction

The structure factor for hard spheres in the Percus-Yevick approximation is given by eq. 9, and is only a function of the volume fraction. Therefore we can simply calculate the bulk modulus for a system of hard spheres when the radius of the spheres and their concentration is given. This was done for the investigated gels, taking into account the densities of the respective aggregates, which were determined experimentally. These calculated values for the bulk moduli K are compared with the experimental shear moduli G in fig. 14. There clearly exists a correlation between the two values. Moreover we have also included in that plot the values for G and K for a gel with decanol as cosurfactant instead of C_{14}DMPO. This value too follows a linear relationship and therefore this scheme of relating structural information to elastic properties seems to be of general applicability. It seems to be possible to predict the mechanical moduli in such systems if structural information from scattering experiments is available.

The mechanical properties of such systems can also be rationalized by means of the interfacial tension of the corresponding system. In our description the gel phase consists of densely packed spheres. These spheres have to develop contact zones if one tries to compress the system one-dimensionally. For such a situation the elasticity modulus E should be given by the interfacial tension γ by eq. 10. Theory of solids gives again a relation between bulk modulus K and elasticity modulus E (eq. 11).

$$E = \frac{2 \cdot \gamma}{r} \tag{10}$$

r : radius of the microemulsion droplet

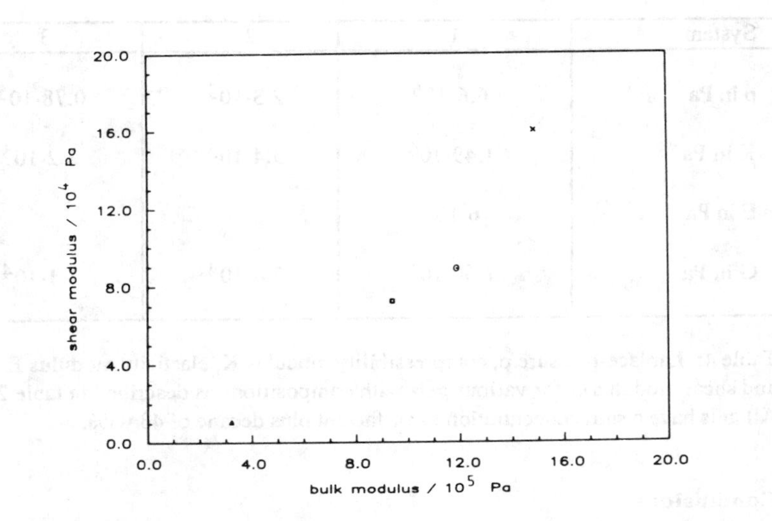

Fig. 14: A comparison of bulk modulus K and shear modulus G for different gels. Molar ratios: Δ: $C_{14}DMAO/C_{14}DMPO$ = 2:8 (system 3); □: $C_{14}DMAO/C_{14}DMPO$ = 6:4 (system 2); x : $C_{14}DMAO/C_{14}DMPO$ = 10:0 (system 1); ⊙: $C_{14}DMAO$/Decanol = 9:1.

$$K = \frac{E}{3\cdot(1 - 2\cdot\mu)} \tag{11}$$

The term $2\cdot\gamma/r$ is identical to the Laplace pressure inside the microemulsion droplets. Lekkerkerker et. al [35] have argued that the Laplace pressure inside such droplets cannot be described by this simple relation but has to be corrected for bending effects of the interface. The correction is the larger the further away the radius of the globules is to the mean radius of curvature under most favourable energetic conditions [36]. But it is conceivable that we are quite close to these conditions in our systems. Therefore we calculated the Laplace pressure p for the individual gel samples investigated from this simple equation, using the experimental values for γ and r. For this calculation we used the interfacial tension γ of a dilute micellar solution as measured against decane. These calculated values for p, as well as the experimental values for the shear modulus G and the calculated values for the bulk modulus K are all given in table 4. Furthermore we have included the experimental value of the elasticity modulus E for the pure aminoxide system, which was determined by a mechanical resonance technique [37]. For all the values good agreement is observed.

System	1	2	3
p in Pa	$6.6\cdot10^5$	$2.3\cdot10^5$	$0.78\cdot10^5$
K in Pa	$1.49\cdot10^6$	$9.4\cdot10^5$	$3.2\cdot10^5$
E in Pa	$6\cdot10^5$	-	-
G in Pa	$1.60\cdot10^5$	$7.4\cdot10^4$	$1.1\cdot10^4$

Table 4: Laplace-pressure p, compressibility modulus K, elasticity modulus E and shear modulus G for various gels with compositions as described in table 2. All gels have a sum concentration of surfactant plus decane of 43 wt%.

3. Conclusions

Ringing gels were found as isotropic, one-phase regions in the ternary phase diagram of the surfactant $C_{14}DMAO$ (or mixtures of it with $C_{14}DMPO$), a hydrocarbon and water. Their location in the phase diagram can be predicted from the intersection of two lines. First the extension of the line of the solubilisation capacity and second a line of constant water content of about 50-60 %.

The experimental work showed that the rodlike micelles being formed by the pure surfactant are transformed into globular aggregates by the solubilisation of hydrocarbon. A sharp rod-sphere transition was observed as a function of the amount of solubilized hydrocarbon. The behavior of these microemulsion droplets can be described by the model of hard sphere particles. From SANS experiments one can conclude that the globules are fairly monodisperse and as a consequence their concentrated solutions show a second scattering maximum in these experiments, besides the normal correlation peak. Scattering experiments also show that the size of the aggregates remains constant for a given surfactant to hydrocarbon ratio when the total concentration is increased. Furthermore it is indicated that the isotropic ringing gel phase is built up from the same spherical particles as the adjacent L_1-phase.

With the structural information obtained by scattering experiments one can calculate bulk moduli under the assumption of a hard sphere model and the use of the Percus-Yevick approximation. These correlate well with experimental values for shear moduli. At the same time the mechanical properties can as well be described by the interfacial tension of the concerning system. The calculated Laplace pressure should be identical to the elasticity modulus of the sample. There is good numerical agreement between calculated and experimental value of the elasticity modulus. Furthermore the calculated Laplace pressures correlate well with calculated bulk moduli and experimental shear moduli. Therefore mechanical properties of such gel systems can be predicted approximately if either structural information from scattering experiments or interfacial tension data is available.

4. References

[1] K. Fontell, Coll. Polym. Sci. in press
[2] H. Hoffmann, G. Ebert, Angew. Chem. 100, 933 (1988)
[3] G. Platz, G. Ebert in Polymer Reaction Engineering (ed. K. H. Reichert, W. Geiseler),Hüthig&Wepf Verlag, Heidelberg 1986
[4] N. Pöllinger, Dissertation Erlangen, 1986
[5] W. Jettka, Dissertation Braunschweig, 1984
[6] F. B. Rosevear, J. Soc. Cosmetic. Chemists 19, 581 (1968)
[7] G. W. Gray, P. A. Winsor, Chapter I in Lyotropic Liquid Crystals (ed. S. Friberg), Advances in Chemistry Series 152, Washington, D. C. 1976
[8] T. Kaufmann, R. Blaser, American Perfumer and Cosmetics 80, 37 (1965)
[9] L. M. Prince (ed.), Microemulsions, theory and practice, Academic Press Inc., New York, San Francisco, London, 1977
[10] H. Kelker, R. Hatz, Chapter 11 in Handbook of Liquid Crystals, Verlag Chemie, Weinheim, 1980
[11] E. Nürnberg, W. Pohler, Progr. Coll. Polym. Sci. 69, 48 (1984)
[12] G. Oetter, Dissertation Bayreuth, 1989
[13] D. W. Osborne, A. J. I. Ward, K. J. O'Neill, in Topical Drug Delivery Formulations (ed. D. W. Osborne, A. H. Amann), Marcel Dekker, New York, in press
[14] K. Fontell, A. Ceglie, B. Lindman, B. Ninham, Acta Chem. Scand. A40, 247 (1986)
[15] K. Fontell, L. Mandell, P. Ekwall, Acta Chem. Scand. 22, 3209 (1969) 16 P. P. A. Spegt, A. E. Skoulios, V. Luzatti, Acta Cryst. 14, 866 (1961)
[17] P. Ekwall, L. Mandell, K. Fontell, Mol. Cryst. Liquid Cryst. 8, 157 (1969)
[18] K. Fontell, K. K. Fox, E. Hansson, Mol. Cryst. Liquid Cryst. Letters 1, 9 (1985)
[19] V. Luzzati, P. A. Spegt, Nature 215, 701 (1967)
[20] A. Tardieu, V. Luzzati, Biochim. Biophys. Acta 219, 11 (1970)
[21] H. Jousma, J. G. H. Joosten, G. S. Gooris, H. E. Junginger, Colloid Polym. Sci. 267, 353 (1989)
[22] G. Oetter, H. Hoffmann, Colloids and Surfaces 38, 225 (1989)
[23] J. L. Burns, Y. Cohen, Y. Talmon, J. Phys. Chem. in press
[24] H. Hoffmann, G. Oetter, B. Schwandner, Progr. Colloid&Polymer Sci. 73, 95 (1987)
[25] K. H. Pospischil, Langmuir 2, 170 (1986)
[26] D. S. Murphy, M. J. Rosen, J. Phys. Chem. 92, 2870 (1988)
[27] J. L. Cayias, R. S. Schechter, W. H. Wade, Adsorption and Interfaces, ACS Symposium Series, Nr. 8, 1975
[28] H. Hoffmann, H. Thurn, in Encyclopedia of Emulsion Technology (ed. P. Becher), 1986
[29] J. K. Percus, G. J. Yevick, Phys. Rev. 110, 1 (1958)
[30] N. W. Ashcroft, J. Lekner, Phys. Rev. 145, 83 (1966)
[31] M. Wertheim, J. Math. Phys. 5, 643 (1964)
[32] M. Gradzielski, H. Hoffmann, G. Oetter, Coll. Polym. Sci. in press
[33] F. H. Ree, W. G. Hoover, J. Chem. Phys. 40, 939 (1967)
[34] R. H. Ottewill, Ber. Bunsenges. Phys. Chem. 89, 517 (1985)
[35] J. Th. G. Overbeek, G. J. Verhoeckx, P. L. de Bruyn, H. N. W. Lekkerkerker, J.

Colloid Interface Sci. 119, 422 (1987)
[36] C. A. Miller, J. Dispersion Sci. Technol. 6, 159 (1985)
[37] J. H. Bartholomäus, C. Führer, Acta Pharm. Technol. 33, 174 (1987)

INDUSTRIAL HYDROCOLLOIDS

E.R. MORRIS
Department of Food Research and Technology,
Cranfield Institute of Technology,
Silsoe College,
Silsoe,
Bedford MK45 4DT
Great Britain

ABSTRACT. Macromolecular shape (conformation) has a dominant effect on hydrocolloid functional properties. Disordered chains give solutions whose viscosity varies in a general, predictable way with concentration and shear rate. Intermolecular ordered structures, stabilised by co-operative arrays of non-covalent bonds, and/or stable aggregates of such structures, can act as 'junction zones' in gels. Networks formed by low-energy interactions between ordered chain segments show gel-like response to small deformation but break down and flow at higher stress, giving solutions that can suspend solid particles or stabilise emulsion droplets. Illustrative examples and investigative techniques are discussed in detail.

1. Introduction

Hydrocolloids are water-soluble polymers that occur naturally, or are obtained from natural polymers by, for example, chemical derivitisation. These materials are widely used in industry, predominantly for generation/control of product rheology [1]. With the notable exception of gelatin (discussed in Section 3.6), the industrial hydrocolloids available at present are all polysaccharides. For the benefit of readers unfamiliar with carbohydrate chemistry, a brief introduction to the primary structure and chain geometry of polysaccharides is given below. In view of the breadth of coverage in this article, most of the references cited are to more specialised previous reviews, rather than to the original literature sources.

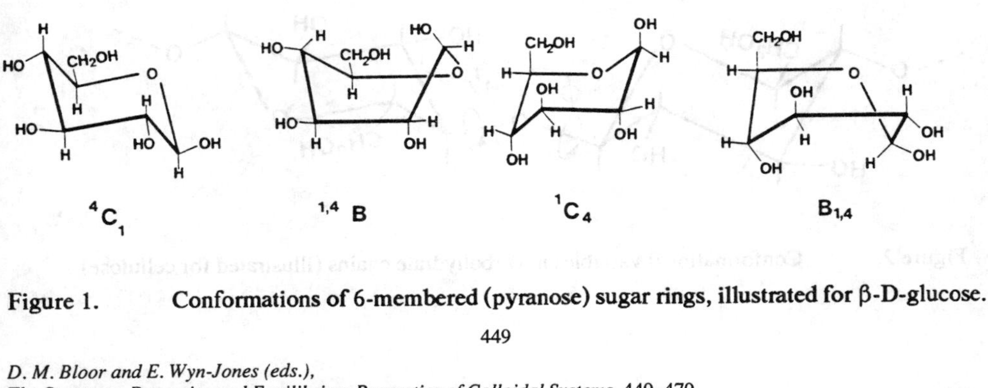

$^4C_1$ $^{1,4}B$ $^1C_4$ $B_{1,4}$

Figure 1. Conformations of 6-membered (pyranose) sugar rings, illustrated for β-D-glucose.

D. M. Bloor and E. Wyn-Jones (eds.),
The Structure, Dynamics and Equilibrium Properties of Colloidal Systems, 449–470.

1.1 CARBOHYDRATE MONOMERS

The basic building block in all industrial polysaccharides is a 6-membered (pyranose) sugar ring (Fig. 1), composed of five carbon atoms and one oxygen. In the projections shown in Fig. 1, carbon atoms are numbered clockwise from the ring oxygen, with C(6) lying outside the ring. The stable conformations of the pyranose ring [2] are chair forms ($^4C_1$ and $^1C_4$) in which all bonds are fully staggered. The boat forms ($^{1,4}B$ and $B_{1,4}$) shown in Fig. 1 are destabilised by eclipsing of bonds around the C(2)-C(3) and C(5)-0(5) linkages, and by van der Waals repulsion between the inward-pointing 'masthead' substituents on C(4) and C(1), and therefore have a very low probability of occurrence.

In the chair conformations, substituents at each carbon atom may be present in either *equatorial* locations, widely spaced around the periphery of the ring, or in crowded *axial* positions above or below the ring. Interconversion between the $^4C_1$ and $^1C_4$ chair forms has the effect of converting all equatorial substituents to axial, and *vice versa*. In general, the stable chair conformation is the one in which the bulky hydroxymethyl substituent at C(6) is equatorial rather than in the sterically crowded axial location.

Hexose monosaccharides (sugars with six carbon atoms, as in Fig. 1) are classified in two groups, according to the steric configuration at C(5), the position of ring closure. In the D series C(6) is equatorial in the $^4C_1$ ring form, while in the mirror-image L series the corresponding stable chair form is $^1C_4$. Within each series sugars are named according to the configuration at C(2), C(3) and C(4), giving eight possible isomers, only some of which occur in nature. In the stable ring form of glucose ($^4C_1$ for D and $^1C_4$ for L) 0(2), 0(3) and 0(4) are all equatorial; in mannose 0(2) is axial; in galactose 0(4) is axial; in gulose 0(3) and 0(4) are axial, and in idose all three are axial. Configuration at C(1) is denoted as α when 0(1) is axial, and β when 0(1) is equatorial. 0(1) is chemically different from the other pendant oxygens of the ring, since it forms part of a hemiacetal group with the ring oxygen, 0(5).

1.2 STRUCTURE AND SHAPE OF POLYSACCHARIDE CHAINS

Linkage of adjacent sugars in carbohydrate chains involves condensation between the hemiacetal OH group at C(1) on one residue and one of the alcohol OH groups of the next residue, with formal elimination of water. Thus, in contrast to proteins and most synthetic polymers, different polysaccharides can be built up from the same monomer unit. For example, polymers of D-glucose include cellulose (β-1,4), amylose (α-1,4), curdlan (β-1,3) and dextran (α-1,6), all of which have very different physical properties.

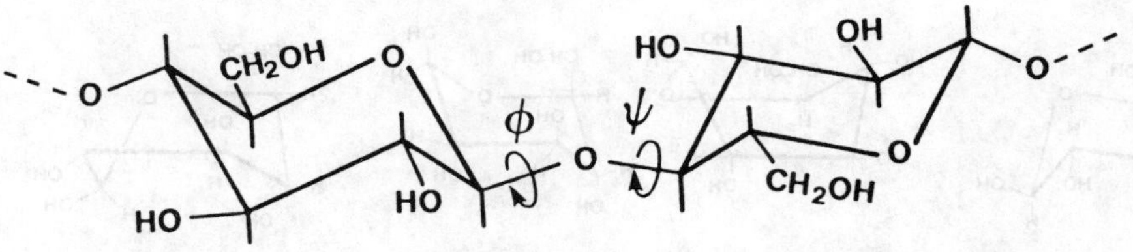

Figure 2. Conformational variables in carbohydrate chains (illustrated for cellulose).

Since, as discussed above, the component monosaccharide units can be regarded as essentially locked in the chair conformation in which C(6) is equatorial, the overall shape of the polysaccharide chain is determined [3] by the dihedral angles between adjacent residues (Fig. 2). These may be either fixed, and have the same values at all positions along the polymer chain, to give ordered chain conformations (as in the solid state), or constantly fluctuating, to give the overall 'random coil' behaviour typical of polymer solutions.

1.3 ORDERED CONFORMATIONS

The nature of the ordered conformations adopted by different polysaccharides depends directly on the relative orientations of the bonds to and from the component monosaccharide residues in the chain [3]. Three bonding patterns (Fig. 3) are of particular importance in understanding the properties of industrial hydrocolloids.

The first pattern is di-equatorial linkage across the pyranose ring (i.e. equatorial bonds at carbons 1 and 4). As outlined above, an equatorial linkage at carbon 1 implies a β configuration, while the oxygen on carbon 4 is equatorial in glucose and mannose (and derivatives such as glucosamine and mannuronic acid) but not in galactose, gulose, idose or their derivatives. Thus 1,4-linked chains of β-D-glucose (cellulose), N-acetyl-β-D-glucosamine (chitin), β-D-mannose (the backbone of the galactomannan family of plant polysaccharides) and β-D-mannuronic acid (from alginate) all share the same linkage geometry. The bonds to and from each residue are parallel and only slightly offset from each other, which gives rise to extended, flat ribbon-like chain geometry in the solid state (Fig. 3a).

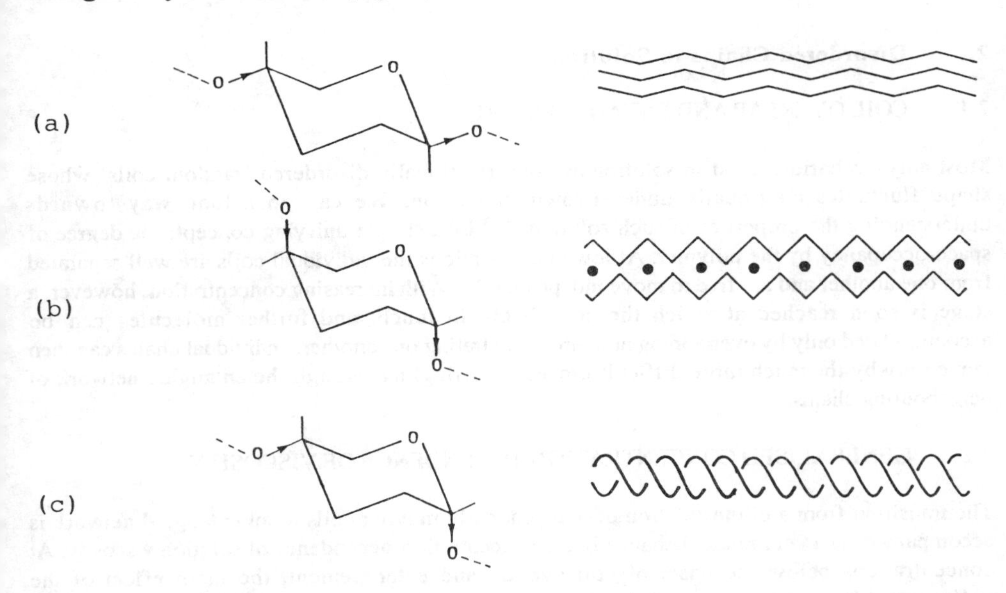

Figure 3. Relationship between the relative orientation of bonds to and from component residues of polysaccharide chains and the nature of the ordered structures adopted in the solid-state. a) Extended ribbons; b) Buckled ribbons; c) Hollow helices.

The second major category is where linkage is again at carbons 1 and 4, but through axial bonds at each position. This occurs for α-linked pyranose residues of galactose, gulose, idose or their derivatives. Once more the bonds to and from each residue are parallel, but in this case they are offset from each other by the full width of the sugar ring, to give rise to a highly buckled ordered chain conformation. When such buckled chains are packed together they form large cavities or interstices, as shown in Fig. 3b. Two important examples of this type of linkage geometry are the 1,4-linked linear chain sequences of α-L-guluronate (from alginate) and α-D-galacturonate (from pectin). In both, divalent metal ions can be incorporated in the interstices between the chains, to reduce intermolecular electrostatic repulsion and stabilise the ordered structure. For obvious reasons this is known as 'egg-box' binding.

In the third class of linkage geometry, the bonds to and from each residue are no longer parallel, which gives rise to helical structures (Fig. 3c). This situation can arise in two different ways. The first is where linkage is again through C(1) and C(4) but one bond is axial and the other equatorial. This is the situation in 1,4-linked α-D-glucose (amylose). The second is where linkage is at positions 1 and 3, rather than diagonally across the ring, for example in 1,3-linked β-D-glucose (curdlan). Normally helical structures of this type pack together coaxially, as in the double helix structure of DNA. For example the solid state structure of amylose is a double helix, while the 1,3-linked β-glucan backbone of curdlan packs as a triple helix. In alternating heteropolysaccharides where one linkage is of the 'ribbon-forming' type and the other is of the 'helix-forming' type, the twist introduced in the latter case again gives rise to overall helix geometry (as in the carrageenan and agar families of algal polysaccharides).

2. Disordered Chains in Solution

2.1 COIL OVERLAP AND ENTANGLEMENT

Most polysaccharides exist in solution as conformationally-disordered 'random coils' whose shape fluctuates continually under Brownian motion. We can go a long way towards understanding the properties of such solutions [4] by a simple unifying concept: the degree of space-occupancy by the polymer. At low concentrations the individual coils are well separated from one another and are free to move independently. With increasing concentration, however, a stage is soon reached at which the coils begin to touch, and further molecules can be accommodated only by overlapping and inter-penetrating one another. Individual chains can then move only by the much more difficult process of wriggling through the entangled network of neighbouring chains.

2.2 GENERALISED CONCENTRATION-DEPENDENCE OF VISCOSITY

The transition from a dilute solution of independently moving coils to an entangled network is accompanied by a very marked change in the concentration-dependence of solution viscosity. At concentrations below the onset of coil overlap and entanglement, the main effect of the polysaccharide coils is to perturb the flow of the solvent by tumbling around and setting up 'countercurrents' which increase the overall viscosity. The magnitude of this effect is roughly proportional to the number of chains present, but second-order effects due to mutual interference of 'countercurrents' from neighbouring molecules increase the concentration-dependence to $\sim c^{1.3}$ (so that doubling concentration increases viscosity by a factor of $\sim$ 2.5). At higher concentrations, where the chains are entangled, viscosity increases much more steeply with increasing concentration, varying as ~ $c^{3.3}$ (so that doubling concentration gives about a ten-fold increase in viscosity). At the onset of entanglement, polysaccharide solution viscosity is about 10 mPa s (i.e. about ten times the viscosity of water), so that almost all practical applications involve entangled networks at higher concentrations.

For any specific polysaccharide sample the onset of coil overlap is determined by two things: the number of chains present (proportional to concentration) and the volume that each occupies. A convenient parameter for characterisation of coil volume is intrinsic viscosity, or 'limiting viscosity number', $[\eta]$. Intrinsic viscosity is the fractional increase in viscosity per unit concentration of isolated chains, and is determined experimentally [4] by measuring 'specific viscosity' [$\eta_{sp} = (\eta - \eta_s)/\eta_s$, where η is solution viscosity and η_s is the viscosity of the solvent] over a range of low polymer concentrations ($\eta_{sp} < 1$) and extrapolating η_{sp}/c to zero concentration, to eliminate the effect of coil-coil interactions through the solvent.

Coil volume can vary widely from sample to sample, with corresponding variation in the critical concentration (c*) at which entanglement begins. However, when measured viscosities are plotted as a function of the extent of space-occupancy by the polymer, characterised by the 'coil overlap parameter' $c[\eta]$, rather than against concentration alone, then results for different 'random coil' polysaccharides superimpose closely [4], irrespective of primary structure and molecular weight, with the onset of entanglement (c* transition) occurring when $c[\eta] \approx 4$.

2.3 'RANDOM COIL' DIMENSIONS

For any polymer/solvent system, intrinsic viscosity increases with molecular weight (M) according to the Mark-Houwink relationship:

$$[\eta] = KM^{\alpha}$$

For a hypothetical 'freely-jointed chain', without steric clashes between chain segments, $\alpha = 0.5$. Certain polymers, under appropriate solvent conditions ('theta conditions') follow this behaviour, but for most polysaccharides in aqueous solution α is substantially higher, often around one, so that intrinsic viscosity (coil volume) increases more or less linearly with increasing chainlength [5].

The K parameter is dependent predominantly on the geometry of the inter-residue linkages within the polymer chain [4]. Polysaccharides in which the bonds to and from each residue are directly opposite each other across the sugar ring (1,4 linkage) and approximately parallel (diequatorial or diaxial) have highly expanded coil dimensions (i.e. high values of K), since the 'sense of direction' of the chain tends to continue more or less unchanged from residue to residue. For the same reason, polysaccharides with this type of linkage geometry normally adopt extended ribbon-like structures in the solid state, as discussed above (Section 1.3).

When the linkages to and from each residue are no longer diagonally opposite one another (e.g. 1,3 linkage) or are no longer parallel (i.e. 1,4 axial-equatorial) a systematic 'twist' in chain direction is introduced, leading to more compact, smaller coils (low values of K) and, as outlined in Section 1.3, to helical structures in the solid state. In 1,6-linked polysaccharides, such as dextran, the residues are linked through three rather than two covalent bonds, which allows considerably greater conformational freedom, again promoting compact coil geometry.

2.4 EFFECT OF CHARGE

In charged polysaccharides, coil dimensions can be substantially altered by changes in ionic strength (I). Electrostatic repulsions between chain segments increase coil volume but can be screened out by addition of salt, allowing the coil to contract towards the dimensions it would have adopted if uncharged. Quantitatively, intrinsic viscosity decreases linearly with $I^{-1/2}$, extrapolating at infinite ionic strength ($I^{-1/2} = 0$) to the intrinsic viscosity of a neutral polysaccharide of the same primary structure and chain length. The slope of $[\eta]$ vs. $I^{-1/2}$ provides [5] an index of chain flexibility, as illustrated later (Section 4.2).

2.5 SHEAR-RATE DEPENDENCE OF VISCOSITY

In dilute solutions, below the onset of entanglement ($c < c^*$), viscosity shows only a slight dependence on shear rate ($\dot{\gamma}$), typically less than 30% over several decades of $\dot{\gamma}$, due to individual coils being stretched out by the flow and offering less resistance to movement. In entangled networks ($c > c^*$), however, the principal mechanism of shear thinning is quite different [4,5]. For the solution to flow, intermolecular entanglements must be pulled apart. At low shear rates, where there is sufficient time for new entanglements to form between different chain-partners, the overall 'crosslink-density' of the network remains constant and the viscosity also remains constant at a fixed, maximum value (the 'zero shear' viscosity). At higher shear-rates, however, where the rate of re-entanglement falls behind the rate of disruption of existing entanglements, the extent of entanglement-coupling decreases progressively with increasing shear-rate and viscosity falls, typically by two or three orders of magnitude over the shear-rate range of practical importance.

Although different entangled 'random coil' polysaccharide solutions can differ widely in their maximum 'zero shear' viscosity, and in the shear rate at which shear thinning begins, the form of shear thinning is entirely general [4] and can be matched, with reasonable precision [6], by the equation:

$$\eta = \eta_o / [1 + (\dot{\gamma} / \dot{\gamma}_{1/2})^p] \qquad (1)$$

where $\dot{\gamma}_{1/2}$ is the shear rate required to reduce viscosity to half the maximum 'zero shear' value (i.e. $\dot{\gamma} = \dot{\gamma}_{1/2}$ when $\eta = \eta_o / 2$) and p is the absolute value of the 'terminal slope' of a double-logarithmic plot of η vs. $\dot{\gamma}$. For samples with a wide polydispersity of chainlength (as in commercial hydrocolloids), p is constant at 0.76. Thus viscosity (η) at any shear rate ($\dot{\gamma}$) can be characterised by two parameters, η_o and $\dot{\gamma}_{1/2}$. Both of these can be obtained from a simple linear plot, as described below.

Equation (1) can be rearranged to the form:

$$\eta = \eta_o - (1/\dot{\gamma}_{1/2})^p \, \eta \, \dot{\gamma}^p \qquad (2)$$

Thus plotting η against $\eta \, \dot{\gamma}^{0.76}$ should give a straight line of intercept η_o and gradient $-(1/\dot{\gamma}_{1/2})^{0.76}$. Figure 4 shows the shear-rate dependence of viscosity plotted in this way [6] for several typical commercial 'random coil' polysaccharides. The plots show acceptable linearity and, in particular, allow the maximum 'zero shear' viscosity to be derived from experimental measurements at higher shear rates.

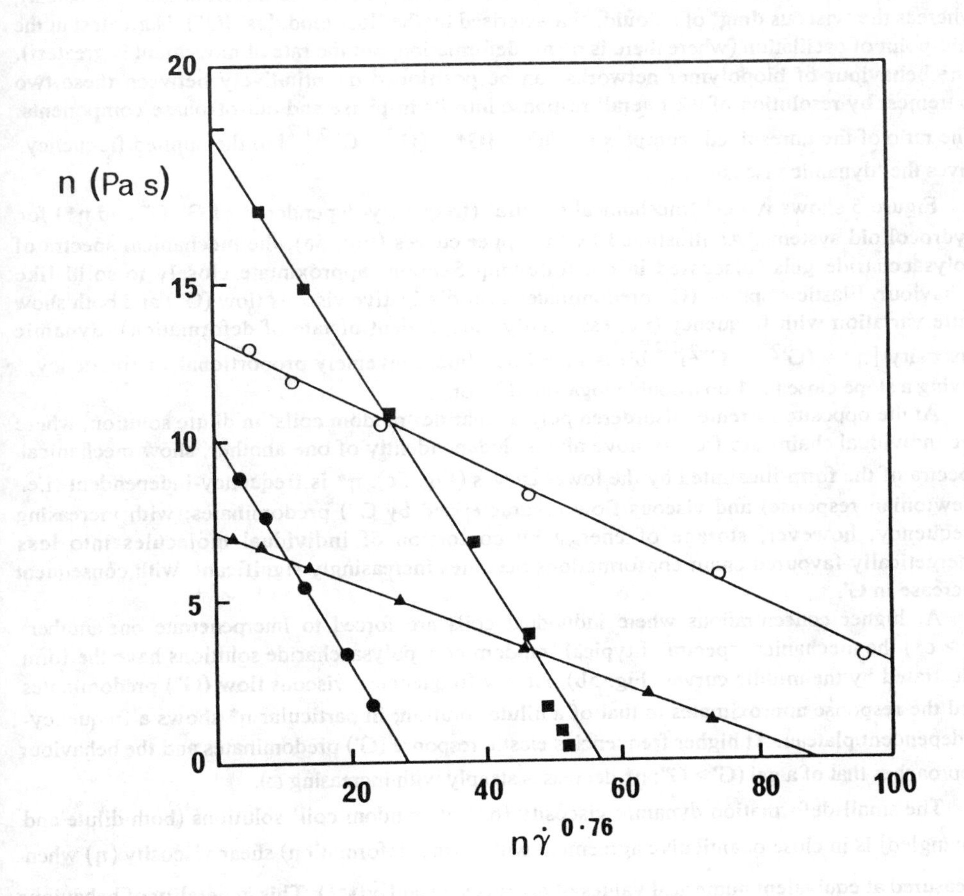

Figure 4. Linear plots (Equation 2) of shear-thinning in typical solutions of entangled 'random coil' polysaccharides. From Morris [6], with permission.

2.6 MECHANICAL SPECTROSCOPY

The discussion so far has considered only 'steady shear' viscosity (η) measured under 'large deformation' conditions (e.g. continuous rotation). Biopolymer networks, including entangled 'random coils', however, also show a substantial element of solid-like response. The degree of solid-like and liquid-like character can be quantified [4,5] by the technique of mechanical spectroscopy, in which the resistance of the sample to a very small oscillatory deformation is measured as a function of frequency (ω). The resistance of a solid, characterised by the 'storage modulus' (G'), is greatest at the extremes of oscillation (where the deformation is greatest), whereas the 'viscous drag' of a liquid, characterised by the 'loss modulus' (G"), is greatest at the mid-point of oscillation (where there is no net deformation, but the rate of movement is greatest). The behaviour of biopolymer networks can be positioned quantitatively between these two extremes, by resolution of the overall response into its in-phase and out-of-phase components. The ratio of the unresolved 'complex modulus' $[G^* = (G'^2 + G''^2)^{1/2}]$ to the applied frequency gives the 'dynamic viscosity', η*.

Figure 5 shows typical 'mechanical spectra' (frequency-dependence of G', G" and η*) for hydrocolloid systems. As illustrated by the upper curves (Fig. 5a), the mechanical spectra of polysaccharide gels (discussed in the following Section) approximate closely to solid-like behaviour. Elastic response (G') predominates over dissipative viscous flow (G") and both show little variation with frequency (i.e. essentially independent of rate of deformation); dynamic viscosity $[\eta^* = (G'^2 + G''^2)^{1/2}/\omega]$ is therefore almost inversely proportional to frequency, giving a slope close to -1 on a double logarithmic plot.

At the opposite extreme, disordered polysaccharide 'random coils' in dilute solution, where the individual chains are free to move almost independently of one another, show mechanical spectra of the form illustrated by the lower curves (Fig. 5c): η* is frequency-independent (i.e. Newtonian response) and viscous flow (characterised by G") predominates; with increasing frequency, however, storage of energy by contortion of individual molecules into less energetically-favoured chain conformations becomes increasingly significant, with consequent increase in G'.

At higher concentrations where individual coils are forced to interpenetrate one another ($c > c^*$) the mechanical spectra of typical 'random coil' polysaccharide solutions have the form illustrated by the middle curves (Fig. 5b). At low frequencies, viscous flow (G") predominates and the response approximates to that of a dilute solution; in particular η* shows a frequency-independent plateau. At higher frequencies elastic response (G') predominates and the behaviour approaches that of a gel ($G' > G''$; η* decreases steeply with increasing ω).

The small-deformation dynamic viscosity (η*) of 'random coil' solutions (both dilute and entangled) is in close quantitative agreement with (large-deformation) shear viscosity (η) when measured at equivalent numerical values of ω (rads s^{-1}) and $\dot{\gamma}$(s^{-1}). This generality of behaviour (the Cox-Merz rule) is well established experimentally over a wide spectrum of conformationally-disordered polymers (natural and synthetic). In particular, the shear-rate dependence of η for entangled polysaccharide coils, discussed above, is closely superimposable on the frequency-dependence of η* illustrated in Fig. 5b.

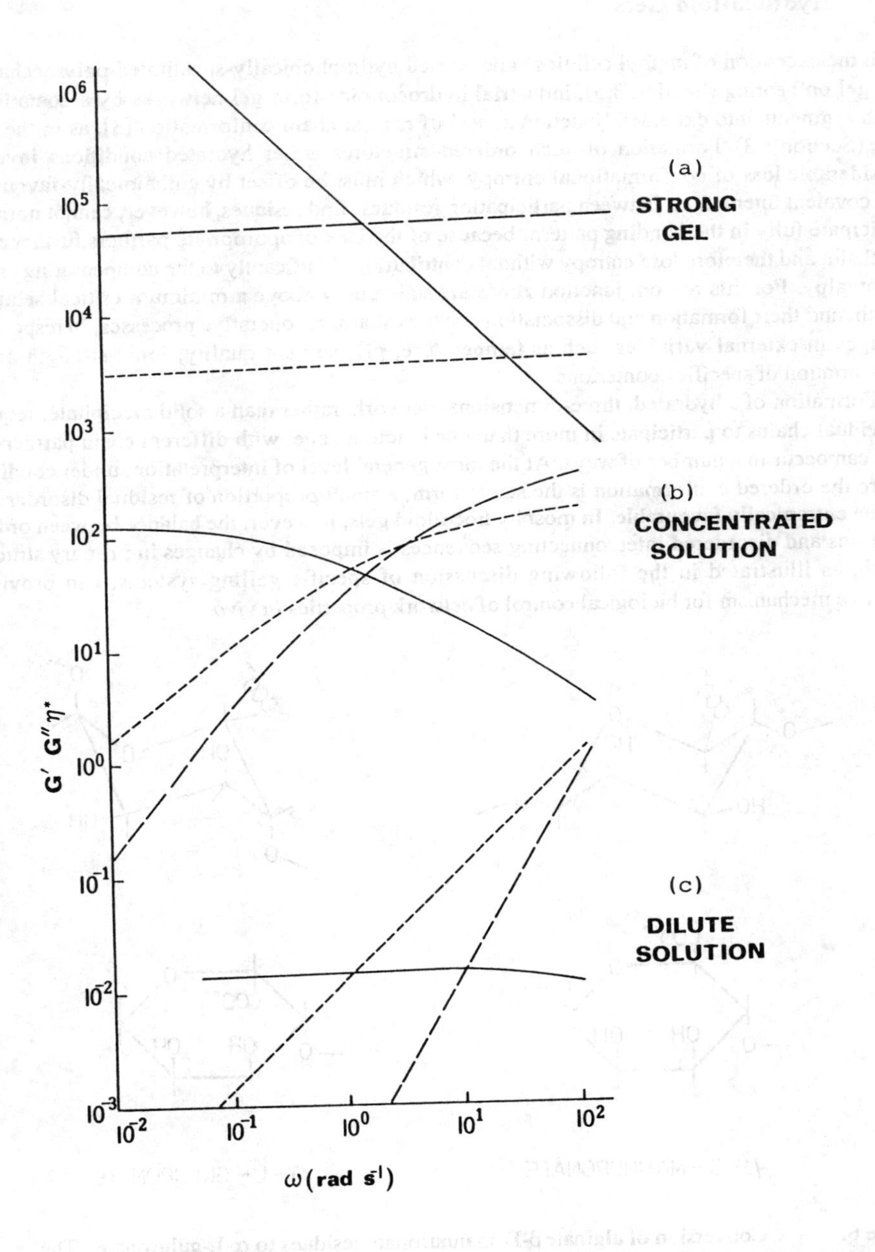

Figure 5. **Frequency (ω) dependence of G' (- - -), G" (-----) and η^* (———) for typical hydrocolloid systems. a) Strong gel; b) Solution of entangled coils; c) Dilute solution of isolated coils.**

3. Hydrocolloid Gels

With the exception of methyl cellulose and related hydrophobically-substituted polysaccharides that gel on heating (Section 3.5), industrial hydrocolloids form gel networks by association of chain segments into extended 'junction zones' of regular chain conformation [3], as in the solid state (Section 1.3). Formation of such ordered structures under hydrated conditions involves considerable loss of conformational entropy, which must be offset by enthalpically-favourable non-covalent interactions between participating residues. End residues, however, cannot normally participate fully in the bonding pattern, because of the lack of appropriate partners further along the chain, and therefore lose entropy without contributing significantly to the compensating change in enthalpy. For this reason, junction zones are stable only above a minimum critical sequence length, and their formation and dissociation occur as sharp, co-operative processes, in response to changes in external variables such as temperature, pH, solvent quality, ionic strength and/or concentration of specific counterions.

Formation of a hydrated, three-dimensional network, rather than a solid precipitate, requires individual chains to participate in more than one junction zone, with different chain partners [3]. This can occur in a number of ways. At the most general level of interpretation, under conditions where the ordered conformation is the stable form, a small proportion of residual disorder may still be entropically favourable. In most hydrocolloid gels, however, the balance between ordered junctions and disordered interconnecting sequences is imposed by changes in primary structure which, as illustrated in the following discussion of specific gelling systems, can provide a sensitive mechanism for biological control of network properties *in vivo.*

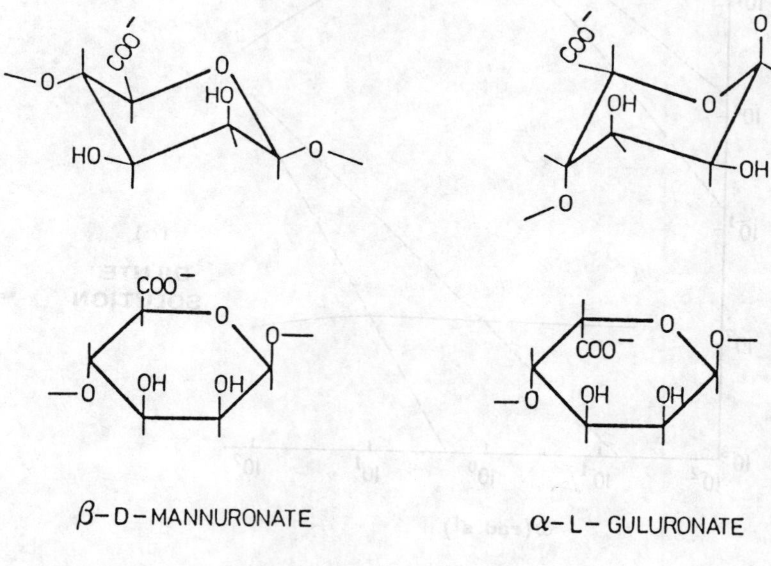

β-D-MANNURONATE α-L-GULURONATE

Figure 6. Conversion of alginate β-D-mannuronate residues to α-L-guluronate. The bottom structures show that change in configuration is confined to epimerisation at C(5). The top structures show the change in ring geometry required to retain the carboxyl group in an equatorial location.

3.1 ALGINATE

The charged polysaccharide alginate, from brown seaweed, forms gels with divalent cations (particularly Ca^{2+}). Alginate [7] is biosynthesised as a 1,4-diequatorially-linked homopolymer of β-D-mannuronate [in which the hydroxymethyl group at C(6) of mannose is replaced by a carboxyl substituent]. As discussed in Section 1.3, the 1,4-diequatorial linkage pattern gives rise to a flat, extended, ribbon-like chain geometry, and in this form the polymer is freely soluble under most ionic conditions. On maturation of the plant, however, the carboxyl group at C(6) of some residues is interchanged enzymically (Fig. 6) with H(5), so that the sugar now becomes α-L-guluronate. To return the C(6) group to the stable, equatorial position, the sugar ring must convert from its original $^4C_1$ conformation to the alternative $^1C_4$ chair form, with consequent conversion of linkage geometry from 1,4-diequatorial to 1,4-diaxial. Thus instead of a flat, extended structure, the ordered conformation of poly-L-guluronate is a highly buckled, two-fold 'zig-zag' and, as illustrated schematically in Fig. 3b, can pack together, with inclusion of an array of site-bound divalent cations, to form the junction zones of the gel network [3].

Calcium ion activity [8] in the presence of mannuronate oligomers of different chain length follows the behaviour expected from polyelectrolyte theory for non-specific 'condensation' of cations onto the polyanion. In the case of polyguluronate, by contrast, there is a sharp reduction in Ca^{2+} activity above a chain length of ~ 20 residues, indicating the onset of a specific co-operative binding process. Polyguluronate chain sequences above, but not below, this critical chain length show large calcium-induced changes in the circular dichroism (CD) of their carboxyl groups, closely similar to the changes observed on formation of calcium alginate gels. It would therefore appear that the critical sequence length for polyguluronate 'egg box' junctions (Fig. 3b) is ~ 20 residues, corresponding to a minimum array of 10 site-bound calcium ions.

Equilibrium dialysis studies of calcium binding to alginates of different composition show that in each case the amount of tightly bound Ca^{2+} resistant to displacement by swamping concentrations of monovalent ions (Na^+) is equivalent to half the stoichiometric requirement of the polyguluronate chain sequences present. A direct explanation of this behaviour is that the most stable structure for calcium polyguluronate is a dimer, in which the interior faces of each of the two-fold chains are involved in cation binding, while the exterior faces are not [3].

Under biological conditions, C(5) epimerisation of β-D-mannuronate to α-L-guluronate (Fig.6) does not go to completion, but confers a block-like structure [7] in which long sequences of fully epimerised poly-L-guluronate are interspersed by residual stretches of unconverted poly-D-mannuronate, and by heteropolymeric regions, neither of which have the geometry necessary for formation of 'egg-box' junctions. The extent of conversion varies throughout the plant, tailoring physical properties to biological function. Thus in the holdfast, where mechanical rigidity is required for anchorage to crevices in the rock, there is a very high content of polyguluronate, whereas fronds, which need flexibility to resist wave damage, have a higher residual mannuronate content, with stipes having alginate of intermediate composition.

The conversion process is also sensitive to calcium. High levels of Ca^{2+} promote formation of heteropolymeric sequences (ranging in structure from almost regular alternation of mannuronate and guluronate to almost random), whereas polyguluronate is formed preferentially at lower concentrations of Ca^{2+}. Thus when the calcium content of the sea-water is low, the alginate is converted to the form that can utilise it most effectively; conversely, at high Ca^{2+} concentrations, where a high content of polyguluronate would make the tissue too brittle, enzymic activity is switched to production of sequences that are less effective in gel formation [7].

3.2 PECTIN

Pectin, from the soft tissue of higher plants, is in many ways the land-based counterpart of alginate in sea plants. Like alginate it is based on a 1,4-diaxially-linked backbone, in this case composed of α-D-galacturonate residues, and again forms gels with Ca^{2+}. Poly-D-galacturonate is the mirror-image of poly-L-guluronate, except in the configuration at C(2), and so an analogous mechanism of cation binding might be anticipated. The calcium-induced CD changes observed for polygalacturonate and polyguluronate are almost exactly equal and opposite, as expected for near mirror-image molecules undergoing the same change in molecular organisation [3]. Calcium ion activity in solutions of galacturonate oligomers [8] also shows a sharp decrease above a minimum critical chain length (in this case ~ 15 residues), and again the concentration of tightly bound Ca^{2+} resistant to displacement with monovalent ions is 50% of the stoichiometric requirement, indicating a closely similar dimeric 'egg-box' structure (Fig. 3b).

Unlike alginate, pectin contains only one type of uronic acid, and inter-chain association is limited in a different way [3]. First, some of the galacturonate residues occur as the methyl ester (Fig. 7a) and therefore can make no electrostatic contribution to calcium binding. There is evidence to suggest that at an early stage of biosynthesis pectin exists in a fully (or almost fully) esterified form which, like the polymannuronate precursor of mature alginate, is incapable of binding calcium. As in alginate, the functionally-active form is then generated by enzymic modification at the polymer level (in this case by de-esterification). The residual ester content varies between different pectin fractions, with corresponding modulation of network properties.

A further, invariant, limitation on junction zone formation is the occurrence in the polymer backbone of occasional residues of 1,2-linked L-rhamnose (Fig. 7b) These are sterically incompatible with incorporation in the 'egg-box' structure, and therefore introduce a 'kink' which terminates inter-chain association. While polyuronates are very resistant to acid hydrolysis, the rhamnosyl linkage is comparatively labile, and on mild hydrolysis the chain is therefore cleaved preferentially at these 'kinking' residues. Chromatographic analysis of the resulting fragments shows a very narrow distribution of molecular weight, corresponding to an uninterrupted sequence length of ~ 25 galacturonate residues between 'kinks'. This is comfortably (but not wastefully) above the minimum sequence length of ~ 15 residues required for co-operative binding of Ca^{2+}. The calcium pectate gel network therefore incorporates two solubilising features (Fig. 7): rhamnosyl 'kinks', which sharply delimit potential junction zones, and esterification, which provides a more subtle mechanism for control of physical properties.

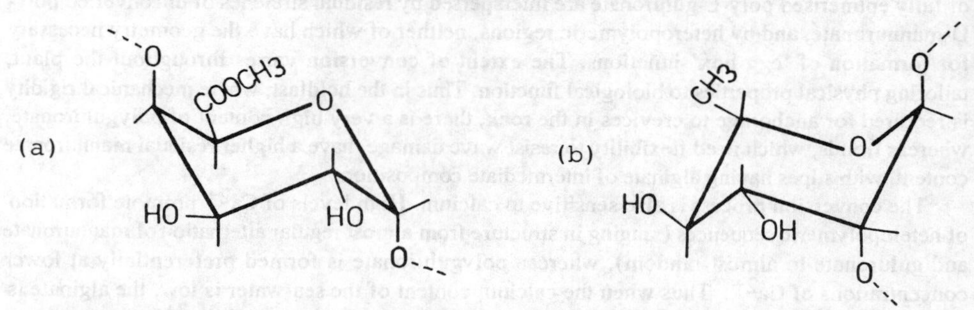

Figure 7. Solubilising features in pectin. a) Methyl esterification of galacturonate residues; b) 1,2-linked L-rhamnose 'kinks'.

3.3 AGAR AND CARRAGEENAN

'Kinking' residues are also important for the gel properties of polysaccharides in the agar and carrageenan families from marine red algae [7]. In both series the ordered conformation [3] is a coaxial double helix, formed by an alternating repeating sequence in which every second residue is a 3,6-anhydride (Fig. 8). The anhydride bridge has the effect of holding the sugar ring in the normally unfavoured chair conformation with C(6) axial. As in the case of alginate, however, these polysaccharides are synthesized initially in a soluble form in which the anhydride bridge is absent, so that the sugar ring adopts the normal chair conformation, which is incompatible with the geometry of the double helix. Subsequent enzymic conversion to the anhydride ring form does not go to completion, and residual unbridged residues act as 'kinks' which limit the extent of helix formation and thus solubilise the network.

Development of a three-dimensional network crosslinked solely through coaxial double helices presents obvious topological problems, and in practice this mechanism of interchain association seems to terminate after the formation of small soluble clusters (or 'domains') consisting typically of about 10 chains, with exchange of partners at 'kinking' residues [3]. Further crosslinking of these 'domains' into a cohesive gel structure involves side-by-side association of double helices from different domains. In the case of highly charged polysaccharides, such as iota carrageenan (Fig. 8b), helix-helix aggregation (and thus gelation) occurs only in the presence of cations (typically K^+ or Ca^{2+}) that can suppress electrostatic repulsion between the participating chains by packing within the aggregate structure.

Aggregation is particularly extensive for the neutral polysaccharide agarose (where no cations are, of course, required), resulting in brittle, turbid gels at an unusually low polymer concentration. Incorporation of helices within the aggregate structure increases their thermal stability, with the result that agarose gels show very pronounced thermal hysteresis, melting at temperatures considerably higher than the setting temperature of the same gel.

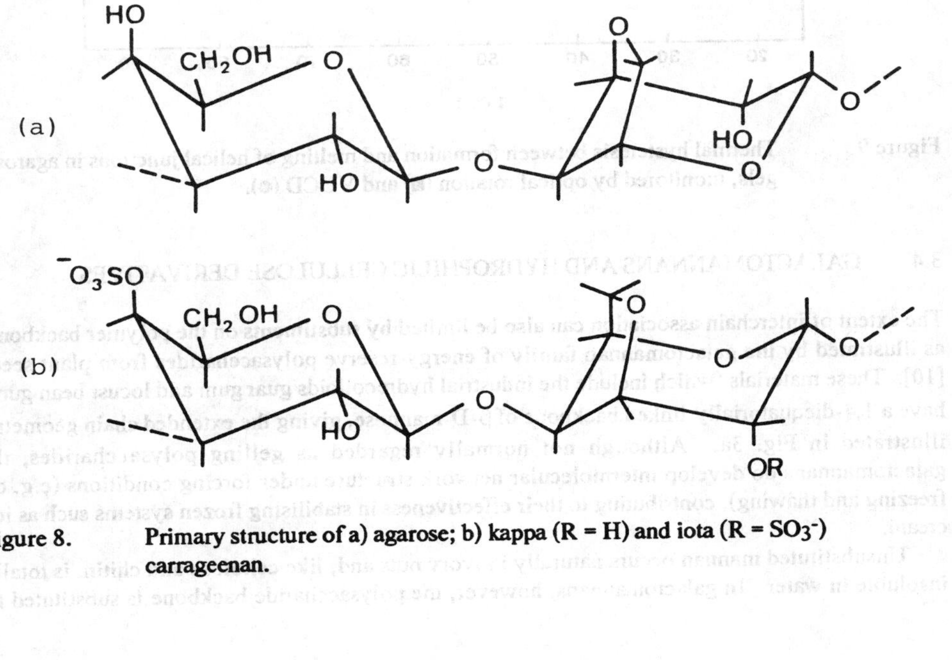

Figure 8. Primary structure of a) agarose; b) kappa (R = H) and iota (R = SO_3^-) carrageenan.

Figure 9 shows the temperature course of helix formation and melting for agarose, as monitored [9] by two different chiroptical methods: optical rotation, which provides a convenient empirical probe of changes in chain geometry, and vacuum ultraviolet circular dichroism (VUCD), a recently developed technique which now allows direct measurement of conformation-sensitive electronic transitions of the polysaccharide backbone, directly analogous to the established application of conventional CD measurements to proteins and polynucleotides.

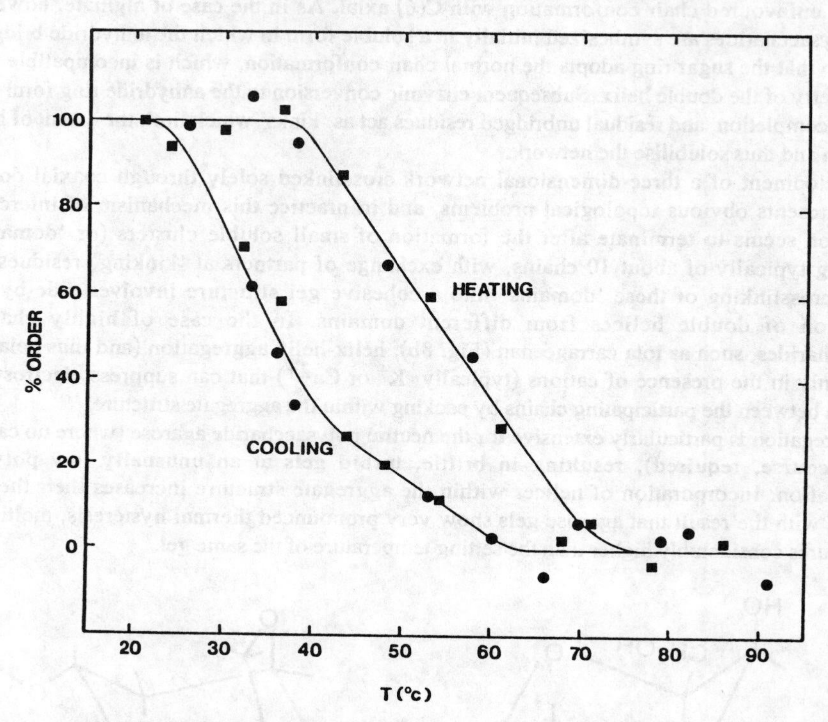

Figure 9. Thermal hysteresis between formation and melting of helical junctions in agarose gels, monitored by optical rotation (■) and VUCD (●).

3.4 GALACTOMANNANS AND HYDROPHILIC CELLULOSE DERIVATIVES

The extent of interchain association can also be limited by substituents on the polymer backbone, as illustrated by the galactomannan family of energy-reserve polysaccharides from plant seeds [10]. These materials (which include the industrial hydrocolloids guar gum and locust bean gum) have a 1,4-diequatorially-linked backbone of β-D-mannose, giving the extended chain geometry illustrated in Fig. 3a. Although not normally regarded as gelling polysaccharides, the galactomannans do develop intermolecular network structure under forcing conditions (e.g. on freezing and thawing), contributing to their effectiveness in stabilising frozen systems such as ice cream.

Unsubstituted mannan occurs naturally in ivory nuts and, like cellulose and chitin, is totally insoluble in water. In galactomannans, however, the polysaccharide backbone is substituted to varying extents by 1,6-linked monosaccharide sidechains of α-D-galactose. These interfere with solid-state packing, and provide an entropic drive to conformational disorder (rotation about the sidechain-backbone linkage), so that highly substituted galactomannans, such as guar gum, are freely soluble. At lower degrees of substitution (as in locust bean gum), however, a delicate balance is achieved, with association and packing of unsubstituted or sparingly substituted regions of the backbone to form a crosslinked network, which is solubilised by more fully substituted chain sequences.

Chemical substituents have a similar solubilising role [11] in cellulose derivatives such as CMC (carboxymethylcellulose). The initial stage in production of these materials [1] usually involves disruption of the native fibrillar structure by swelling in alkali, to permit penetration of the reagents required for chemical derivitisation, with the reaction proceeding preferentially from the sites of initial attack to give a block-like distribution of substituents. As in the galactomannans, unsubstituted or sparingly substituted regions of the polymer backbone have a tendency to re-associate, giving rise to a thixotropic network.

3.5 HYDROPHOBICALLY-SUBSTITUTED CELLULOSE DERIVATIVES

As well as the hydrophilic (e.g. carboxymethyl) substituents mentioned above, commercial hydrocolloids are also produced by derivitisation of cellulose with hydrophobic substituents, in particular methyl and/or hydroxypropyl groups. An unusual feature of these materials is that they form gels on heating, by clustering of the hydrophobic substituents into a micellar structure. In the sol state the hydrophobic groups are believed to be stabilised by a 'cage' of structured water molecules, which is disrupted on heating [11]. The resulting gels are highly turbid or opaque and dissociate on cooling, usually at temperatures substantially lower than those at which they form.

3.6 GELATIN

Gelatin [12] is a denatured form of collagen, the structural protein of animal connective tissue. In contrast to normal globular proteins, collagen has extended regions of quasi-repeating structure, every third residue being glycine and about half of the remaining residues being proline or hydroxyproline, both of which have a ring structure incompatible with the α-helix and β-sheet conformations commonly adopted by protein chains. Instead, collagen forms a coaxial triple helix, which provides the crosslinks in gelatin gels. Current evidence [13] suggests that the junction zones are formed by two strands from one molecule, folded back into a hairpin structure, together with a third strand from another chain.

Although in principle the peptide units of which proteins are composed can exist in two isomeric forms, *cis* and *trans*, the *cis* form normally has much higher energy and therefore a very low probability of occurrence. In proline and hydroxyproline, however, the cyclic structure reduces the energy difference, so that in gelatin, in the high-temperature sol state, a substantial proportion of these residues exist in the *cis* form. Only the *trans* conformation, however, can be incorporated in the triple helix structure. The activation-energy barrier between the two forms is very high (~ 85 kJ mol^{-1}), so that isomerisation of *cis* residues limits the rate of helix growth [14] relative to the competing back reaction (helix unwinding). Hence, in comparison with the polysaccharide systems discussed above, gelation of gelatin is extremely slow.

4. Rheological Properties of 'weak gels'

As described in Section 3, solutions of conformationally-disordered ('random coil') polysaccharides interacting only by physical entanglement show the following striking generalities of behaviour, irrespective of their chemical composition or chain length:

1) Their shear-thinning behaviour has a general form which can be quantified by:

   $\eta = \eta_o / [1 + (\dot{\gamma}/\dot{\gamma}_{1/2})^{0.76}]$

2) Zero-shear viscosity (η_o) shows a generalised dependence on c[η], with a marked change in concentration-dependence at $c^* \approx 4/[\eta]$.
3) Their mechanical spectra (frequency-dependence of G' and G") have a characteristic form, with G" > G' at low frequency and G' > G" at high frequency.
4) The frequency (ω) dependence of η* from 'small deformation' oscillatory measurements is superimposable on the shear-rate ($\dot{\gamma}$) dependence of η from 'large deformation' unidirectional measurements.

Certain polysaccharide solutions, however, violate all the above generalities. The best characterised example is xanthan (the extracellular bacterial polysaccharide from *Xanthomonas campestris*), which has achieved rapid commercial success due to its unusual, and technologically valuable, rheological properties.

4.1 XANTHAN

Although solutions of xanthan flow freely, they show mechanical spectra similar to those of gels: G' exceeds G" at all frequencies, with little frequency-dependence in either modulus [4]. In practical terms [1] this gel-like character is reflected in the ability of xanthan to hold particles in suspension or stabilise emulsions over long periods. The shear-thinning behaviour of xanthan is also unusual; the gradient of double-logarithmic plots of η *vs.* $\dot{\gamma}$ is usually substantially higher than the maximum value of -0.76 observed for entangled solutions and, in normal commercial salt forms, there is no evidence of a 'Newtonian plateau' at low shear-rates. Xanthan solutions also violate the Cox-Merz rule: double logarithmic plots of η *vs.* $\dot{\gamma}$ and η* *vs.* ω are approximately parallel, but the absolute values of η* are substantially higher than those of η at equivalent frequencies and shear rates [4], the divergence increasing with increasing concentration of polymer.

The origin of these unusual solution properties can be traced to the conformational behaviour of xanthan. At high temperature and/or low ionic strength, xanthan exists in solution in a fluctuating, disordered form, and displays the normal properties of a 'random coil'. On cooling, or on addition of salt, however, it undergoes [15] a sharp, co-operative transition to a rigid, ordered conformation and in this form displays the unusual 'weak gel' properties outlined above. The detailed nature of the ordered structure remains the subject of some controversy.

Xanthan has a cellulose backbone, with alternate residues carrying charged trisaccharide sidechains. Solid-state x-ray evidence is equally consistent with two possible models, a co-axial double helix and a single helix stabilised by ordered packing of sidechains along the polymer backbone [3]. Arguments in favour of both models have been advanced by different investigators from solution studies by various techniques. For reasons discussed in detail elsewhere [15], the present author believes that the ordered structure is a single helix which, under certain salt conditions, can form side-by-side dimers, indistinguishable by many criteria from coaxial double helices.

Despite this uncertainty, however, there seems little doubt that the rheological properties of xanthan arise from weak association of ordered chain segments to form a continuous three-dimensional network. In contrast to 'true' gels, however, the xanthan 'weak gel' network is not strong enough to be self-supporting against gravity, but can resist smaller stresses such as those imposed by small particles in suspension. It will also remain intact under small strains, but is broken down by larger deformations. Hence values of η^* from small deformation oscillatory measurements (where the network is conserved) are higher than the corresponding values of η from rotational measurements, where the network is destroyed.

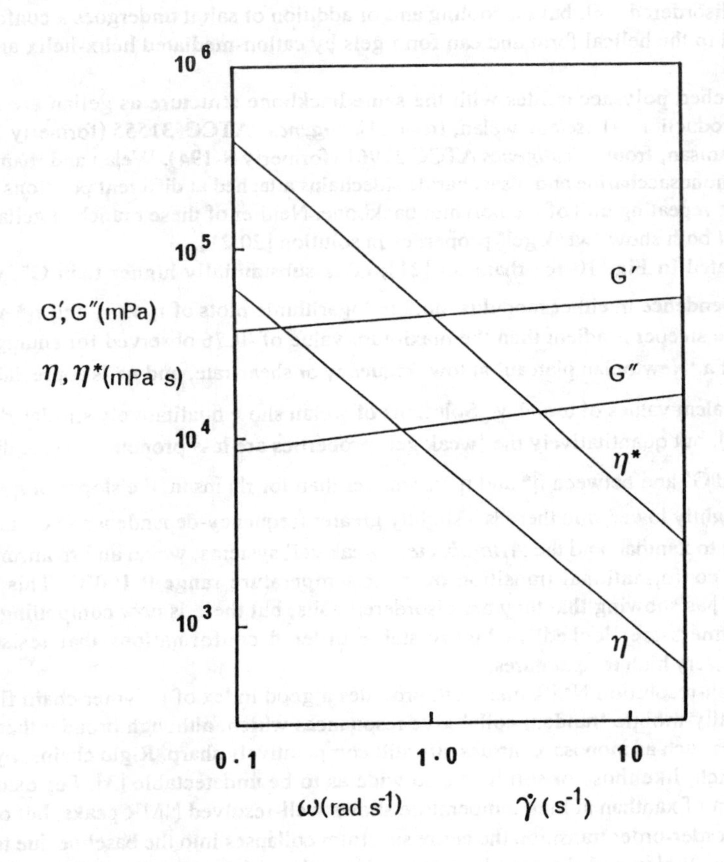

Figure 10. Solution properties [21] of rhamsan (1% w/v; Na^+ salt form). Compare with the mechanical spectra shown in Fig. 5 for typical polysaccharide solutions and gels.

4.2 OTHER BACTERIAL POLYSACCHARIDE 'WEAK GELS'

Other polysaccharides which, like xanthan, adopt an ordered conformation in solution also show analogous rheological properties. The bacterial screening programme which led to the discovery of xanthan also found similar properties in polysaccharides from certain *Arthrobacter* species. Subsequent studies showed these materials to be conformationally-ordered [16]. More recently the Kelco Division of Merck has developed a series of bacterial polysaccharides [17] that share the same polymer backbone and show unusual rheological behaviour. The parent member of the series is gellan gum (formerly known as S-60), produced as an extracellular polysaccharide by *Auromonas elodea*(ATCC 31461).

Gellan has a linear tetrasaccharide repeating sequence and adopts a three-fold, parallel double-helix structure in the solid state [18]. At high temperature and low ionic strength gellan exists in solution as a disordered coil, but on cooling and/or addition of salt it undergoes a conformational transition [17] to the helical form and can form gels by cation-mediated helix-helix aggregation [19,20].

Two branched polysaccharides with the same backbone structure as gellan are already in commercial production. These are welan, from *Alcaligenes* ATCC 31555 (formerly known as S-130) and rhamsan, from *Alcaligenes* ATCC 31961 (formerly S-194). Welan and rhamsan have, respectively, monosaccharide and disaccharide sidechains attached at different positions within the tetrasaccharide repeating unit of the polymer backbone. Neither of these branched gellan variants forms gels, but both show 'weak gel' properties in solution [20,21].

As illustrated in Fig. 10 for rhamsan [21], G' is substantially higher than G", with little frequency-dependence in either modulus, double-logarithmic plots of η *vs.* $\dot{\gamma}$ and η^* *vs.* ω are linear and have steeper gradient than the maximum value of -0.76 observed for entangled coils, with no sign of a 'Newtonian plateau' at low frequency or shear-rate, and η^* is appreciably higher than η at equivalent values of ω and $\dot{\gamma}$. Solutions of welan show qualitatively similar rheological behaviour [20], but quantitatively the 'weak gel' properties are less pronounced (the differences between G' and G" and between η^* and η are smaller than for rhamsan, the slopes of η *vs.* $\dot{\gamma}$ and η^* *vs.* ω are slightly lower, and there is a slightly greater frequency-dependence in G' and G").

In contrast to xanthan and the *Arthrobacter* 'weak gel' systems, welan and rhamsan show no evidence of a conformational transition over the temperature range 0-100°C. This has been interpreted [17] as showing that they are disordered coils, but there is now compelling evidence that both polymers are 'locked' in highly stable ordered conformations that resist thermal denaturation to very high temperatures.

Firstly, high-resolution NMR linewidth provides a good index of polymer chain flexibility. Conformationally-mobile 'random coils' give resonances which, although broader than those of small molecules such as monosaccharides, are still comparatively sharp. Rigid chains, by contrast, give peaks which, like those of solids, are so wide as to be undetectable [3]. For example, the disordered form of xanthan at high temperature shows well-resolved NMR peaks, but on cooling through the disorder-order transition the entire spectrum collapses into the baseline due to extreme line-broadening. Welan and rhamsan both show featureless high-resolution NMR spectra at all accessible temperatures, indicative of rigid chain geometry [21].

As outlined previously (Section 2.4), the response of polyelectrolyte intrinsic viscosity, $[\eta]$, to changes in ionic strength (I) also gives an empirical index of macromolecular flexibility. Figure 11 shows the dependence of $[\eta]$ on $I^{-1/2}$ for welan, rhamsan and one of the least flexible disordered polysaccharides (sodium alginate). As expected from polyelectrolyte theory, the plots are linear. Empirically, the ratio of the slope of such plots to the measured intrinsic viscosity at fixed ionic strength (0.1; $I^{-1/2} = 3.16$) correlates directly [5] with chain flexibility (i.e. the extent of molecular expansion in response to increased intramolecular electrostatic repulsion on decreasing ionic strength). By this criterion also, welan [20,21] and rhamsan [21] are far less flexible than the stiffest known disordered polysaccharides, again indicating that both are conformationally ordered.

The intrinsic viscosity of rhamsan (~ 150 dl g^{-1}) is remarkably high. By comparison, commercial xanthans normally have intrinsic viscosities around 40-50 dl g^{-1}, while values for 'random coil' polysaccharides of high molecular weight (~ 10^6 or greater) are typically in the range 5-25 dl g^{-1}, depending on backbone geometry and chain flexibility [5].

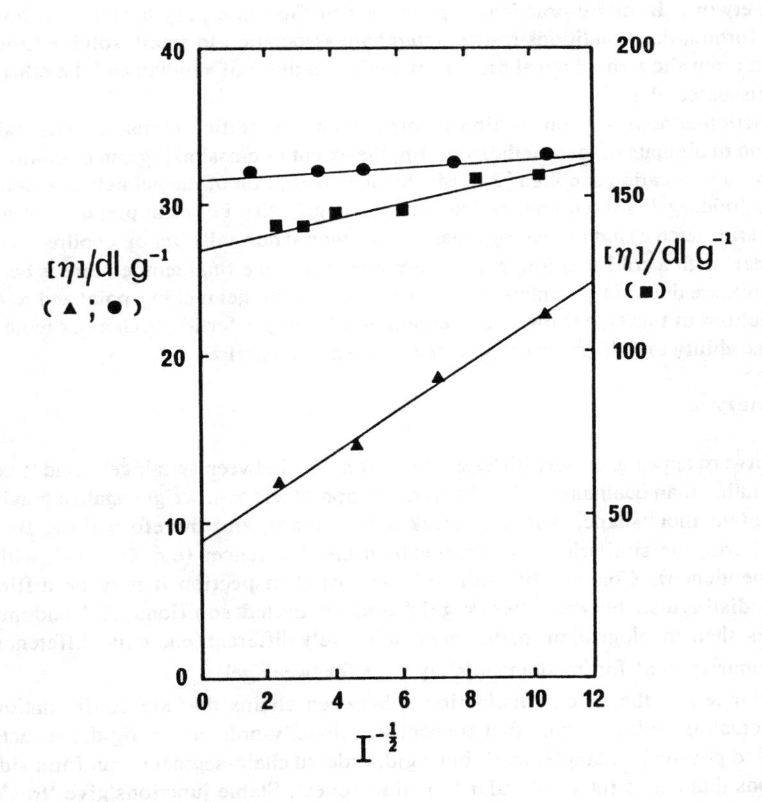

Figure 11. Variation [21] of intrinsic viscosity, [η], with ionic strength (I) for welan (●) and rhamsan (■) in comparison with a stiff 'random coil' polysaccharide, alginate (▲). Note that the intrinsic viscosities of rhamsan (right-hand axis) are much higher than those of welan and alginate (left-hand axis).

In contrast to the parent polysaccharide, gellan [19], the rheological properties of welan and rhamsan show little, if any, dependence on the nature and concentration of the counterions present [21]. A possible interpretation of this behaviour, and of their conformational stability at high temperature, is that both adopt the double-helical structure characterised for gellan, but that the sidechains fold down along the polymer backbone, giving additional stability to the ordered structure but inhibiting the cation-mediated aggregation process [19,20] necessary for gel formation.

4.3 DISCONTINUOUS NETWORKS OF GELLING POLYSACCHARIDES

As discussed in Section 3, formation of 'true' polysaccharide gels normally involves crosslinking of chains through conformationally-ordered 'junction zones' to develop a continuous three-dimensional network. In many practical applications of the same polysaccharides, however, processing or formulation conditions restrict interchain association to small, soluble 'microgel' clusters, which often show rheological properties similar to those of xanthan and the other 'weak gel' systems discussed above.

Where junction-zone formation requires incorporation of specific cations, as in the calcium-induced gelation of alginate or low-methoxy pectin, the extent of crosslinking can be controlled by limiting the amount of cation supplied [22]. Mechanical disruption of the gel network as it forms can also give a 'microgel' solution rather than a cohesive gel [23]. For example, if a hot solution of a polysaccharide such as agar or carrageenan, which would normally gel on cooling, is mixed, under high shear, with a cold solution (e.g. of sucrose) so that the final temperature is below the gel point, the mixture does not gel unless heated above the normal gel melting-point and re-cooled. 'Microgel' solutions of this type show characteristic 'weak gel' properties, such as extreme shear-thinning and the ability to hold particles in suspension over long periods.

5. Summary

Although to outward appearance very different, the distinction between 'weak gels' and 'true gels' is quantitative rather than qualitative. 'True' gels can support their own weight against gravity, and therefore maintain their shape, whereas 'weak gels' cannot, and therefore flow. By many rheological criteria, the similarities are greater than the differences (e.g. $G' > G''$, with little frequency-dependence). Conversely, although on visual inspection it may be difficult or impossible to distinguish between 'weak gels' and entangled solutions of 'random coil' polysaccharides, their rheological properties are qualitatively different (e.g. quite different shear-thinning behaviour; $\eta = \eta^*$ for 'random coils', $\eta < \eta^*$ for 'weak gel').

In molecular terms, the crucial distinction is between chains that are conformationally-disordered, fluctuating coils and those that are conformationally-ordered and rigid. Interaction of coils is limited to physical 'entanglement', but rigid, ordered chain-segments can form side-by-side associations that act as intermolecular 'junction zones'. Stable junctions give 'true' gels; junctions of lower binding energy give 'weak gels', but the difference is one of degree rather than of kind. Conversion of a fluctuating, disordered coil to a rigid, ordered structure involves a substantial loss of conformational entropy and will therefore occur only if the enthalpic contributions from non-covalent interactions within the ordered structure are strong (e.g. co-operative arrays of hydrogen bonds or anion-cation interactions). This is often the case in 'true' gels. Side-by-side association of pre-existing ordered structures, by contrast, involves little loss of entropy and may therefore be induced by weak attractions such as van der Waals forces or dipolar interactions. This is the situation in 'weak gels'.

ACKNOWLEDGEMENTS

I thank Dr G. Robinson for unpublished results (Figures 10 and 11) from his PhD research [ref. 21].

REFERENCES

[1] Whistler, R.L. and BeMiller, J.N. (1973) 'Industrial Gums', Academic Press, New York.
[2] Stoddart, J.F. (1971) 'Stereochemistry of Carbohydrates', Wiley, New York.
[3] Rees, D.A., Morris, E.R., Thom, D. and Madden, J.K. (1982) 'Shapes and interactions of carbohydrate chains', in G.O. Aspinall (ed.), The Polysaccharides, Vol. 1, Academic Press, New York, pp. 195-290.
[4] Morris, E.R. (1984) 'Rheology of hydrocolloids', in G.O. Phillips, D.J. Wedlock and P.A. Williams (eds.), Gums and Stabilisers for the Food Industry 2, Pergamon Press, Oxford, pp. 57-78.
[5] Morris, E.R. and Ross-Murphy, S.B. (1981) 'Chain flexibility of polysaccharides and glycoproteins from viscosity measurements', Techniques in Carbohydrate Metabolism, B310, Elsevier, London, pp. 1-46.
[6] Morris, E.R. (1990) 'Shear thinning of random coil polysaccharides: characterisation by two parameters from a simple linear plot', Carbohydrate Polymers, 13, 85-96.
[7] Painter, T.J. (1983) 'Algal polysaccharides', in G.O. Aspinall (ed)., The Polysaccharides, Vol. 2, Academic Press, Orlando, Florida, pp. 195-285.
[8] Kohn, R. (1975) 'Ion binding of polyuronates - alginate and pectin', Pure and Applied Chemistry, 42, 371-397.
[9] Morris, E.R., Stevens, E.S., Frangou, S.A. and Rees, D.A. (1986) 'Total optical rotation of agarose: relation to observable transitions in the vacuum ultraviolet', Biopolymers, 25, 959-973.
[10] Dea, I.C.M. and Morrison, A. (1975) 'Chemistry and interactions of seed galactomannans', Advances in Carbohydrate Chemistry and Biochemistry, 31, 241-312.
[11] Rees, D.A. (1972) 'Polysaccharide gels: a molecular view', Chemistry and Industry, 630-636.
[12] Ward, A.G. and Courts, A. (1977) 'The Science and Technology of Gelatin', Academic Press, London.
[13] Busnel, J.P., Morris, E.R. and Ross-Murphy, S.B. (1989) 'Interpretation of the renaturation kinetics of gelatin solutions', International Journal of Biological Macromolecules, 11, 119-125.
[14] Bächinger, H.P., Bruckner, P., Timpl, R. and Engel, J. (1978) 'The role of cis-trans isomerisation of peptide bonds in the coil-triple helix conversion of collagen', European Journal of Biochemistry, 90, 605-614.
[15] Norton, I.T., Goodall, D.M., Frangou, S.A., Morris, E.R. and Rees, D.A. (1984) 'Mechanism and dynamics of conformational ordering in xanthan polysaccharide', Journal of Molecular Biology, 175, 371-394.
[16] Darke, A., Morris, E.R., Rees, D.A. and Welsh, E.J. (1978) 'Spectroscopic characterisation of order-disorder transitions for extracellular polysaccharides of *Arthrobacter* species', Carbohydrate Research, 66, 133-144.
[17] Crescenzi, V., Dentini, M. and Dea, I.C.M. (1986) 'The influence of sidechains on the

dilute solution properties of three structurally related bacterial anionic polysaccharides', Carbohydrate Research, 160, 283-302.

[18] Chandrasekaran, R., Millane, R.P., Arnott, S. and Atkins, E.D.T. (1988) 'The crystal structure of gellan', Carbohydrate Research, 175, 1-15.

[19] Grasdalen, H. and Smidsrod, O. (1987) 'Gelation of gellan gum', Carbohydrate Polymers, 7, 371-393.

[20] Robinson, G., Manning, C.E., Morris, E.R. and Dea, I.C.M. (1988) 'Sidechain-mainchain interactions in bacterial polysaccharides', in G.O. Phillips, D.J. Wedlock and P.A. Williams (eds.), Gums and Stabilisers for the Food Industry 4, IRL Press, Oxford, pp. 173-181.

[21] Robinson, G. (1990) 'Functional properties of industrial polysaccharides', PhD Thesis, Cranfield Institute of Technology, UK.

[22] Sime, W.J. (1984) 'The practical utilisation of alginates in food gelling systems', in G.O. Phillips, D.J. Wedlock, and P.A. Williams (eds.), Gums and Stabilisers for the Food Industry 2, Pergamon Press, Oxford, pp. 177-188.

[23] Harris, P. and Pointer, S.J. (1986) 'Edible gums', UK Patent GB 2128871B.

KINETICS OF THE INTERACTION OF THE POTENTIAL-SENSITIVE DYE OXONOL V WITH LIPID VESICLES

R.J. CLARKE[1] AND H.-J. APELL[2]
1) School of Chemical Sciences, University of East Anglia, Norwich NR4 7TJ, U.K.
2) Department of Biology, University of Konstanz, D-7750 Konstanz, F.R.G.

ABSTRACT. The interaction of the dye oxonol V with unilamellar dioleoylphosphatidylcholine vesicles was investigated using a fluorescence stopped-flow technique. At low dye/lipid concentration ratios, on mixing with the vesicles, the dye exhibits an increase in fluorescence, which occurs in two phases. According to the dependence of the reciprocal relaxation time on vesicle concentration, the rapid phase appears to be due to a second-order binding of the dye to the lipid membrane, which is very close to being diffusion-controlled. The slow phase is almost independent of vesicle concentration, and it is suggested that this may be due to a change in dye conformation or position within the membrane, possibly diffusion across the membrane to the internal monolayer. As one proceeds to higher dye/lipid concentration ratios a third intermediate phase characterised by a decrease in fluorescence appears, which can be attributed to fluorescence quenching via an inner-filter effect as the local concentration of dye in the membrane rises. A kinetic model is presented which successfully predicts the time course of the fluorescence change due to the binding reaction over the complete range of dye and vesicle concentrations.

1. Introduction

Membrane potential is an important parameter in many biological processes, e.g., nerve impulse transmission [1,2], vision [1] and kidney function [3]. Changes of the membrane potential during such processes often come about by the pumping of ions across the cell membrane by enzymes embedded in it, e.g., the Na^+,K^+-ATPase [1-3]. Therefore, in order to investigate the mechanism of such pump enzymes and to gain an insight into biological processes with which they are involved, a means of rapidly monitoring the membrane potential is required. A direct method of doing this is to use microelectrodes [1,2], however this requires the piercing of the membrane and is hence quite an invasive technique and its application is limited to relatively large cells. An alternative technique is to determine the potential indirectly by using potential-sensitive dyes [4-8], which respond to changing potential with a change in their absorption or fluorescence spectrum.

Although the application of potential-sensitive dyes in biochemical research may not necessarily require a complete understanding of the mechanism of the dyes' response, knowledge of their mechanisms may yield information useful in designing better probe molecules. Inthemselves they are also interesting in that they provide a model system for studying the interaction of small hydrophobic species with lipid membranes. Over the last 10 years many

D. M. Bloor and E. Wyn-Jones (eds.),
The Structure, Dynamics and Equilibrium Properties of Colloidal Systems, 471–492.

different dye molecules have been used and a number of different mechanisms have been proposed to explain their response to changing membrane potential. In general, the change in membrane potential must cause a perturbation of the dyes' electronic system or a change in its chemical environment. Possible mechanisms for this include:

(1) a change in concentration of membrane-bound dye [9-17]
(2) a change in position or orientation of the dye in the membrane [10,15,18]
(3) a change in the state of aggregation of the dye [9,10,15,18-20]
(4) an electrochromic effect [21]
(5) a combination of the above

In order to understand how the dye responds to membrane potential, however, its interaction with the lipid membrane in the absence of a potential must be understood.

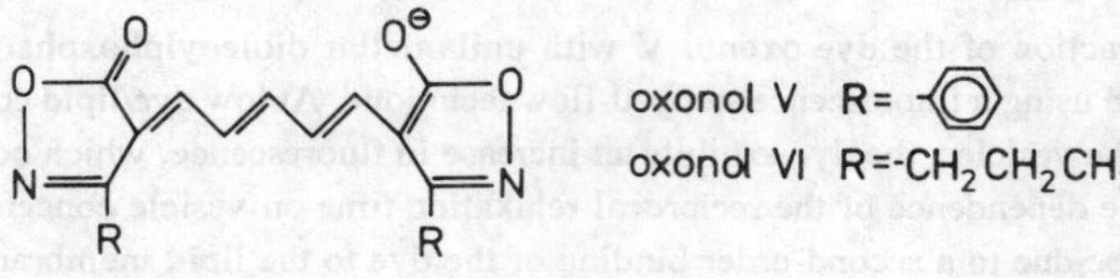

oxonol V R=
oxonol VI R=$-CH_2CH_2CH_3$

Figure 1. Structures of oxonol V and VI.

Oxonol dyes, such as oxonol V and VI (see Fig. 1), have been widely used as potential-sensitive dyes in membrane experiments [22]. They have a pK_a near 4.2 [23] and are thus anionic at physiological pH values, but the charge is delocalized over the molecule so that they are still membrane permeable. Because of their hydrophobicity they bind strongly to lipid membranes and consequent changes in their absorbance and fluorescence spectra are observed. Here equilibrium and kinetic data are presented for the interaction of oxonol V with unilamellar dioleoylphosphatidylcholine vesicles, and a kinetic model is described to explain the time course of the dyes' fluorescence change in a stopped-flow experiment.

2. Materials and Methods

2.1 MATERIALS

Dioleoylphosphatidylcholine was obtained from Avanti Polar Lipids, Birmingham, AL, U.S.A.; oxonol V (bis (3-phenyl-5-oxoisooxazol-4-yl) pentamethine oxonol) was from Molecular Probes, Junction City, OR, U.S.A. The phospholipid contents of the vesicle suspensions were determined by the phospholipid B test from Wako Pure Chemical Industries, Osaka, Japan. Sodium cholate and imidazole were from Serva, Heidelberg, and from Sigma, respectively. All other reagents were obtained from Merck or BDH (analytical grade). Dialysis tubing was purchased from Serva and Medicell International, London.

2.2. VESICLE PREPARATION

Lipid vesicles were prepared from synthetic dioleoylphosphatidylcholine as described previously by a detergent dialysis method with sodium cholate, producing homogeneous unilamellar vesicles with an average outer diameter of 72 nm [24,25]. All vesicle suspensions were prepared in a buffer containing 30 mM imidazole and 75 mM K_2SO_4, adjusted to pH 7.2 with H_2SO_4. Vesicle concentrations were estimated by dividing the lipid concentration by the average number of lipid molecules per vesicle, which was calculated as described previously [22] from the vesicle radius and the partial specific volume of lipid to be 45200.

2.3. STOPPED-FLOW MEASUREMENTS

Stopped-flow experiments were carried out using a stopped-flow unit (ZWS II) of Sigma Instruments, Berlin, with fluorescence detection at right angles to the incident light beam. By the use of appropriate filters the excitation wavelength was maintained at 580 (±5) nm and the emission was observed at wavelengths ≥ 630 nm. The solutions in the drive and reservoir syringes were equilibrated to a temperature of 22° C prior to each experiment. The drive syringes were driven by compressed air, and the mixing time of the stopped-flow unit was determined to be approximately 7 milliseconds.

The interaction of the dyes with phosphatidylcholine vesicles was investigated by mixing a dye solution with an equal volume of vesicle suspension. Experiments were performed at varying dye/lipid concentration ratios. All solutions contained 75 mM K_2SO_4 to prevent any undesired osmotic effects. The relaxation times obtained are the average of at least four transients.

Computer simulations of the stopped-flow transients were performed using a Digital VAX computer.

2.4 STATIC FLUORESCENCE AND ABSORBANCE MEASUREMENTS

Fluorescence measurements were carried out in a Perkin-Elmer LS 50 luminescence spectrometer with a thermostatically controlled cuvette holder. The excitation wavelength was set to 580 nm (slit width 15 nm) and the emission was recorded at wavelengths ≥ 600 nm (slit width 20 nm). The oxonol V stock solution contained 0.27 mM dye in ethanol. 2 µl of this solution was added to 2 ml of buffer in the cuvette to obtain a final (total) oxonol concentration of 270 nM. The fluorescence spectrum was then recorded before and after the addition of varying aliquots of vesicle suspension.

Absorbance measurements were carried out in a Varian Cary 219 spectrophotometer with a thermostatically controlled cuvette holder. Spectrophotometric titrations were performed at constant dye concentration and varying vesicle concentration as well as constant vesicle concentration and varying dye concentration.

3. Equilibrium Results

On addition of vesicles to a dye solution the dye exhibits a red shift of its absorbance maximum of approximately 20 nm and a slight increase in its molar absorptivity at λ_{max} (see Fig. 2). An isosbestic point is observed at approximately 613 nm. Titration of vesicles with dye shows that as the dye concentration increases there is a decrease in the magnitude of the absorbance change due to the vesicles, indicating that the vesicles are saturable with dye (see

Fig. 3). Thus, it seems that the most appropriate description of the dye-vesicle interaction is a binding mechanism, whereby dye binds to vesicles possessing a discrete number of binding sites [26-28]. A partition mechanism involving the transfer of dye from an aqueous to a lipid phase can be excluded, since it does not allow for the possibility of saturation. Accordingly, the spectral data have been interpreted using the Langmuir adsorption isotherm [29-31]:

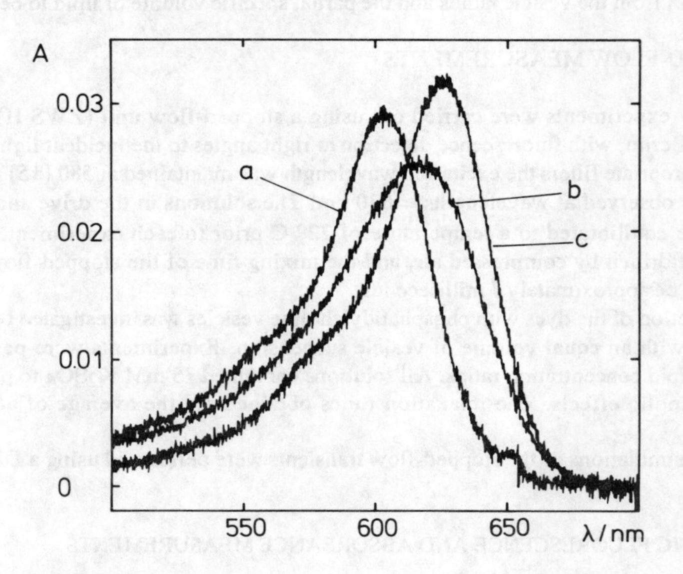

Figure 2. Variation of the oxonol V absorbance spectrum in the presence of dioleoylphosphatidylcholine vesicles; T = 22° C, pH 7.2. a: 266 nM oxonol V alone, b: 266 nM oxonol V + 3,080 nM lipid, c: 266 nM oxonol V + 30,800 nM lipid.

$$\bar{r} = \frac{nK\bar{c}_D}{1 + K\bar{c}_D} \tag{1}$$

where r is the equilibrium concentration of bound dye per unit concentration of vesicles, n is the number of binding sites per vesicle, c_D is the equilibrium concentration of free dye, and K is the association constant for the reaction:

$$\text{Dye} + \text{Free binding site} \overset{K}{\rightleftarrows} \text{Occupied site} \tag{2}$$

Thus, the association constant is defined by

$$K = \frac{[\text{Occupied sites}]_{eq}}{\bar{c}_D\,[\text{Free sites}]_{eq}} \qquad (3)$$

Combining the absorbance data from dye and vesicle titrations, values of $\bar{r}$ and $\bar{c}_D$ can be calculated and plotted according to the Scatchard relationship (see Fig. 4), from which n and K can be determined from the intercept and slope, respectively [32]. The value of r has been expressed as the concentration of bound dye per unit concentration of lipid, so that the intercept, y, represents the number of binding sites per lipid molecule rather than the number of sites per vesicle. The values found are:

$K = 3.31\ (\pm 0.22) \cdot 10^{6}\ M^{-1}$

$y = 0.161\ (\pm 0.003)$ sites per lipid molecule

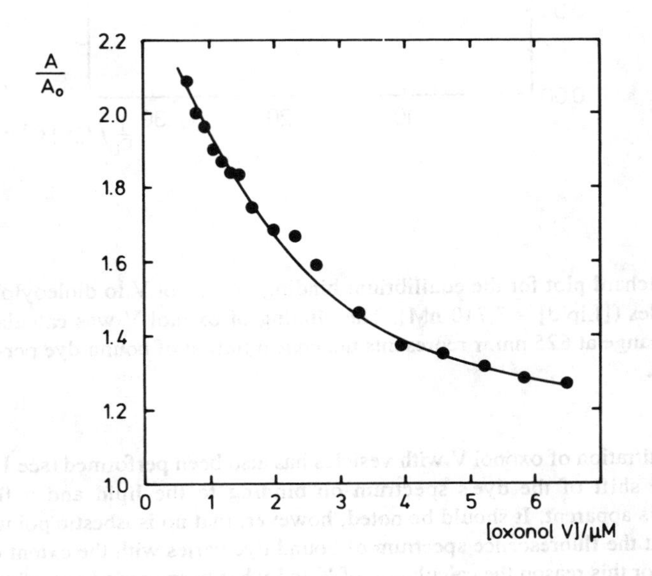

Figure 3. Absorbance A of oxonol V at 625 nm as a function of dye concentration at a constant total lipid concentration of 7,710 nM. A is referred to the absorbance, A_o, at zero lipid concentration.

The value of y indicates that in a fully saturated vesicle there are 1/y ≈ 6 lipid molecules for every bound dye molecule. These values are comparable to those reported elsewhere [31] for the binding of oxonol V to soybean lipid and dipalmitoyl lecithin vesicles.

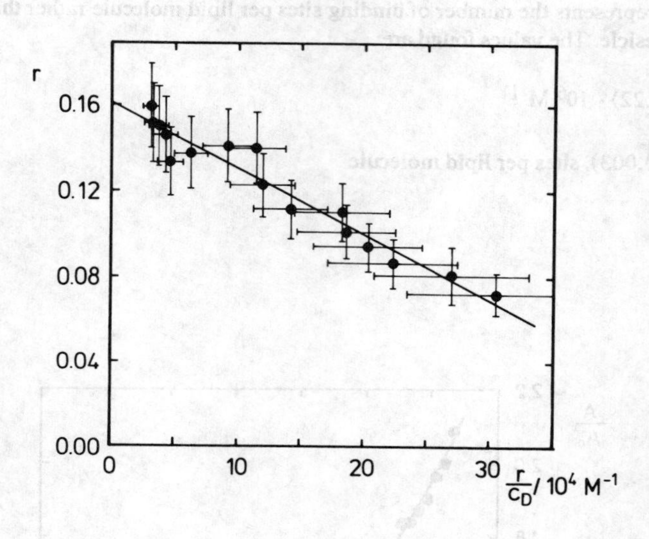

Figure 4. Scatchard plot for the equilibrium binding of oxonol V to dioleoylphosphatidylcholine vesicles ([Lipid] = 7,710 nM). The binding of oxonol V was calculated from the absorbance change at 625 nm. r represents the concentration of bound dye per unit concentration of lipid.

Fluorescence titration of oxonol V with vesicles has also been performed (see Fig. 5). Again there is a red shift of the dye's spectrum on binding to the lipid and a fluorescence enhancement is apparent. It should be noted, however, that no isosbestic point is observed, suggesting that the fluorescence spectrum of bound dye varies with the extent of binding to the vesicles. For this reason the calculation of K and y has been carried out using absorbance measurements alone.

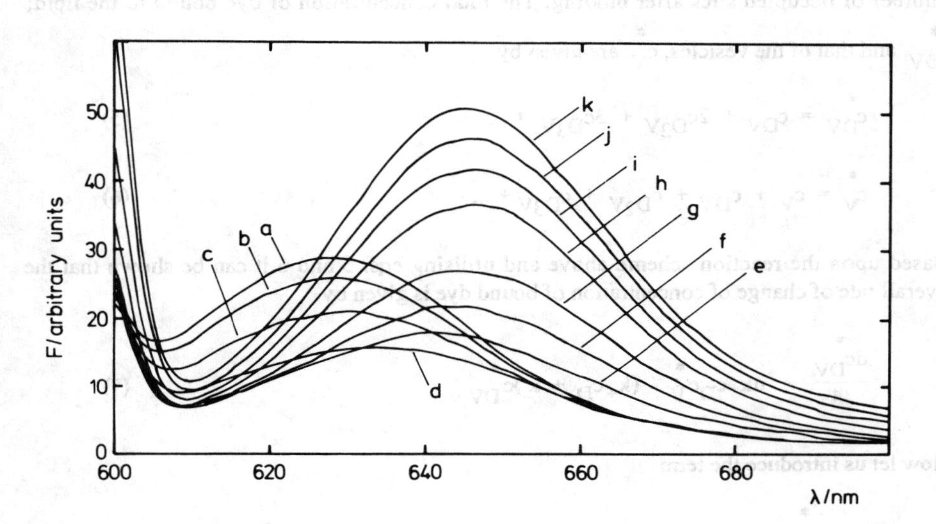

Figure 5. Variation of the oxonol V fluorescence spectrum in the presence of dioleoylphosphatidylcholine vesicles; T = 22°C, pH 7.2. a: 266 nM dye alone. b - k: After the addition of vesicles to final lipid concentrations in the range 125 - 30,900 nM.

4. Kinetic Theory

Now let us consider the kinetics of the dye-vesicle binding process in terms of the occupation of binding sites. First let us define the rate constants k_+ and k_-, which refer to the association and dissociation reactions of dye with a given binding site (see reaction (2)), so that $K = k_+/k_-$. For the binding of dye, D, to a vesicle, V, with n binding sites one can write the following reaction scheme:

$$D + V \underset{k_-}{\overset{nk_+}{\rightleftarrows}} DV$$

$$D + DV \underset{2k_-}{\overset{(n-1)k_+}{\rightleftarrows}} D_2V \qquad (4)$$

$$D + D_2V \underset{3k_-}{\overset{(n-2)k_+}{\rightleftarrows}} D_3V$$

$$\cdots$$

Implicit in these relations is the assumption that there are no interactions between the dye molecules in the lipid membrane, so that the dye association rate constant for each step changes merely by a statistical factor corresponding to the initial number of empty sites on the vesicle. The backward rate constant changes by a statistical factor corresponding to the number of occupied sites after binding. The total concentration of dye bound to the lipid, c^*_{DV}, and that of the vesicles, c^*_V, are given by

$$c^*_{DV} = c_{DV} + 2c_{D_2V} + 3c_{D_3V} + ... \quad (5)$$

$$c^*_V = c_V + c_{DV} + c_{D_2V} + c_{D_3V} + ... \quad (6)$$

Based upon the reaction scheme above and utilising eqs. 5 and 6 it can be shown that the overall rate of change of concentration of bound dye is given by

$$\frac{dc^*_{DV}}{dt} = nk_+c_Dc^*_V - (k_+c_D + k_-)c^*_{DV} \quad (7)$$

Now let us introduce the term

$$r = \frac{c^*_{DV}}{c^*_V} \quad (8)$$

which represents the ratio of bound dye to the total vesicle concentration at any time during the reaction, and the total dye concentration, c^*_D, which is given by

$$c^*_D = c_D + c^*_{DV} \quad (9)$$

Substituting for c_{DV} from eq. 8 in eq. 7 and combining this with the definitions of c_D and K given above in eqs. 9 and 3, respectively, eventually yields

$$\frac{dr}{dt} = -[(n-r)k_+c^*_V + \frac{n}{n - \bar{r}} \cdot k_-](r - \bar{r}) \quad (10)$$

$$\bar{r} = \frac{K(n - \bar{r})c^*_D}{K(n - \bar{r})c^*_V + 1} \quad (11)$$

$\bar{r}$ designates the equilibrium value of r. In terms of the dye binding sites, the values of r and (n - r) correspond to the number of occupied sites per vesicle and the number of free sites per vesicle, respectively. Thus, under conditions of great excess of vesicles, i.e., n >> r, eqs. (10) and (11) reduce to the forms previously derived [22] for the case where saturation of binding sites can be neglected. Integration of eq. (10) yields the following expression for r as a function of time:

$$r = \frac{b\bar{r}(1 - e^{at})}{\bar{r}k_+c^*_V - be^{at}} \quad (12)$$

where,

$$a = (n - \bar{r})k_+c^*_V + \frac{n}{n - \bar{r}} \cdot k_- \quad (13)$$

$$b = \frac{n[(n - \bar{r})k_+c^*_V + k_-]}{n - \bar{r}} \quad (14)$$

In the stopped-flow experiments the binding of dye to the vesicles has been followed by monitoring the change in dye fluorescence. The fluorescence, F, at any time, t, can be defined by the following equation

$$F = c_Df_w + c^*_{DV} f_l \quad (15)$$

where f_w and f_l are the values of fluorescence intensity per mole of dye in water and bound to the lipid, respectively. If f_w and f_l are both considered to be constants, it can be shown [22] that the fluorescence signal should follow a time course governed by a relaxation time, τ, given by

$$1/\tau = (n - r)k_+c^*_V + \frac{n}{n - \bar{r}} \cdot k_- \quad (16)$$

If experiments are performed at great excess of vesicle binding sites over dye molecules such that n >> r, eq. (16) reduces to

$$1/\tau = nk_+c^*_V + k_- \quad (17)$$

which is of the same form as the equation derived in a previous paper [22] for conditions of excess vesicle concentrations. In this case a single exponential relaxation would be expected, and a plot of the reciprocal relaxation time versus the total vesicle concentration should yield a straight line, from which the rate constants nk_+ and k_- can be determined from the slope and intercept, respectively. The expression nk_+ represents the rate constant for binding of a dye molecule to a vesicle with n binding sites.

The equations (10-14) derived above are in fact quite general, irrespective of the means of detecting the binding reaction, and eq. (16) could equally well be applied to absorbance measurements. In the case of fluorescence detection, however, the actual course of the fluorescence change during the binding process may become more complicated at high dye/lipid concentration ratios, because the fluorescence of a dye molecule in the lipid may be affected by the proximity of neighbouring dye molecules. Thus, in eq. (15), f_l may not in fact be a constant, but instead it may vary with the value of r, e.g. a dye molecule binding to a lipid binding site may be expected to have a higher fluorescence if it binds in a region free of other dye molecules than if it binds next to an already bound dye molecule, because of the likelihood of self-reabsorption of the light quanta. As an approximation let us assume that there are only two possible fluorescence states of dye within the lipid:

1) dye isolated, with fluorescence intensity f_l^i

2) dye bound next to another dye molecule, with fluorescence intensity f_l^q (i.e. quenched).

Let us now introduce the quantity P_i, the probability that a given dye molecule in the lipid is isolated, which is a function of r. The fluorescence of dye in the lipid, f_l, at any given value of r is then given by

$$f_l = f_l^i P_i + f_l^q (1 - P_i) \tag{18}$$

Now substituting this expression for f_l into eq. 15 and replacing c_D by $c_D = c^*_D - c^*_{DV}$ as well as c^*_{DV} by $c^*_{DV} = rc^*_V$, one obtains that the fluorescence change during the course of the reaction is described by

$$F - F_o = c^*_V r[f_l^q + P_i(f_l^i - f_l^q) - f_w] \tag{19}$$

where F_o $(= c^*_D f_w)$ is the initial fluorescence of the dye solution before any reaction has occurred. Now in order to evaluate this expression for varying values of r, the dependence of P_i on r is required.

Consider the binding sites to be arranged on a two-dimensional lattice over the vesicle, where each binding site is surrounded by z nearest neighbours. The probability that a certain binding site is occupied is given by r/n. Therefore, the probability that a given binding site is free is expressed by (1 - r/n). The probability that for a given bound dye molecule all the neighbouring binding sites are unoccupied is then given by

$$P_i = (1 - \frac{r}{n})^z \tag{20}$$

For computer simulations of stopped-flow fluorescence experiments the following procedure has been carried out:

1) For given values of c^*_D, c^*_V, n, k_+ and k_- solve eq. (11) for $\bar{r}$.

2) Calculate r as a function of time from eqs. (12-14).
3) From the r values calculate P_i from eq. (20) using an assumed value of z.
4) Calculate F as a function of time from eq. (19) using the previously calculated values of r and P_i.

5. Kinetic Results

When dye is rapidly mixed with vesicles in the stopped-flow apparatus, as described in section 2, two relaxation processes are observed, both of which are characterized by an increase in fluorescence (see Fig. 6). The two processes are widely separated in terms of their time scales, the faster process occurring in the range of tens of milliseconds and the slower in the range of tens of seconds. If the reciprocal relaxation time of the faster process is plotted vs. vesicle concentration (see Fig. 7A), one can see that a straight line is obtained, suggesting that the process being observed is due to the binding of dye to the vesicle membrane (cf. eq. 17). From the slope of the plot the second-order rate constant for the binding step can be estimated as:

$$nk_+ = 1.18\,(\pm 0.05) \cdot 10^{11}\ M^{-1}\ s^{-1}$$

This is the rate constant for binding of dye to a vesicle. The rate constant for binding to a particular binding site, k_+, is obtained by dividing by the no. of sites per vesicle. The no. of sites per lipid molecule was calculated to be 0.161 (see section 3), and for the 72 nm diameter vesicles used for the kinetic experiments the number of lipid molecules per vesicle is 45,200 (see section 2). Thus, for the vesicles used here, n = 45,200 • 0.161 = 7,280 binding sites per vesicle. Thus,

$$k_+ = \frac{1.18 \cdot 10^{11}}{7.280} = 1.62\,(\pm 0.10) \cdot 10^{7}\ M^{-1}\ s^{-1}$$

Theoretically it should be possible to determine the rate constant, k_-, for dissociation of dye from the membrane from the intercept of the plot. However, the value is too small to allow accurate determination from the experimental data. Nevertheless, k_- can be indirectly estimated by using the previously determined value of the association constant, K (see section 3). Thus,

$$k_- = \frac{k_+}{K} = \frac{1.62 \cdot 10^{7}}{3.31 \cdot 10^{6}} = 4.9(\pm 0.6)\ s^{-1}$$

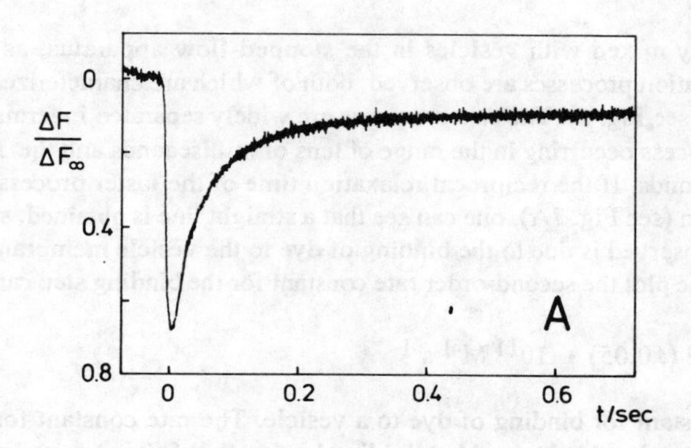

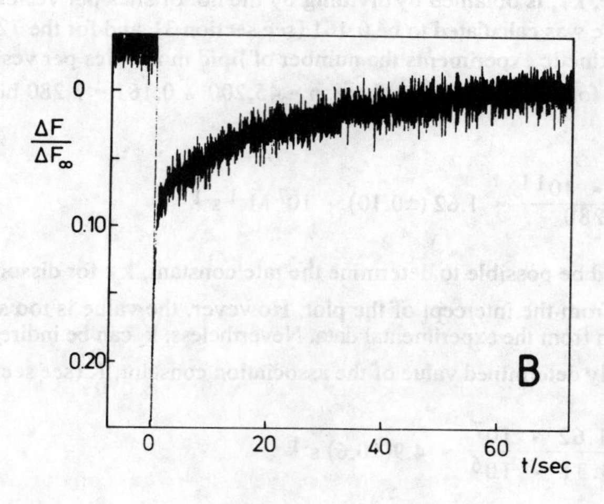

Figure 6. **Stopped-flow traces. (A) Oxonol V in buffer containing 75 mM K_2SO_4 and vesicles in buffer containing 75 mM K_2SO_4 were mixed to final concentrations of 75 nM dye and 0.14 nM vesicles (lipid concentration 6.3 μM); T = 22° C, pH 7.2. (B) As in panel A but over a longer time range.**

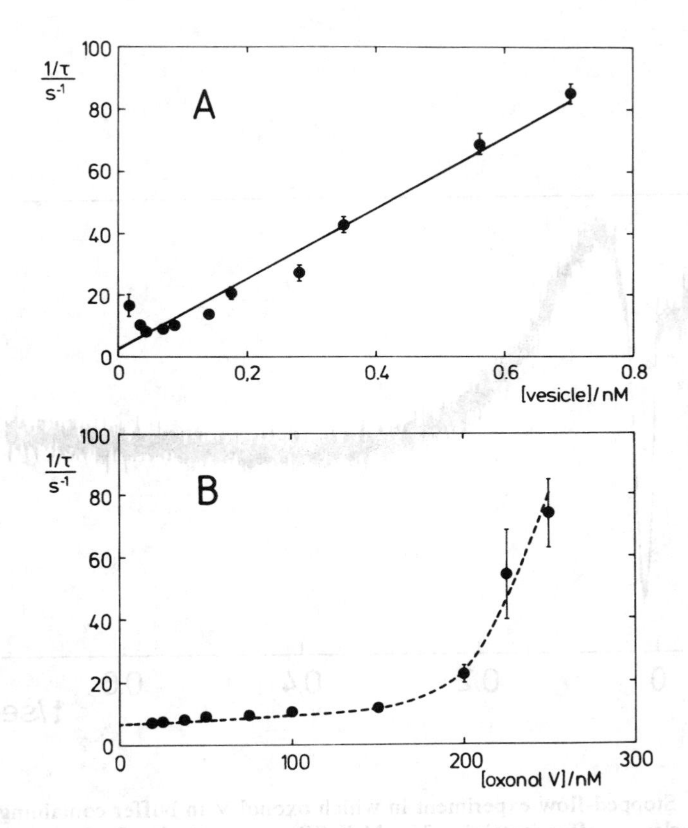

Figure 7. Kinetic plots for the faster process of oxonol V. (A) Dependence of the reciprocal relaxation time on vesicle concentration. [Oxonol V] = 75 nM. (B) Dependence of the reciprocal relaxation time on oxonol V concentration. [Vesicle] = 0.070 nM.

The dependence of the reciprocal relaxation time of the faster process on dye concentration is depicted in Fig. 7B. At low dye concentrations, the reciprocal relaxation time appears to be practically independent of dye concentration, as expected from eq. (17). At high dye concentrations (i.e., at dye/lipid concentration ratios ≥ 0.05), however, there is an apparent marked increase in the reciprocal relaxation time. This will be further discussed shortly.

If one proceeds to higher dye concentrations at constant vesicle concentration, at approximately 225 nM oxonol V, a fluorescence decrease is observed after the initial rapid rise (see Fig. 8). Subsequent to this decrease the slower rise in fluorescence is still observed over tens

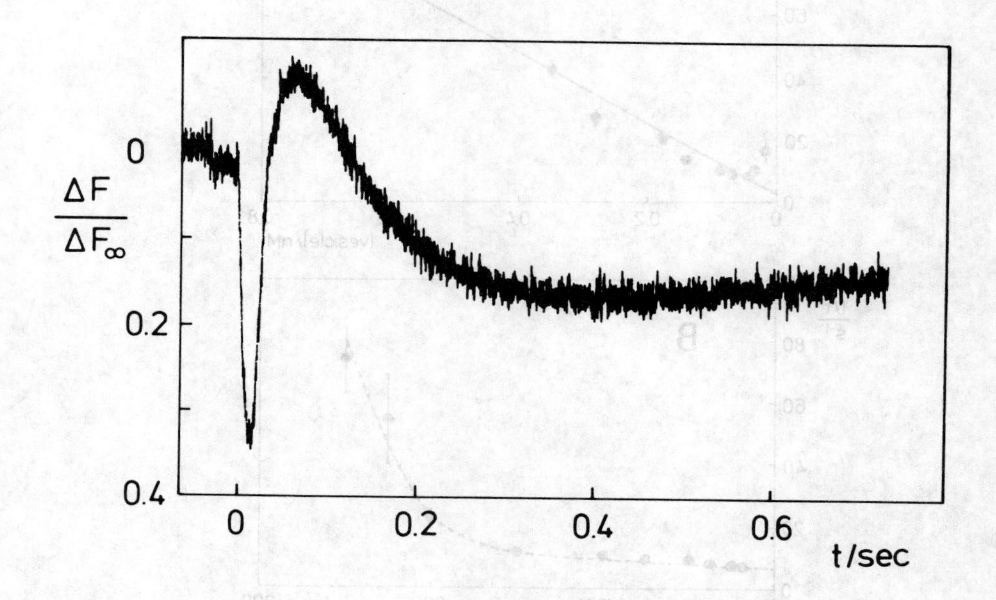

Figure 8. Stopped-flow experiment in which oxonol V in buffer containing 75 mM K_2SO_4 and vesicles in buffer containing 75 mM K_2SO_4 were mixed to final concentrations of 250 nM dye and 0.07 nM vesicles; T = 22° C, pH 7.2.

of seconds. With 225 nM oxonol V and 3.2 µM lipid in the buffer a dye/lipid concentration ratio of 0.07 is obtained, corresponding to one dye molecule per 14 lipid molecules. A computer simulation of this stopped-flow trace using the method described at end of section 4 is shown in Fig. 9. The values of the parameters used for the simulation are as follows: $c_D^* = 2.5 \cdot 10^{-7}$ M, $c_V^* = 7.0 \cdot 10^{-11}$ M, $k_+ = 1.6 \cdot 10^7$ M^{-1} s^{-1}, $k_- = 4.9$ s^{-1}, n = $7.3 \cdot 10^3$ sites per vesicle, $f_w = 7.15 \cdot 10^7$ (arbitrary units) M^{-1}, $f_1^i = 19.0\,8 \cdot 10^7$ (arb. units) M^{-1}, $f_1^q = 0.$ (arb. units) M^{-1}, and z = 3. The values of k_+, k_- and n are as calculated

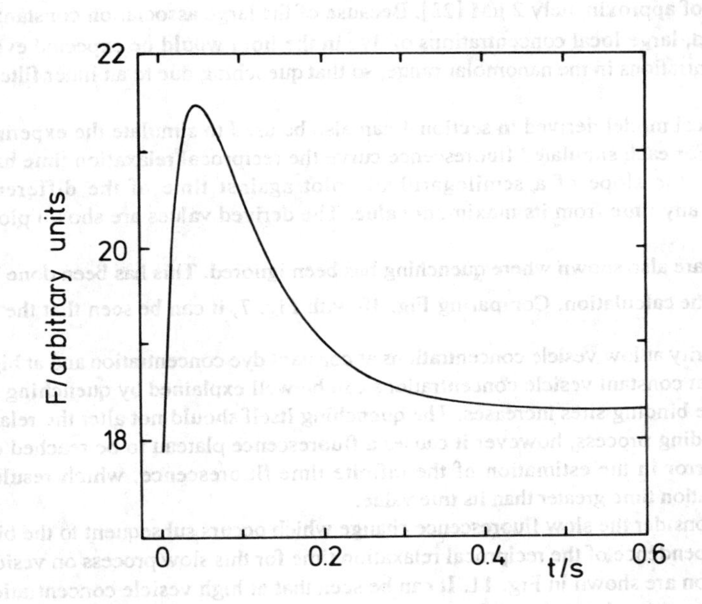

Figure 9. **Computer simulation of a fluorescence stopped-flow trace using eqs. (11 - 14, 19 and 20). Values of the parameters used are: $c^*_D = 2.5 \cdot 10^{-7}$ M, $c^*_V = 7.0 \cdot 10^{-11}$ M, $k_+ = 1.6 \cdot 10^7$ M^{-1} s^{-1}, $k_- = 4.9$ s^{-1}, $n = 7.3 \cdot 10^3$ sites per vesicle, $f_W = 7.15 \cdot 10^7$ (arbitrary units) M^{-1}, $f^i_1 = 19.08 \cdot 10^7$ (arb. units) M^{-1}, $f^q_1 = 0.$ (arb. units) M^{-1} and $z = 3$.**

previously. The value of f_W was estimated from Fig. 5 at 645 nm (the λ_{max} for lipid-bound dye). The value of f^i_1 was likewise estimated from Fig. 5 after extrapolating to infinite vesicle concentration. It is assumed that neighbouring dye molecules in the lipid phase are completely quenched, so that $f^q_1 = 0$. The value of z was chosen after trying several small integral values, with 3 seeming to give the best correlation between the experimental and simulated data. As can be seen, the theory described in section 4 is able to successfully explain the form of the observed fluorescence stopped-flow trace. The initial rapid rise in fluorescence can be

attributed to the binding of dye to the lipid, where it undergoes a fluorescence enhancement due to its transfer to the nonpolar environment of the lipid. As the binding reaction proceeds, however, and the vesicles become more saturated with dye, the probability that a dye molecule will bind near an already bound dye increases. Thus, a fluorescence maximum is reached and then a slower decay of fluorescence occurs, which can be attributed to the increase in local dye concentration in the lipid and quenching via an inner filter effect. It should be noted that for oxonol V there is a significant degree of overlap between the absorbance and fluorescence spectra, and in ethanolic solution the fluorescence of the dye alone is only linear up to concentrations of approximately 2 μM [23]. Because of the large association constant of the dye for the lipid, large local concentrations of dye in the lipid would be expected even with total dye concentrations in the nanomolar range, so that quenching due to an inner filter effect is most likely.

The theoretical model derived in section 4 can also be used to simulate the experimental data in Fig. 7. For each simulated fluorescence curve the reciprocal relaxation time has been estimated from the slope of a semilogarithmic plot against time of the difference in fluorescence at any time from its maximum value. The derived values are shown plotted in Fig. 10.

Simulations are also shown where quenching has been ignored. This has been done by setting $f_1^q = f_1^i$ in the calculation. Comparing Fig. 10 with Fig. 7, it can be seen that the deviations from linearity at low vesicle concentrations at constant dye concentration and at high dye concentrations at constant vesicle concentrations can be well explained by quenching as the saturation of the binding sites increases. The quenching itself should not alter the relaxation time for the binding process, however it causes a fluorescence plateau to be reached earlier and hence an error in the estimation of the infinite time fluorescence, which results in a reciprocal relaxation time greater than its true value.

Now let us consider the slow fluorescence change which occurs subsequent to the binding process. The dependence of the reciprocal relaxation time for this slow process on vesicle and dye concentration are shown in Fig. 11. It can be seen that at high vesicle concentration and low dye concentrations the reciprocal relaxation time is relatively insensitive to changes in concentration. If one proceeds to high dye/lipid concentration ratios, however (i.e. high dye concentrations at constant vesicle concentration or low vesicle concentrations at constant dye concentration), a significant increase in the rate is apparently observed. The origin of this reaction is less clear than that of the fast phase. However, since an increase in $1/\tau$ with vesicle concentration is not observed, a binding reaction can be eliminated. A likely explanation is that the slow phase is due to a change in dye conformation or position within the membrane, possibly diffusion across the membrane to the internal monolayer, a process whose rate would be expected to depend on the structure of the dye and the membrane fluidity. In this case it is again possible that the apparent increase in $1/\tau$ at high dye/lipid concentration ratios

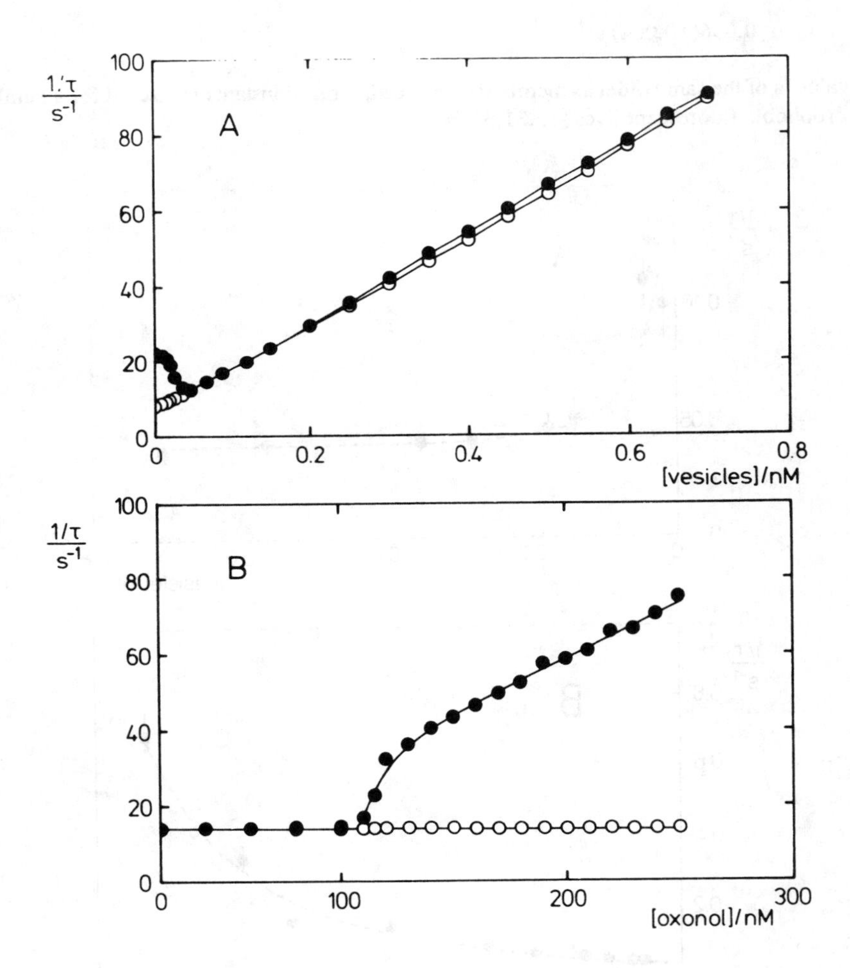

Figure 10. **Simulated kinetic plots for the binding reaction of oxonol V with vesicles. (A) Dependence of the reciprocal relaxation time on vesicle concentration. [Oxonol V] = 75 nM. (B) Dependence of the reciprocal relaxation time on oxonol V concentration. [Vesicle] = 0.070 nM. The filled data points are derived from simulations where quenching is included, the open data points refer to simulations where quenching is neglected ($f_1^q = f_1^i$). The values parameters, except for the concentrations, are as given in Fig. 9.**

could be explained by quenching as the binding sites in the internal monolayer become saturated, causing the calculation of an artificially high $1/\tau$ value. If one extrapolates from the graphs to zero dye concentration (see Fig. 11B) or infinite vesicle concentration (see Fig. 11A), where no quenching of dye fluorescence would be expected, the apparent rate constant of the slow process, k_S, can be estimated to be

$$k_S = 0.036(\pm 0.004)\ s^{-1}.$$

This value is of the same order as membrane permeation rate constants measured for a number of hydrophobic fluorescent dyes [15,31,33,34].

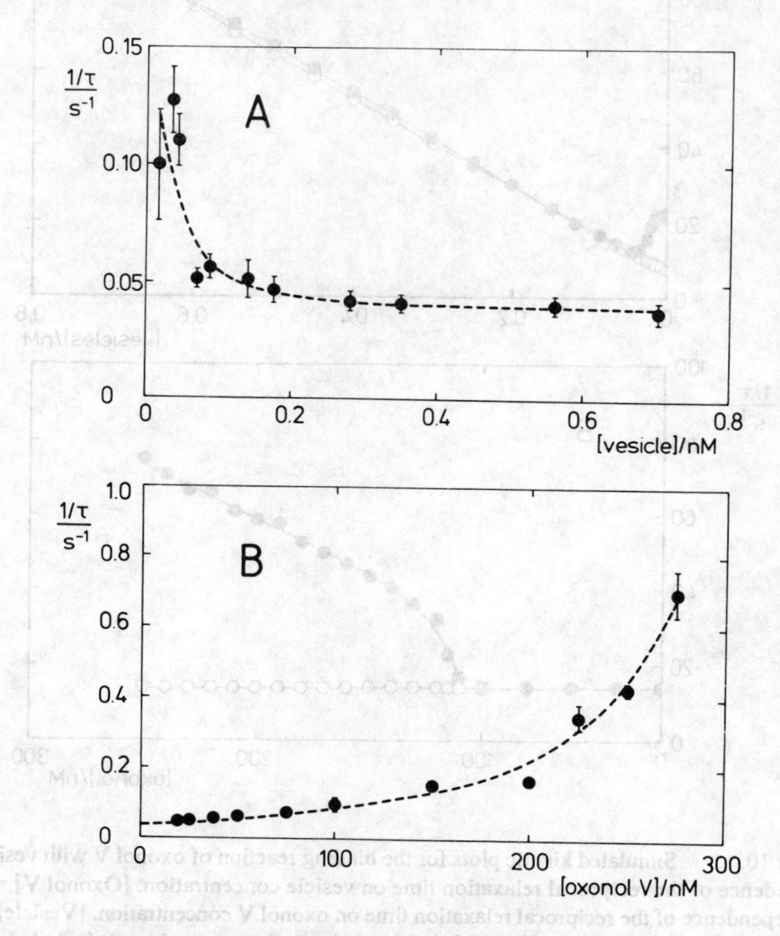

Figure 11. Kinetic plots for the slower process of oxonol V. (A) Dependence of the reciprocal relaxation time on vesicle concentration. [Oxonol V] = 75 nM. (B) Dependence of the reciprocal relaxation time on oxonol V concentration. [Vesicle] = 0.070 nM.

6. Discussion

Oxonol V has been found to interact with lipid vesicles via a two-step process. Based on the dependence of the reciprocal relaxation times on vesicle concentration, the two steps can be interpreted as being due to rapid binding of the dye to the external lipid monolayer of the vesicle, followed by a slow change in the dye's environment within the membrane, possibly due to diffusion across the membrane to the internal monolayer. In the case of the binding reaction it has been found that both the equilibrium and kinetic results are best described by assuming the presence of saturable binding sites on the vesicle membrane. A kinetic model incorporating saturation and fluorescence quenching has been described from which the time course of a fluorescence stopped-flow experiment can be simulated for the complete range of dye and vesicle concentrations.

The nature of the binding sites should briefly be discussed. In fact it is inconceivable that a synthetic lipid membrane should have pre-formed binding sites for the dye oxonol V. The binding probably proceeds with the dye pushing its way into the lipid bilayer with hydrophobic interactions as the driving force. Nevertheless, it seems that the vesicles are not able to accept an indefinite number of dye molecules. The binding of a dye molecule to the lipid probably causes an expansion of the surface of the bilayer, and due to this process a limit may be reached where the binding of further dye molecules becomes thermodynamically unfavourable. Thus, although the lipid membrane has no true pre-formed binding sites for the dye, the assumption of saturable binding sites provides the best mathematical description of the dye-vesicle interaction.

The rate constant for the binding of dye to a vesicle, nk_+, can now be compared with the theoretical diffusion-controlled value of the rate constant, calculated from the Smoluchowski equation [35]. Since the vesicles are much larger than the dye molecules, they can be considered as remaining stationary during the course of the reaction and their diffusion coefficient may therefore be neglected. The radius of collision can also be considered to equal the vesicle radius. Thus, the diffusion-controlled rate constant is given by

$$k_{diff} = 4\pi LDr \tag{21}$$

where D denotes the dye diffusion coefficient, r the vesicle radius and L Avogadro's number. According to the calculations of Smith et al. [12], the diffusion coefficient of oxonol V in aqueous solution is equal to $6.4 \cdot 10^{-6}$ cm^2 s^{-1}. Using this value and an average vesicle radius of 36 nm, the theoretical diffusion-controlled rate constant k_{diff} for oxonol V is calculated to be $1.74 \cdot 10^{11}$ M^{-1} s^{-1}. The value of k_{diff} is thus comparable with the experimentally determined value of nk_+: $1.18 \cdot 10^{11}$ M^{-1} s^{-1}. It seems then that the binding reaction is very close to being diffusion-controlled. Diffusion control has previously been reported for the binding of the fluorescent dyes N-phenyl-naphthylamine [35] and 1-anilino-8-naphthalenesulphonate [34,36] to phospholipid vesicles.

The interaction of oxonol V with azolectin vesicles [12,31] and chromaffin granules [37] have been previously investigated. Two kinetic phases were also observed in these studies, and they have similarly been attributed to binding and diffusion across the membrane. The values of the association rate constants obtained, however, were within the range 10^6- 10^7 M^{-1} s^{-1}, i.e., much lower than the value reported here for dye-vesicle binding. However, the method which has been used in order to obtain these values involves measuring the dependence of the observed reaction rate on dye concentration rather than vesicle or lipid concentration, which according to the kinetic theory presented earlier is more appropriate.

Thus, the physical significance of these values is uncertain. In the case of the dissociation rate constant, k_-, and the observed rate constant for the slower process, k_S, the values given here are comparable to those reported by Bashford et al. [31] for the interaction of oxonol V with azolectin vesicles. The small differences can be attributed to the different lipids used for the vesicle preparation.

Finally, it should be noted that the kinetic equations developed here to explain the fluorescence stopped-flow kinetics of oxonol V-vesicle interaction are of quite general application for the saturable binding of small hydrophobic species to lipid vesicles.

Acknowledgements

The authors would like to thank Professor P. Läuger for many valuable discussions, correspondence, and suggestions concerning this work and for his help in preparing the manuscript. This work has been financially supported by the Deutsche Forschungsgemeinschaft (Sonderforschungsbereich 156). R.J.C. acknowledges with gratitude financial support from the Alexander-von-Humboldt-Stiftung and from the Leverhulme Trust.

References

1 Stryer, L. (1988) Biochemistry, 3rd edn., W.H. Freeman and Company, New York, chapters 37 and 39.

2 Adam, G., Läuger, P. and Stark, G. (1988) Physikalische Chemie und Biophysik, 2nd edn., Springer-Verlag, Heidelberg, chapter 9.

3 Stein, W.D. (1967) The Movement of Molecules across Cell Membranes, Academic Press, New York, chapter 7.

4 Bashford, C.L. and Smith, J.C. (1978) Methods Enzymol. 55, 569.

5 Cohen, L.B. and Salzberg, B.M. (1978) Rev. Physiol. Biochem. Pharmacol. 83, 33.

6 Waggoner, A.S. (1979) Annu. Rev. Biophys. Bioeng. 8, 47

7 Freeman, J.C. and Laris, P.C. (1981) Int. Rev. Cytol. Suppl. 12, 177.

8 Waggoner, A.S. (1985) in: The enzymes of biological membranes, 2nd. edn., ed. A.N. Martonosi, Plenum, New York, vol. 3, p. 313

9 Waggoner, A.S., Wang, C.H. and Tolles, R.L. (1977) J. Membr. Biol. 33, 109

10 Dragsten, P.R. and Webb, W.W. (1978) Biochem. 17, 5228.

11 Smith, J.C. and Chance, B. (1979) J. Membr. Biol. 46, 255.

12 Smith, J.C., Frank, S.J., Bashford, C.L., Chance, B. and Rudkin, B. (1980) J. Membr. Biol. 54, 127.

13 Smith, J.C., Hallidy, L. and Topp, M.R. (1981) J. Membr. Biol. 60, 173.

14 Smith, J.C., Graves, J.M. and Williamson, M. (1984) Arch. Biochem. Biophys. 231, 430.

15 Cabrini, G. and Verkman, A.S. (1986) J. Membr. Biol. 9, 171

16 Apell, H.-J. and Bersch, B. (1987) Biochim. Biophys. Acta 903, 480

17 George, E.B., Nyirjesy, P., Basson, M., Ernst, L.A., Pratap, P.R., Freedman, J.C. and Waggoner, A.S. (1988) J. Membr. Biol. 103, 245

18 Ross, W.N., Salzberg, B.M., Cohen, L.B., Grinvald, A., Davila, H.V., Waggoner, A.S. and Wang, C.H. (1977) J. Membr. Biol. 33, 141

19 Sims, P.J., Waggoner, A.S., Wang, C.H. and Hoffman, J.F. (1974) Biochem. 13, 3315

20 Singh, A.P. and Nicholls, P. (1985) J. Biochem. Biophys. Methods 11, 95

21 Loew, L.M., Bonneville, G.W. and Surow, J. (1978) Biochem. 17, 4065

22 Clarke, R.J. and Apell, H.-J. Biophys. Chem. 34, 225

23 Smith, J.C., Russ, P., Cooperman, B.S. and Chance, B. (1976) Biochem. 15, 5094

24 Apell, H.-J., Marcus, M.M., Anner, B.M., Oetliker, H. and Läuger, P. (1985) J. Membr. Biol. 85, 49

25 Marcus, M.M., Apell, H.-J., Roudna, M., Schwendener, R.A., Weder, H.G. and Läuger, P. (1986) Biochim. Biophys. Acta 854, 270

26 Brocklehurst, J.R., Freedman, R.B., Hancock, D.J. and Radda, G.K. (1970) Biochem. J. 116, 721

27 Haigh, E.A., Thulborn, K.R., Nichol, L.W. and Sawyer, W.H. (1978) Aust. J. Biol. Sci. 31, 447

28 Blatt, E., Chatelier, R.C. and Sawyer, W.H. (1984) Chem. Phys. Lett. 108, 397

29 Tanford, C. (1961) Physical Chemistry of Macromolecules, Wiley, New York, p. 533

30 McLaughlin, S. and Harary, H. (1976) Biochemistry 15, 1941

31 Bashford, C.L., Chance, B., Smith, J.C. and Yoshida, T. (1979) Biophys. J. 25, 63

32 Scatchard, G. (1949) Ann. N.Y. Acad. Sci. 51, 660

33 Bashford, C.L., Chance, B. and Prince, R.C. (1979) Biochim. Biophys. Acta 545, 46

34 Haynes, D.H. and Simkowitz, P. (1977) J. Membr. Biol. 33, 63

35 Woolley, P. and Diebler, H. (1979) Biophys. Chem. 10, 305

36 Haynes, D.H. and Staerk, H. (1974) J. Membr. Biol. 17, 313

37 Scherman, D. and Henry, J.P. (1980) Biochim. Biophys. Acta 599, 150

EQUILIBRIUM AND DYNAMIC INVESTIGATION ON THE MAIN PHASE TRANSITION OF DIPALMYTOYLPHOSPHATIDYLCHOLINE VESICLES CONTAINING POLYPEPTIDES: A DSC AND IODINE LASER T-JUMP STUDY

A. GENZ*, T.Y. TSONG& and J.F. HOLZWARTH*+

Fritz-Haber-Institut der Max-Planck-Gesellschaft, Faradayweg 4-6, D-1000 Berlin 33, West Germany

& Department of Biochemistry, University of Minnesota, St. Paul, Minnesota 55104, USA

+ to whom all correspondence should be submitted.

ABSTRACT. The influence of the linear polypeptide gramicidin A' (GA') on the dynamics of the main phase transition in single shell DPPC vesicles was investigated in equilibrium and kinetic measurements. Additional experiments were carried out with vesicles containing low amounts of the synthetic polypeptide lys_2-gly-leu_{24}-lys_2-ala-amide. Turbidity measurements showed a broadening of the phase transition for both polypeptides and a hysteresis in the transition curves. DSC peaks indicated that at least two types of lipids coexist for DPPC-GA' vesicles in the concentration range $0.02 \leq X(GA') \leq 0.04$. The transition enthalpy was decreasing linearly for both polypeptides in the investigated concentration range. Kinetic experiments were performed on our iodine-laser T-jump arrangement (ILTJ) using turbidity detection. In the region of the main phase transition an increasing percentage of the overall turbidity change was shifted from the ms to the µs time range when increasing amounts of polypeptides were present in the phospholipid bilayer. We conclude that domains containing lipids of different order (clusters) as they are assumed to coexist in pure DPPC bilayers become much less important if polypeptides are incorporated.

<u>Abbreviations</u>

DMPC	Dimyristoylphosphatidylcholine
DPPC	Dipalmytoylphosphatidylcholine
GA'	Gramicidin A'
PP	Lys_2-Gly-Leu_{24}-Lys_2-Ala-amide
ILTJ	Iodine-Laser-Temperarure-Jump
DSC	Differential Scanning Calorimetry
NMR	Nuclear Magnetic Resonance
ΔH	Enthalpy of the Phase Transition
CD	Circular Dichroism
T_m	Temperature of the Main Phase Transition Midpoint
Ala	Alanine
Gly	Glycine

D. M. Bloor and E. Wyn-Jones (eds.),
The Structure, Dynamics and Equilibrium Properties of Colloidal Systems, 493–515.

Leu	Leucine
Lys	Lysine
Trp	Tryptothan
Val	Valine

1. Introduction

Gramicidin A from bacillus brevis is a linear polypeptide antibiotic of 15 hydrophobic amino acids of altering L and D conformation (MW = 1879). The amino acid sequence has been determined:

HCO-L-VAL_1-Gly_2-L-Ala_3-D-Leu_4-L-Ala_5-D-Val_6-L-Val_7-D-Val_8-L-Trp_9-D-Leu_{10}-L-Trp_{11}-D-Leu_{12}-L-Trp_{13}-D-Leu_{14}-L-Trp_{15}-$NHCH_2CH_2OH$ [1].

Commercial mixtures (GA') contain gramicidin A, B and C, approximately 70%, 10% and 20% respectively, which differ in their amino acid composition in position 11. Gramicidin B carries L-phenylalanine and gramicidin C L-tyrosine. Recently a new type, Gramicidin K was isolated [2].

Gramicidin A' forms passive channels for the transport of small monovalent cations across phospholipid bilayer membranes (for a review see [4]). The gramicidin induced ion permeability was studied in conductivity experiments [5, 6] and simultaneous conductance fluorescence studies [7]. These experiments indicated that the conducting channel is a dimer. NMR work using specifically incorporated ^{13}C and ^{19}F nuclei could show that the CH_2OH-terminus neighbouring Trp_{15} is located near the surface of the bilayer and the NH_2-terminus is buried deep within the hydrophobic part of the membrane [9]. The two antiparallel monomers are hydrogen bonded at their NH_2-termini forming a left handed β-helical dimer. This model for the gramicidin channel was first proposed by Urry [10] and is now generally accepted [11, 12, 13].

A 5 Å Fourier map of gramicidin A' has been published [2]. These hydrogen deuterium difference neutron diffraction experiments showed the gramicidin dimer to be a 32 Å long cylinder with a diameter of 6-10 Å. The ion channel has a diameter of 4Å [14]. The conformation of gramicidin A' in different phospholipid bilayers was studied by CD spectroscopy [15]. The spectra showed significant differences for gramicidin in DMPC or DPPC bilayers, indicating a difference in the dimerisation constant for gramicidin in the two lipids, which may be caused by destabilisation of the dimer in thicker bilayers. No change in the CD spectra occurred at the phase transition of DMPC bilayers, and no influence of monovalent ions was detected. Also a change in the peptide/lipid ratio between 1/45 and 1/15 did not alter the spectra. ^{31}P NMR measurements showed an immobilisation of the lipid phosphate headgroups above the main phase transition if high amounts of gramicidin (>50 weight %) were incorporated [18]. Chapman et al. [19] used a variety of physical techniques to study the interaction of gramicidin A with lipid bilayers of DMPC and DPPC. DSC measurements showed a broadening of the phase transition and a linear decrease of the transition enthalpy with increasing amounts of gramicidin up to a GA' mole fraction of 0.06 inside the DPPC bilayer. Above this concentration the decrease in enthalpy became smaller.

The authors concluded from their calorimetric data that each gramicidin removed 6 DPPC molecules from the cooperative melting process; this number is equivalent to the first layer of lipids around the polypeptide. They attributed this effect to a reduction of gauche isomers for the "fluid" lipid chains in direct contact with gramicidin, while in the crystalline phase GA' prevented the neighbouring fatty acid chains from "crystallisation". A detailed difference infrared study [20] on the DMPC/GA' system showed that below T_m gramicidin increases the proportion of gauche conformers and induces an increase in liberal motion of the acyl chains. The authors reported that above T_m gramicidin produced an increase in chain order at low concentrations and an increase in chain disorder with respect to the pure lipid at high concentrations. ESR experiments [19] showed a fraction of immobilised lipid molecules for very high GA'/lipid rations (≥ 1/5), which is attributed to a trapping of the lipids between gramicidin molecules [19,21].

As illustrated above numerous studies have been carried out to characterise the gramicidin channel in phospholipid bilayers and the influence on the neighbouring lipid molecules. But so far no work on the question how gramicidin influences the kinetics of the main phase transition in phospholipid membranes was published.

As simple protein analogue to study lipid-protein interactions in model membranes Davis et al. [22] synthesised amphophilic polypeptides with hydrophilic amino acid residues at each end and a central core of hydrophobic amino acids. The linear dimension of the peptide lys$_2$-gly-leu$_{24}$-lys$_2$-ala-amide is designed to span DPPC bilayers. The authors could show by CD measurements that the peptide has an α-helical conformation in methanol. The DPPC-peptide system was further characterised with DSC, X-ray diffraction and NMR measurements; a strong influence of the peptide on the behaviour of the lipid molecules could be shown [22]. A detailed DSC study on this system was carried out by Morrow et al. [23]. The authors adapted regular solution theory to explain the observed phase behaviour. They tried to show that the peptide does not remove lipid molecules from the main transition, like it is assumed in the "annulus" or "boundary lipid" concept, but rather influences the transition thermodynamics for the bilayer as a whole. ^{2}H NMR experiments on this system [24] showed that only one type of lipid exists in the time range of 10^{-4}s, a result which is in disagreement with most other observations and should be due to problems with this technique.

In our earlier work we have investigated the kinetics of the main phase transition in single shell phospholipid vesicles [25, 26, 27, 28] using the highly flexible iodine laser T-jump technique (ILTJ) developed by Holzwarth and coworkers [25, 29]. In the work presented here we would like to focus our attention on the question how the polypeptide gramicidin A' influences the main phase transition of single shell DPPC vesicles. Additional measurements were performed with vesicles containing small amounts of lys$_2$--gly-leu$_{24}$-lys$_2$-ala-amide.

2. Materials and Methods

DPPC was purchased from Fluka (Switzerland). Gramicidin A' a mixture of Gramicidin A, B and C (approx. 70%, 10% and 20%) [3] was delivered by Sigma (USA) under the commercial name Gramicidin D. The synthetic polypeptide lys$_2$-gly-leu$_{24}$-lys$_2$-ala-amide (PP) was a generous gift of R.S. Hodges and M. Bloom (Edmonton and Vancouver, Canada). The buffer pH=7.4 consisted of 0.1 M NaCl (Merck, Germany) and 0.01 M Tris (hydroxymethyl) aminomethan (Fluka) in triply distilled water. All chemicals were of the highest commercially available grade and used without further purification. The vesicles were prepared by the modified injection method.

Kremer et al. [30] described this preparation first for unilamellar DMPC and DPPC vesicles. We applied their technique to prepare vesicles of pure DPPC and polypeptide containing lipid samples in the following way: A mixture of DPPC and GA' in ethanol was slowly (0.02 ml/min) injected into Tris-buffer at 54°C and then dialysed against pure buffer for at least 12 h. Vesicles of high GA'/DPPC ratio (≥ 1/20) tended to aggregate and were mildly sonified for 5 min after dialysis. The final lipid concentration was 2.7 mM and the DPPC/GA' ratio was checked by centrifuging a part of the solution at 1500 g for 90 min. The pellet was then suspended in ethanol and the GA' concentration calculated from the absorption band of tryptophan (ε = 22800 as determined in control experiments). More than 90% of the gramicidin was incorporated into the vesicles for DPPC/GA' ratios ≥ 20.

The vesicles containing PP were prepared according to the procedure described above. Only vesicles with small amounts of the polypeptide could be prepared because at DPPC/PP ratios <60 the preparations were no longer stable. We attribute this to the relatively high curvature of the vesicles from our preparations.

Vesicles of pure DPPC and preparations containing GA' or PP were checked by electron microscopy using the negative staining technique (figure 1b and ref. 28).

The samples containing polypeptides showed slightly increased diameters (size ≈ 62 nm) and a broader size distribution, compared to vesicles of pure DPPC (size ≈ 42 nm ref. 28).

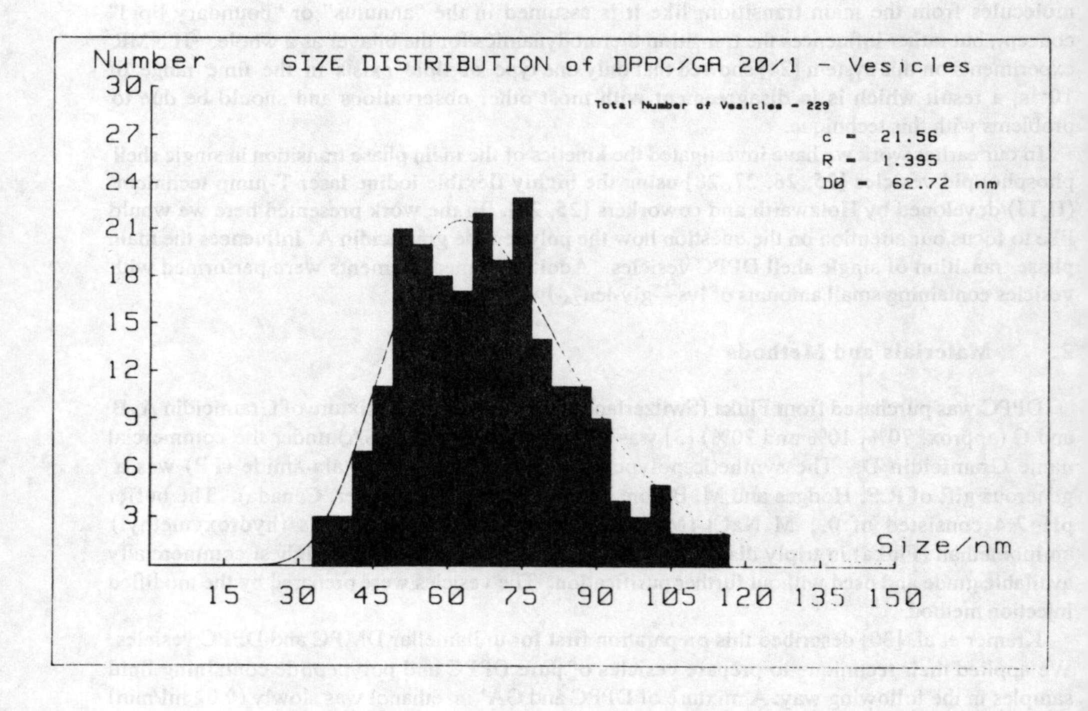

Fig. 1a) Size distribution of vesicles containing DPPC + GA at a ratio of 20/1 from electron micrographs achieved by negative staining with uranylacetate 1 (b).

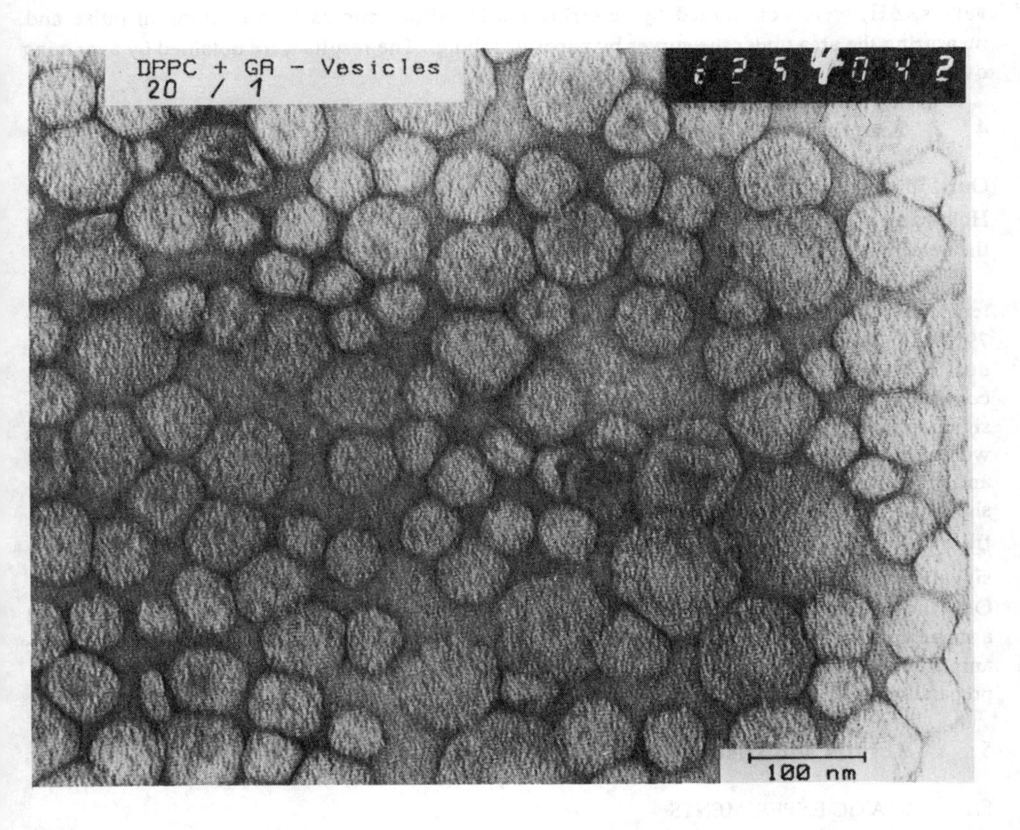

Fig. 1b) Electronmicrograph of DPPC vesicles containing gramicidin at a ratio of DPPC/GA = 20 achieved from samples which were negatively stained by uranylacetate.

3. Static Measurements

Turbidity transition curves were measured on a UV/VIS spectrometer (Perkin Elmer 555) at 300 nm for DPPC and DPPC+PP vesicles and at 360 nm for DPPC+GA' vesicles. The temperature inside the cuvette (Hellma, QS, Germany) was scanned with a rate of 18°C/h on a Haake (Germany) thermostat equipped with a PG11 temperature control unit and determined in control experiments with a digital thermometer. The phase transition is represented by the temperature dependence of the order parameter OP=$(S_c\text{-}S) / (S_c\text{-}S_f)$ using linear extrapolations for the turbidities of the crystalline S_c and fluid states S_f [31].

DSC measurements were performed on a MC-1 microcalorimeter (Amherst, Mass.) using 0.9 ml of solution for reference as well as sample cell and a scan rate of 21.7°C/h. The enthalpy values, ΔH, were determined by referring the transition curves to a calibration pulse and measuring the area under the curves by paper weighing. The results were obtained by averaging over at least two different vesicle preparations.

4. Kinetic Experiments

Our Iodine Laser T-jump apparatus (ILTJ) with turbidity detection is described in detail by Holzwarth et al. [25,29]. The laser was used in the oscillator mode (2.4 μs pulse relaxation) and the detection wavelength was 360 nm.

Relaxation processes in three different time ranges were monitored during one single temperature jump. Signals in the μs time range were recorded with a Tektronix transient digitiser 7912AD equipped with a 7A22 amplifier plug in. A Biomation 1010 waveform recorder in the dual time base mode was used to register the two ms relaxation signals. The signals were computer fitted on a HP 9845B. The programme allows sampling of the signals to improve the signal to noise ratio. Single relaxation times were calculated with their error values. The results were averaged over two (for GA') or three (for PP) different vesicle preparations. For low GA' and PP concentrations the relaxation process around 1 ms (Signal 4) was superimposed by a faster signal (about 5 times) with a very weak amplitude. The amplitudes A_4 given in the figures result from the superposition of both signals. The relaxation times (τ_4) are those of the slower main signal. All amplitudes were normalised to $\Delta T = 1$ K and can be compared among each other. Only the measurements for the highest GA' concentration were done with a different geometrical arrangement for the detection system, because of the higher turbidity of this sample. Therefore the amplitudes for this probe are not exactly comparable with those obtained from the other preparations.

5. Results and Discussion

5.1 STATIC EXPERIMENTS

The amount of light scattered by the phospholipid vesicles changes during the main phase transition, due to a change of the optical density of the bilayer. For a detailed theoretical discussion see [32]. Consequently temperature dependent phase changes can be derived from turbidity measurements. A simple order parameter (see Materials and Methods) is used to monitor the main phase transition in vesicles prepared from pure DPPC or the more complex phase changes in DPPC vesicles containing GA' or PP (figures 2a, b). The curves indicate broadening of the transition with increasing GA' and PP content. A second effect of peptide incorporation is an increasing hysteresis effect in the transition behaviour, which reaches 1.5 K for DPPC/GA' = 25 and 2 K for DPPC/PP = 60 (figures 2c, 2d).

The phase transition temperature T_m which is an average between the up and down going temperature/turbidity curves where the order parameter reaches the value 0.5. T_m is slightly reduced by incorporation of GA', in agreement with [19] and our DSC measurements. For PP, T_m is slightly increased in agreement with measurements of the first moment of ^2H NMR spectra [22] while the peak maximum in DSC measurements remains unchanged by PP incorporation.

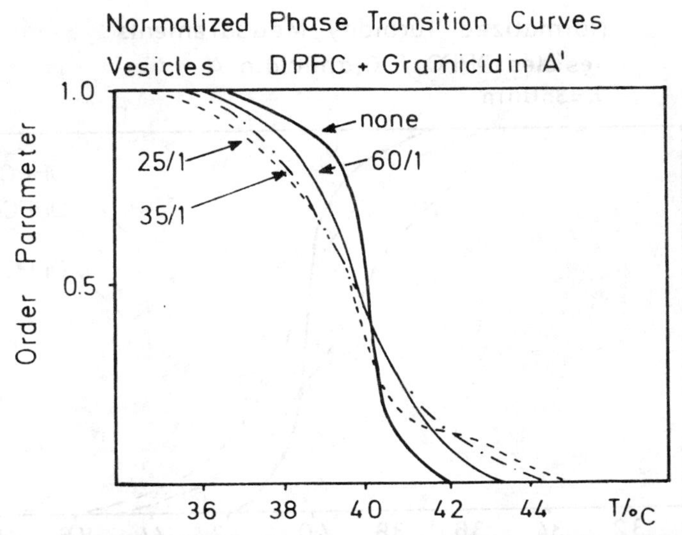

Fig. 2a: **Phase transition curves for DPPC vesicles containing GA' at different ratios DPPC/GA. The order parameter is calculated from static turbidity measurements (C_{DPPC} = 2.7 mM).**

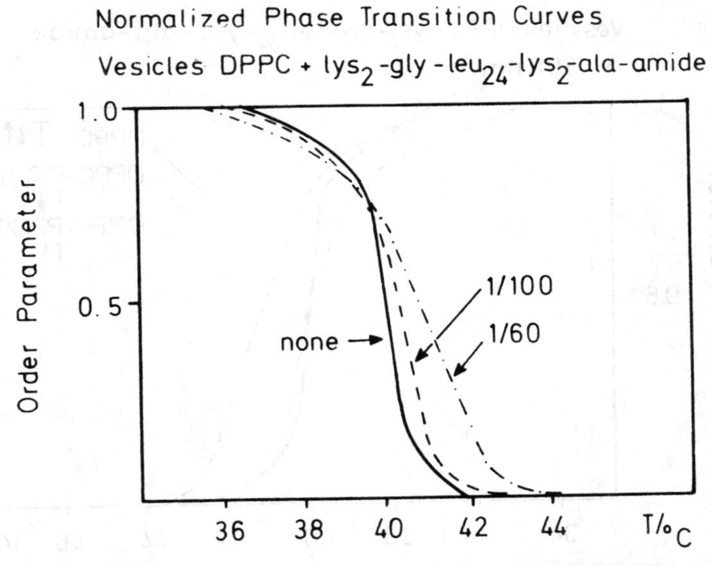

Fig. 2b: **Phase transition curves for DPPC vesicles containing PP at different ratios PP/DPPC. The order parameter is calculated from static turbidity measurements (C_{DPPC} = 2.7 mM).**

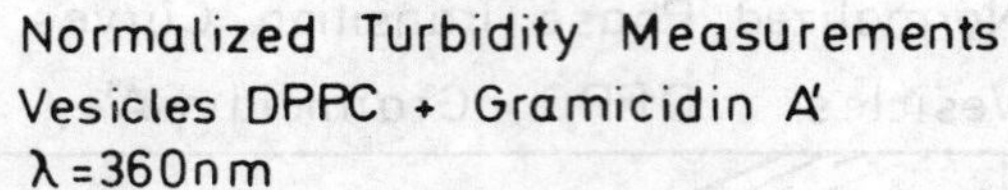

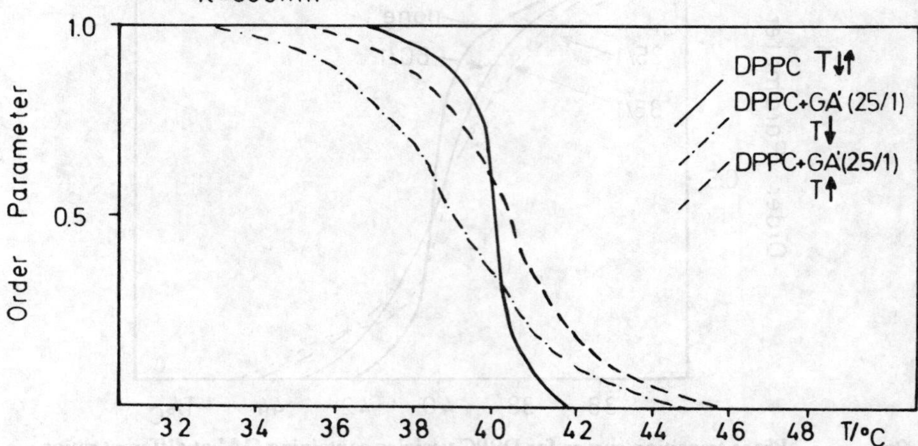

Fig. 2c: **Phase transition curves for DPPC vesicles containing GA' at a ratio DPPC/GA of 25/1 for up and down scanning of the temperature T showing hysteresis only with GA.**

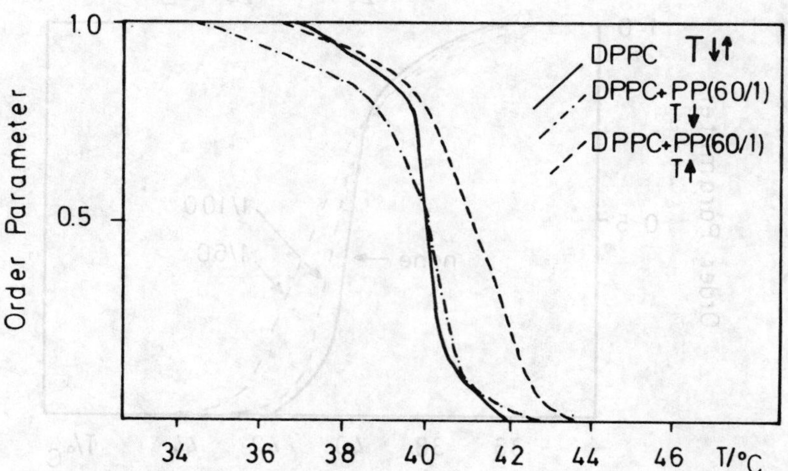

Fig. 2d: **Phase transition curves for DPPC vesicles containing PP at a ratio of DPPC/PP = 60/1 for up and down scanning of temperature T showing hysteresis only with PP.**

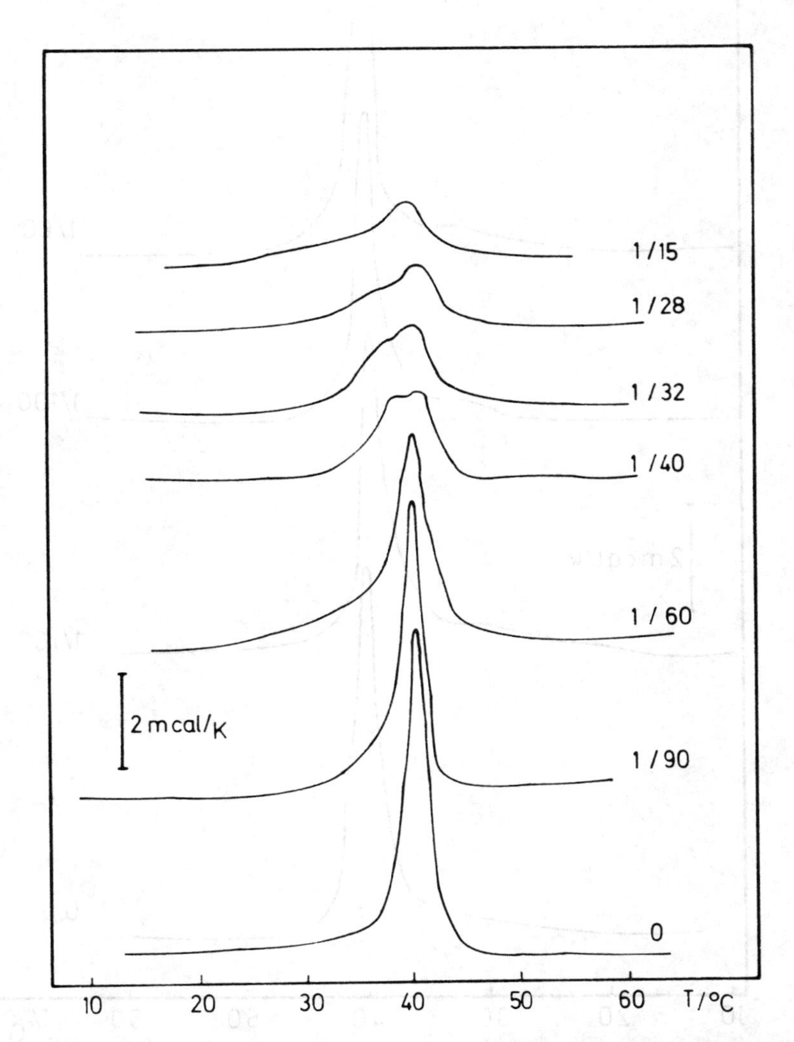

Fig. 3a: **Differential scanning calorimetric results of DPPC vesicles containing different amounts of GA' at a ratio GA/DPPC (C_{DPPC} = 2.7 mM; scan rate = 21.7°C/h).**

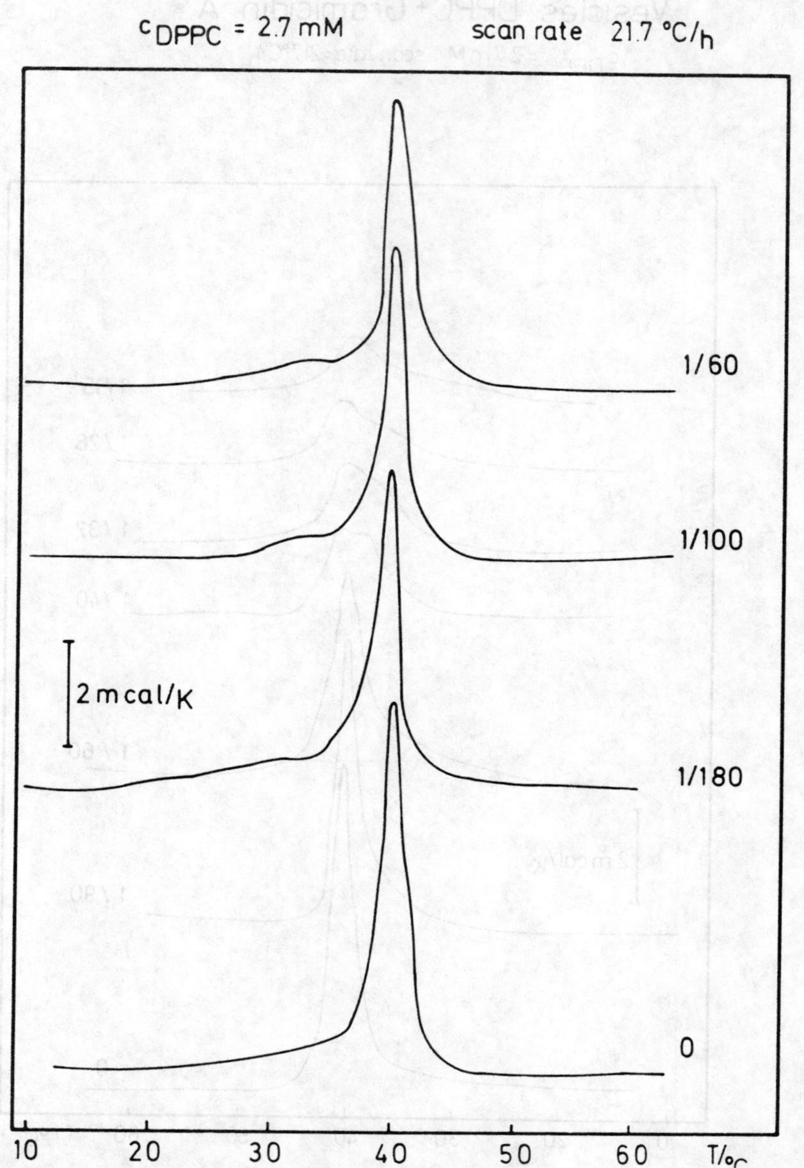

Fig. 3b: **Differential scanning calorimetric results of DPPC vesicles containing different amounts of PP at a ratio PP/DPPC (C_{DPPC} = 2.7 mM; scan rate = 21.7°C/h).**

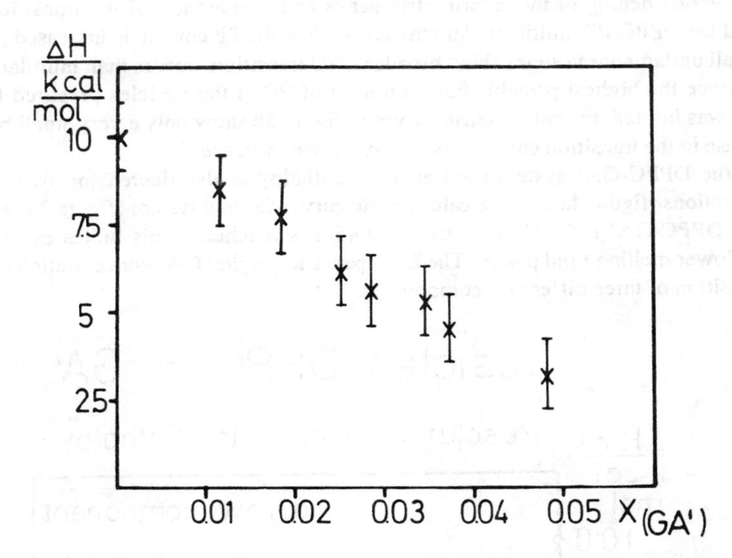

Fig. 4a: **The variation with GA' (a) content of the transition enthalpies ΔH of DPPC Vesicles.**

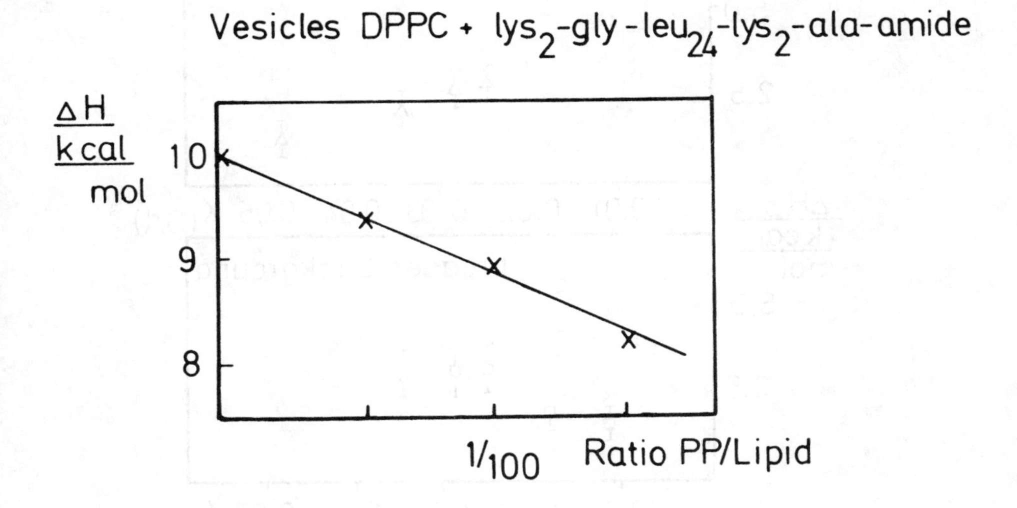

Fig. 4b: **The variation with PP (b) content of the transition enthalpies ΔH of DPPC Vesicles.**

The DSC measurements were performed for vesicles of different GA' and PP contents (figures 3a, b). A broadening of the calorimetric peaks and a reduction of the transition enthalpy was reported for DPPC-PP multilamellar structures when the PP content is increased [23]. It is known that small unilamellar vesicles show broader DSC transition curves than multilamellar structures [33]. Since the highest possible concentration of PP in the vesicles prepared by the injection method was limited, the calorimetric curves in figure 3b show only a very small broadening while a decrease in the transition enthalpy is clearly shown in figure 4b.

For the DPPC-GA' system the transition enthalpy is also decreasing with increasing GA' concentrations (figure 4a). In the calorimetric curves for this system (figure 3a) a shoulder occurs when a DPPC/GA' ratio of approximately 45/1 is reached. This indicates the existence of a second lower melting lipid phase. The DSC peaks for higher GA' concentrations may even be the superposition of three difference components.

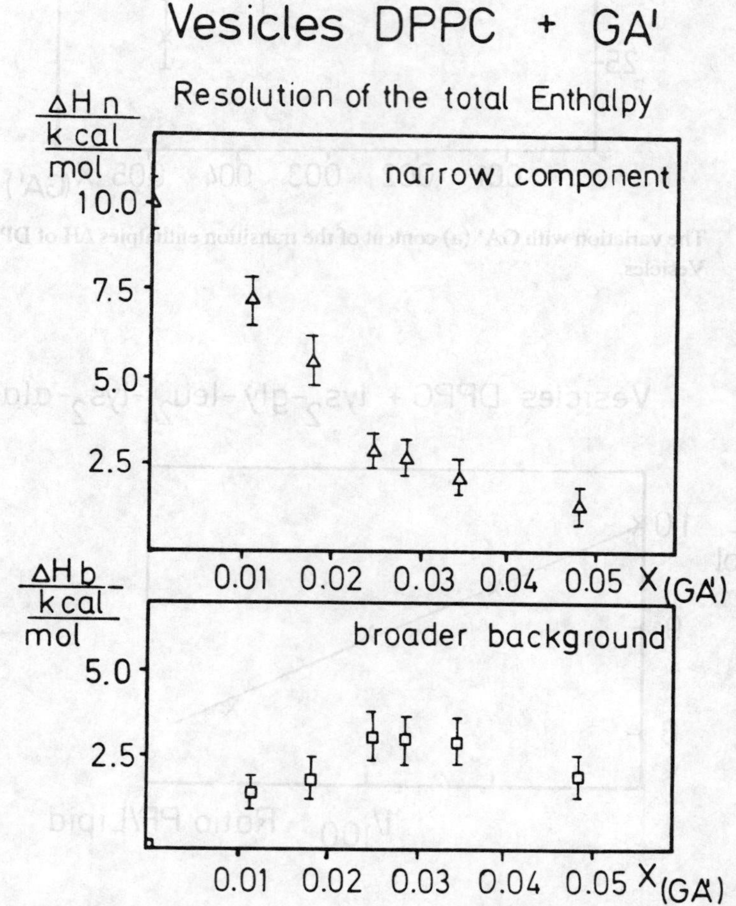

Fig. 5: The deconvolution of the total enthalpy into a narrow component (Δ) and a broader background (□) for GA' in DPPC vesicles.

In figure 5 it is tried to separate the calorimetric curves into a narrow component which is attributed to the melting of a lipid phase not influenced by GA' and a broader background. The decrease of the narrow component is not linear, a break occurs around X(GA') = 0.022, which is the concentration where the background reaches some kind of plateau. An extrapolation of the decreasing narrow component shows that it should vanish around X(GA') = 0.07 which is the concentration where a linear extrapolation for ΔH from figure 3a reaches the zero line. The broader background is increasing until X(GA') = 0.022 and decreasing again for X(GA) >0.04.

A calculation for the simple assumption that only the first layer of lipids around each gramicidin is influenced in its phase transition behaviour [34] should show a maximum for the amount of influenced lipids if the gramicidin concentration reached 0.125; this is clearly not in agreement with our experimental data. Therefore the ideas of an exponential decay of the polypeptide's influence on the lipids described by a coherence length [35] or a more general influence on the whole bilayer structure [23] seems more likely than the assumption of a single layer of boundary lipids around the polypeptide.

Moreover the influence of the curvature [36] on the phase behaviour of the DPPC/GA' vesicles should not be neglected. The GA' monomers are likely to prefer the outer monolayer caused by the higher package density in the head group region of the inner monolayer. In our case the size of the vesicles is increased (fig. 2) so that no marked difference exists between the inside and the outside of the vesicles; this is due to our preparation method.

An interesting new idea to describe the phase behaviour of polypeptide/phospholipid systems is based on the mismatch between the hydrophobic parts of both components [37]. In the case of the DPPC/GA' system the hydrophobic interior of the lipid is longer than the hydrophobic outside of a gramicidin dimer and therefore a special lipid arrangement near the gramicidin molecules seems necessary.

6. Kinetic Experiments

ILTJ measurements were performed for pure DPPC vesicles, vesicles containing different amounts of gramicidin (1/60, 1/35 and 1/15) and low amounts of the synthetic polypeptide lys_2-gly-leu_{24}-lys_{24}-lys_2-ala-amide (figure 6). T is the temperature at which the T-jump started. The experiments with pure DPPC vesicles were carried out using a laser energy of 0.85 J, while for vesicles containing polypeptides a higher energy was used to achieve greater relaxation amplitudes. For better comparison the relaxation amplitudes (A) were normalised to a temperature jump ΔT = 1K, assuming that A is proportional to ΔT, which is a good approximation for 0.07 K $\leq \Delta T \leq$ 1.8 K as verified in control experiments.

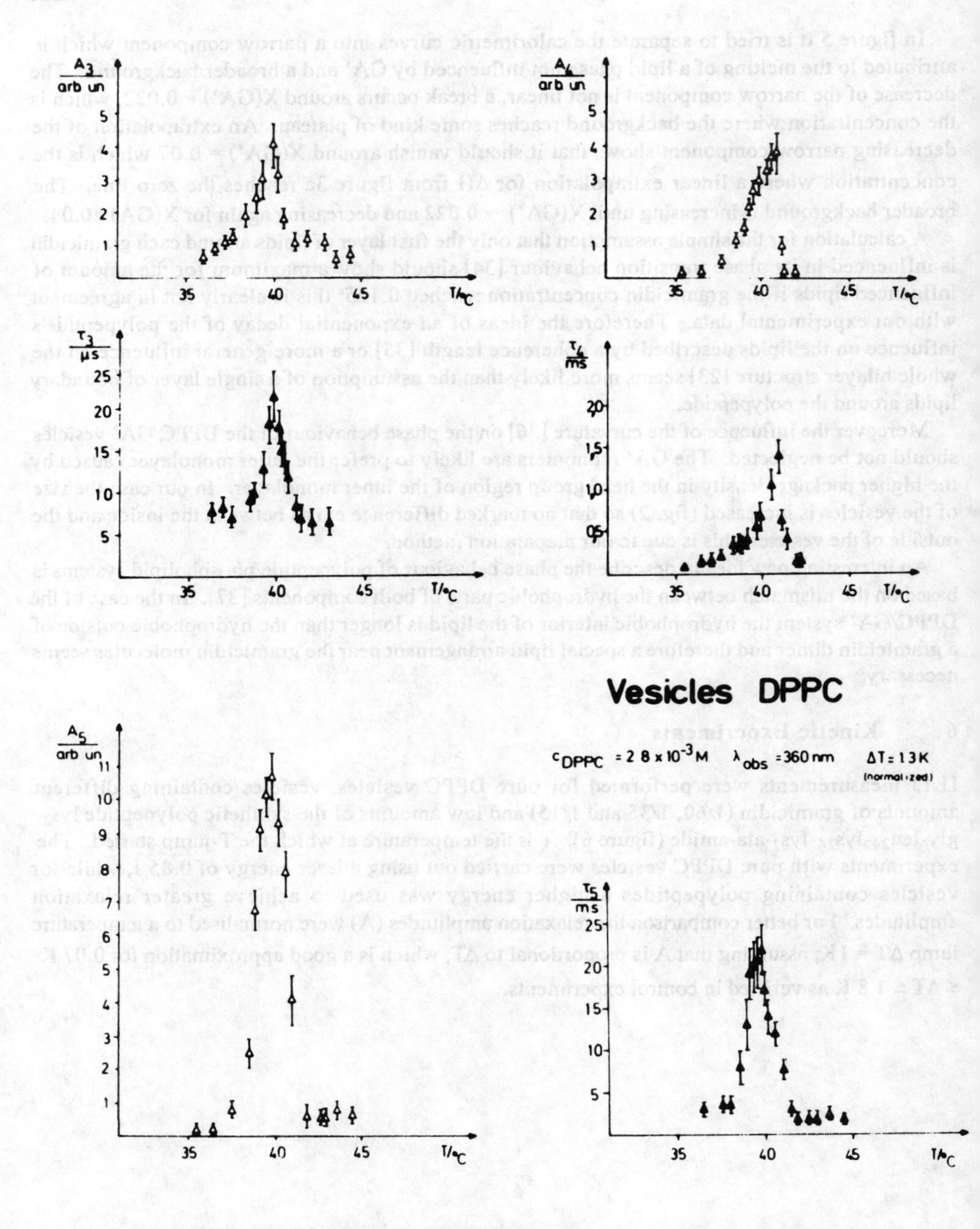

Fig. 6a: **Relaxation times and their corresponding amplitudes A as a function of the temperature where the T-jump started (a) pure DPPC vesicles.**

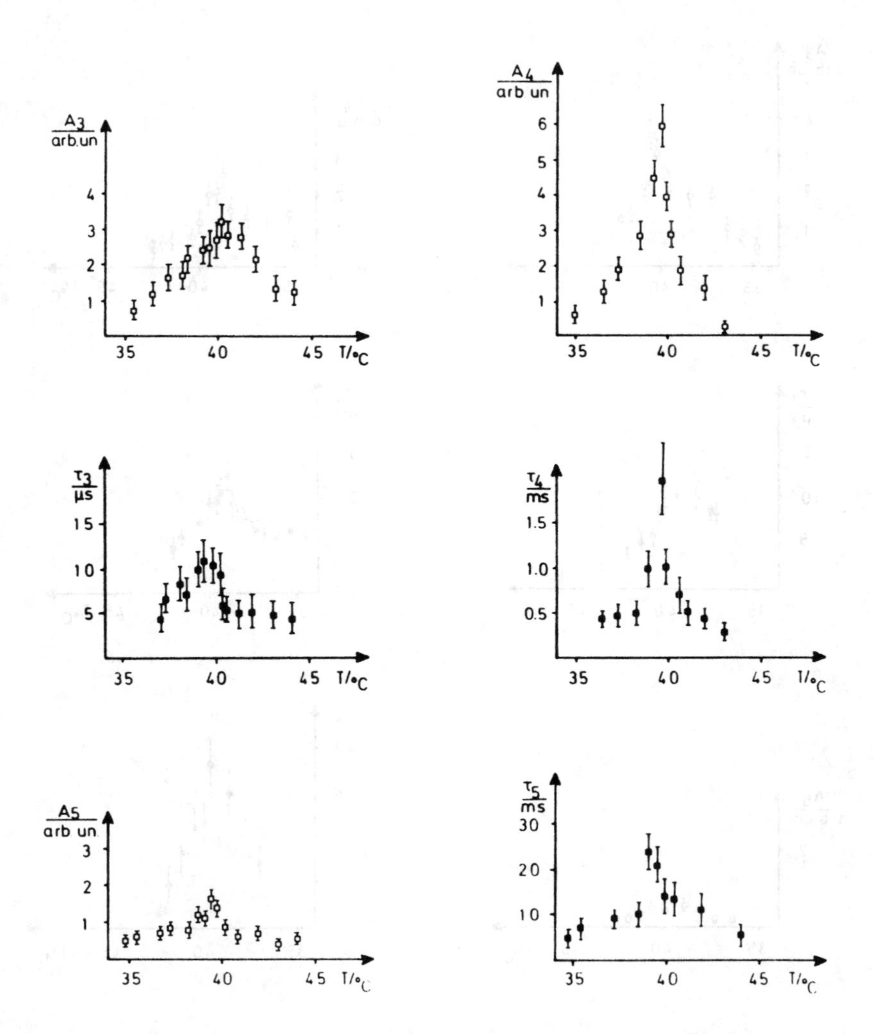

Fig. 6b: Relaxation times and their corresponding amplitudes A as a function of the temperature where the T-jump started (b) vesicles with a DPPC/GA' ratio of 60/1.

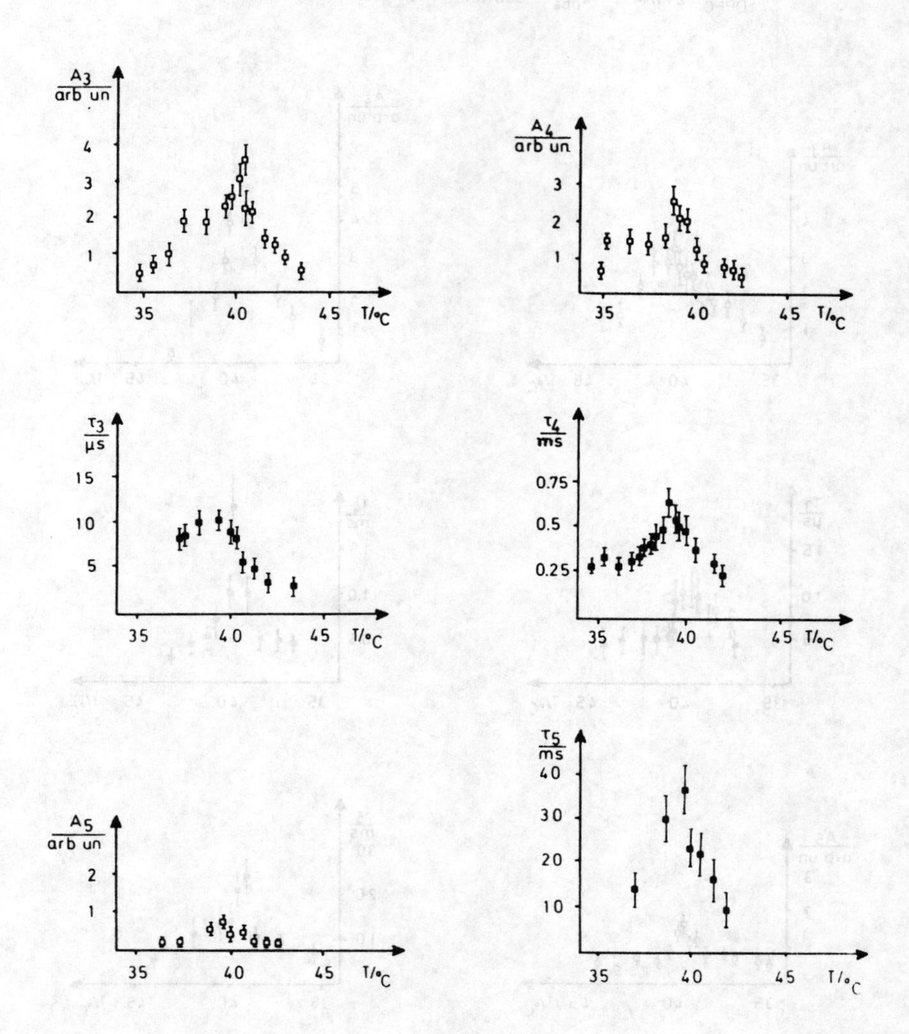

Fig. 6c: Relaxation times and their corresponding amplitudes A as a function of the temperature where the T-jump started (c) vesicles with a DPPC/GA' ratio of 35/1.

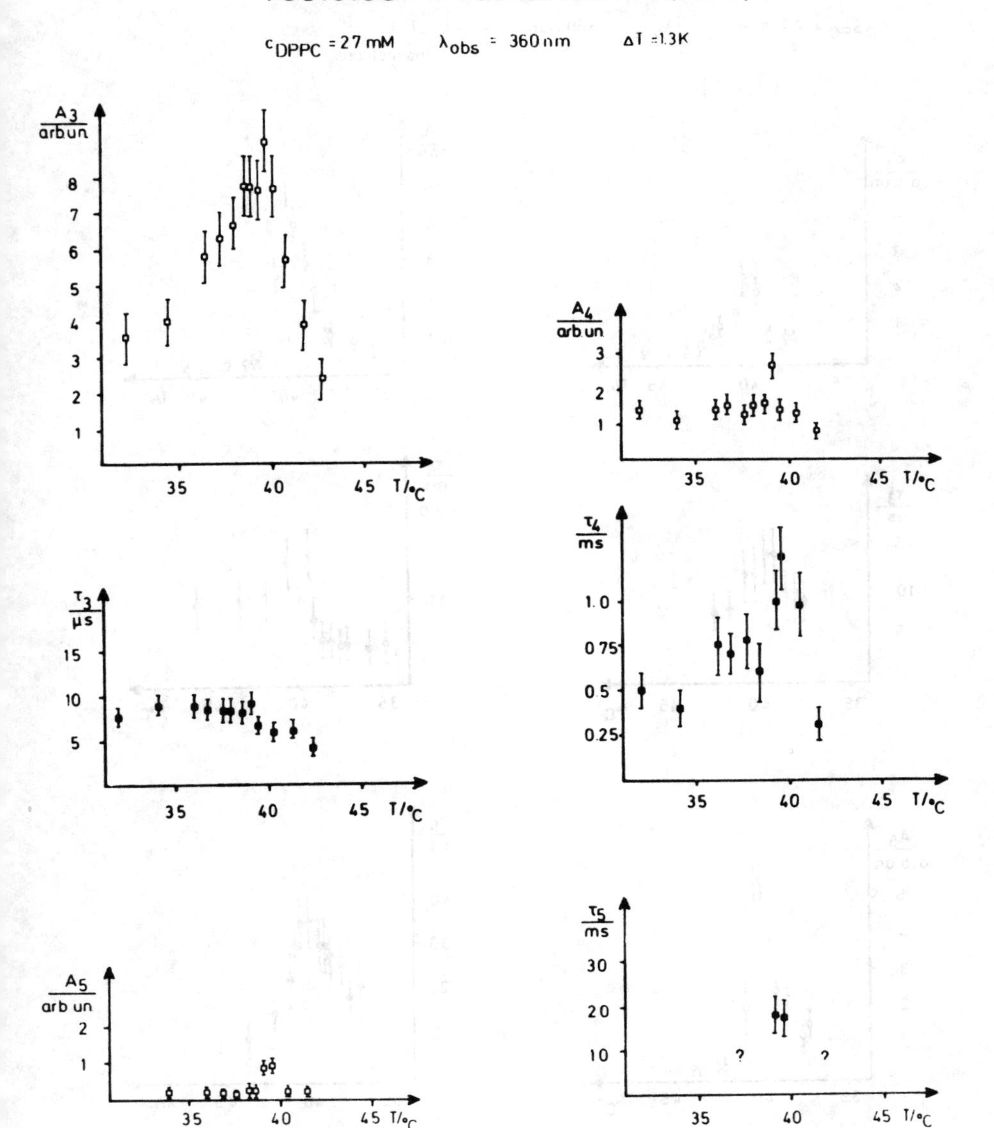

Fig. 6d: **Relaxation times and their corresponding amplitudes A as a function of the temperature where the T-jump started (d) vesicles with a DPPC/GA' ratio of 15/1.**

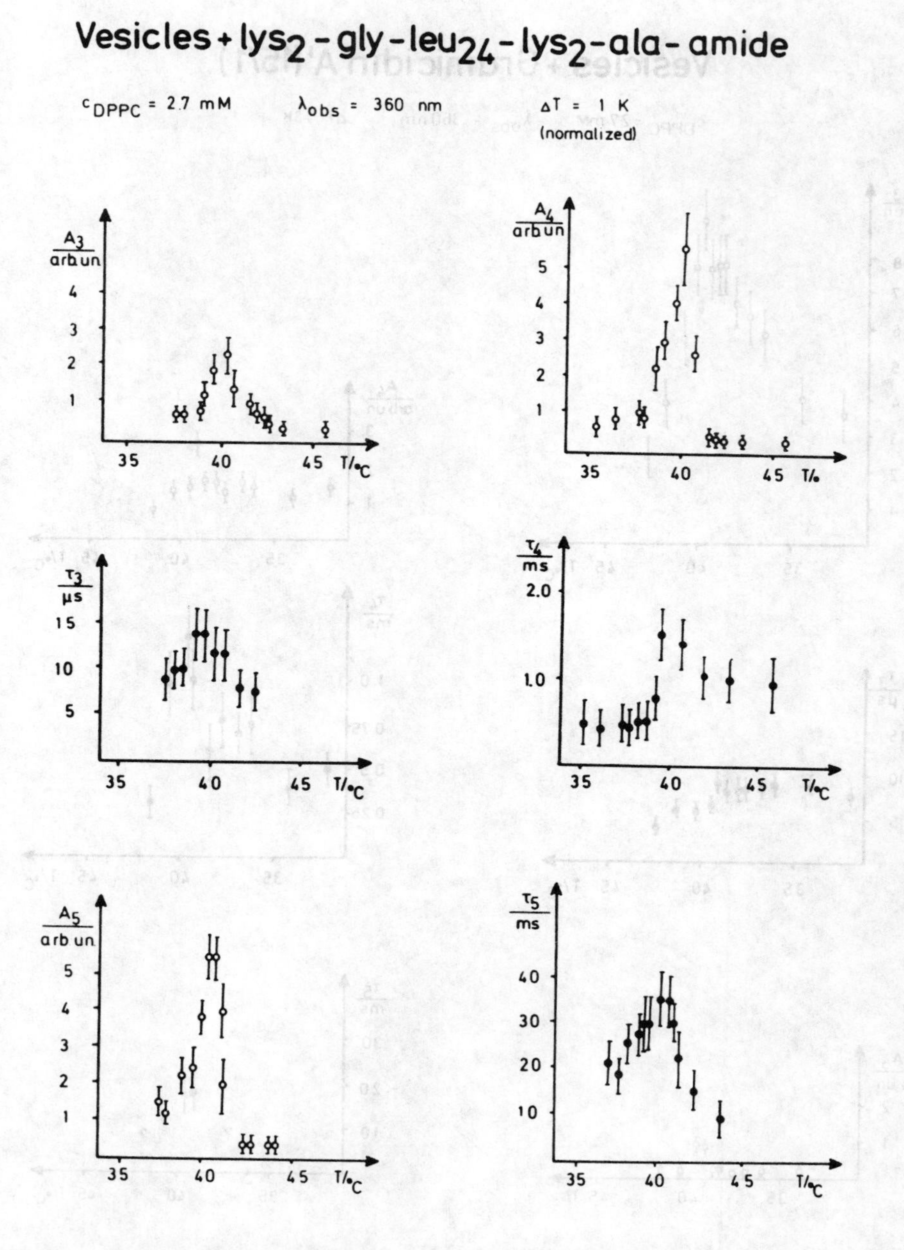

Fig. 6e: Relaxation times and their corresponding amplitudes A as a function of the temperature where the T-jump started (e) vesicles with a DPPC/PP ratio of 100/1.

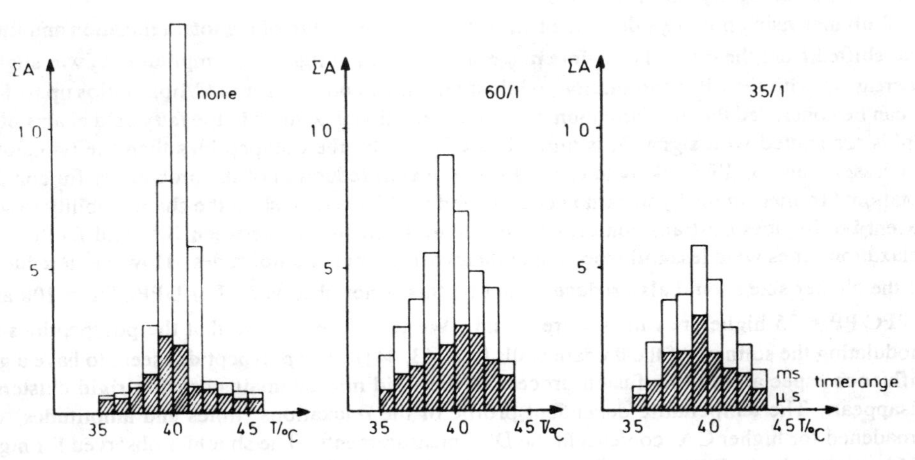

Fig. 7: **Superposition of the relaxation amplitudes in the micro- and millisecond range for vesicles containing different amounts of GA' (C_{DPPC} = 2.7 mM).**

Figure 7 allows a comparison of the overall relaxation amplitude ΣA for vesicles containing different GA' concentrations. Vesicles with a DPPC/GA' ratio of 15/1 showed a higher turbidity and had to be measured using smaller cuvettes and a different geometrical arrangement of the detection system. Therefore the relaxation amplitudes are not exactly comparable with those observed for smaller GA' concentrations.

The kinetic results for pure DPPC vesicles are in agreement with ILTJ investigations of the phase transition applying fluorescence [27] or absorption [28] as detection parameters. A mode locked iodine laser was used to produce temperature jumps for investigations of kinetics in the ns time range [25, 38]. Two relaxation processes were found. An explanation for the underlying molecular events is given in the papers cited above. Both processes are mainly attributed to changes of single molecules, while the relaxation processes in the μs and ms time range, which were investigated in the work presented here, are strongly cooperative. Their relaxation times and amplitudes show strong maxima near T_m. The processes 4 and 5 were associated with the coexistence of clusters of different lipid order which are fluctuating on the ms time scale. This interpretation [25, 27] is based on ideas first introduced by Adam [39] and further developed by Tsong and Kanehisa [40] More recent Monte-Carlo simulations [41, 42, 43] using a microscopic interaction model resulted in a better understanding of the cluster distribution or more general in the density fluctuations during the phase transition [42]. The results of these new theoretical simulations are in contrast to mean field calculations where the important density fluctuations are not included because of the long range energy exchange and the use of too few different states of order of the lipids.

The kinetic experiments with polypeptides incorporated in the DPPC vesicles indicated that the lipid clusters are strongly influenced by the peptides. Small amounts of GA' or PP (figures 6b and 6e) caused a dramatic reduction of the relaxation amplitude A_5 which is the biggest one in pure DPPC vesicles. A further increase of the GA' content also decreased the amplitude A_4. Gramicidin creates lipids in an intermediate state of order; this restricts the fluid as well as the crystalline microdomains to smaller sizes at higher gramicidin concentrations. These phenomena causes the reduction of amplitude A_4 which is associated with the growing of the fluid areas and the shrinking of the crystalline clusters.

With increasing polypeptide content an increasing percentage of the total relaxation amplitude was shifted from the ms to the μs time range as illustrated in figure 7. Amplitude A_3 was slightly decreasing with the PP incorporation, while it remained constant for GA'/lipid ratios up to 1/35. It can be concluded that the formation of complex rotational isomers in the fatty acid chains of the lipids associated with signal 3, is much less effected by the polypeptides than the two slower processes 4 and 5. PP is likely to cause a slight overall reduction of the probability for complex rotational isomers in the hydrocarbon chains, while GA' should restrict the chain mobility to some extent but induces cis-trans conformations due to the mismatch between GA' and DPPC. The relaxation times were less influenced than their corresponding amplitudes. However a reduction of the cluster size should also reduce τ_4 and τ_5; this is not observed. For DPPC/PP = 100 and DPPC/PP = 35 higher τ_5 values were found. We therefore assume that the polypeptides are modulating the softness of the domain walls (see [43, 44]). The polypeptides seem to have a great influence especially on the fusion process of the fluid microdomains causing rigid clusters to disappear. The temperature dependent profile of the relaxations (times and amplitudes) was broadened for higher GA' contents in the DSC measurements. The shoulder observed for higher GA' contents in the DSC measurements was also found in the time and amplitude profiles of the kinetic experiments while the PP vesicles showed only very little overall broadening like in the corresponding DSC peaks.

We are still far from giving a detailed molecular model of the kinetic processes and specific interactions during the phase change in phospholipid vesicles containing polypeptides. But further experiments and computer models using the kinetic data presented here should give more insight into this very complex and exciting field.

7. Acknowledgements

We would like to thank F. Meyer who drew the figures, E. Lerch for the computer programme and R. Groll who prepared the electron micrographs. The polypeptide lys$_2$-gly-leu$_{24}$-lys$_2$-ala-amide was a generous gift of R.S. Hodges (Edmonton) and M. Bloom (Vancouver). A travel grant for A. Genz (NIH M28795) is gratefully acknowledged.

8. References

1. R. Sages and B. Witkop, J. Am. Chem. Soc. 87, 2011-2020, (1965).

2. R.E. Koeppe, D.A. Paczkowski and W.L. Whaley, Biochem. 24, 2822-2826 (1985).

3. J.D. Glickson, D.F. Mayers, J.M. Settine and D.W. Urry, Biochem. 11, 477-486 (1972).

4. D.S. Andersen, Ann. Rev. Physiol. 46, 531-548 (1984).

5. E. Bamberg and P. Läuger, J. Membr. Biol. 11, 177 (1973).

6. E. Bamberg and P. Läuger, BBA 367, 127-133 (1974).

7. W.R. Veatch, R. Mathies, M. Eisenberg and L. Stryer, J. Mol. Bio. 99, 75-92 (1975).

8. W. Veatch and L. Stryer, J. Mol. Biol. 113, 89-102 (1977).

9. S. Weinstein, B.A. Wallace, E.R. Blout, J.S. Morrow and W. Veatch, PNAS 76, 4230-4234 (1979).

10. D.W. Urry, PNAS 68, 672-679 (1971).

11. S. Weinstein, B.A. Wallace, J.S. Morrow and W.R. Veatch, J. Mol. Bio. 143, 1-19 (1980).

12. E. Bamberg and H. Janko, BA 465, 486-499 (1977).

13. N.R. Clement and J.M. Gold, Biochem, 20, 1544-1548 (1981).

14. A. Finkelstein and O.S. Andersen, J. Membr. Biol. 59, 155-171 (1981).

15. B.A. Wallace, W.R. Veatch and E.R. Blout, Biochem. 20, 5754-5760 (1981).

16. E. Oldfield, R. Gilmore, M. Glaser, H.S. Gutowsky, J.C. Hshung, S.J. Kang, T. King, M. Meadows and R. Rice, J. Bio. Chem. 256, 1160-1166 (1978).

17. B.A. Cornell and M. Kenivy, BBA 732, 705-710 (1983).

18. S. Rajan, S.Y. Kang, H.S. Gutowsky and E. Oldfield, J. Bio. Chem. 256, 1160-1166 (1981).

19. D. Chapman, B.A. Cornell, A.W. Eliasz and A. Perry, J. Mol. Biol. 13, 517-538 (1977).

20. D.C. Lee, A. Aziz and D. Chapman, BBA 769, 49-56 (1984).

21. D.A. Pink, A. Georgallas and D. Chapman, Biochem. 20, 7152-7157 (1981).

22. J.H. Davis, D.M. Clare, R.S. Hodges and M. Bloom, Biochem. 22, 5298-5305 (1983).

23. M.R. Morrow, J.C. Huchschilt and J.H. Davis, Biochem. 24, 5369-5406 (1985).

24. J.C. Huchschilt, R.S. Hodges and J.H. Davis, Biochem. 24, 1377-1386 (1985).

25. J.F. Holzwarth, V. Eck and A. Genz, in "Spectroscopy and Dynamics of Biological Systems", eds. P.M. Bayley and R. Dale, 351-377, Acad. Press, London (1985).

26. A. Genz and J.F. Holzwarth, Colloid & Polymer Sci. 263, 484-493 (1985).

27. A. Genz and J.F. Holzwarth, Europ. Biophys. J., 13, 323 (1986).

28. J.F. Holzwarth, Faraday Discussion No. 81, Lipid Vesicles and Membranes, 353-366 (1986).

29. a) J.F. Holzwarth, A. Schmidt, H. Wolff and R. Volk, J. Phys. Chem. 81, 2300 (1977)
b) J.F. Holzwarth, Laser Temperature Jump, pp. 47-76, in: Techniques and Applications of Fast Reactions in Solution, W.J. Gettins and E. Wyn-Jones, eds., D. Reidel Publ. Comp. Dordrecht, Holland (1979).

30. J.M. Kremer, M.W. Eskey, C. Pathmamanoharan and P.H. Wiersema, Biochem. 16, 3932-3925 (1977).

31. B. Gruenewald, S. Stankowski and A. Blume, FEBS Lett. 102, 227 (1979).

32. C.S. Chong and K. Colbow, BBA 436, 260-282 (1976).

33. P.W.M. Van Dijck, B. De Kruyff, P.A. Aarts, A.J. Verkleij and J. De Gier, BBA 506, 183-191 (1978).

34. B.A. Cornell, M.M. Sacre, W.E. Peel and D. Chapman, FEBS Lett. 90, 29 (1978).

35. F. Jähnig, Biophys. J. 36, 329-345 (1981).

36. D. Lichtenberg, E. Freire, C.F. Schmidt, Y. Barenholz, P. L. Felgner and T.E. Thompson, Biochem. 20, 3462-3466 (1981).

37. O.G. Mouritsen and M. Bloom, Biophys. J. 46, 141-153 (1984).

38. B. Gruenewald, W. Frisch and J.F. Holzwarth, BBA 641, 311-319 (1981).

39. a) G. Adam, Zeutschrift f. Naturforschung 23b, 181-197 (1968);
b) G. Adam, Cooperative Transitions in Biological Membranes, pp. 220-231, in: Synergistics: Cooperative Processes in Multi-Component Systems, H. Haken ed., B.G. Teubner Verlag, Stuttgart (1973).

40. M.I. Kanehisa and T.Y. Tsong, J. Am. Chem. Soc. 100, 424-432 (1978).

41. O.G. Mourtisen, BBA 731, 217-221 (1983).

42. O.G. Mourtisen, A. Boothroyd, R. Harris, N. Jan, T. Lookman, L. Mac Donald, D.A. Pink and M.J. Zuckermann, J. Chem. Phys. 79, 2027-2041 (1983).

43. O.G. Mouritsen and M.J. Zuckermann, Europ. Biphys. J. 12, 75-86 (1985).

44. O.G. Mouritssen, Phys. Rev. B 31, 2613-2616 (1985).

AN INVESTIGATION ON THE PENETRATION OF LIPIDS IN THE BILAYER OF STRATUM CORNEUM

H. SUHAIMI
Chemistry Department
University of Agriculture Malaysia
43400 UPM Serdang Selangor
MALAYSIA

S.E. FRIBERG
Chemistry Department
Clarkson University
Postdam, N.Y. 13676
USA

ABSTRACT. A simplified mixture of model stratum corneum was analyzed to make an estimation on the penetration of lipids on the stratum corneum lipid structure. Small x-ray diffractometry and optical microscopy were employed for these purposes.

The small angle x-ray diffractograms showed the lipids to be partitioned in the hydrated acid/soap host layered structure between the methyl group layers. Cholesterol, however, which was entirely located in the acid/soap chains, attracted other compounds from the location between the methyl group layers.

1. Introduction

Water is an essential component of the animal body and the prevention of its escape into the surrounding atmosphere is one of the important functions of the skin. The main barrier to water transport through the skin is located in the outermost layer, the stratum corneum (1,2). This fact is clearly demonstrated by an experiment in which the stratum corneum was removed by repeated stripping using adhesive tape (1).

Taking this fact into consideration, in addition to the observation (3) that the stratum corneum of living and dead skin show identical barrier properties and finally that the lipids in the stratum corneum are the essential compounds for the water barrier (4,5), the intense research interest in the composition and organization of the stratum corneum lipids (6-12) appears well justified.

Many suggestions have been made for the structural organization of the stratum corneum lipids. The layered structure originally proposed by Elias (13) is the basis for this article. In this structure the hydrocarbon chains are mirroring each other over the methyl groups while the polar groups are separated by an aqueous layer. A recent analysis (14) of the calcium content in the epidermis showed high concentrations of calcium in the lamellar bodies and after their secretion into the interstices of the granular-cornified layer interface. This result was interpreted as indicating a role for calcium for the initial bilayer formation in the lower stratum corneum.

The results prompted Downing and collaborators to extend their study on bilayers (15) at pH

D. M. Bloor and E. Wyn-Jones (eds.),
The Structure, Dynamics and Equilibrium Properties of Colloidal Systems, 517–528.

7.5 to include also the influence of calcium ions (16). The results showed the presence of calcium ions changed the liposomes to aggregates of parallel sheets, supporting the earlier suggested role of calcium (14).

However, the upper part of the stratum corneum contains but little calcium (14) and the lipid structure in these parts, with a pH in the range 4.5-6 still consists of broad bilayer (17). It seems that an alternate explanation must be found for the bilayer structure in the upper part of stratum corneum.

With this article we would like to draw attention to the role of the free fatty acids in the formation of a multilayer structure. A combination of a free fatty acid and its corresponding soap spontaneously form a multilayered structure with water provided unsaturated fatty acids constitute a fraction of them (18). We have used this soap/acid lamellar structure as a basis for a layered structure containing the lipids of the stratum corneum in a recent article (19).

This layered structure is stable also at the low pH values found in the upper part of stratum corneum and should be a useful complement to the structures suggested by Downing and collaborators (14,15). With the present contribution we demonstrate the potential of the model to give information about the organisation of the different lipids in the layered structure.

Small angle x-ray diffractometry is used to determine layer spacing in the layered structure and optical microscopy with polarised light is applied to give information about liquid crystal/crystal variations.

2. Experimental

2.1 MATERIALS

The materials used to model the stratum corneum lipids are given in Table I according to Elias et al (12). The materials were all of the highest purity and were used without further purification. The water was twice distiled.

2.2 PREPARATION OF SAMPLE

The sample for the x-ray analysis was prepared by adding the lipids (Table I) according to the sequence shown below:

1. Unsaturated FFA (oleic, linoleic and palmitoleic acids)
   followed by:
2. myristic acid
3. palmitic acid
4. stearic acid
5. phosphatidylethanolamine
6. cholesterol
7. ceramide
8. oleic acid palmityl ester
9. squalene

TABLE I - Composition of Model Epidermal Lipid

Component	Source	Purity	Wt% in Mixture
Phosphatidyl-enthanolamine	Avanti Polar Lipids	99%	5%
Cholesteryl Sulfate	Research Plus	98%	2%
Cholesterol	Fisher		14%
Triolein	Sigma	99%	25%
Free Fatty Acids	Sigma		19%
Myristic		99%	3.8%
Linoleic		99%	12.5%
Oleic		99%	33.1%
Palmitic		99%	36.8%
Palmitoleic		99%	3.6%
Stearic		99%	9.9%
Oleic Acid Palmityl Ester	Sigma	98%	6%
Squalene	Aldrich	98%	7%
Pristane	Aldrich	96%	4%
Ceramides	Sigma	99%	18%

The water content for each composition was varied in each series of samples from 30% to 40% water. The samples were mixed by centrifuging repeatedly through a construction in a sealed 7 mm glass tube. The samples were then allowed to equilibrate at 30°C for 24 hours. Oxygen sensitive compounds i.e. phosphatidylethanolamine and ceramides were handled in a nitrogen atmosphere. The mixed samples were analyzed by optical microscopy and low angle x-ray diffraction. All glassware was rinsed with ether to eliminate the possibility of external lipid contamination. Care was taken to keep the model lipid mixture frozen when not being used for the preparation of samples.

2.3 X-RAY DIFFRACTION

Low angle x-ray diffraction measurements were obtained by use of a Kiessig low-angle camera from Richard Seifert. Ni filtered Cu radiation was used and the reflection determined by a Tennelec position sensitive detector system (Model PSD-1100).

2.4 PHOTOMICROGRAPHY

An Olympus-BH polarizing microscope, attached to an automatic exposure Olympus camera (Model C-35A) was used for photomicrography. Pre-cleaned microscope slides and covers were selected, and then buffed with lint-free tissue immediately before use. A small amount of the sample was transferred from the sample tube onto the glass slide and was immediately covered with the slide cover. The sample was then sheared between the slide and the cover to thickness of about 5 to 10 microns and was left for a few minutes for equilibration. The appearance of the sample was then observed between crossed polarizers. A representative region was then selected and photographed at a magnification of 200.

3. Results

The unsaturated fatty acid/soap combination gave a lamellar liquid crystal with water contents in the range 30-40% by weight. The optical appearance, Fig. 1, is typical of a lamellar liquid crystal. Addition of the saturated fatty acids, Figure 2, show a typical optical pattern of a distorted lamellar liquid crystal due to the presence of crystalline lipidic material. The addition of cholesterol obviously returned the structure to the liquid crystalline state, Fig. 3; the optical pattern is unambiguous. The ceramide and the ester changed the system to a two-phase structure for which one was a well-defined liquid crystal (Fig. 4 and 5). The final addition of hydrocarbon gave an optical pattern of a liquid crystalline phase with a liquid dispersed in it (Fig. 6).

The results of small angle x-ray diffractograms are given in Figure 7 for water contents in the range 30.3 - 40.2% corresponding to water ratios of 0.43 - 0.67. The interlayer spacing of the unsaturated fatty/acid increased from 45.8Å to 49.0Å in the interval used. Addition of myristic acid increased the interlayer spacing by 2.3Å retaining the change with water content. The subsequent addition of palmitic and stearic acid caused no significant change in the interlayer spacing nor in the dependence on the water content. Addition of phosphatidylethanolamine gave an increase of approximately 3Å with retained dependence on the water content.

The behaviour after addition of cholesterol was entirely different and an extremely strong dependence on the water content was now found with an increase of 4Å at the lowest water content but an increase of 10Å at the highest water content. Addition of the ceramide, the oleic acid palmityl ester and the squalene each gave approximately a 2Å increase in the interlayer spacing, restoring the dependence of the water content to the original value.

The curves of interlayer spacings versus water content were extrapolated to zero water content

Figure 1. The unsaturated fatty acid/soaps give a lamellar liquid crystal (32% water).

Figure 2. With addition of the saturated fatty acid/soaps, the optical pattern indicates microcrystals embedded into a liquid crystalline matrix (32% water).

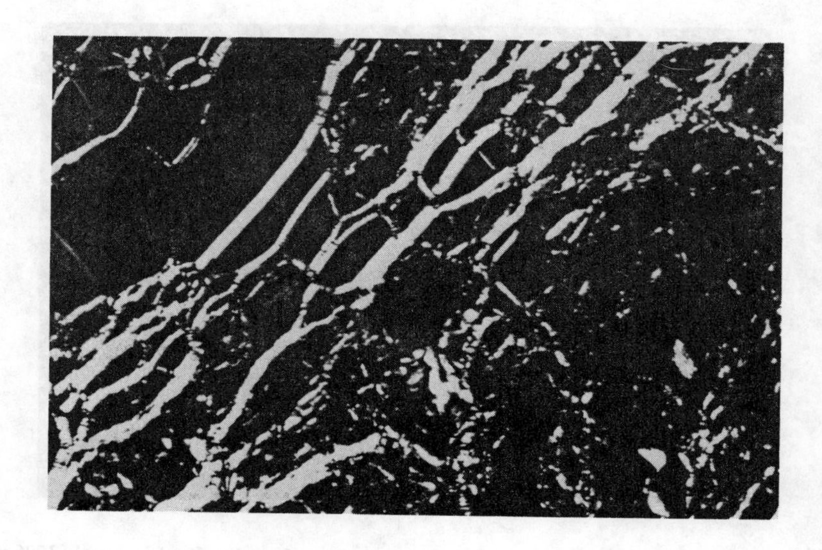

Figure 3. Addition of cholesterol gave optical pattern typical of a lamellar liquid crystal with well-defined oil streaks (32% water).

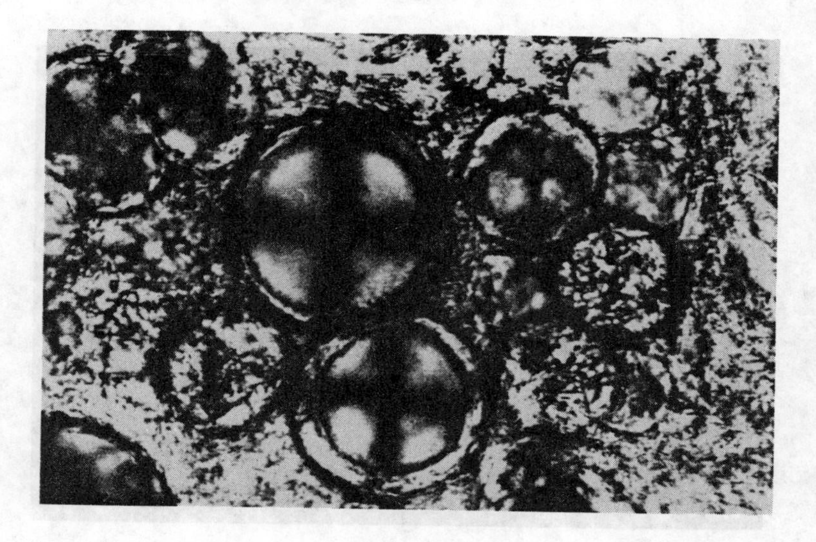

Figure 4. Addition of ceramide gave an optical pattern typical of a two-phase system (32% water).

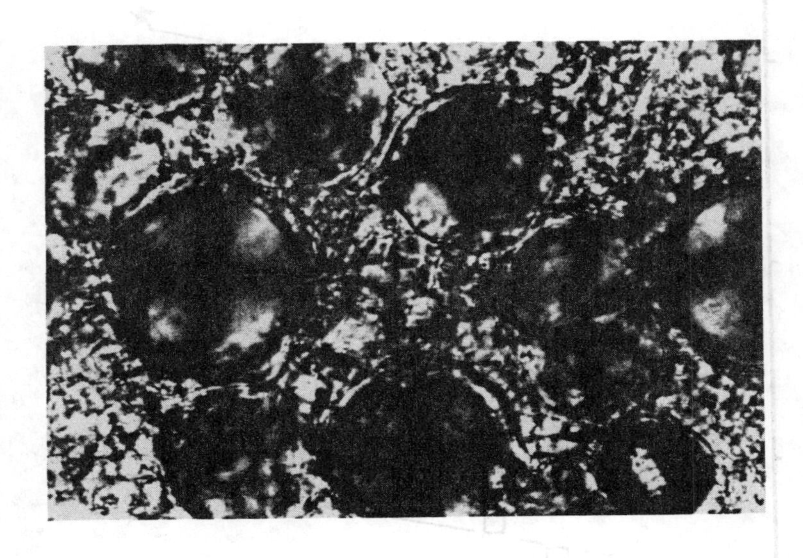

Figure 5. Addition of ester retained the optical pattern obtained after addition of the ceramide (32% water).

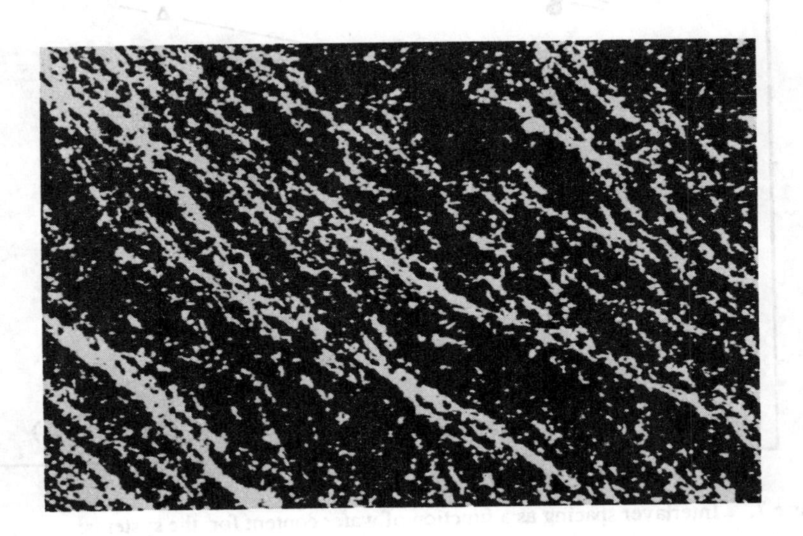

Figure 6. Addition of hydrocarbon gave a two-phase system well dispersed.

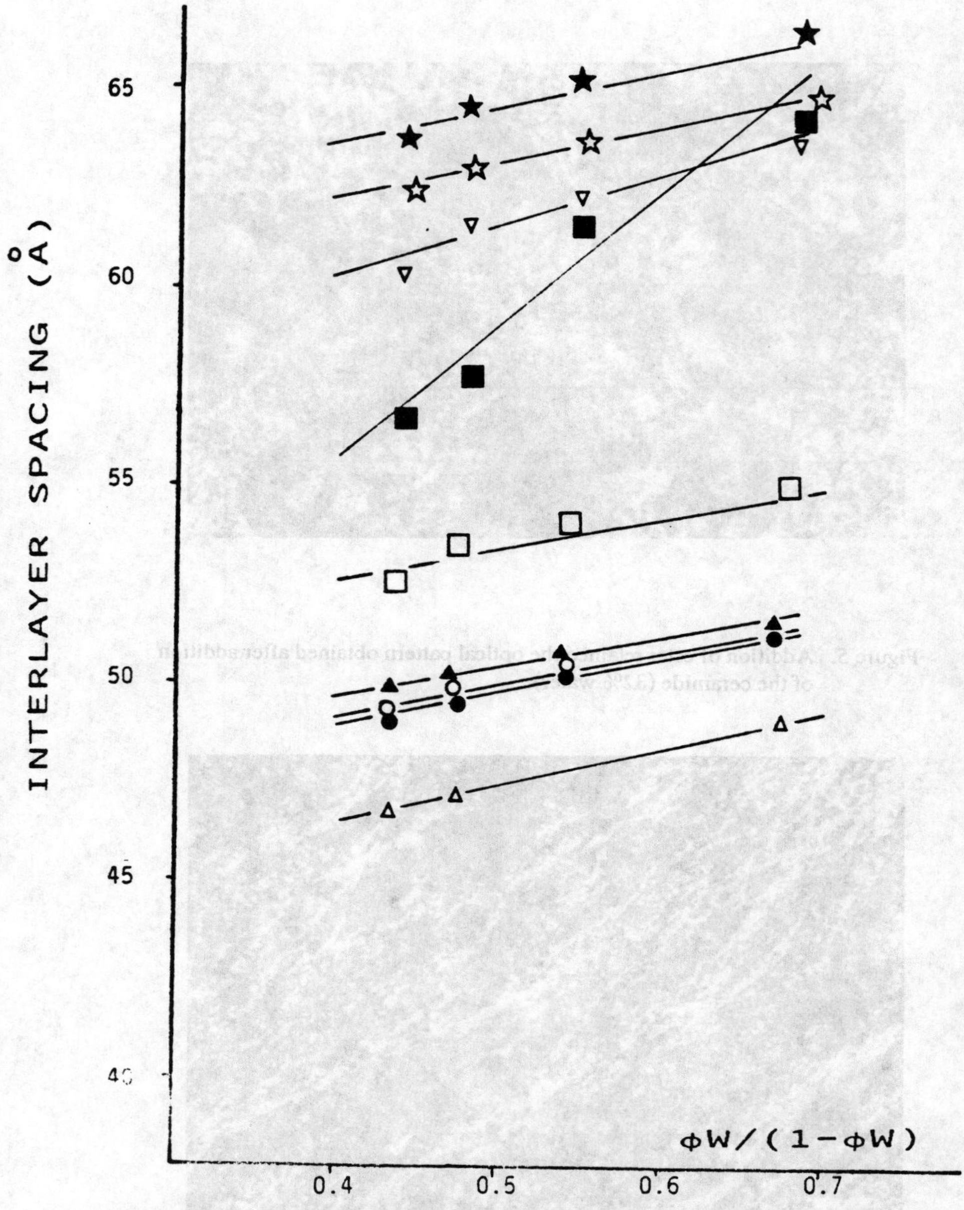

Figure 7. Interlayer spacing as a function of water content for the systems: Δ, unsaturated FFA; 0, myristic + Δ; ●, palmitic + 0; ▲, stearic + ●; □, phosphatidylethanolamine (PE) + ▲; ■, cholesterol + □; ∇, ceramide + ■;☆, oleic acid palmityl ester + ∇;★, squalene +☆

in order to estimate the thickness of the amphiphilic layer. The values are given in Table II together with values of the slope of the straight lines and in Table III indicating molecular weight and density of each compound.

The table shows a regular increase of the extrapolated values of interlayer spacings, d_o, similar to those found in the investigated range of water content except for the composition into which the cholesterol has been added. The d_o values for that composition was reduced to that of unsaturated fatty acids e.g. the lowest possible.

Table II. Values of interlayer spacing extrapolated to zero water content and the slope of the straight line

Compound	d_o	$\frac{\partial d}{\partial(w/1-w)}$
1 Unsaturated fatty acids	42.6	9.6
2 = 1 + Myristic acid	45.9	8.3
3 = 2 + Palmitic acid	45.7	8.4
4 = 3 + Stearic acid	46.8	7.3
5 = 4 + PE	49.7	7.5
6 = 5 + Cholesterol	42.6	33.3
7 = 6 + Ceramide	55.0	13.1
8 = 7 + Ester	58.8	8.4
9 = 8 + Squalene	60.1	8.8

4. Discussion

The results demonstrate that layered structures are formed by the stratum corneum lipids also at the low pH values in the upper part of the layer. As a matter of fact the pH value is the fundamental condition for the layered structure to form as is directly evident from the extensive analysis by Ekwall and co-workers (18).

In the present investigations the amounts of saturated fatty acids exactly corresponded to the distribution given by Elias and co-workers (12). In our earlier report (19) their fraction was reduced in order to retain a liquid crystalline structure of the host. Such an arrangement was advantageous in order to demonstrate the applicability of the method; our present more realistic approach is more useful for obtaining information about the conditions in stratum corneum.

Unsaturated fatty acid/soap combinations form liquid crystals at room temperature. The combination with saturated fatty acids on the other hand form crystals. The presence of crystalline lipid compounds in the stratum corneum does not appear reasonable and the problem of accommodating the saturated fatty acids in a pliable structure is a major one for the stratum corneum lipids. Two of the present results refer to this problem.

In the first instance the small angle x-ray data, Fig. 7, reveal that the saturated fatty acids do penetrate the layered structure of the lamellar liquid crystal; the interlayer spacing is significantly changed by their presence. The photographs of the optical patterns, on the other hand, demonstrate that their presence perturbed the dislocation pattern of the liquid crystal (20). Figure 2 is typical of a lamellar liquid crystal with a low content of a non-homogeneous material. The course of this perturbation was not due to crystals of acid soaps like those found in water-free systems. Our DSC analysis of acid soaps (21) has shown that the hydrated forms of the saturated acid soaps display a strongly reduced melting point in comparison with the water-free system.

Table III. Some properties of the lipids

Component	MW. g/mole	Density, g/ml
Oleic Acid	282.47	0.8935
Linoleic Acid	280.45	0.8686
Palmitoleic Acid	254.42	0.9000 (Assumed Value)
Myristic Acid	228.38	0.8439
Palmitic Acid	256.43	0.8527
Stearic Acid	284.50	0.9408
Phosphatidylethanolamine	743.06	1.0000
Cholesterol	386.66	1.0520
Ceramide	650.10	0.9000 (Assumed Value)
Ester	506.90	0.9000 (Assumed Value)
Squalene	410.73	0.8584
Pristane	268.53	0.7850
Triolein	885.50	0.8988

Hence, it appears justified to conclude that the unsaturated and saturated fatty acid/soaps are well mixed in the layered structure. The nature of the perturbation by the saturated fatty acids was not investigated because the results after addition of cholesterol made such an analysis superfluous. The addition of cholesterol is unambiguously described by the optical pattern and the small angle x-ray results. Both show the structure to be returned to the original liquid crystalline state of the unsaturated fatty acid/soap combination with identical interlayer spacing of the lipid of the layers. Presumably disordering caused by the addition of cholesterol was sufficient to cause those fractions of the fatty acids and the phosphatidylethanolamine which resided between the methyl group layers to penetrate the amphiphilic layer anchored at the water layers. The action of cholesterol in the bilayer part of the biomembrane is an interesting analogous phenomenon

(22,23).

The results lend themselves to a preliminary estimation of the location of the individual components from the change of interlayer spacing after their addition. It is obvious that the increase of interlayer spacing would be proportional to added volume provided all the added compound were localised between the methyl groups in the layered structure and that no change in interlayer spacing would occur if the added compound to its entire extent penetrated between the hydrocarbon chains in the host structure*. Those conclusions assume no change in the order parameter of the host compounds nor any change in the tilt of the hydrocarbon chains.

These assumptions make the present estimations of the location of added compounds artificial to a degree, but the exercise may have some value in giving a quantitative - albeit purely formal - value to the influence of cholesterol on the structural organization of the lipid components.

With the assumption of retained degree of order and no change of tilt of the hydrocarbon chains formal values for the percentage of each compound penetrating into the host hydrocarbon chains are given in Table IV.

An interesting point is the strong influence of cholesterol on the structure. With low water content the addition of cholesterol brings all the polar compounds into the space, penetrating the hydrocarbon chains of the free fatty acids; as a matter of fact, the interlayer distance without water is reduced to the same value as was found for the liquid crystalline phase with the lamellar liquid crystal of the unsaturated fatty acid/soaps alone. This interlayer distance is identical to the length of fully extended hydrocarbon chains of compounds in the C_{16}-C_{18} range.

The second influence of cholesterol was to reduce the water penetration into the lipid part of the structure. For the fatty acid/soap liquid crystal, a formal calculation of the water penetration gives 80% while the composition with cholesterol gives 17%; a drastic difference.

Table IV. Fraction of each lipid penetrating into the bilayer

Lipid	$d_{i,calc}$	$d_{i,exp}$	$d_{i,calc}$ - $d_{i,exp}$, Δd	$\Delta d/d_{i,calc}$
Myristic	3.452	3.272	0.180	0.05
Palmitic	32.900	-0.174	33.074	1.01
Stearic	4.667	1.080	3.587	0.77
PE	10.80	2.94	7.860	0.73
Cholesterol	24.845	-7.14	31.985	1.29

It is essential to realize that the large increase of interlayer spacing (≅10Å) experienced at 40% (water ratio 0.67, Fig. 6) when cholesterol was added to the mixture is not due to the cholesterol being localised between the methyl group layers. The huge value of interlayer spacing is due to the fact that the water did not penetrate into the lipid space.

The values for compositions to which ceramide and the remaining components were added cannot be analysed in a corresponding manner. The optical patterns display their two-phase nature and an analysis of the x-ray results are meaningless without a separation and analysis of the individual phases.

* Corresponding amount of water was added to the structure with each addition of lipid.

Such an investigation would have insignificant relevance for the examination of the lipid structure in stratum corneum and was considered unwarranted.

5. References

1. A.M. Kligman, The Epidermis, (W. Montagna, Ed.), Academic Press: New York-London, Chapter 20 (1964).
2. P.M. Elias and D.S. Friend, J. Cell. Biol. 65, 185 (1975).
3. G.E. Burch and P. Wisnor, Arch. Internal. Med. 74, 437 (1944).
4. P.M. Elias, N.S. McNutt and D.S. Friend. Anat. Rec. 189, 577.
5. W.P. Smith, M.S. Christenson, S. Nachet and E.H. Gans, J. Invest. Dermatol, 78, 7 (1982).
6. P.M. Elias, J. Goerke and D.S. Friend, J. Invest. Dermatol. 68, 535, (1977).
7. G.M. Gray and R.J. White, Ibid 70, 336 (1978).
8. G.M. Gray and H.J. Yardley, J. Lipid Research 16, 435 (1975).
9. E.G. Bligh and W.J. Dyer, Can. J. Biochem. Physiol.
10. W. Abraham, P.W. Wertz and D.T. Downing, J. Lipid Research 26, 761 (1985).
11. A.W. Ranasinghe, P.W. Wertz, D.T. Downing and J.C. Mackenzie, J. Invest. Dermatol. 86, 187 (1986).
12. P.M. Elias, Ibid 80, 44 (1983).
13. P.M. Elias, J. Dermatol. 20, 1 (1981).
14. G.K. Menon, S. Grayson and P.M. Elias, J. Invest. Dermatol. 84, 508 (1985).
15. P.W. Wertz, W. Abraham, L. Landmann and D.T. Downing, J. Invest. Dermatol. 87, 582 (1986).
16. W. Abraham, P.M. Wertz, L. Landmann and D.T. Downing, J. Lipid Research 26, 761 (1985).
17. P.M. Elias and D.S. Friend, J. Cell Biol. 65, 501 (1975).
18. P. Ekwall, Advances in Liquid Crystals, (G.H. Brown. Ed.), Academic Press: New York (1975) p.1.
19. S.E. Friberg and D.W. Osborne, J. Disp. Sci. & Techn. 6(4), 485 (1985).
20. F.B. Rosevear, J. Am. Oil Chem. Soc. 31, 628 (1954).
21. S.E. Friberg and L.B. Goldsmith, Unpublished results.
22. Biological Membranes (D. Chapman, Ed.), Academic Press: New York, (1968).
23. B.P. Schoenborn, Chem. Eng. News 55, 31 (1977).

FOAM STABILITY IN NON-AQUEOUS MULTI-PHASE SYSTEMS

S. E. FRIBERG and WEI-MEI SUN
Chemistry Department
Clarkson University
Potsdam, NY 13676

ABSTRACT. Foam stability in non-aqueous systems is a different phenomenon from that of aqueous foams, because no solid monomolecular film of surfactant is found and the surface free energy stabilization is not present.

Foams with excellent stability are obtained if the condensed part contains a rigid phase in addition to the liquid one. Examples of such foams are presented.

1. Introduction

Foam stability is a fascinating subject and a great number of studies have been published [1-4]. Most of these treatments are concerned with aqueous systems in which an approach combining surface and colloidal forces [5] provide an excellent agreement with experimental results [6,7].

Unfortunately, this avenue is not available for hydrocarbon systems, because the hydrocarbon based surfactants do not adsorb to the surface and are, hence, not able to serve as stabilizers. This is a serious disadvantage because foams from hydrocarbons serve in a manyfold of capacities. So for example, is carbon dioxide flooding the preferred method for extended oil recovery in the continental U.S. [8,9], a great number of food products are in aerated form [10], and foamed polymers are ubiquitous in all aspects of life. Foams from organic solvents may also be a strong disadvantage; so is the case in distillation and extraction processes.

The fundamental knowledge about these foam systems is rudimentary compared to that of aqueous systems. However, some progress has been made in recent years, and the following sections emphasize a few phenomena, which in the future may serve as a nucleus from which to establish a more complete structure.

The treatment has been divided into three sections dealing with evanescent foams, hydrocarbon foams and foams from polar organic solvents.

2. Evanescent Foams

The unstable foams formed when a solution of two organic solvents are separated by tower distillation or extraction, are a pronounced problem which may lead to temporary shut-down of the operation. They were early treated in the engineering literature [11] relating the stability to

D. M. Bloor and E. Wyn-Jones (eds.),
The Structure, Dynamics and Equilibrium Properties of Colloidal Systems, 529–539.

Marangoni flow. This flow from the bulk solution to the bubble cap was assumed to be induced because enhanced evaporation from the bubble compared to that of the bulk solution. The difference in evaporation rate means a relative increase of the less volatile component in the bubble resulting in an increase of surface tension. The increased surface tension in turn causes liquid to be drawn from the bulk and the bubble cap retains its thickness; i.e., it remains stable.

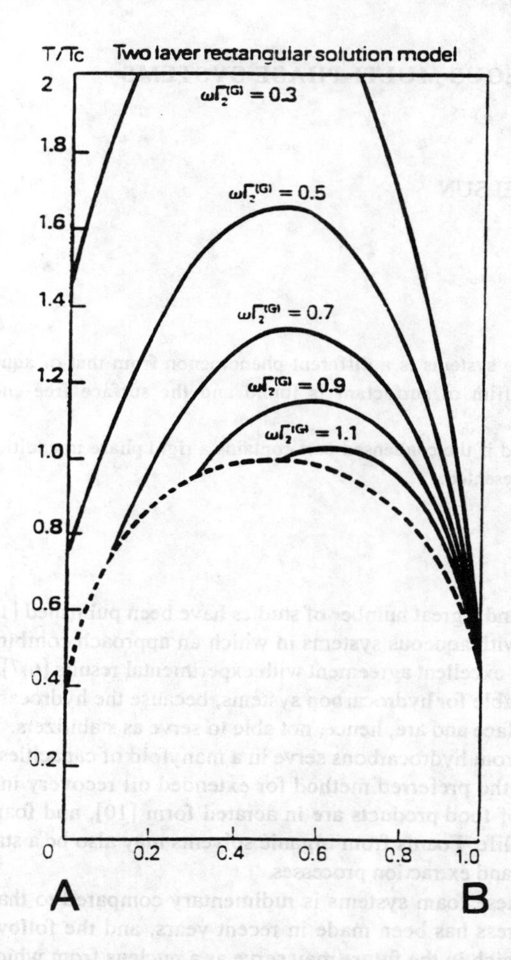

Figure 1. Surface excess in the foam of the component with lower surface tension B in the one phase region of two organic solvents with an upper convolute point according to calculations using regular solution theory [13].

Ross [12] has pointed out that such an explanation is too inclusive. There are many systems of this kind, which show no foaming under evaporation. Hence, foaminess should be a more specific feature and Ross focused his attention on structural changes in the solution prior to phase separation [13]. The excess surface concentration of one component in a binary system can be calculated by means of the "two surface layer" model of regular solutions [14].

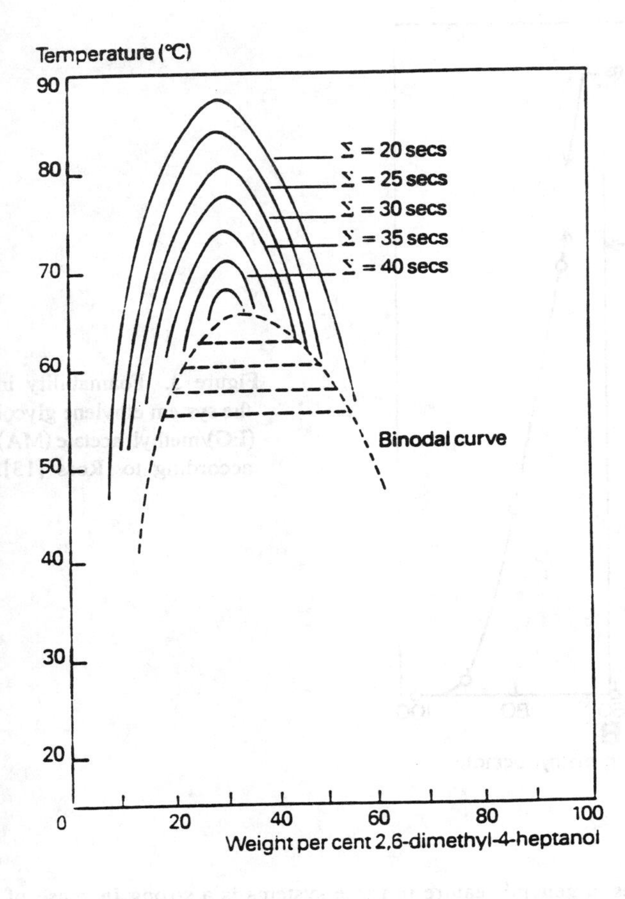

Figure 2. Experimental foamabilities in the system ethylene glycol/2.6-dimethyl-4-heptanol [13].

The results [13], Fig. 1, showed rather large surface excess of the component with the lower surface tension and that this excess was not maximal at the critical point but shifted toward the component with lower surface tension. Experimental determinations [13] of foamability in the system ethylene glycol/2.6-dimethyl-4-heptanol gave the opposite result, Fig. 2. The maximum foam stability was now found towards the component with higher surface tension, the ethylene glycol. So far the difference has not been explained. One factor that may be useful to evaluate is the fact that the stability of these foams, less than a minute, may be dependent also on drainage rate. Hence, the increased viscosity contribution from the compound with higher surface tension would reduce drainage and the foamability would be enhanced with more of the high surface tension compound.

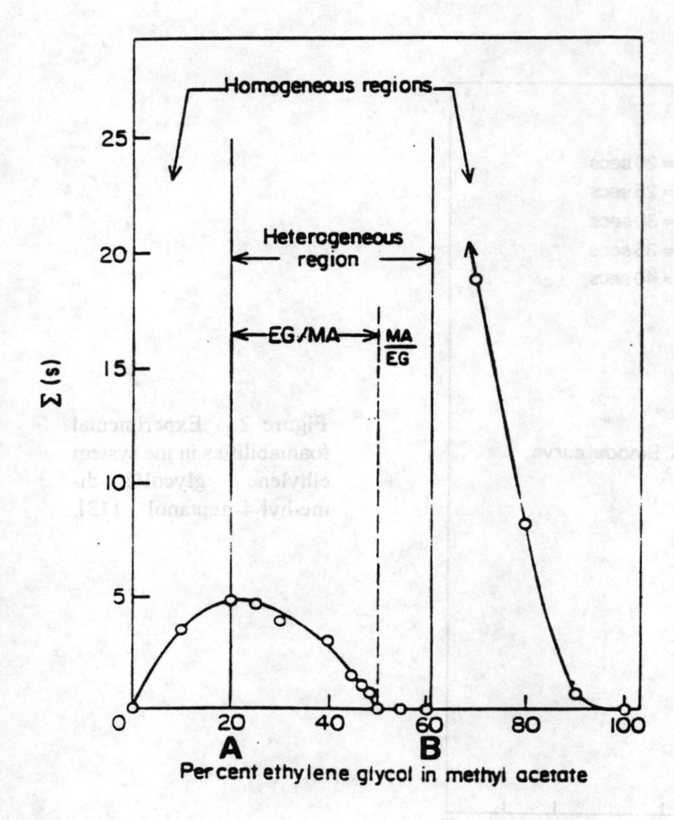

Figure 3. Foamability in the system ethylene glycol (EG)/methyl acetate (MA) according to Ross [13].

Aside from these discrepancies, a general feature in these systems is a strong increase of foaminess close to a phase limit. An excellent example is provided by the combination of ethylene glycol and methyl acetate, Fig. 3 [13]. The increase of foaminess is conspicuous close to the phase limit when methyl acetate is added to the ethylene glycol. In the two-phase region there is no foaming as long as the ethylene glycol is the continuous phase, but the foaminess increases steadily after the inversion in the two-phase region. The methyl acetate solution gave reduced foaminess with reduced amount of ethylene glycol. The pure methyl acetate, of course, gave no foam.

The absence of foaming for the two-phase region which was continuous in ethylene glycol MA/EG was explained by wetting [13]. The surface tensions of the two conjugate solutions were EG(MA) (B Fig. 3). = 26.2 nM/m and MA(EG) (A, Fig. 3) = 25.4 mN/m. The interfacial tension between B and A is negligible and hence, the spreading pressure of A on B is positive and that of B on A is negative. B will be defoamed by A while the reverse does not happen, all in agreement with the experimental results, Fig. 3.

It is interesting to note that these influences are also found in binary solutions of polymers [15,16].

These investigations have all been concerned with foamability; the ratio of steady foam height to gas flow rate. Long term stability of foams cannot be expected in systems of this kind, because they lack a rigid surface film of surfactant.

Such stability is instead achieved, when the second condensed phase in the system shows rigidity and, in addition, spreads on the liquid phase. Behaviour of this kind is found in hydrocarbon solutions in equilibrium with a liquid crystalline phase.

3. Stable Foams from Hydrocarbons

Foams with excellent stability were found a few years ago [17] in a system containing only hydrocarbon and a hydrocarbon chain surfactant. These foams were without change in the foam height for more than 100 hours.

The reason for this enhanced stability is obvious from microscopy photos of the foams, Fig. 4 [17]. The liquid crystal is preferentially located at the surface to air, forming thick layers of adsorbed phase. With this information, the stabilization mechanism is immediately evident; the interface toward air is immobilized by a thick layer of viscous or even rigid liquid crystal and, in addition, fragments of liquid crystal are blocking the drainage from the foam.

Figure 4. The liquid crystal (radiant) is adsorbed at the air bubble surface (black spheres).

This result explains the fact that the foams are stable, but leaves one query, the fundamental reason for the liquid crystal being preferentially located at the surface toward the air. This question has been answered later [18]. The solution was based on the contributions from different molecular groups toward surface free energy. Zisman [19] found that a surface of methyl groups showed a surface tension of approximately 23 mN/m while that of methylene groups was 30 mN/m.

This information was applied to a system of water, hydrocarbon and nonionic surfactant [18] in which the change from an inverse micellar solution L_2, Fig. 5, to a lamellar liquid crystal was accompanied by an onset of foam stability. The inverse micelle, Fig. 5, is characterized by chain disorder allowing the hydrocarbon to penetrate the structure freely. The increase of water in the total composition force an increase of the area of the polar groups and the inverse micelles are transformed to a lamellar liquid crystal, Fig. 5. The surface of this structure consists mostly of methyl groups [20]. The methyl groups give a lower surface energy than the methylene groups and, hence, the lamellar liquid crystal has a lower surface tension than the solution. With this fact in mind, its location at the surface toward the air is an expected phenomenon.

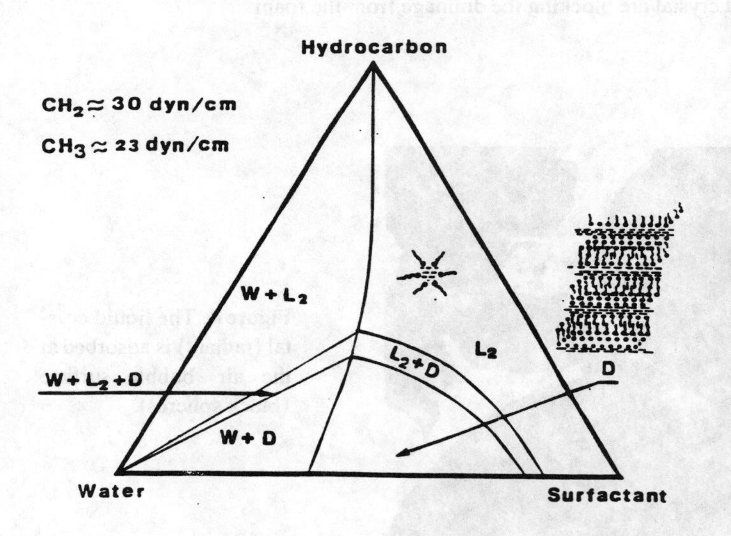

Figure 5. The isotropic solution L_2 contains inverse micelles of the surfactant and water. The solution is in equilibrium with a lamellar liquid crystal D in a two-phase L_2+D.

A direct proof of the difference in surface tension between a solution and a liquid crystal was considered difficult due to the experimental problems of determining surface free energy of a liquid crystal. Hence, the proof was obtained in an indirect manner [18], by selecting two hydrocarbons with different surface tension for foam stability measurements. The liquid crystal would not be expected to spread on the hydrocarbon with low surface free energy and no foam stability should be expected for that combination. The results [18] confirmed the hypothesis. The system with high surface tension hydrocarbon gave stable foams for all hydrocarbon contents provided the total composition remained within the two-phase region liquid/liquid crystal, Fig. 5. On the other hand, the system in which the hydrocarbon had a lower surface tension than the liquid crystal gave no foam stability when the hydrocarbon was part of the total composition.

The results are interesting not only because they provide a simple method to obtain stable foams from hydrocarbons, but because the mechanism is a case of induced surface activity. The surfactant is per se not surface active, but increased water content causes it to change into a new association structure in which the low energy methyl groups are preferentially exposed at the surface.

The liquid crystals have also been shown to be instrumental in the stabilization of foams from polar organic solvents, but with a different mechanism.

4. Stable Foams from Polar Organic Solvents

These investigations were made on two systems [21]. In the first one, the glycerol was combined with sodium octanoate and decanol; in the second system the surfactant was sodium dodecyl sulfate.

The difference between the two combinations is illustrated by their phase diagrams, Fig. 6A,B. With the short surfactant a continuous isotropic solubility region was found from glycerol to decanol while the longer surfactant gave a lamellar liquid crystal in the central part of the system.

The stability of foams in the system of sodium octanoate varied mostly with the glycerol content, Table 1, while the influence of the surfactant content was limited to changes close to the plait point. The values in Table 1 are averaged for different surfactant concentrations.

Table 1. Stability of foams from the isotropic solution in Fig. 6,A.

Glycerol Decanol (Wt ratio)	$t_{1/2}$(s)
1/9	5
2/8	10
4/6	16
5/5	35
6/4	34
8/2	208

These foams show limited stability and the main factor appears to be the viscosity. With high glycerol content the foams were more stable.

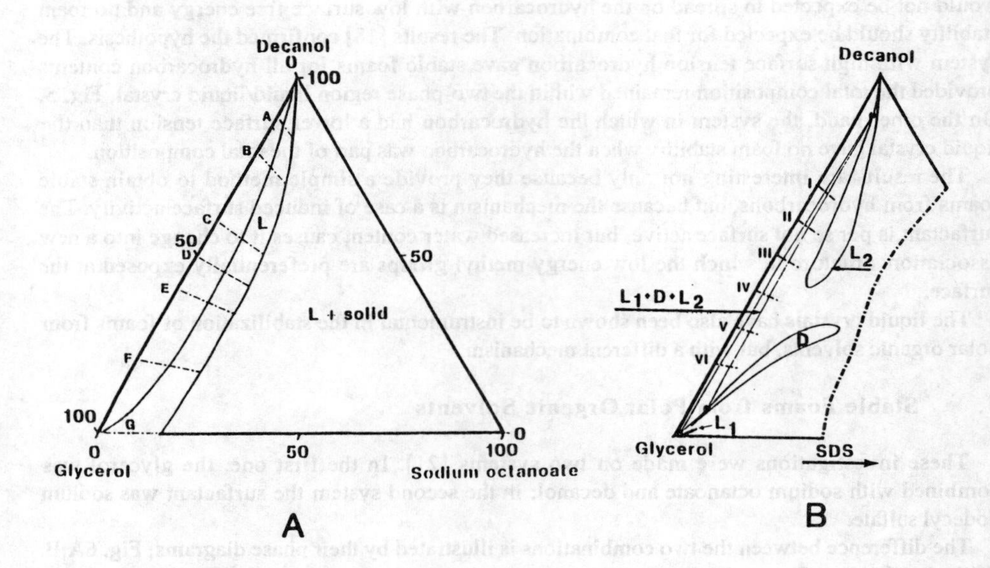

Figure 6. The phase diagram of water plus decanol with sodium octanoate (A) and with sodium dodecyl sulfate (B).

The system with sodium dodecyl sulfate (SDS), Fig. 6B, gave entirely different results. Foams along the tie-line of maximum SDS content in the two-phase region to the left in the diagram, Fig. 6B, were extremely unstable with life times of seconds only. The situation was drastically changed when the SDS concentration was increased to move the total composition into the three phase region $L_1 + L_2 + D$, Fig. 6B. With as little as 0.5% by weight of the liquid crystal the foam lasted for 50 hours and more, Fig. 7.

This result is remarkable because such a small amount of liquid crystal had the most pronounced effect and because the foam height varied with time in an unusual manner, Fig. 7. The foam height remained constant until breakdown began, when it was rapidly reduced to zero. Initial breakage obviously lead to all the foam bubbles breaking.

Microphotographs [21] showed the liquid crystal in this case not to be localized at the surface to the air, but rather evenly distributed in the condensed part of the foam. This means that its main function in this case was not to stabilize the interface, but to prevent drainage by increasing the viscosity and also by direct clogging of the junctions in the foam.

Such a function is in accordance with the shape of the curves of foam height versus time, Fig. 7. With no significant drainage the destabilizing action is restricted to transport of liquid from the films to the junctions by the action of the La Place pressure difference.

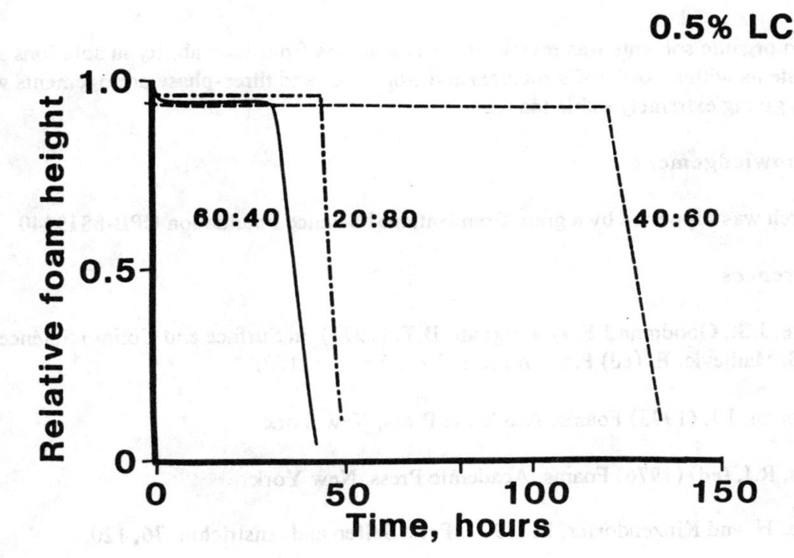

Figure 7. Foam height versus time for compositions within the three-phase region, Fig. 6,B, with 0.5% liquid crystal present. The numbers show ratios between the amounts of glycerol and decanol solution, Fig. 6,B.

The traditional colloidal forces have no significant effect at film thicknesses of the order of microns. The absence of drainage means that the thinning of the films takes place uniformly in the entire foam.

This is opposite to the usual trend in which the films thin mostly in the top part of the foam. The lower parts of the foam are then not thinned to the same extent because here the drained liquid is replenished by liquid draining from the top of the foam. Consequently the lower part of the foam retains its stability and the energy pulses from bursting bubbles at the top are accommodated without destabilization.

In contrast, a foam in which the films are uniformly thinned approaches a state in which the entire structure becomes very sensitive to perturbation. As a result the energy from the first breaking film is not accommodated by increasingly thick films in the lower part of the foam. Instead the breakage of one film leads to a subsequent breakage of adjacent ones. The energy released is increased with each film breaking and a catastrophic event is induced. The foam height versus time for such a foam corresponds to the one in Fig. 7.

5. Summary

Foaming in organic solvents was reviewed with examples from foamability in solutions and two-phase systems without ordered structures and from two- and three-phase arrangements with liquid crystals giving extremely stable foams.

6. Acknowledgement

This research was supported by a grant from National Science Foundation CPB-8819140.

7. References

1. Clunie, J.S., Goodman J.F. and Ingram, B.T. (1971) in Surface and Colloid Science Vol. 3, Matijević, E. (ed) Plenum Press, New York p. 167.

2. Bikerman, J.J. (1973) Foams, Academic Press, New York

3. Akers, R.J. (ed) (1976) Foams, Academic Press, New York

4. Lange, H. and Kinzendorfer, B. (1974) Fette Seifen and Anstrichm. 76, 120.

5. Vrij, A. and Overbeek, J.Th.G. (1968) J. Am. Chem. Soc. 90, 3074.

6. de Feijter, J.A. and Vrij, A. (1978) J. Colloid Interface Sci. 64, 269.

7. Radoev, B.P., Scheludko, A.D. and Manev, E.D. (1983) J. Colloid Interface Sci. 95, 254.

8. Enhanced Oil Recovery (1984) National Petroleum Council, Washington, DC.

9. Surfactant-Based Mobility Control, Smith D.H. (ed) (1988) ACS Symposium Series 373, American Chemical Society, Washington DC.

10. Food Emulsions, Friberg, S.E. (ed) (1976) Marcel Dekker, New York.

11. Zuiderweg, F.J. and Harmens, A. (1958) Chem. Eng. Sci. 9, 89.

12. Ross, S. and Nishioka, G. (1981) Chemistry and Industry p. 47.

13. Ross, S. and Patterson, R.E. (1979) J. Phys. Chem. 83, 2226.

14. Defay, R. and Prigogine, I, (1950) Trans. Faraday Soc. 46, 199.

15. Prigorodov, V.N. (1970) Kolloid Zh. 32, 793.

16. Ross, S. and Nishioka, G. (1977) Colloid Polymer Sci. 255, 560.

17. Friberg, S.E., Wohn, Ch.-S. Greene, B. and Gilder, R.V. (1984) J. Colloid Interface Sci. 101, 593.

18. Friberg, S.E. , Blute, I., Kunieda, H. and Stenius, P. (1986) Langmuir 2, 121.

19. Zisman, W. in Contact Angle, Wettability and Adhesion, Gould, R.F. (ed) (1964) `Advances in Chemistry 43, American Chemical Society, Washington, DC, p. 7.

20. Gruen, D.W.R. (1982) Chem. Phys. Lipids 30, 105.

21. Friberg, S.E., Blute, I. and Stenius, P. (1989) J. Colloid Interface Sci. 127, 573.

THE PROFILE OF A PLATEAU BORDER NEAR A VERTICAL FOAM FILM

J.B.M. HUDALES and H.N. STEIN,
Laboratory of Colloid Chemistry,
Eindhoven University of Technology,
P.O. Box 513, 5600 MB EINDHOVEN,
The Netherlands.

ABSTRACT. **The pressure in a Plateau border is calculated from the latter's profile measured by light interference, with the objective of better understanding the phenomenon of marginal regeneration. The local pressure in the border, at small heights above the bulk liquid, unexpectedly turns out to be lower than the hydrostatic equilibrium pressure. This can be understood through flows in the border caused by a surface tension gradient at the border/film transition. Such flows in the Plateau/border (upwards near the film, downwards far away from the film) have indeed been observed, by flow visualization through addition of solid particles.**

1. Introduction

The problem at the background of the present investigation is the lack of understanding of marginal regeneration. This phenomenon was observed by Mysels c.s. [1] as an exchange of liquid near the border of a thin liquid film, against liquid from the surrounding border. It was established as the most important cause of drainage of vertical thin liquid films with mobile surfaces, which are frequently encountered in foams formed from surfactant solutions at concentrations higher than the cmc; it causes drainage to occur predominantly in a horizontal direction to the nearest border rather than in vertical direction. In spite of this insight dating back to 1959, a basic understanding of the causes of marginal regeneration is still lacking.

One of the unknown parameters in this phenomena is the local pressure in the Plateau border connecting a vertical thin liquid film with the supporting frame. Originally this was assumed [1] to be at least equal to the hydrostatic equilibrium pressure at the height concerned, larger values than the equilibrium ones being expected if the border offers a non-negligible resistance to downward flow. The present investigation was started with the objective of ascertaining whether such is the case, in the special case of films drawn from a 0.02 M cetyl-trimethylammonium bromide solution. Experimentally the method developed by Sheludko c.s. [2-5] and by Haydon c.s. [6-7] for measuring contact angles between a horizontal film and its Plateau border, was adapted to vertical films supported by a glass frame. By extending the measurement of interference fringes to larger distances from the film into the Plateau border, the curvature of the film-near region of the

D. M. Bloor and E. Wyn-Jones (eds.),
The Structure, Dynamics and Equilibrium Properties of Colloidal Systems, 541–549.

border can be measured, and through it the local pressure in the border can be calculated.

2. Principle of the Experimental Method

According to the Laplace equation, the pressure difference p between two phases separated by a surface with principal radii of curvature R and R_1 and R_2:

$$\Delta p = \gamma \left(\frac{1}{R_1} + \frac{1}{R_2} \right) \tag{1}$$

where γ = surface (or interfacial) tension.

In the case of a Plateau border near a vertical liquid film contained in a rectangular frame, one of the radii is much larger than the other hence its influence can be neglected. If we consider a horizontal section through film and border, the curvature with the large radius is connected with a vertical section through the border. Equation (1) becomes:

$$\Delta p = \frac{\gamma}{R_1} \tag{2}$$

with R_1 = the radius of curvature of the horizontal section concerned. This can be calculated if the half border thickness (y) is known as a function of the horizontal distance from the border into the film (x), through

$$R_1 = \frac{\left\{ 1 + \left(\frac{dy}{dx}\right)^2 \right\}^{\frac{3}{2}}}{\frac{d^2y}{dx^2}} \tag{3}$$

(see fig. 1 for the coordinate system employed).

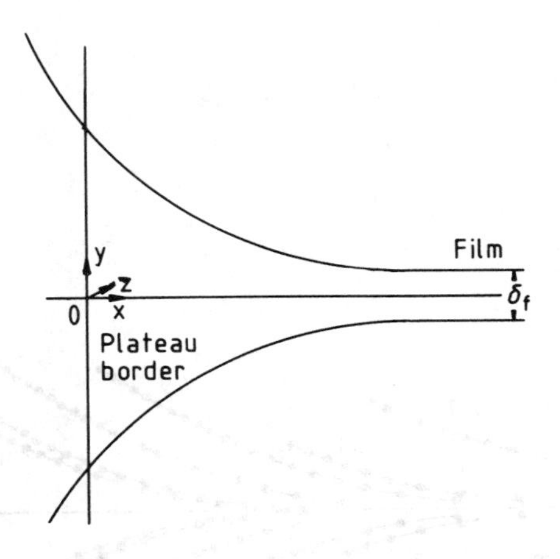

Figure 1. Coordinate system employed.

In equation (3), a minus sign is employed because of the curvature concerned being concave (when seen from the surrounding gas phase).

y(x) is calculated from the interference fringes between light reflected at the front and back surfaces. These were measured by means of a Leitz Metalloplan microscope, using a 50 W super pressure mercury lamp (Osram HB 050). Details of the experimental set-up can be found in the references [8] and [9].

The experimental arrangement for the observation of the interference fringes in the border differ in two essential respects from the traditional one employed in the case of films [1, 10]:

a. illumination is conical in the Leitz microscope rather than plane parallel as in the conventional apparatus;
b. the border front and back surfaces are curved, both concave when seen from the surrounding gas phase, and therefore not parallel.

However, a detailed calculation of the effects mentioned [8, 9] showed that they introduced, at not too large distances from the film, only negligible errors into the thickness calculated by application of formulae developed for plane parallel film surfaces and plane parallel illumination.

3. Results

Fig. 2 shows some typical profiles of the film/border transition region, with the film to the left and the border to the right, for increasing times after formation of the film. Subsequent profiles are referred to the same (arbitrarily chosen) origin with regard to x = 0; y = 0 is chosen in the film midplane.

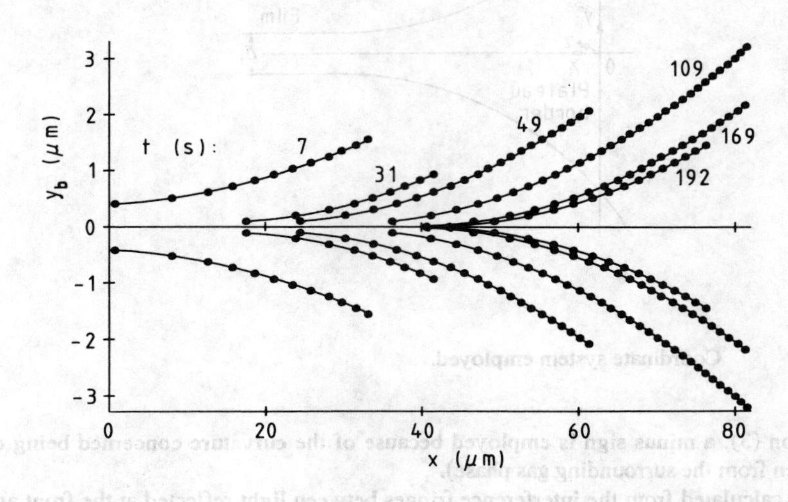

Figure 2. Profiles Plateau Border near a vertical free liquid film (0.02 M CTAB), at various times after film formation. Height above bulk liquid: 6.3 mm.

With increasing time, the border thins and withdraws (alternatively stated the film expands) at the level concerned. Profiles were measured at two heights: 6.3 and 11.3 mm above the bulk liquid, respectively.

In the context of the present paper, especially the radius of curvature is interesting. This was found to be independent of x over the distances covered by the experiments.

Fig. 3 shows a comparison of the radii of curvature calculated from the experimental profiles, with the radii corresponding to the hydrodynamic equilibrium pressures. It is seen that, while there is no distinct difference between the experimental and theoretical radii at a height of 11.3 mm above the bulk liquid, at 6.3 mm above the bulk liquid an experimental radius of curvature is found smaller than the theoretical one. This means that at that height there is a smaller pressure than corresponds with hydrodynamic equilibrium. This is particularly interesting, because theoretically the pressure at any height should be at least equal to the hydrodynamic equilibrium one as long as there is downward liquid motion in the border. A smaller pressure than the

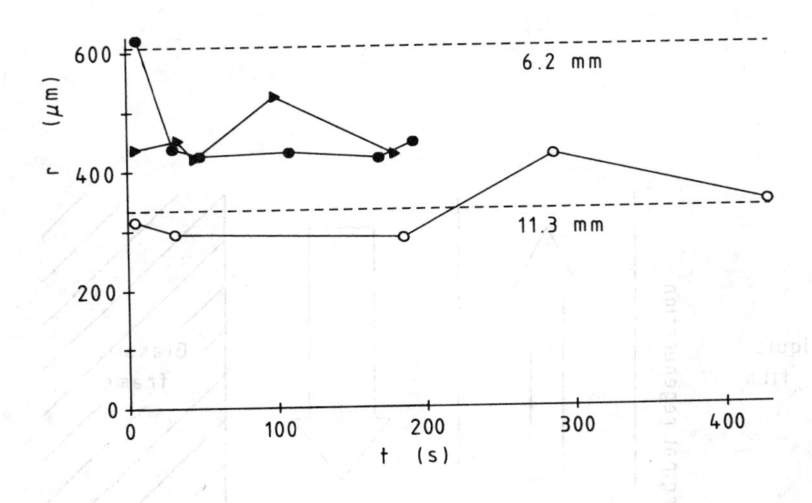

Figure 3. Experimental and theoretical radii of curvature of Plateau Border. The theoretical values were calculated on the assumption of hydrostatic equilibrium pressure.

hydrodynamic equilibrium one at small heights above the bulk liquid should lead to upward motion. However, the withdrawal of the border (or expansion of the film) at that height with increasing time (fig. 2) indicates clearly that there must be, at least in a net sense, a downward liquid motion in the border.

Flow visualization on addition of small glass particles (< 5 µm, density 2.47 g.cm^{-3}) showed that, in the border, two flow regions can be discerned: in the film - near region, the liquid flows upward (fig. 4), while near the frame the liquid flows downward. In between, particles show no, or only a restricted, more or less circular motion. Other horizontal flow in the border between the two regions of upward and downward flow is absent.

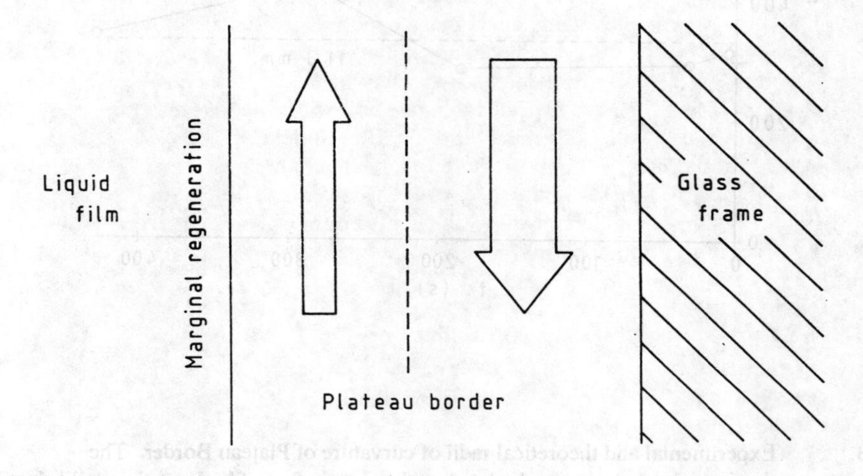

Figure 4. Flows in plateau border seen on addition of glass particles.

4. Discussion

Any mechanism proposed must first of all account for the fact, that in the Plateau border two regions are seen on flow visualization, one of upward and one of downward flow, with no net horizontal liquid transport between these regions. This indicates that any tendency to flow horizontally, from the frame-near to the film-near region of the border, must be counteracted by some effect. The most probable factor, in the field of thin films of surfactant solutions, is a surface tension gradient, directed such as to suppress the tendency of horizontal pressure equalization. In the case at hand this means that the surface tension near the film must be lower than that near the frame, in one horizontal section; this would make a flow from the frame-near to the film-near region of the border an anti-Marangoni one.

Such a surface tension gradient may arise in the following way:

In the film, the surface tension increases with increasing height [11]. The fundamental reason for this is, that a horizontal section line through the front or back surface of the film must support a greater weight at large height than at low height. By marginal regeneration, volume elements of the film are incorporated into the film-near region of the border.

If we assume that in this process there is no significant transport of surfactant or solvent molecules to or from the surface, then the new border volume elements coming from the film will impart their relatively high surface tension to the film-near region of the border.

Thus, a surface tension gradient in the following direction arises (fig. 5):

$$\gamma_{bulk} < \gamma_A < \gamma_B < \gamma_C$$

This leads to a Marangoni flow from A to C, becoming stationary only when it is counteracted either by viscous drag, or by a pressure gradient in the opposite direction. Viscous drag occurs for the upward flow, only at the border / film boundary and at the transition between the upward and downward flow regions, because front and back surfaces of the border are subject to the same surface tension gradient and thus are expected to move with the same velocity, minimizing any velocity gradient in the direction perpendicular to the film midplane. Thus, the local pressure differs from the hydrostatic equilibrium pressure to an extent which increases with increasing height: at low heights above the bulk liquid the pressure is lower than the equilibrium pressure; at 11.3 mm height, it is approximately equal to the equilibrium pressure; at larger heights it is expected to be larger than the equilibrium pressure. The former two statements are confirmed by experiment (fig. 3).

By this process, the border at larger heights becomes too bulky in its film-near region (at C in fig. 5). This leads to horizontally levelling off: a flow from C to D, entails a flow downwards (D → E → F). This flow is then no more a Marangoni one, and the surface tension may be supposed to be equal to that at C because the liquid entrains its own surface.

Thus, the surface tension sequence in the border becomes:

$$\gamma_{bulk} < \gamma_A < \gamma_B < \gamma_C = \gamma_D = \gamma_E = \gamma_F$$

and this leads to a suppression of horizontal pressure equalization because a flow, from E to B say, would be an anti-Marangoni one.

The question may arise why this flow is established rather than horizontal surface tension equalization, through flow from E to B. Such a flow, however, would lead to a slightly increased surface tension at E (compared with the surface tension of the bulk solution), because at E there is only a finite amount of liquid available for surface renewal. At A, on the other hand, bulk liquid is sucked up. This increases the driving force for a flow from A to C, viz. the surface tension difference between A and C, and makes the upward flow from A to C self-reinforcing. A similar self-reinforcement is assumed in the establishment of Marangoni instabilities due to temperature gradients [12].

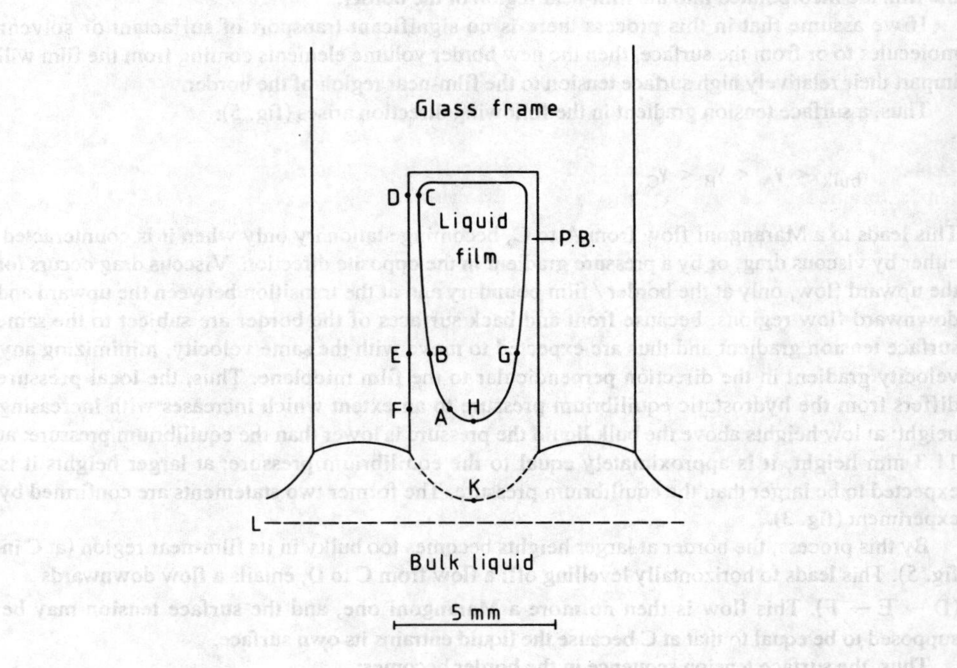

Figure 5. Schematic drawing of film and Plateau border in a frame. P.B. = Plateau border; L: bulk liquid meniscus. See text for remaining symbols.

The assumed sequence of surface tensions leads to a surface tension difference at F, between Plateau border and bulk liquid, which counteracts the transport of liquid from the border to the bulk liquid. It follows from the gradual thinning of the border (fig. 1) that this transport is not completely suppressed, presumably because of the relatively large thickness of the border at the point concerned.

The mechanism is a tentative one, since its most important starting point, the surface tension differences in the Plateau border, cannot be measured independently. However, it accounts for the experiments.

5. Conclusions

Two experimental effects are reported which contradict established ideas:

a. The pressure in the film-near region of the Plateau border, at small heights above the bulk meniscus, is lower than the hydrostatic equilibrium pressure;

b. In the Plateau border both upward and downward flows occur, the former in its film-near region.

Both phenomena can be explained by the hypothesis that marginal regeneration creates a surface tension gradient in the film-near region of the Plateau border.

6. References

1. Mysels, K.J., Shinoda, K. and Frankel, S., 'Soap Films: Studies of their Thinning and a Bibliography', Pergamon Press 1959.

2. Scheludko, A., Kolloid-Z, 155, 39-44 (1957)

3. Scheludko, A., Platikanov, D. and Manev, E., Disc.Far.Soc. 40, 253- 265 (1965).

4. Scheludko, A., Radoev, B. and Kolarov, T., Trans.Far.Soc. 64, 2213- 2220 (1968).

5. Scheludko, A., Radoev, B. and Kolarov, T., Trans.Far.Soc. 64, 2864- 2873 (1968).

6. Haydon, D.A. and Taylor, J.L., Nature 217 , 739-740 (1968).

7. Requena, J.L., Billet, D.F. and Haydon, D.A., Proc.Roy.Soc. London A-347, 141-159 (1975).

8. Hudales, J.B.M., Foam Films in the Absence and Presence of Solid Particles, Ph.D. Thesis, Eindhoven 1989.

9. Hudales, J.B.M. and Stein, H.N., J.Coll.Int.Sci., accepted for publication.

10. Lyklema, J., Rec.Trav.Chim. 81, 890-897 (1962).

11. Lucassen, J., in: Surfactant Series 11: Anionic Surfactans, Physical Chemistry of Surfactant Action (E.H. Lucassen-Reinders, ed.), Marcel Dekker 1981, p. 218-265.

12. Tritton, D.J., Physical Fluid Dynamics, Clarendon Press 2nd ed., Oxford 1988, p. 47.

THE ANAMOLOUS EFFECT OF ELECTROLYTES ON SURFACTANT MONOLAYER SURFACE PRESSURE-AREA ISOTHERMS

M.S. ASTON
BP Research Centre
Sunbury-on-Thames
MIDDLESEX TW16 7LN

ABSTRACT. Quantitative discussions given previously have shown that comparison of the surface pressure-area isotherms of insoluble surfactants is misleading, when the subphase contains varying concentrations of electrolyte. This is confirmed in the present paper, which gives further qualitative arguments, showing that surface tension-area per molecule isotherms may be more appropriate in interpreting surfactant-electrolyte interactions. In addition, surface laser light scattering measurements confirm that concentrated ammonium nitrate has no significant effect on the surface tension of the close packed monolayer of a block copolymer surfactant.

1. Introduction

The interaction between surfactants and electrolytes is important in a wide range of systems ranging from commercial emulsion formulations to biological membranes. In a recent paper [1], the interactions between concentrated electrolytes (to 8 molar) and nonionic "insoluble" surfactant monolayers were discussed. It was shown using a detailed thermodynamic argument that the comparison of surface pressure-area (π-A) isotherms when electrolyte is present in the subphase is fundamentally unsound. The π-A isotherm becomes more expanded with increasing electrolyte concentration, regardless of any surfactant-electrolyte interaction. The effect was observed with all three nonionic surfactants studied, which were of differing chemical type; i.e., a poly(12-hydroxystearic acid)-polyethylene oxide-poly(12-hydroxystearic acid) block copolymer known as B246, an alcohol ethoxylate ($C_{14}E_2$), and sorbitan monooleate. The main electrolyte studied was ammonium nitrate. It was suggested that the surface tension-area per molecule (γ-A) isotherm should be examined to obtain a true picture of the surfactant-electrolyte interaction.

An important observation was that the surface tension of the surfactant close-packed monolayer was independent of the ammonium nitrate concentration.

The present paper provides additional understanding of the observed effects from the qualitative discussions given. In addition, surface laser light scattering(SLLS) measurements are performed on the B246 monolayer. This technique was used as a non-invasive method of determining surface tension, thus avoiding any difficulties with wetting/contact angle effects. This is in contrast with the Wilhelmy plate method used previously [1,2]. The aim was to use the SLLS method to confirm that the surface tension of the close-packed monolayer was independent of the electrolyte concentration.

D. M. Bloor and E. Wyn-Jones (eds.),
The Structure, Dynamics and Equilibrium Properties of Colloidal Systems, 551–555.

2. Background

In a typical insoluble monolayer study, the surface pressure (π), due to the surfactant monolayer, is measured as a function of the area per surfactant molecule (A) at the surface. π is defined by:

$$\pi = \gamma_o - \gamma \quad (1)$$

where γ_o is the surface tension of the clean (surfactant-free) surface and γ is the surface tension in the presence of the surfactant monolayer.

Calculation of the area per molecule is straightforward from the mass of surfactant added, its molar mass, and the total area of surface available. A is varied either by compressing the monolayer using a movable barrier, or by successive addition of surfactant.

For most systems, including mixtures of insoluble surfactants, equation 1 is acceptable. However, the situation becomes more complex when a combination of a soluble and an insoluble surface active component is involved. Such a case is an insoluble surfactant monolayer at the electrolyte solution/air interface. Whilst the surfactant reduces the surface tension, the electrolyte raises it; both components are surface active. With electrolyte in the subphase, equation 1 becomes:

$$\pi' = \gamma_o' - \gamma' \quad (2)$$

in which γ' is the measured surface tension and γ_o' is the surface tension of the electrolyte solution in the absence of the surfactant.

Effectively, equation 2 assumes the effect of the electrolyte is independent of the surfactant coverage at the surface; ie, γ_o' is a constant even though π' and A are changing. This is an incorrect assumption. The electrolyte raises the surface tension of pure water, but (as observed) may have no effect on the surface tension of the surfactant close packed monolayer. This leads to the observed apparent expansion of the π-A isotherm.

3. Experimental

3.1 MATERIALS

The surfactant used was B246, a block copolymer (ICI Speciality Chemicals) having the structure PHS-PEO-PHS, where PHS is poly(12-hydroxystearic acid) and PEO is polyethylene oxide. The number average molar mass is 3543 g mol^{-1}, determined by vapour pressure osmometry. Each PHS chain has been shown to contain approximately 5 HS units [3]. The surfactant was used as received, without further purification.

The water used was deionised and then doubly distilled. Analytical grade hexane and ethanol were used for the mixed spreading solvent (10:1 mixture), for both the Langmuir trough work and the SLLS measurements. Ammonium nitrate was obtained as pure recrystallised material from Frederick Allen and Sons, Ltd, London.

3.2 THE LANGMUIR TROUGH

Details of the Langmuir trough used, and its operation, have been given previously [1]. The surfactant was spread on the surface of the aqueous phase as a solution in the spreading solvent, containing 300 mg dm^{-3} surfactant. Aliquots of the spreading solution were successively applied at constant total trough area, until a surface pressure of approximately 15 mN m^{-1} was reached. To obtain surface pressures above this, the monolayer was compressed using a moving barrier (rate of area change: 1.0 cm^2 s^{-1}). The Wilhelmy plate method was used to measure surface tension, from which the surface pressure was calculated using equations 1 or 2.

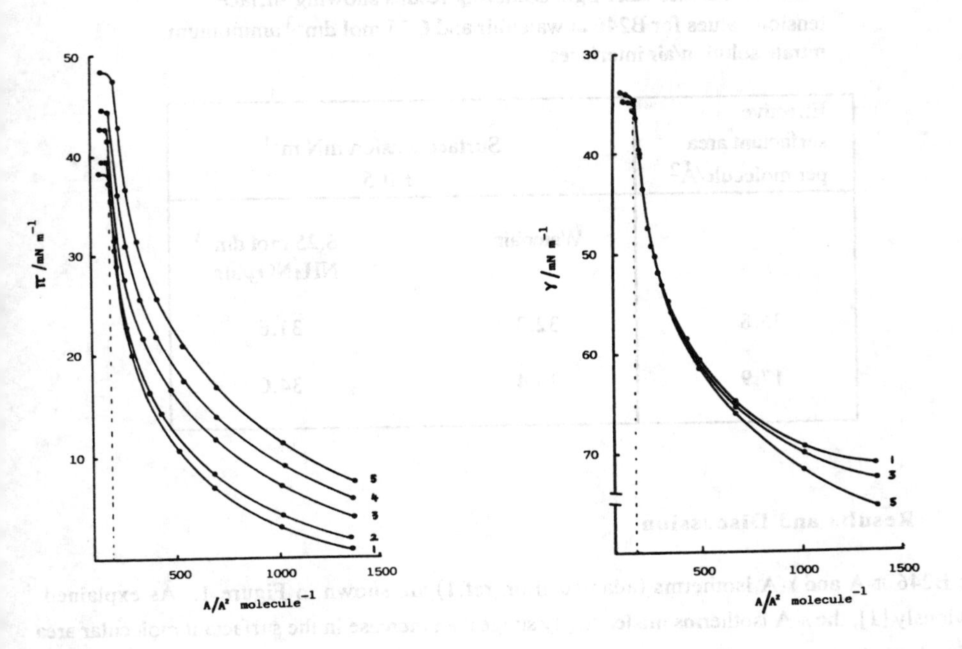

Figure 1. Surface pressure-area (π-A) and surface tension-area (γ-A) isotherms for the polymeric surfactant B246 at various ammonium nitrate concentrations (25°C). The numbered curves refer to the following ammonium nitrate concentrations (mol dm^{-3}): 1: 0.00, 2: 1.25, 3: 4.00, 4: 6.25, 5: 8.25.

3.3 SURFACE LASER LIGHT SCATTERING

The theory of SLLS is complex and will not be discussed here. A large number of papers have appeared in the literature on the subject and the reader is referred to these for details [4,5,6]. The essence of the technique is to look at light scattered at small angles from the main reflected beam, when light from a laser is incident on a liquid surface. Small shifts in wavelength occur as a result of capillary waves present at the surface. Scattering was analysed using a photomultiplier detector

and Malvern correlator; time domain analysis of the resulting correlation function was performed using the analytical procedures of Earnshaw as referred to in model D of his theoretical treatments [6]. Although the technique can give information on surface and bulk rheological terms [5], the present measurements were limited to the determination of the surface tension. An argon-ion laser (wavelength = 488 nm) was used as the light source.

B246 Spreading solution was added from a microsyringe to the water/air or 6.25 mol dm^{-3} ammonium nitrate solution/air surface, in a cell having a total liquid surface area of 66.48 cm^2. Sufficient spreading solution was added to ensure a close packed surfactant monolayer at the surface.

Table 1. Surface laser light scattering results showing surface tension values for B246 at water/air and 6.25 mol dm^{-3} ammonium nitrate solution/air interfaces.

Effective surfactant area per molecule/Å^2	Surface tension/mN m^{-1} ± 0.5	
	Water/air	6.25 mol dm^{-3} NH_4NO_3/air
35.8	32.3	31.6
17.9	33.4	34.0

4. Results and Discussion

The B246 π-A and γ-A isotherms (adapted from ref.1) are shown in Figure 1. As explained previously [1], the π-A isotherms misleadingly suggest an increase in the surfactant molecular area as the electrolyte concentration is increased. On the other hand, the γ-A isotherms converge at the onset of close packing of the surfactant monolayer (300 ± 50 Å^2 molecule^{-1}), indicating no surfactant-electrolyte interaction. Following the arguments in Section 2, it appears that the electrolyte 'sees' the surface only when the surfactant monolayer is less than close packed. Thus, electrolyte raises the surface tension at relatively low surfactant coverage, but at the onset of close packing of the monolayer, the electrolyte ceases to have access to the surface and so no longer affects the tension. The inference must be that the electrolyte has no significant effect upon the area occupied by the surfactant molecule at the surface.

Further evidence that the surface tension of the close-packed monolayer is unaffected by the electrolyte is given in Table 1. The data show the surface tension at effective area per molecule values of 17.9 and 35.8 Å^2 molecule^{-1}, for water/air and 6.25 mol dm^{-3} ammonium nitrate/air interfaces. These area per molecule values are below the minimum close packed area of 100 Å^2 molecule^{-1} obtained by vertical extrapolation of the π-A or γ-A isotherm (see Figure 1). Thus, they represent true close packed surfactant systems, in the presence of excess surfactant. Clearly, from Table 1, the presence of the electrolyte does not significantly affect the measured surface tension, in agreement with the γ-A data of Figure 1. The slight increase in surface tension on decreasing the effective area per molecule from 35.8 to 17.0 $Å^2$ may be associated with collapse processes in the monolayer, but it is not considered to be significant, and in any case does not alter the findings with regard to the effects of the electrolyte.

5. Conclusions

At electrolyte concentrations which significantly raise the surface tension of water, surface pressure-area (π-A) isotherms of insoluble surfactants can be misleading.

Surface tension-area per molecule isotherms probably give a better indication of the true effect of the electrolyte on the surfactant close packed area.

Surface laser light scattering data adds confirmatory evidence for the arguments given. In particular, it has been shown for B246 polymeric surfactant, that the surface tension of the close packed monolayer is unaffected by the presence of high concentrations of ammonium nitrate in the subphase.

6. References

1. Aston, M.S. and Herrington, T.M. Submitted to J. Colloid Interface Sci.

2. Aston, M.S., Herrington, T.M. and Tadros, Th.F. To be published.

3. Aston, M.S., Bowden, C.J., Herrington, T.M. and Tadros, Th.F. 'Polymer Association Structures', Ed. M.A. El-Nokaly, ACS Symposium Series, **384**, 338 (1989).

4. Hard, S. and Neuman, R.D. J. Colloid Interface Sci. **83**, 315 (1981).

5. Crilly, J.F. and Earnshaw, J.C. J. Physique **48**, 485 (1987).

6. Earnshaw, J.C. and McGivern, R.C. J. Colloid Interface Sci. **123**, 36 (1988).

SURFACTANT MOLECULAR GEOMETRY WITHIN PLANAR AND CURVED MONOLAYERS IN RELATION TO MICROEMULSION PHASE BEHAVIOUR

R. AVEYARD, B.P. BINKS and P. D.I. FLETCHER
Surfactant Science Group, School of Chemistry,
University of Hull, Hull, England HU6 7RX

ABSTRACT. The phase behaviour of oil + water + surfactant mixtures is strongly dependent on the molecular geometry of the surfactant *within the monolayer.* In this article we show how surfactant molecular geometrical parameters in both curved and flat monolayers may be estimated. We describe examples for which the variation of surfactant geometry with electrolyte concentration, temperature, alkane chain length and the concentration of a second surface active species have been measured quantitatively. Finally, we describe a method whereby the extent of oil mixing with a surfactant monolayer at the air-water interface may be determined.

1. Introduction

Surfactants are widely used throughout the chemical industry for a wide range of applications including detergency, formulation of agricultural products, food manufacture, mineral processing, crude oil recovery and many others [1]. Surfactants are normally classified as "hydrophilic" or "lipophilic" and a number of hydrophilic-lipophilic balance (HLB) scales have been devised to aid the formulation chemist to optimise surfactant performance for a particular application [2-4]. The HLB behaviour of a surfactant within a particular system containing oil + water may be investigated and characterised in terms of the microemulsion phase behaviour [5,6] and such studies over the past few decades have highlighted the role of the effective surfactant molecular geometry in determining surfactant phase behaviour [7,8] and hence HLB. The purpose of this review is to show how relevant surfactant molecular geometrical parameters may be determined experimentally and to illustrate some underlying principles of the relationship between the observed phase behaviour and the molecular structure of the surfactant. Rather than give an exhaustive catalogue of empirical findings, we have selected results drawn mainly from our own recent work.

2. Microemulsion phases formed in mixtures containing oil + water + surfactant

Microemulsions are thermodynamically stable, isotropic dispersions of oil in water (o/w) or water in oil (w/o) stabilised by a surfactant. At a microscopic level, microemulsions consist of oil and water domains separated by a monolayer of surfactant. The domain sizes are typically in the range 5 - 50 nm and hence the systems scatter light weakly giving the microemulsions

D. M. Bloor and E. Wyn-Jones (eds.),
The Structure, Dynamics and Equilibrium Properties of Colloidal Systems, 557–581.

their transparent or bluish appearance. Microemulsions are generally of low viscosity. Reviews of studies of microemulsion phase structures may be found in references [9-13].

When a surfactant is added to a two-phase oil + water system at a moderate concentration in excess of a critical value, denoted as the critical microemulsion concentration (cmc), one may obtain either a Winsor I, Winsor III or Winsor II system [14]. A Winsor I system is one in which an o/w microemulsion coexists with an excess oil phase. A Winsor II system consists of a w/o microemulsion with a conjugate phase of excess water. In the Winsor III system, a microemulsion phase generally containing comparable amounts of oil and water coexists with both excess oil and water phases. The phase system progression Winsor I - III - II (shown in Figure 1) may be obtained by systematically changing a number of variables which include electrolyte concentration, temperature, the nature of the oil component and the type and concentration of a second surface active material ("cosurfactant") [15].

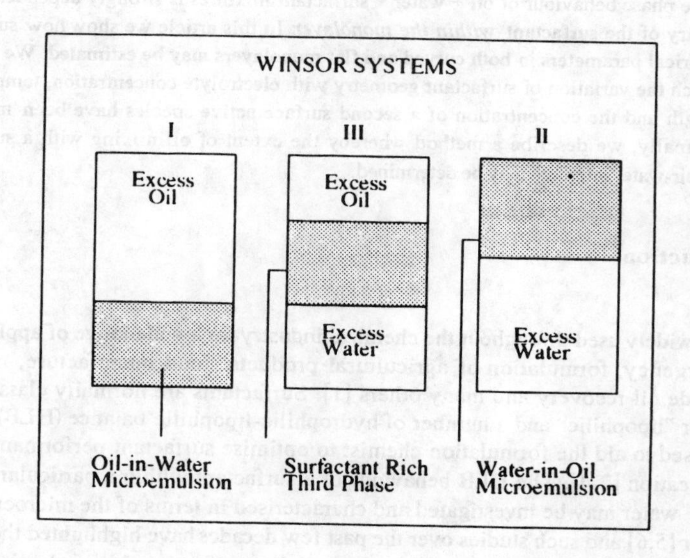

Figure 1. Schematic diagram of Winsor I, III and II equilibrium multiphase systems.

The Winsor systems described above contain a monolayer of surfactant at the planar interface separating the bulk phases which is in equilibrium with the curved monolayer coating the microemulsion aggregates [16,17]. The curvature of the latter monolayer changes progressively from positive in the Winsor I system to negative in the Winsor II case. The microemulsion phase of the Winsor III system is thought to consist of a "sponge - like" structure of interconnected oil and water domains in which the average net curvature of the monolayer is close to zero [18,19]. We are primarily concerned with Winsor I and II systems here in which the microemulsion aggregates are spherical droplets and will consider these further.

For the Winsor I and II systems the microemulsion droplets coexist with a separate phase of the excess dispersed component (i.e. oil for Winsor I and water for Winsor II). The droplets are free to solubilise more or less dispersed component (and hence swell or shrink in size) until the system free energy is minimised. This generally occurs when the droplet size is close to the "natural" size, that is, the inverse of the "natural" curvature of the surfactant monolayer. Hence the formation of microemulsion phases is strongly associated with the tendency of the monolayer to curve which, in turn, is determined by the molecular geometry of the surfactant. For a simple geometrical model in which it is assumed the droplets are spherical and monodisperse, the radius of the droplet core r_c is related to the area of the surfactant headgroup A_h, tailgroup area A_t and the length of the surfactant molecule δ according to the appropriate form of equation 1.

$$r_c = \delta/((A_t/A_h)^{1/2} - 1) \text{ for w/o droplets}$$
$$r_c = \delta/((A_h/A_t)^{1/2} - 1) \text{ for o/w droplets} \qquad (1)$$

The relative magnitudes of A_t and A_h determine the preferred monolayer curvature (and hence microemulsion type) as illustrated schematically below (Figure 2). This type of geometrical

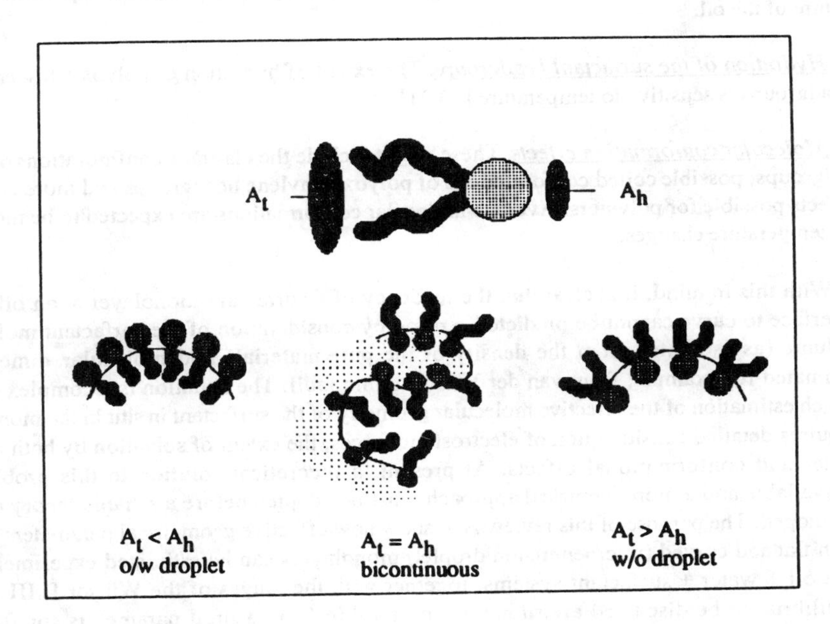

Figure 2. Representation of a surfactant molecule showing the cross sectional head and tail areas A_h and A_t. The relative magnitudes of these areas determine the preferred monolayer curvature.

discussion was first given by Israelachvili, Mitchell and Ninham who introduced the packing factor P (= surfactant chain group volume/headgroup area x surfactant chain length) [7,8] which is approximately equivalent to the ratio A_t/A_h.

If the surfactant geometry was solely determined by the surfactant molecular structure then a particular surfactant would yield a single preferred monolayer curvature (and hence a single microemulsion droplet size in an equilibrated two-phase system) independent of prevailing conditions of temperature, electrolyte concentration and the nature of the oil. It is well known that microemulsion droplet size and type (o/w or w/o) is sensitive to all these variables [15] and it follows that the tendency of the monolayer to curve is also affected. Hence it is the effective molecular geometry in situ in the monolayer that must be considered. This effective geometry differs from that of the isolated surfactant molecule due to a number of effects:

1. Electrostatic repulsion between neighbouring ionic headgroups. This has the effect of increasing the effective area of the headgroup above its "bare" size. The effective size is decreased by the addition of electrolyte which leads to a shielding of the charge.

2. Solvation of the surfactant tailgroups by the apolar oil solvent. Penetration of the tail region by the oil swells the effective tail area and the extent to which this occurs is dependent on the nature of the oil.

3. Hydration of the surfactant headgroups. The extent of hydration of polyoxyethylene type headgroups is sensitive to temperature [20,21].

4. Molecular conformation effects. These might include the cis-trans configurations of alkyl tailgroups, possible coiled conformations of polyoxyethylene headgroups and more complex effects possible for polymers. Average molecular conformations are expected to be modified by temperature changes.

With this in mind, it is clear that the tendency of a surfactant monolayer at an oil-water interface to curve cannot be predicted *a priori* by consideration of the surfactant molecular volume (as estimated from the density of the pure material) and molecular dimensions estimated for example, from van der Waals atomic radii. The situation is a complex one in which estimation of the effective molecular geometry of the surfactant in situ in the monolayer requires detailed consideration of electrostatic forces, the extent of solvation by both oil and water and conformational effects. At present, a theoretical solution to this problem is unavailable and a more empirical approach must be adopted before a reliable theory can be developed. The purpose of this review is to show how effective geometrical parameters within both flat and curved (microemulsion droplet) monolayers can be estimated experimentally. The oil + water + surfactant systems, together with the ranges of the Winsor I, III and II equilibria, to be discussed are summarised in Table 1. Measured parameters for flat and curved monolayers of the same surfactant system will be compared. Finally, we will show that surface tension measurements at the air-water surface can give quantitative information concerning the extent of penetration of oils into surfactant monolayers although such penetration may or may not be the same as at an oil-water interface.

3. Curved (i.e. microemulsion droplet) monolayers at the oil-water interface

Information on surfactant geometry in curved monolayers may be obtained by measuring microemulsion droplet sizes. In this regard it is important to distinguish between single-phase, "made-up" microemulsions and those which form at equilibrium with excess dispersed phase (i.e. two-phase Winsor I or Winsor II systems). Single phase microemulsions are stable only over a range of conditions of composition, temperature, electrolyte concentration etc. The phase boundaries are of two types. At the so-called solubilisation phase boundary excess dispersed phase separates giving either a Winsor I or II two phase system. Here the droplets

Table 1. Summary of the surfactant + water + oil systems referred to in this paper including the ranges of the Winsor I, III and II regions.

System	Variable	Winsor I	Winsor III	Winsor II	Figures
AOT/heptane 25°C	[NaCl]/M	0-0.045	0.045-0.55	>0.055	3-6,11
AOT/dodecane 25°C	[NaCl]/M	0-0.08	0.08-0.12	>0.12	11
AOT/heptane 0.05 M NaCl	T/°C	>26	24-26	10-24	13
AOT/0.1 M NaCl 25°C	N	>12	12	7-12	12
AOT/heptane 0.017 M NaCl/ 30°C	$X_{dodecanol}$	0-0.25	0.25-0.27	>0.27	15
AOT/heptane 0.1 M NaCl/ 25°C	X_{SDS}	>0.11	0.08-0.11	0-0.08	9,16
$C_{12}E_5$/heptane	T/°C	10-26	26-30	>30	7,8,10,14
$C_{12}E_5$/tetra-decane	T/°C	30-44	44-52	>52	7

AOT = sodium bis 2-ethylhexylsulphosuccinate
SDS = sodium dodecyl sulphate
$C_{12}E_5$ = dodecyl pentaethylene glycol ether
N is the alkane chain length
X refers to the mole fraction of the subscripted species in the surface

are free to swell or shrink by solubilising more or less dispersed component as already discussed and hence the size corresponds to the "natural" one determined by the tendency of the monolayer to curve. It is this "natural" size (and, of course, the type of microemulsion formed) that is related to the HLB of the surfactant.

In single phase microemulsion samples not at the solubilisation phase boundary the droplet size is mainly determined by geometrical constraints imposed by the sample composition. Changing a variable such as temperature or composition may cause a second type of phase boundary to be reached (called the haze curve for w/o microemulsions and the cloud point curve for o/w microemulsions) at which a surfactant-rich phase separates. This phase separation is driven by attractive inter-aggregate interactions which increase as the single phase microemulsion region is traversed from the solubilisation boundary to the haze or cloud point curve. Changing the variable so that one moves from the solubilisation to the haze or cloud boundary generally causes only relatively small changes in the microemulsion droplet size [22-25] and hence the fixed droplet size becomes increasingly different from the natural size.

For a microemulsion consisting of spherical, monodisperse droplets coated by a surfactant monolayer, the droplet core radius is proportional to the molar ratio of dispersed component to surfactant within the droplets (R), i.e.

$$r_c = (3\,V/A_h)\,R \text{ for w/o droplets}$$

$$r_c = (3\,V/A_t)\,R \text{ for o/w droplets} \qquad (2)$$

where V is the molecular volume of the dispersed component (oil or water as appropriate).

Different droplet radii are determined depending on the experimental technique used. For example, many measurements have been made using dynamic light scattering in which the droplet hydrodynamic radius r_h is obtained [26-28]. In this case, r_h is the sum of the core radius and the thickness of the surfactant monolayer including entrapped solvent (δ). If the surfactant area (A_h or A_t) and δ are independent of R, their values may be obtained from the slope and intercept of the linear plot of r_h versus R. Since microemulsion droplets will, in general, have a finite polydispersity of sizes the surfactant areas obtained will be average values. Such a plot for w/o microemulsions stabilised by AOT is shown in Figure 3. A linear least squares fit of the data gives $\delta = 1.4 \pm 0.1$ nm and $A_h = 0.50 \pm 0.03$ nm^2 [28,29]. In this study the radius - R relationship was found to be identical for both "made-up" systems and for equilibrated two phase systems (Winsor II) for which R values were measured by chemical analysis [29].

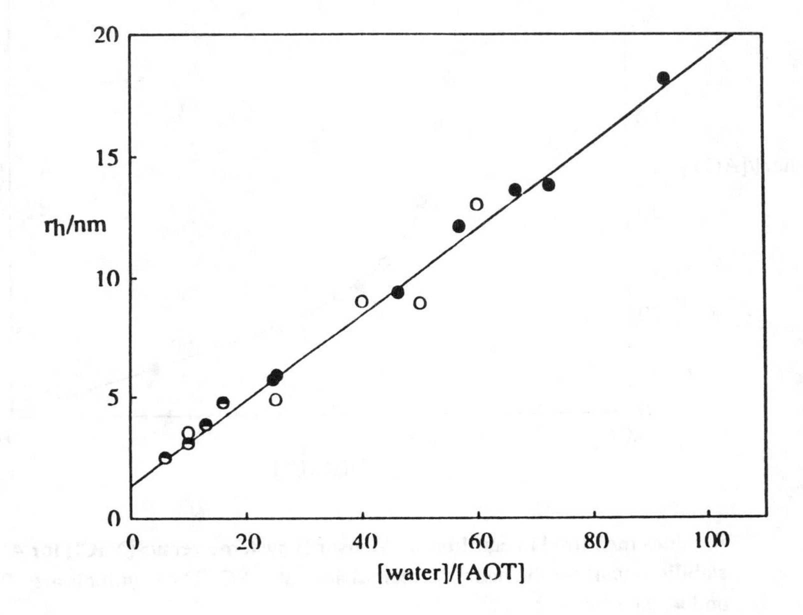

Figure 3. **Hydrodynamic radius versus R for AOT stabilised water-in-heptane microemulsions at 25°C. Data symbols : ● and 0, "made-up" microemulsions [28,29], • equilibrated microemulsions [29].**

For the example of AOT stabilised w/o microemulsions formed in equilibrium with excess water phases containing varying concentrations of NaCl, the R value decreases with increasing salt concentration [29,30] as shown in Figure 4. Combination of this data with the relationship between radius and R (established earlier) enables us to calculate the natural droplet radius as a function of the salt concentration. Since the linearity of the r_h versus R plot implies that A_h is independent of R (within 20% or so) over this range (note that this is not the case for low R values corresponding to reverse micelles [31,32]), it appears that the changing tendency of the monolayer to curve with salt concentration is a consequence of the changing value of A_t. This rather surprising result is, at present, unexplained. This is shown in Figure 5 which depicts the variation of the "natural" values of $A_h$and A_t (calculated using equation 1) with salt concentration. For the o/w microemulsions A_h decreases with increasing [NaCl] whereas A_t (shown as the solid line) is constant. For the w/o case, the constant value of A_h is indicated as the dashed line.

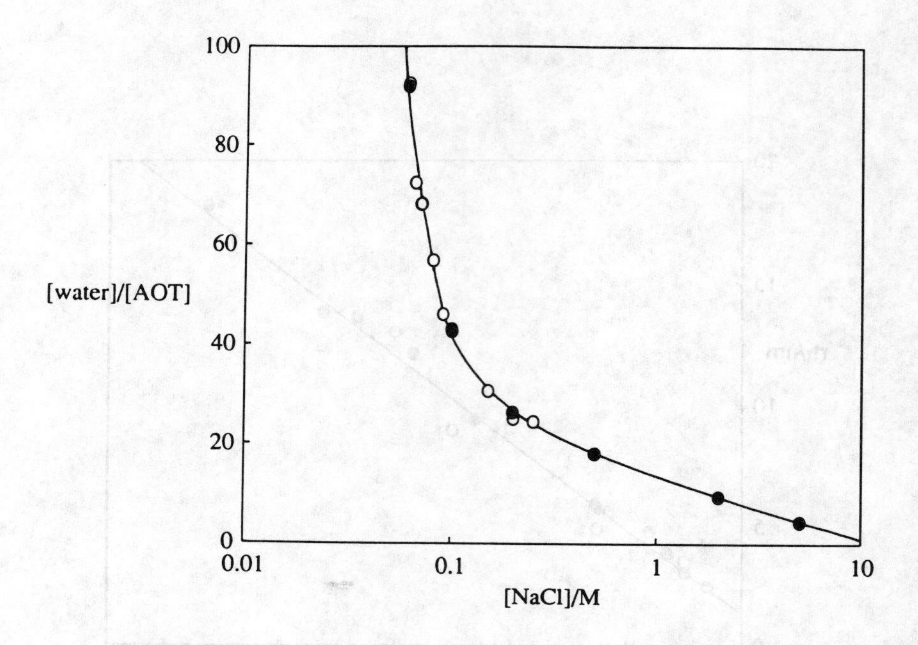

Figure 4. R values measured in equilibrium Winsor II systems versus [NaCl] for AOT stabilised water-in-heptane microemulsions at 25°C. Data symbols are : 0 and •, data from [29,30].

For the case of o/w microemulsions stabilised by AOT, the electrically charged droplets interact very strongly. Hence, it is more difficult to obtain reliable droplet sizes using scattering methods such as dynamic light scattering which are sensitive to interactions. A technique which is insensitive to droplet interactions is time resolved fluorescence which yields values for the number of surfactant molecules within a droplet, i.e. the aggregation number N_{agg} [33-36]. Assuming spherical, monodisperse droplets with A_t independent of R (as was done in deriving equation 2), N_{agg} is related to R for o/w droplets as follows:

$$N_{agg} = \{(36\pi V^2)/A_t^2\} R^2 \quad (3)$$

Hence plots of N_{agg} versus R^2 should be linear and A_t may be obtained from the slopes. Note that δ is not obtained using this method. Plots for o/w microemulsions stabilised by AOT and containing either heptane or nonane are shown in Figure 6. The derived value of A_t is 0.52 ± 0.04 nm^2 for both alkanes. Combination of this data with measurements of R versus salt concentration allows the calculation of the natural radii as a function of [salt] [37]. The changing value of A_h can be estimated using equation 1 if δ is assumed to be equal to the value measured for the w/o case.

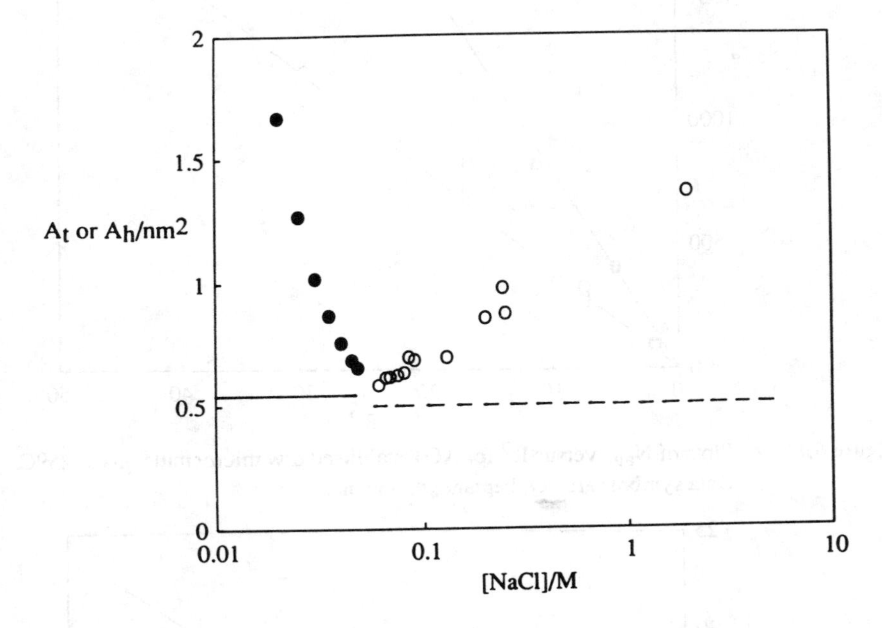

Figure 5. A_t and A_h values for AOT monolayers at their natural curvature (i.e. at the microemulsion droplet surfaces in the Winsor I and II systems) versus [NaCl] at 25°C. Data symbols are : ●, A_h for o/w systems; O, A_t for w/o systems.

This then gives us the complete variation of the surfactant geometrical parameters within the natural curved monolayers as the system is phase inverted from a Winsor I to a Winsor II system by the addition of NaCl. Figure 5 shows the variation of A_h and A_t for both the o/w and the w/o cases. It can be seen that the effective packing parameter (= A_t/A_h) passes through unity close to 0.05 M NaCl where Winsor III systems are formed. At this NaCl concentration in this particular system the oil-water interfacial tension reaches an ultralow minimum value [29] which generally occurs when the monolayer has a preferred curvature close to zero (P = 1). Discussion of the values is reserved for a later section.

The effects of *temperature* on the surfactant geometry within curved monolayers of the nonionic surfactant $C_{12}E_5$ have been derived as follows. Values of A_t, A_h and δ have been obtained from the plots of r_h (measured using dynamic light scattering [38]) versus R shown in Figure 7. The values are :

$A_t = 0.39 \pm 0.03$ nm^2, $\delta = 3.3 \pm 0.2$ nm for heptane-in-water droplets.

$A_t = 0.29 \pm 0.03$ nm^2, $\delta = 3.0 \pm 0.2$ nm for tetradecane-in-water droplets.

$A_h = 0.35 \pm 0.06$ nm^2, $\delta = 1.6 \pm 2$ nm for water-in-heptane droplets.

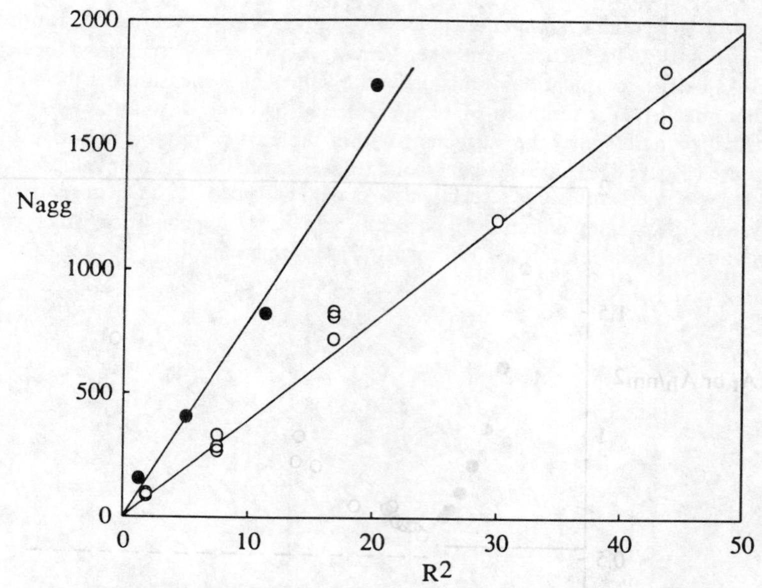

Figure 6. **Plots of N_{agg} versus R^2 for AOT stabilised o/w microemulsions at 25°C. Data symbols are : O, heptane; ●, nonane.**

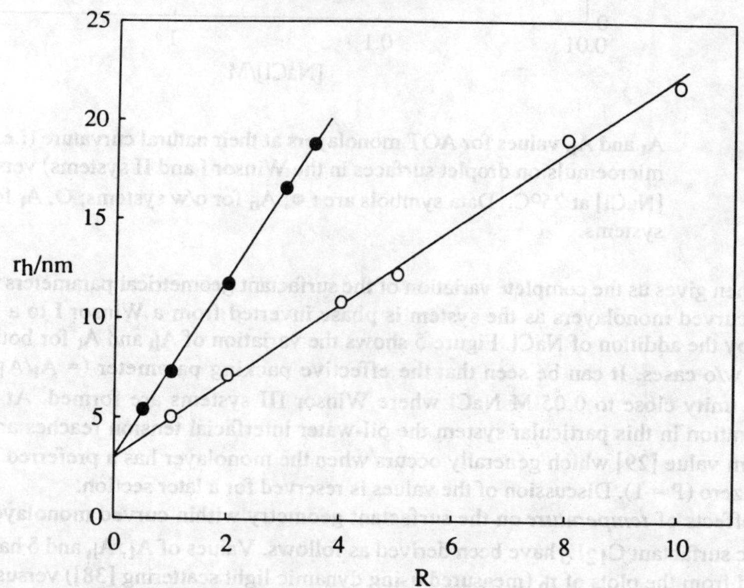

Figure 7. **Plots of r_h versus R for $C_{12}E_5$ stabilised o/w microemulsions at their solubilisation phase boundaries. Data symbols are : O, heptane; ●, tetradecane. Reprinted with permission from R. Aveyard et al., Langmuir (1989), 5, 1210. Copyright (1989) American Chemical Society.**

It is noteworthy here that a longer chain length oil gives a lower value of A_t for this system. This is in contrast to the AOT system where similar A_t values were measured for heptane and nonane. As before combination of this data with values of R measured at the solubilisation phase boundaries as a function of temperature allows the derivation of the effective geometrical parameters of the surfactant within the curved monolayers as a function of temperature (Figure 8). The dashed and solid lines indicate the constant values of A_t for o/w and A_h for w/o microemulsions respectively. As with variation of NaCl concentration for the AOT system, the packing parameter P passes through unity at a condition which corresponds to minimum interfacial tension and Winsor III phase formation [38].

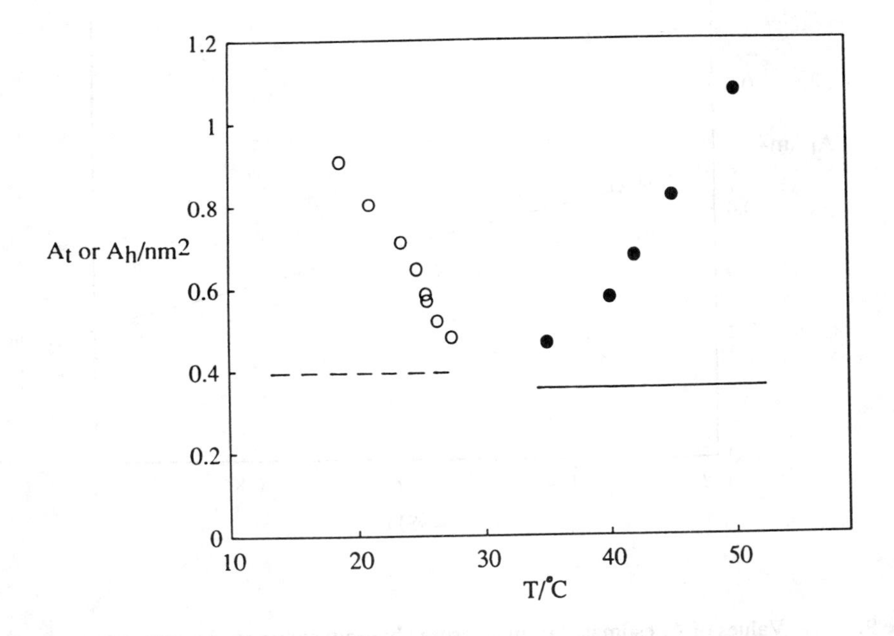

Figure 8. A_t and A_h values for $C_{12}E_5$ monolayers at their natural curvature versus temperature with heptane as oil. Data symbols : O, A_h for o/w systems; •, A_t for w/o systems.

Areas per surfactant molecule in curved monolayers have been estimated for mixed monolayers of AOT with SDS. At the high salt concentration used in this study (0.1 M NaCl) AOT has a preferred negative curvature (i.e. $A_t > A_h$) whereas SDS has $A_t < A_h$. Hence, addition of SDS to the system causes phase inversion from w/o to o/w [39]. Average values of A_t were obtained from measurements of r_h of o/w droplets (at infinite dilution) using equation 2 with δ taken as 1.9 nm being the estimated mean of the lengths of two surfactants for various values of the mole fraction X = [SDS]/([AOT] + [SDS]). Values of A_t for o/w microemulsion droplets versus X are shown in Figure 9. As expected, the mean area decreases with increasing X. The solid line shows the area expected if A_t varies linearly with

X from 0.52 nm^2 for pure AOT (see above) to 0.22 nm^2 estimated to be the value for a close packed monolayer of the single tailed SDS with no penetration by the oil solvent. This example serves to illustrate an approach to obtaining geometrical parameters within mixed monolayers. Currently, insufficient systematic data is available to give a complete description of the type shown in Figures 5 and 8 for mixed monolayers.

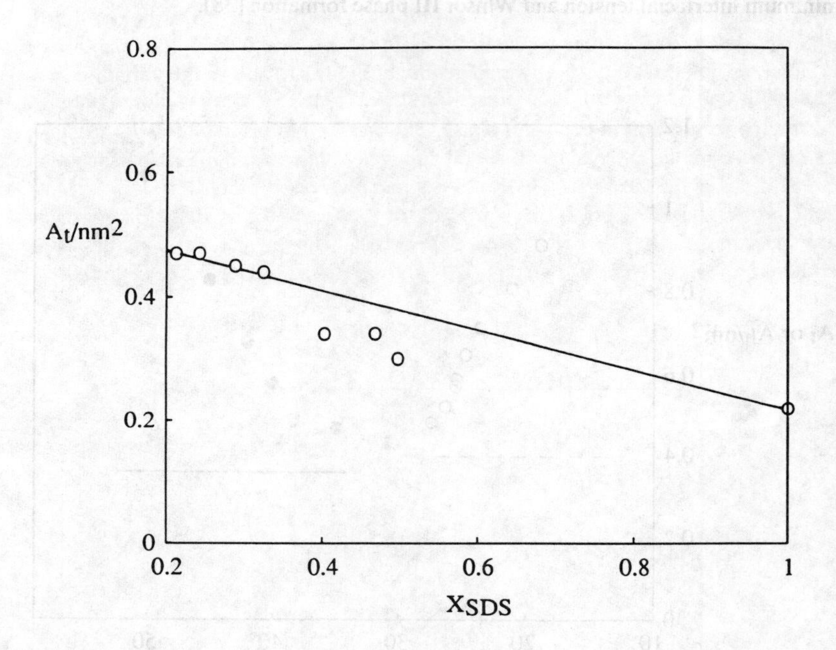

Figure 9. Values of A_t estimated from measured hydrodynamic radii versus mole fraction of SDS in the interface for o/w microemulsions stabilised by mixtures of AOT and SDS at [NaCl] = 0.1 M and 25°C.

4. Flat monolayers at oil-water interfaces

In the preceding section we showed how quantitative information concerning the effective surfactant geometry within monolayers at their "natural" curvature could be obtained by measuring microemulsion droplet sizes within Winsor multiphase microemulsion systems. In such systems a surfactant monolayer at the planar interface separating the bulk phases co-exists at equilibrium with the curved monolayer. The area per surfactant molecule in this planar interface may be obtained by measuring the oil-water interfacial tension as a function of surfactant concentration.

The general behaviour is as follows. The interfacial tension of the oil-water interface in the absence of surfactant is generally 30 - 50 mN m^{-1}. For low surfactant concentrations (below the cmc) all the surfactant is present as monomeric species and will, in general, reach an equilibrium distribution between the oil and the water phases. For ionic surfactants such as AOT, the distribution of the monomer lies heavily in favour of the water phase and virtually no AOT is present in the oil phase. For nonionics such as $C_{12}E_5$ the monomer distribution lies heavily in favour of the oil. Increasing the monomer concentration causes the interfacial tension to drop to a value which may vary from a few mN m^{-1} to an ultralow value of 10^{-3} - 10^{-4} mN m^{-1} depending on the system. Further surfactant addition at concentrations greater than the cmc leads to the formation of aggregates in either the oil, water or a third phase. Above the cmc the monomer concentrations in the oil and water phases and the oil-water interfacial tension do not generally change much. Figure 10 shows the variation of interfacial tension with $C_{12}E_5$ concentration present in the oil phase. A similar plot is obtained for the surfactant concentration in the water phase. The concentrations at which the tension reaches a constant value are the cmc values in the oil and water phases, and the ratio of the oil and water concentrations gives the surfactant monomer partition coefficient [40]. (As a digression, we note here that the surfactant HLB is not related to the monomer partition coefficient but is related to whether the microemulsion aggregates form in the oil or water. For example, mixtures of $C_{12}E_5$ + heptane + water at 20 °C form o/w emulsions when emulsified and o/w

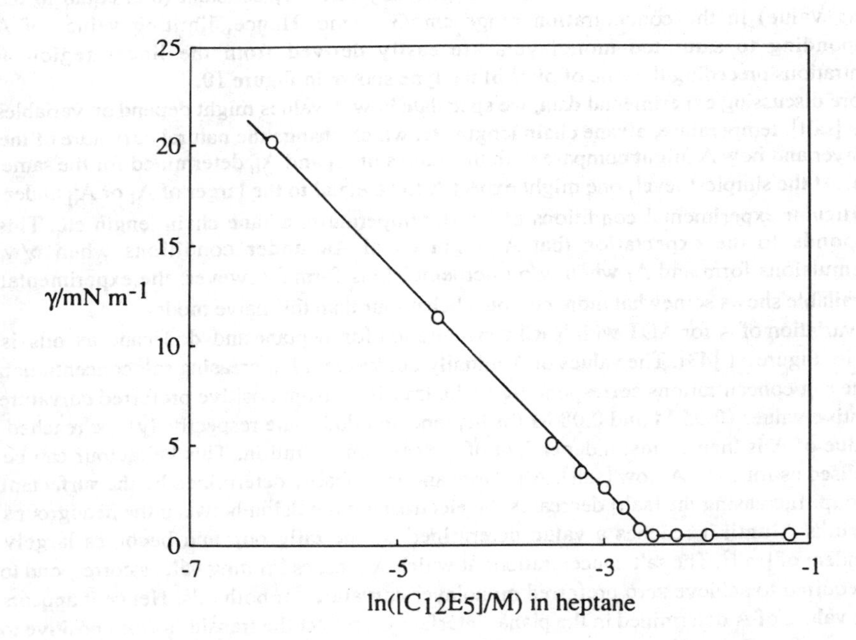

Figure 10. Heptane-water interfacial tension versus $C_{12}E_5$ concentration in the oil phase at 50°C.

microemulsions (i.e. Winsor I systems are formed) when allowed to equilibrate although the monomer distributes heavily in favour of the oil with monomer partition coefficient $[C_{12}E_5]_{oil}/[C_{12}E_5]_{water}$ = of 480±80 [40].)

We denote the area of the plane interface occupied by a single surfactant molecule as A (without subscript to distinguish the value from A_t and A_h measured in the curved monolayers). For a nonionic surfactant A is related to the change in interfacial tension with surfactant concentration according to the appropriate Gibbs equation [41].

$$A = - kT/(d\gamma/d\ln f[\text{surfactant}]) \qquad (4)$$

In equation 4, f is the activity coefficient of the surfactant. For systems containing ionic surfactants and added salt, A is obtained using a modified form of equation (4) due to Matijevic and Pethica [42]:

$$A = - kT \{1 + ([\text{surfactant}]/([\text{salt}] + [\text{surfactant}]))/(d\gamma/d\ln f[\text{surfactant}])\} \qquad (5)$$

The value of A decreases with increasing surfactant concentration as the surfactant adsorbs until a limiting value is reached corresponding to the formation of a surfactant monolayer of the maximum possible surface density of surfactant. Since surfactant monolayers of this type are rather incompressible, the value of A is generally virtually constant (and equal to the limiting value) in the concentration range cmc/3 - cmc. Hence, limiting values of A corresponding to saturated monolayers are easily derived from the linear region at concentrations preceding the cmc of plots of the type shown in Figure 10.

Before discussing experimental data, we speculate how A values might depend on variables such as [salt], temperature, alkane chain length etc. which change the natural curvature of the monolayer and how A might compare with the values of A_t and A_h determined for the same system. At the simplest level, one might expect A to be equal to the larger of A_t or A_h under the particular experimental conditions of [salt], temperature, alkane chain length etc. This corresponds to the expectation that A might equal A_h under conditions when o/w microemulsions form and A_t when w/o microemulsions form. However, the experimental data available shows somewhat more complex behaviour than this naive model.

The variation of A for AOT with NaCl concentration for heptane and dodecane as oils is shown in Figure 11 [43]. The values of A initially decrease with increasing salt concentration until the salt concentrations corresponding to the transition from positive preferred curvature to negative values (0.05 M and 0.08 M for heptane and dodecane respectively) are reached. The value of A is then almost independent of the salt concentration. This behaviour can be rationalised as follows. At low [salt], A is large and is probably determined by the surfactant headgroup. Increasing the [salt] decreases the electrostatic repulsion between the headgroups and shrinks A until it reaches a value determined by the tailgroup and becomes largely independent of [salt]. The salt concentrations at which A reaches limiting values correspond to those required to achieve zero preferred monolayer curvature for both oils. Hence it appears that the values of A determined in the planar interface do reflect the transition from positive to negative preferred monolayer curvature. However, the naive expectation that A might equal the larger of $A_h$or A_t for the curved monolayer is not borne out by a quantitative comparison of the values in Figures 11 and 5.

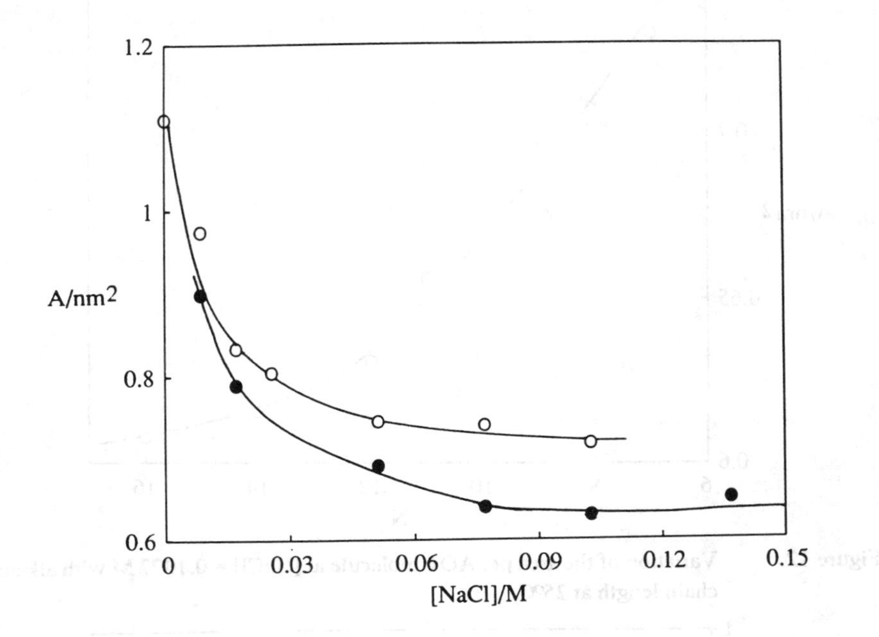

Figure 11. Area per molecule versus NaCl concentration for saturated monolayers of AOT adsorbed at the heptane-solution interface (O) and dodecane-solution interface (●) at 25°C.

The limiting values of A at high salt concentration, which might be expected to be determined mainly by the tailgroup area, are different for heptane and dodecane. This effect is illustrated in more detail in Figure 12 which shows the variation of A with alkane chain length N at a constant, high [salt] at which all these systems have a negative preferred monolayer curvature. The areas decrease with increasing chain length implying that longer chain length alkanes penetrate and swell the tail region of the AOT monolayer less than short chain alkanes. This observation is in accord with measurements of the variation of the preferred curvature of monolayers with N [44,45] and with theory [46]. We will return to this point in the last section of this review where we will show how the extent of alkane penetration into surfactant monolayers (at air-water surfaces) may be quantitatively measured.

We now consider the effect of temperature. For Winsor systems containing AOT at constant [NaCl], low temperatures lead to the formation of w/o microemulsions (a Winsor II system) whereas o/w droplets form at high temperatures in a Winsor I system. Hence a plot of A versus temperature, shown in Figure 13, might be expected to be dominated by the tail area at low temperatures and the head area at high temperatures and to show some type of break point at the temperature corresponding to zero preferred monolayer curvature. At low temperatures A increases only slightly with temperature but increases more sharply above approximately

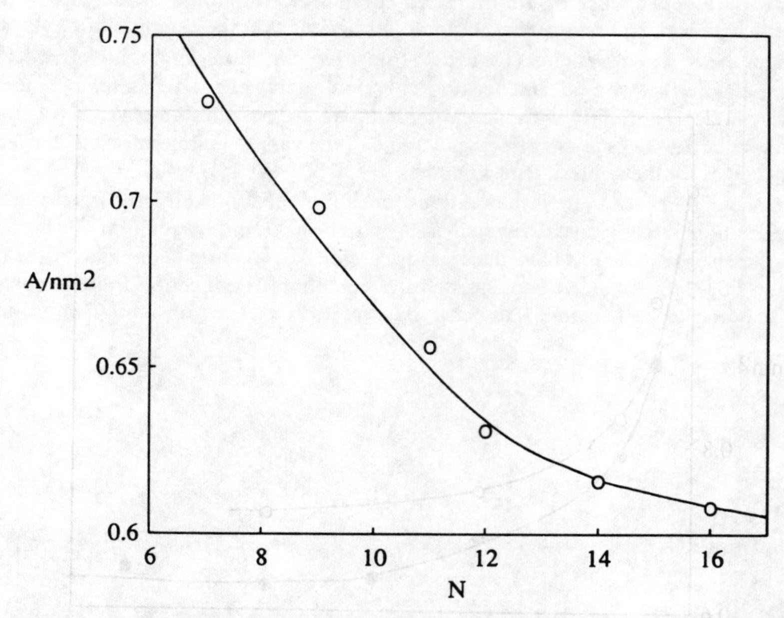

Figure 12. **Variation of the area per AOT molecule at [NaCl] = 0.1027 M with alkane chain length at 25°C.**

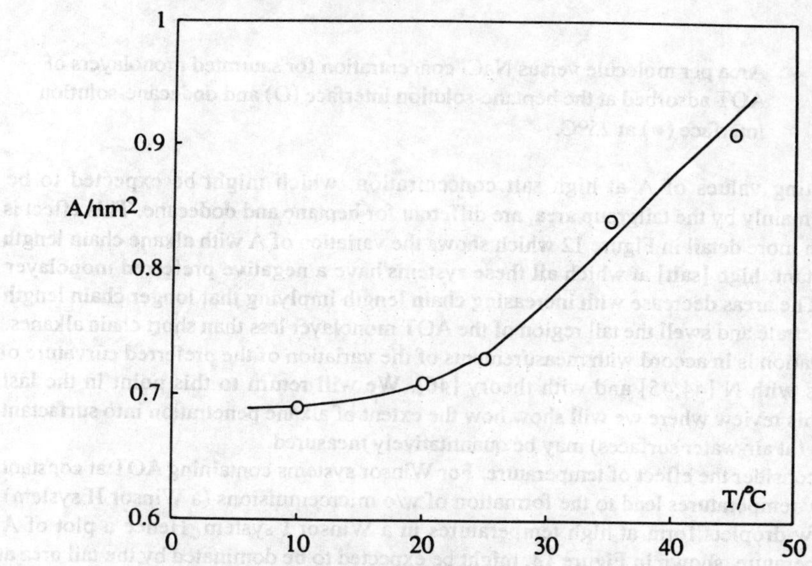

Figure 13. **Area per AOT molecule at the heptane-0.0512 M aqueous NaCl interface versus temperature. For this system the PIT occurs at 25°C.**

The temperature dependence of the preferred curvature of nonionic surfactants of the polyoxethylene type is the reverse of that for ionic surfactants. For the example of $C_{12}E_5$ in two-phase heptane + water systems (Figure 14) positive curvature aggregates (o/w) are formed at temperatures below 27°C and negative curvature aggregates at higher temperatures. The value of A increases slightly with increasing temperature but with no obvious break at 27°C [40]. As with the AOT system effected by [NaCl], the values are intermediate between the A_t and A_h values for the same system (Figure 8).

We have discussed how A varies when microemulsion inversion is effected by changing the salt concentration or temperature. Inversion may also be achieved by addition of a second surface-active species which must have the opposite preferred curvature to monolayers of the first surfactant. For AOT + water + heptane mixtures containing a salt concentration equal to 0.017 M it has been shown (Figure 5) that A_h is larger than A_t (i.e. o/w microemulsions

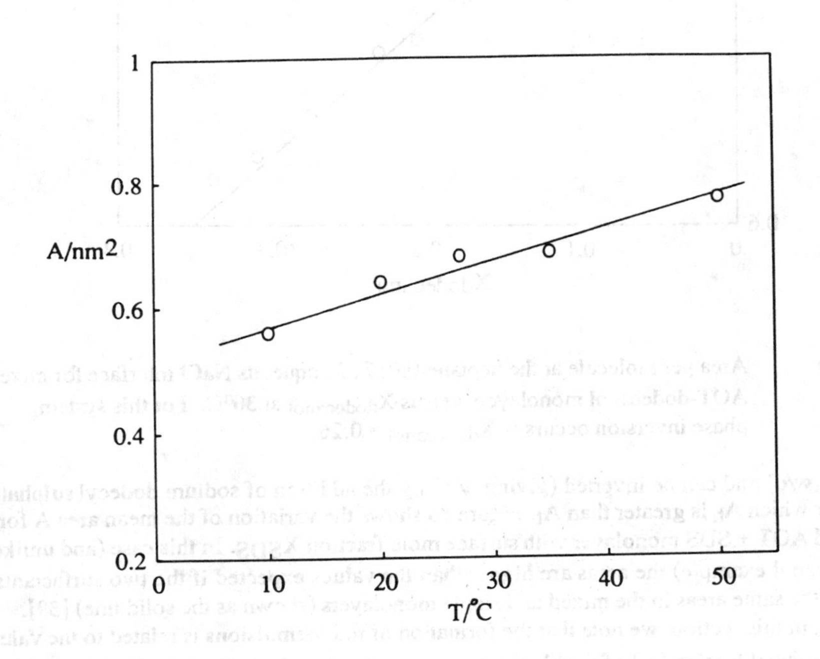

Figure 14. Area per molecule for $C_{12}E_5$ adsorbed at the heptane-water interface versus temperature.

form). Addition of dodecanol, for which A_t is greater than A_h, causes inversion at a mole fraction of dodecanol $X_{dodecanol}$ in the planar monolayer of 0.26. Figure 15 shows that the mean area per surfactant species decreases monotonically with no obvious break point at the inversion composition [43]. The solid line is drawn assuming a linear variation in A with $X_{dodecanol}$, from A = 0.79 nm^2 at $X_{dodecanol}$ = 0 to 0.25 nm^2 at $X_{dodecanol}$ = 1. In contrast, for the AOT system at [NaCl] equal to 0.1 M, the preferred monolayer curvature is

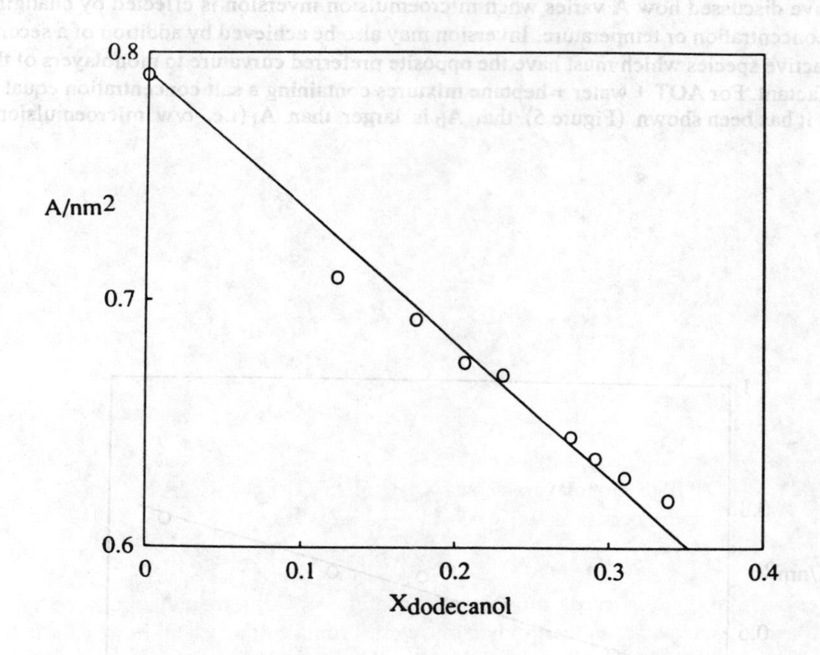

Figure 15. Area per molecule at the heptane-0.0171 M aqueous NaCl interface for mixed AOT-dodecanol monolayers versus $X_{dodecanol}$ at 30°C. For this system phase inversion occurs at $X_{dodecanol}$ = 0.26.

negative (w/o) and can be inverted (giving o/w) by the addition of sodium dodecyl sulphate (SDS) for which A_h is greater than A_t. Figure 16 shows the variation of the mean area A for the mixed AOT + SDS monolayer with surface mole fraction X_{SDS}. In this case (and unlike the dodecanol example) the areas are higher than the values expected if the two surfactants occupied the same areas in the mixed and single monolayers (shown as the solid line) [39].

Finally, in this section, we note that the formation of microemulsions is related to the value of the interfacial tension (γ_c) of the planar interface separating the bulk phases in Winsor I or II systems measured at surfactant concentrations in excess of the cmc. In general, larger microemulsion droplets are formed at lower γ_c and a minimum γ_c is observed under conditions when the preferred surfactant curvature is zero. The surfactant present in the planar

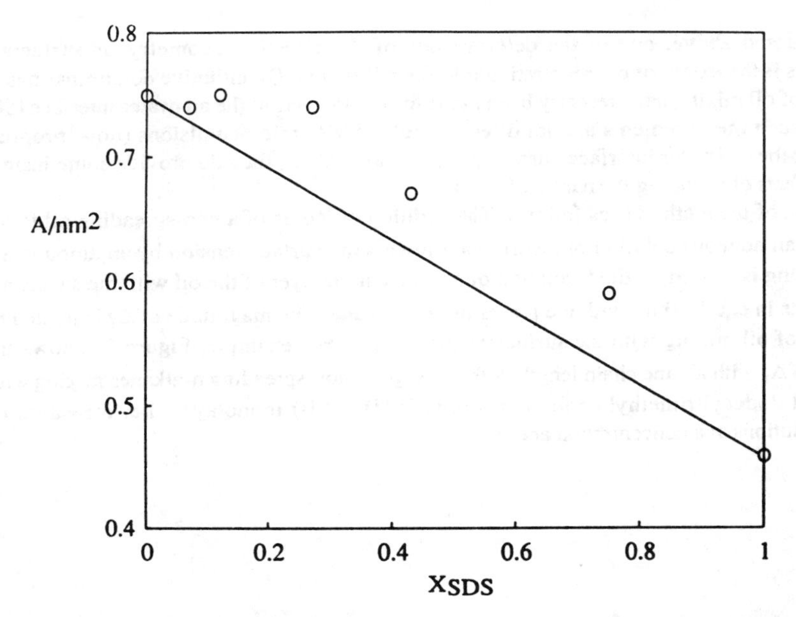

Figure 16. **Area per molecule at the heptane-0.1 M aqueous NaCl interface for mixed AOT-SDS monolayers versus X_{SDS} at 25°C. For this system phase inversion occurs at X_{SDS} = 0.1.**

monolayer and in the curved monolayers coating the microemulsion droplets in Winsor I and II systems are in thermodynamic equilibrium. For a system in which the free energies associated with droplet dispersion and inter-droplet interactions can be neglected, the tension γ_c is related to the free energy required to "unbend" the surfactant monolayer coating the droplets to give a planar film between the bulk phases. If it is assumed that the droplet interface has a tension of zero and the monolayer has negligible thickness, then γ_c is expected to scale with droplet radius according to:

$$\gamma_c = (2K + \bar{K})/r^2 \qquad (6)$$

where K and $\bar{K}$ are the rigidity and Gaussian curvature elastic modulii respectively [47-49]. This equation also assumes that the surfactant monolayer is incompressible, i.e. the surface concentrations of surfactant (and areas per molecule) in the curved and planar monolayers are the same. Clearly, this approach is only approximate; the surfactant monolayer thickness is not negligible in relation to the microemulsion droplet radii and a more realistic theory must also incorporate differences in surfactant area between the plane and curved monolayers of the type discussed here.

5. Flat monolayers at air-water surfaces

As discussed above, one of the determinants of the effective geometry of surfactant monolayers is the extent of oil penetration into the tail region. Quantitative determinations of the extent of oil mixing have recently been made for monolayers at the air-water interface [50]. Although such measurements are not directly applicable to microemulsions (no appropriate method for the oil-water interface currently appears available), they do provide some insight into the effects of changing the nature of the oil.

The basis of the method is as follows. The addition of lenses of a non-spreading oil to the surface of an aqueous solution of a surfactant lowers the surface tension by an amount $\Delta\gamma$. This lowering is caused by the formation of a mixed monolayer of the oil with the surfactant which exists in equilibrium with the lenses of bulk alkane. The magnitude of $\Delta\gamma$ is related to the extent of oil mixing with the surfactant monolayer. For example, Figure 17 shows the variation of $\Delta\gamma$ with alkane chain length N for a range of non-spreading n-alkanes mixing with a saturated dodecyltrimethylammonium bromide (DoTAB) monolayer (i.e. measured on DoTAB solutions at a concentration above the cmc).

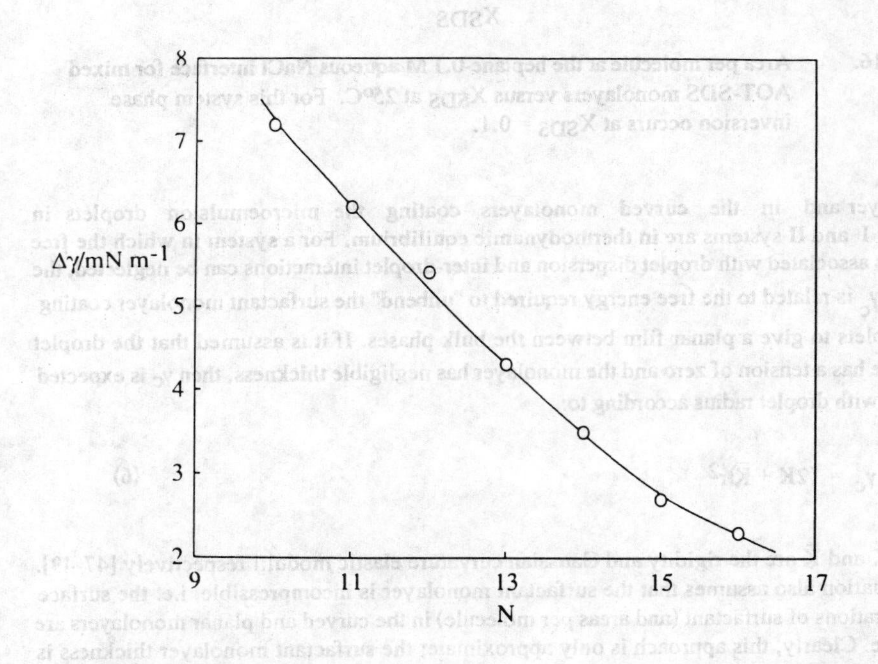

Figure 17. **Tension lowering versus N for the addition of non-spreading alkanes at the air-0.02 M aqueous DoTAB solution interface at 25° C.**

As expected, the extent of mixing decreases as N increases. In order to determine the mixed film composition quantitatively, the surface excess concentrations of both the surfactant (Γ_s) and the oil (Γ_o) must be measured. The value of Γ_s in the mixed film is determined from measurements of γ as a function of surfactant concentration in the presence of the oil using the Gibbs equation, and Figure 18 shows a plot for DoTAB.

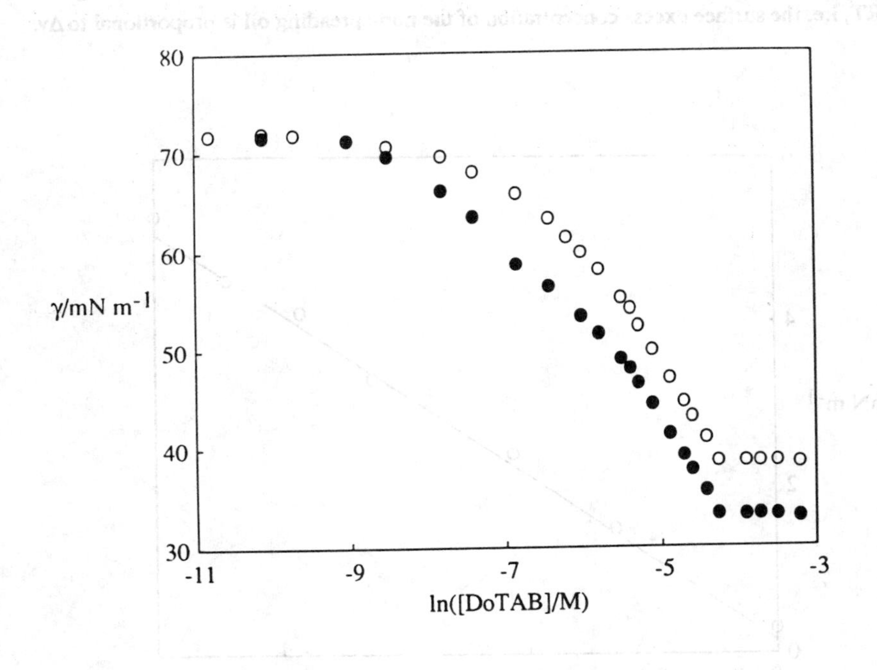

Figure 18. Variation of surface tension with DoTAB concentration in the absence (O) and presence (●) of dodecane at 25°C.

We have obtained Γ_o for the same system at the cmc as follows. The liquid hydrocarbon squalane (2,6,10,15,19,23-hexamethyltetracosane) does not spread on DoTAB solutions and it lowers γ only very slightly (approximately 0.3 mN m^{-1}), indicating that it does not mix significantly with the DoTAB monolayer. Dodecane and squalane are miscible in all proportions and we assume the mixtures to be ideal. Values of $\Delta\gamma$ obtained using mixtures with mole fraction of dodecane ($X_{dodecane}$) are shown in Figure 19. The values of Γ_o may be obtained using

$$\Gamma_o = (X_{dodecane}/RT)(d\Delta\gamma/dX_{dodecane}) \quad \text{at constant [surfactant]} \qquad (7)$$

For pure dodecane ($X_{dodecane}$ = 1) Γ_o is 1.20 x 10^{18} molecules m^{-2} and the corresponding Γ_s (derived from the data of Figure 18) is 1.52 x 10^{18}molecules m^{-2}. Thus, the mixed film formed with a saturated DoTAB monolayer contains approximately 8 molecules of dodecane per 10 molecules of DoTAB when in equilibrium with lenses of pure dodecane. It should be noted that, if the linearity of the $\Delta\gamma$ versus $X_{dodecane}$ plot is general, then Γ_o is simply given by $\Delta\gamma$/RT, i.e. the surface excess concentration of the non-spreading oil is proportional to $\Delta\gamma$.

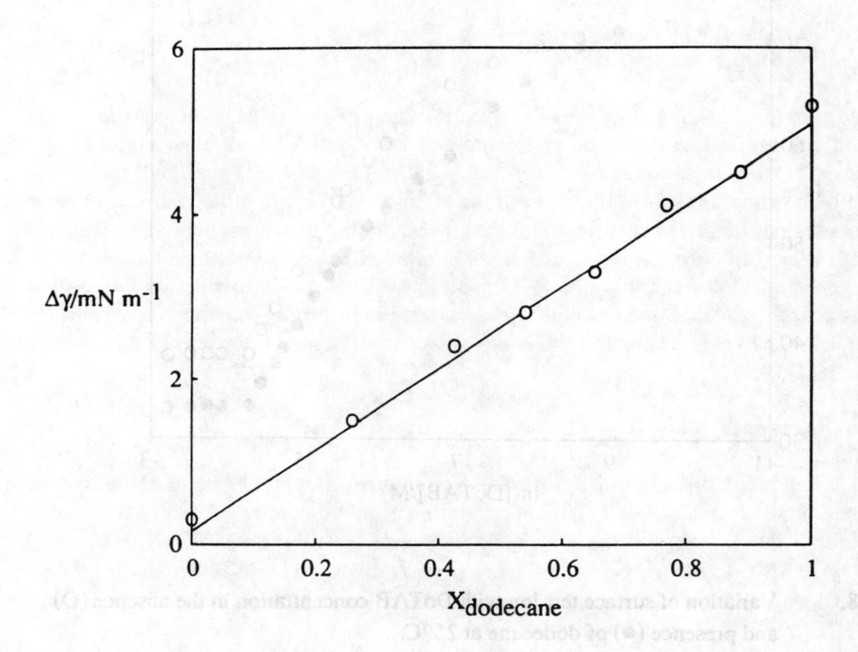

Figure 19. Tension lowering of DoTAB solutions at 25°C caused by (squalene-dodecane) mixtures, [DoTAB] = 19.7 mM.

Using the approach described above, we have carried out a quantitative investigation of the effects of surfactant and hydrocarbon chain length on the extent of mixing [51]. Some of the main findings are listed below:

1. **For the mixing of n-alkanes with a particular surfactant monolayer, increasing the alkane chain length leads to decreased extent of mixing.**

2. For a homologous series of n-alkyltrimethylammonium bromides, increasing the chain length of the surfactant leads to increased mixing for a given alkane. Effects 1 and 2 have been predicted theoretically for microemulsion droplet surfaces [46].

3. For saturated surfactant monolayers, the area occupied per surfactant molecule is little affected following the incorporation of the oil.

4. For a range of saturated surfactant monolayers and oils, the difference between the tension of the surface containing the mixed film and the interfacial tension between the oil and the surfactant solution is generally close to the surface tension of the oil. Thus the tail side of the mixed film at the air-water surface is energetically similar to the surface of the pure oil.

6. Concluding remarks

We have attempted to show how the effective molecular geometry of surfactants *in situ* in curved and flat monolayers may be estimated and have collected together some of the systematic data available. One feature relevant to microemulsions which emerges is that there appears to be no simple relationship between the effective surfactant molecular geometry in monolayers at their preferred curvature and the corresponding planar monolayers. Although the methods described here do give some insight into the effects of solution conditions on the effective geometry within monolayers, it is hoped that this discussion will stimulate further detailed structural investigation, including characterisation of these systems using recently developed techniques such as neutron and X-ray reflectivity [52] and fluorescence microscopy [53].

Acknowledgements

We wish to thank British Petroleum Research (Sunbury), Bevaloid Ltd. and the Science and Engineering Research Council (U.K.) for funding of the research discussed here. One of the authors (P.D.I.F.) thanks the Organising Committee of the NATO ASI Meeting (Aberystwyth, Sept. 1989) for being invited to present this work. We also thank the Royal Society of Chemistry and the American Chemical Society for permission to reproduce Figures 3, 4, 6, 11, 12, 16, 18, 19 and 7 respectively.

References

1. Myers, D. (1988) Surfactant Science and Technology, VCH, New York.
2. Griffin, W.C. (1949) J. Soc. Cosmet. Chem., 1, 311-326.
3. Davies, J.T. and Rideal, E.K. (1963) Interfacial Phenomena, Academic Press, London.
4. Winsor, P.A. (1954) Solvent Properties of Amphiphilic Compounds, Butterworths, London.
5. Arai, H. and Shinoda, K. (1978) J. Colloid Interf. Sci., 65, 589-591.
6. Shinoda, K. and Friberg, S. (1986) Emulsions and Solubilisation, Wiley, New York.

7. Israelachvili, J.N., Mitchell, D.J. and Ninham, B.W. (1976) J. Chem. Soc. Faraday Trans. 2, 72, 1525-1568.
8. Mitchell, D.J. and Ninham, B.W. (1981) J. Chem. Soc. Faraday Trans. 2, 77, 601-629.
9. Langevin, D. (1988) Acc. Chem. Res., 21, 255-260.
10. Bourrel, M. and Schechter, R.S. (1988) Microemulsions and Related Systems, Dekker, New York.
11. Martellucci, S. and Chester, A.N. Eds. (1989) Progress in Microemulsions, Plenum, New York.
12. Pileni, M.P. Ed. (1989) Structure and Reactivity in Reverse Micelles, Elsevier, Amsterdam.
13. Meunier, J., Langevin, D. and Boccara, N. Eds. (1987) Physics of Amphiphilic Layers, Springer Verlag, Berlin.
14. Winsor, P.A. (1948) Trans. Faraday Soc. 44, 376-398.
15. Aveyard, R. (1987) Chem. and Ind., July, 474-478.
16. Cazabat, A.M., Langevin, D., Meunier, J. and Pouchelon, A. (1982) Adv. Colloid Interf. Sci., 16, 175-199.
17. Aveyard, R., Binks, B.P., Lawless, T.A. and Mead, J. (1988) Can. J. Chem., 66, 3031-3037.
18. Scriven, L.E. (1976) Nature 263, 123-125.
19. de Geyer, A. and Tabony, J. (1985) Chem. Phys. Lett., 113, 83-88.
20. Rendall, K. and Tiddy, G.J.T. (1984) J. Chem. Soc. Faraday Trans. 1, 80, 3339-3357.
21. Lindman, B., Shinoda, K., Olsson, U., Anderson, D., Karlstrom, G. and Wennerstrom, H. (1989) Colloids and Surfaces, 38, 205-224.
22. Kotlarchyk, M., Chen, S-H and Huang, J.S. (1982) J. Phys. Chem., 86, 3273-3276.
23. Lang. J., Jada, A. and Malliaris, A. (1988) J. Phys. Chem., 92, 1946-1953.
24. Clark, S., Fletcher, P.D.I. and Ye, X., (1990), Langmuir, in press.
25. Jada, A., Lang, J., Zana, R., Makhloufi, R., Hirsch, E. and Candau, S.J. (1990) J. Phys. Chem., 94, 387-395.
26. Eicke, H.F. and Rehak, J. (1976) Helv. Chim. Acta, 59, 2883-2891.
27. Day, R.A., Robinson, B.H., Clarke, J.H.R. and Doherty, J.V. (1979) J. Chem. Soc. Faraday Trans. 1, 75, 132-139.
28. Nicholson, J.D. and Clarke, J.H.R. (1984) Eds. Mittal, K.L. and Lindman, B., Surfactants in Solution, volume 3, Plenum, New York, p.1663-1675.
29. Aveyard, R., Binks, B.P., Clark, S. and Mead, J. (1986) J. Chem. Soc. Faraday Trans. 1, 82, 125-142.
30. Fletcher, P.D.I. (1986) J. Chem. Soc. Faraday Trans. 1, 82, 2651-2664.
31. Bridge, N.J. and Fletcher, P.D.I. (1983) J. Chem. Soc. Faraday Trans. 1, 79, 2161-2169.
32. Zulauf, M. and Eicke, H.F. (1979) J. Phys. Chem., 83, 480-486.
33. Infelta, P.P., Gratzel, M. and Thomas, J.K. (1974) J. Phys. Chem., 78, 190-195.
34. Almgren, M., Lofroth, J-E., and van Stam, J. (1986) J. Phys. Chem., 90, 4431-4437.
35. Lianos, P., Lang, J., Strazielle, C. and Zana, R. (1982) J. Phys. Chem., 86, 1019-1025.
36. Tachiya, M. (1975) Chem. Phys. Lett., 33, 289-292.

37. Fletcher, P.D.I. (1987) J. Chem. Soc. Faraday Trans. 1, 83, 1493-1506.
38. Aveyard, R., Binks, B.P. and Fletcher, P.D.I. (1989), Langmuir, 5, 1210-1217.
39. Aveyard, R., Binks, B.P., Mead, J. and Clint, J.H. (1988) J. Chem. Soc. Faraday Trans. 1, 84, 675-686.
40. Aveyard, R., Binks, B.P., Clark, S. and Fletcher, P.D.I. (1990) to be published.
41. Aveyard, R. and Haydon, D.A. (1973) An Introduction to the Principles of Surface Chemistry, Cambridge University Press.
42. Matijevic, E. and Pethica, B.A. (1958) Trans. Faraday Soc. 54, 1382-1389.
43. Aveyard, R., Binks, B.P. and Mead, J. (1986) J. Chem. Soc. Faraday Trans. 1, 82, 1755-1770.
44. Kahlweit, M. and Strey, R. (1985) Angew. Chem. Int. Ed. Engl., 24, 654-668.
45. Blum, F.D., Pickup, S., Ninham, B.W., Chen, S.J. and Evans, D.F. (1985) J. Phys. Chem., 89, 711-713.
46. Mukherjee, S., Miller, C.A. and Fort, T. (1983) J. Colloid Interf. Sci., 91, 223-243.
47. de Gennes, P.G. and Taupin, C. (1982) J. Phys. Chem. 86, 2294-2304.
48. Binks, B.P., Meunier, J., Abillon, O. and Langevin, D. (1989) Langmuir, 5, 415-421.
49. Cates, M.E., Andelman, D., Safran, S.A. and Roux, D. (1988) Langmuir, 4, 802-806.
50. Aveyard, R., Cooper, P. and Fletcher, P.D.I. (1990) J. Chem. Soc. Faraday Trans., 86, 211-212.
51. Aveyard, R., Cooper, P. and Fletcher, P.D.I. (1990) to be published.
52. Lee, E.M., Thomas, R.K., Penfold, J. and Ward, R.C. (1989) J. Phys. Chem., 93, 381-388.
53. Suresh, K.A., Nittmann, J. and Rondelez, F. (1989) Prog. Colloid Polym. Sci., 79, 184-193.

LIGHT SCATTERING EXPERIMENTS ON ANISOTROPIC LATEX PARTICLES

V. DEGIORGIO
Dipartimento di Elettronica
Università di Pavia
27100 Pavia, Italy

ABSTRACT. Static and dynamic light scattering experiments on crystalline polymer colloids are reviewed. It is shown, in particular, that such experiments allow one to derive the intrinsic optical anisotropy of the particles, and give both translational and rotational diffusion coefficients.

1. Introduction

Static and dynamic light scattering has been widely used for the characterization of suspended particles, such as macromolecules, inorganic and polymeric colloids, micelles and other aggregating species [1-6]. Little attention was paid, however, to particles possessing an intrinsic optical anisotropy. Recently, latex particles with a crystalline structure and with small polydispersity have been produced by emulsion polymerization methods and studied by light scattering [7,8] and electric birefringence [9] experiments. I will review in this article the light scattering experiments which gave information on the intrinsic anisotropy of the material, on the particle size, and on the translational and rotational diffusion coefficients. As I will show, the particles represent a very interesting model system for fundamental studies of colloidal solutions. At the same time, the measurement of the degree of crystallinity performed by light scattering may give useful information for the industrial utilization of the particles.

Two different kinds of polymer colloids have been studied: rod-like polytetrafluoroethylene (PTFE) latex particles and spheres of a PTFE copolymer, both supplied as stable water-dispersions. We have measured the degree of polarization of scattered light, the average scattered intensity and the intensity correlation function versus the scattering angle θ. The most striking property of the dispersion is that it is impossible to suppress scattering even by index-matching the particles with an appropriate solvent (a mixture of water and glycerol, in our case). In the index-matching situation, the scattered light is strongly depolarized, so that the intensity correlation measurements allow one to easily derive both the translational and the rotational diffusion coefficients of the latex particles.

2. Experimental

The samples used were aqueous solutions of latex particles obtained by emulsion polymerization. Two different kinds of particles were studied: polytetrafluoroethylene (PTFE) rods [7], and spheres made of a melt processable PTFE copolymer [8]. Both are products from Montefluos, S.p.A., Milano, Italy, and were kindly prepared for us by Dr. D. Lenti and Dr. M. Visca. As

D. M. Bloor and E. Wyn-Jones (eds.),
The Structure, Dynamics and Equilibrium Properties of Colloidal Systems, 583–596.

observed with the electron microscope (Fig. 1), the particles are fairly monodisperse. The original latex was purified by dialysis which is very efficient in removing weakly adsorbed

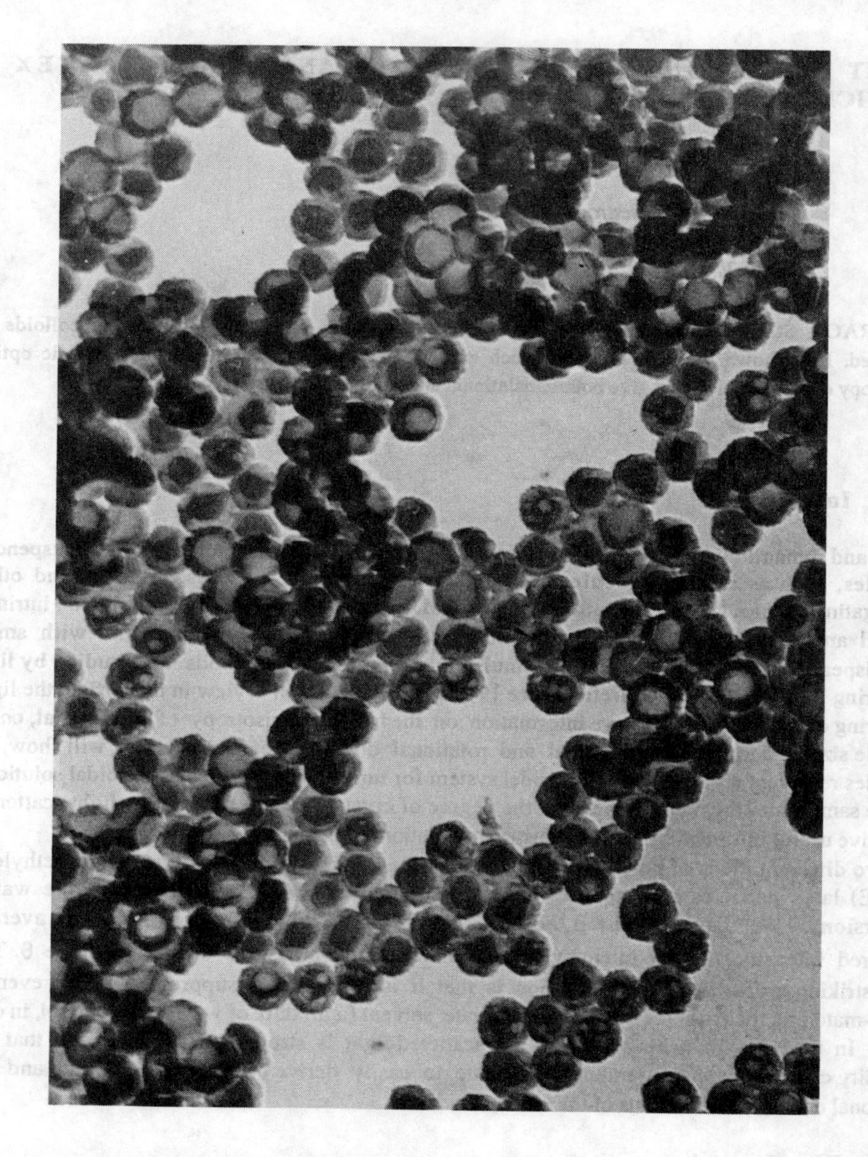

Fig. 1. Transmission electron microscope photograph of spherical latex particles. The average particle radius is 87 nm.

molecules and ions. It should be noted that the dialysis process only partially removes the strongly adsorbed fluorinated surfactant used in the polymerization procedure. Dialysis was carried out for over 12 weeks, until a stable conductivity value of about 100 mS was reached. The light scattering samples were prepared by diluting the latex solution with BDH AnalaR water, with a small amount of nonionic surfactant added to prevent slow coagulation. All the used solutions contained about 10 mM NaCl, which was added to partially screen electrostatic interactions, easing the interpretation of the experimental results. The PTFE rods have length a of about 0.33 mm and diameter of about 0.15 mm. Two different kinds of spheres, with the same copolymer composition, but differing in the "melt flow index" (MFI), (the higher MFI corresponding to a lower molecular weight of the polymeric chains), were used for the static measurements. The average radius is 70 nm for the spheres having MFI = 0.6 (henceforth called copolymer 1), and 90 nm for the spheres having MFI = 5.5 (copolymer 2). Spheres coming from a preparation similar to that of copolymer 2, but having a slightly different size (radius ≈ 85 nm) were used for the intensity correlation measurements.

The light scattering apparatus includes an argon laser operating at $\lambda = 514.5$ nm, a cylindrical scattering cell, a photomultiplier tube mounted on a rotating arm, and a digital correlator for the measurement of the intensity correlation function of scattered light. All measurements were performed at room temperature. The index of refraction of the solvent, n_s, was varied by mixing water with glycerol. In order to keep the turbidity of the sample approximately constant, when the optical mismatch between particle and solvent was varied the particle volume fraction was changed accordingly. The used volume fractions were in the range $2.5*10^{-3}$ -$4*10^{-2}$ for spheres, and even smaller for rods. The values of n_s reported on the abscissa axis of Figs.2 and 3 are relative to $\lambda = 589$ nm (sodium line).

By measuring the ratio P_i/P_t between the incident and transmitted power of the laser beam as a function of n_s, we derive the turbidity T as $T = (1/L)\ln(P_i/P_t)$, where L is the pathlength of the beam in the scattering cell. The intensity and polarization of scattered light were measured at a scattering angle $\theta = 90°$ as functions of the refractive index of the solvent. The geometry of the experiment is the following: the incident beam is linearly polarized in the vertical direction, the scattering plane is horizontal. We call I_{VV} (I_{VH}) the intensity of scattered light with vertical (horizontal) polarization.

The experimental data for PTFE rods at a volume fraction $\Phi = 10^{-3}$ are shown in Fig.2. We see that the turbidity never goes to 0, that is, it is impossible to optically match particles and solvent. The minimum turbidity corresponds to a weight fraction of glycerol of 35%. The behaviour of I_{VV} is similar to that of the turbidity, whereas I_{VH} is independent of n_s. The experimental data for copolymer 1 are shown in Fig.3, where we have reported the specific turbidity T/Φ, I_{VV} and I_{VH}. The minimum turbidity corresponds to a weight fraction of glycerol of 20.5%. The minimum of I_{VV} occurs in the same position as the minimum of T. At the minimum, the ratio I_{VV}/I_{VH} is equal to 1.85. Fig.4 presents the comparison between the I_{VV} curves obtained with copolymer 1 and 2. We see that both have a parabolic shape, but the two minima are in different positions (the minimum turbidity for copolymer 2 corresponds to a glycerol weight fraction of 17.5%), and the widths of the two parabolas are also different. The ratio I_{VV}/I_{VH} for copolymer 2 is the same as that of copolymer 1.

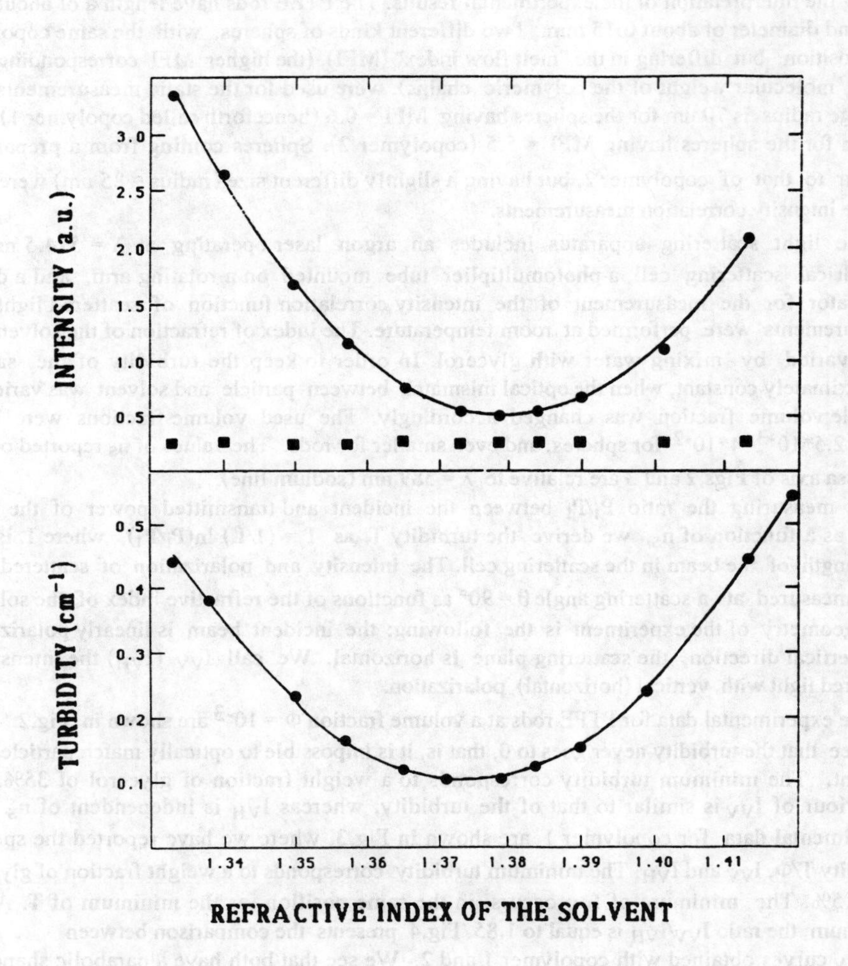

Fig.2. a) intensity of scattered light I_{VV}(●) and I_{VH}(■) for a dispersion of PTFE latex as a function of the refractive index of the solvent n_s. b) turbidity T as a function of n_s.

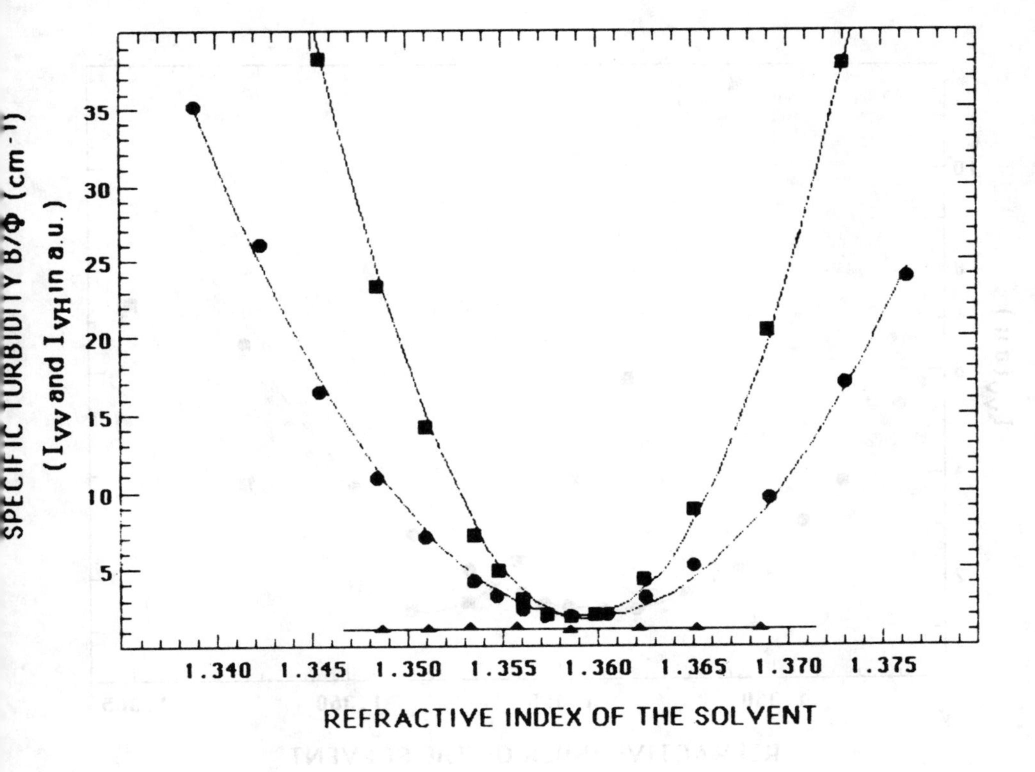

Fig.3. Specific turbidity (●), and intensity of scattered light I_{VV} (■) and I_{VH} (▲) as a function of the refractive index of the solvent n_S.

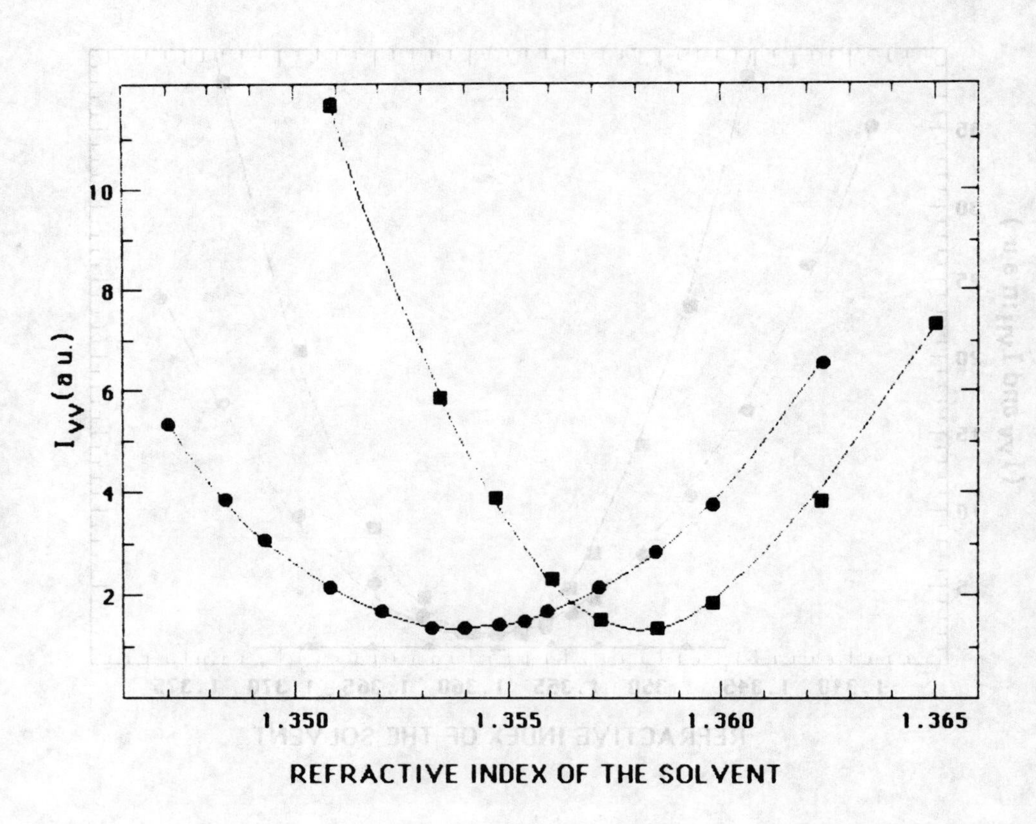

Fig.4. I_{VV} as a function of the refractive index of the solvent for copolymer 1 (■) and 2 (●).

The intensity of scattered light was measured, for PTFE rods in an approximately index-matched solvent, as a function of the modulus of the scattering vector k. We recall that $k = (4\pi n/\lambda)\sin(\theta/2)$, where λ is the wavelength of incident light, and n is the index of refraction of the scattering medium. The results are shown in Fig.5. We see that the two curves have a similar shape, apart from a scaling factor.

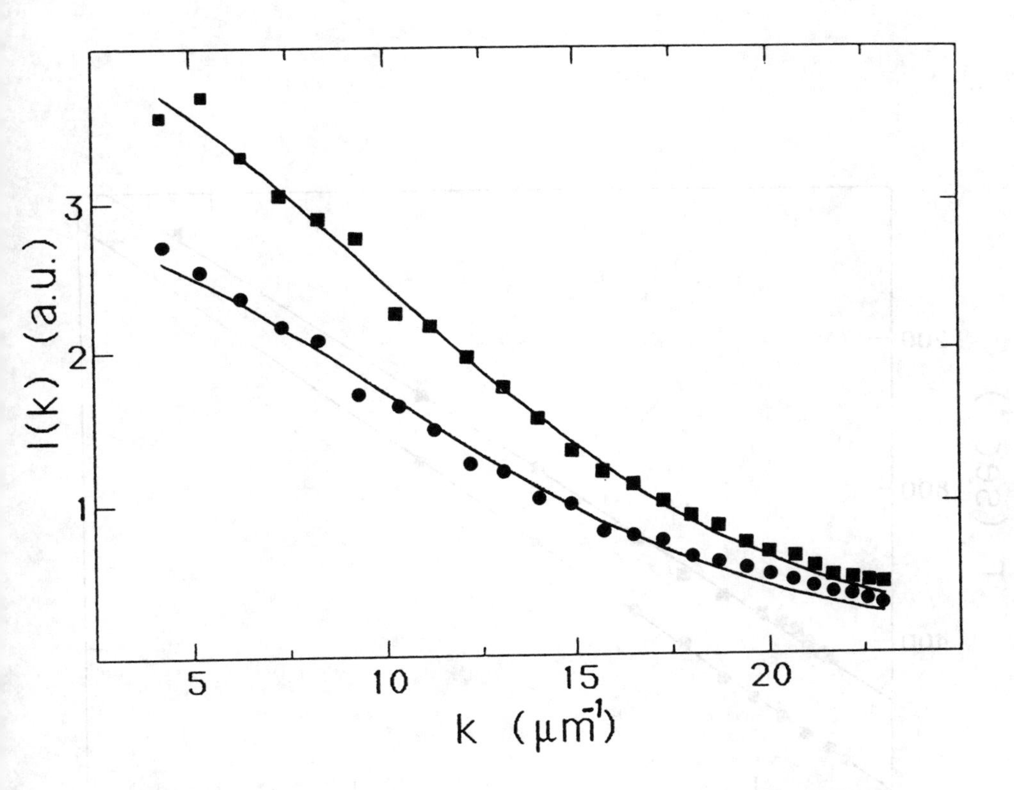

Fig.5. Intensity of scattered light, vertically (■) and horizontally (●) polarized, plotted as a function of k.

Intensity-correlation-function measurements were performed at various scattering angles. Both the correlation functions of vertically and horizontally polarized light show a nearly exponential behaviour. We have analyzed the data with the usual cumulant fit. The behaviour of the first cumulant versus k^2 is reported in Fig.6 for PTFE rods and Fig.7 for the spherical latex.

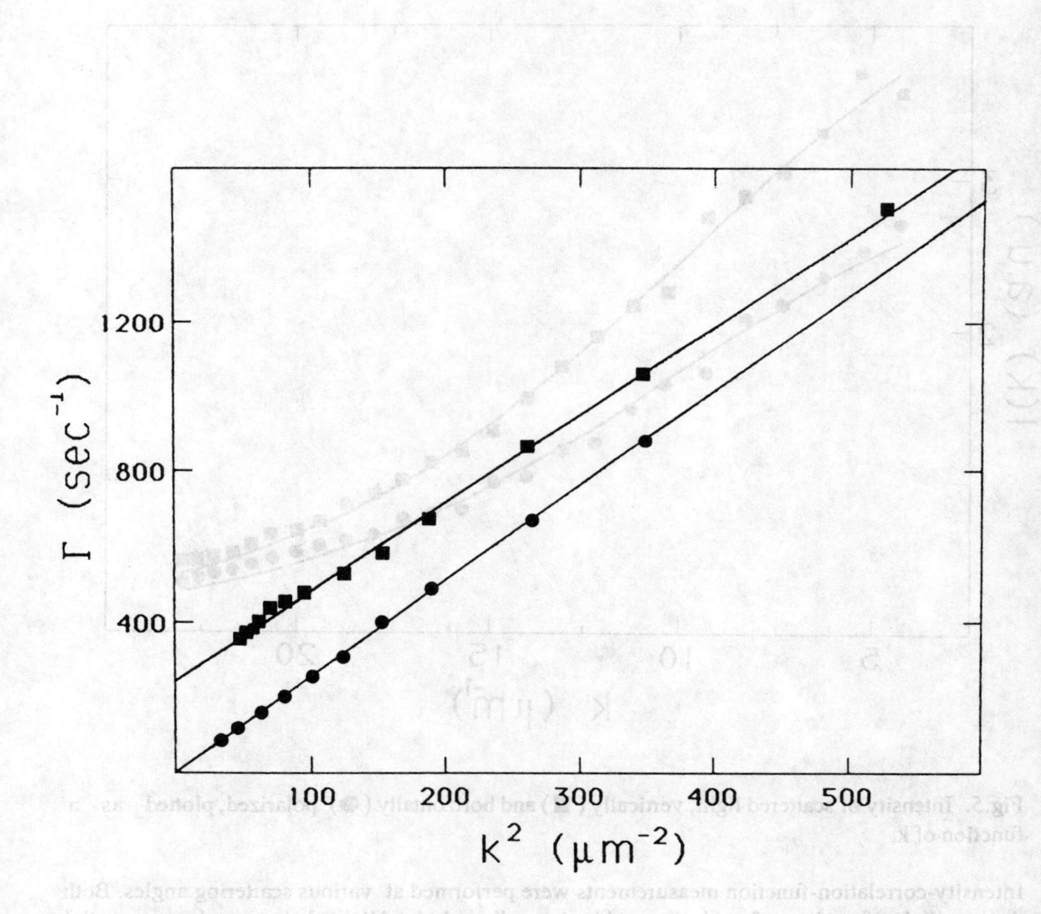

Fig.6. Reciprocal of the decay time of the correlation function of scattered light for vertical (■) and horizontal (●) polarization plotted as a function of k^2.

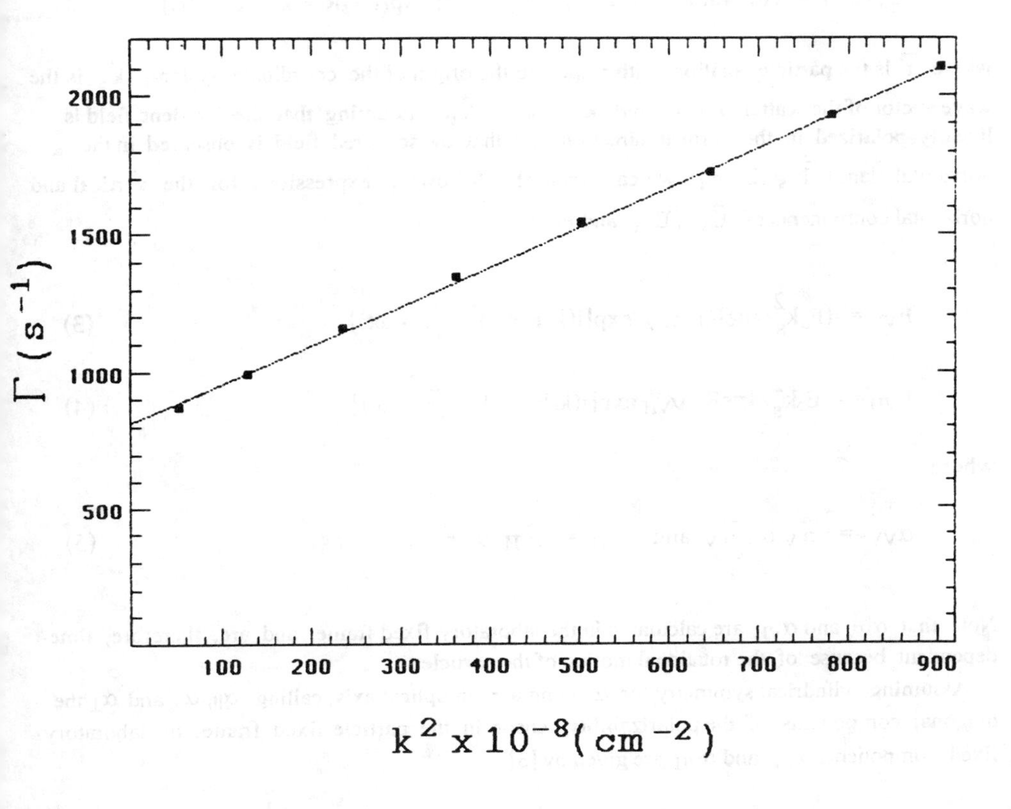

Fig.7. Reciprocal of the decay time of the correlation function of I_{VH} plotted as a function of k^2.

3. Discussion

The general treatment of static and dynamic light scattering from large anisotropic particles is rather complex [1-3]. In our case we can take advantage of the fact that the latex particles are suspended in a nearly index-matched solvent, so that we can interpret the experimental results by using the Rayleigh-Debye approximation [1], even if the particle size is comparable to λ.
We first recall the expression of the field scattered by a small anisotropic particle characterized by a polarizability tensor $\underline{\underline{\alpha}}$. If the incident field is given as

$$\vec{E}_i(\vec{r},t) = \vec{n}_i \; E_o \exp[i(\vec{k}_i \cdot \vec{r} - \omega t)] \tag{1}$$

where $\vec{n}_i$ is a unit vector in the direction of the incident field, the scattered field $\vec{E}_s$ observed at distance R in the far field is given by

$$\vec{E}_s(R,t) = (E_o/4\pi\varepsilon R)\, \vec{k}_s \times [\vec{k}_s \times (\underline{\underline{\alpha}} \cdot \vec{n}_i)]\, \exp[i(k_s R + \vec{k} \cdot \vec{r} - \omega t)] \tag{2}$$

where $\vec{r}$ is the particle position with respect to the origin of the coordinate system, $\vec{k}_s$ is the wave vector of the scattered field, and $\vec{k} = \vec{k}_s - \vec{k}_i$. Assuming that the incident field is linearly polarized in the vertical direction and that the scattered field is observed in the horizontal plane ($\vec{k}_s \perp \vec{n}_i$), we can write the following expressions for the vertical and horizontal components of $\vec{E}_s$, E_{VV} and E_{VH}:

$$E_{VV} = -(E_o k_s^2/4\pi\varepsilon R)\, \alpha_{VV}\, \exp[i(k_s R + \vec{k} \cdot \vec{r} - \omega t)] \tag{3}$$

$$E_{VH} = -(E_o k_s^2/4\pi\varepsilon R)\, \alpha_{VH}\, \exp[i(k_s R + \vec{k} \cdot \vec{r} - \omega t)] \tag{4}$$

where

$$\alpha_{VV} = \vec{n}_V \cdot \alpha \cdot \vec{n}_V \text{ and } \alpha_{VH} = \vec{n}_H \cdot \alpha \cdot \vec{n}_V . \tag{5}$$

Note that α_{VV} and α_{VH} are calculated in the laboratory-fixed frame, and are, therefore, time-dependent because of the rotational motion of the particle.

Assuming cylindrical symmetry for $\underline{\underline{\alpha}}$ along a main optical axis, calling $\alpha_{||}$, $\alpha_{\perp}$ and $\alpha_{\perp}$ the diagonal components of the polarizability tensor in the particle-fixed frame, the laboratory-fixed components α_{VV} and α_{VH} are given by [3]

$$\alpha_{VV} = \alpha + (16\pi/45)\beta Y_{2,0}(\delta,\phi) \tag{6}$$

$$\alpha_{VH} = i(2\pi/15)\beta[Y_{2,1}(\delta,\phi) + Y_{2,-1}(\delta,\phi)] \tag{7}$$

where $\alpha = (\alpha_{||} + 2\alpha_{\perp})/3 - V(n_s^2 - 1)$ is the average excess polarizability of the particle with respect to the solvent, V is the particle volume, $\beta = \alpha_{||} - \alpha_{\perp}$ is the anisotropy of the particle polarizability, $Y_{2,m}$ is the second order spherical harmonic of index m, and δ and ϕ are the angles describing the optical axis direction relative to the laboratory frame. By decoupling rotations from translations, assuming that the solution is sufficiently dilute to make interactions negligible, and averaging over all particle orientations, we can calculate from Eqs.3 and 4 the field correlation functions

$$|G_{VV}(\tau)| = C\,[\alpha^2 \exp(-D_T k^2 \tau) + (4/45)\beta^2 \exp(-D_T k^2 \tau - 6D_R \tau)] \tag{8}$$

$$|G_{VH}(\tau)| = C(\beta^2/15)\exp(-D_T k^2 \tau - 6D_R \tau) \tag{9}$$

where D_T and D_R are, respectively, the translational and rotational diffusion coefficients of the particle, and C is a constant. Note that the average scattered intensities are simply derived from Eqs.8 and 9 as: $I_{VV} = G_{VV}(0)$, $I_{VH} = G_{VH}(0)$.

The turbidity T is given by

$$T = N\sigma \tag{10}$$

where N is the concentration of scatterers and σ is the total scattering cross-section which can be expressed as [1,3]:

$$\sigma = A[\alpha^2 + (2/9)\beta^2] \tag{11}$$

where A is a constant. Since, in our case, both α and β are much smaller than $(\alpha_{||} + 2\alpha_{\perp})/3$, we can write approximately σ and the scattered intensities as:

$$\sigma = A'[(\bar{n} - n_s)^2 + (2/9)(n_{||} - n_{\perp})^2] \tag{12}$$

$$I_{VV} = C'[(\bar{n} - n_s)^2 + (4/45)(n_{||} - n_{\perp})^2] \tag{13}$$

$$I_{VH} = C'(n_{||} - n_{\perp})^2/15 \tag{14}$$

where $\bar{n} = [(n_{||}^2 + 2n_{\perp}^2)/3]$ is the average index of refraction of the particles, and A' and C' are constants. As discussed previously [7-9], since the particles are almost index-matched by the solvent, we can treat light scattering in the Rayleigh-Gans approximation: if the particle is homogeneous, this is equivalent to saying that the field scattered by the particle is obtained by multiplying the small particle results (Eqs.3 and 4) by the particle form factor. This will affect only the angular distribution of the scattered intensity, leaving unchanged the dependence on n_s, and the temporal behaviour of the correlation functions. We can still use, therefore, Eqs.12-14 and Eqs.8-9, respectively, for the interpretation of the static and dynamic light scattering data.

I will first discuss the data relative to PTFE rods. It is seen from Fig.2b that the measured turbidity is indeed a parabolic function of the refractive index of the solvent. The full line in the figure represents the fit with Eq.12. The fit parameters are $n_{||} - n_{\perp} = 0.046 \pm 0.005$, and $\bar{n} = 1.374 \pm 0.002$, in good agreement with the values given in Ref.9. A similar analysis was applied to the plot of I_{VV} versus n_s presented in Fig.2a. The fit parameters are: $n_{||} - n_{\perp} = 0.064 \pm 0.005$ and $\bar{n} = 1.378 \pm 0.002$. The discrepancy between the two anisotropy values derived from curves a and b of Fig.1 is somewhat larger than the experimental uncertainty. Presently, we have no explanation for the observed discrepancy. A point to be checked is whether crystalline PTFE is indeed uniaxial.

We have performed a comparison of the measured angular dependence of the scattered intensity with the form factor calculated for a prolate ellipsoid having semi-axes a,b,b, with 2a = 0.33 µm, and 2b = 0.15 µm. The calculated P(k), multiplied by an appropriate factor in order to fit the experimental data, is shown by the full curves in Fig.5. We see that the fit confirms the values of the height and diameter of the particles estimated from the electron microscope observation.

I will now discuss the dynamic light scattering data concerning the PTFE rods. As it is well-known, the measured quantity is the intensity correlation function which, under the hypothesis that the scattered field is a Gaussian process, is related to the modulus of the field correlation function by a simple relation [3]. The obtained field correlation functions were nearly exponential. According to Eq.9, the first cumulant of G_{VH} is given by

$$\Gamma = D_T k^2 + 6 D_R \tag{15}$$

We see, indeed, from the G_{VH} data of Fig.6 that Γ is a linear function of k^2. The fit with Eq.15 gives the diffusion coefficients

$$D_T = (2.3 \pm 0.1) \times 10^{-8} \quad cm^2/s$$

and

$$D_R = (43 \pm 5)\ s^{-1}$$

The values of the diffusion coefficients calculated from the known dimensions of the cylindrical particles [3] are: $D_T = 2.34 \times 10^{-8}$ cm^2/s and $D_R = 55\ s^{-1}$. The agreement is satisfactory if we take into account that the non-negligible polydispersity of the sample affects D_R much more than D_T.

In the case of G_{VV}, Eq.8 predicts a superposition of two exponentials. If we analyze G_{VV} in terms of a cumulant expansion, we find

$$\Gamma = D_T k^2 + 6[1/(1+H)]\, D_R \tag{16}$$

where H is the ratio between the amplitudes of the two exponentials. In our case, $H \approx 20$ which means that the contribution of the second exponential is very small. Eq.16 predicts for Γ a linear dependence on k^2, with the same slope as that predicted by Eq.15. The slope derived from the G_{VV} data is $(2.5 \pm 0.1) \times 10^{-8}$ cm^2/s, in good agreement with the value of D_T given above. Since H is much larger than 1, the intercept with the vertical axis is very small for the G_{VV} data, so that it is not possible to reliably derive the value of the rotational diffusion coefficient in this latter case.

We can conclude that all the data are consistent with the hypothesis that the PTFE rod is a single crystal. The fact that PTFE rods have a crystalline structure was already known previously [10], but no measurement of the optical anisotropy was available.

I will now discuss the data relative to the spherical latex. Also in this case (see Fig.3) the measured turbidity is a parabolic function of the refractive index of the solvent. The full line in Fig.3 represents the fit with Eq.12. The fit parameters are $n_{||} - n_{\perp} = 0.0086 \pm 0.0008$, and $\bar{n} = 1.3581 \pm 0.0002$. A similar analysis is applied to the plot of I_{VV} presented in Fig.3. The fit with Eq.13 gives: $n_{||} - n_{\perp} = 0.0087 \pm 0.0008$ and $\bar{n} = 1.3581 \pm 0.0002$, in very good agreement with the values derived from the turbidity curve.

We note that the ratio I_{VV}/I_{VH}, as calculated from Eqs.13 and 14 for $\bar{n} = n_s$, is equal to 4/3. The fact that the experimental ratio is larger than 4/3 indicates that the particle cannot be treated as a single crystal, but has a more complicated internal structure which includes, perhaps, both amorphous and crystalline regions. In order to understand the effect of a possible non-uniformity of the local polarizability, we have calculated the scattered intensity for polycrystalline spheres, by assuming a completely random distribution for the orientation of the optical axis of different crystalline regions. If r is the number of cystalline regions, we find that the scattered intensity can still be expressed by Eqs.13 and 14 with the only difference that, instead of the anisotropy $n_{||} - n_{\perp}$, we have an "effective" anisotropy $(n_{||} + n_{\perp})/r$. This suggests that the large difference in the measured anisotropy between the PTFE latex studied in Ref.1 and the PTFE copolymers studied in this paper is mainly due to the fact that the pure PTFE particle is essentially a single crystal, whereas the latex particles obtained with the PTFE copolymer are polycrystalline. Furthermore, the difference in the measured anisotropy between copolymers 1 and 2 is probably due to the difference in the number r of crystalline regions.

In the case of copolymer 2, the fit of the I_w curve of Fig.4 gives $n_{||} - n_{\perp} = 0.0160 \pm 0.0008$ and $\bar{n} = 1.3540 \pm 0.0005$. It is interesting to note that different values of the MFI correspond to different values of the optical anisotropy of the latex particles.

Finally, we shall discuss the dynamic light scattering data. The data of Fig.7 show that the measured Γ is a linear function of k^2. The fit with Eq.15 gives the diffusion coefficients $D_T = (1.43 \pm 0.05) \times 10^{-8}$ cm ^{2}/s and $D_R = (135 \pm 5)$ s^{-1}. By taking T = 22°C and a solvent viscosity of 1.73 cp, we derive from the experimental value of D_T a particle radius of 87 nm, and from the experimental D_R a particle radius of 88 nm. The internal agreement between the two data is excellent, and the consistency with the electron microscope evaluation is satisfactory.

To summarize the discussion about the PTFE copolymer, the light scattering data clearly show that the particles are optically anisotropic. The sensitivity of the technique is remarkable, because anisotropies $n_{||} - n_{\perp}$ smaller than 0.01 are easily measured.

As a conclusion, we have given a rather complete characterization of PTFE rodlike particles and of spherical latex particles made of a PTFE copolymer. In particular, the former particles are crystalline whereas the latter are only partially crystalline, with a degree of crystallinity showing a correlation with the melt flow index of the polymer. Our data indicate that the light scattering technique can easily be used as a method to detect the crystalline structure of colloids. From a more fundamental point of view, the latex we have studied represents a very interesting model system for optically anisotropic particles. As far as we know, the data reported in this paper represent the first light-scattering measurement of the rotational diffusion constant of a spherical colloid particle.

4. References

1. van de Hulst H.C. (1981), Light Scattering by Small Particles, Dover, New York.
2. Kerker M. (1969), The Scattering of Light and Other Electromagnetic Radiation, Academic Press, New York.
3. Berne B.and Pecora R."Dynamic Light Scattering", Wiley, New York, 1975.
4. Degiorgio V., Corti M. and Giglio M. (Eds.) (1980), Light Scattering in Liquids and Macromolecular Solutions, Plenum, New York.
5. Dahneke B.E. (Ed.) (1983), Measurement of Suspended Particles by Quasi-Elastic Light Scattering, Wiley, New York.
6. Pecora R. (Ed.) (1985), Dynamic Light Scattering. Applications of Photon Correlation Spectroscopy, Plenum, New York.
7. Piazza R., Stavans J., Bellini T. and Degiorgio V.(1989), Opt. Commun. 73, 263.
8. Piazza R., Stavans J., Bellini T., Degiorgio V., Lenti D. and Visca M. (1990), Progr. Colloid Polym. Sci. (to appear).

9. Bellini T., Piazza R., Sozzi C. and Degiorgio V.(1988), Europhys. Lett. 7, 561.
10. Ottewill R.H. and Rance D.G.(1986), Coll. Polymer Sci. 264, 982

ELECTRIC BIREFRINGENCE STUDIES OF MICELLAR AND COLLOIDAL DISPERSIONS

V. DEGIORGIO
Dipartimento di Elettronica
Universita' di Pavia
27100 Pavia, Italy

ABSTRACT. Static and dynamic measurements of the Kerr effect in three different systems are reviewed. The first system is an aqueous dispersion of polytetrafluoroethylene (PTFE) polymer colloid having a crystalline structure. The second is a polydisperse solution of rod-like ionic micelles. The third is a nonionic micellar solution near the cloud point.

1. Introduction

When an electric field is applied to a solution of macromolecules, the macromolecules orientate and the solution exhibits a macroscopic optical anisotropy. When the applied field is removed, the orientational order is randomized by rotational diffusion. Static and dynamic electric birefringence studies of macromolecules or other colloidal particles are helpful for determining the size and shape of the particles [1]. In this article the discussion will be confined to three specific examples which are illustrative of the variety of applications of the technique. The first example concerns measurements of electric birefringence (also called Kerr effect) on a dispersion of an electrically charged rod-like polymer colloid having a crystalline structure [2]. It is found that the Kerr constant of such a system is exceptionally large because the particles possess an intrinsic optical anisotropy. Furthermore, by exploiting the fact that the index of refraction of the particles is almost matched to that of the solvent, the Kerr constant was measured in situations in which interparticle interactions are not negligible. The data show a considerable enhancement of the Kerr constant in this collective regime. The second example deals with a polydisperse solution of rod-like ionic micelles [3]. It has been shown that the relaxation of electric birefringence is asymptotically characterized by a stretched exponential behaviour. Such a result suggests a rather general and powerful method to characterize relaxation phenomena in polydisperse systems: in particular, the value of the stretching exponent can be directly related to the shape of the size distribution of the particles. The third example refers to solutions of nonionic micelles near the cloud point [4]. The Kerr constant of such a system is found to increase dramatically as the temperature of the solution approaches the cloud point temperature. This effect is connected with the existence of attractive intermicellar interactions, and does not necessarily imply the formation of strongly elongated micelles.

2. The Technique

In presence of an applied electric field E, the solution becomes optically anisotropic in the sense that a light beam travelling in a direction perpendicular to E has a different velocity depending whether its direction of polarization is parallel or perpendicular to E. If E is sufficiently small, the

D. M. Bloor and E. Wyn-Jones (eds.),
The Structure, Dynamics and Equilibrium Properties of Colloidal Systems, 597–612.

induced birefringence is $\Delta n = B\lambda E^2$, where λ is the wavelength of the light beam and B is called the Kerr constant of the medium. All the experiments described in this article are performed in the regime in which Δn is proportional to E^2 (Kerr regime). The transient electric birefringence (TEB) experiment consists of applying a rectangular pulse of electric field to the liquid mixture and observing the transient as well as the steady-state induced birefringence. A general description of the technique can be found in Ref.5. A scheme of the used apparatus is shown in Fig.1. We recall that the insertion of a quarter-wave plate between the Kerr cell and the analyzer allows one to establish a linear relation between Δn and the output of the photodetector. The cell itself is made from an optical glass cuvette which was selected for its low residual stress birefringence. The optical pathlength is 60 mm, and the electrodes have a separation of 1-3 mm.

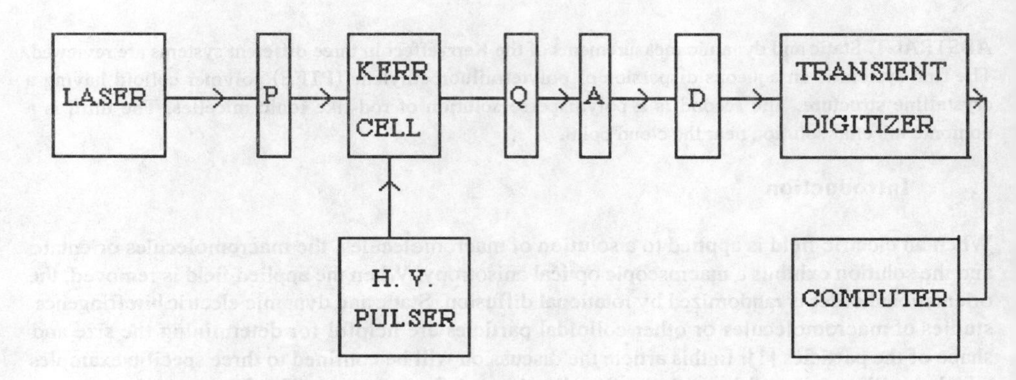

Fig.1. Scheme of the electric birefringence experiment. P: polarizer, Q: quarter-wave plate, A: analyzer, D: photodetector.

The output of the photodetector is sent to a transient digitizer and averager (Data 6000, Data Precision). The decay is sampled over 1000 points. We use, typically, the last 100 points for the evaluation of the baseline. The accuracy in the determination of the baseline is better than 0.2%. The overall response time of the apparatus is less than 1 μs. A more detailed description of the apparatus can be found in [6].

3. Crystalline Polymer Colloids

We have studied dilute dispersions in water of polytetrafluoroethylene (PTFE) particles [2]. The material was a gift from M. Visca and D. Lenti, Montefluos, Spinetta Marengo, Italy. PTFE was prepared by a dispersion polymerization process which in the presence of emulsifier yields stable aqueous latices. The latex particles thus formed are highly crystalline [7]. As observed with the electron microscope (see Fig.2), the particles have a rodlike shape with length of about 0.33 μm and diameter of about 0.15 μm. The particles are fairly monodisperse: the estimated standard deviation in volume is about 16% [8].

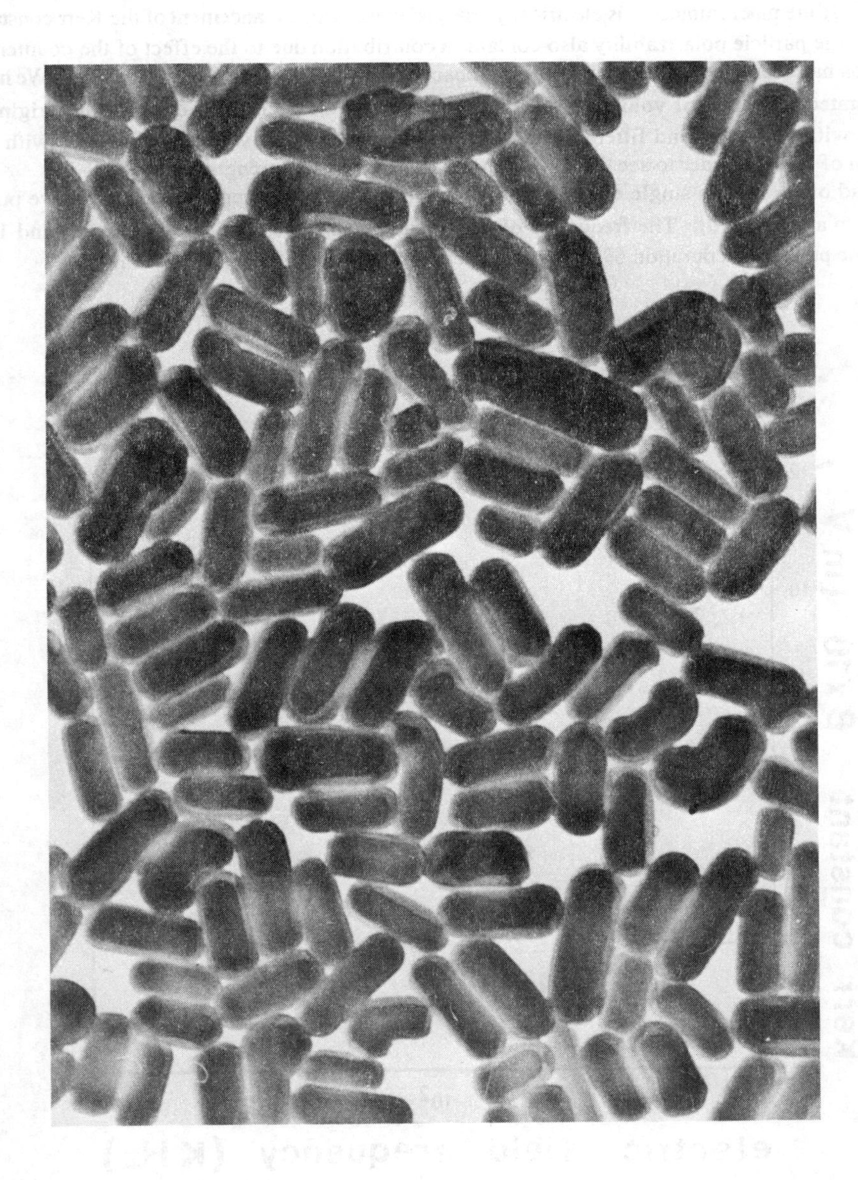

Fig.2. Transmission electron microscope photograph of rod-like PTFE latex particles.

Because of the preparation procedure, it is known that the particles are negatively charged. The measurement of the surface charge density by conductometric titration [8] gave a value of 0.3

$\mu C/cm^2$. If the macromolecule is electrically charged there is an enhancement of the Kerr constant because the particle polarizability also contains a contribution due to the effect of the counterion cloud, as has been shown by experiments on tobacco mosaic virus (TMV) solutions [9]. We have investigated the range of volume fractions Φ between 10^{-4} and 10^{-2} by diluting the original sample with deionized and filtered water. Some measurements have been performed with the addition of NaCl in order to see the dependence of B on the ionic strength.

Instead of applying a single electric-field pulse, in this work we applied a square-wave pulse with zero average [10]. The frequency of the square wave υ was varied between 0.3 and 10^3 kHz. The pulse had a duration of 10 ms and a peak-to-peak amplitude of 10-30 V.

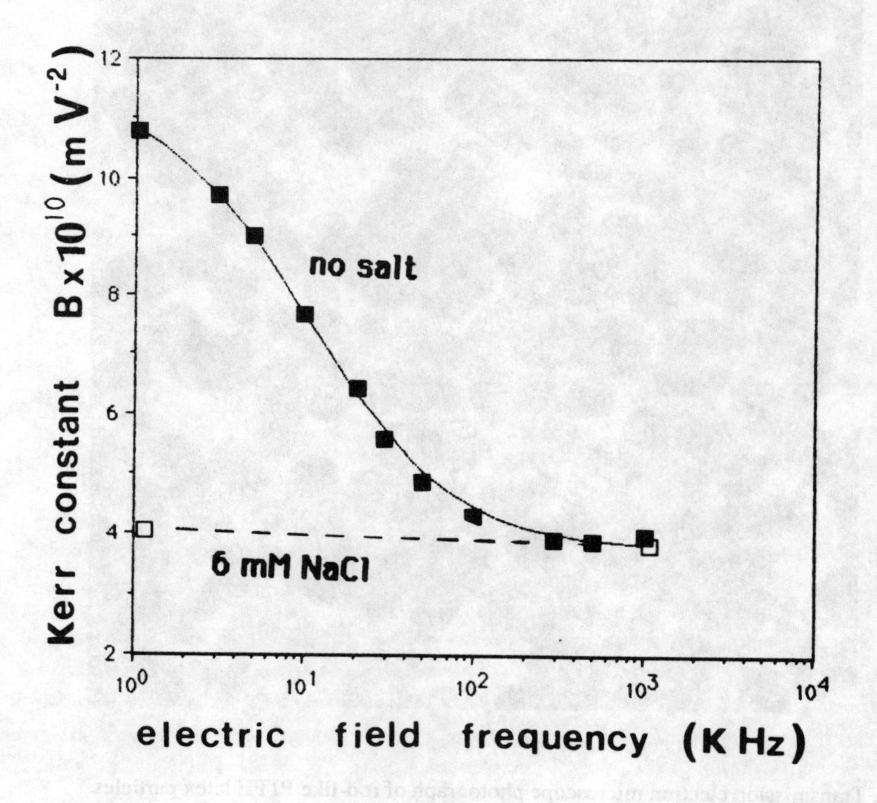

Fig.3. Kerr constant of a dispersion of PTFE rods as a function of the frequency of the applied square wave. The volume fraction is $4x10^{-4}$. ■: no salt, □: 6 mM NaCl.

We report in Fig.3 the behaviour of the Kerr constant B as a function of the frequency of the square wave at the volume fraction $4x10^{-4}$. Fig.3 shows that B decreases with υ, and reaches a plateau when $\upsilon > 300$ kHz. The ratio between the low-and high-frequency values is r = 2.7. If we define a cut-off frequency υ_c as the frequency at which r is reduced by a factor 2, we find from Fig.3 that υ_c = 10 kHz. We have studied the frequency behaviour of B at various volume fractions: the results are qualitatively similar to those reported in Fig.3. It should be noted that the lowest applied frequency, 1 kHz, is still much larger than τ_R^{-1}, where τ_R is the relaxation time due to rotational diffusion. Since the particles do not possess a permanent dipole moment, the orienting torque is a result of an induced polarization which can be reversed, in periods short compared with the relaxation time, by reversing the applied field. The torque remains unchanged upon field reversal. The results of Fig.3 indicate that an important contribution to the induced polarization comes from the counterion atmosphere which surrounds the charged particles. The observed cut-off frequency is the frequency above which the counterions are unable to follow the variation of the applied field. We have found that the contribution of the ion atmosphere becomes less and less important as the ionic strength of the solution is increased. As shown in Fig.3, the addition of 6 mM NaCl is sufficient to cancel completely the counterion contribution. This finding is in qualitative agreement with the theoretical model by O'Konski and Krause[11].

The high-frequency value of B shown in Fig.3 is $4x10^{-10}$ mV^{-2}. Notwithstanding the fact that the solution is very dilute, such a value is three orders of magnitude larger than the Kerr constant of nitrobenzene. If we define a specific Kerr constant B/Φ, we can compare the experimental value with the value calculated from the known shape, size, dielectric constant and refractive index of the particle [1]. We find that the calculated value is about 50 times smaller than the experimental result. The only possible explanation for this discrepancy is that the particles, being crystalline, possess an intrinsic optical anisotropy. For symmetry reasons, the structure will be uniaxial with the optical axis coincident with the rod axis. Since the observed birefringence is positive, $n_{||} > n_{\perp}$, where $n_{||}$ ($n_{\perp}$) is the index of refraction of PTFE for polarization parallel (perpendicular) to the optical axis. From the experimental B/Φ, we derive $n_{||} - n_{\perp} = 0.04 \pm 0.005$.

In order to get an independent evaluation of the intrinsic optical anisotropy of the particles we have also measured the turbidity of the solution as a function of the refractive index of the solvent n_s. (These results are described in the accompanying article by the same author in this publication).

We have discussed so far the properties of very dilute dispersions in terms of single particle parameters. Since the particles are charged, the electrostatic interactions might play some role in determining the properties of the dispersion even at rather small volume fractions. To get some information on this aspect, we performed measurements of B as a function of the volume fraction. Some results are reported in Fig.4 where the low-frequency specific Kerr constant, measured at various ionic strengths, is plotted versus Φ. We see that there is an important collective effect when no salt is added to the dispersion (curve a). By adding NaCl, the collective effect is progressively washed out. At high ionic strength, the low-frequency B/Φ coincides with the high-frequency value which does not show any dependence on the volume fraction (curve e). Curves b, c, and d are well described by a linear relation,

$$B/\Phi = B_o/\Phi\,(1 + k\,\Phi), \qquad (1)$$

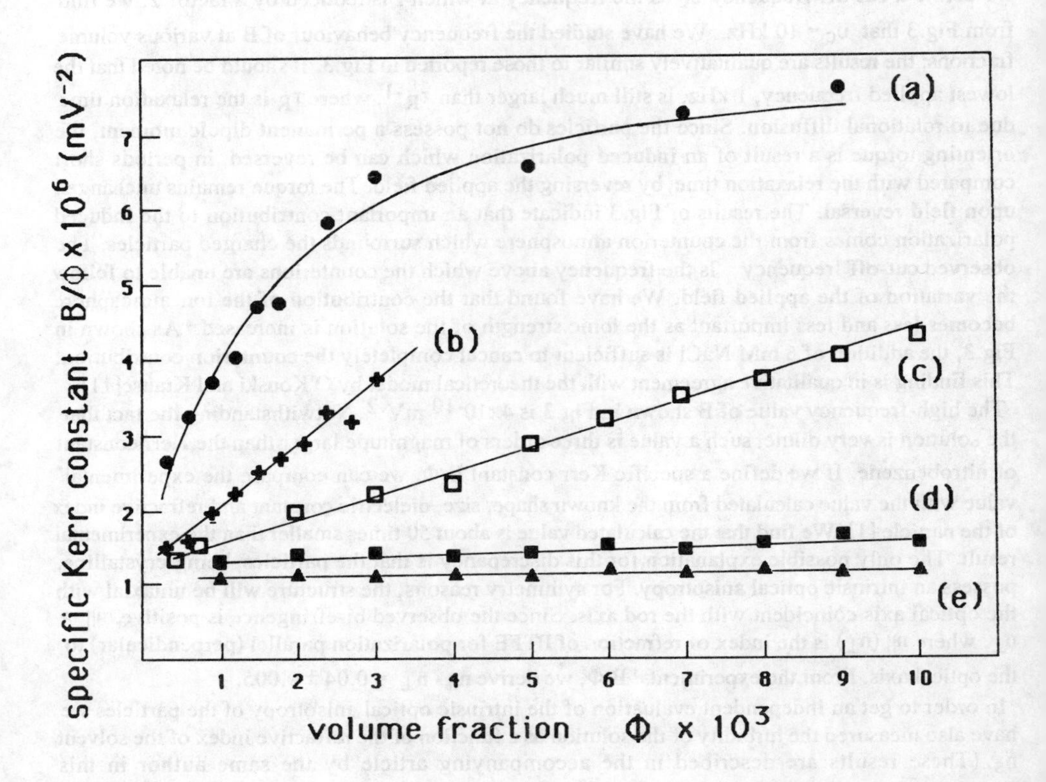

Fig.4. Specific Kerr constant versus volume fraction. Curve a): no added salt; curve b): 0.2 mM NaCl; curve c): 1 mM NaCl; curve d): 3 mM NaCl. Curves a-d: low-frequency response; curve e): high-frequency response, any ionic strength.

where k is a kind of second virial coefficient which strongly depends on the added salt concentration. B_O seems to be very weakly dependent on the ionic strength. Curve a cannot be fitted as a straight line because in that case the ionic strength is itself a function of the particle concentration. The fact that the measured k is positive indicates that the collective effect consists in an enhancement of the average particle orientation in the direction of the applied electric field. The effect is due to the interaction among the polarized counterion clouds surrounding each particle. In fact, when the frequency of the field is too high to distort the ion atmosphere, no collective contribution is found.

The last data to discuss concern the relaxation of the electric birefringence. We have found that both rise and decay of the birefringence pulse follow the same time behaviour. Such a behaviour is not exponential because the sample is polydisperse. From the initial logarithmic derivative of the EB signal, we derive a relaxation time of 5.7 ms which is in good agreement with that expected from a rotational diffusion constant $D_R = 55\ s^{-1}$, provided that the effect of polydispersity is included [12]. In the investigated range of volume fractions, we found no appreciable dependence of D_R on Φ.

As a final comment, the existence of a rather strong intrinsic optical anisotropy implies that dispersions of PTFE-rods are a good medium for the observation of the optical Kerr effect. In fact a successful experiment has been recently performed with a sample having the same origin as our sample [13].

4. Rod-Like Micelles

Here the discussion will concern the measurement of the time dependence of electric birefringence in a polydisperse solution of rod-like ionic micelles which were first reported in Ref.3. Since the conductivity of the sample must be very low in an electric birefringence experiment, it is possible to study elongated micelles only if they are formed at very low concentrations (below a few mM) of both ionic amphiphile and salt. This limitation restricts considerably the choice of possible systems. Our samples consisted of water solutions of an ionic amphiphile, cetylpyridinium chloride (CPyCl), and a univalent salt, sodium salicylate (NaSal). Both CPyCl and NaSal were purchased from Merck Chemie. The system we have chosen was previously investigated by Hoffmann and coworkers [14,15] by various techniques, and found to form elongated and polydisperse micelles at very low concentrations of both salt and amphiphile. The same system forms, at concentrations larger than those we have used, a viscoelastic gel. We have performed measurements at 25°C and 3 mM NaCl, in the amphiphile concentration range 0.2 mM < c < 3 mM. Voltage pulses with heights of 300-1500 V/cm and durations of 1-5 ms were used. We show in the insert of Fig.5 a typical birefringence pulse (averaged over 50 shots). In this work we have focussed our attention on the decay of the pulse which is simpler to study and to interpret. With B(t) the observed birefringence at time t, and t = 0 the time at which the applied electric- field pulse terminates, we define a normalized decay relaxation function as $R(t) = B(t)/B(0)$. In the dilute regime the decay of the induced birefringence is connected with the rotational diffusion time of the elongated particles. Since the solution is polydisperse (different rod-lengths are present), the relaxation process is a superposition of many exponential contributions. Fig.5 shows a log-log plot of $-\ln[R(t)]$ versus the scaled time t/τ for three distinct concentrations. We find that linear behaviour is obeyed asymptotically over more than one decade in the reduced variable t/τ. This means that the asymptotic behaviour of the relaxation can be described by a stretched- exponential (SE) function:

$$R(t) \approx \exp[-(t/\tau)^{\alpha}]. \tag{2}$$

The slope of the best-fit straight line to the linear portions of the data gives the stretching exponent α.

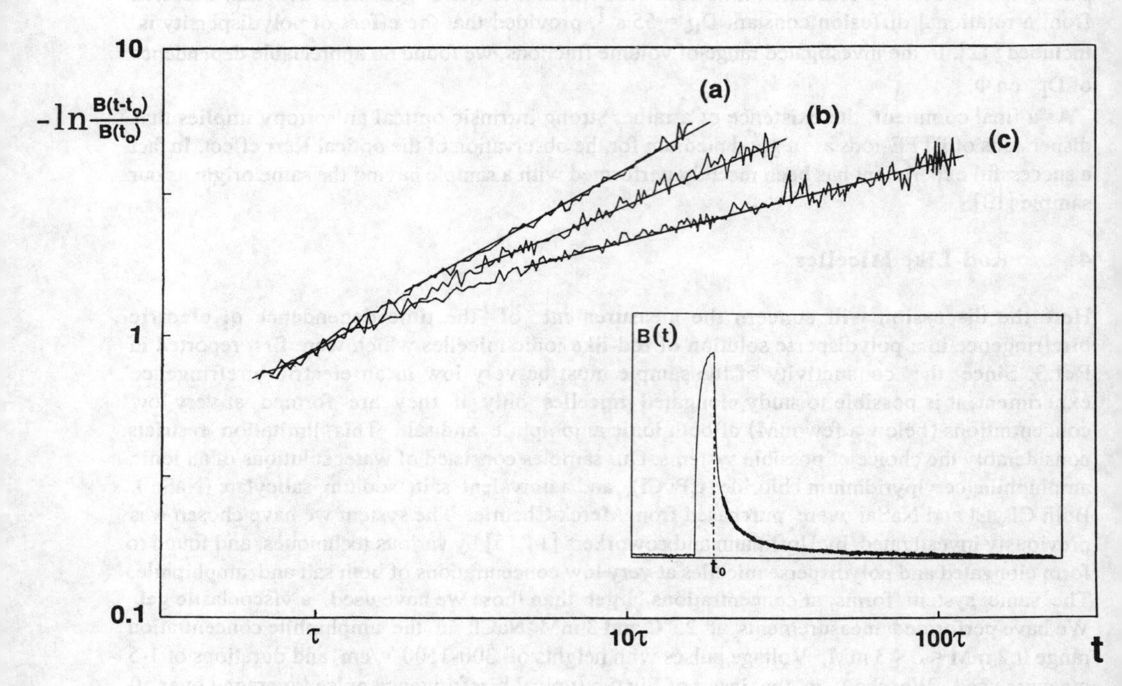

Fig.5. The logarithm of normalized electric-birefringence decay vs. scaled time, presented in a log-log plot to demonstrate stretched-exponential relaxation. The system consists of a polydisperse solution of cylindrical micelles of cetylpyridinium chloride in 3mM NaCl. a): 3 mM CPyCl, α = 0.53; b): 1 mM CPyCl, α = 0.41; c): 0.6 mM CPyCl, α = 0.26. The insert shows the full birefringence pulse whose thickened portion appears as curve a) in the log-log plot.

The behaviour of α as function of the amphiphile concentration is reported in Fig.6. We see that α grows from about 0.25 at amphiphile concentrations below 0.5 mM to about 0.5 for c larger than 1 mM. We have calculated the critical concentration for entanglement c* by using for the average rod length the experimental data of Ref.14. We find c* ≈ 0.6 mM.

A simple theoretical model [3,16] explains the dilute-regime results. If we make the assumption that intermicellar interactions play a negligible role, B(t) can be written as a sum of individual-particle responses, that is:

$$B(t) = \int_0^\infty P(m)S(m)\exp[-t/\tau(m)]dm \qquad (3)$$

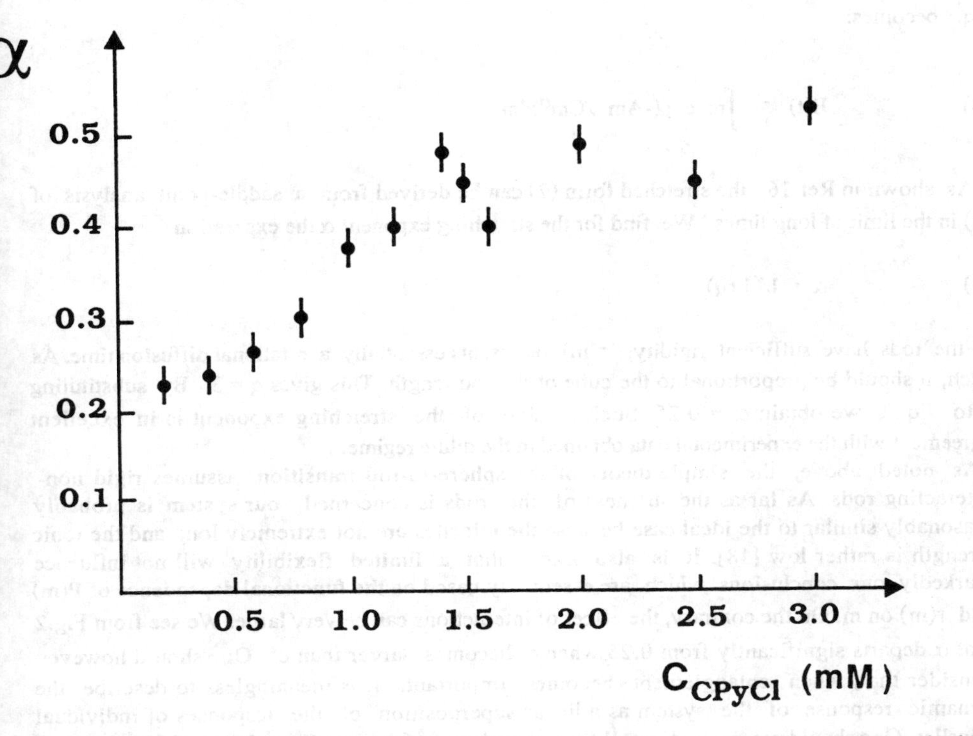

Fig.6. The exponent a of the stretched-exponential decay plotted as a function of the amphiphile concentration.

where m is the aggregation number of the micelles, P(m) is the probability distribution of aggregation numbers, S(m) is a signal function describing the contribution of each process to the observed signal, and $\tau(m)$ is the time constant characterizing the relaxation of birefringence for micelles of aggregation number m. In order to assign the functional dependence of P(m) and S(m) on the aggregation number m, we make use of the model of the sphere-to-rod transition which was developed in detail by Benedek and co-workers [17]. Such a model yields directly the probability distribution P(m) as P(m) = Aexp(-Am), where the constant A depends on amphiphile concentration, ionic strength and temperature [17]. The dependence of S(m) on m is unknown a priori because the magnitude of the Kerr effect is influenced by several factors such as the form anisotropy, the intrinsic anisotropy of counterions, and the ionic strength of the solution. Our results [3] indicate that, at low c, $B \sim c^2$. If we consider that B coincides with $B(0) = \int P(m)S(m)dm$, and that the theory of the sphere-to-rod transition of ionic micelles assigns a concentration dependence of the constant A of the type $A \sim \sqrt{c}$, we can derive $S(m) \sim m^r$, with r = 3. We assume, in addition, that $\tau(m)$ behaves as a power-law of m,

$\tau(m) = Cm^q$, where C is a constant. By using the explicit expressions for S(m), $\tau(m)$ and P(m), Eq.3 becomes:

$$(4) \qquad B(t) \sim \int m^r \exp(-Am - t/Cm^q) dm$$

As shown in Ref.16, the stretched form (2) can be derived from a saddle-point analysis of (4) in the limit of long times. We find for the stretching exponent α the expression

$$(5) \qquad \alpha = 1/(1+q)$$

If the rods have sufficient rigidity, $\tau(m)$ represents essentially a rotational diffusion time. As such, it should be proportional to the cube of the rod length. This gives q = 3. By substituting into Eq.5, we obtain α = 0.25. Such a value of the stretching exponent is in excellent agreement with the experimental data obtained in the dilute regime.

As noted above, the simple theory of the sphere-to-rod transition assumes rigid non-interacting rods. As far as the stiffness of the rods is concerned, our system is probably reasonably similar to the ideal case because the micelles are not extremely long and the ionic strength is rather low [18]. It is also likely that a limited flexibility will not influence markedly our conclusions which are essentially based on the functional dependence of P(m) and $\tau(m)$ on m. On the contrary, the effect of interactions can be very large. We see from Fig.2 that α departs significantly from 0.25 when c becomes larger than c*. One should however consider that, when entanglement becomes important, it is meaningless to describe the dynamic response of the system as a linear superposition of the responses of individual micelles. One should use, instead, a collective description of the micellar solution: the existence of a stretched exponential decay must be associated with the simultaneous excitation of many collective modes.

It should be mentioned that in the past few years a great variety of experimental data for dielectric, magnetic, NMR and mechanical relaxation phenomena in complex random systems have been shown to obey a SE behaviour [3]. The similarity of relaxation processes in rather different random systems, such as glasses, spin-glasses, polymers, viscous fluids, disordered dielectrics, and critical binary mixtures, is striking, and indicates a common origin of the universal behaviour. From this point of view, the polydisperse micellar solution can represent an interesting model system.

As a conclusion, we have seen that the relaxation of electric birefringence in a polydisperse solution of rod-like micelles follows asymptotically a stretched-exponential behaviour. The exponent α is found to depend on the concentration of micelles, ranging from about 0.25 for very dilute solutions (weakly interacting particles) to about 0.50 for semi-dilute systems (particle concentration larger than the critical concentration for entanglement c*). At variance with most of the previously reported observations of SE decay, we can give for our system (in the dilute regime) a "microscopic" theory of SE relaxation. Our results indicate that the analysis in terms of SE behaviour can represent a rather general and powerful method to characterize relaxation phenomena in polydisperse systems, and that the value of α contains direct information on the shape of the size distribution in the polydisperse system.

In the concentration range where the interparticle interactions are small, the observed value of the stretching exponent is in very good agreement with the theoretical prediction. When interactions become important, α grows, as expected from qualitative arguments.

5. Nonionic Micellar Solutions

This section will describe electric birefringence measurements performed on solutions of polyoxyethylene nonionic amphiphiles from the family $C_iH_{2i+1}(OCH_2CH_2)_jOH$, hereafter called C_iE_j [4]. All systems have been studied as a function of the temperature at fixed concentration. The system C_iE_j-H_2O is known to present a lower consolute curve (called cloud curve in the literature) with a minimum at a critical temperature T_c and a critical concentration c_c. The value of T_c depends on the hydrophilic-lipophilic balance of the amphiphile. The value of c_c is usually low (below 5%) except for short-chain amphiphiles. It has been found that B shows a power-law divergence in all systems, in agreement with the behaviour expected for a critical binary mixture. The measured exponent is however larger than predicted by existing theories. In the case of the system $C_{12}E_6$ -H_2O which was studied in a wide temperature region the data are consistent with the hypothesis of a moderate micellar growth.

There have been previous EB studies of solutions of nonionic amphiphiles. Hoffmann et al.[19] have studied solutions of C_8E_4, $C_{10}E_4$ and $C_{12}E_4$ at low concentration below T_c. They find no electric birefringence for C_8E_4 and $C_{10}E_4$, and some for $C_{12}E_4$ solutions. Neeson et al.[20] have studied $C_{12}E_6$ and $C_{12}E_8$ solutions at high concentration near the isotropic-hexagonal phase boundary. They attribute the observed electric birefringence to the existence of nonspherical micelles which grow in size near the boundary of the liquid crystalline phase. In the experiment of Ref.21 the Kerr coefficient of $C_{12}E_6$ and $C_{12}E_8$ solutions at the critical concentration was measured as a function of the temperature T as T approaches T_c. It was found for both systems that B grows considerably as the temperature distance T_c -T is reduced. The experimental data are consistent with a power-law behaviour of the type $B \sim (T_c - T)^{-\Psi}$. It was suggested in Ref.21 that the effects observed near T_c are connected with the existence of critical concentration fluctuations and do not imply the formation of very elongated micelles.

The used nonionic amphiphiles are high-purity products prepared by the group of Dr. Platone (Eniricerche, S.Donato, Milano, Italy). In order to lower the sample conductivity, the compounds were furtherly purified by means of repeated extraction with organic solvents to reduce the amount of residual ionic impurities. Typically the resistivity of our samples was 300 kΩcm at 20°C.

The sample volume was chosen to be rather large (about 20 cm^3) in order to reduce the degradation effects due to the metallic electrodes. It is known that such a degradation can become appreciable when the sample is kept for many hours in the cell at high temperature. The cell temperature was controlled within 0.01°C. Voltage pulses had height of 0.3-1 kV and duration of 10-300 μs. The transition temperature of the solution is carefully determined in the cell itself by monitoring the cell turbidity as the temperature is increased in steps of 0.01°C, allowing to the sample a sufficiently long equilibration time between one step and another.

B was measured as function of the temperature distance from the critical point for aqueous solutions of C_6E_3, C_8E_4, $C_{10}E_5$, and $C_{12}E_6$ (the latter in both H_2O and D_2O), all prepared at the critical concentration [4]. We always found positive values for B. All the investigated solutions present, far from the cloud point, a very low Kerr coefficient, comparable to the value expected for pure water, $B = 3\times10^{-14}$ m/V^2. All systems show a considerable increase of B as the temperature is raised toward the critical point. In a double logarithmic plot B is found to depend linearly on the temperature distance T_c -T when the data are taken sufficiently close to T_c, as shown in Fig.7 for the system C_6E_3-H_2O. Far from T_c the behaviour of B may be more complex as shown in Fig.8 for the system $C_{12}E_6$ - H_2O which was studied in a wide range of temperatures (8-50°C).

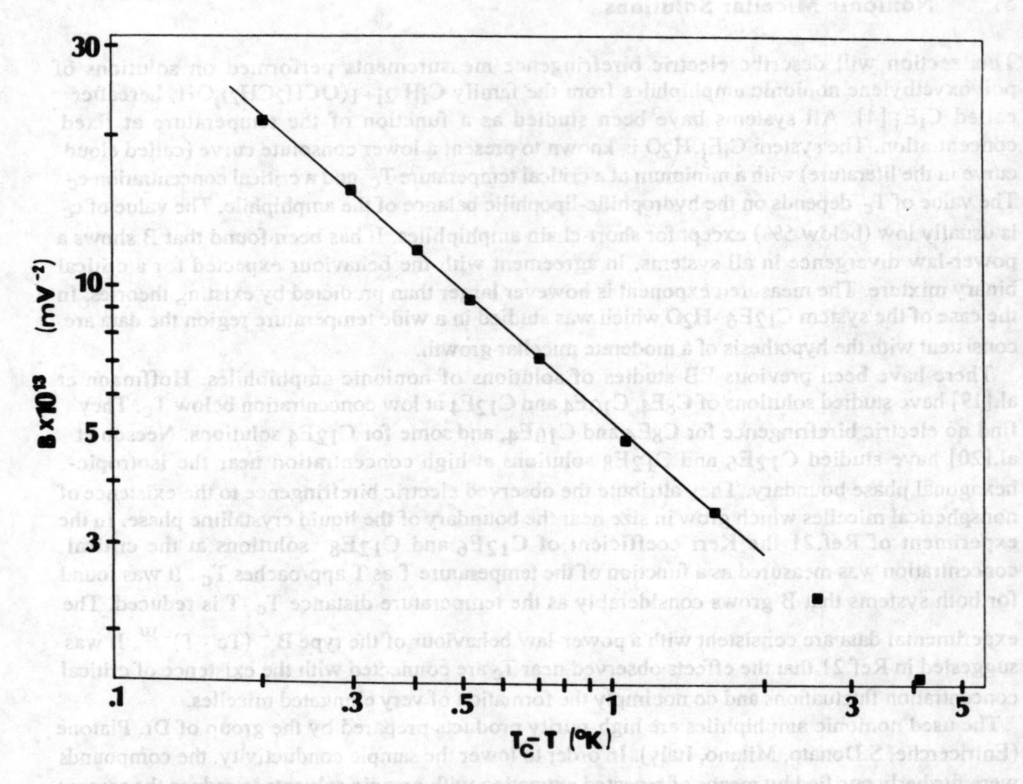

Fig.7. Kerr constant vs. the temperature distance from the critical point for a 13% solution of C_6E_3 in H_2O (T_c = 46.05°C)

Electric birefringence data on colloidal solutions can be simply interpreted in terms of the optical anisotropy of the individual colloidal particles only if the solution is sufficiently dilute to allow the neglection of interparticle interactions. It is important to stress that interactions may lead to a Kerr effect even if the colloidal particles are spherical. A clear example of this effect is represented by the experiments performed on microemulsions consisting of a dispersion of water-in-oil droplets [22]. In such a system electric birefringence may arise because the attractive interactions among the droplets generate statistical clusters which have a nonspherical shape. Of course, this effect

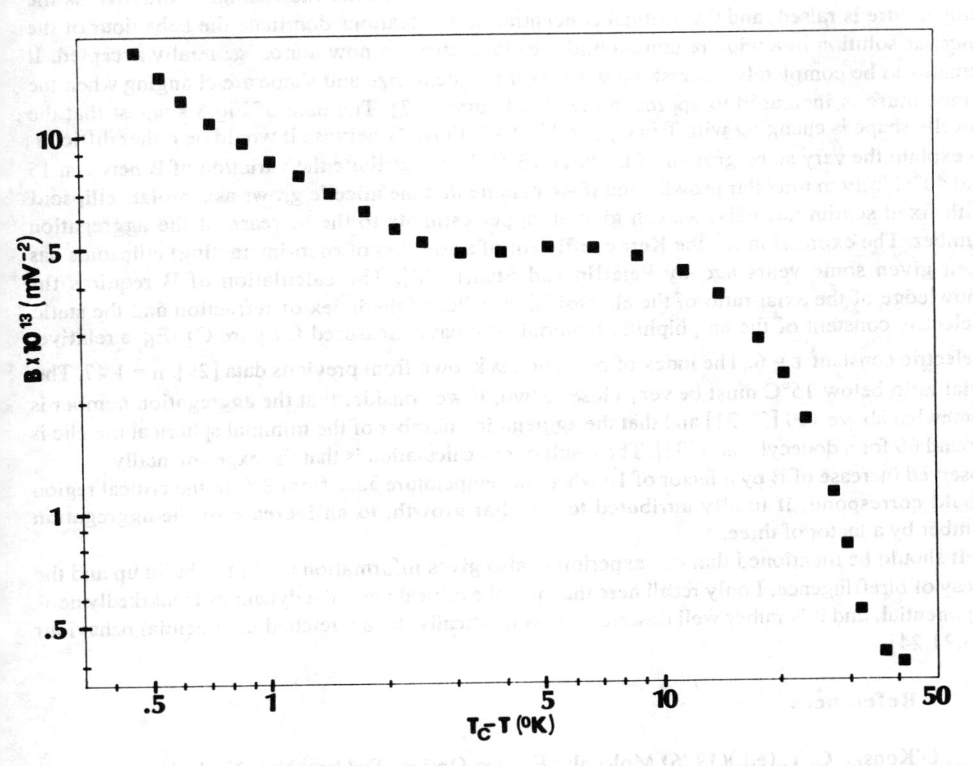

Fig.8. Kerr constant vs. the temperature distance from the critical point for a 2.2% solution of $C_{12}E_6$ in H_2O ($T_c = 51.49°C$)

becomes particularly significant near phase transition lines, such as the percolation threshold. A case which is particularly interesting for our discussion is that of a binary liquid mixture near a critical consolute point. It has been shown indeed that B may become very large in the critical region, even for liquid mixtures which present negligible electric birefringence far from the critical point [23,16,24]. The origin of the effect observed in critical systems is not in the anisotropy of individual molecules but is rather in the anisotropy of the spontaneous concentration fluctuations which are oriented (or deformed?) by the applied electric field. Theory [25] predicts that B should grow as the critical point is approached, following a power-law behaviour as a function of the reduced temperature with a critical exponent that is close to the one associated with the divergence of the correlation range ($\Psi \approx 0.6$).

The value of Ψ for the system C_6E_3 -H_2O, as derived from the data of Fig.7, is 0.88 ± 0.04. A similar value of the critical exponent is found for the other critical micellar solutions [4], and also for critical mixtures of non-associating liquids [16,24].

Aqueous solutions of nonionic amphiphiles have been studied in the last few years by a variety of techniques. The point of view [26] that the cloud curve is a consolute curve at which the micellar solution separates into two micellar solutions with distinct concentrations, that the phase separation is driven by intermicellar interactions which become increasingly attractive as the temperature is raised, and that critical concentration fluctuations dominate the behaviour of the micellar solution in a wide region around the cloud curve is now almost generally accepted. It remains to be completely understood whether the micelle size and shape are changing when the temperature is increased to approach the cloud curve [27]. The data of Fig.8 suggest that the micelle shape is changing with T in $C_{12}E_6$-H_2O solutions. Otherwise it would be rather difficult to explain the very steep growth of B above 15°C. If we attribute the variation of B between 15 and 40°C fully to micellar growth, and if we assume that the micelle grows as a prolate ellipsoid with fixed semiminor axis, we can give an upper estimate to the increase of the aggregation number. The expression for the Kerr coefficient of a solution of (non-interacting) ellipsoids has been given some years ago by Peterlin and Stuart [28]. The calculation of B requires the knowledge of the axial ratio of the ellipsoidal micelle, of the index of refraction and the static dielectric constant of the amphiphilic material. We have measured for pure $C_{12}E_6$ a relative dielectric constant $\varepsilon = 6$. The index of refraction is known from previous data [29], $n = 1.47$. The axial ratio below 15°C must be very close to two, if we consider that the aggregation number is somewhat above 100 [30,31] and that the aggregation number of the minimal spherical micelle is around 60 for a dodecyl chain [32]. The result of our calculation is that the experimentally observed increase of B by a factor of 10 when the temperature goes from 8°C to the critical region would correspond, if totally attributed to micellar growth, to an increase of the aggregation number by a factor of three.

It should be mentioned that our experiment also gives information about the build up and the decay of birefringence. I only recall here that near the critical point the dynamics is markedly non-exponential, and it is rather well described, asymptotically, by a stretched-exponential behaviour [16,21,24].

6. References

1. O'Konski C. T. (ed.)(1976),Molecular Electro-Optics, Dekker, New York; Jennings B. R.(ed.) (1979), Electro-Optics and Dielectrics of Macromolecules and Colloids, Plenum, New York.

2. Bellini T., Piazza R., Sozzi C. and Degiorgio V. (1988), Europhys. Lett. 7, 561.

3. Bellini T., Mantegazza F., Piazza R. and Degiorgio V. (1989), Europhys. Lett. 10, 499.

4. Degiorgio V. and Piazza R. (1987), Progr. Colloid Polym. Sci. 73, 76.

5. Fredericq E.and Houssier C. (1973), Electric Dichroism and Electric Birefringence, Clarendon Press, Oxford; see also: Eden D.and Elias J.G.(1983), in Dahneke B. (ed.), Measurement of Suspended Particles by Quasi-Elastic Light Scattering , J. Wiley, New York, p.401.

6. Piazza R., Degiorgio V. and Bellini T.(1986), J. Opt. Soc. Am. B3,1642; Opt. Commun. 58,400.

7. Ottewill R.H. and Rance D.G.(1986), Coll. Polymer Sci. 264, 982

8. Visca M., unpublished.

9. O'Konski C.T. and Haltner A.J.(1957), J. Am. Chem. Soc. 79, 5634.

10. Wijnenga S.S., van der Touw F. and Mandel M.(1985), Polym. Commun. 26, 172 .

11. O'Konski C.T. and Krause S. (1970), J. Phys. Chem. 74, 3243

12. Jennings B.R. and Oakley D.M.(1982), Appl. Opt. 21, 1519.

13. Pizzoferrato R., Marinelli M., Zammit U., Scudieri F., Martellucci S. and Romagnoli M. (1988), Optics Commun. 68, 231.

14. Rehage H. and Hoffmann H. (1988), J. Phys. Chem. 92, 4712 .

15. Hoffmann H., Platz G., Rehage H. and Schorr W.(1982), Adv. Colloid Interface Sci. 17, 275 ; Hoffmann H. (1985), in Degiorgio V. and Corti M. (Eds.),Physics of Amphiphiles: Micelles, Vesicles, and Microemulsions, North-Holland, Amsterdam, p.160.

16. Piazza R., Bellini T., Degiorgio V., Goldstein R.E., Leibler S. and Lipowsky R. (1988), Phys. Rev.B 38, 7223 .

17. Benedek G.B. (1985),in: Degiorgio V. and Corti M. (eds.), Physics of Amphiphiles: Micelles, Vesicles and Microemulsions, North-Holland, Amsterdam , p.223.

18. For a discussion of flexibility of rod-like micelles see: Porte G. and Appell J. (1981), J. Phys.Chem.85, 2511 .

19. Hoffmann H., Kielman H. S., Pavlovic D., Platz G. and Ulbricht W.(1981), J. Colloid Interface Sci. 80, 237 .

20. Neeson P. G., Jennings B. R. and Tiddy G. J. T.(1983), Faraday Disc. Chem. Soc. 76, 353 .

21. Degiorgio V. and Piazza R.(1985), Phys. Rev. Lett. 55, 288 .

22. Guering P. and Cazabat A. M.(1983), J. Phys. Lett. (Paris) 44, 601 ; Eicke H-F., Hilfiker R. and Thomas H.(1985), Chem. Phys. Lett. 120, 272 .

23. Pyzuk W.(1980), Chem. Phys. 50, 281; Pyzuk W., Majgier-Baranowska H. and Ziolo J. (1981),Chem. Phys. 59, 111.

24. Bellini T. and Degiorgio V.(1989), Phys. Rev. B 39, 7263.

25. Goulon J., Greffe J. L. and Oxtoby D. W.(1979), J. Chem. Phys. 70, 4742 ; Hoye J. S. and Stell G.(1984), J. Chem. Phys. 81, 3200 .

26. Degiorgio V. (1985) in: Degiorgio V. and Corti M. (eds.), Physics of Amphiphiles: Micelles, Vesicles and Microemulsions , North-Holland, Amsterdam , p.303.

27. There is a very wide literature about this point: see, besides Ref.26, the Proceedings of the Conferences "Surfactants in Solution" edited by Mittal and Lindman, and by Mittal and Bothorel.

28. Peterlin V. and Stuart H. A.(1939), Z. Physik 112, 129 ; see also O'Konski C. T.and Krause S. (1976), in: O'Konski C. T.(ed.),Molecular Electro-Optics , Dekker, New York, p.63.

29. Corti M. and Degiorgio V.(1981), J. Phys. Chem. 85, 1442 ; Corti M., Minero C. and Degiorgio V. (1984), J. Phys. Chem. 88, 309.

30. Zulauf M., Weckström K., Hayter J. B., Degiorgio V.and Corti M.(1985), J. Phys. Chem. 89, 3411.

31. Zana R. and Weill C.(1985), J. Phys. Lett. (Paris) 46, 953 .

32. Israelachvili J. N., Mitchell D. J. and Ninham B. W.(1976), J. Chem. Soc. Faraday Trans. II 72, 1425.

POLYMERS AT INTERFACES: STATICS, DYNAMICS AND EFFECTS ON COLLOIDAL STABILITY

M.A. COHEN STUART,
Department of Physical and Colloid Chemistry,
Dreijenplein 6,
6703 HB Wageningen,
The Netherlands.

ABSTRACT. This contribution is a short resumé of the lectures given in the ASI course. It was felt by the author that the interested reader would be better served with such a resumé and a list of central references giving access to the field, than with a lengthy tutorial text.

It has been known for a long time that polymers may adsorb on the particles of a colloidal dispersion and, if they do so, affect the stability of the dispersion. We will discuss here **physisorption** (i.e. attachment by weak forces, no chemical bonds) of polymers. Accounts of chemisorption and its effects can be found in a monograph by Napper [1].

The forces generated by the adsorbed polymers can be either **attractive**, namely when the chains form bridges between two neighbouring particles, or **repulsive** when the polymer forms a protective sheath around the particles which precludes any bridge formation [2].

1 Statics

In order to develop an understanding of these forces, adsorption as such was first studied empirically. Already as early as 1951 [3] a qualitative picture was proposed for the structure of an adsorbed chain molecule in terms of **trains** (segments in contact with the substrate), **loops** (subchains starting from the surface and returning there) and tails (freely dangling chain ends). Attempts were also made to put the theory of polymer adsorption on a quantitative footing [4,5] but the problem of many interacting chains accumulated at a surface at first seemed formidable and hardly tractable. However, during the seventies much progress was made [6,7] and a very versatile lattice model was developed by Scheutjens and Fleer [8]. This model is already explained in some detail in the contribution by Lyklema (see this volume). Here we emphasize only that a central assumption in the theory (apart from it being a lattice model) is that the segment density is only a function of the distance normal to the interface; in directions parallel to the interface the density is homogeneous. This 'mean field' assumption has been criticized by polymer physicists, notably by de Gennes who developed an alternative scaling theory [9] but for most experimental systems the assumption is very good [10]. In a recent review, [11] the predictions of the Scheutjens Fleer theory have been compared with a vast amount of experimental data, selected from careful studies on well-defined systems. The main conclusions for homopolymers (adsorpting from dilute solution) are :

D. M. Bloor and E. Wyn-Jones (eds.),
The Structure, Dynamics and Equilibrium Properties of Colloidal Systems, 613–617.

(1) for not too small segmental binding energy the adsorption isotherm has a very steep initial rise and a well-defined plateau; ('high affinity' isotherm).

(2) the plateau adsorption increases with chain length; in good solvents the increase levels off at high chain length, in θ-solvents the plateau adsorbed amount becomes proportional to log M.

(3) the plateau adsorption increases with increasing segmental binding strength but levels off when this energy exceeds 3 kT. A sharp transition from adsorption to desorption occurs when the binding strength is outweighed by the elastic (conformational) free energy [11,12].

(4) The segment density falls monotonously with distance from the surface showing a dense contribution at short distance due to loops and a more dilute contribution but of larger extension due to tails; for long chains, the mass adsorbed in tails may be 10-15% of the total adsorbed mass. Very long tails are possible.

(5) The thickness of the adsorbed layer is largely a function of the adsorbed mass; unsaturated layers are thin although near saturation the thickness increases very steeply, especially in good solvents. When the thickness is measured by a hydrodynamic method (tangential solvent flow), the effect is largely due to tails.

Essentially all these predictions were rather well corroborated by experiment. In addition, mixtures of polymers with different chain lengths (polydisperse polymers) were studied both theoretically [13] and experimentally [11,14] and the important consequences of preference and exchange were laid bare.

The Scheutjens-Fleer theory is a numerical theory. The quest for analytical expressions was also pursued, but at first the approximations made were too crude. The original recursion formula can be rewritten into a second order differential equation, which was further simplified by assuming (i) low densities, (ii) very ('infinitely') long chains and (iii) a negligible contribution of tails [15]. Especially the latter approximation is a poor one : two eigenfunctions are important, one of which gives rise to tails [16]. A more recent calculation takes this into account [17].

Scheutjens and Fleer went on to apply their formalism to polymer between two surfaces, thus describing the situation of two particles interacting through their polymer layers. They considered two cases : (i) all molecules (solvent and polymer) can leave the gap between the two walls when these are moved together; there is, so to say, an infinite reservoir of molecules and the chemical potentials of all components remain fixed during the approach. (ii) only solvent can leave the gap; the amount of polymer between the plates remains constant.

For case (i) there is attraction at all distances; this does certainly not correspond to experimental results where both attraction and stability can be found. Therefore this was considered an irrelevant case.

Case (ii) is interesting. Here one finds monotonously repulsive situations when the adsorbed amount is high, but when the surface is not saturated with polymer an attractive minimum develops, which first deepens with decreasing adsorption and then disappears again when the surfaces become bare. This is the case of **bridging attraction** which is obviously most effective when the adsorbed amount is somewhat below saturation ('starved layer') [15,18].

This finding is in good agreement with experimental data which show optimum flocculation at some intermediate polymer dose [2]. It seems that the situation is simple and well understood : dispersed particles are repulsive or attractive depending on polymer dose; since an equilibrium (thermodynamic) theory is capable to describe it we need not consider temporal effects.
However, reality appeared to be more complex.

2 Dynamics

Under the apparent agreement between theory and experiment (as far as the effect of the polymer dose on flocculation is concerned) lies a little paradox. Most flocculation experiments are conducted with dispersions which are initially stable due to electrostatic repulsion. The particles in such dispersions have little tendency to approach each other closely. However, the attraction due to polymer bridging develops only at very short distances since unsaturated polymer layers tend to be thin. We can only solve this paradox if we assume that the polymer molecules must be in an extended state in order to initiate bridging. But since the extended state cannot persist for starved polymer layers, temporal effects must come into play.

Such effects were showing up soon in an interesting study by Pelssers [19]. For permanently attractive particles (such as particles subjected to Van der Waals forces and little or no electrical repulsion to counteract this) one expects aggregation to occur as described by the Von Smoluchowsky theory (or a refinement thereof). According to this theory, the number of singlets, N_1, decreases in time(t) with an inverse square law :

$$N_1(t) = N_0 (1 + k_f N_0 t)^{-2}$$

where N_0 is the initial particle concentration, and k_f the flocculation rate constant.

Pelssers carried out studies with polystyrene latex particles which were flocculated with polyethylene oxide (PEO). He measured N_1 as a function of both time and particle concentration, at the optimum PEO dose. Much to his surprise, the singlet number did **not** decrease monotonously in time but came to a complete standstill after some initial rapid drop. Also the initial rapid drop was not a smooth function of N_0. Rather, there was almost no flocculation below a certain threshold value, after which N_1 dropped steeply to almost zero. In addition, it was found that a certain minimum shear rate was necessary to bring about flocculation [19].

The scenario which finally proved useful in understanding the results is the following. The polymer adsorption process has three steps : (1) mass transfer in the bulk solution (2) attachment (3) conformational relaxation. Step (1) can be conveniently described by classical expressions for collision rates in (sheared) dispersions, and step (2) is, for uncharged flexible polymers too rapid to be rate determining. Therefore, the number of attached molecules per colloidal particle follows a first order rate law. In order to form bridges, these attached molecules must extend far enough to reach the surface of a second particle. However, the electrostatic repulsion between the particles maintains a minimum distance of approach which is still fairly large. Therefore it is conceivable that freshly attached, extended molecules can form a bridge, but relaxed, flattened ones cannot. Hence, it is the number of extended ('active') molecules per particle which counts, and this number will **increase** due to attachment, but **decrease** due to unfolding. Assuming the unfolding rate to be first order (too), one finds that the number of active molecules goes through a maximum depending on concentrations and rate constants.

The idea is now that a certain minimum number of active polymer molecules per particle is needed before bridging starts. If unfolding is too rapid in comparison with attachment, no flocs are formed. If the attachment rate is enhanced, e.g. by shearing or by increasing the particle and polymer concentrations, bridging may be induced. The schedule in fig. 1 illustrates this scenario. A quantitative model was developed which was found to describe the data very well [20].

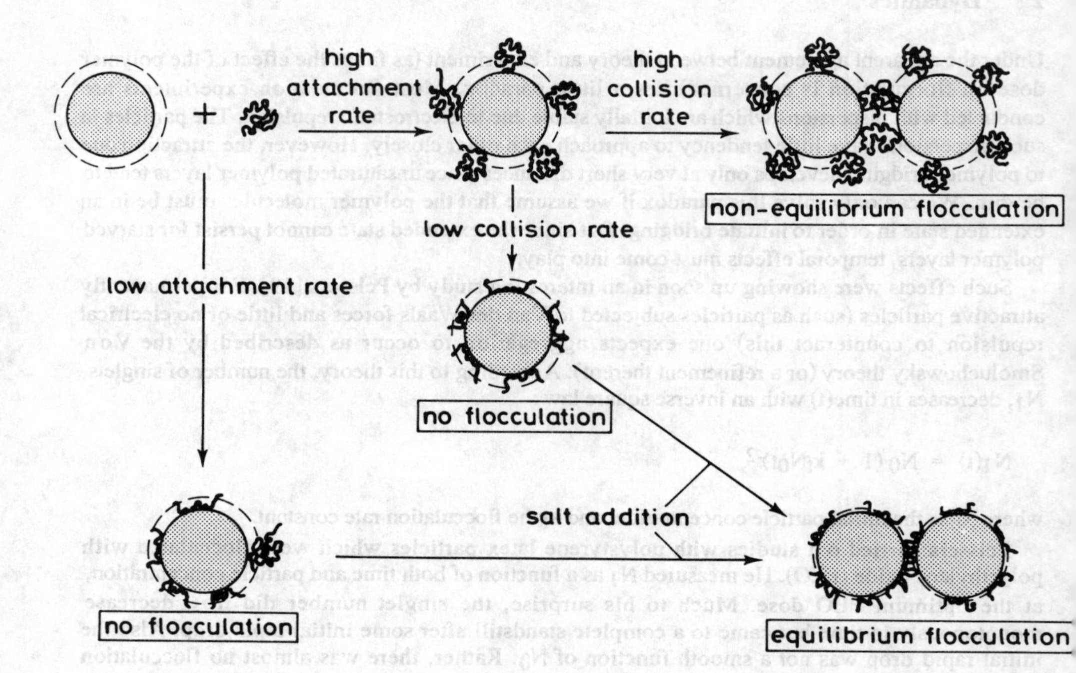

Figure 1. Pathways in bridging flocculation

References

1. D.H. Napper. Polymeric Stabilisation of Colloidal Dispersions, AC Press, London (1983)
2. B. Vincent, Adv. Colloid Interface Sci. 4, 193 (1974).
3. E. Jenkel, B.Rumbach, Z. Electrochem. 55, 612 (1951).
4. C.A.J. Hoeve, J. Chem. Phys. 42, 2558 (1965); J. Chem. Phys. 43, 3007 (1965); J. Chem. Phys. 44, 1505 (1966).
5. A. Silberberg, J. Chem. Phys. 46, 1105 (1967); J. Chem. Phys. 48, 2835 (1968).
6. E.A. DiMarzio, R.J. Rubin, J. Chem. Phys. 55, 4318 (1971).
7. R.J. Roe, J. Chem. Phys. 60, 4192 (1974).
8. J.M.H.M. Scheutjens, G.J. Fleer, J. Phys. Chem. 83, 1619 (1979); J. Phys. Chem. 84, 178 (1980).
9. P.G. de Gennes, Macromolecules 14, 1637 (1981).
10. G.J. Fleer, J.M.H.M. Scheutjens, M.A. Cohen Stuart, Colloids Surfaces 31, 1 (1988).
11. M.A. Cohen Stuart, T. Cosgrove, B. Vincent, Adv. Colloid Interface Sci. 24, 143 (1986).
12. M.A. Cohen Stuart, G.J. Fleer, J.M.H.M. Scheutjens, J. Colloid Interface Sci. 97, 515 (1984); J. Colloid Interface Sci. 97, 526 (1984).
13. J.M.H.M. Scheutjens, G.J. Fleer in 'The Effect of Polymers on Dispersion Properties',

Tadros T.F., Ed. Ac. Press, London (1982), 145.

14. M.A. Cohen Stuart, J.M.H.M. Scheutjens, G.J. Fleer, J. Pol. Sci. Pol. Phys. Ed. **18**, 559 (1980).
15. M.A. Cohen Stuart, in 'Future Directions in Polymer Colloids', M.S. El-Aasser, R. Fitch, Eds., NATO ASI Series, E 138, p. 229 (1987).
16. J.M.H.M. Scheutjens, G.J. Fleer, M.A. Cohen Stuart, Colloids Surfaces **21**, 285 (1986).
17. H.J. Ploehn, W.B. Russel, Macromolecules **22**, 266 (1989).
18. J.M.H.M. Scheutjens, G.J. Fleer, Macromolecules **18**, 1882 (1985).
19. E.G.M. Pelssers, M.A. Cohen Stuart, G.J. Fleer, Colloids Surfaces **38**, 15 (1989).
20. E.G.M. Pelssers, M.A. Cohen Stuart, G.J. Fleer, J. Chem. Soc. Faraday Transaction I (1990, in press).

EQUILIBRIUM STRUCTURE AND PROPERTIES OF COLLOIDAL DISPERSIONS

R. RAJAGOPALAN
Department of Chemical Engineering
University of Houston
Houston, Texas 77204-4792, USA

C. S. HIRTZEL
Department of Chemical Engineering and Materials Science
Syracuse University
Syracuse, New York 13244-1190, USA

ABSTRACT. Tailored colloidal dispersions serve as excellent model many-body systems and, therefore, offer an attractive opportunity to study a number of problems in condensed matter physics, including phase transitions and critical phenomena in systems interacting through soft, long-ranged potentials, glass transitions in atomic systems, and two-dimensional phase transitions. Consequently, self-organization in colloidal dispersions has attracted the attention of chemists and physicists in recent years. Here, we present an introduction to order/disorder transitions in model colloidal dispersions. After a brief review of the physical basis of self-organization and crystallization in colloids, some recent attempts to predict the equilibrium structure and properties of dispersions using effective hard-sphere models, statistical mechanical perturbation theories and integral equation approaches are reviewed, with emphasis on the last method in view of the popularity of integral equation methods in analyzing the structure of globular proteins and micelles.

1. Introduction

Traditionally, one of the major thrusts of classical colloid science has been the study of the surface chemistry and its effect on the stability of dispersions against coagulation (Verwey and Overbeek 1948). Since the colloidal particles in most cases carry charges and interact over relatively large distances, the chemical aspects of stability were understandably the first to attract attention and were, for a long time, considered the most immediate for a proper understanding of colloidal phenomena. Moreover, until recently, much of the literature in this area has been concerned with interaction effects in *dilute* dispersions. The term '*dilute dispersion*' in this context stands for dispersions in which the phenomenon of interest is almost exclusively influenced by interactions between at most two particles at a time. This restriction, however, seldom holds in dispersions of practical interest, either because the concentrations of the particles in the dispersions are large or because the Debye screening length is of the order of the average interparticle separation or larger (Hirtzel and Rajagopalan 1985). A striking example of this can be seen in the solid-like properties

D. M. Bloor and E. Wyn-Jones (eds.),
The Structure, Dynamics and Equilibrium Properties of Colloidal Systems, 619–633.

displayed by turnip yellow mosaic virus solutions at very low concentrations (at volume fractions of the order of 0.01) at room temperatures (Hastings 1978).

The spatial ordering and the topological disorder that result from the many-body interactions and from the incessant Brownian motion of the colloidal particles imply that the task of relating macroscopic properties to microscopic forces and motion must take a statistical route. In fact, the microstructural details of colloidal dispersions, both from equilibrium and nonequilibrium points of view, resemble those of molecular and atomic fluids, and it is often profitable, conceptually as well as analytically, to treat colloidal dispersions as macroscopic analogues of simple atomic fluids. Thus, in the case of interacting colloidal dispersions, micellar solutions and microemulsions, the term '*supramolecular fluids*' conveniently defines both the conceptual basis and the methods of analysis needed for understanding the hierarchy of events that contribute to the observed macroscale behavior.

The purpose of this chapter is to provide a broad overview of the equilibrium structure and properties of colloidal dispersions. We have taken a tutorial approach in presenting this overview and, in addition, have kept the theoretical and mathematical details to a minimum since they may be obtained from the numerous references cited throughout the text. For the same reason, we have also avoided reproducing figures and tabulations of data here. In order to keep the general concepts tractable, the discussion here will focus primarily on monodispersed (i.e., one-component) systems. Some of the recent work on interactions, phase transitions, and critical phenomena in micellar solutions and microemulsions are reported in Chen and Rajagopalan (1990); this compendium also has a review of Monte Carlo and Brownian dynamics computer experiments applied to colloidal dispersions.

2. Many-Body Interactions in Dispersed Systems

The influence of many-body interactions on the formation of local structure in dispersions has been recognized for a long time, although systematic attempts to relate such structures (and structural transitions) to interaction forces are of recent origin. [One exception in this regard is the investigation of Kirkwood and Mazur (1952), which traces the "liquid-like" short-range order observed in charged protein suspensions to colloidal interaction effects using a statistical mechanical integral equation technique; see Section 3.3.] We shall not attempt to provide a comprehensive historical account of such studies, experimental or theoretical, in this section; instead, our objective here will be to merely illustrate, using some of the experimental studies reported in the literature, some examples of short-range and long-range ordering in colloidal dispersions and the implications of such ordering to the macroscopic properties and phenomena.

2.1. SPATIAL ORDERING IN COLLOIDS AND SOME IMPLICATIONS

A historical review of the observation of ordering in colloidal dispersions (along with a phenomenological, albeit now outdated, interpretation) has been presented by Efremov (1976). The iridescence caused by spatial ordering in latex dispersions of monodispersed particles is now well known (for some early reports, see Hiltner and Krieger 1969, Vanderhoff *et al.* 1970, and Krieger and Hiltner 1971); a result of the Bragg scattering caused by crystal-like ordering of the particles, this has been noted and studied by numerous investigators since then. Despite the fact that this interest in periodic colloid structures may seem to be of relatively recent origin, the formation of order in colloidal systems has been recorded and studied for over fifty years. For instance, Langmuir (1938) attributed the unusual mechanical properties of bentonite sols and the optical properties of Schiller layers in iron oxide sols to the formation of order in these dispersions of nearly-monosized particles. The biological literature offers a fascinating variety of ordered structures of supramolecular dimensions, including the body-centered cubic structures observed in bushy stunt virus (see Bernal and Fankuchen 1941) and the face-centered cubic structures observed in solutions of tipula virus (see Williams and Smith 1957; because of the opal-like iridescence of the resulting crystals, this virus is often called tipula iridescent virus). An extensive review of experimental observations on ordering in ionic micelles, macroions, proteins and virus particles is also available in Ise and Okubo (1980). [Of these, ordering in latex dispersions has been a popular example since it leads to visually striking results (because of the iridescence, which can be observed with naked eyes) and since the ordered structures themselves can be seen directly through microscopes. Phase diagrams of crystallization in latex dispersions have been constructed by Hachisu, Kobayashi and Kose (1973) through direct optical observations and have been the subject of considerable study since then; see, for example, Hirtzel and Rajagopalan 1985; Safran and Clark 1987; and Ise and Sogami 1988.]

Finally, it is important to note that, in addition to their importance in colloid science, order/disorder transitions in colloids offer an excellent opportunity for studying some of the fundamental problems in condensed matter physics at convenient length and time scales; for instance, shear-induced phase transitions, ordering in two-dimensions, influence of defects and grain boundaries in melting of solids, phase transitions in fluids interacting through 'soft' long-range interactions, and the dynamics of glass transitions can be studied conveniently by using tailored colloids as model many-body systems; see Section 4.

The equilibrium thermodynamics of dispersions has a direct influence on some of the dynamic properties as well. The (semi-empirical) description advanced by Hastings (1978) to explain the observed yield stresses of low density virus solids offers an instructive illustration of the effects of equilibrium structural transitions on transport properties. Following an idea originally used by Williams, Crandall and Wojtowics (1976) to estimate the heat of melting and the entropy change during the melting of crystals of polystyrene spheres, Hastings relates the yield stresses of crystalline structures in turnip yellow mosaic virus solutions to the Gibbs free energy change accompanying the shearing of the crystals. The colloidal interaction effects enter these calculations through the assumed (and highly simplified) models for the energies of the ordered and disordered dispersions. Although Hastings's analysis is not free of empirical parameters, the general success of his description attests to the validity of the basic hypothesis that the observed yield stresses must be related to the structural changes and the interaction effects in the virus solution.

With some restrictions, the thermodynamic contributions can be identified separately also in the case of diffusion coefficients. As well known, the Stokes-Einstein equation defines such a relation between the collective diffusion coefficient, D_c, and the osmotic pressure gradient, $(\partial p/\partial \rho)$:

$$D_c = (\partial p/\partial \rho)_T \, [f(\rho) \, . \, (1-\phi)]^{-1} \quad (1)$$

where ϕ is the volume fraction of the particles and $f(\rho)$ the density-dependent friction coefficient. [The factor, $(1-\phi)^{-1}$, appears in Equation (1) so that D_c and $f(\rho)$ are expressed in terms of a volume-fixed frame of reference.] The osmotic pressure is related to the pair interaction potential, $u(r)$, and the equilibrium structure through the virial equation:

$$p = k_B T\rho - (2\pi\rho^2/3) \int_0^\infty r^3(du/dr)\, g(r;\rho)dr \quad (2)$$

where k_B is the Boltzmann constant and T the absolute temperature. (See Section 3 below. The number density ρ is explicitly shown in the notation for the radial distribution function $g(r;\rho)$ to emphasize the dependence of equilibrium structure on concentration.) The diffusion coefficient D_c, which can be measured at the low-scattering-vector limit in, say, quasi-elastic light-scattering spectroscopy, is a measure of the rate of relaxation of concentration fluctuations and provides the same information obtained in macroscopic gradient diffusion experiments. In view of the relation between the low-scattering limit of the structure factor and the osmotic pressure gradient, D_c can also be written as

$$D_c = [k_B T/S(0)]\,[f(\rho) \,.\, (1 - \phi)]^{-1} \quad (3)$$

In contrast to D_c, the self-diffusion coefficient, D_s, usually measured at large scattering vectors, is affected directly by colloidal interactions only at large times. On the other hand, in the short-time limit, only the hydrodynamic interaction is important, but in this case the effects of colloidal forces enter indirectly through the resulting rearrangement of the local geometric distribution of the particles. That is, the extent of hydrodynamic interactions is influenced by the local equilibrium structure and this influence can be incorporated easily in the determination of D_s by averaging the hydrodynamic mobility of the diffusing particle over the equilibrium distribution (see, for example, Hirtzel and Rajagopalan 1990).

3. Equilibrium Structure and Properties of Interacting Dispersions

3.1 THE REPRESENTATION OF SPATIAL DISTRIBUTION

The most direct piece of information on the equilibrium state (of an isotropic dispersion) that can be obtained theoretically is the radial distribution function, g(r), which is a measure of the local density variation in the dispersion as a function of distance from the center of an arbitrary particle. The radial distribution function is also directly accessible experimentally as structure factor S(q) through suitable radiation scattering techniques. The structure factor can be expressed in the terms of g(r) as follows:

$$S(q) = 1 + (4\pi\rho/q) \int_0^\infty [g(r) - 1]\, r \sin qr\, dr \quad (4)$$

where q is the magnitude of the scattering vector, which is related to the scattering angle θ and wavelength λ (in vacuo) of the radiation used in the measurement through

$$q = (4\pi n/\lambda) \sin(\theta/2) \quad (5)$$

The quantity n in Equation (5) is the refractive index of the medium in the case of light scattering and is replaced by unity in the case of X-ray or neutron scattering techniques. In view of Equation

(4) one can also write g(r) in terms of S(q) by

$$g(r) = 1 + (2r\pi^2\rho)^{-1} \int_0^\infty [S(q) - 1]\, q \sin qr\, dq \tag{6}$$

The radial distribution function, in combination with the pair-potential of interaction, provides the information necessary for obtaining the equilibrium properties of the dispersion such as free energies, chemical potential, osmotic pressures and isothermal compressibilities.

3.2 THE ESSENTIAL PHYSICS: HARD-SPHERE MODELS AND PERTURBATION THEORIES

It is instructive to begin the discussion of spatial distribution of the particles with a consideration of a simple forerunner of more rigourous statistical mechanical approaches. For instance, the simplest possible analysis of phase transitions in colloids is to treat the dispersion as a suspension of hard, mutually-impenetrable spheres. This is of course equivalent to assuming the simplest possible pair interaction between the particles, namely,

$$u(r) = \begin{cases} 0 & r > d \\ \infty & r \leq d \end{cases} \tag{7}$$

where d is the hard-sphere diameter. The adequacy of this approximation and some of the needed (empirical and theoretical) modifications to this approximation have been discussed extensively in the literature. We shall consider this briefly since there is sufficient merit in the approaches based on hard-sphere models under certain conditions (for instance, in the case of some microemulsions and for polymethylmethacrylate spheres in nonpolar liquids such as benzene; see Nieuwenhuis, Pathmamanoharan and Vrij 1981; and Cebula, Myers and Ottewill 1982). More importantly, hard-sphere approximations frequently form a good reference point for subsequent refinements to accommodate softer repulsions or attractive forces.

It was pointed out by Kirkwood (1939) over fifty years ago that a system of particles interacting via hard-sphere potentials would undergo a phase transition from a disordered, fluid-like structure to an ordered, solid-like structure at a volume fraction well below the one corresponding to the closest packing. The molecular dynamics "experiments" of Alder and Wainwright (1957, 1962) eventually established the validity of Kirkwood's prediction, and subsequent statistical mechanical extensions and refinements have placed the range of volume fractions where the two phases coexist at roughly 0.5 to 0.55 (see, for example, Ziman 1979). This hard-sphere phase transition is generally known as the Kirkwood-Alder transition and offers, at least in retrospect, a conceptual basis for understanding phase transitions in colloids. First suggested by Stigter (1954), this correspondence between colloidal phase transitions and Kirkwood-Alder transition was resurrected about twenty years later by Wadati and Toda (1972); see, also, van Megen and Snook (1975a,b). As mentioned earlier, using this correspondence to understand formation of order in colloids is equivalent to assuming the pair interaction between the particles to be of the hard-sphere type -- a simplification that is acceptable in the case of uncharged particles but becomes a rough approximation even in the case of sterically-stabilized particles. The colloidal forces modify the hard-sphere interaction in two ways: first, the repulsive interaction between the overlapping electrical double layers increases the 'effective' diameter of the particles; secondly, since the electrostatic repulsion decays over a finite region and is reduced in intensity by the attractive force, the combined forces soften the interaction potential. Many of the subsequent attempts on casting colloidal order/disorder transformations in terms of Kirkwood-Alder transitions have been concerned with devising simple methods for defining effective hard-sphere diameters in terms of empirical observations of known pair-potentials; see Castillo *et al.* (1984) for references. This approach is, in fact, implicit in the Wadati-Toda proposal, but it is now known that more rigourous corrections to deviations from effective hard-sphere approximations may be necessary in most cases. Similarly, the 'melting' of the ordered states can be often sufficiently accurately approximated using empirical rules such as the Lindemann's criterion (which states that an ordered structure will melt when the root-mean-square displacement of the particles from their lattice sites is a characteristic fraction of lattice spacing); this has, for instance, been used by Forsyth *et al.* (1978), who also used Onsager's classical work on ordering in nematic fluids (Onsager 1949) to explain ordering in tobacco mosaic virus.

Although effective-hard-sphere approximations seem sufficient for predicting the freezing and melting boundaries of the phase diagram, they generally fall far short of the required accuracy for estimating equilibrium properties of fluid-like dispersions. The radial distribution functions of effective-hard-sphere fluids can indeed mimic the general features of the local structure of the actual fluid. Nevertheless, the differences are usually substantial enough to affect the properties calculated from the approximate g(r) and the effective-hard-sphere potential.

A rigorous estimation of two-phase coexistence regions requires the minimization of relevant thermodynamic potentials. This implies that corrections to the hard-sphere approximations mentioned above are necessary, and one usually resorts to statistical mechanical perturbation theories for this purpose. The perturbation approaches are based on the premise that the essential features of the equilibrium structure and properties of dense fluids are determined by the harsh, repulsive core of the pair-potential and that the differences between these and the actual structure or properties can be expanded in terms of a suitable perturbation to the core of the potential. (The core is usually approximated by an effective hard-sphere interaction, and this establishes the conceptual link between perturbation theories and the above-mentioned hard-sphere approximation.) A basic outline of the perturbation theories is discussed in another chapter in this book (in the context of inversion of static structure factor to obtain effective pair-potentials; see Rajagopalan 1990) and therefore we will not address this method in any more detail here. Details of applications of these concepts to colloids may be found in Snook and van Megen (1976), van Megen and Snook (1976), and Castillo *et al.* (1984). The perturbation theories are usually used in combination with similar approximations for the crystalline phase; for the latter, approximations based on so-called 'cell models', usually sufficient near the melting region, are easier to employ, as described in the above references.

3.3. INTEGRAL EQUATION THEORIES

The integral equation methods have been particularly popular in the literature for analyzing static structure factors in micellar solutions and microemulsions, since some of the integral equation formalisms lead to analytical solutions for the structure factor (for screened coulombic potentials). In view of this, we discuss this class of approach in some detail.

The integral equation theories can be classified under two classes depending on the starting point and the method of derivation. One of these starts from the definition of distribution functions in terms of the potential of interaction and, if pair-additivity of interaction potentials is assumed, leads to two equivalent integral equation hierarchies, known as the Born-Green-Yvon hierarchy and the Kirkwood hierarchy, respectively. Although these are exact, they both require "closure" approximations to break the hierarchy; that is, the nth member of the hierarchy leads to the (n+1)-body distribution function in terms of the *(n+2)-body* distribution function, and unless the latter can be expressed in terms of the former through a suitable approximation, the sequence cannot be terminated. The simplest closure device is the Kirkwood superposition (KS) approximation, which equates (in the simplest case) the three-body distribution function to a product of the appropriate collection of two-body distribution functions. Additional details on the physical significance of the KS approximation may be found in Watts (1973).

One of the earliest attempts to apply statistical mechanical theories to dispersions (of proteins or macromolecules, in this instance) seems to have been made by Kirkwood and Mazur (1952), who used the Born-Green-Yvon formulation of the integral equation for g(r) with the KS approximations. If one defines a function $\zeta(r)$ by

$$g(r) = \exp\left[-\beta u(r) + \zeta(r)/r\right] \tag{8}$$

the integral equation can be written as

$$\zeta(r) = \int_0^\infty \left[K(|r-s|) - K(r+s)\right]\left[g(s) - 1\right] s\, ds \tag{9}$$

where the Born-Green kernel K is given by

$$K(t) = \pi\beta\rho \int_t^\infty (s^2 - t^2)\,(du/ds)\, g(s)\, ds \tag{10}$$

Kirkwood and Mazur solved Equations (9) and (10), using an iterative numerical procedure, for a Yukawa potential (i.e., screened-coulombic potential). The resulting radial distribution functions, which can be obtained as functions of particle size, volume fraction, Debye-Hückel length, etc., can be used to demonstrate the onset of liquid-like local ordering and long-range (crystalline) ordering as the volume fraction approaches and exceeds a critical value. These observations are qualitatively correct in general, but are quantitatively restricted to the case of thick double layers because of the form of the pair potential chosen. The above paper of Kirkwood and Mazur is of only historical interest presently since the numerical results presented there are limited. However, Schaefer (1977) has made an attempt to compare some experimental structure factor data (on polystyrene latex dispersions) with the solution of the Born-Green integral equation under the superposition approximation. For classical fluids, comparisons of the results based on the KS-approximation with those of the "exact" Monte Carlo method have demonstrated that the superposition approximation is poor for both hard and soft potentials. However, as noted by Croxton (1974), the Born-Green-Yvon equation in the superposition approximation fails to yield convergent or physically acceptable solutions at densities that correspond to densities at which phase transitions occur. Schaefer, therefore, associates the conditions under which such numerical instabilities occur in the case of colloidal fluids with conditions at which the melting transition may be expected. However, the agreement of the KS-estimates with Schaefer's experimentally-observed melting temperatures is very poor.

An alternative formulation, which is based on writing the *total correlation function*

$$h(r) = g(r) - 1 \tag{11}$$

in terms of a *direct correlation function*, $c(r)$, and an indirect part, is more useful. The indirect part is assumed to result from the influence of particle 1 on particle 3, which in turn affects the position of particle 2, either directly or indirectly through other particles. When this effect is weighted by the density ρ and averaged over all possible positions of particle 3, one arrives at the well-known *Ornstein-Zernike equation:*

$$h(r_{12}) = c(r_{12}) + \rho \int c(r_{13})\, h(r_{23})\, dr_3 \tag{12}$$

where r_{ij} is the distance between particle i and particle j. This equation has been used, particularly for Yukawa fluids (i.e., screened-coulombic fluids), along with the *Percus-Yevick* (PY) and *Hypernetted Chain* (HNC) closure conditions

$$c(r) = g(r)\,\{1 - \exp[\beta u(r)]\} \qquad : \text{PY} \tag{13}$$

$$c(r) = g(r) - 1 - \ln g(r) - \beta u(r) \qquad : \text{HNC} \tag{14}$$

to obtain the structure factors. The Ornstein-Zernike equation has been also solved for the thin double layer case by Keavey and Richmond (1976) under the Percus-Yevick approximation. Schaefer (1977) has in fact used numerical solutions of the Ornstein-Zernike equation (for both PY and HNC closures) to compare the resulting structure factors with his experimental data. The HNC closure scheme seems to lead to better agreement than the PY condition (consistent with what is known in the case of coulombic potentials). However, the differences in $S(q)$ are large enough to cause substantial deviations in the thermodynamic properties that one may like to derive from the computed values of $S(q)$. Since it is difficult to measure the effective surface potential in an interacting dispersion, the computed structure factors may be fitted to experimental data using adjustable parameters such as effective charge or potential of the particles. Although the accuracy of such a procedure has not been examined in detail in the case of PY and HNC closures, this procedure has been used frequently with another closure approximation known as the *mean spherical approximation*(MSA) as discussed below. One difficulty in using the PY and the HNC approximations is that the solutions can be obtained only numerically. An exception in this regard is provided by the work of Hayter and Penfold (1981a), which leads to closed-form *analytical* solutions to the Ornstein-Zernike equation in the MSA for the Yukawa potential. The mean spherical approximation sets the closure relation as follows:

$$c(r) = -\beta u(r) \qquad r > d \tag{15}$$

$$h(r) = -1 \qquad r \leq d \tag{16}$$

The Hayter-Penfold solution is equivalent to the earlier solutions of Waisman (1973) and Pastore *et al.* (1980) and has been used by Hayter and Penfold (1981b) in their investigation of the charge and aggregation numbers of sodium dodecyl sulfate micelles. The above solution places no restriction on the sign of the Yukawa tail in the potential, so that the effect of attractive interactions can also be examined; see Hayter and Zulauf (1982). However, large values of (attractive) contact potentials lead to numerical problems in the evaluation of the closed-form solution. Nevertheless, as long as the contact potentials remain low, the Hayter-Penfold solution provides an opportunity to examine long-range clustering due to the attractive tail and the resulting phase change.

Two limitations of the above solution deserve mention. One of these is related to the form of the potential used and the other to the range of validity of the mean spherical approximation itself.

The first of these refers to the restriction of the solution to the Yukawa potential; i.e., closed-form analytical solutions do not seem possible for non-Yukawa forms of the potentials. It appears that this restriction can be circumvented empirically by equating the actual structure factor to that of a hard-sphere dispersion with the same second virial coefficient (Bendedouch and Chen 1983), at least for dilute systems. The values of S(q) calculated in this manner seem nearly identical to the Hayter-Penfold results for low micellar concentrations. One reason for this equivalence is that the Yukawa potential and the logarithmic form of the Derjaguin-Landau-Verwey-Overbeek potential (Verwey and Overbeek 1948) are very nearly the same for separations larger than the equivalent hard-sphere diameter (see Bendedouch and Chen 1983; additional details on the experimental systems used by Chen and coworkers are given in Bendedouch, Chen and Koehler 1983a,b; Chen and Sheu 1990). In addition, if the general features of more complicated pair-potential functions can be approximated by multiple Yukawa tails, then the essential physics of the interactions can be studied in the MSA for such potentials since analytical solutions can be obtained for multiple Yukawa potentials.

The second limitation of the Hayter-Penfold solution is that the mean spherical approximation yields reliable results only for sufficiently high particle concentrations ($\phi > 0.2$) when the contact potential (known as the coulombic coupling, in the case of screened coulombic potentials) is large. [A lucid explanation of this is given by Hayter (1988); see also Chapter 14 of Hunter (1989).] While this restriction is satisfied in the case of the micellar systems studied by Hayter and Penfold (1981b), dispersions of charge colloidal particles such as the ones used by Brown *et al.* (1975) interact strongly at very low concentrations. Under the latter conditions, the mean spherical approximation can lead to unphysical, negative contact values, g(d), for the radial distribution functions. To avoid this, a *rescaled mean spherical approximation* (RMSA) can be formulated (Hansen and Hayter 1982) following a similar prescription used in the study of one-component plasmas. The RMSA takes advantage of the fact that contact configurations are extremely unlikely in very strongly interacting dispersions; that is, g(r) becomes practically zero for

$$r \sim (3/4\pi\rho)^{1/3} \tag{17}$$

since the effect of the actual hard-core (i.e., r = d) on the structure is overshadowed by the effect of the large magnitude of the Yukawa tail near r = d. Therefore, the actual diameter can be rescaled to an effective diameter such that the interaction potential remains the same beyond the rescaled diameter. As long as the effective contact potential at the rescaled diameter is large, the physics of the problem remains unaffected. Moreover, since the rescaled diameter increases the effective volume fraction, the mean spherical approximation becomes more accurate. The magnitude of scaling is determined by taking the contact value of g(r) in the mean spherical approximation and adjusting the effective diameter such that the new contact value is zero; see Hansen and Hayter (1982) for details. The results of RMSA agree very well with the HNC solution, which is generally more accurate than unscaled MSA for long-range coulombic interactions. In fact, Monte Carlo calculations (Svensson and Jönsson 1983) also agree well with the RMSA results for weakly-charged colloids. Comparison of RMSA results with experimental data of Brown *et al.* (1975) is also favourable (Hansen and Hayter 1982), but the compressibility limit of the structure factor [i.e., S(0)] differs significantly from experimental data (see also Hess and Klein 1983). To what extent this is caused by the polydispersity of the sample used by Brown *et al.* is not known.

The success of MSA and RMSA in the range of parameters where the approximations hold is at least partly attributable to the form of the interaction potential used by Hayter and Penfold (1981a) and Hansen and Hayter (1982). Chen and Sheu (1990) have shown recently that the solution of a contracted version of the multicomponent Ornstein-Zernike equation for an asymmetric electrolyte solution (in the primitive, coulombic model for the charge interactions) leads to a direct correlation function of the Yukawa form for the macroion correlations. Consequently, one would expect the one-component model of macroionic solutions used in the Hayter-Penfold formulation (with the Yukawa form of interaction) to lead to good results for the structure factor, as has been the case. Some additional details and related problems are discussed in Chen and Sheu 1990.

3.4 OTHER APPROACHES

A number other techniques such as molecular dynamics and Brownian dynamics computer experiments (in combination with empirical criteria such as Lindemann criterion for melting and Hansen-Verlet criterion for freezing), density functional theories, and self-consistent phonon theories have also been used in recent years to determine the phase diagrams of electrostatically stabilized dispersions. These fall outside the scope of this chapter, but details on these may be found in Rosenberg and Thirumalai (1986, 1987), Kremer *et al.* (1987), Robbins *et al.* (1988), and Hirtzel and Rajagopalan (1990).

4. Concluding Remarks

The cooperative effects seen in dispersions when the particles are influenced by strong many-body interactions are in many instances straightforward analogues of what are observed in atomic and molecular systems. As a consequence, the interpretation of the equilibrium structure and properties of colloidal systems is made considerably easier by the developments already well-known in solid-state and liquid-state physics. These analogies, however, are much more easily applied in the case of colloids with rigid, spherical particles (with minor deviations in sizes and charges from the ideal monodisperse case). Even in this restricted situation, the available experimental data are limited, and initiation of systematic studies have begun only in the last five years. One may also note here that inclusion of broad polydispersity in sizes and charges in the theoretical analyses poses special challenges. Moreover, the need to include nonspherical and deformable colloidal particles makes the task of predicting the structure and properties from first principles considerably more difficult and challenging in the case of many practical systems.

Despite the above limitations, the study of even the restricted case of spherically symmetric monodispersed colloids (or, colloids with controlled polydispersity) has important applications. As mentioned earlier, such colloidal dispersions can be 'tailored' to mimic atomic systems so that a number of problems in condensed matter physics can be studied at very convenient length and time scales. These include mechanisms of ordering and melting in two-dimensional systems, phase transition and critical phenomena in materials with soft, long-ranged potentials, and glass transition in molecular and supramolecular fluids, among others.

For example, in the case two-dimensional systems, the use of tailored colloids offers a number of advantages that are difficult to achieve or not possible in atomic level experiments (Murray and Van Winkle 1988; Van Winkle and Murray 1988). In the case of the latter, one usually employs rare-gas atoms adsorbed on substrates (such as graphite) which themselves introduce significant modulation of the spatial and orientational order of the two-dimensional adsorbed layer. In contrast, it is possible to use relatively inert and smooth substrates on a colloidal scale if one uses model colloids confined between suitably treated glass plates. (However, if necessary, modulation of any symmetry and interaction strength can also be introduced on the glass plates to study the effects of modulation.)

Since a relatively high level of contrast-matching can be achieved by coating the colloidal particles suitably and by adjusting the chemical composition of the carrier fluid, one now has the opportunity to study interactions in dense systems directly using suitable scattering techniques (Pusey and van Megen 1987; Sirota *et al.* 1989). The opportunity for studying glass transitions using model systems is particularly appealing. The prediction of glass transitions using kinetic and hydrodynamic theories of liquids which incorporate nonlinear feedback mechanisms has received renewed attention in the literature recently (Leutheusser 1984; Bengtzelius *et al.* 1884; Das *et al.* 1985; Das and Mazenko 1986; Kirkpatrick 1985; and Geszti 1983). The glass transition is characterized by sharp divergence of the shear viscosity, vanishing self-diffusion coefficients and diminishing density correlations. Colloidal dispersions pass through similar transitions at high volume fractions (beyond the values at which crystallization occurs), and changes in properties similar to the ones seen in atomic systems occur. Since the time-scale of the glass transition in the case of colloids is in the range of hours, one can study the dynamics much more easily. The structure and properties of colloidal systems thus have a much broader impact than may be realized from the point of view of colloid chemistry alone.

5. Acknowledgments

We would like to thank the National Science Foundation and the Petroleum Research Funds, administered by the American Chemical Society, for partial support.

6. References

Alder, B. J. and Wainwright, T. E. (1957) J. Chem. Phys., 27, 1208.

Alder, B. J. and Wainwright, T. E. (1962) Phys. Rev., 127, 359.

Bendedouch, D., and Chen, S.-H. (1983) J. Phys. Chem., 87, 1653.

Bendedouch, D., Chen, S.-H., and Koehler, W.C. (1983a) J. Phys. Chem., 87, 153.

Bendedouch, D., Chen, S.-H., and Koehler, W.C. (1983b) J. Phys. Chem. 87, 2621.

Bengtzelius, U., Götze, W., and Sjölander, A. (1984) J. Phys. C, 17, 5915.

Bernal, J.D. and Fankuchen, I. (1941) J. Gen. Physiol., 25, 111.

Brown, J.C., Pusey, P.N., Goodwin, J.W. and Ottewill, R.H. (1975) J. Phys. A: Math. Gen., 8, 664.

Castillo, C. A., Rajagopalan, R. and Hirtzel, C. S. (1984) Rev. in Chem. Eng., 2, 237.

Cebula, D.J., Myers, D.Y. and Ottewill, R.H. (1982) Colloid Polym. Sci., 260, 96.

Chen, S.-H. and Rajagopalan, R., Eds. (1990) Micellar Solutions and Microemulsions: Structure, Dynamics and Statistical Thermodynamics, Springer-Verlag, NY.

Chen, S.-H. and Sheu, E. (1990) pp. 1-25 in Micellar Solutions and Microemulsions: Structure, Dynamics and Statistical Thermodynamics, S.-H. Chen and R. Rajagopalan, Eds., Springer-Verlag, New York.

Croxton, C.A., (1974) Liquid State Physics - A Statistical Mechanical Introduction, Cambridge University Pr., Cambridge, U.K.

Das, S. P. and Mazenko, G. F. (1986) Phys. Rev. A, 34, 2265.

Das, S. P., Mazenko, G. F., Ramaswamy, S. and Toner, J. J. (1985), Phys. Rev. Lett., 54, 118.

Efremov, I. F. (1976) pp. 85 in Periodic Colloid Structures, pp. 85-192 in "Surface and Colloid Science", Vol. 8, E. Matijevic, Ed., Wiley-Interscience, New York.

Forsyth, P.A., Marcelja, S., Mitchell, D.J. and Ninham, B.W. (1978) Adv. Colloid Interface Sci., 9, 37.

Geszti, T. (1983) J. Phys. C., 16, 5805.

Hachisu, S., Kobayashi, Y. and Kose, A. (1973) J. Colloid Interface Sci., 42, 342.

Hansen, J. P. and Hayter, J. B. (1982) Mol. Phys., 46, 651.

Hastings, R. (1978) Phys. Lett., 67A, 316.

Hayter, J. B. (1988) pp. 500-511 in Ordering and Organisation in Ionic Solutions, N. Ise and I. Sogami, Eds., World Scientific, Singapore.

Hayter, J. B. and Penfold, J. (1981a) Mol. Phys., 42, 109.

Hayter, J. B. and Penfold, J. (1981b) J. Chem. Soc., Faraday Trans. I, 77, 1851.

Hayter, J. B. and Zulauf, M. (1982) Colloid Polym. Sci., 260, 1023.

Hess, W. and Klein, R. (1983) Adv. in Phys., 32, 173.

Hiltner, P. A. and Krieger, I. M. (1969) J. Phys. Chem., 73, 2386.

Hirtzel, C. S. and Rajagopalan, R. (1985) Colloidal Phenomena: Advanced Topics, Noyes Publ., Park Ridge, New Jersey.

Hirtzel, C. S. and Rajagopalan, R. (1990) pp. 111-142 in Micellar Solutions and Microemulsions: Structure, Dynamics and Statistical Thermodynamics, S.-H. Chen and R. Rajagopalan, Eds., Springer-Verlag, New York.

Hunter, R. J. (1989) Foundations of Colloid Science, Vol. II, Clarendon Pr., Oxford, England.

Ise, N. and Okubo, T. (1980) Acct. Chem. Res., 13, 303.

Ise, N. and Sogami, I., Eds. (1988) Ordering and Organisation in Ionic Solutions, World Scientific, Singapore.

Keavey, R. P. and Richmond, P. (1976) J. Chem. Soc., Faraday Trans.II, 72, 773.

Kirkpatrick, T. R. (1985) Phys. Rev. A, 31, 939.

Kirkwood, J. G. (1939) J. Chem. Phys., 7, 919.

Kirkwood, J. G. and Mazur, J., (1952) J. Polym. Sci., 9, 519.

Kremer, K., Grest, G. R. and Robbins, M. O. (1987) J. Phys. A: Math. Gen., 20, L181.

Krieger, I. M. and Hiltner, P. A. (1971) pp. 63-72 in Polymer Colloids, R.M. Fitch, Ed., Plenum, New York.

Langmuir, I. (1938) J. Chem. Phys., 6, 873.

Leutheusser, E. (1984) Phys. Rev. A, 29, 2765.

Murray, C. A. and Van Winkle, D. H. (1988) Phys. Rev. Lett., 58, 1200.

Nieuwenhuis, E. A., Pathmamanoharan, C. and Vrij, A. (1981) J. Colloid Interface Sci., 81, 196.

Onsager, L. (1949) Ann. N.Y. Acad. Sci., 51, 627.

Pastore, G., Nappi, C., DeAngelis, U. and Forlani, A. (1980) Phys. Lett., 78A, 75.

Rajagopalan, R. (1990), this volume.

Robbins, M. O., Kremer, K. and Grest, G. S. (1988) J. Chem. Phys., 88, 3286.

Rosenberg, R. O. and Thirumalai, D. (1986) Phys. Rev. A, 33, 4473.

Rosenberg, R. O. and Thirumalai, D. (1987) Phys. Rev. A, 36, 5690.

Safran, S. A. and Clark, N. A., Eds. (1987) Physics of Complex and Supermolecular Fluids, Wiley, New York.

Schaefer, D.W. (1977) J. Chem. Phys., 66, 3980.

Snook, I. and van Megen, W. (1976) J. Colloid Interface Sci., 57, 47.

Svensson, B. and Jönsson, B. (1983) Mol. Phys., 50, 489.

Vanderhoff, J. W., van den Hul, H. J., Tausk, R.J., and Overbeek, J.Th.G., (1970) pp. 15-44 in Clean Surfaces, G. Goldfinger, Ed., Marcel Dekker, New York .

van Megen, W. and Snook, I. (1975a) Chem. Phys. Lett., 35, 399.

van Megen, W. and Snook, I. (1975b) J. Colloid Interface Sci., 53, 172.

van Megen, W. and Snook, I. (1976) J. Colloid Interface Sci., 57, 40.

Van Winkle, D. H. and Murray, C. A. (1988) J. Chem. Phys. 89, 3885.

Verwey, E. J. W. and Overbeek, J. Th. G. (1948) Theory of the Stability of Lyophobic Colloids, Elsevier, Amsterdam, The Netherlands.

Waisman, I. (1973) Mol. Phys., 25, 45.

Wadati, M. and Toda, M. (1972) J. Phys. Soc. Japan, 32, 1147.

Watts, R. O. (1973) in A Specialist Periodical Report: Statistical Mechanics, Vol. 1, K. Singer, Ed., The Chemical Society, London, U.K.

Williams, R., Crandall, R. S. and Wojtowics, P. J. (1976) Phys. Rev. Lett., 37, 348.

Williams, R. D. and Smith, K. (1957) Nature, 179, 119.

Ziman, J. M. (1979) Models of Disorder, Cambridge Univ. Pr., Cambridge, U.K.

STRUCTURE, DYNAMICS AND EQUILIBRIUM PROPERTIES OF INORGANIC COLLOIDS

J. D. F. RAMSAY
Chemistry Division
AEA Technology, Harwell Laboratory
Oxfordshire, OX11 0RA, UK

ABSTRACT. Recent applications of neutron scattering techniques in the determination of the structure, dynamics and equilibrium properties of inorganic colloids are described, with particular reference to metal oxide sols and gels and clay colloids dispersed in water. Incoherent quasielastic neutron scattering (IQENS) is particularly valuable in the study of the dynamics and structure of protonated systems such as polynuclear metal cations, which form on hydrolysis of Al(III) and Zr(IV) in aqueous solution. Details of the structure, short-range order and interactions in concentrated oxide sols can be obtained from small angle neutron scattering (SANS). The non-equilibrium structure and alignment of rod-shaped particles under shear can be determined from anisotropic analysis of SANS measurements made in situ, as illustrated with sepiolite clay dispersions.

1. Introduction

Colloidal dispersions of inorganic particles in liquids have applications in a wide range of commercial products such as paints, pigments, cosmetics, pharmaceuticals, printing inks, ceramics, ferro-fluids and paper coatings for example. Inorganic dispersions also occur in the environment as sediments, flocs and suspended colloidal particulates in natural waters [1]. Oxide and clay colloids in aqueous media are of widespread importance. Such systems are charge stabilised and when concentrated may develop an equilibrium structure, due to strong interparticle interactions [2]. Under shear these dispersions frequently exhibit marked non-Newtonian and viscoelastic properties which are related to a reversible disruption of the structure.

In this brief review we will illustrate the interrelationships between the structure, dynamics and equilibrium properties of oxide and clay colloids and highlight the particular versatility of neutron scattering techniques which have recently been used extensively to probe such systems.

2. Techniques for Studying Structure, Interactions and Equilibrium Properties

Some of the techniques which have been used to characterise oxide and clay colloids and the information obtainable are summarised in Table 1. Many of these are familiar techniques which have been described extensively elsewhere [3] and need not be dealt with here. Radiation

D. M. Bloor and E. Wyn-Jones (eds.),
The Structure, Dynamics and Equilibrium Properties of Colloidal Systems, 635–651.

TABLE 1. Techniques for characterisation of structure, interactions and equilibrium properties of inorganic colloids.

	Technique for Study	Information
Colloid Interactions	Rheology; measurements under steady and oscillatory shear (viscoelastic behaviour)	Flow behaviour of dispersions - particle interactions at high volume fraction
	Electrophoresis - electrophorectic light scattering	Electrophoretic mobility - surface charge, colloid stability
	Inelastic and quasi-elastic neutron scattering I.r. spectroscopy neutron diffraction	Properties of interfacial water layers - particle interaction, polynuclear hydroxy ions
Colloid Structure	Static light scattering	Particle size of sols M.W. and size of sol aggregates
	Quasielastic light scattering (photon correlation spectroscopy)	Diffusional and rotational motion of colloid particles (surface/solvent interactions)
	Ultracentrifugation	Sedimentation coefficient
	Small angle X-ray scattering Small angle neutron scattering	Particle size/shape of sol; particle ordering (radial) distribution function); inter-particle forces

scattering techniques (light, X-rays, neutrons) have been employed extensively and more recently neutron scattering has proved very powerful [4].

Neutron scattering can arise through interaction with atomic nuclei [5,6]. The scattering from a single rigidly fixed atom can be defined in terms of its cross section, σ, where $\sigma = 4\pi b^2$. Here b is defined as the scattering length of the bound atom. However, when scattering occurs from matter which is composed of an assembly of non-rigidly bound atoms, there will be two distinct contributions to the total cross section, which arise from coherent and incoherent effects. The first of these, σ_{coh}, results in interference between the neutron waves scattered by the nuclei, and is associated with a coherent cross section given by $\sigma_{coh} = 4\pi b^2_{coh}$. The second, σ_{inc}, is due to interactions between the spin states of the neutron and the nucleus, and gives rise to isotropic scattering; it does not exist for nuclei having zero spin, eg. ^{12}C and ^{16}O.

Values of b_{coh}, σ_{inc}, the neutron absorption cross section, σ_a, and the scattering lengths of X-rays, $10^{12}b_{(X\text{-}ray)}/cm$ for different nuclei, many of which occur in inorganic colloids such as oxides and clays, are given in Table 2.

Table 2. Neutron coherent scattering lengths, b_{coh}, incoherent and absorption cross-sections, σ_{inc} and σ_a, for different elements compared with scattering lengths for X-rays, $b_{(X\text{-}ray,Q=0)}$

	$10^{12}b_{coh}/cm$	$10^{24}\sigma_{inc}/cm^2$	$10^{24}\sigma_a/cm^2$	$10^{12}b_{(X\text{-}ray)}/cm$
H	-0.374	79.7	0.33	0.281
D	0.667	2.0	0.0005	0.281
C	0.665	0.0	0.0035	1.686
N	0.94	0.3	1.9	1.967
O	0.58	0.0	0.00019	2.249
Al	0.35	0.0	0.23	3.654
Si	0.42	0.0	0.17	3.935
Ti	-0.34	3.0	6.1	6.2
Fe	0.95	0.4	2.6	7.308
Zr	0.72	0.3	0.18	11.3
Ce	0.48	0.0	0.63	16.3
Th	0.98	0.0	7.4	25.3
U	0.842	0.0	7.5	25.9

This shows that b_{coh} varies erratically from element to element, and even for different isotopes - a feature which can be exploited in contrast variation studies. Another important feature is the large value of σ_{inc} for the proton, which dominates that for other nuclei; this makes incoherent scattering measurements particularly suitable for the study of hydrogenous materials (eg. water, polymers, etc.) [7-10], in situations where other spectroscopic techniques are unsuited because of absorption problems for example.

One of the most important features of neutron scattering compared to other radiation techniques is its ability to probe both the structure and dynamic properties of materials over a wide spatial range extending from 1 to 10^3 Å as illustrated in Figure 1.

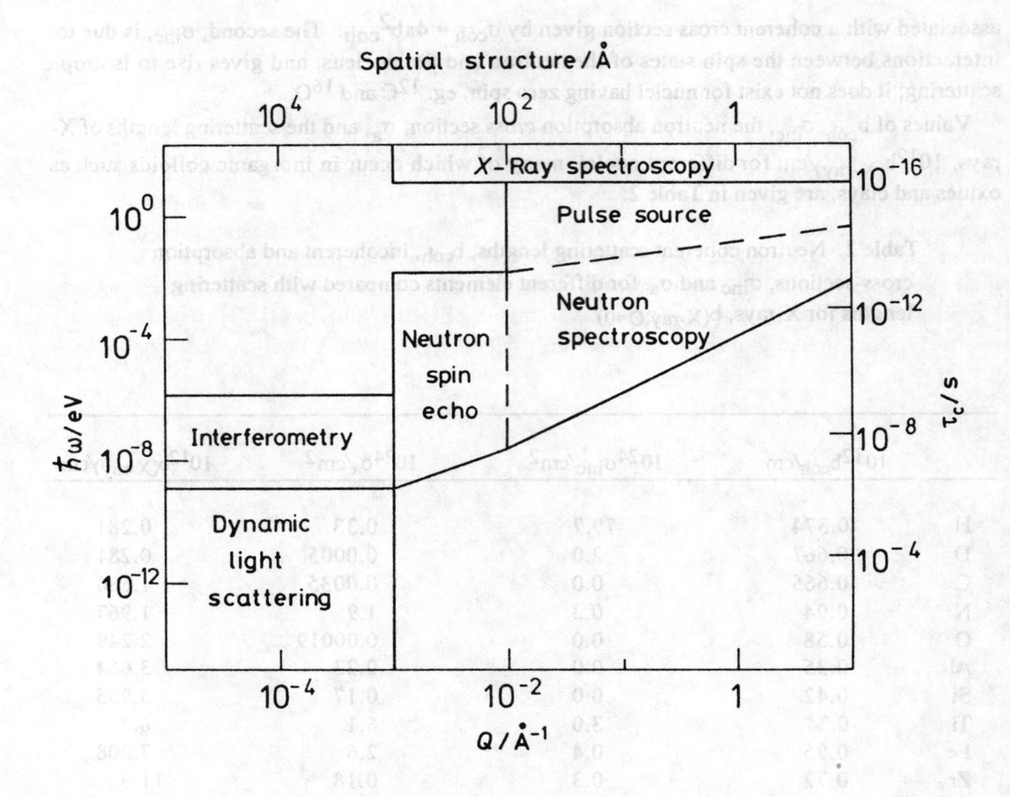

Figure 1. Energy and spatial resolution of different scattering techniques.

(Here Q, the momentum transfer is given by $4\pi\sin\theta/\lambda$, where λ is the wavelength and 2θ the scattering angle). For structural studies of colloids the upper part of this range is particularly important and overlaps with that covered by small angle X-ray scattering (SAXS) and static light scattering (SLS). Furthermore, because of the high energy resolution which is now attainable in neutron scattering, the energy, $\hbar\omega$, and corresponding timescale of dynamic processes, τ_c, that can be studied extends from approximately 10^{-14}s to the region of 1 μs which overlaps with that covered by dynamic light scattering (DLS). This makes the study of diffusion processes in both molecular and colloidal systems possible.

This wide range of timescale, which is equivalent to a correspondingly large energy transfer range, $\hbar\omega$, as illustrated in Figure 1, has been particularly valuable in the study of incoherent scattering processes. This will be illustrated here with reference to polynuclear cations in aqueous solution. Studies of dynamic processes in colloid systems using quasielastic coherent neutron scattering have however been more restricted. These have included measurements of the diffusion of latex particles using the recently developed high resolution spin-echo technique [11].

3. Investigations of the Dynamics and Structure of Polynuclear Ions by Incoherent Quasielastic Neutron Scattering (IQENS)

Polynuclear ions are formed during the partial hydrolysis of aqueous solutions of polyvalent metal cations, such as Al(III), Fe(III), Zr(IV), Th(IV) and Pu(IV) [12]. The nature of these species is of fundamental interest in inorganic solution chemistry and of importance in many areas ranging from nuclear technology to environmental chemistry [1]. Such species are also frequently the precursors in the formation of larger colloidal oxide particles (Figure 2). Despite their importance investigations of polynuclear ions in aqueous solution have been restricted, possibly because of their complexity and ill defined nature. Novel information has however recently been obtained from incoherent neutron scattering measurements on reasonably well defined solutions of relatively concentrated solutions (>0.1 mol dm^{-3}) of Al(III) and Zr(IV) polynuclear ions [13].

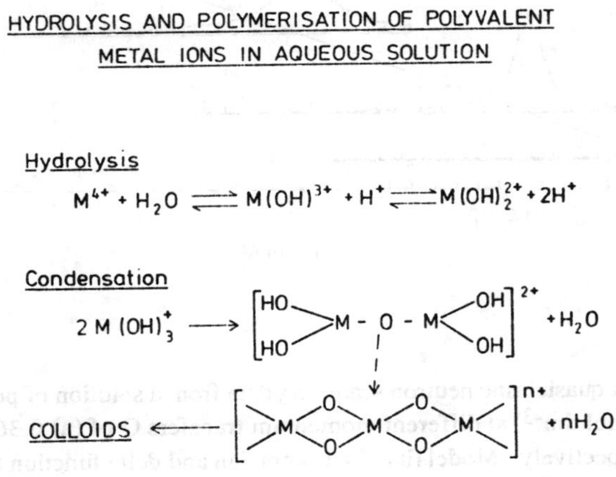

HYDROLYSIS AND POLYMERISATION OF POLYVALENT METAL IONS IN AQUEOUS SOLUTION

Hydrolysis

$$M^{4+} + H_2O \rightleftharpoons M(OH)^{3+} + H^+ \rightleftharpoons M(OH)_2^{2+} + 2H^+$$

Condensation

$$2\,M(OH)_3^+ \longrightarrow \left[(HO)_2M-O-M(OH)_2\right]^{2+} + H_2O$$

$$\downarrow$$

$$\text{COLLOIDS} \longleftarrow \left[\,M(O)_2M(O)_2M\,\right]^{n+} + nH_2O$$

Figure 2. Hydrolysis and polymerisation of polyvalent metal ions in aqueous solution.

Here we will illustrate the results obtained from quasielastic scattering (IQENS) measurements at medium (24 μev) and high (~1 μev) resolution with time-of-flight and back-scattering spectrometers at the ILL Grenoble. Here we note again that IQENS is highly suited to probe the dynamic behaviour of protons in a variety of different systems [14,15]. In this particular situation scattering arises from molecular water in the aqueous solution and protons associated with the hydrated ions.

This feature is demonstrated by the scattering at medium resolution for a solution of Zr(IV) polynuclear ions at a concentration of 3.6 mol dm^{-3}, in Figure 3.

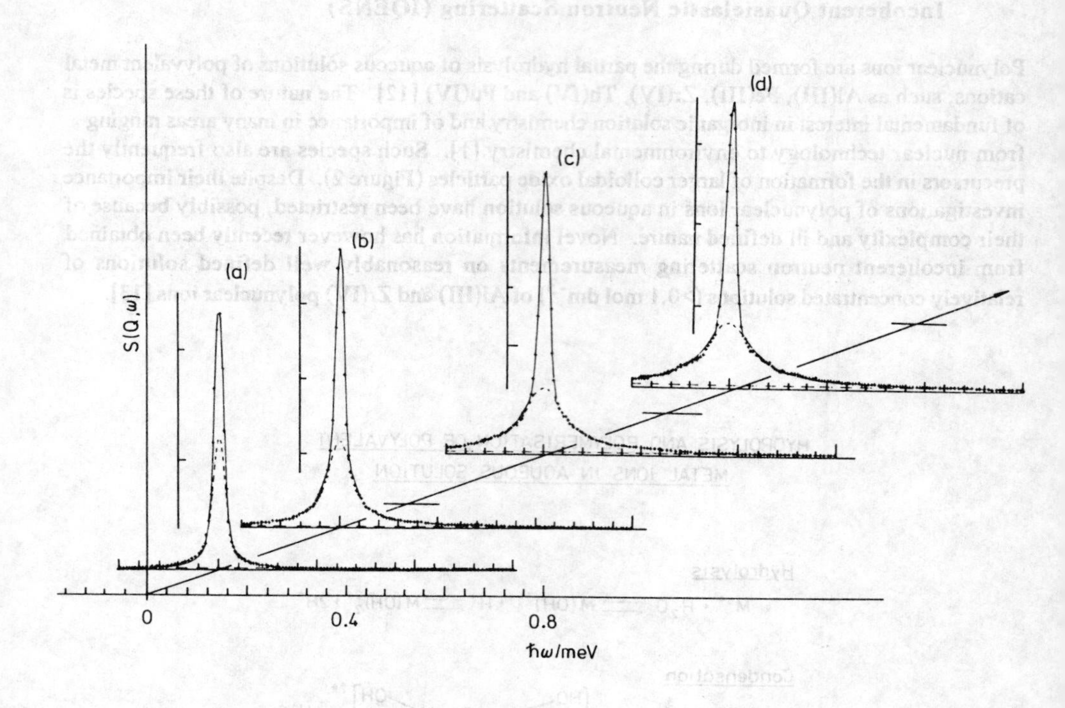

Figure 3. Incoherent quasielastic neutron scattering data from a solution of polynuclear zirconium ions (Zr(IV) = 3.6 mol dm^{-3}) at different momentum transfers Q of (a) 0.36, (b) 0.61, (c) 0.87, and (d) 1.06 Å^{-1} respectively. Model fits of a Lorentzian and delta function to experimental data correspond to the free and bound proton fractions in the sample.

Here the scattering law, S(Q,ω) can be fitted with a single Lorentzian, folded with a delta function. The weightings of the two components are the same over the total range of Q (0.1 to 1.1 Å^{-1}) and arise from two different fractions of protons in "free", γ_F, and bound, γ_B, states. The bound fractions, γ_B, have an almost linear dependence on the Zr(IV) and Al(III) concentration of the solutions, as illustrated in Figure 4. This feature indicates that the bound proton fraction is due to molecular water associated with the ion, together with a smaller proportion of structural OH groups. For Zr(IV) ion solutions the values of γ_B correspond closely to a species having a molecular formula [$ZrO(OH).4H_2O$]. For Al(III) a molecular formula of [$Al_2O(OH)_3.3H_2O$] was derived. These two anion deficient systems have metal:anion ratios of 1:1 and 1:0.5 respectively.

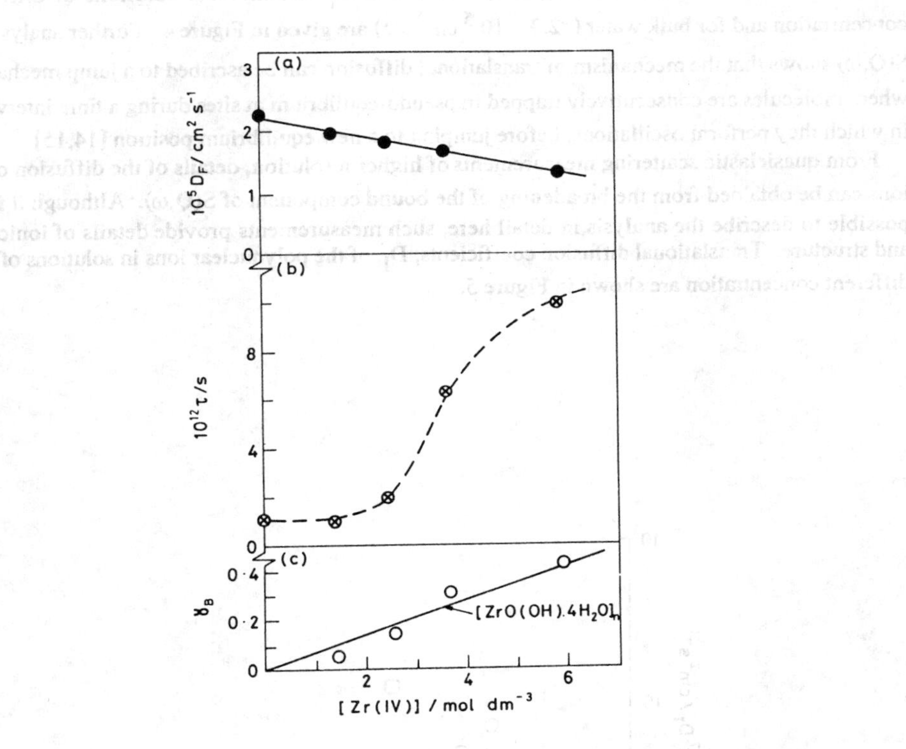

Figure 4. Incoherent quasielastic neutron scattering data is obtained from medium resolution (IN5) measurements on Zr(IV) polynuclear ions solutions of different concentration: (a) D_F, translational diffusion constant of "free water", (b) "jump" time, (c) γ_B, "bound" proton fraction.

From a further analysis of the broadened Lorentzian component of S(Q,ω) corresponding to the free fraction, the dynamics of the water in the solutions can be obtained. Thus in the lower limit of Q ($Q^2/Å^{-2}$ <0.2) the diffusion behaviour of the free water is continuous and the following simple scattering law is obeyed:

$$S_{inc}(Q,\omega) = \frac{1}{\pi}\left[\gamma_F \left\{\frac{D_F Q^2}{\omega^2+(D_F Q^2)^2}\right\} + \gamma_B \left\{\frac{D_B Q^2}{\omega^2+(D_B Q^2)^2}\right\}\right] \quad (1)$$

where Q and ω are the momentum and energy transfers respectively and D_F and D_B are the respective translational diffusion constants of the free and bound fractions. Here $D_B << D_F$ and at medium resolution the second term can be represented by a delta function corresponding to the instrumental resolution, namely $\gamma_B\ \delta(\omega)$. Values of D_F obtained for solutions of different concentration and for bulk water (~2.3×10^{-5} $cm^2\ s^{-1}$) are given in Figure 4. Further analysis of S(Q,ω) shows that the mechanism of translational diffusion can be ascribed to a jump mechanism where molecules are consecutively trapped in pseudo-equilbrium at sites during a time interval τ_o in which they perform oscillations, before jumping to a new equilibrium position [14,15].

From quasielastic scattering measurements of higher resolution, details of the diffusion of the ions can be obtained from the broadening of the bound component of S(Q,ω). Although it is not possible to describe the analysis in detail here, such measurements provide details of ionic size and structure. Translational diffusion coefficients, D_I, of the polynuclear ions in solutions of different concentration are shown in Figure 5.

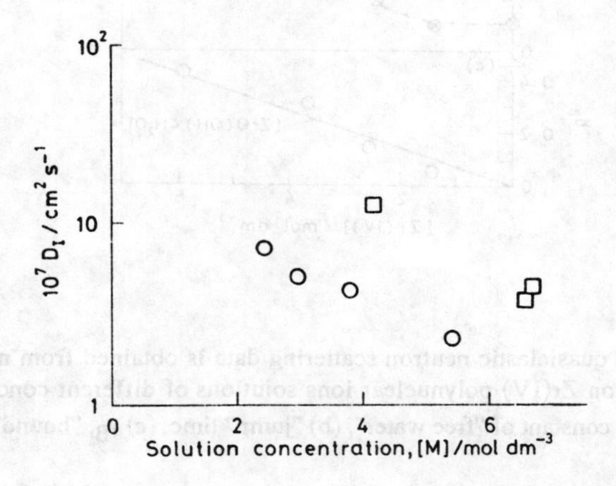

Figure 5. Translational diffusion of polynuclear ions, D_I, obtained from high resolution IQENS. o, Zr(IV); □, A1(III).

The markedly smaller value for the Zr(IV) species compared with Al(III) reflects the considerably larger size of the Zr(IV) species (~2 nm). Furthermore, the values of D_I for the Zr(IV) ion are in accord with those measured by dynamic light scattering [13] and show a similar decreasing trend as the ionic strength is increased. This reduction in D_I reflects considerable ion-ion interaction possibly due to hydrodynamic effects. This eventually leads to an effective immobilisation of the ions in the glassy gel, which is finally provided from the highly viscous concentrated solutions.

4. Investigations of Structure and Interactions in Concentrated Oxide Sols by Small Angle Neutron Scattering (SANS)

Neutron coherent scattering has its counterpart in small angle X-ray scattering (SAXS) and diffraction, for which the theory is similar [16], although the possibilities afforded by contrast variation are unique to neutron scattering [17]. Small angle scattering, SAS, arises from variations of scattering length density which occur over distances d_{SAS} (where $d_{SAS} \sim \lambda/2\theta$) corresponding to a scattering angle 2θ for radiation with a wavelength (λ) exceeding the normal interatomic spacings in solids and liquids. The application of SANS to the study of concentrated colloidal dispersions has recently had a considerable impact on our understanding of these systems [4,18,19]. Typical systems where SANS studies can provide information on structure and interactions are illustrated schematically in Figure 6.

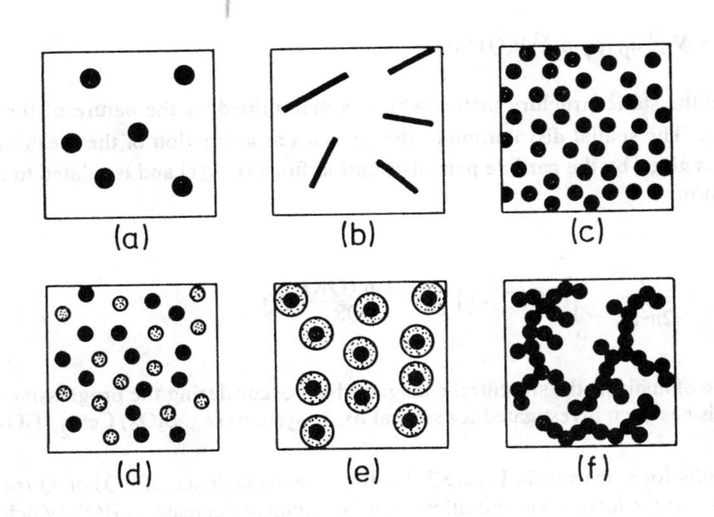

Figure 6. Schematic illustration of different types of colloidal system which may be studied by small angle neutron scattering: (a) and (b), dilute colloid dispersions of regularly shaped particles; (c) and (d), concentrated dispersions of interacting particles - in (d) there is a two component mixture; (e) dispersion of particles having an adsorbed layer of different scattering length density, (eg surfactant or polymer molecules); (f) aggregates of particles.

We will consider in particular those typified by (a) and (c) which represent dilute and concentrated oxide sols composed of small relatively monodispersed isotropic particles, such as silica.

For a dilute colloidal dispersion of non-interacting colloidal particles the scattered intensity I(Q) is given by [16]:

$$I(Q) = V_p^2 n_p (\rho_p - \rho_s)^2 P(Q) \tag{2}$$

where V_p is the particle volume, n_p the number density and ρ_p and ρ_s are respectively the scattering length density of the particles and solvent. P(Q) is a form factor which depends on the particle shape and also orientation for anisotropic particles. For spheres of radius, R, it is given by:

$$P(Q) = \left[\frac{3\,[\sin(QR) - QR\cos(QR)]}{Q^3R^3}\right]^2 \tag{3}$$

Equation 3 is valid for dilute dispersions containing widely separated non-interacting particles. In more concentrated dispersions, the intensity distribution is modified by the effects of interference, which depend on the spatial ordering of the particles. The scattered intensity is then

$$I(Q) = V_p^2 n_p (\rho_p - \rho_s)^2 P(Q)\, S(Q) \tag{4}$$

where S(Q) is the static structure factor, which is determined by the nature of the interaction potential, $\Phi(r)$. The spatial distribution of the particles as a function of the mean interparticle separation, r, is given by the particle pair-distribution function g(r) and is related to S(Q) by the Fourier transform

$$g(r) - 1 = \frac{1}{2\pi^2 n_p} \int_0^\infty [S(Q) - 1]\; Q^2\; \frac{\sin(QR)}{QR}\, dQ \tag{5}$$

Using this type of analysis the structural changes which occur during the progressive conversion of sols into gels has been investigated for several oxide systems (eg. SiO_2, CeO_2, TiO_2, FeOOH) [4,20,21].

Typical results for silica sols in Figure 7 illustrate the dependence of I(Q) on Q for a range of concentrations. At the lowest concentration there is a gradual decrease in I(Q), which has a form expected (cf equation (2)) for discrete non-interacting particles with a diameter of ~16nm. The development of the maxima at higher concentrations is caused by interference and indicates that the particles are not arranged at random but have some short-range ordering due to interparticle repulsion. The corresponding S(Q) derived from these results is shown in Figure 8.

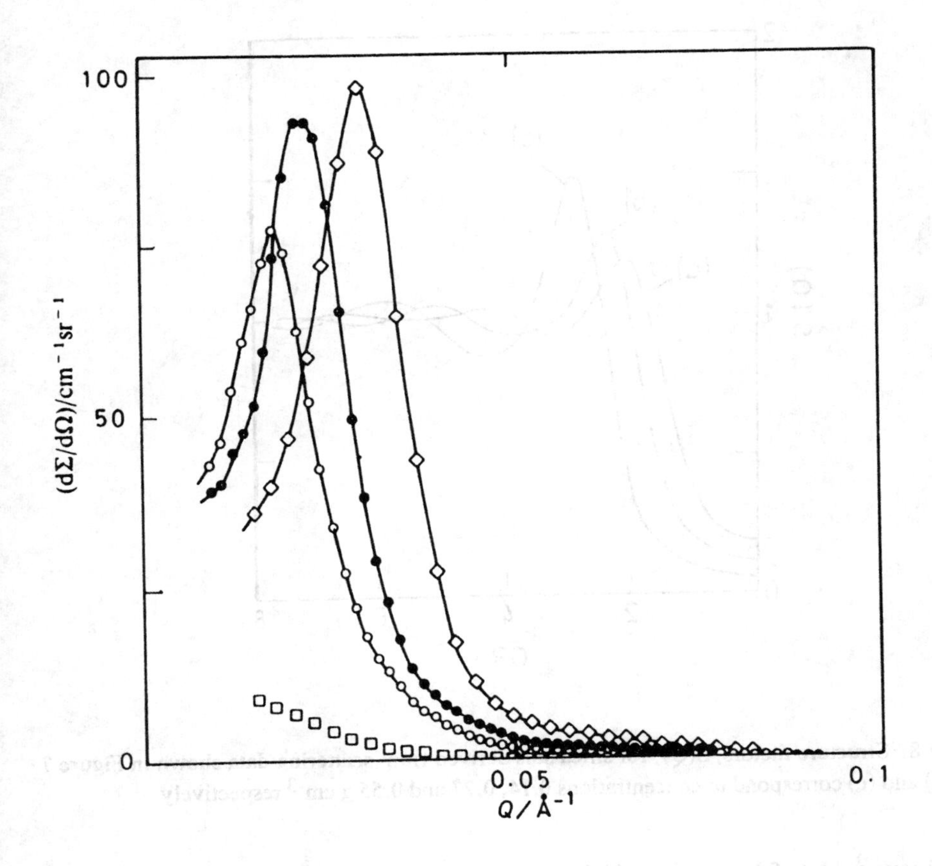

Figure 7. Small angle neutron scattering from silica sols of different concentration, c: □, 0.014; o, 0.14; ●, 0.27; and ◇, 0.55 g cm^{-3}. Particle diameter is ~ 16 nm.

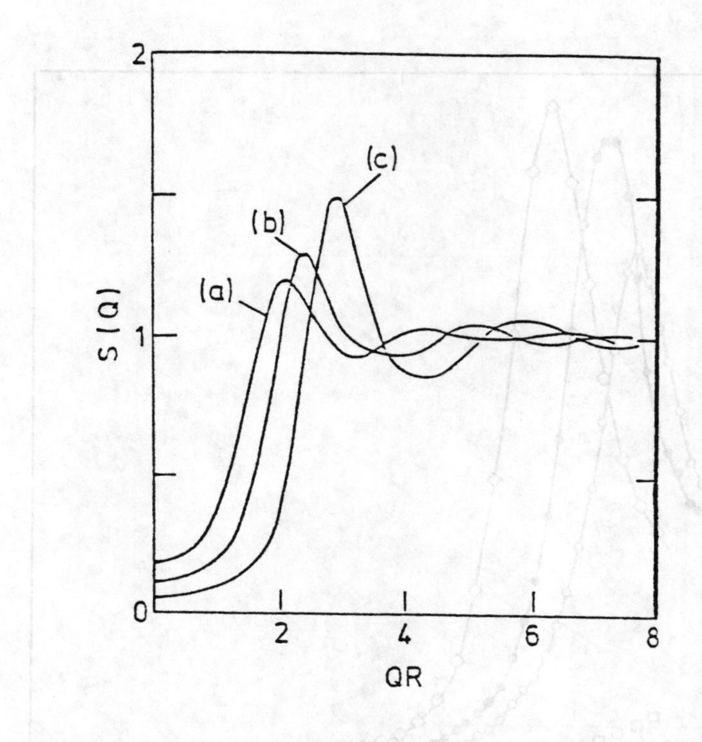

Figure 8. Structure factors, S(Q), for silica sols derived from scattering data shown in Figure 7. (a), (b) and (c) correspond to concentrations 0.14, 0.27 and 0.55 g cm^{-3} respectively.

Thus the movement of the maxima to higher values of Q with increasing sol concentration reflects a reduction in the equilibrium separation distance $r_{g(r)max}$, ($=r^*$), between the particles as derived from equation (5). An insight into the structural changes which occur on converting sols into gels can be obtained from the dependence of r^* on the sol concentration, c. In general, for these discrete particle systems it is found that the interparticle separation decreases as $c^{1/3}$ and approaches that of the particle diameter, 2R, as the solid gel is formed. This feature, which is illustrated by results for ceria and silica sols of different particle size in Figure 9, indicates that there is little change in the relative arrangement of particles when the volume fraction is increased by over an order of magnitude. This maintenance of local order during the sol to gel conversion process is a result of the repulsive double-layer interaction between the particles which forces them eventually to adopt an efficiently packed structure in the gel [22].

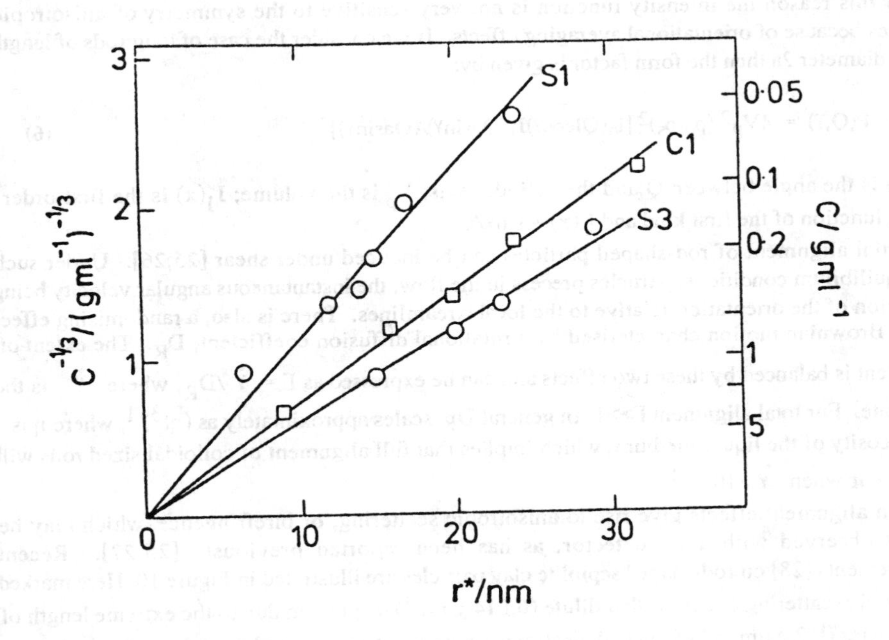

Figure 9. Dependence of r* on the concentration, c, of sols and gels.

These interparticle interactions are of fundamental importance because they also control many of the macroscopic properties of concentrated sols, such as the rheology and stability. A direct insight into the interaction behaviour can be obtained from the form of S(Q). In general the experimental S(Q) is simulated by that obtained from a model potential, $\Phi(r)$. The rescaled mean spherical approximation (RMSA) has proved particularly successful [19,23] and allows us to obtain the effective surface charge and surface potential, Ψ_o, of the oxide colloid [24].

5. Investigations of Anisotropic Particles under In-Situ Shear by SANS

In the two previous situations we have considered colloidal systems under equilibrium conditions. Thus in the case of solutions of polynuclear ions we were able to derive details of the diffusion behaviour arising from dynamic fluctuations. Although such effects produce temporal decreases or increases of entropy, these are time-averaged about an equilibrium condition. In IQENS such fluctuations are evident from the quasielastic broadening about the incident neutron energy. Equivalent information may be derived from quasielastic light scattering by directly measuring the correlation function of fluctuations in intensity with time. In the case of SANS we measure a time-averaged or "static" intensity.

For this reason the intensity function is not very sensitive to the symmetry of anisotropic particles because of orientational averaging effects. If we consider the case of long rods of length 2l and diameter 2a then the form factor is given by:

$$P(Q,\gamma) = 4V_p^2 (\rho_p-\rho_s)^2 [j_o(Ql\cos\gamma)J_1(Qa\sin\gamma)/(Qa\sin\gamma)]^2 \quad (6)$$

Here γ is the angle between Q and the cylinder axis; V_p is the volume; $J_1(x)$ is the first-order Bessel function of the first kind and $j_o(x) = \sin x/x$.

Partial alignment of rod-shaped particles can be induced under shear [25,26]. Under such non-equilibrium conditions, particles precess in the flow, the instantaneous angular velocity being a function of the orientation relative to the local streamlines. There is also, a randomising effect due to Brownian motion characterised by a rotational diffusion coefficient, D_R. The extent of alignment is balanced by these two effects and can be expressed as $\Gamma = \dot{\gamma}/D_R$, where $\dot{\gamma}$ is the shear rate. For total alignment $\Gamma >> 1$. In general D_R scales approximately as $(\eta l^3)^{-1}$, where η is the viscosity of the liquid medium, which implies that full alignment of colloidal sized rods will only occur when $\dot{\gamma} > 10^3\ s^{-1}$.

Such alignment effects give rise to anisotropic scattering, or birefringence, which may be readily observed with a 2D-detector, as has been reported previously [25-27]. Recent measurements [28] on rod-shaped sepiolite clay particles are illustrated in Figure 10. Here marked anisotropic scattering occurs with a dilute (0.014 g ml^{-1}) dispersion due to the extreme length of the particles (l ~2.5 µm; a ~15 nm). Anisotropic scattering was indeed observed even at the lowest shear rates studied ($\dot{\gamma}$ ~ 25 s^{-1}) and persisted for more than ten minutes after shear. Such measurements highlight the potential value of SANS for investigating relaxation processes in clay dispersions and gaining insight into thixotropic phenomena.

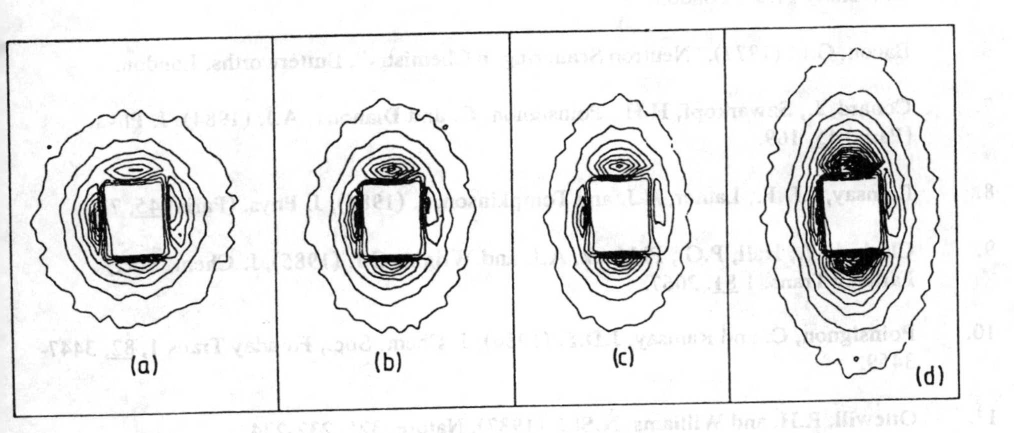

Figure 10. Intensity contour plots for 1.4% sepiolite dispersion at different shear rates, γ/s^{-1}: (a) 0; (b) 2.5 x 10^2; (c) 2.5 x 10^3; (d) 1.25 x 10^4.

6. Acknowledgements

The work described was undertaken as part of the Underlying Research Programme of the UKAEA. Contributions of many collaborators, in particular, Drs. C Poinsignon, R M Richardson, S W Swanton, P Lindner and Mr R G Avery are especially acknowledged. Provision of experimental facilities and support at the Institut Laue Langevin are also gratefully acknowledged.

7. References

1. Stumm, W. and Morgan, J.J. (1981), "Aquatic Chemistry", Wiley Interscience, New York.

2. van Olphen, H. (1977), "Clay Colloid Chemistry", Wiley Interscience, New York.

3. see e.g. "Colloidal Dispersions" (1982), in J.W. Goodwin (ed.), Special Publication No.43, Royal Society of Chemistry, London.

4. Ramsay, J.D.F. (1986), "Recent developments in the characterisation of oxide sols using small angle neutron scattering techniques", Chem. Soc. Rev., 15, 335-371.

5. Marshall, W and Lovesey, S.W. (1971), "Theory of thermal neutron scattering", Oxford University Press, London.

6. Bacon, G.E. (1977), "Neutron Scattering in Chemistry", Butterworths, London.

7. Conard, J., Szwarkopf, H.H., Poinsignon, C. and Dianoux, A.J. (1984), J. Phys. (Paris) 45, 169.

8. Ramsay, J.D.F., Lauter, H.J. and Tompkinson, J. (1984), J. Phys. (Paris) 45, 73

9. Clark, J.W., Hall, P.G., Pidduck, A.J. and Wright, C.J. (1985), J. Chem. Soc., Faraday Trans. 1 81, 2067.

10. Poinsignon, C. and Ramsay, J.D.F. (1986), J. Chem. Soc., Faraday Trans 1, 82, 3447-3459.

11. Ottewill, R.H. and Williams, N.St.J. (1987), Nature, 325, 232-234.

12. Baes, C.F. and Mesmer, R.E. (1976), "The Hydrolysis of Cations", Wiley Interscience, New York.

13. Ramsay, J.D.F. (1986), "Neutron and light scattering studies of aqueous solutions of polynuclear ions", in G W Neilson and J.E. Enderby (eds.), "Water and Aqueous Solutions", Adam Hilger, Bristol. pp. 207-218.

14. Springer, T. (1972), in "Quasielastic neutron scattering for the investigation of diffusive motions in solids and liquids", Springer Tracts in Modern Physics, Vol. 64., Springer-Verlag, Berlin, N.Y.

15. Springer, T. (1978), in "Dynamics of solids and liquids by neutron scattering", S W Lovesey and T Springer (eds.), Springer-Verlag, Berlin, N.Y., pp 255-300.

16. Guinier, A. and Fournet, G. (1955), "Small angle scattering of X-rays", Wiley, New York.

17. Jacrot, B. (1976), Rep. Prog. Phys., 39, 911.

18. Ottewill, R.H. (1982), "Colloidal Dispersions", J.W. Goodwin (ed.), Special Publication No.43, Royal Society of Chemistry, London, pp143-164.

19. Hayter, J.B. (1983), Faraday Discuss. Chem. Soc., 76, 7-17.

20. Ramsay, J.D.F., Avery, R.G. and Benest, L. (1983) Faraday Discuss. Chem. Soc., 76, 53-64.

21. Ramsay, J.D.F. and Booth, B.O. (1983), J. Chem. Soc., Faraday Trans. 1, 79, 173.

22. Ramsay, J.D.F. (1989), Pure and Appl. Chem., 61, 1963-1968.

23. Hayter, J.B. and Penfold, J. (1981), J. Chem. Soc., Faraday Trans. 1, 77, 1851.

24. Penfold, J. and Ramsay, J.D.F. (1985), J. Chem. Soc., Faraday Trans. 1, 81, 117.

25. Hayter, J.B. and Penfold, J. (1983), J. Colloid Polym. Sci., 261, 1022.

26. Hayter, J.B. and Penfold, J. (1984), J. Phys. Chem., 88, 4589.

27. Oberthur, R.C. (1984), Revue Phys. Appl., 19, 663.

28. Lindner, P., Ramsay, J.D.F. and Swanton, S.W. (1990), Instit. Laue Langevin Annual Rep., (to be published).

THE RHEOLOGY OF CONCENTRATED DISPERSIONS OF AGGREGATED PARTICLES

R. BUSCALL
ICI Corporate Colloid Science Group
PO Box 11, The Heath
Runcorn, Cheshire
WA7 4QE

ABSTRACT. Recent work on the rheology of gels formed by the irreversible aggregation of spherical colloidal particles is reviewed briefly with an emphasis on scaling. The properties considered include the gel modulus and strength, non-linear viscoelasticity, shear-flow and the effect of flow on structure.

1. Introduction

This paper is concerned with the rheology and structure of particulate gels. The discussion is centred around available data pertaining to concentrated aqueous dispersions of colloidal particles subjected to irreversible aggregation, i.e. fast coagulation in the primary minimum. It should however be borne in mind that there are a whole range of systems that are related in a structural and rheological sense. Examples are aerogels, coagulated proteins (and other systems part-way between simple particulate-gels and conventional cross-linked polymers) and powder compacts.

In shear flow particulate gels formed by irreversible aggregation show a yield stress below which they show solid-like viscoelasticity and above which they show non-Newtonian flow typified by strong shear-thinning and, often, irreversible changes in structure with time. Particulate gels of this type also show substantial compressive strength, this for example can be maintained on drying allowing the formation of low-density aerogels.

The stucture of aggregates formed in dilute dispersions has been extensively studied of late [1-4]. It has been shown that aggregates formed by irreversible aggregation have a fractal structure and scattering experiments and quantitative microscopy have yielded fractal dimensions in fair agreement with those expected from simulation [5]. In concentrated dispersions gelation occurs by overcrowding of the growing aggregates. There have been rather few studies of the structure of the gelled state, however scattering experiments on for example silica gels do appear to confirm the existence of a fractal sub-structure [6,7]. Knowing that the structure is fractal on certain scales does not alone allow gel-point to be predicted since fractal aggregates can fill space at any particle concentration depending upon their size. However, above the gel-point the requirement to fill the space available suggests that the size (radius) of a typical fractal cluster ξ is such that

D. M. Bloor and E. Wyn-Jones (eds.),
The Structure, Dynamics and Equilibrium Properties of Colloidal Systems, 653–658.

$$\phi_c(\xi) \approx \phi_o \qquad [1]$$

where $\phi_c(\xi)$ is the mean volume-fraction of particles inside the typical cluster and ϕ_o is the overall volume-fraction in the system. The former is given by

$$\phi_c(\xi) \approx \left(\frac{\xi}{a}\right)^{d_f - 3} \qquad [2]$$

where a is some cut-off length of order the particle-size. Thus equation 1 becomes

$$\xi \approx a\,\phi_o^{1/(d_f - 3)} \qquad [3]$$

These relationships can be used to predict the properties of the gel given a knowledge of the properties of an individual cluster and how these depend upon its size and fractal dimension. This has been done by Ball and Brown [8,9] for the elastic moduli (G,K,E) and the DC conductivity (C), with the results

$$G,K,E \sim \phi_o^{\mu_{el}} \; ; \quad C \sim \phi_o^{\mu_{cond}} \qquad [4]$$

with the exponents given by

$$\mu_{el} = \frac{3 + d_{chem}}{3 - d_f} \; ; \quad \mu_{cond} = \frac{1 + d_{chem}}{3 - d_f} \qquad [5]$$

where d_{chem}, the chemical-length exponent measures the length of the shortest path through the cluster. d_{chem} is not independent of the fractal dimension d_f, however for fractal dimensions of order 2 and above it is close enough to unity for this to be approximately the case. For diffusion-limited cluster-cluster aggregation the two exponents are predicted to have values of about 1.8 (conductivity) and 3.6 (elasticity). To the best of the authors knowledge the conductivity exponent has not been measured experimentally. However the elasticity exponent has and the experimental results agree rather well with this prediction as will be discussed below. Given a fractal sub-structure one might also expect other and more complex properties to scale similarly.

2. Gelation

The critical concentration or gel-point can be determined by detecting the onset of elastic rigidity or, in the case of conducting particles, DC conductance. The gel-point is found to depend upon the mechanism of aggregation and upon whether the samples have subsequently been subjected to flow [10-12]. However for spherical particles the gel-point is found to be equal or greater than 0.05, values of about 0.05 - 0.07 being typical of fast, irreversible aggregation in the absence of flow, higher values being observed for orthokinetic aggregation, reversible aggregation and for samples subjected to subsequent shearing. The gel-points for anisometric particles (rods and plates) tend to be lower, everything else being equal.

3. Gel Strength and Modulus

The shear modulus (in the linear viscoelastic regime) has been measured for gels formed from spherical particles (polymer latices and silica particles) over a wide range of concentration and particle size [10-13]. The modulus is found to rise rapidly from zero at the gel-point and then turn over to the anticipated power-law form. The experimental exponents lie in the range 3.5 to 4.2 in fair agreement with the predictions of Ball and Brown. Another readily measured property is the compactive or compressive strength, this being defined as the minimum uniaxial stress required to cause irreversible compaction or consolidation of the gel, this being measured in a pressure filter [13] or a centrifuge [14,15] . The compactive strength is found to scale like the modulus and to have values similar in magnitude. A possible reason for this correspondence is as follows: The compaction process probably necessitates the breaking of few if any bonds in the structure, rather the structure probably folds in on itself with new bonds forming at the new contact points. That the strength should then scale with the elasticity follows. (It is found that the shear strength does not, inevitably bonds need to be broken in order to facilitate shear flow).

The mechanical properties of silica aerogels have been studied by Woignier and Phalippou [16]. They found that the moduli E,G and K scaled like those in the wet state with an exponent of 3.7. The flexural strength and toughness were also measured and were found to scale with exponents of 2.6 and 1.6 respectively. The exponent for the strength is similar to the exponent found for the shear strength of wet gels (see below).
A similar elasticity exponent has been reported for coagulated protein gels and for powder compacts/ceramic green-bodies [17] .

In summary then there appears to be limited but rather compelling evidence for a universal elasticity exponent of between 3.5 and 4, in good agreement with the overcrowded fractal-aggregate model of Ball and Brown. This agreement however should not be taken to suggest that matters are as straightforward as they appear. The model requires the existence of fractal sub-clusters of typical size ξ given by equation 3. However for concentrations much above the gel-point and more particularly of order 0.2 or so ξ ceases to be large compared to the cut-off length and in this sense the model breaks down. An alternative model that might be considered more appropriate at high-densities has been given by Seng and Fen [18]. They examined the effect of removing particles at random from a close-packed lattice of bonded particles. This is not strictly a good model of the aggregated state since the procedure eventually cuts the gel to give unbonded clusters and particles, however this feature is probably unimportant in the high density limit, and furthermore it can be corrected for by focussing upon percolating cluster only. The exponent found is again about 4. The essential difference between the two models is the way the correlation-length decreases with increasing concentration, what actually happens could however be determined by scattering and this would be well-worth doing. It is probably reasonable to assume that the overcrowded fractal cluster model has to be essentially correct at $\phi \sim \phi_G$ and the depleted-lattice model at $\phi \sim \phi_{max}$. The nature of the cross-over remains to be determined . However it would appear that it is not noticed in experimental elasticity measurements because of the similarity of the limiting exponents. That the exponent does not change has been confirmed experimentally by making measurements over a wide density range for aerogels [16] and aggregate-networks formed from latex particles [19].

4. Non-Linear Viscoelasticity and Yield in Shear

Creep-compliance measurements have shown that a non-linear viscoelastic response is elicited at very small stresses and strains in the case of gels formed from spherical particles [10-12]. The critical stress depends upon density but the critical strain is found to be of order 0.0005 to 0.001 typically. The instantaneous modulus is found to decay slowly but exponentially with increasing instantaneous strain and data for different concentrations and particle-sizes scale if the modulus is normalised on its limiting value and the strain on the critical value. Eventually yield occurs at a strain of order 0.2. The difference between the magnitude of the critical strain and the yield strain of two decades or so is remarkable. There are two obvious sources of non-linearity, finite stretching of the inter-particle bonds eliciting an anharmonic response, and finite deformation of the chains of particles. The former can be ruled out as a source of the extended non-linear compliance as the forces are short-range on the scale of a particle diameter. That the structure becomes less rigid with increasing strain or stress as is implied by the decay in modulus is also surprising per-se since at these volume-fractions (0.1 to 0.25) the stress-carrying paths would be expected to be rather direct. A possible explanation is that bonds are in fact broken in these paths but that the stress is then taken up by more convoluted paths that do not support the stress initially. Support for this hypothesis is provided by elastic recovery measurements which showed that recovery was incomplete in the non-linear region [10,11]. Data for the yield stress itself were found to be consistent with

$$\sigma_y \quad \propto \quad \phi^3 a^{-2} \qquad [6]$$

The dependence on density was thus weaker than that of the compressive strength (as was that on particle size) and the shear strength was typically two orders of magnitude smaller than the compressive strength. It was also found that the yield stress showed a systematic statistical scatter, implying that the yielding process in shear is rather inhomogenous.

5. Effect of Shear on Structure

Characterisation of the flow properties of aggregated suspensions is complicated by the fact that both reversible and irreversible changes in the viscosity, and thus by implication the structure, can occur with time. Thus a common expedient is to subject samples to prolonged shear at high shear-rates prior to measurement at lower shear-rates, this procedure giving reproducible flow curves [20]. This causes the clusters to become more compact and for the gel, if the concentration is high enough for one still to exist, to comprise a network of tight "floccules", this type of structure being the basis of the rather successful Elastic Floc model of Hunter et.al. [20]. Irreversible changes of structure with time appear to be particularly favoured and more pronounced when the shearing is carried out at constant stress rather than constant shear-rate [21-23]. Thus Mills [21,23] has demonstrated irreversible reductions in viscosity of two orders of magnitude or more by this means. It was shown by freeze-etch electron microscopy that the drop in viscosity was accompanied by a transformation from a ramified structure in which the average co-ordination was about three to a pauci-disperse suspension of roughly-spherical clusters of about 10 μm diameter with an internal density estimated to be in excess of 0.5. That controlling the stress rather than the shear-rate might favour this type of structural change does not seem unreasonable. Shearing at constant (high) shear-rate causes gross breakdown of the flocs and consequently they might be expected to reform in roughly the same way when the shearing is stopped. In the controlled-stress experiments the shear-rate, which is typically very low initially, rises gradually with time as the viscosity drops and high shear-rates are only experienced at the latter stages when the clusters are compact and strong. The rearrangement is envisaged as a "self-comminution" process whereby looser,weaker parts of the clusters are abraded by collision, with the denser stronger parts growing by aggregation. Mills also performed some measurements of the elasticity and compressive strength of the conditioned gels. The gel-points were found to be somewhat higher and the magnitudes of the modulus and the strength were found to be reduced, everything else being equal. The exponent characterising the concentration dependence was however found to remain unchanged (the data were limited but consistent with this idea as far as they go). This implies that in scaling terms it is irrelevant whether the network is formed from primary particles or dense spherical clusters, further work is however required in order to establish this unambiguously. The complexity of the shear flow behaviour of aggregated dispersions in the concentrated regime and the inter-dependence of structure and flow means that existing studies have only scratched the surface. Much more remains to be done and parallel studies of structure and rheology will be essential if a lucid picture is to be constructed.

6. Role of Computer Simulation

The complexity of the shear-flow behaviour renders the prospects for theoretical models poor, except perhaps for limiting cases [20]. There are however encouraging signs that computer simulation has better prospects. In a recent paper Chen and Doi [24] describe a simulation of the shear flow of adhesive spheres. The adhesion was characterised by a contact force f_c and the simulations were carried out at reduced shear-rates between about 0.01 and 1, the reduced shear-rate $\dot{\gamma}_R$ being given by

$$\dot{\gamma}_R = \frac{6\pi a \eta_o \dot{\gamma}}{f_c} \qquad [7]$$

where η_o is the medium viscosity. Brownian motion was not included and it was assumed that the flocs were free-draining. The particles were placed at random initially and the aggregation was allowed to take place under the influence of the imposed overall velocity gradient. It was found that gelation in the sense of the formation of a percolating cluster under the imposed shear was first detected at a volume fraction of close to 0.1. At lower volume fractions the flocs were found to have a fractal dimension of about 2 and the size of the flocs was found to reduce like the inverse 1/3 power of the shear-rate. The viscosity was found to decrease with increasing shear-rate with an exponent of -0.56. Bearing in mind the simplifying assumptions and the fact that the shearing times were rather short as a result of the computer time required, these results are very encouraging. The effects and trends seen are in qualitative agreement with experiments and the insight they provide is considerable. This pioneering study bodes well for the future.

7. Conclusions and Comments

The static rheological properties of particulate gels formed by rapid, irreversible aggregation appear to show scaling behaviour consistent with the idea of a fractal sub-structure. There remains however a need for complimentary structural studies. The flow properties of the degraded gels are in general complex and poorly defined and understood, there being an inextricable link between flow, properties and structure. However it is clear from the study by Chen and Doi that simulation should prove an invaluable aid to understanding in this area.

A more complete set of references to earlier work is given in ref. 10.

The paper by Ramsay in this volume also refers to further relevant work on particulate gels formed from inorganic particles.

8. References

1. D A Weitz and M Oliveria, Phys. Rev. Lett., 52, 1433 (1984)
2. M Matsushita, K Sumida Y.Sawada, J.Phys. Soc. Jpn, 54, 2786 (1985)
3. P N Pusey and J G Rarity, Mol. Phys., 62, 411 (1987)
4. J P Wilcoxon, J E Martin and D W Schaefer, Phys. Rev. A, 39, 2675 (1989)
5. P Meakin, Adv. Colloid Interface Sci., 28, 249 (1988)
6. D W Schaefer, Mat. Res. Soc. Symp. Proc., 79, 47 (1987)
7. D W Schaefer, Mat. Res. Soc. Bull., 13, 22 (1988)
8. R C Ball and W D Brown, private communication
9. W D Brown, Ph.D. Thesis, Dept. of Physics, University of Cambridge (1987)
10. R Buscall, P D A Mills and G E Yates, Colloids Surf., 18, 341 (1986)
11. R Buscall, I J McGowan, P D A Mills, R F Stewart, D Sutton, L R White and G E Yates, J. non-Newtonian Fluid Mech., 24, 183 (1987)
12. R Buscall, P D A Mills, J W Goodwin and D W Lawson, J. Chem.soc. Faraday Trans. 1, 84, 4249 (1988)
13. D N Sutherland, Ph.D. Thesis, Dept. of Chemical Engineering, University of Cambridge (1964)
14. R Buscall, Colloids Surf., 5, 269 (1982)
15. R Buscall and L R White, J Chem. Soc. Faraday Trans. 1, 83, 873 (1987)
16. T Woignier and J Phalippou, Rev. Physique Appliquee, 24, C4- 179 (1989)
17. K Kendall, N McN Alford and J D Birchall, Proc. R. Soc. Lond. A 412, 269 (1987)
18. S Feng and P N Sen, Phys. Rev. Lett., 52, 216 (1984)
19. R Buscall and A J Morton-Jones, Unpublished Work
20. R J Hunter, in "Modern Trends of Colloid Science" (H-F Eicke ed.) Birkhauser Verlag Basel 1985, and references therein
21. P D A Mills, paper presented at RSC conference on "Aggregation in Colloidal Dispersions", London, December 1988
22. B Grover, M.Sc. Thesis, School of Chemistry, University of Bristol (1986)
23. J W Goodwin, B Grover and P D A Mills, paper in preparation.
24. S Chen and M Doi, J. Chem. Phys., 91, 2656 (1989)

RHEOLOGICAL PROPERTIES, INTERPARTICLE FORCES AND SUSPENSION STRUCTURE

J.W. GOODWIN
Department of Physical Chemistry
University of Bristol
Cantock's Close
Bristol BS8 1TS
U.K.

1. Introduction

Control of the rheological properties of colloidal dispersions is often a main feature of formulation work; indeed the colloidal state is often the best way of achieving the best rheology. It is only infrequently that dilute systems are used and the rheological properties in concentrated systems are the result of the interplay between the volume fraction and the forces acting between the particles reinforced when necessary by the addition of "thickeners" such as soluble polymers or further colloidal components.

The division between "dilute" and "concentrated" is not only somewhat arbitrary but often dependent on tradition. It can be convenient to define dilute dispersions as ones in which the particle motion is uncorrelated. This means that the average nearest neighbour distance is in excess of the range of the interparticle interactions (electrostatic, van der Waals' and hydrodynamic). The implication of this lack of correlation is that there is no continuous structure.

At first sight, dispersions of hard spheres are the simplest to study from a theoretical standpoint. In the absence of Brownian motion, i.e. large particles and low temperatures, a continuous structure would not be expected until the volume fraction $\phi > 0.5$ when a phase transition is predicted from computer simulations [1]. Of course the excluded volume of the particles is increased by the presence of the long range repulsion that can arise with charged particles as well as Brownian motion. These effects both reduce the volume fraction at which the phase transition occurs. Polymeric stabiliser layers move with the particles and must be included in any calculation of the effective volume fraction of the dispersion. However, experiments [2] on systems of hard spheres with Brownian motion show the systems to be viscous liquids at $\phi < 0.45$ with Newtonian or slightly shear-thinning behaviour.

The viscosity of a suspension of hard spheres has been calculated from a rigourous hydrodynamic analysis to be [3,4]

$$\eta(o) = \eta_o[1 + 2.5\phi + 6.2\phi^2 + O(\phi^3)] \qquad (1)$$

where η_o is the viscosity of the continuous phase and low shear rate limit of the viscosity, $\eta(o)$, is used as a uniform (random) particle distribution was assumed in the derivation. Equation 1 is only useful for $\phi < 0.1$ and an approximate analysis is required for higher volume fractions as the

D. M. Bloor and E. Wyn-Jones (eds.),
The Structure, Dynamics and Equilibrium Properties of Colloidal Systems, 659–679.

multibody hydrodynamics cannot be solved. A particularly useful description of the viscosity is due to Krieger [3]:

$$\eta(o) = \eta_o\left(1 - \frac{\phi}{\phi_m}\right)^{-2.5\,\phi_m} \tag{2}$$

where ϕ_m is the maximum packing fraction at which flow can occur. Values of $0.60 < \phi_m < 0.64$ corresponding to dense random packing often give good fits to the data. However ϕ_m is frequently used as a fitting parameter as factors such as polydispersity, shape, and particle or layer flexibility result in values of ϕ_m outside this range being required.

2. The State of Dispersion

Whether a colloidal dispersion is stable or whether the particles are aggregated depends on the interplay between the forces of repulsion (electrostatic or steric) and those of attraction (van der Waals' and depletion). The interaction between particles is usually expressed in terms of the potential energy or the pair potential , V(R), as a function of the distance between particle centres, R. Typical pair potentials are sketched in Figure 1. 1a, is the typical long range repulsive tail

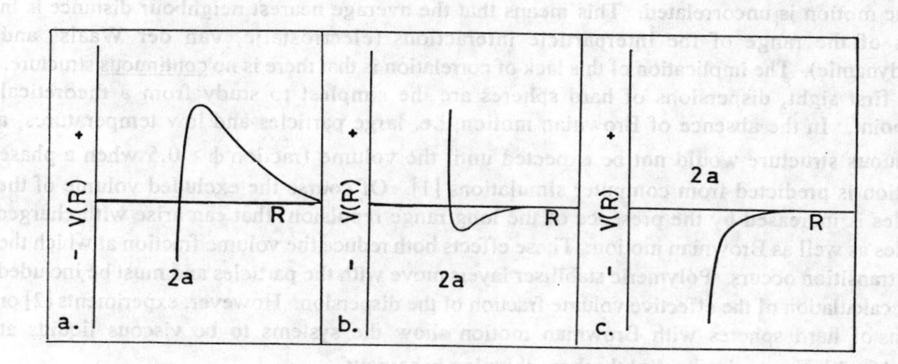

Figure 1. Pair potential curves for colloidal particles; a) electrostatically stabilised, b) weakly flocculated, c) coagulated.

DISPERSION STATE	CONCENTRATION INCREASING →				
STABLE	SINGLETS	SINGLETS	SINGLETS	ORDERED LIQUID-LIKE STRUCTURE	GLASS/ CRYSTAL
WEAKLY FLOCCULATED	SINGLETS AND DOUBLETS	SINGLETS PLUS COMPACT CLUSTERS	SINGLETS PLUS LARGE CLUSTERS	CONTINUOUS NETWORK	→
COAGULATED	DOUBLETS, TRIPLETS, ETC.	FRACTAL CLUSTERS	CONTINUOUS NETWORK	→	
RHEOLOGY	VISCOUS FLUID		VISCOELASTICITY		

Figure 2. **Summary of the effect of dispersion state on the rheology of dispersions.**

giving electrostatically stable systems. 1b, is more typical of sterically stabilised systems in which the shallow minimum gives a weakly attractive or flocculated system. The strong attraction in 1c, gives a coagulated system. At low concentrations colloids in all three states of dispersion behave as viscous liquids although pseudoplasticity or shear thinning may be observed. As the volume fractions are increased viscoelastic responses will be observed with each system although the onset will be found at different volume fractions. Generally the smaller the particle size, the lower the volume fraction at which a viscoelastic response can be measured. The continuous structures produced at moderate to high volume fractions from systems with any of these types of interaction are capable of the storage of energy when subjected to small strains i.e. elasticity can be observed. This stored energy can be dissipated by a structural rearrangement relieving the stress.

3. Linear Viscoelasticity

Only linear viscoelastic responses will be discussed below and it is important to remember that the conclusions drawn are only reliable if sufficiently small strain and strain rates are used for the measured responses to be either linear or at least very close to linear. In addition, only the shear components of the stress and strain tensors will be utilised so that normal stresses and extensional flows will not be discussed here.

For linear behaviour the elastic response obeys Hooke's Law so that the shear stress, σ, is directly proportional to the shear strain, γ, with the constant of proportionality being the elastic modulus G:-

$$\sigma = G\gamma \tag{3}$$

The units of the modulus are force per unit area, as is stress since the strain is a relative deformation. The work done in producing a finite deformation gives the energy stored, E so that:

$$E = \int_0^{\gamma} \sigma d\gamma = G \int_0^{\gamma} \gamma d\gamma$$

i.e.
$$E = \frac{1}{2} G \gamma^2 \qquad (4)$$

The viscous component gives the dissipation of energy (as heat) and is given by Newton's Law:

$$\sigma = \eta \frac{d\gamma}{dt} \qquad (5)$$

Writing d/γ as $\dot{\gamma}$, the rate of energy dissipation is:

$$\dot{E} = \sigma \dot{\gamma}$$

i.e.
$$\dot{E} = \eta \dot{\gamma}^2 \qquad (6)$$

At this point it is useful to carry out the thought experiment implied by Maxwell in considering viscosity as the decay of elastically stored energy.

i) Apply a small strain very rapidly to a material. The components of the fluid (molecules, macromolecules or colloidal particles) are moved from their equilibrium positions in energy minima.

ii) The consequence is that a restoring force is produced as the structure is distorted i.e. work has been done and the energy is stored elastically.

iii) The components diffuse to produce a rearrangement of the new structure to accommodate the strains and allow the components to move into new energy minima. That is the stress has relaxed and a permanent or viscous strain has been produced.

The most important feature of this relaxation process is the time-scale. This is a characteristic for the particular system and is known as the Stress Relaxation Time, τ. It is of course related to the diffusivity of the components.

The timescale of our experimental measurements is a matter of choice and equipment, but we can define an observation time as t_o. A useful material classification can now be made using the dimensionless group known as the Deborah Number, D_e:

$$D_e = \frac{\tau}{t_o} \qquad (7)$$

so that as:

$D_e \rightarrow \infty$ we have an elastic solid;
$D_e \rightarrow 0$ a viscous liquid but as
$D_e \rightarrow O(1)$ we have a viscoelastic material.

Our sensitivity covers a range of approximately 10^{-3} s to 10^{3} s and hence it should be no surprise that most of the colloidal systems that we design to have "interesting rheology" have relaxation times that lie in this range. Examples can be chosen from ceramic slips, paints and inks, pharmaceutical products, foodstuffs and detergents.

4. Experimental Procedures

4.1 STRESS RELAXATION

This experiment is the practical application of the thought experiment described above. The sample is placed between a cone and plate or in concentric cylinders and the strain is applied by rapidly moving one surface with respect to the other. The strain growth profile should be linear and as rapid as mechanically possible up to a preset value. After this, the strain is maintained at this constant value. The stress profile is that shown in Figure 3 and shows a growth portion followed by the relaxation portion. Provided that , $t_a \ll \tau$, where t_a is the strain rise time, all the dynamic information can be obtained from the relaxation curve.

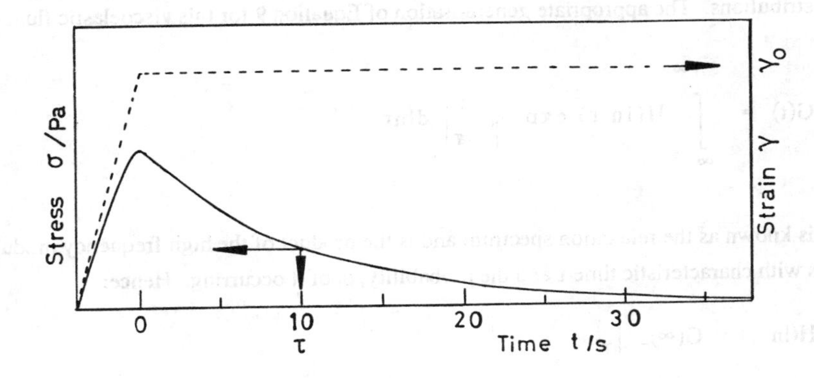

Figure 3. The stress relaxation experiment.

For example, for the simplest case of a single relaxation process occurring in the viscoelastic fluid, the change in stress with time follows an exponential decay:

$$\sigma(t) = \sigma(o) \exp\left(-\frac{t}{\tau}\right) \tag{8}$$

where $\sigma(t)$ is the measured stress at time t and $\sigma(o)$ is the peak stress measured as the deformation reaches the plateau value γ_o. The result is directly analogous to chemical relaxation through first order reaction rates. The relaxation modulus, G(t), is defined:

$$G(t) = \frac{\sigma(t)}{\gamma_o}$$

and hence:

$$G(t) \rightarrow G(\infty) \text{ as } t \rightarrow O;$$

$$G(t) = G(\infty) \exp\left(-\frac{t}{\tau}\right) \tag{9}$$

$G(\infty)$ is the high frequency limit of the storage modulus determined in the forced oscillation experiments described in the section following this one.

It is unusual for a process to show a single relaxation time and a distribution about a mean value is to be expected from any diffusive process. Rather than attempting to sum a large number of exponentials, as would of course be required if discrete times were chosen from the distribution, it is more useful to think in terms of a continuous distribution or spectrum of processes. In addition, a logarithmic scale is generally used and this is capable of handling very broad distributions. The appropriate generalisation of Equation 9 for this viscoelastic fluid is:-

$$G(t) = \int_{-\infty}^{+\infty} H(\ln \tau) \exp\left(-\frac{t}{\tau}\right) d\ln\tau \tag{10}$$

$H(\ln \tau)$ is known as the relaxation spectrum and is the product of the high frequency modulus for a process with characteristic time τ and the probability, ρ, of it occurring. Hence:

$$H(\ln \tau) = G(\infty)\rho \big|_{\tau} \tag{11}$$

The limiting behaviour of the fluid described by Equation 10 is:

$$t \rightarrow O \quad G(t) \rightarrow G(\infty) = \int_{-\infty}^{+\infty} H(\ln \tau) d\ln\tau \tag{12}$$

i.e. this is just the sum of all the elastic contributions which subsequently relax independently of each other, i.e. the processes are assumed to be non-coupled.

Complete relaxation may not have occurred during the timescale of the experiment and then the material would be displaying a solid-like response. Experimentally we observe the stress to fall from the peak value of $\sigma(o)$ to a long time (or "equilibrium") value σ_e giving a non-relaxing component to the modulus G_e so that for a solid component:

$$G(t) = G_e + \int_{-\infty}^{+\infty} H(\ln \tau) \exp\left(-\frac{t}{\tau}\right) d\ln\tau \tag{13}$$

As $t \rightarrow o$,

$$G(t) \rightarrow G(\infty) = G_e + \int_{-\infty}^{+\infty} H(\ln \tau) d\ln\tau \tag{14}$$

and as $t \rightarrow \infty$,

$$G(t) \rightarrow G_e$$

The inversion of Equations 10 and 13 to obtain the relaxation spectrum is usually done by an approximate method. The simplest, and therefore a frequently used approximation is that known as Alfrey's Rule [7]. In this approximation the exponential function in the integral is replaced by a step function varying between zero and unity at $t = \tau$. Equation 10 now reduced to

$$G(t) = \int_{\ln t}^{\infty} H(\ln \tau) d\ln\tau$$

If we write the integral of $H(\ln \tau)$ with respect to $\ln \tau$ as $\hat{H}(\ln \tau)$ so that:

$$\frac{d\hat{H}(\ln \tau)}{d \ln \tau} = H(\ln \tau)$$

then

$$G(t) = \int_{\ln t}^{\infty} H(\ln \tau) \, d \ln = [\hat{H}(\ln \tau)]_{\ln t}^{\infty},$$

i.e. $$G(t) = \hat{H}(\ln \tau)\big|_{\tau = \infty} - \hat{H}(\ln \tau)\big|_{\tau = t}$$

differentiating G(t) with respect to ln t yields as the first term on the right-hand side is a constant:

$$\frac{dG(t)}{d\ \ln\ t} = -\frac{d\,\hat{H}(\ln\ \tau)}{d\ \ln\ t}\bigg|_{\tau = t}$$

that is:

$$-\frac{dG(t)}{d\ \ln\ t}\bigg|_{t=\tau} = H(\ln\ \tau) \qquad (15)$$

This is just the simplest example of Leibniz' Rule for the differentiation of integrals and gives the relaxation spectrum as the negative slope of the curve of relaxation modulus plotted against the logarithm of the time.

4.2 FORCED OSCILLATION

During a forced oscillation test a sample is subjected to a continuous oscillating strain of small amplitude and over a set of fixed frequencies. A sinusoidal deformation is usually employed and the stress and strain profiles are shown schematically in Figure 4. The measured quantities are the peak values of stress and strain, σ_o and γ_o respectively, and the difference in phase of the oscillations expressed as δ radians. With oscillation frequency as ω rad s^{-1} and Δt as the difference between the two wave forms on the time base,

$$\delta = \frac{\Delta 2\pi}{\omega}$$

Any sinusoidal waveform can be expressed as sum of sine and cosine components in complex number notation:-

$$\gamma^*(\omega) = \gamma_o(\cos \omega t + i \sin \omega t)$$

$$\sigma^*(\omega) = \sigma_o(\cos(\omega t + \delta) + i \sin(\omega t + \delta))$$

The ratio of stress to strain is a modulus:

$$G^* = \frac{\sigma^*(\omega)}{\gamma^*(\omega)} ,$$

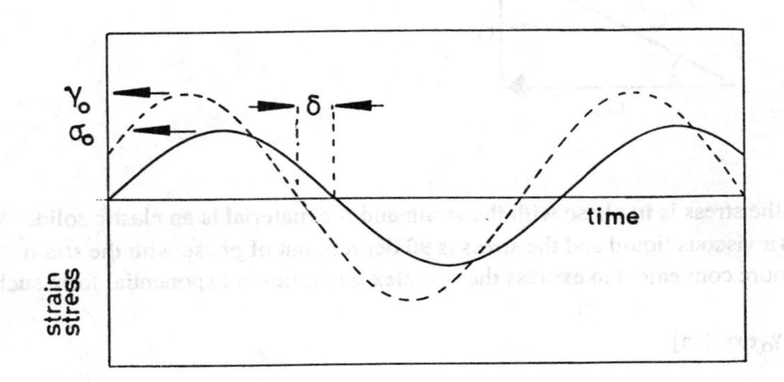

Figure 4. The forced oscillation experiment.

where G^* is known as the complex modulus. It follows that:

$$G^* = \frac{\sigma_o}{\gamma_o}\left\{\frac{\cos(\omega t + \delta) + i \sin(\omega t + \delta)}{\cos \omega t + i \sin \omega t}\right\}$$

so that

$$G^* = \frac{\sigma_o}{\gamma_o}\{\cos \delta + i \sin \delta\}.$$

This equation is normally written in terms of a storage modulus, G', and a loss modulus, G'':-

$$G^* = G' + i\,G'' \tag{16a}$$

so that

$$G' = \frac{\sigma_o}{\gamma_o}\cos \delta \tag{16b}$$

$$G'' = \frac{\sigma_o}{\gamma_o}\sin \delta; \tag{16c}$$

i.e.

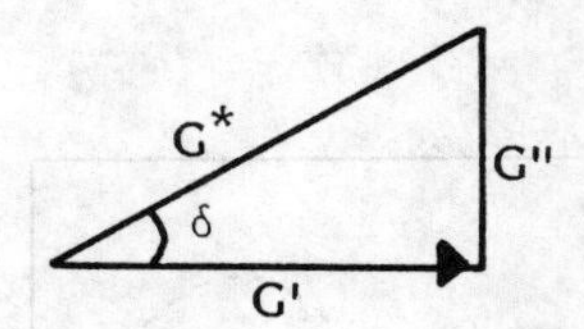

Within $\delta = 0$, the stress is in phase with the strain and the material is an elastic solid. At $\delta = \pi/2$ the material is a viscous liquid and the stress is 90 degrees out of phase with the strain.

It is often more convenient to express the complex quantities in exponential form such that

$$\gamma^* = \gamma_o \exp[i\omega t]; \tag{17a}$$

$$\sigma^* = \sigma_o \exp[i(\omega t + \delta)] \tag{17b}$$

i.e. $$G^* = \frac{\sigma_o}{\gamma_o} \exp i\,\delta , \tag{17c}$$

The strain rates is $\dot{\gamma}^* = i\,\omega\gamma_o \exp[i\omega t]$

$$\frac{\dot{\gamma}^*}{i\omega} = \gamma^* \tag{17d}$$

From Equation 16 and 17:

$$\frac{\sigma^*}{\gamma^*} = G' + iG''$$

and

$$\frac{\sigma^*}{\dot{\gamma}^*} = \frac{-iG'}{\omega} + \frac{G''}{\omega}$$

giving the complex viscosity

$$\frac{\sigma^*}{\dot{\gamma}^*} = \eta^* = \eta' - i\,\eta'' \tag{18}$$

where $\eta' = \frac{G''}{\omega}$ and $\eta'' = \frac{G'}{\omega}$

The Cox-Mertz rule equates the viscosities measured by continuous shear thus:

$$\eta^*_{\omega \to o} = \eta(o)$$

although this rule is not always obeyed in practice.

For the simplest viscoelastic fluid when subjected to a stress σ will deform at a rate $\dot{\gamma}$ which can be calculated from the sum of the elastic and viscous contributions. The constitutive equation for a Maxwell fluid is then:

$$\dot{\gamma} = \dot{\gamma}_{el.} + \dot{\gamma}_{vis.} = \frac{\dot{\sigma}}{G(\infty)} + \frac{\sigma}{\eta(o)} \tag{19}$$

The material constants $\eta(o)$ and $G(\infty)$ define the relaxation time from

$$\tau = \frac{\eta(o)}{G(\infty)} \tag{20}$$

By substitution of the expressions for the oscillating stress, stress rate and strain rates in Equation 19 from Equation 16 gives the frequency dependence of the storage and loss modulus:

$$G'(\omega) = G(\infty) \frac{(\omega\tau)^2}{1 + (\omega\tau)^2} ; \tag{21a}$$

$$G''(\omega) = G(\infty) \frac{\omega\tau}{1 + (\omega\tau)^2} \tag{21b}$$

and

$$\eta'(\omega) = \eta(o) \frac{1}{1 + (\omega\tau)^2} \tag{21c}$$

where use has been made of Equation 20. Figure 5 shows the graphs of the functions in Equation 21 against reduced frequency.

Although Equation 21 describes the behaviour of a material with a single relaxation time, it is more usual to observe a distribution of times about a mean value for each process. This is best described with integral equations in a similar manner to the stress relaxation. Hence for a spectral distribution:

$$G'(\omega) = H(\ln \tau) \frac{(\omega\tau)^2}{1 + (\omega\tau)^2} \, d \ln \tau; \tag{22a}$$

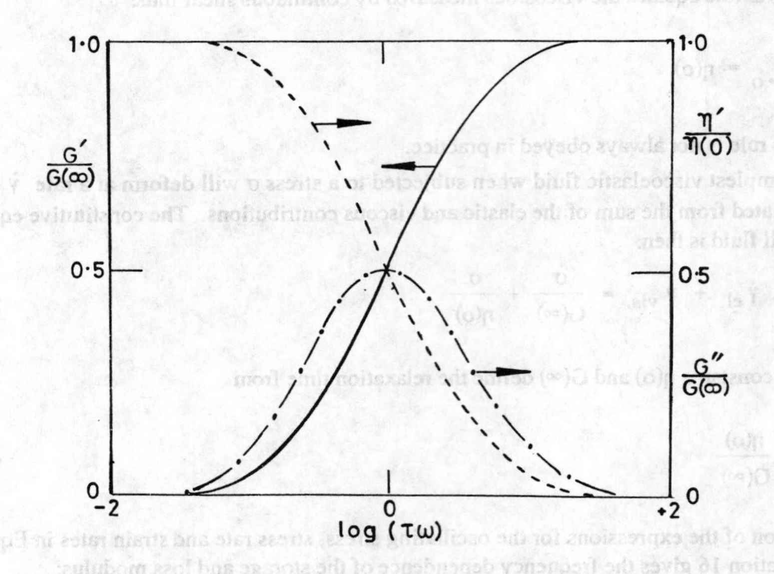

Figure 5. The dynamic response of a Maxwell fluid.

$$G''(\omega) = H(\ln\ \tau)\ \frac{\omega\tau}{1 + (\omega\tau)^2}\ d\ln\tau; \tag{22b}$$

and
$$\eta(\omega) = \int_{-\infty}^{+\infty} H(\ln\ \tau)\ \frac{\tau}{(1 + \omega\tau)^2}\ d\ln\tau \tag{22c}$$

$H(\ln\tau)$ is the same relaxation spectrum introduced earlier in Equation 10. It is useful to use Equations 22a and 22c to look at the response in the high and low frequency limits.

As $\omega \to \infty$, $(\omega\tau)^2 \sim 1 + (\omega\tau)^2$ and

$$\underset{\omega \to \infty}{G'(\omega) \to G(\infty)} = \int_{-\infty}^{+\infty} H(\ln\ \tau)\ d\ln\tau; \tag{23a}$$

and as $\omega \rightarrow 0$, $1 + (\omega\tau)^2 \rightarrow 1$ and

$$\eta'(\omega) \rightarrow \eta(o) = \int_{-\infty}^{+\infty} H(\ln \tau)\, d\ln\tau \; ; \tag{23b}$$

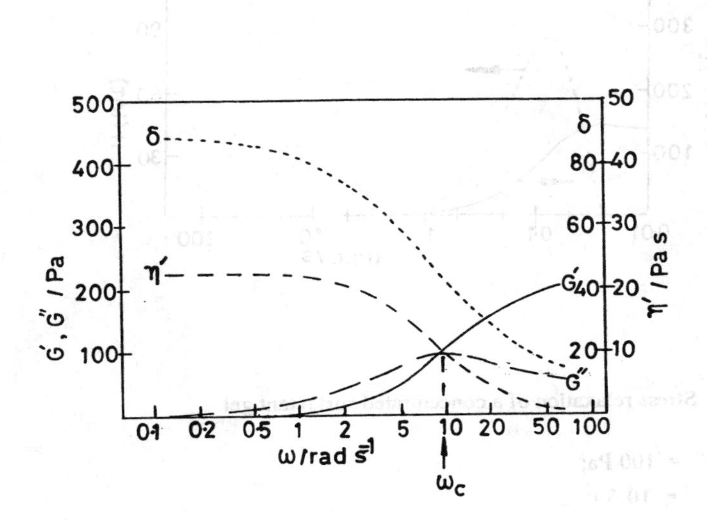

Figure 6. Dynamic modulii obtained from a concentrated surfactant gel using a Bohlin VOR rheometer.

The experimental data shown in Figures 6 and 7 were obtained using a Bohlin VOR rheometer with a concentrated surfactant system containing rod-like micellar units. It was commercial shower gel preparation. The form of the curves in Figure 6 are very similar to a Maxwell fluid, i.e. a system with a single relaxation time. The half width of the response is almost two orders of magnitude in frequency which is only very slightly broader than the half width in Figure 5. The following material quantities were obtained:

ω_c = 9 rads^{-1} ≡ τ = 0.11 s
$(\delta = \pi)$

η' = 22.5 Pa s
$\omega \rightarrow o$

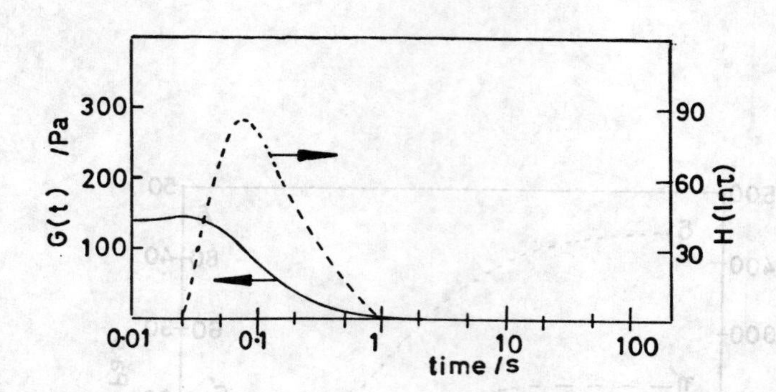

Figure 7. **Stress relaxation of a concentrated surfactant gel.**

at ω_c, $G' = G'' = 100$ Pa;

$\eta' = 10.5$ Pa s

and $G(\infty) = \dfrac{22.5}{0.11} = 202$ Pa.

The stress relaxation curve shown in Figure 7 for the same material shows a $G(t)_{t\rightarrow o} \sim 150$ Pa which is close to the $G(\infty)$ estimate from Figure 6. The strain was applied over 20 ms and some relaxation will have occurred during this time as the characteristic relaxation time was only 0.11 s. The relaxation spectrum peaks at 0.09 s and has fallen to zero at 1 s. This compared favourably with Figure 6 where the storage modulus has fallen to zero at a frequency of 0.9 rads s^{-1} and integration of the spectrum gives a value of $G(\infty)$ of 170 Pa.

It is always important to check that the material being investigated is behaving in a linear fashion if the theory of linear viscoelasticity is being utilised. A simple procedure is to carry out an oscillation test at a single frequency whilst progressively increasing the amplitude of the peak strain. In the linear region the stress will increase in direct proportion to the strain giving constant values for the modulii. A good example is given in Figure 8 which shows data from a concentrated latex [8] at a volume fraction of 0.38 in 10^{-2} dm^{-3} sodium chloride solution. The viscoelasticity is due to the interparticle forces in this example as the material is simply well

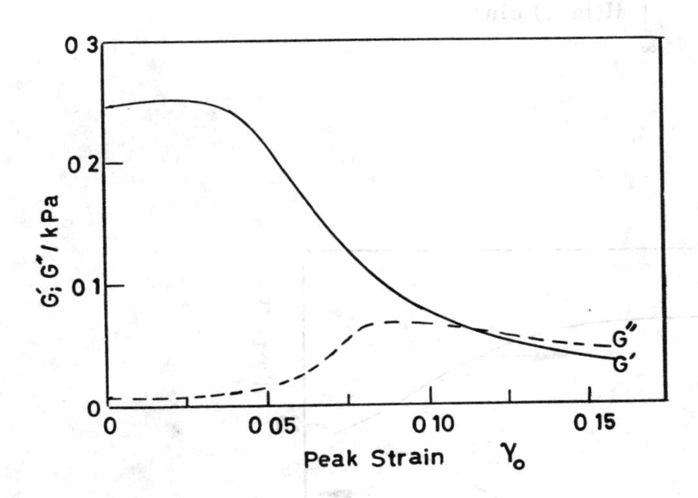

Figure 8. Strain dependence of the dynamic moduli for a concentrated polymer latex ($\phi = 0.38$) in 10^{-2} mol dm^{-3} sodium chloride. [8]

dispersed, charged, spherical particles in a salt solution. The response is linear within the experimental uncertainty up to a peak strain of 0.04. A test frequency of 31.4 rad s^{-1} was chosen and at low strains the material is showing an elastic solid-like response with G'>G"; i.e. tan $\delta \to 0$. The behaviour rapidly changes with increasing strain until at strains in excess of 0.12 the loss modulus becomes greater than the storage and material response is fluid-like. This process is a shear-induced melting and has strong implications with regard to the suspension structure. Under quiescent conditions these latex systems show marked iridescence at moderate volume fractions indicating a high degree of order. Over most of the concentrated (i.e. structured) suspension range there is faced centred cubic symmetry [9]. If a strain of 0.04 is taken as the point where melting occurs, a yield stress can be defined from the modulus in the linear regime as

$$\sigma_y \sim 250 \ . \ 0.04 = 10 \text{ Pa}$$

This is the simplest approximation and means that complex modulus is being approximated by a step function which drops to zero at $\gamma_o > 0.04$. A slightly better estimate could be obtained if integration under the experimental curve is taken to the strain where the most rapid change in viscosity occurs. However this increased the yield stress by less than factor of two in the example given in Figure 7. The important point to note is that plastic flow could be expected with this latex system at stress levels of $\sigma > 10$ Pa is applied for long periods of time. Figure 9 shows the stress relaxation curve obtained for this material and shows that only partial relaxation of the stress occurs; i.e. the material has become solid-like and Equation 23a should be:

$$G(\infty) = G_e \int_{-\infty}^{+\infty} H(\ln \tau)\, d\ln\tau \tag{24}$$

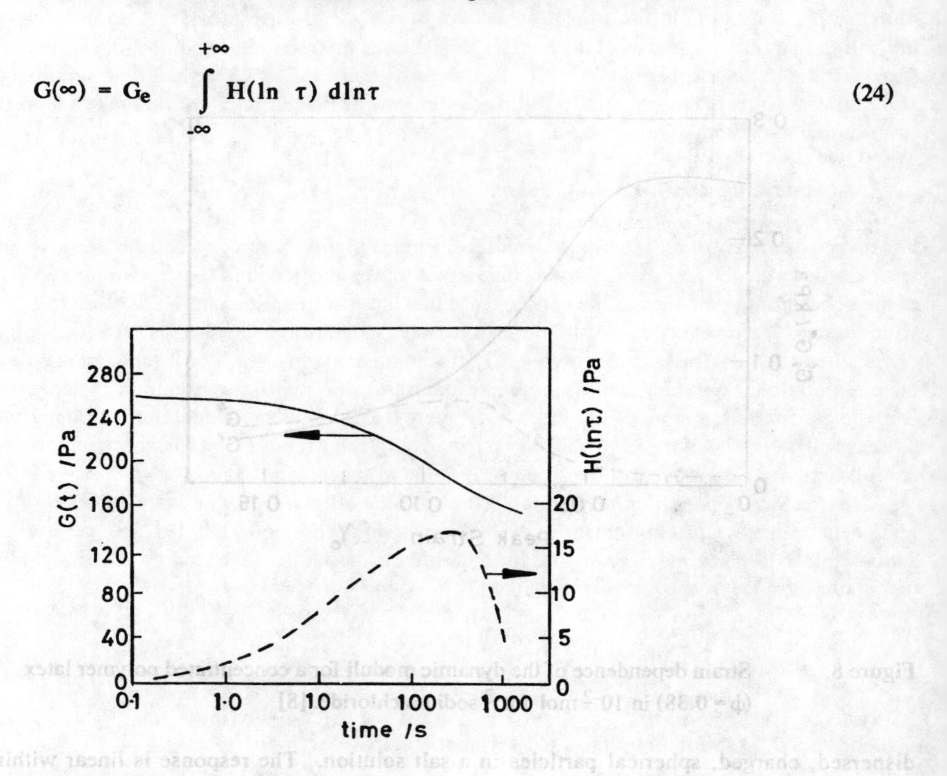

Figure 9. Stress relaxation for a polymer latex at $\phi = 0.38$.

where G_e is the long time residual or equilibrium modulus. The implication for the viscosity is of course that $\eta(o) \rightarrow \infty$. At $\phi < 0.36$ at 10^{-2} mol dm^{-3} electrolyte, this latex showed complete relaxation and liquid-like behaviour. The liquid-like phase transition was dependent on electrolyte concentration so that:

[NaCl]	10^{-4}	10^{-3}	10^{-2}
ϕ at transition	0.06	0.18	0.37
$\kappa(r - 2a)$	4	4.5	6.5

It is interesting to note that the transition from a liquid-like response to a solid-like response with this particle size latex occurred at surface to surface particle separations of several diffuse layer dimensions, for example the transition occurs at 6.5 κ^{-1} at 10^{-2} mol dm^{-3} sodium chloride indicating that only quite weak interparticle interactions are required to control the structure and dominate the rheological behaviour. In all of these systems the structure had f.c.c symmetry and at the lower electrolyte concentrations the counter ions of the particles would make a significant contribution to the total ionic strength.

5. Modelling the Limiting Behaviour

For disperse systems with long range repulsive forces, highly ordered structures are formed and viscoelasticity can be readily measured. The structure has face-centred cubic symmetry over most of the concentration range with the consequence that the co-ordination number is close to 12 over a wide volume fraction range whilst the mean particle separation decreases. This is in contrast to dispersions in which the particles are weakly flocculated due to attractive forces. In this case the mean separation is constant at the energy minimum position whilst the packing density increases (i.e. the co-ordination number increases). When systems are coagulated due to strong attractive forces, a fractal structure is produced [10] and this requires a different approach from the equilibrium structure models of the first two dispersion states.

A lattice model is particularly suitable for systems with strong repulsive forces due to the regularity of the structures produced as evident from diffraction of light [9] and neutrons [11]. The simplest lattice model gives the high frequency modulus in terms of the mean nearest neighbour distance, r, and the pair potential [12]:

$$G(\infty) = \frac{0.83}{R} \frac{\partial^2 V(R)}{\partial R^2} \Big|_{\bar{R}} \tag{25a}$$

The constant 0.83 results from the f.c.c. symmetry and:

$$\bar{R} = 2a \left[\frac{\phi_m}{\phi}\right]^{1/3} \tag{25b}$$

The pair potential V(R) can be calculated from the electrostatic repulsion between the particles although account of the total ion concentration should be made [13,14].

When the suspension structure is produced by weak attractive forces, as for example can occur with large sterically stabilised particles or systems in which depletion flocculation is important [15], lattice models are inappropriate and a better statistical description of the structure is required. This may be done by using a model for the pair distribution function g(R), and then summing the pair interactions using this description of the structure in a similar manner to the calculation for molecular liquids [16]. This technique has been used successfully for large latex particles in high electrolyte concentrations [17]. The resulting prediction of the high frequency modulus was:

$$G(\infty) = \frac{2\pi}{15} \rho^2 \int_0^\infty g(R) \frac{d}{dR} \frac{r^4 V(R)}{dR} dR \tag{26}$$

where ρ is the particle number density. (Of course is a lattice model is assumed, and only the first nearest neighbour interaction is considered, Equation 26 reduced to a similar form to Equation 25a). Figure 10A illustrates schematically the type of pair distribution calculated for the weakly attractive case with corresponding co-ordination number N_1 calculated from the distribution. The

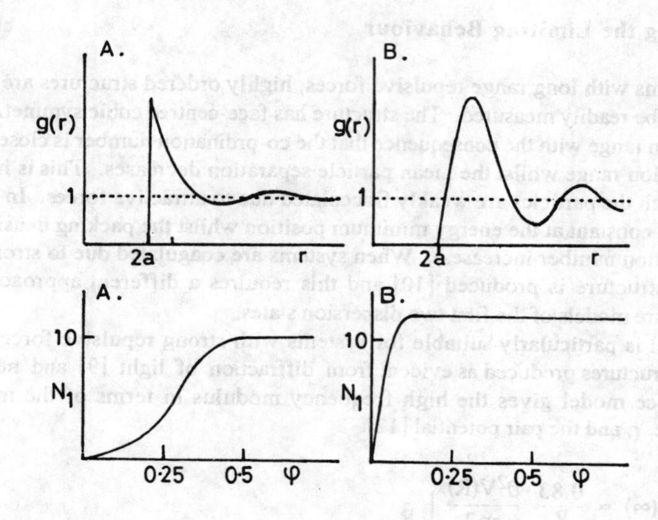

Figure 10. The pair distribution function and co-ordination number dependence on volume fraction for:- A. a weakly flocculated dispersion, B. an electrostatically stabilised dispersion.

marked densification of the structure occurs at c.a. $\rho \sim 0.3$ which means experimentally that a space-filling system would be observed in this region. For contrast, the pair distribution function and co-ordination number profile are shown schematically in Figure 10B for a system with strong repulsion dominating the pair interactions. Figure 11 shows the results calculated from Equations 25 and 26 that gave good fits to experimental data [17, 18].

6. Measurement of the High Frequency Plateau Modulus

When studying the more solid-like samples of concentrated dispersions, relaxation times in excess of 10^2 s can be observed and measurements using forced oscillation or stress relaxation are capable of sufficiently high frequencies (i.e. up to *ca* 50 rad s^{-1}) that the limiting values of G' and G(t) respectively yield good estimates of $G(\infty)$. However, the more fluid systems have much shorter relaxation times and then a different technique is required to obtain an estimate of $G(\infty)$. For many of the weak colloidal gels showing viscoelasticity wave propagation provides a rapid

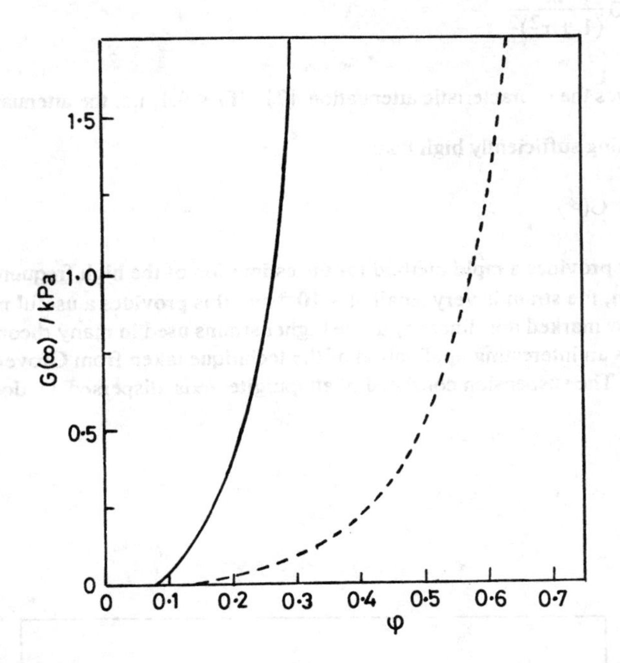

Figure 11. The dependence of G(∞) on volume fraction for polymer latices which were: ——— electrostatically stabilised [18], - - - - - weakly flocculated [17].

assessment of the wave rigidity modulus $\tilde{G}$. In an apparatus such as the Rank Shearometer (Rank Bros., Bottisham, Cambridge), the time taken for a shear wave to travel between two parallel discus is recorded for fixed disc separations. The slope of the disc separation-time plot gives the wave velocity v. This is then used with the suspension density to give the wave rigidity modulus, $\tilde{G}$.

$$\tilde{G} = \rho v^2 \tag{27}$$

The attenuation of the wave with propagation distance enables the wave rigidity modulus to be split into storage and loss components at the frequency of the wave. It is important that the variation in the height of the chosen wave peak is measured over a range of disc separations to give the attenuation as the decay in peak height with time is often dominated by the friction in the instrument bearing (and this will not of course vary with disc separation). The decay of the wave is experimental with distance and decreases with a characteristic decay distance of x_o. Both the velocity v ms^{-1} and the frequency of the travelling wave ω rad s^{-1} are easily measured and give the dynamic modulii as:

$$G'\big|_{\omega} = \tilde{G}\,\frac{(1-r^2)}{(1+r^2)^2}\;; \tag{28a}$$

$$G''_{\omega} = \tilde{G}\frac{2r}{(1+r^2)^2} \tag{28b}$$

where $r = \dfrac{v}{x_o\,\omega}$ gives the characteristic attenuation [7]. If $r < 0.1$, i.e. the attenuation is small, the frequency is becoming sufficiently high that:

$$G' = \tilde{G} \sim G(\infty)$$

and the experiment provides a rapid method for the estimation of the high frequency behaviour of the gel. In addition, the strain is very small at $< 10^{-4}$ and this provides a useful methodology for systems which show marked non-linearity at the higher strains used in many rheometers.

Figure 12 shows an interesting application of the technique taken from Grover's work [19] on an oleophilic clay. The suspension consisted of attapulgite rods dispersed in dodecane using a

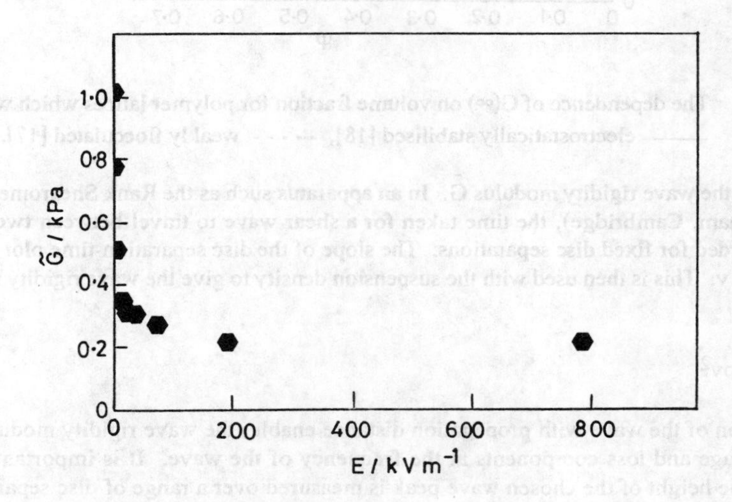

Figure 12. The effect of applied field on the wave rigidity modulus of an organophilic attapulgite clay dispersion [19].

cationic surfactant. The wave rigidity modulus was measured as a function of an D.C. field applied between the plates. It is interesting to note that the modulus dropped rapidly with only moderate applied fields as the structure of the suspension became linearised due to the field. The work of Hoffman described in this volume showed an alignment of rod-like micellar units perpendicular to an applied field using both optical and neutron scattering techniques. This type of structural arrangement would result in a low modulus system. The experiment provides a graphic illustration of the interplay between interparticle forces and suspension structure in the determination of the viscoelastic properties of suspensions.

References

1. J.G. Kirkwood, J. Chem. Phys., 919-925, 7, (1939).

2. J.C. van der Werff and C.G. de Kruif, J. Rheol., 421, 33, (1989).

3. A. Einstein, "Investigations on the Theory of the Brownian Motion", Dover, New York, 1956.

4. G.K. Batchelor, J. Fluid. Mech., 97, 83, (1977).

5. I.M. Kreiger, Adv. Coll. and Interface Sci., 111, 3, (1972).

6. M. Reiner, Physics Today, 17, 62, (1969).

7. J.D. Ferry, "Viscoelastic Properties of Polymers", Wiley, New York, 1980.

8. S. Keeping, Ph.D. thesis, Bristol 1989.

9. P.A. Hiltner and I.M. Krieger, J. Phys. Chem., 2386, 73, (1969).

10. R. Buscall, P.D. Mills, J.W. Goodwin and D. Lawson, J. Chem. Soc., Faraday Trans. I, 4249, 89, (1988).

11. D.J. Cebula, J.W. Goodwin, G.C. Jeffrey, R.H. Ottewill, A. Parentich and R.A. Richardson, Faraday Discuss. Chem. Soc., 37, 76, (1983).

12. R. Buscall, J.W. Goodwin, M.W. Hawkins and R.H. Ottewill, J. Chem. Soc. Faraday Trans. I, 2869, 78, (1982).

13. D.W. Benzing and W.B. Russel, J. Coll. and Interface Sci., 163, 83, (1981).

14. B. Beresford-Smith and D.Y.C. Chan, Chem. Phys. Lett., 474, 92, (1982).

15. D.H. Napper, "Polymeric Stabilisation of Colloidal Particles", Academic Press, 1983.

16. R.W. Zwanzig and R.D. Mountain, J. Chem. Phys., 4464, 43, (1965).

17. J.W. Goodwin, R.W. Hughes, S.J. Partridge and C.F. Zukoski, J. Chem. Phys., 559, 85, (1986).

18. I.C. Callaghan, J.W. Goodwin and S. Keeping, submitted.

19. B. Grover, Ph.D. thesis, Bristol 1990.

INTERFACIAL RHEOLOGY AND ITS APPLICATION TO INDUSTRIAL PROCESSES

J.H. CLINT
Colloid Science Branch
BP Research Centre
Chertsey Road
Sunbury-on-Thames
MIDDLESEX TW16 7LN

ABSTRACT. In practical systems containing liquids, the interfaces are subjected to disturbances from equilibrium. Changes in shape and area of an interface lead to the concepts of interfacial shear and dilatational rheology and these are explained. Methods for measuring the appropriate interfacial rheological parameters are described and the relaxation mechanisms responsible for this behaviour are discussed. An important aspect of interfacial rheology is the Marangoni effect in which monolayer material moves in response to local differences in surface pressure. The motion is coupled to the adjacent liquid in the bulk phases and this has important consequences in such processes as film thinning, foam stability, droplet coalescence and wave damping, of all which are illustrated.

1. Introduction

A great many industrial processes involve liquids. Where more than one liquid phase is involved (e.g. 2 immiscible liquids) or where a liquid and a gas phase are present, then a complete description of the process must take account of the properties of the interface. In most systems of practical interest the interface is not static but is subject to changes in area or shape which in some cases can be taking place on a very rapid time scale. Because of this it is usually inadequate to use only equilibrium properties of the interface in our description of the system.

One reason for this is that interfacially active molecules diffuse to and adsorb at interfaces at finite rates. Where the changes in interface geometry are occuring on a time scale comparable to that of the diffusion and adsorption processes, then it is the dynamic properties of the interface which become important. The science of how interfaces respond to imposed stresses is known as interfacial rheology. It is the intention of this article to outline the principles of this subject in sufficient detail to explain how they can be applied to processes such as emulsion stabilisation and resolution, and foam stabilisation and breaking.

2. Coalescence of Fluid Droplets

An important effect, which plays a significant role in most of the processes to be considered, is that of coalescence between fluid droplets. When two approaching droplets are forced together, even momentarily during a collision, their surfaces become flattened at the contact point and a thin film of the continuous phase is formed between them. If this film remains intact during the whole encounter, then the droplets cold reseparate in more or less their original state. If the film ruptures

D. M. Bloor and E. Wyn-Jones (eds.),
The Structure, Dynamics and Equilibrium Properties of Colloidal Systems, 681–694.

then irreversible coalescence of the droplets will occur. In a sheared system the larger droplet may then be broken down again into smaller droplets. The important point is that the stability of the system is dependent to a large extent on the stability of this thin film.

Once the film has formed and the relative motion between the approaching droplets has stopped, the film continues to thin under the influence of capillary forces. Because of the pressure difference across a curved surface, the liquid of the continuous phase within the film, where the interfaces are almost planar, is at a higher pressure than that in the bulk. This capillary pressure is the main driving force causing thinning of the film. Only when the film becomes extremely thin will dispersion (van der Waals) forces make a significant additional contribution. Thinning is opposed by viscous effects due to the bulk viscosity of the film and other effects due to interfacial rheology as we shall see later. A useful model study involves measurement of film thickness in systems thinning under controlled capillary pressure as described in the next section.

3. Film Thinning

Optical reflectance methods can be used to measure the thickness of a thin film during drainage. The technique can be applied equally well to liquid films in gas or to liquid in liquid films. The measurements are most conveniently done in a horizontal cell mounted on a microscope stage where the liquid film is held on a small orifice [1].

The usefulness of this type of study is illustrated by the measurement of the kinetics of drainage of crude oil films in water. Figure 1 shows results for two different crude oils. To understand the differences between the two oils it is helpful to recall the Stephan-Reynolds equation for drainage of liquid of viscosity η from between two rigid discs of radius r, forced together by an applied pressure p.

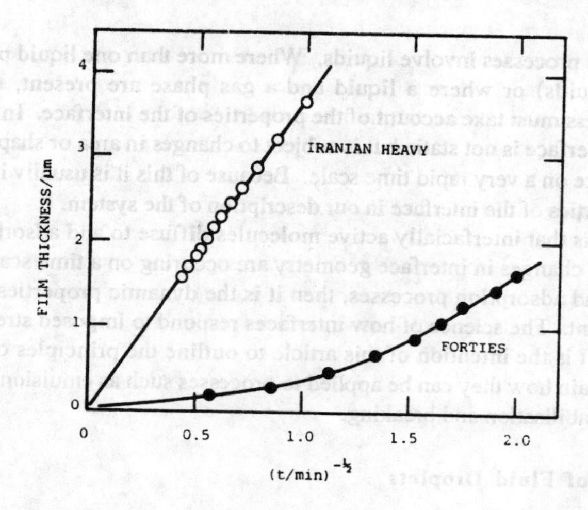

Figure 1. Kinetics of drainage of crude oil films in water at 25°C.

The film thickness can be written in the form

$$h = (3\eta r^2/4p)^{1/2}.t^{-1/2} \quad (1)$$

This relationship implies that, for different oils draining in water, plots of thickness against $t^{-1/2}$ should be straight lines whose gradients reflect differences in oil viscosity and capillary pressure (proportional to oil/water interfacial tension).

The straight line obtained with Iranian Heavy oil suggests that the Stephan-Reynolds equation does indeed apply. In contrast the rate of drainage of a film of Forties crude oil in water is much more rapid than can be accounted for by differences in viscosity and interfacial tension. Also the relationship does not follow equation 1. The implication is that the Iranian Heavy/water interface is behaving sufficiently like a rigid surface that it can be regarded as a solid boundary. With Forties oil the interface is much more fluid and more complex momentum transfer is taking place.

This example illustrates well the significance of interfacial rheology in systems of great practical importance. Crude oils are produced with considerable quantities of water which becomes emulsified in the oil as it passes through the region of very high shear at the well-head choke. The water is removed, preferably before transmission along pipelines, by sedimentation which is greatly speeded up by adding chemical demulsifiers to promote coalescence to form larger water droplets. Forties crude oil is much easier than Iranian Heavy to demulsify in practice; a fact which is consistent with the data in Figure 1.

Clearly the study of interfacial rheology in such systems, and the factors affecting it, should lead to better understanding and control of processes such as demulsification.

4. Interfacial Rheology

In any study of rheology we are concerned with the response of our system to applied stress. For bulk rheology the stress is most frequency a shear stress. With interfaces, the area of the portion being studied can be changed just as easily as the shape. As a result dilatational (sometimes called dilational) stresses, which tend to expand an interface uniformly in two dimensions, are used as well as shear stresses. Since the experimental methods for these two processes are different, we shall consider them separately.

4.1 INTERFACIAL SHEAR RHEOLOGY

When a section of interface is subjected to a tangential stress this results in a shear strain. A tangential force F acting on a length y of surface is equivalent to a shear stress F/y. If this produces a rate of strain dv/dx (i.e. a velocity gradient over the interface) then the stress and rate of strain are related by

$$F/y = \eta_s\,(dv/dx) \quad (2)$$

The coefficient in Equation 2 is the surface or interfacial shear viscosity, η_s. The viscous dissipation can be considered as acting within a surface film of thickness z, perhaps that of a monomolecular layer. Then the equivalent bulk viscosity of the interfacial material would be

$$\eta = \eta_s/z \quad (3)$$

Typical values for η_s are in the range 1 μPa.s.m, to 1 mPa.s.m, for example for condensed films of stearic acid on water at low pH. The equivalent bulk viscosity, assuming a monolayer thickness of 2.5 nm, is 400 to 4 x 10^5 Pa.s which is approximately that for a substance like butter [2]. Such viscosities indicate that monolayers in the so-called solid phase are indeed rather solid-like confirming the orientation of the hydrocarbon chains and the strong lateral interactions between adjacent molecules.

4.2 MEASUREMENT OF INTERFACIAL SHEAR VISCOSITY

The most direct method for measuring surface shear viscosity for insoluble monolayers is the "canal method", often used with a Langmuir trough. The trough is divided into two compartments connected by a narrow canal of width W and length l between two floating barriers as shown in Figure 2.

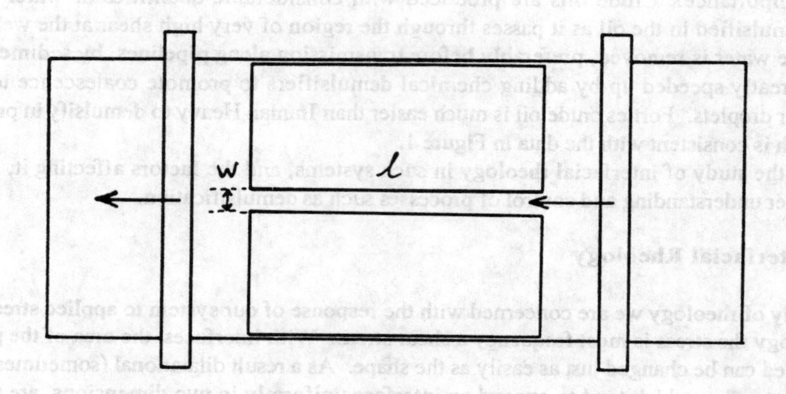

Figure 2. Canal method for surface shear viscosity measurement. Barriers are moved so as to maintain constant surface pressure difference between ends of canal of length l and width w.

The two normal barriers spanning the trough are moved together in such a way as to force the monolayer through the canal whilst maintaining a constant surface pressure difference $\Delta\pi$ between the two halves of the trough. The flow rate through the canal Q is derived in the same way as is done for Poiseuille's equation but in two dimensions. The result is

$$Q = \Delta\pi . W^3 / (12\eta_s l) \tag{4}$$

from which η_s can be calculated directly.

Equation 4 is derived with the assumption that there is perfect slippage between the monolayer and the immediately adjacent liquid. In reality momentum transfer between the monolayer and the subphase means that a thin layer of liquid is dragged along by the monolayer, thereby increasing the resistance to flow. This coupling of motion between the monolayer and the adjacent liquid is usually referred to as the Marangoni effect [3,4] and is quite general. It was originally discussed in terms of the self-healing effect in thin liquid films but it is an equally important effect in the mechanisms of stability and instability in foams and emulsions. We shall return to this later.

4.2.1 *The Biconical Bob Rheometer.* For films or monolayers at the oil/water interface, a very useful method is the biconical bob rheometer. Figure 3 shows a general drawing of the equipment.

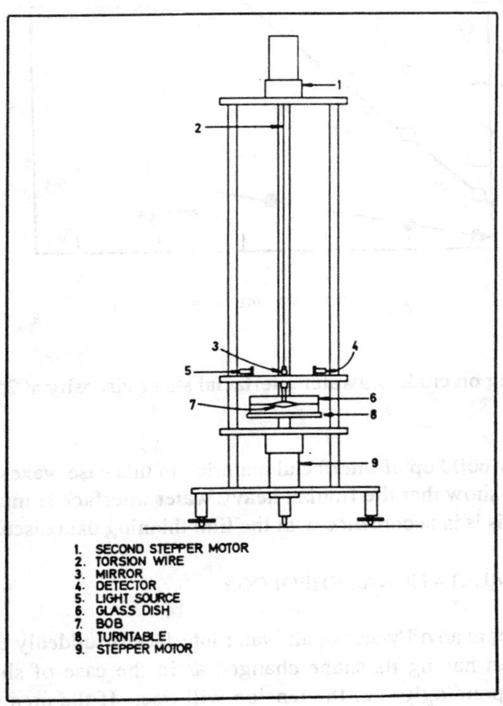

Figure 3. Biconical bob rheometer for interfacial shear viscosity.

The sharp-edged bicone, suspended from a sensitive torsion wire, is located at the oil/water interface for a system contained in a cylindrical dish. The dish can be rotated at different speeds using a stepper motor. For interfaces with comparatively low viscosity, a constant rotation mode may be used where the dish is rotated at various speeds and the steady state deflection of the bob observed. For this technique a "blank" experiment also has to be performed in which oil and water phases have identical bulk viscosity to those of interest but without an interfacial film being present; i.e. pure oils and surfactant free aqueous phase. Details of the experimental technique and calculations can be found in the literature [5]. For higher viscosities a damped oscillation method can be used and for very high viscosities the instrument can be operated in a creep compliance mode [6].

Typical results for oil/water interfaces are shown in Figure 4 for the two crude oils already discussed. The data show how the interfacial shear viscosity in such systems tends to increase

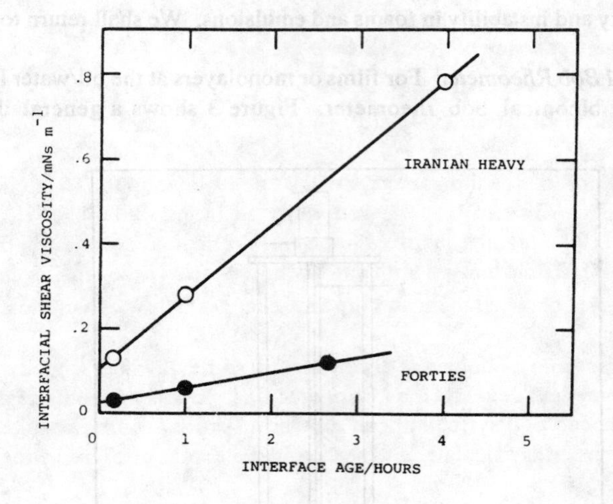

Figure 4. Effect of ageing on crude oil/water interfacial shear viscosity at 25°C.

with time due to the slow build up of interfacial material, in this case waxes and asphaltenes from the oil. The results also show that the Iranian heavy/water interface is much more viscous than that for Forties oil and this is in accordance with the film thinning data discussed in Section 3.

4.3 INTERFACIAL DILATATIONAL RHEOLOGY

If the adsorbed monolayer at an oil/water or air/water interface is suddenly expanded uniformly in all directions, rather than having its shape changed as in the case of shear, then the surface pressure will fall correspondingly, i.e. the tension will rise. If the area is initially A and the change in area is dA, then the resulting change in tension can be expressed as

$$d\gamma = \varepsilon^*.dA/A \qquad (5)$$

The coefficient ε^* is called the interfacial dilatational modulus and the star is used to denote that it is a complex number. The modulus is complex because in cyclic changes in area, for example, there is a phase difference between the area and tension changes. Like all complex numbers the dilatational modulus can be split into real and imaginary parts

$$\varepsilon^* = |\varepsilon|\,(\cos\phi + i.\sin\phi) \qquad (6)$$

where $|\varepsilon|$ is the amplitude of the modulus, ϕ is the phase difference between tension and area changes, and i is $\sqrt{-1}$. The equation can also be written

$$\varepsilon^* = \varepsilon_d + i.\omega\eta_d \qquad (7)$$

where ω = the angular frequency of the cyclic area change,

ε_d = the interfacial dilatational elasticity, and

η_d = the interfacial dilatational viscosity.

For any interface the dilatational parameters will be dependent on the frequency of the cyclic area change. This behaviour can be followed using an interfacial film balance fitted with oscillatory barriers [7]. The method involves imposing cyclic area changes on the interface and measuring changes in interfacial tension with a Wilhelmy plate. These tension changes, together with the phase differences between them and the area changes, allow calculation of ε_d and η_d at each frequency.

The technique suffers from a number of disadvantages some of which can be overcome by the use of the pulsed drop method developed by Clint et al. [8]. In this method a small drop of oil is formed in water on the end of a syringe tip, as shown in Figure 5. Area changes are calculated from drop diameters, tension is calculated by measuring the excess pressure inside the drop with a sensitive pressure transducer.

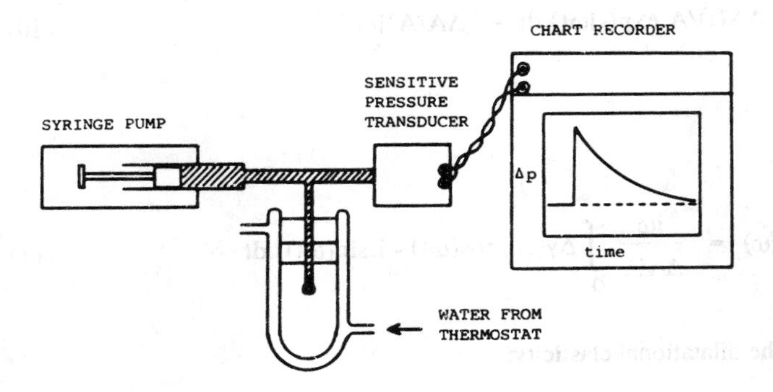

Figure 5. Pulsed drop method for interfacial dilatational rheology.

Instead of the conventional oscillatory method for dilatational modulus measurement, the single pulse Fourier transform method is used [9]. When the cell containing the aqueous solution has reached thermal equilibrium, the drop radius r_1 is measured. A fixed volume pulse is then injected from the syringe pump over a short time period (preferably <1 second) which increases the radius to r_2. The variation of pressure with time is then followed on a chart recorder which itself has a very short time constant. Alternatively the pressure transducer can be interfaced to a microcomputer which assists subsequent data handling.

All of the pressure change after the rapid rise is assumed to take place at a constant drop radius, the final radius r_2. Then the interfacial tension at any time, $\gamma(t)$ is equal to $\Delta p(t).r_2/2$. The interfacial modulus is usually written

$$\varepsilon^* = d\gamma/d\ln A = \varepsilon' + i\varepsilon'' \tag{8}$$

Taking Fourier transforms of the numerator and denominator converts the perturbation time function $\Delta A(t)/A$ and the response time function $\gamma(t)$, to the frequency function:

$$\varepsilon^*(\omega) = \frac{\int_{-\infty}^{\infty} \Delta\gamma(t).\exp(-i\omega t).dt}{\int_{-\infty}^{\infty} \Delta A(t)/A.\exp(-i\omega t).dt} \tag{9}$$

For a perfect step function (instantaneous area change),

$$\int_{-\infty}^{\infty} \Delta A(t)/A.\exp(-i\omega t).dt = (\Delta A/A)i\omega \tag{10}$$

Therefore

$$\varepsilon^*(\omega) = \frac{i\omega}{\Delta A/A} \int_0^{\infty} \Delta\gamma(t).[\cos(\omega t) - i.\sin(\omega t)].dt \tag{11}$$

The real part gives the dilatational elasticity:

$$\varepsilon' = \varepsilon_d(\omega) = \frac{\omega}{\Delta A/A} \int_0^{\infty} \Delta\gamma(t).\sin(\omega t).dt \tag{12}$$

The imaginary part gives the dilatational viscosity:

$$\varepsilon'' = \omega\eta_d(\omega) = \frac{\omega}{\Delta A/A} \int_0^{\infty} \Delta\gamma(t).\cos(\omega t).dt \tag{13}$$

where ω = angular frequency in radians per second. Equations 12 and 13 can be used to calculate ε_d and η_d at any frequency from the decay curve using a microcomputer. It is adequate to take approximately 100 points along the decay curve for these calculations.

Examples of results are shown in Figures 6 and 7 for a model system of 10 ppm stearic acid dissolved in n-decane against distilled water at pH 2.5 to prevent ionisation of the acid. Figure 6

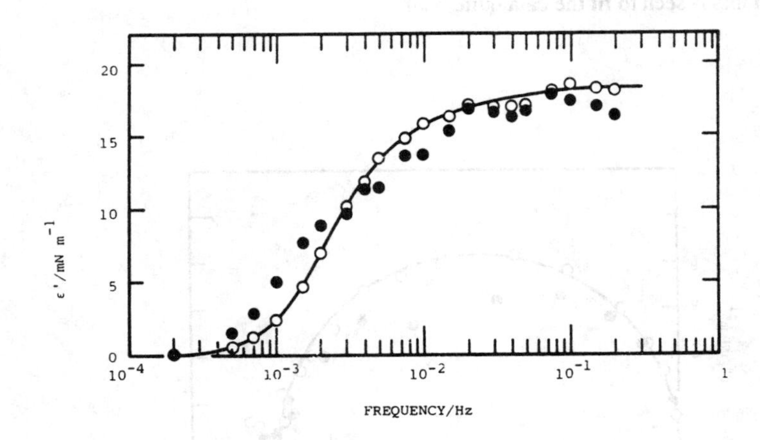

Figure 6. Real part of interfacial dilatational modulus for 10 ppm stearic acid in n-decane/distilled water, pH 2.5 at 25ºC. Open circles-trough method; filled circles - drop method.

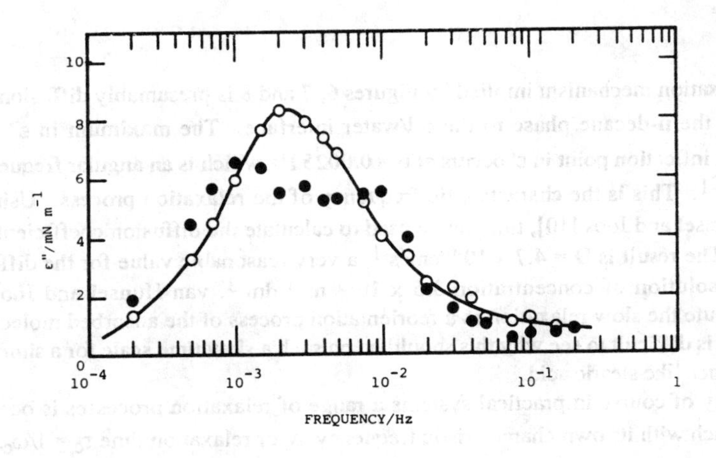

Figure 7. Imaginary part of interfacial dilatational modulus. (Systems and symbols as for Figure 6).

shows the real (elastic) component and Figure 7 the imaginary (frequency x viscosity) component of the modulus. The results are shown in comparison with data obtained using the trough technique, also using the Fourier transform method. The shapes of the curves for ε' and ε" are very close to those expected for a single relaxation mechanism. This is illustrated more strikingly

in Figure 8 where a Cole-Cole plot is shown. A single relaxation mechanism has a semi-circular Cole-Cole plot and this is seen to fit the data quite well.

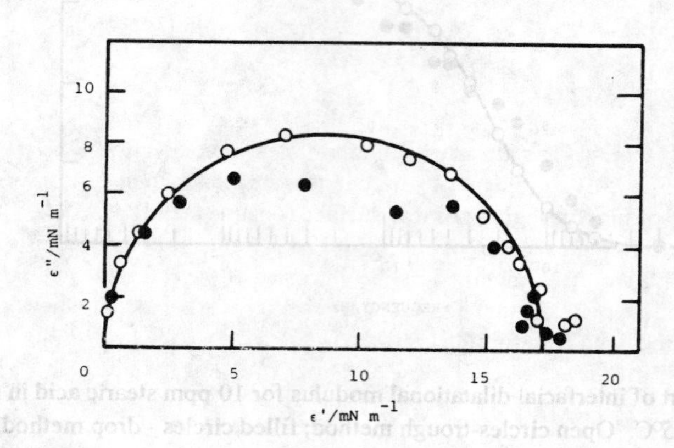

Figure 8. Cole-Cole plot for interfacial dilatational modulus. (Systems and symbols as for Figure 6).

The single relaxation mechanism implied by Figures 6, 7 and 8 is presumably diffusion of the stearic acid from the n-decane phase to the oil/water interface. The maximum in ε" which corresponds to the inflection point in ε' occurs at $\upsilon = 0.0025$ Hz which is an angular frequency $\omega = 2\pi\upsilon = 0.0157\ s^{-1}$. This is the characteristic frequency of the relaxation process. Using the method of van Hunsel and Joos [10], this can be used to calculate the diffusion coefficient in the n-decane phase. The result is $D = 4.7 \times 10^{-6}\ cm^2s^{-1}$, a very reasonable value for the diffusion coefficient for a solution of concentration 3.5×10^{-5} mol dm^{-3}. van Hunsel and Joos [10] disagree and attribute the slow relaxation to a reorientation process of the adsorbed molecules at the interface but it is difficult to see why this should be on such a slow time scale for a simple low molar mass substance like stearic acid.

More frequently of course in practical systems a range of relaxation processes is occurring simultaneously, each with its own characteristic frequency ω_c or relaxation time $\tau_c = 1/\omega_c$. The curves of ε' and ε" against log(ω) then become broadened out. If only two or three separate relaxation processes are occuring with widely different relaxation times, then they can often be assigned to specific mechanisms. Common mechanisms at interfaces are

(a) Diffusion of one or more components of the monolayer into either or both of the bulk phases, depending on their relative solubilities.
(b) Slow collapse of insoluble monolayers.
(c) Molecular reorientation or rearrangements within the monolayer.

All of these processes might have characteristic frequencies in the region 10^{-4} to 1 Hz ($\omega = 6.3 \times 10^{-4}$ to $6.3\ s^{-1}$) which is the frequency range accessible to the experimental method described. More recently surface laser light scattering techniques have been used to explore additional relaxation processes with much higher characteristic frequencies up to about 100 kHz [11,12]. These processes may be associated with vertical motion of the interface during the collective motion of spontaneous ripples (ripplons) at the interface, although this is disputed by Hard and Neuman[12].

5. Marangoni Effects

The Marangoni effect [3,4] has already been mentioned in connection with the canal method for surface shear viscosity. If a gradient of surface pressure exists, then monolayer material tends to flow from regions of high to regions of low surface (or interfacial) pressure. This motion by the interfacial monolayer creates a drag on the liquid adjacent to it in the bulk phases. The speed of flow of monolayer material is governed by the viscosity of the interface and the bulk viscosity of the adjacent fluids. Many important processes are strongly influenced by the Marangoni effect and some of these are included here for illustration.

5.1 TEARDROP FORMATION

A well known example is that of teardrop formation on the inside of glasses containing strong alcoholic drinks like sherry. Alcohol lowers the surface tension of water by being adsorbed at the air/water interface. In addition the mixture tends to spread up the inside of the glass in the form of a thin wetting film. Localised air currents or eddies cause localised evaporation of some ethanol which raises the surface tension. The gradient of surface tension thereby set up causes movement of the monolayer at the air-water interface towards the spot where evaporation occurred. Drag on the underlying sherry film results in a build up of liquid at this point beyond the thickness of the normal wetting film. The result is the formation of a teardrop which runs down the glass.

5.2 FOAM STABILITY AND ANTIFOAMS

Foams are assemblies of thin films formed from solutions of surface active agents. Surfactant monolayers stabilise the films against drainage below a certain thickness by providing electrical or steric repulsion barriers. By a kind of inverse Marangoni effect, drainage of liquid from within the film is hindered if the interfacial viscosity of the adsorbed monolayers is high. As a result the foam lifetime is increased by high surface viscosity. For example, addition of dodecyl alcohol to aqueous solutions of sodium dodecyl sulphate causes a very marked increase in surface viscosity and an equally large increase in foam lifetime.

In addition to this simple drainage argument, the ability of the films to resist localised shocks (e.g. mechanical or thermal) is critically dependent on the rheology of the interface. Suppose a local disturbance results in sudden thinning at a point in the film. The size of the surface pressure gradient created will depend on the dilatational elasticity expressed at a frequency corresponding to the speed of creation of the disturbance. For very slow disturbance of the interface, the pseudostatic compressibility is the important parameter. For rapid expansion the dilatational elasticity at high frequencies is the governing factor. The pressure gradient created will then relax by a number of mechanisms. A soluble surfactant could diffuse to the rarified interfaces and equalise the surface pressure; Marangoni flow of the interface could also even out the surface pressure. The latter could drag liquid into the thinned region whereas the former would not. This liquid flow acts to oppose the original thinning. Since the relative importance of restoration by diffusion will increase with increasing surfactant concentration, the lower foam persistence often observed at higher detergent concentration is explicable.

Antifoams and foam breakers work mainly via the Marangoni effect. They are highly surface active materials which find their way to the interface and spread rapidly to displace the indigeneous surfactant. This rapid movement of the monolayer drags with it enough of the underlying fluid to cause local thinning of the foam film, to such an extent that film rupture becomes more likely. A good example from the oil industry is the use of silicone fluids as antifoams for crude oil. They are injected at very low concentration into the crude oil at a point before the pressure is lowered to allow dissolved gas to separate. Natural gas in the crude oil can then be released without the formation of excessive foam.

5.3 EMULSION COALESCENCE

Similar considerations can often explain the various phenomena observed during coalesence of emulsion droplets. When two oil drops collide a thin film of water is formed between them and this must thin rapidly and rupture before coalescence can occur. Similarly in a water-in-oil emulsion an oil film forms and ruptures. High interfacial viscosity for example due to adsorbed protein monolayers, will reduce the rate of thinning of films and hence slow down the rate of coalescence. High interfacial dilatational elasticity will create large tension gradients during collisions and then the Marangoni flow of liquid into the film will oppose thinning and improve stability.

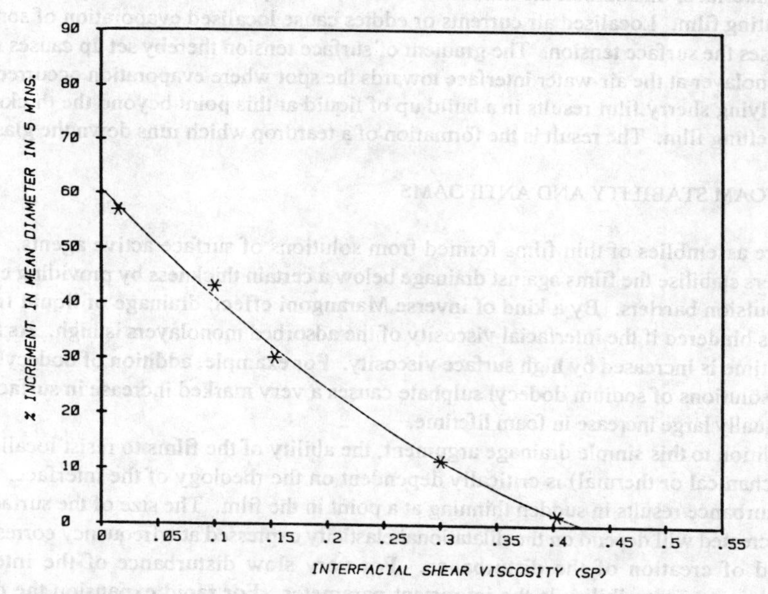

Figure 9. Effect of interfacial shear viscosity (varied by changing electrolyte concentration) on rate of coalesence of Salem crude oil droplets in water. (Wasan et al., [13]).

Figure 9 shows data for the rate of coalescence of crude oil droplets in aqueous electrolyte solution as a function of the interfacial shear viscosity of the system. The correlation is quite clear: high interfacial viscosity greatly impedes coalescence.

5.4 WAVE DAMPING

Waves on liquid surfaces can be of many types. These cover a wide spectrum from low frequency longitudinal and gravity waves, via capillary waves to high frequency thermal ripples (ripplons). The approximate frequency range covered is 0.1 to 10^5 Hz. As waves pass a specific location the interface is periodically expanded and contracted. Any viscoelastic behaviour of the interface causes a resistance to this periodic motion which dissipates energy and results in damping of the waves. In Benjamin Franklin's historic experiment where he spread a "teaspoonful" of oil on the pond at Clapham Common the most striking effect was the damping of surface ripples over an area of "half an acre". A simple calculation tells us that the polar oil, presumably something similar to olive oil, had spread to a thickness of around 2.5 nm, i,.e. a monomolecular layer.

6. References

1. Callaghan, I.C. and Neustadter, E.L., Chemistry & Industry, 17.1.81, p53 (1981).

2. Davies, J.T. and Rideal, E.K., "Interfacial Phenomena", Academic Press, Second edition, p252 (1963).

3. Marangoni, Nuovo Cim., 2, 5-6, 239 (1871).

4. Marangoni, Nuovo Cim., 3, 97 (1878).

5. Grist, D.M., Neustatdter, E.L. and Whittingham, K.P., J. Canadian Pet. Tech., 20(2), 74 (1982).

6. Cairns, R.J.R., Grist, D.M. and Neustadter, E.L. "Theory and Practice of Emulsion Technology", Brunel Symposium, 16-18 September (1974).

7. Graham, D.E., Jones, T.J., Neustadter, E.L. and Whittingham, K.P., "Interfacial Rheological Properties of Crude Oil Water Systems", 3rd International Conference on Surface and Colloid Science, Stockholm, Plenum Press (1979).

8. Clint, J.H., Neustadter, E.L. and Jones, T.J., Rev. Pet. Sci., 13, 135 (1981).

9. Loglio, G., Tesei, U. and Cini, R., J. Colloid Interface Sci., 71, 316 (1979).

10. Van Hunsel, J. and Joos, P., Colloids & Surfaces, 25, 251 (1987).

11. Langevin, D., J. Colloid Interface Sci., 80, 412 (1981).

12. Hard, S. and Neuman, R.D., J. Colloid Interface Sci., 120, 15 (1987).

13. Wasan, D.T., Shah, S.M., Aderangi, N., Chan, M.S. and McNamara, J.J., Soc. Petroleum Engineers J., 409, December (1978).

EFFECTIVE INTERACTION POTENTIALS OF COLLOIDS FROM STRUCTURAL DATA: THE INVERSE PROBLEM

R. RAJAGOPALAN
Department of Chemical Engineering
University of Houston
Houston, Texas 77204-4792, USA

ABSTRACT. We present here an examination of a method based on perturbation theory for extracting effective interaction potentials from static structure factors of monodispersed colloids and simple fluids. Experience with *forward* theories, namely, statistical mechanical theories that attempt to predict the structure from known pair-potentials, appears to indicate that the resulting structure is not very sensitive to the details of the pair-potential. However, it is also known that both the radial distribution function and the static structure factor must be examined together to determine the effect of the pair-potential on the structure. Although the past attempts on the development of solutions to the inverse problem have been largely unsuccessful, the inverse problem has attracted considerable attention in view of the potential benefits that can be derived from successful theories. Using a perturbation approach, which assumes that the large-scale structure is determined by the excluded-volume effect of the core of the potential and that the finer details are supplied by the rest of the potential, we show that the essential details of the effective pair-potential can be extracted from the static structure factor. The performance of this approach is examined for two model potentials for which the structure factors are first obtained using an independent theory and computer simulations.

1. Introduction

Systematic investigations of the physical chemistry and the chemical physics of formation (and structural transitions) of supramolecular assemblies in solution are central to the understanding of self-organization and cooperative phenomena in colloidal, macromolecular and biological dispersions. Such studies, which lie at the interface of solution chemistry, colloid science, and condensed-matter physics, have for a long time relied almost entirely on (classical) thermodynamic descriptions since detailed structural information and interaction forces needed for a closer scrutiny has been largely inaccessible experimentally. A notable example of such a case (although not the only one) is the study of self-association phenomena in surfactant solutions, namely, the formation of association colloids (such as micelles and microemulsions) and phase transitions and critical phenomena in such solutions (Chen and Rajagopalan 1990). However, in recent years, advances in (non-intrusive) radiation scattering techniques for the study of supramolecular solutions have now opened up significant new opportunities for obtaining structural data

D. M. Bloor and E. Wyn-Jones (eds.),
The Structure, Dynamics and Equilibrium Properties of Colloidal Systems, 695–706.

experimentally (including short-range and long-range order and *intra*-species structure) so that proper statistical mechanical tools and appropriate condensed matter theories can be applied profitably to study phase transitions and cooperative phenomena in these systems. Historical and emerging perspectives on these and related problems and current research advances and needs are discussed in recent monographs and compendia (see, for example, Degiorgio and Corti 1985; Hirtzel and Rajagopalan 1985; and Chen and Rajagopalan 1990).

The specific focus of this lecture will be on an issue of central relevance to the above topics, namely, the extraction of effective potentials of interactions in strongly interacting dispersions of charged species from static structural data. Not only is this information important for the determination of the near-field and far-field features of the interaction forces and for the determination of basic parameters like the effective charge on the particles, but it is also the most important element of any rigorous statistical mechanical formulation of the complex hierarchy of phenomena in such systems. It is important to note that the results discussed here also have direct implications in the study of interactions in some 'classical' systems such as rare-gas liquids and liquid metals.

2. Inversion of Structural Data: Background

Traditionally, statistical physics has concerned itself primarily with the description of the microscopic and macroscopic properties of the system under consideration from *given* interatomic or intermolecular pair potentials. In contrast, the inverse problem of extracting effective interaction potentials from observed structural data has received limited (and, to this day, incomplete) attention. The inverse problem is especially important in the case of interactions in dispersions with 'liquid-like' ordering since under such circumstances potentials (or forces) based on the dilute-limit may not be rigorously correct. [The term 'liquid-like ordering' alludes to the fact that the dispersed species can be treated, conceptually and analytically, as elementary constituents of a fluid-like collection of 'particles'. The host medium in which they are dispersed may be assumed to be structureless and treated as an effective dielectric. Correspondingly, the motion of the species is described by Brownian dynamics (with appropriate hydrodynamic contributions) rather than by molecular dynamics; see Castillo *et al.* 1984; Hirtzel and Rajagopalan 1990.] The interaction potentials in the dilute limit are based on the assumption that the interacting species are imbedded in an infinite reservoir of counterions and that they have surfaces in equilibrium with the bulk - conditions that do not always hold in strongly interacting dispersions. Even in the case of classical liquids, the effective potentials contain a wealth of fundamental information on density dependence, possible long-range oscillatory behaviour of the interaction forces and cooperative effects (March and Senatore 1984).

Recent attempts to extract effective potentials from structural data can be traced to a method suggested by Johnson and March (1963), for rare-gas liquids and liquid metals, based on certain integral equation theories developed originally for predicting the equilibrium structure of liquids from known potentials [with the Percus-Yevick (PY) and the hypernetted chain (HNC) closure schemes (Friedman 1985)], and the limited efforts in the last twenty years have been largely unsuccessful. The PY and the HNC effective potentials can be obtained in a straightforward manner from the known static structure factor S(q), where q is the magnitude of the scattering vector q, through

$$\beta\, u_{PY}(r) \quad = \quad ln[1 - c(r)/g(r)] \qquad (1)$$

and

$$\beta\, u_{HNC}(r) \quad = \quad g(r) - c(r) - 1 - ln\, g(r) \qquad (2)$$

where g(r) and c(r) are, respectively, the radial distribution function and the direct correlation function, which are related to S(q) through Fourier transforms [see Eqns. (4) and (5)].

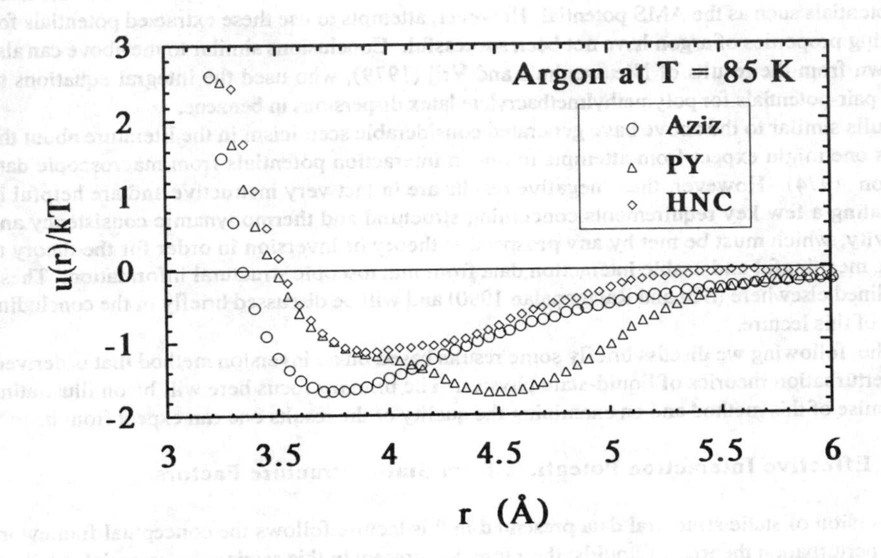

Figure 1. Inverted effective pair-potentials for argon based on the Percus-Yevick (PY) closure and the hypernetted chain (HNC) closure. The potential marked 'Aziz' is the AMS pair-potential mentioned in the text.

However, the integral equation approximations break down at high densities (or, in the case of supramolecular fluids, when liquid-like ordering develops). Consequently, the extracted potentials bear little resemblance to the ones obtained by fitting thermodynamic and transport properties to assumed forms of potentials. An example of such a case is illustrated in Figure 1, which shows the effective pair potential in liquid argon (at 85 K and at the reduced density, $\rho\sigma^3$ = 0.84) obtained through the PY and the HNC integral equation theories. This particular example of liquid argon has been chosen here as a model system since very good experimental scattering data are available in the literature (Yarnell *et al.* 1973) and since the interaction potentials in argon have been studied extensively by computer simulations in the past. Figure 1 also shows a three-parameter potential known as the Aziz-Maitland-Smith potential, parametrized for argon by

Aziz, and the properties of argon derived from it match experimental data very well for argon at *low* densities (Fender and Halsey 1962; Maitland *et al.* 1981). The Aziz-Maitland-Smith (AMS) potential is shown primarily for comparison, and this potential is strictly valid only in the dilute-limit. (Three-body corrections are needed for predicting at least some of the properties of argon at liquid-densities.) It is clear from the figure that both the PY and the HNC pair-potentials [obtained from the data of Yarnell *et al.* using Eqs. (1) and (2)] deviate substantially, in both the location and the magnitude of the minimum in the potential. Equally important is the fact that the hard core of the potential as well as the attractive tail also differ significantly. The PY and HNC potentials shown are actually *effective* pair-potentials and one would expect them to deviate from *pair*-potentials such as the AMS potential. However, attempts to use these extracted potentials for predicting properties of argon have not been successful. Conclusions similar to the above can also be drawn from the results of Nieuwenhuis and Vrij (1979), who used the integral equations to extract pair-potentials for polymethylmethacrylate latex dispersions in benzene.

Results similar to the above have generated considerable scepticism in the literature about the success one might expect from attempts to obtain interaction potentials from macroscopic data (Croxton 1974). However, these negative results are in fact very instructive and are helpful in formulating a few key requirements concerning structural and thermodynamic consistency and sensitivity, which must be met by any prospective theory of inversion in order for the theory to provide meaningful and usable interaction data from macroscopic structural information. These are outlined elsewhere (Sen and Rajagopalan 1990) and will be discussed briefly in the concluding section of this lecture.

In the following we discuss briefly some results based on an inversion method that is derived from perturbation theories of liquid-state physics. The primary focus here will be on illustrating the promise of this method and on examining the quality of the results one can expect from it.

3. Effective Interaction Potentials from Static Structure Factors

The inversion of static structural data presented in this lecture follows the conceptual framework used in perturbation theories of liquids; therefore, we present in this section the essential details of the forward formalism necessary to follow the logical structure of the inversion.

In the perturbation theories, the configurational integral is expanded in a series, relative to a convenient reference potential, in terms of a perturbation of the pair-potential (McQuarrie 1976). Since excellent descriptions of various forms of perturbation theories and approximations, including the optimized random-phase approximation that will be used here, are readily available in the literature, the present section will be restricted to a basic outline of the details necessary for understanding the approach used in the inversion method. Results of applications of Barker-Henderson and Chandler-Weeks-Andersen perturbation schemes to colloidal fluids are presented in Castillo *et al.* 1984; the high level of accuracy one can expect for dense systems from such perturbation schemes is demonstrated through Monte Carlo computer experiments in Hirtzel and Rajagopalan (1990).

Both in the forward perturbation theories and in the inversion scheme, it is assumed that the pair potential, u(r), can be written as

$$u(r) = u_o(r) + u_p(r) \tag{3}$$

where $u_o(r)$ is a suitable reference potential (usually an appropriate representation of the hard core) and $u_p(r)$ is the perturbation (the soft attractive tail). In the inversion scheme one begins with the static structure factor, S(q), at known intervals of the scattering vector, q. The radial distribution function, g(r), and the direct correlation function, c(r), are related to S(q) through

$$g(r) \quad = \quad [\rho\,(2\pi)^3]^{-1} \int dq\ [S(q)-1]\ \exp\,(iq{\cdot}r) \qquad (4)$$

and

$$c(r) \quad = \quad [\rho\,(2\pi)^3]^{-1} \int dq\ \{[S(q)-1]/S(q)\}\ \exp\,(iq{\cdot}r) \qquad (5)$$

Eqs. (4) and (5) can be used to obtain the radial distribution function and the direct correlation function for the fluid from the observed structure factor. In the case of isotropic systems, the above equations can be simplified and can be written in terms of the scalar variables q and r. Details may be found in Hansen and McDonald (1986) or in standard books on scattering theory or statistical mechanics.

The (forward) perturbation theories begin with the decomposition of the pair-potential as in Eqn. (1). One then usually chooses an effective hard-sphere diameter d, which defines a hard-sphere potential $u_d(r)$ that replaces $u_o(r)$ in Eqn. (1) and defines the trial potential, $u_T(r)$, i.e.,

$$u_T(r) = u_d(r) + u_p(r). \qquad (6)$$

A number of options exists for the selection of d; the use of the following criterion due to Lado is known to lead to better predictions of the thermodynamic properties of the fluid (Lado 1984; Hansen and McDonald 1986):

$$\int [\partial y_d(r)/\partial d]\ \{\exp\,[-\beta u_o(r)] - \exp\,[-\beta u_d(r)]\}\ dr \quad = \quad 0, \qquad (7)$$

where y_d is the well-known cavity function corresponding to u_d. Finally, the trial structure factor can be written as (Andersen *et al.* 1976)

$$S_T(q) \quad = \quad S_d(q)\,/\,[1 + S_d(q)\,\rho\,\beta\,u_p(q)]\,, \qquad (8)$$

where r is the number density, $u_p(q)$ is the Fourier transform of an optimized $u_p(r)$ which is redefined inside the hard core in terms of a polynomial obtained by a well-known optimization prescription (Andersen *et al.* 1976), and $S_d(q)$ is the structure factor of a hard-sphere fluid of diameter d. The radial distribution function in the so-called EXP-approximation can be written as

$$g_{EXP}(r) \quad = \quad y_d(r)\ \exp\,[C_L(r) - u_o(r)], \qquad (9)$$

where $C_L(r)$ is a renormalized potential (Andersen *et al.* 1976).

In the case of inversion, we will begin with an estimate of the effective hard-sphere diameter, d, from the compressibility limit of the structure factor, namely, S(0). This is then used in combination with Eqn. (8) and the given structure factor data to obtain the perturbative part of the effective potential and subsequently the renormalized potential. The core can now be 'softened' using Eqn. (9) under the assumption that the experimental radial distribution function can be approximated by $g_{EXP}(r)$. The resulting effective potential is refined further through additional iterations by returning to the above scheme with a new hard-sphere diameter obtained from the extracted potential. The details of this inversion scheme and the implications of the assumptions involved in the procedure will be presented elsewhere (Sen and Rajagopalan 1990); here, we restrict ourselves to a brief discussion of some sample results.

4. Results

We now present some results based on this *predictor-corrector* method. Two different classes of potentials, one a simple potential with a single minimum (a Lennard-Jones potential) and another with a more complex shape with two minima (a weakly charged model colloidal dispersion), have been chosen for examining the method. In order to avoid the complications arising from possible effects of three-body potentials on structure, we restrict our attention here to data generated from *known* pair-potentials using an *independent* theoretical method. For this purpose, we have chosen the reference hypernetted chain (RHNC) theory. We have independently verified, using computer simulations, that the structure factors computed using RHNC for the above Lennard-Jones potential and the colloid potential are sufficiently accurate for testing the inversion method.

4.1 A LENNARD-JONES FLUID

The parameters of the Lennard-Jones potential considered are taken to be $\sigma = 1$ and $(\varepsilon/k_B T) = -0.1$, where σ is the so-called collision diameter and ε is the minimum in the pair-potential:

$$u(r) = 4\,\varepsilon\,[\,(\sigma/r)^{12} - (\sigma/r)^{6}\,] \qquad (10)$$

The structure factor data for this potential were generated for a fluid at a reduced density, $\rho\sigma^3$, of 0.382 [corresponding to a 'volume fraction', $(\pi\rho\sigma^3/6)$, of 0.2] using RHNC. This moderately low value for the density was taken so that the structure factor generated was sufficiently accurate. Although computer simulations could have been used for this purpose, we chose to use the RHNC theory in order to avoid the need for smoothing the computer-generated data (to eliminate the statistical noise) and to avoid the need for extending the data sets further in both r and q spaces. However, the accuracy of the data generated from the RHNC theory was verified using simulations. (The use of an analytical theory for generating structural data allows the choice of arbitrarily fine intervals in r and q, without any concern for statistical noises.) The resulting structure factor was used in the inversion scheme and the extracted potential is shown in Figure 2 along with the known potential. The excellent agreement between the two is evident from this figure. An enlargement of the results near the minimum further illustrates the accuracy of the results.

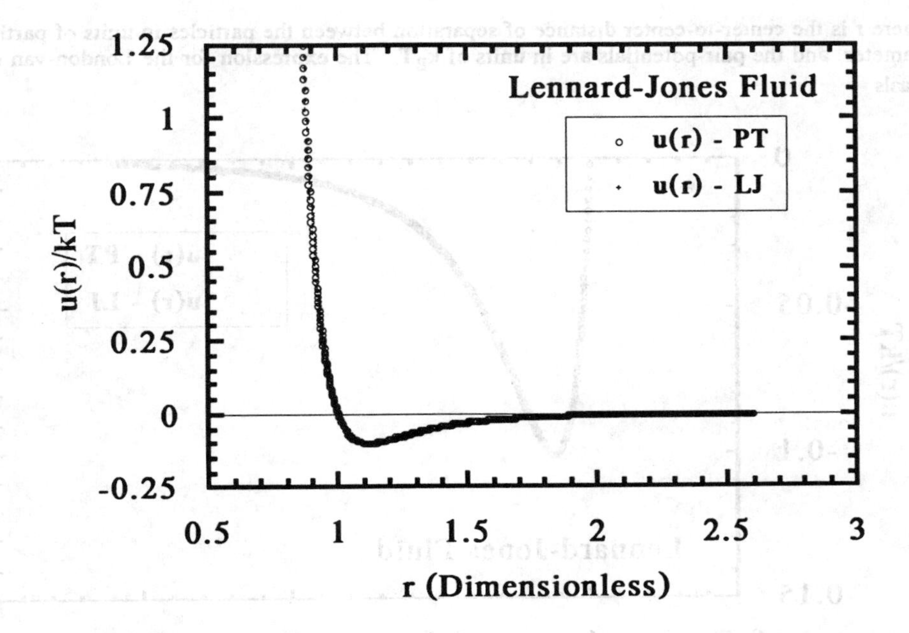

Figure 2. Inverted potential based on the perturbation method for an LJ fluid. Please see the text for details.

4.2 A MODEL COLLOIDAL FLUID

It is now instructive to examine a case corresponding to colloidal interactions. A model colloidal potential, which is described below in some detail, is chosen for this purpose. The structure factors and the radial distribution functions were generated for a colloidal fluid interacting through this potential for a number of densities using the RHNC theory. Salgi (1990) has found that the RHNC results compare very well with the results of Monte Carlo simulations (for the model colloidal potential used here) for low densities (e.g., volume fractions of about 0.15 or lower). At larger volume fractions (especially around 0.4 or above), even though the radial distribution function appears to be quite accurate, the structure factor at low q [particularly the S(0) value, which is related to the isothermal compressibility] is in error by about 30% or more. Since the S(q) at low q's will have significant contributions to the extracted potential for large r, the larger density has been chosen here so that the resulting error in the extracted u(r) could be examined.

The hypothetical, but sufficiently realistic, model colloidal potential used here has a hard core (of unit diameter), a strong London-van der Waals attraction, u_A, a weak electrostatic repulsion, u_E, and a steep Born-type repulsion near the core, u_B:

$$u(r) = u_A(r) + u_E(r) + u_B(r) , \tag{11}$$

where r is the center-to-center distance of separation between the particles in units of particle diameter, and the pair-potentials are in units of k_BT. The expression for the London-van der Waals

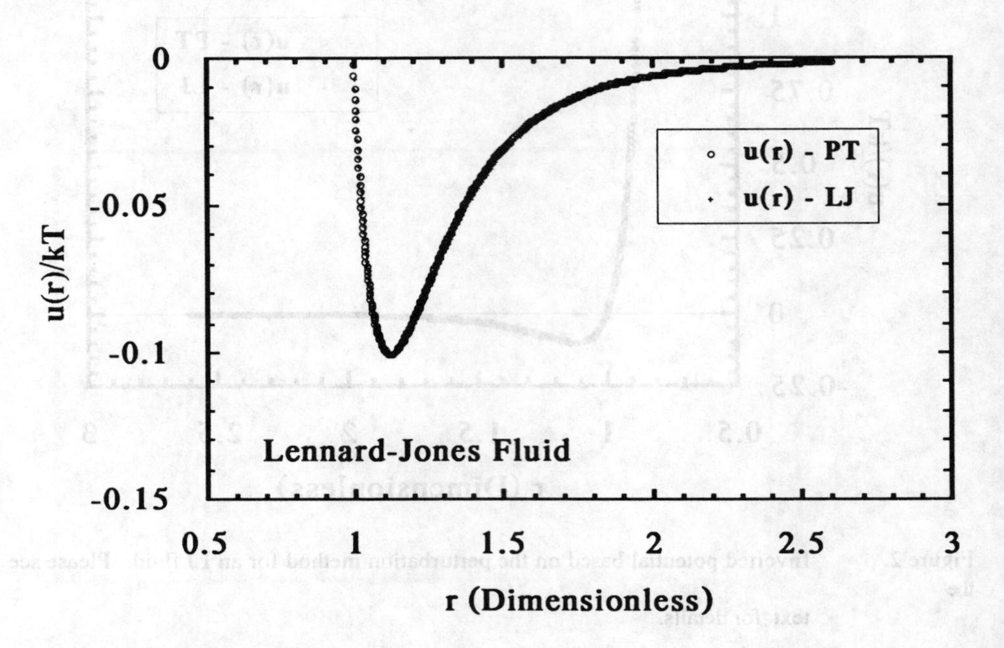

Figure 3. Enlargement of Figure 2 near the minimum in the extracted and the known potentials.

interaction u_A between two identical spheres is given by (Vold and Vold 1983; Castillo *et al.* 1984) :

$$u_A(r) = - N_{LO} \{ 1/(2r^2) + 1/[2(r^2-1)] + \mathit{ln}[(r^2-1)/r^2] \} , \qquad (12)$$

where $N_{LO} = H/(6k_BT)$.

The Hamaker constant, H, in the above equation is a material property. The repulsive contribution, u_E, in Eqn. (11) arises from the electrostatic interaction between the diffuse portions of the electrical double layers surrounding each particle. The general expressions for electrostatic interaction energies are rather complicated and depend on whether the interacting surfaces keep their potentials constant or their charge densities constant. We have used an expression that reproduces the essential features well for systems with constant surface-charge densities and thick electrical double layers, i.e., small inverse-screening lengths, $\kappa a \geq 2.5$ (Vold and Vold 1983;

Verwey and Overbeek 1948):

$$u_E(r) \quad = \quad N_E\,[\,\exp(\,-\,\kappa a\,(r-1)\,)\,]\,/\,r, \qquad (13)$$

where a is the radius of the particles (equal to 0.5 in our dimensionless units) and the

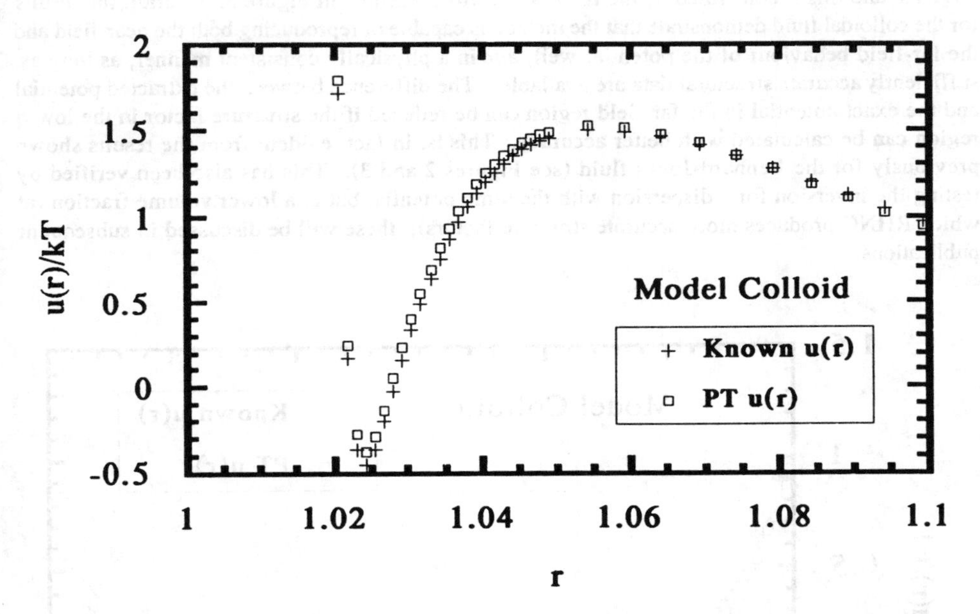

Figure 4. The inverted and the known potentials for the model colloid near the primary minimum in the potentials.

premultiplying factor N_E is related to the contact potential. The other repulsive contribution, u_B, despite its very short range, has to be taken into consideration for dispersions with moderate or low electrostatic interactions, since the probability of finding two particles at very close separations can be far from negligible at sufficiently high particle concentrations. An expression of the following form was used for the Born repulsion u_B so that the appropriate shape of the pair-potential could be generated:

$$u_B(r) \quad = \quad N_B\,(r-1)^{-12} \qquad (14)$$

with N_B is a scale factor that we call the Born parameter. The numerical values that were used for the dimensionless parameters are $N_{LO} = 1.025$, $N_E = 10.0$, $\kappa a = 3.75$, and $N_B = 2.275 \times 10^{-20}$. The resulting pair-potential has both a *primary* minimum and a *secondary* minimum, with a moderate energy barrier in between. The attractive force immediately beyond the primary minimum is significant.

Figures 4 and 5 show that the inverted pair-potential for volume fraction of 0.4 is in good agreement with the known potential. All the major features of the potential and the shape are reproduced well. As mentioned earlier, the structure factor generated by the RHNC theory for volume fraction equal to 0.4 is not very accurate in the compressibility limit; consequently, one should expect to see disagreements between the model potential and the extracted potential for large r's, and this is confirmed by the far-field results presented in Figure 5. Overall, the results for the colloidal fluid demonstrate that the method is capable of reproducing both the near-field and the far-field behaviour of the potential well, *and* in a physically consistent manner, as long as sufficiently accurate structural data are available. The difference between the extracted potential and the exact potential in the far-field region can be reduced if the structure factor in the low-q region can be calculated with better accuracy. This is, in fact, evident from the results shown previously for the Lennard-Jones fluid (see Figures 2 and 3). This has also been verified by testing the inversion for a dispersion with the same potential but at a lower volume fraction (at which RHNC produces more accurate structure factors); these will be discussed in subsequent publications.

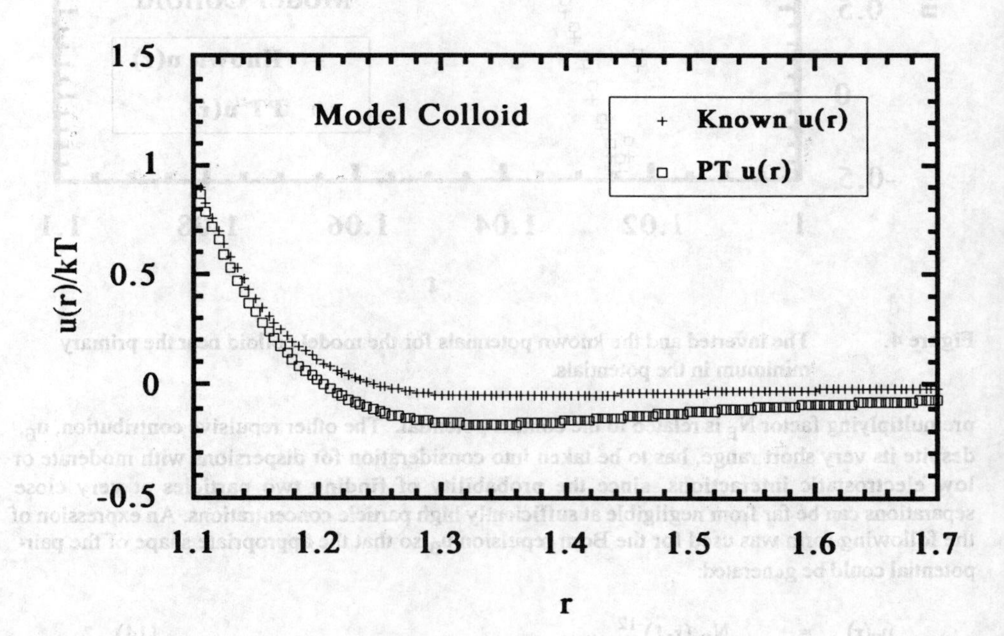

Figure 5. The inverted and the known potentials for the model colloid near the secondary minimum. The difference between the two is due to the large inaccuracy in the structure factor data near $q = 0$.

5. Concluding Remarks and Caveats

The primary objectives of this lecture have been to demonstrate that static structure factors retain a considerable amount of information on the microscopic details of pair interaction potentials and that such details can be recovered to a large extent if the inversion method places proper emphasis on the crucial features of the interactions. The results shown above support these arguments. The method presented here is also the first successful analytical scheme that does not require any computer simulations in the inversion process. Nevertheless, a number of outstanding issues remain and require systematic examination. These issues include (i) the influence of errors in the structure factor (i.e., the sensitivity of the inverted potential to these errors), (ii) effects of cutoffs in the range of q's over which the data are available, and (iii) the influence of smoothing and extension procedures that are used for refining and extending the range over which the data are available. It should also be emphasized here that the inverted potentials actually represent effective pair-potentials when three-body interactions and other non-additive interactions are present in the system. In view of this, additional constraints (such as consistency between experimentally measured thermodynamic or transport properties and the ones derived from the extracted potential) may need to be imposed. These and related issues will be addressed in future publications.

6. Acknowledgments

We would like to acknowledge the support from the National Science Foundation (through grant CPE 83-19862) which helped us to initiate studies in this area. We also thank the Petroleum Research Funds, administered by the American Chemical Society, for partial support. Some of the work presented in this lecture was done in collaboration with Dr. A. K. Sen, presently at the Saha Institute of Nuclear Physics in Culcutta, India, during his tenure at the University of Houston as a PRF Fellow. The assistance of Mr. Jean-François Guérin and Mr. Paul Salgi, graduate students at the University of Houston, is also appreciated.

7. References

Andersen, H. C. and Chandler, D. (1970) J. Chem. Phys. 53, 547.

Andersen, H. C., Chandler, D. and Weeks, J. D. (1972) J. Chem. Phys. 56, 3812.

Castillo, C.A., Rajagopalan, R. and Hirtzel, C.S. (1984) Rev. in Chem. Eng. 2, 237.

Chen, S.-H. and Rajagopalan, R., Eds., (1990) Micellar Solutions and Microemulsions: Structure, Dynamics, and Statistical Thermodynamics, Springer-Verlag, New York.

Croxton, C. A. (1974) Liquid State Physics - A Statistical Mechanical Introduction, Cambridge University Press, Cambridge, England.

Degiorgio, V. and Corti, M., Eds., (1985) Physics of Amphiphiles: Micelles, Vesicles and Microemulsions, North-Holland, Amsterdam.

Fender, B. E. F., and Halsey, Jr., G. D. (1962) J. Chem. Phys. 36, 1881.

Friedman, H. L. (1985) A Course in Statistical Mechanics, Prentice-Hall, Englewood Cliffs, NJ.

Hansen, J. P. and McDonald, I. R. (1986) Theory of Simple Liquids, 2nd. Edn., Academic Pr., New York.

Hirtzel, C. S. and Rajagopalan, R. (1985) Colloidal Phenomena: Advanced Topics, Noyes Publ., Park Ridge, NJ.

Hirtzel, C. S. and Rajagopalan, R. (1990) 'Computer Experiments for Structure and Thermodynamic and Transport Properties of Colloidal Fluids', in Micellar Solutions and Microemulsions: Structure, Dynamics, and Statistical Thermodynamics, Chen, S-H. and Rajagopalan , R., Eds., Springer-Verlag, New York.

Johnson, M. D. and March, N. H. (1963) Phys. Lett. 3, 313.

Lado, F. (1984) Mol. Phys. 52, 871.

Maitland, G. C., Rigby, M., Smith, E. B. and Wakeham, W. A. (1981) Intermolecular Forces, Clarendon Press, Oxford.

March, N.H. and Senatore, G. (1984) Phys. Chem. Liq. 13, 285.

McQuarrie, D. A. (1976) Statistical Mechanics, Harper & Row, New York.

Nieuwenhuis, E. A. and Vrij, A. (1979) J. Colloid Interface Sci. 72, 321.

Salgi, P. (1990) Ph. D. Dissertation, University of Houston, in progress.

Sen, A. K. and Rajagopalan, R. (1990) in preparation.

Verwey, E. J. W. and Overbeek, J. Th. G. (1948) Theory of the Stability of Lyophobic Colloids, Elsevier, Amsterdam.

Vold, R. D. and Vold, M. J. (1983) Colloid and Interface Chemistry, Addison-Wesley, Reading, MA.

Yarnell, J. L., Katz, M. J., Wenzel, R. G. and Koenig, S. H. (1973) Phys. Rev. A 7, 2130.

COMPUTER SIMULATION OF THE COAGULATION AND FLOCCULATION OF COLLOIDAL PARTICLES

ERIC DICKINSON

Procter Department of Food Science,
University of Leeds,
Leeds LS2 9JT, U.K.

ABSTRACT. Understanding how structures form from interacting particles is one of the central issues of colloid science. Computer simulation is a powerful tool in the systematic study of the various physico-chemical factors influencing the morphology of colloidal structures such as aggregates, gels and sediments. This article outlines the basic principles and methodology of two computational methods, Monte Carlo and Brownian Dynamics, which are particularly suited to the simulation of colloidal systems. The use of these techniques is illustrated for DLVO-type particles by recent studies of floc dissociation, and the structures of aggregates produced by reversible flocculation and irreversible coagulation. The main factors considered are Brownian motion, external fields, and the roles of interparticle hydrodynamic and colloidal interactions.

1. INTRODUCTION

Computer simulation provides a valuable bridge between theory and experiment, and its applications in colloid science range from models of soot [1] to models of food [2]. With complex problems which are difficult or impossible to solve analytically, computer simulation provides the link between the microscopic details and the macroscopic behaviour. That is, it enables us better to understand why soot or food have the structures they have, and what are the implications of those structures for the properties of the materials.

Numerical data from simulation studies may be used in two ways. On the one hand, they may be compared with results from real laboratory experiments, thereby testing the simplifying assumptions made in the simulation model, especially those relating to the nature of the interparticle forces. Alternatively, the data may be compared with predictions from an approximate analytic theory incorporating the same assumed set of interparticle forces. In both cases, we can think of the simulations as being 'computer experiments'. The great advantage of a computer experiment over a real laboratory experiment, however, is

D. M. Bloor and E. Wyn-Jones (eds.),
The Structure, Dynamics and Equilibrium Properties of Colloidal Systems, 707–727.

that, whereas particle interactions in the latter are limited by the restricted chemical nature of the available materials, the nature and strength of particle interactions in a computer simulation can be varied at will. This enables increased insight to be gained into what are the crucial factors influencing the observed structure and properties.

In recent years, progress in understanding colloidal aggregation phenomena has been greatly assisted by the use of computer simulation. A notable stimulus has been the analysis of large-scale aggregate structures using the language of fractal dimensions [3,4]. In this paper, however, the emphasis is more on the short-range aggregate structure which is more sensitively dependent on the microscopic details than is the long-range structure. We need to distinguish clearly between two kinds of processes: coagulation and flocculation. The word 'coagulation' (from the Latin 'cogere', meaning 'to drive together') is used to describe the irreversible coming together of colloidal particles to form rigid aggregates with interparticle distances of the order of atomic dimensions. The word 'flocculation' (from the Latin 'floccus', meaning 'tuft of wool') is used to describe the reversible formation of loose aggregates (flocs) in which the particles may be held relatively far apart.

Two kinds of computer simulation technique are particularly suited to colloidal systems: Monte Carlo and Brownian Dynamics. The Monte Carlo technique gives time-averaged equilibrium properties of systems of weakly interacting particles, and is therefore suitable for studying the structure of flocculated colloids. Brownian Dynamics describes the motion of Brownian particles in a hydrodynamic medium, and is therefore suitable for the simulation of the dynamic processes of colloidal coagulation, aggregate growth, and floc dissociation. A third kind of simulation technique, molecular dynamics, is also useful to the colloid scientist, although its primary application is in the computer simulation of liquids [5]. Molecular dynamics has been used to study the structure of polydisperse colloids in the vicinity of the order—disorder transition [6] and to model the dynamics of collective particle motion in shear flow [7,8]. Recent advances in the methodology and applications of molecular dynamics are well documented elsewhere [5,9].

2. MONTE CARLO SIMULATION

2.1. Principles and Methodology

The Monte Carlo method is a numerical technique for estimating the value of multidimensional integrals based on the use of random numbers. The name 'Monte Carlo' appears first to have been coined by Metropolis and Ulam [10] in the title of a paper published in 1949. Configurations of particles in a Monte Carlo simulation are not followed in time, but are generated randomly according to a prescribed set of rules. By averaging over a very large number of configurations typical of the equilibrium condition, structural and thermodynamic properties are obtained.

At equilibrium, the probability of a system being in a state of energy U is proportional to the Boltzmann factor, $\exp(-\beta U)$, where β is the reciprocal of kT (T is temperature, and k is Boltzmann's constant). According to the ergodic theorem of statistical mechanics, the amount of time that the system spends in the state of energy U is also directly proportional to $\exp(-\beta U)$. In the Monte Carlo method, particles are moved from an initial configuration according to a criterion which guarantees that the subsequently generated configurations occur with probabilities proportional to $\exp(-\beta U)$ for those configurations. By assigning equal weights to all the configurations, and averaging the appropriate quantities over them, various equilibrium properties of the system are evaluated.

Let us consider a property ξ which has a value $\xi(s)$ for a state s of energy U(s). The expected value of ξ is

$$\langle\xi\rangle = \int \xi(s)\exp[-\beta U(s)]\,ds \Big/ \int \exp[-\beta U(s)]\,ds$$

$$\approx \sum_s \xi(s)\exp[-\beta U(s)] \Big/ \sum_s \exp[-\beta U(s)] \quad , \qquad (1)$$

where the integrals are approximated by summations over a very large number of discrete states ($1 \le s \le s_{max}$). In principle, we could try to evaluate the ensemble average $\langle\xi\rangle$ in equation (1) by generating the configurations totally at random. It turns out, however, that such a crude Monte Carlo approach, while theoretically acceptable, is extremely inefficient for non-dilute systems of impenetrable particles. The reason for this is that, because of the exponential term, the major contribution to the integral comes from just a small proportion of the generated configurations. Imagine, for instance, trying to fill a box to high volume fraction with hard spheres of diameter d introduced sequentially at random positions within the box. Just one single pair overlap, due to two sphere centres finding themselves less than d apart, would lead to any 'trial' being rejected, since $\exp(-\beta U) \rightarrow 0$ as $U \rightarrow \infty$. On this basis, the vast majority of totally randomly generated 'trials' would be rejected. A solution to the problem is to generate 'trials' such that they lie close to previously accepted states; because of the structural integrity of the particles, these 'trials' are much more likely to make a contribution to the ensemble average. The problem then is to generate this sequence of states in such a way that each state occurs with the correct probability, so that the average over all the generated states is indeed the ensemble average.

The Monte Carlo technique usually used in the simulation of liquids or colloidal suspensions is the so-called 'importance sampling' method of Metropolis and coworkers [11]. In an importance sampling procedure, configurations are generated with a probability density

$$P(s) = \exp[-\beta U(s)] \Big/ \int \exp[-\beta U(s)]\,ds \quad . \qquad (2)$$

This enables a sequence of states to be chosen which makes an important contribution to the ensemble average, i.e., avoiding the high-overlap states. One way of doing this is the Metropolis method [11] which utilizes the theory of Markov chains [12]. A Markov chain is a random sequence of states (s_1, s_2, ..., s_{n-1}, s_n, s_{n+1}, ...) where state s_{n+1} depends only on the 'present' state s_n and not on any of the preceding states (s_1, s_2, ..., s_{n-1}). So, using an underlying stochastic matrix, a sequence of states is generated by reference only to the immediately preceding state [5]. In practice, this means that, for a dense system of model colloidal particles existing in state s_n, what we do to generate state s_{n+1} is to choose at random one particle j, and move it with uniform probability to any point in a small region Ξ around j by making movements along each of the coordinate directions. Typically, in a three-dimensional simulation, the region Ξ is a cube of side-length 2L centred at the position of particle j in state s_n. The value of the maximum possible displacement in any one direction, L, is then an adjustable parameter which controls the convergence of the Markov chain.

Once a possible new state has been generated, it may or may not be accepted. Let us suppose that the trial movement of particle j changes the total energy of the system from U(old) to U(new). The rule for accepting or rejecting this new configuration as state s_{n+1} depends on the magnitude of the energy change ΔU = U(new) - U(old). There are two possibilities:

(i) if $U \leq 0$, the move is immediately accepted;

(ii) if $U > 0$, the move is accepted with a probability $\mathbf{P} = \exp(-\beta\Delta U)$ by comparing the value of $\mathbf{P}$ with that of a random number which has been generated uniformly over the interval (0, 1).

If the move is accepted, state s_{n+1} is the new configuration with particle j in its new position. If the move is rejected, however, state s_{n+1} is identical to the old configuration (i.e., state s_n), and so the old configuration is counted again in the evaluation of the ensemble average from equation (1).

In any particular simulation, the magnitude of the maximum particle displacement L is normally obtained by trial and error. Choosing too large a value of L leads to most moves being rejected, and so sampling of phase space proceeds rather slowly. Similarly, there is slow movement through phase space if L is too small, since in this case the positions of particles in sequential states will tend to be very close together. It is standard practice to adjust L to give an acceptance ratio of about 50%, though there is evidence [13,14] to indicate that an acceptance ratio of 10-20% maximizes the (root) mean-square displacement of particles with respect to computer time. We note, however, that the changes in position of individual particles during a Monte Carlo simulation do not correspond to real particle motion, and it is quite incorrect to postulate, for a sequence of evolving Monte Carlo states, that the number of steps is proportional to time. The temptation to extract dynamic information from a Monte Carlo simulation should always be resisted!

Computationally, the most expensive part of a Monte Carlo simulation is the repeated calculation of pair interaction energies as individual particle positions change throughout the run. Fortunately, the new determination of ΔU at each step does not involve a complete N-particle recalculation of the configurational energy of state s_{n+1}, but only the change associated with moving the single particle j. Nevertheless, in a simulation of several million configurations, this is still a large computational exercise. A common way to save computer time is to count only interactions out to some specified 'cut-off' distance beyond which it is assumed that the pair energy is exactly zero. Systems of spheres interacting with hard-sphere potentials are particularly suited to the Monte Carlo technique, since ΔU is either $+\infty$ (overlap), in which case the move is immediately rejected, or 0 (no overlap), in which case it is immediately accepted.

2.2. Concentrated Suspensions

Monte Carlo simulation has been used successfully to determine the structure and equation of state of concentrated suspensions of colloidal particles interacting with pairwise-additive electrostatic and London—van der Waals forces (i.e., DLVO-type potentials) [15,16].

Equilibrium structure in stable concentrated colloidal systems is expressed in terms of the radial distribution function g(r). This measures how much the local density $\rho g(r)$, a distance r from the centre of a given particle, differs on average from the bulk density $\rho = N/V$, where N is the number of particles and V is the volume. To evaluate g(r), we imagine ourselves 'sitting' on a particle and repeatedly counting the number of neighbours with centres lying in an infinitesimal element of volume dV a distance r away; the mean number is then $\rho g(r)dV$. In a simulation at constant N, V and T (the canonical ensemble), the radial distribution function is given by

$$g(r) = \langle N(r, r+\delta r)\rangle / 2\pi\rho r^2 \,\delta r \quad , \tag{3}$$

where $N(r, r+\delta r)$ is the number of particles between r and $r+\delta r$ from a given particle, and the angular brackets denote the canonical average. The equation of state is given by

$$pV/NkT = 1 - (6NkT)^{-1} \left\langle \sum_{i \neq j} r_{ij} \left[du(r_{ij})/dr_{ij} \right] \right\rangle \quad , \tag{4}$$

where p is the osmotic pressure, and $u(r_{ij})$ is the interaction energy between particles i and j at centre-to-centre separation r_{ij}.

A surprisingly small number of particles is sufficient for the reliable simulation of bulk colloidal systems. One or two hundred are usually enough so long as periodic boundary conditions are employed to eliminate surface effects. That is, the real 'infinite' system is represented in the computer by a basic simulation cell—usually cubic in shape—replicated periodically throughout space. When, during a Monte Carlo move, a particle leaves the basic cell through one of its faces, a replica particle moves in through the opposite face to take its place. Replica particles in neighbouring cells are also included in the new calculation of the configurational energy change ΔU. It is assumed that particle i interacts only with the representation of particle j (real particle or image) which lies nearest to it. This is known as the nearest image distance convention. The consequence of periodicity in Monte Carlo simulations of bulk systems is that meaningful structure, as measured by g(r), is obtainable only over distances of the order of half the box size.

The equilibrium properties of stable colloidal dispersions of rigid spherical particles are now reasonably well understood in terms of the theories of liquid state physics, and the role of Monte Carlo simulation in helping to bring this about has been adequately documented in earlier reviews [15-17]. The statistical description of deformable particles (droplets, proteins, biological cells, etc.) is still, however, very much in its infancy [2]. One approach to the problem which has been made recently [18,19] is a Monte Carlo simulation of a concentrated suspension of deformable particles modelled as a two-dimensional assembly of cyclic lattice chains. In this model, each deformable particle is a 'necklace' of identical segments, and its average size and shape in the suspension is determined by the particle concentration and the strength of interactions between non-bonded segments on the same or on different necklaces. The model exhibits variable polydispersity, with individual sloppy particles having a character intermediate between conventional rigid particles and highly flexible polymer molecules. The radial distribution function and the equation of state of the concentrated suspension have been determined as a function of the intersegmental interactions. Adsorption behaviour at a planar interface has also been investigated [20] with a view to developing a simple statistical model of protein adsorption.

2.3. Flocculated Systems

Monte Carlo simulation is ideally suited to the study of the structure of aggregates of reversibly flocculated particles in a concentrated colloidal suspension. We have recently reported [21] results for pairwise-additive potentials of the square-well form:

$$u(r) = \begin{cases} \infty , & (r < d) \\ -\epsilon , & (d \leq r \leq D) \\ 0 . & (r > D) \end{cases} \qquad (5)$$

The potential in equation (5) has a hard-sphere repulsion of range d, and an attraction of small finite range D and constant strength ϵ. In contrast to more realistic continuous potentials (e.g., DLVO), for which u(r) is never exactly zero at any finite separation r, there is no difficulty with the square-well potential of distinguishing precisely between flocculated and unflocculated arrangements. At any instant, if a single path can be traced between a group of particles in steps of

$r \leq D$, then the group forms part of the same floc. Any particle is unflocculated if its centre lies further than the distance D from any other particle.

Monte Carlo results have been obtained [21] for an assembly of 512 square-well particles in a cubic box with periodic boundary conditions. The well width is set at 0.1 d (i.e., $D/d = 1.1$). Table 1 gives the computed average fractions of particles existing as monomers, dimers and trimers as ϵ is increased from 0.5 kT to 2.25 kT at constant particle

TABLE 1. Fractions of particles existing as monomers (f_1), dimers (f_2), trimers (f_3) and larger flocs (f_{4+}) as a function of the well-depth ϵ at volume fraction $\phi = 0.05$

ϵ/kT	f_1(%)	f_2(%)	f_3(%)	f_{4+}(%)
0.5	77	17	4	2
1.0	68	20	7	5
1.5	55	22	10	13
2.0	40	19	12	29
2.25	24	11	7	58

volume fraction $\phi = 0.05$. We note that flocculated dimers are most predominant for $\epsilon \sim 1.5$ kT, and flocculated trimers for $\epsilon \sim 2$ kT. More than half the particles are flocculated for $\epsilon \gtrsim 2$ kT. Phase separation into coexisting gas-like and liquid-like colloidal phases occurs for $\epsilon \gtrsim 2.5$ kT. An analysis has been made [21] of flocculated trimer shape distributions in terms of a single bond angle, and of flocculated tetramer shape distributions in terms of the proportions of 3-bonded, 4-bonded, 5-bonded and 6-bonded structures. With simulated aggregates containing more than four particles, structures are conveniently described by the scaling relationship

$$R_G \sim n^{1/d_f} \quad , \tag{6}$$

where R_G is the radius of gyration of an n-particle floc, and d_f is the fractal dimension. A plot of $\ln R_G$ against $\ln n$ ($5 \leq n \leq 15$) for $\phi = 0.05$ and $\epsilon = 1.5$ kT is consistent with $d_f \approx 1.8$, but a slightly larger fractal dimension ($1.8 < d_f < 2.1$) gives a better fit to the numerical data at larger well-depths ($1.5 < \epsilon/kT < 2.5$). These values are rather similar to the fractal dimensions reported [3,4] for large-scale irreversibly coagulated aggregates simulated by diffusion-limited ($d_f = 1.75$) and reaction-limited aggregation ($d_f = 2.05$). In the one-phase region, it would appear that flocculated colloidal aggregates take up the same sort of disordered fractal-type structure as is found with coagulated aggregates.

The square-well potential has also been used [22] to model the well-known phenomena of bridging flocculation and depletion flocculation. A

TABLE 2. Effect of polymer—particle interaction energy ϵ on flocculation by bridging in square-well mixture model (see text for details)

ϵ/kT	s_{all}	s_{max}	$f_1(\%)$
0.0	1.05	2.1	92
0.5	1.15	2.8	77
1.0	1.32	3.9	59
1.3	1.58	5.4	42

binary mixture of large spheres (colloidal particles) + small spheres (polymer molecules) is simulated using the Monte Carlo technique. The polymer—particle interaction is represented by a square-well potential $u_{12}(r)$ with variable well-depth ϵ. Table 2 gives some calculated average properties of a three-dimensional system of 2025 small spheres ($\phi_1 = 0.020$) + 50 large spheres ($\phi_2 = 0.062$) exhibiting flocculation by bridging. Characteristic size parameters are set at $d_2/d_1 = 5$, $d_{12}/d_1 = 3$, and $D_{12}/d_1 = 3.5$. It is assumed that two colloidal particles are bridged together in a floc if they both lie within a distance D_{12} of the same polymer molecule. The three quantities listed as a function of ϵ in table 2 are the mean floc size (s_{all}), the average size of the largest floc in the simulated system (s_{max}), and the fraction of the colloidal particles existing as monomers (f_1). We note that, for a well-depth $\epsilon \gtrsim 1\,kT$, over half the colloidal particles are involved in bridging flocculation. Visualization of instantaneous configurations of the same model system in two dimensions shows the formation of stringy aggregates of bridging flocs. With negative values of ϵ (i.e., square-well repulsion), depletion flocculation becomes evident with the appearance of stringy flocs separated by regions of high local polymer concentration.

3. BROWNIAN DYNAMICS SIMULATION

3.1 Principles and Methodology

To simulate the time-dependent behaviour of a system containing solute particles and a very much larger number of solvent molecules (plus other small solute molecules, ions, etc.), where one is interested only in the motion of the large solute particles, the type of computational method used is called stochastic dynamics. Where inertial terms are quite negligible, and the motion is overwhelmingly diffusive, the technique is called Brownian Dynamics [23]. Each colloidal particle in a Brownian Dynamics simulation diffuses in a force field caused by the presence of neighbouring particles, with spatial correlations in the motion determined by position-dependent hydrodynamic interactions acting between the particles.

The underlying mathematics of Brownian movement was set out at the turn of the century by a Frenchman, Louis Batchelier, in the form of a statistical treatment of share prices on the Paris stock exchange [24]. The essential physics of Brownian movement was described by Einstein a few years later. The key feature is that a particle which executes a random walk has an average displacement which is proportional to the square root of time, unlike classical force-free motion which is linear in time. So, for motion in one dimension, the average displacement x_{av} in time t is given by

$$x_{av} = (\langle x^2 \rangle)^{1/2} = (2Dt)^{1/2} \quad , \tag{7}$$

where D is the diffusion coefficient, and $\langle x^2 \rangle$ is the mean-square displacement in the x-direction. Implicit in equation (7) is the idea of a diffusive time interval τ which is small compared with t, but is nevertheless of such magnitude that movements executed by the particle in two successive time intervals τ are mutually independent.

A fascinating facet of the pure Brownian walk is its interminably mazy motion. The relative tortuosity of the path is not diminished by any change in spatial or temporal scale. It is a self-similar object with a fractal dimension of $d_w = 2$. The Brownian motion of real colloidal particles, even in very dilute suspension, differs from this ideal behaviour at very short times due to molecular correlations, and at long times due to interactions with objects (surfaces, particles, etc.) or fields (electrical, gravitational, etc.). For an isolated particle diffusing in an external field, equation (7) may be rewritten as the sum of a stochastic term and a systematic term,

$$x_{av} = (2Dt)^{1/2} + Kt \quad , \tag{8}$$

where K is a constant. The particle trajectory is essentially the same as for pure Brownian motion for $t \ll 2D/K^2$, and pure classical linear motion for $t \gg 2D/K^2$. When the stochastic and systematic terms are both significant, which is often the situation, the particle executes a biased random walk with d_w somewhere between 1 and 2.

The translational diffusive motion of a group of N particles in an incompressible hydrodynamic medium is described by the Langevin equation [25]

$$m_i(dv_i/dt) = \sum_j \zeta_{ij}v_j + F_i + \sum_j \alpha_{ij}q_j \quad , \quad (1 \leqslant i,j \leqslant 3N) \tag{9}$$

where m_i is the particle mass associated with index i, v_i is the velocity component in direction i, F_i is the sum of external and all interparticle (colloidal) forces acting in direction i, and the sum is over all 3N translational degrees of freedom. Equation (9) is really nothing more than an elaborate form of Newton's equation of motion. The left-hand side is mass X acceleration, and the right-hand side is a sum of three forces: frictional, systematic, and stochastic. The frictional force depends on the components of the configuration-dependent friction tensor ζ_{ij}. The coefficients α_{ij} in the stochastic term are related to ζ_{ij} by

$$\zeta_{ij} = (kT)^{-1} \sum_k \alpha_{ik}\alpha_{jk} \quad . \tag{10}$$

The set of random numbers $\{q_j\}$ are described by a Gaussian distribution

$$\langle q_i(0)q_j(t)\rangle = 2\delta_{ij}\delta(t) \quad , \tag{11}$$

where δ_{ij} is the Kronecker delta and $\delta(t)$ is the Dirac function.

A Brownian Dynamics algorithm based on equations (9) to (11) was devised by Ermak and McCammon [26]. In terms of the 3N X 3N diffusion tensor,

$$D_{ij} = kT\,(\zeta_{ij})^{-1} \quad , \tag{12}$$

the change in particle coordinate i during time-step Δt is given by

$$\Delta r_i = \sum_j (\partial D^\circ_{ij}/\partial r_j)\Delta t + (kT)^{-1}\sum_j D^\circ_{ij}F^\circ_j\Delta t + \mathbf{R}_i(D^\circ_{ij},\Delta t) \quad ,$$

$$(1 \leq i,j \leq 3N) \tag{13}$$

where the symbol $^\circ$ denotes that the quantity is evaluated at the start of the time-step. The chosen value of the interval Δt should be long compared with the characteristic time associated with solvent molecular motion, but sufficiently short for the quantities F_j and D_{ij} to remain sensibly constant over the simulation time-step. The stochastic term $\mathbf{R}_i$ has first and second moments given by

$$\langle \mathbf{R}_i(\Delta t)\rangle = 0, \tag{14}$$

$$\langle \mathbf{R}_i(\Delta t)\mathbf{R}_j(\Delta t)\rangle = 2D^\circ_{ij}\Delta t \quad . \tag{15}$$

The evaluation of the set of stochastic displacements $\{\mathbf{R}_i\}$ involves taking the square root of the diffusion tensor; this is the most time-consuming part of the implementation of the algorithm. Notice that the stochastic term is not truely random, but is related to the hydrodynamic interactions via the diffusion tensor. The resulting set of stochastic displacements from equations (14) and (15) for all N particles in the three coordinate directions (x, y and z) ensures thermal equilibration of the system by reflecting the balance between frictional forces and solvent collisional forces as prescribed in the fluctuation—dissipation relationship [equation (10)].

The Brownian Dynamics algorithm of equation (13) is different from a deterministic molecular dynamics algorithm [9] in that it neither has a definite solution, nor does it conserve energy. Another difference is that, while the diffusion coefficient D is an output result from many a molecular dynamics simulation, the diffusion tensor D_{ij} forms part of the input information in Brownian Dynamics. As written, equation (13) applies only to the translational Brownian motion of the particles, but the algorithm has also been generalized [27] to include rotational diffusion, as well as the coupling of translational and rotational motions. Examples of the implementation of this generalized algorithm have recently been described [28,29].

Hydrodynamic interactions are included in the Brownian Dynamics simulation through the mobility tensor $b_{ij} = D_{ij}/kT$. At low Reynolds number, the velocities of two macroscopic spheres acted upon by external forces $\underline{F}_1$ and $\underline{F}_2$ can be written as

$$\underline{v}_1 = b_{11}\underline{F}_1 + b_{12}\underline{F}_2 \quad , \tag{16}$$

$$\underline{v}_2 = b_{21}\underline{F}_1 + b_{22}\underline{F}_2 \quad , \tag{17}$$

where the torque on each sphere is zero. For identical particles of radius a, the mobility tensor has the general form [30]

$$b_{ij} = (6\pi\eta a)^{-1}\Big\{A_{ij}(r)(\underline{r}_{12}\underline{r}_{12}/r^2) + B_{ij}(r)\big[\mathbf{I} - (\underline{r}_{12}\underline{r}_{12}/r^2)\big]\Big\} \tag{18}$$

where $\underline{r}_{12}$ is the vector between the centres of particles 1 and 2, $A_{ij}(r)$ and $B_{ij}(r)$ are distance-dependent mobility coefficients, $r = |\underline{r}_{12}|$ is the scalar separation, and η is the Newtonian viscosity of the dispersion medium. Relative translational motion of the pair of spheres along the line of centres is determined by $A_{11} - A_{12}$, and rotation about the centre of friction by $B_{11} - B_{12}$. Explicit expressions for A_{11}, A_{12}, B_{11} and B_{12} in powers of a/r are available [31] to order r^{-12}. The mobility tensor to order $(a/r)^3$ has the form

$$b_{ij} = (6\pi\eta a)^{-1}\Big\{\delta_{ij}\mathbf{I} + \tfrac{3}{4}(a/r)(1-\delta_{ij})\big[\mathbf{I} + (\underline{r}_{12}\underline{r}_{12}/r^2)\big] - \tfrac{1}{2}(a/r)^3(1-\delta_{ij})\big[3(\underline{r}_{12}\underline{r}_{12}/r^2) - \mathbf{I}\big]\Big\} \quad , \tag{19}$$

where $\mathbf{I}$ is the unit tensor. Equation (19) is known as the Rotne—Prager tensor [32]. It is a convenient approximation from the computational viewpoint because it is algebraically simple, it always gives a positive definite 3N X 3N diffusion matrix for r > 2a in multiparticle systems, and the gradient term in equation (13) can be omitted since $\underline{\nabla}\cdot b_{ij} = 0$.

Computer simulators have come to take for granted the validity of assuming that particle—particle interactions are pairwise-additive. While this assumption is very often appropriate for intermolecular pair interactions and colloidal potentials of mean force, it is certainly not so for hydrodynamic interactions, except at high dilution. Adopting pairwise additivity in concentrated systems where pairs of particles are able to approach closely ($r \rightarrow 2a$) can lead to physically absurd theoretical predictions (e.g., negative diffusion coefficients [33]) and computational problems due to non-positive-definite diffusion matrices

[34]. The dominant contribution to translation from a group of n spheres has been shown [35] to be of order $(a/r)^{3n-5}$. This means that the neglect of multibody interactions is consistent with the use of the Rotne—Prager tensor [equation (19)], that three-body interactions first appear at order $(a/r)^4$, and that four-body interactions first appear at order $(a/r)^7$. Unfortunately, these higher order terms are expensive to include in simulations because of their mathematical complexity. One compromise solution [34] is to make an allowance for hydrodynamic screening by including only the short-range pairwise-additive mobilities within a hydrodynamic cut-off distance of the order of a few particle radii. An alternative approach [36] is to adopt an effective pair tensor which allows for multibody hydrodynamic interactions via a set of empirical screening constants which depend on the local concentration. We note, however, that the concept of hydrodynamic screening in a colloidal suspension is only an approximation, since there always remains a long-range term in r^{-1} however high the particle concentration. Without some sort of hydrodynamic cut-off, it is difficult to simulate bulk systems by Brownian Dynamics due to the problem of reconciling the long-range term with the use of periodic boundary conditions. One possible solution is through an Ewald summation [37] like that used to handle long-ranged electrostatic interactions in molecular dynamics [5], although whether equations such as (18) or (19) are rigorously correct in a simulation with periodic boundaries has been questioned [38].

In practice, for many systems of colloidal interest, the complication of multibody hydrodynamics is of little consequence in relation to the phenomena being investigated. This is particularly so for the case of aggregation processes. A justification for the use of simplified hydrodynamic expressions—or the neglect of hydrodynamic interactions altogether—in some Brownian Dynamics simulations is that the results are found to be weakly dependent on the form of the hydrodynamic equations (see below). Only when particles spend most of their time very close together (i.e., $r - 2a \ll a$) does one need to be especially careful about the exact form of the hydrodynamic expressions [23].

3.2. Floc Dissociation

The Brownian Dynamics algorithm described above has been used to study the dissociation of small aggregates of secondary-minimum flocculated DLVO-type particles in the absence of any external field [39,40], in the presence of gravity [34,41], and in the presence of simple shear flow [42,43]. Isolated flocs containing from 2 to 57 particles have been considered over a range of conditions.

We can test the validity of the computational technique for the case of a doublet of secondary-minimum flocculated particles, since the hydrodynamic interaction for an isolated pair of spheres is known exactly, and an analytic theory is available [44] to test the kinetic results against. Repeated simulation runs have been carried out [39,42] for doublets of 2 μm diameter particles, starting at a pair separation $r = r_{min}$ corresponding to the bottom of the secondary minimum, and ending at $r = r_{max}$ corresponding to an attractive energy of less than a few per cent of kT. Three different DLVO-type potentials are considered with well-depth and position parameters, ϵ_{min} and r_{min}, as indicated in table 3. Values of the mean dissociation time $\bar{t}$, based on N_{sim} discrete dissociation events, are compared with values of $\bar{t}$ calculated by Chan

TABLE 3. Mean doublet dissociation time $\bar{t}$ from Brownian Dynamics simulation and theory for three DLVO-type potentials with secondary-minimum well-depth ϵ_{min} and well-position r_{min}

ϵ_{min}/kT	$r_{min}/\mu m$	N_{sim}	$\bar{t}/s$	
			simulation	theory[c]
0.87	2.024	543[a]	0.258 ± 0.015	0.253
4.7	2.008	185[a]	1.49 ± 0.12	1.38
10.7	2.005	20[b]	140 ± 40	164

[a] ref. 39; [b] ref. 42; [c] ref. 44

and Halle [44] using a 'first passage time' method. We see that the two sets of mean dissociation times are in agreement within the estimated statistical uncertainty of the simulation data. The results indicate that a time-step of $\Delta t \sim 10\ \mu s$ is satisfactory for simulating colloidal systems of this type.

The results in table 3 are a stringest test of the Brownian Dynamics simulation scheme, since, for spheres that get as close together as these ($2.003 \lesssim r/a \leq 2.2$), it is essential that very accurate numerical expressions be used for the hydrodynamic interactions. With colloidal interactions that are of longer range, however, the nature of the hydrodynamic approximation is not so important. Table 4 shows some results for simulations of trimers of 1 μm diameter particles with pair

TABLE 4. Mean dissociation time $\bar{t}$ from Brownian Dynamics simulation for linear and equilateral trimers up to terms of order $(a/r)^n$ in the hydrodynamics

n	$\bar{t}/s$	
	linear	equilateral
1	0.49	0.37
3	0.42	0.32
4	0.51	0.42
6	0.50	0.35
7	0.53	0.39

potential having a shallow secondary minimum of ϵ_{min} = 0.8 kT at r_{min} = 2.35 a. Initial pair separations are set at $r_{12} = r_{23} = r_{min}$ and $r_{13} = 2r_{min}$ for linear trimers, and $r_{12} = r_{23} = r_{13} = r_{min}$ for equilateral trimers, and a trimer is deemed to have dissociated when each of the pair separations exceeds 1.5 μm (= 3a). Table 4 lists values of $\bar{t}$ obtained with Δt = 20 μs and N_{sim} = 300 by truncating the hydrodynamic interaction series expansion at various values of $(a/r)^n$. It is evident that the Rotne–Prager tensor (n = 3) gives faster dissociation than the leading 3-body term (n = 4), but the other approximations (n = 1, 6 and 7) give essentially the same results within the statistical uncertainty (± 0.03 s). These results, together with others not reported here, go some way towards justifying the neglect of multibody hydrodynamics in simulations of the coagulation of DLVO-type particles (see below), especially when one is primarily interested in aggregate structure and size distributions, as opposed to absolute rate constants [40].

In the presence of an externally applied flow field, the algorithm may be written in the form [42]

$$\Delta r_i = \Delta r_i \{\text{eqn (13)}\} + v_i \Delta t + \mathbf{S}_i^{\circ} \rho \Delta t \quad , \tag{20}$$

where ρ is the rate of strain tensor, $\mathbf{S}_i$ is a third-order mobility tensor (the shear tensor), and v_i is the local velocity associated with the tidal flow. The first term on the right-hand side of equation (20) is simply that of equation (13). The second and third terms, derived from the analysis of Batchelor and Green [45] for non-Brownian spheres, allow for the additional effect of fluid flow on the particle motion. The quantity $v_i \Delta t$ is the absolute displacement arising from the bulk motion of the unperturbed fluid, and $\mathbf{S}_i^{\circ} \rho \Delta t$ represents the influence of interparticle hydrodynamics on the shear-induced motion.

Flocculated doublets of DLVO-type particles have been simulated in simple shear, and aggregate life-times have been computed [42] as a function of shear-rate, well-depth, and doublet orientation with respect to the flow field. Particles interacting with secondary minima of several kT may rotate many revolutions before dissociating. For example for the potential in table 3 with ϵ_{min} = 10.7 kT, it was found that the application of a shear-rate of 10 reciprocal seconds leads to a lowering of the mean life-time from 140 ± 40 s to 95 ± 35 s. With one orbit every 2 seconds, this corresponds to an average of about 40 doublet rotations prior to eventual dissociation. With larger colloidal flocs in shear flow, the dissociation behaviour is much more complicated [43].

3.3. Irreversible Aggregation

Brownian Dynamics simulation has been used to study the irreversible coagulation of DLVO-type particles [40,46–48] and their deposition into sediments [40,49]. Emphasis is placed here on discussing the short-range structural features which are not adequately described by the simplified simulation models based on idealized trajectories or simple sticking probabilities [50]. Using the Brownian Dynamics approach, it is possible to evaluate systematically how colloidal structure is affected by such factors as particle size, dispersed-phase volume fraction, external field strength, hydrodynamic interactions, and the nature of the interparticle colloidal forces.

In a low-density sediment or gel formed from irreversibly aggregated spherical colloidal particles, three spatial scales of structure may be identified: (i) short-range order from packing and excluded volume effects, (ii) medium-range disorder associated with the fractal-type characteristics of diffusion-limited aggregation processes, and (iii) long-range uniformity for a material that is homogeneous macroscopically. This description is summarized by defining a normalized particle–particle distribution function [51]

$$G(r) = \begin{cases} g(r) , & (2a \leq r \leq \gamma) \\ (r/\lambda)^{d_f - 3} , & (\gamma < r \leq \lambda) \\ 1 . & (r > \lambda) \end{cases} \qquad (21)$$

In equation (21), g(r) is equivalent to the short-range liquid-like radial distribution function which extends out to $r = \gamma$, d_f is the fractal dimension, and λ is the characteristic network length of the connected gel or sediment.

Once particles become rigidly stuck together, the rotational motion of the non-spherical aggregates becomes a relevant factor in the dynamical behaviour and in the evolution of the subsequent aggregation. This makes the simulation more complicated than with freely moving independent spheres. With relatively small m-particle aggregates (i.e., $m \lesssim 10$), it is feasible to simulate aggregate Brownian motion by means of a constraints algorithm [52]. An iterative constraints procedure that takes into account hydrodynamic interactions between all the particles has been described by Allison and McCammon [53]. This kind of algorithm was used to simulate coagulation in a concentrated dispersion [46] and isolated encounters of monomers + dimers, monomers + trimers, and dimers + dimers [48]. As the application of constraints becomes increasingly expensive in terms of computer time as aggregates become larger, a simpler alternative approach [47] is to neglect rotational diffusion altogether, and represent translational diffusion of the m-particle aggregates using a size-dependent scalar diffusion parameter

$$D_m = (kT/6\pi\eta a)\, m^c \, , \qquad (22)$$

where c is a scaling exponent. A cluster–cluster aggregation model [54] based on hydrodynamics at the Rotne–Prager level gives $c = -0.54$.

To explore the influence of colloidal interactions on the structure of aggregates formed by irreversible coagulation, we have considered [47] two different unstable DLVO-type potentials (corresponding to different electrolyte concentrations). Potential I is attractive at all separations, and strongly so as $r \to 2a$. Potential II is also strongly attractive at close surface-to-surface separations, and weakly so at large separations, but it has a repulsive region at intermediate values ($2.15 \lesssim r/a \lesssim 2.3$). So, while potential I just has a primary minimum, potential II has a small primary maximum of height $\sim 3\,kT$ above a small secondary minimum of depth $\sim 2\,kT$. Intuitively, one might expect the secondary minimum in potential II to cause pairs to associate loosely at separations $r \approx 2.3\,a$ before thermal motion either induces dissociation (as in section 3.2) or causes irreversible aggregation into the primary minimum ($r \rightarrow 2a$) after jumping over the small primary maximum. This interpretation is consistent with the plots of g(r) obtained for systems containing the final single aggregate produced from a Brownian Dynamics simulation of 512 particles in a cubic box with periodic boundary conditions. At volume fraction $\phi = 0.15$, g(r) for potential II has a substantial peak at $r = 2.3\,a$ corresponding to non-bonded pairs in the secondary minimum, and a broader peak at $r \simeq 4a$ corresponding to the second 'shell' of nearest neighbours, as in the liquid-like structure characteristic of stable concentrated suspensions [16]. On the other hand, g(r) for potential I was found to have little structure. The effective fractal dimensions calculated from plots of $\ln R_G$ against $\ln n$ [see equation (6)] are, however, similar for both potentials at the same value of ϕ. This shows that the form of the DLVO-type potential during coagulation affects the short-range aggregate structure, as measured by the particle—particle correlations ($r \leq \gamma$), but not the larger scale structure, as measured by the fractal dimension.

4. CONCLUDING REMARKS

We have seen that the numerical statistical mechanical techniques of Monte Carlo and Brownian Dynamics can be used successfully to simulate the coagulation and flocculation behaviour of model colloidal systems. These simulation methods are essential for understanding properly the factors controlling aggregate structure, since analytic theory in this field is still at a relatively early stage of development [55]. The advantage of the Monte Carlo and Brownian Dynamics techniques is that they can handle colloidal interactions of any arbitrary degree of complexity, and are therefore able to approach closely to the situation in real colloids. Sometimes, however, such complexity is not required, in which case more idealized, and computationally cheaper, modelling procedures may be utilized, e.g., in the study of the packing of hard discs [2,56-58], or the generation of very large-scale self-similar structures [3,4]. In most of these cases, the intelligent use of modern computer graphics provides great insight into the structure of real particulate materials as seen under the optical or electron microscope [59]. What computer simulation tells us is that even the most simple set of rules, when repeated over and over again, can lead to structures of such interconnected order and disorder that they can easily be confused with many of highly complicated structures found in nature.

Looking ahead, one topic which is ripe for computer simulation study is the role of adsorbing and non-adsorbing polymers in the flocculation behaviour of colloidal particles. This is an area in which computer simulation lags behind both theory and experiment [60-62]. Some recent progress has been made in simulating the heteroflocculation of spherical particles [22], the aggregation of associating polymers and surfactants [63,64], and the competitive adsorption of polymers and surfactants [65]. What needs to be done now, using Monte Carlo or Brownian Dynamics, is the bringing together within the same computer simulation model of a mixture of both solid particles and flexible polymer molecules. With such a simulation model, one would be able to study how the nature of polymer—particle interactions affects the mechanism of polymer-induced flocculation and phase separation.

REFERENCES

1. Richter, R., Sander, L. M. and Cheng, Z. (1984) 'Computer simulation of soot aggregation', J. Colloid Interface Sci., 100, 203-209.

2. Barker, G. C. and Grimson, M. J. (1989) 'Food colloid science and the art of computer simulation', Food Hydrocolloids, 3, 345-363.

3. Jullien, R. and Botet, R. (1987) Aggregation and Fractal Aggregates, World Scientific, Singapore.

4. Meakin, P. (1988) 'Fractal aggregates and their fractal measures', in C. Domb and J. L. Lebowitz (eds), Phase Transitions and Critical Phenomena, Academic Press, London, vol. 12, chap. 3.

5. Allen, M. P. and Tildesley, D. J. (1987) Computer Simulation of Liquids, Oxford University Press.

6. Dickinson, E., Parker, R. and Lal, M. (1981) 'Polydispersity and the colloidal order—disorder transition', Chem. Phys. Lett., 79, 578-582.

7. Woodcock, L. V. (1984) 'Origin of shear dilatancy and shear thickening phenomena', Chem. Phys. Lett., 111, 455-461.

8. Heyes, D. M., Morriss, G. P. and Evans, D. J. (1985) 'Nonequilibrium molecular dynamics study of shear flow in soft disks', J. Chem. Phys., 83, 4760-4766.

9. Fincham, D. and Heyes, D. M. (1985) 'Recent advances in molecular dynamics computer simulation', Adv. Chem. Phys., 63, 493-575.

10. Metropolis, N. and Ulam, S. (1949) 'The Monte Carlo method', J. Amer. Stat. Assoc., 44, 335-341.

11. Metropolis, N., Rosenbluth, A. W., Rosenbluth, M. N., Teller, A. H. and Teller, E. (1953) 'Equation of state calculations by fast computing machines', J. Chem. Phys., 21, 1087-1092.

12. Hammersley, J. M. and Handscomb, D. C. (1964) Monte Carlo Methods, Chapman and Hall, London.

13. Wood, W. W. and Jacobsen, J. D. (1959) 'Monte Carlo calculations in statistical mechanics', Proceedings of the Western Joint Computer Conference, San Francisco, pp. 261-269.

14. Groot, R. D., van der Eerden, J. P. and Faber, N. M. (1987) 'The direct correlation function in hard sphere fluids', J. Chem. Phys., 87, 2263-2270.

15. Dickinson, E. (1983) 'Statistical mechanics of colloidal suspensions', in D. H. Everett (ed.), Colloid Science, Specialist Periodical Report, Royal Society of Chemistry, London, vol. 4, pp. 150-179.

16. van Megen, W. and Snook, I. (1984) 'Equilibrium properties of suspensions', Adv. Colloid Interface Sci., 21, 119-194.

17. Hirtzel, C. S. and Rajagopalan, R. (1985) Advanced Topics in Colloidal Phenomena, Noyes Publications, Park Ridge, NJ.

18. Dickinson, E. (1984) 'Statistical model of a suspension of deformable particles', Phys. Rev. Lett., 53, 728-731.

19. Dickinson, E. and Euston, S. R. (1989) 'Statistical study of a concentrated dispersion of deformable particles modelled as an assembly of cyclic lattice chains', Molec. Phys., 66, 865-886.

20. Dickinson, E. and Euston, S. R. (1990) 'Simulation of adsorption of deformable particles modelled as cyclic lattice chains: a simple statistical model of protein adsorption', J. Chem. Soc., Faraday Trans., 86, in press.

21. Dickinson, E., Elvingson, C. and Euston, S. R. (1989) 'Structure of simulated aggregates formed by reversible flocculation', J. Chem. Soc., Faraday Trans. 2, 85, 891-900.

22. Dickinson, E. (1989) 'A model of a concentrated dispersion exhibiting bridging flocculation and depletion flocculation', J. Colloid Interface Sci., 132, 274-278.

23. Dickinson, E. (1985) 'Brownian dynamics with hydrodynamic interactions: the application to protein diffusional problems', Chem. Soc. Rev., 14, 421-455.

24. Dickinson, E. (1986) 'Brownian dynamics and aggregation: from hard spheres to proteins', Chem. Industry, 158-163.

25. Deutch, J. M. and Oppenheim, I. (1971) 'Molecular theory of Brownian motion for several particles', J. Chem. Phys., 54, 3547-3555.

26. Ermak, D. L. and McCammon, J. A. (1978) 'Brownian dynamics with hydrodynamic interactions', J. Chem. Phys., 69, 1352-1360.

27. Dickinson, E., Allison, S. A. and McCammon, J. A. (1985) 'Brownian dynamics with rotation—translation coupling', J. Chem. Soc., Faraday Trans. 2, 81, 591-601.

28. Dickinson, E. and Honary, F. (1986) 'A Brownian dynamics simulation of enzyme—substrate encounters at the surface of a colloidal particle', J. Chem. Soc., Faraday Trans. 2, 82, 719-727.

29. Ying, R. and Peters, M. H. (1989) 'Torque constraints for modeling the behaviour of rigid and semirigid macromolecules', J. Chem. Phys., 91, 1287-1293.

30. Batchelor, G. K. (1976) 'Brownian diffusion of particles with hydrodynamic interaction', J. Fluid Mech., 74, 1-29.

31. Schmitz, R. and Felderhof, B. U. (1983) 'Mobility matrix for two spherical particles with hydrodynamic interaction', Physica, A116, 163-177.

32. Rotne, J. and Prager, S. (1969) 'Variational treatment of hydrodynamic interaction in polymers', J. Chem. Phys., 50, 4831-4838.

33. Glendinning, A. B. and Russel, W. B. (1982) 'A pairwise additive description of sedimentation and diffusion in concentrated suspensions of hard spheres', J. Colloid Interface Sci., 89, 124-143.

34. Bacon, J., Dickinson, E. and Parker, R. (1983) 'Simulation of particle motion and stability in concentrated dispersions', Faraday Discuss. Chem. Soc., 76, 165-178.

35. Mazur, P. and van Saarloos, W. (1982) 'Many-sphere hydrodynamic interactions and mobilities in a suspension', Physica, A115, 21-57.

36. Snook, I., van Megen, W. and Tough, R. J. A. (1983) 'Diffusion in concentrated hard sphere dispersions: effective two-particle mobility tensors', J. Chem. Phys., 78, 5825-5836.

37. Beenakker, C. W. J. (1986) 'Ewald sum of the Rotne—Prager tensor', J. Chem. Phys., 85, 1581-1582.

38. Smith, E. R. (1987) 'Boundary conditions on hydrodynamics in simulations of dense suspensions', Faraday Discuss. Chem. Soc., 83, 193-198.

39. Bacon, J., Dickinson, E., Parker, R., Anastasiou, N. and Lal, M. (1983) 'Motion of flocs of two or three interacting colloidal particles in a hydrodynamic medium', J. Chem. Soc., Faraday Trans. 2, 79, 91-109.

40. Ansell, G. C. and Dickinson, E. (1987) 'Brownian dynamics simulation of the formation of colloidal aggregate and sediment structure', Faraday Discuss. Chem. Soc., 83, 167-177.

41. Dickinson, E. and Parker, R. (1984) 'Brownian encounters in a polydisperse sedimenting system of interacting colloidal particles', J. Colloid Interface Sci., 97, 220-231.

42. Ansell, G. C., Dickinson, E. and Ludvigsen, M. (1985) 'Brownian dynamics of colloidal aggregate rotation and dissociation in shear flow', J. Chem. Soc., Faraday Trans. 2, 81, 1269-1284.

43. Ansell, G. C. and Dickinson, E. (1986) 'Brownian dynamics simulation of the fragmentation of a large colloidal floc in simple shear flow', J. Colloid Interface Sci., 110, 73-81.

44. Chan, D. Y. C. and Halle, B. (1984) 'Dissociation kinetics of secondary-minimum flocculated colloidal particles', J. Colloid Interface Sci., 102, 400-409.

45. Batchelor, G. K. and Green, J. T. (1972) 'The hydrodynamic interaction of two small freely-moving spheres in a linear flow field', J. Fluid Mech., 56, 375-400.

46. Ansell, G. C. and Dickinson, E. (1985) 'Aggregate structure and coagulation kinetics in a concentrated dispersion of interacting colloidal particles', Chem. Phys. Lett., 122, 594-598.

47. Ansell, G. C. and Dickinson, E. (1987) 'Short-range structure of simulated colloidal aggregates', Phys. Rev., A35, 2349-2352.

48. Dickinson, E. and Elvingson, C. (1988) 'The structure of aggregates formed during the very early stages of colloidal coagulation', J. Chem. Soc., Faraday Trans. 2, 84, 775-789.

49. Ansell, G. C. and Dickinson, E. (1986) 'Sediment formation by Brownian dynamics simulation: effect of colloidal and hydrodynamic interactions on the sediment structure', J. Chem. Phys., 85, 4079-4086.

50. Dickinson, E. (1989) 'Structure of simulated colloidal deposits', Colloids Surf., 39, 143-159.

51. Dickinson, E. (1987) 'Short-range structure in aggregates, gels and sediments', J. Colloid Interface Sci., 118, 286-289.

52. van Gunsteren, W. F. and Berendsen, H. J. C. (1982) 'Algorithms for Brownian dynamics', Molec. Phys., 45, 637-647.

53. Allison, S. A. and McCammon, J. A. (1984) 'Transport properties of rigid and flexible macromolecules by Brownian dynamics simulation', Biopolymers, 23, 167-187.

54. Meakin, P., Chen, Z. Y. and Deutch, J. M. (1985) 'The translational friction coefficient and time dependent cluster size distribution of three dimensional cluster—cluster aggregation', J. Chem. Phys., 82, 3786-3789.

55. Cohen, R. D. (1989) 'The probability of capture and its impact on floc structure', J. Chem. Soc., Faraday Trans. 2, 85, 1487-1503.

56. Rubinstein, M. and Nelson, D. R. (1982) 'Order and deterministic chaos in hard-disk arrays', Phys. Rev., B26, 6254-6275.

57. Bideau, D., Gervois, A., Oger, L. and Troadec, J. P. (1986) 'Geometrical properties of disordered packings of hard disks', J. Physique, 47, 1697-1707.

58. Dickinson, E., Milne, S. J. and Patel, M. (1989) 'Ordering in simulated packed beds formed from binary mixtures of particles in two dimensions: implications for ceramic processing', Powder Technol., 59, 11-24.

59. Kaye, B. H. (1989) A Random Walk Through Fractal Dimensions, VCH Publishers, Weinheim.

60. Wong, K., Cabane, B. and Somasundaran, P. (1988) 'Highly ordered microstructure of flocculated aggregates', Colloids Surf., 30, 355-360.

61. Vincent, B., Edwards, J., Emmett, S. and Croot, R. (1988) 'Phase separation in dispersions of weakly-interacting particles in solutions of non-adsorbing polymer', Colloids Surf., 31, 267-298.

62. Canessa, E., Grimson, M. J. and Silbert, M. (1989) 'Theory of phase equilibria in polymer stabilized colloidal suspensions', Molec. Phys., 67, 1153-1166.

63. Balazs, A. C., Anderson, C. and Muthukumar, M. (1987) 'A computer simulation for the aggregation of associating polymers', Macromolecules, 20, 1999-2003.

64. Chakrabarti, A. and Toral, R. (1989) 'Computer simulation of the aggregation process in self-associating polymer and surfactant systems', J. Chem. Phys., 91, 5687-5693.

65. Dickinson, E. and Euston, S. R. (1989) 'Computer simulation of the competitive adsorption between polymers and small displacer molecules', Molec. Phys., 68, 407-421.

COMPUTER SIMULATIONS OF FLUIDS AND SOLUTIONS OF ORGANIC MOLECULES

S. TOXVAERD
Chemistry Laboratory III
H.C. Ørsted Institute
University of Copenhagen
DK-2100 Copenhagen 0, Denmark

ABSTRACT. Computer simulations of organic molecules can either be performed by the molecular dynamics technique, which gives the classical mechanical solution of the molecular motions, or by stochastic methods like the Monte Carlo technique. The different methods are analysed and their results are critically reviewed by comparing them with corresponding data for n-alkanes.

1. Introduction

The physics of fluids consisting of complex molecules is one of the fascinating and challenging problems of modern condensed-matter physics. Because of the topological interactions, these systems display a rich and complex thermodynamic and viscoelastic behaviour. Though experiments have been very important in elucidating many of these properties, they are often unable to investigate the microscopic or molecular origin directly. For this reason, computer simulation of the molecular motions can play an important role in understanding fluids and solutions. Although the capability of computers has exponentially increased in the past few decades with a doubling of capacity every five years, the present generation of (parallel) computers are only able to simulate relatively small systems compared to eg. colloidal suspensions, biomolecules and polymers.

The computer systems consist of models of the molecules and are often treated by classical mechanics. But, under all circumstances they are a drastic reduction and simplification of the real systems. What kind of simplifications are possible depend, however, entirely on the phenomena of interest. It is quite often possible to leave out molecular details, which are of importance for the specific - but not for the qualitative behaviour. A well-known example of such a general law for fluids is the law of corresponding states, which allows us to calculate the equation of state for simple fluids by a computer simulation from only a few molecular parameters such as the Van der Waals diameter and energy etc. For more complex systems such as a fluid of n-alkanes there also exists a corresponding state mapping of the individual alkanes into an equation of state [1-2], and it will be demonstrated that the present generation of computers allows us to simulate such systems and to obtain quantitative agreement with experimentally obtained equation of state data, but only after modelling the alkanes in a quite detailed manner with the correct bond lengths and bond angles, energies etc. However, if we are only interested in the qualitative behaviour, such as the dependence of the end-to-end distance with number of chain-units, a much simpler molecular model for alkanes will give the general dependence by scaling in a correct manner.

The scaling concept is a very important tool for understanding complex systems [3]. It is the

D. M. Bloor and E. Wyn-Jones (eds.),
The Structure, Dynamics and Equilibrium Properties of Colloidal Systems, 729–739.

first and most important test a molecular model must pass, and often this gives all the information needed at present by giving the qualitative behaviour and the key to adjust or extend the model in order to describe a specific substance or solution quantitatively. How crude a model can be might be surprising for a chemist. Often simple lattice systems such as the Ising model will give the correct description of the general phenomena [3-4]. There are of course many examples, where these coarse grained models do not help us much, and where the detailed structure is essential for gaining a molecular insight. Such an example is a model for an enzyme. Here the conformational changes, and the bioactive site(s) depend on small free energy changes in the molecules, and the water or membrane environment play a crucial role for its activity. There exist computer programs for such systems, but it is fair to say that the programs and the models must be developed further before we can benefit from their molecular description.

The computer programs fall into two categories, one is the so-called molecular dynamics (MD) method, and the other category contains different statistical based methods like the Monte Carlo (MC) technique. The following section presents the two methods together with some comments about their usefulness as models of macromolecules.

2. Computer Models for Simulating Systems of Macromolecules

Classical molecular dynamics (MD) solve Hamiltons equations for a set of generalised coordinates q_i and p_i. In practice for macromolecules we will usually take the cartesian coordinates of the atomic centre $r_i = q_i$ or even consider groups of atoms, e.g. $-CH_2$, phenyl or a polystyrene sphere as one unit. Let $U(r^N)$ be the energy of the system of N subsystems. The equation of motion is then in the most simple form given by:

$$r_i(t+h) = 2r_i(t) - r_i(t-h) - \frac{\partial U(r^N)}{\partial r_i} h^2/m_i \qquad (1)$$

where one, in the simplest possible way constructs the positions at time t+h from the positions at time t and t-h. The time increment h is typically of the order 1% of the mean collision time or fastest vibration time. There is one very important fact to notice at this point. The coupled set of equations are (in the best case) chaotic which means that nearby trajectories separate exponentially in time so any numerical inaccuracy will rapidly erode the solution of the individual trajectories. A simple MD calculation, however gives the correct linear response regime [5]. In other words, to consider a single trajectory may not give any meaning, but the mean behaviour obtained from many trajectories give e.g. the correct self-diffusion coefficient. It is therefore more important that a calculation procedure is stable, which in the present case means that it conserves energy. The simple scheme (1) is time-symmetrical, reversible and is observed to conserve the energy in accordance with the thermodynamic laws. There exists a series of higher order predictor-corrector schemes to calculate the trajectories more accurately. If not used for some special purpose there is, however, no need for this, if we are only interested in the thermodynamical behaviour and the behaviour in the linear response regime.

In statistical simulations one often generates a Markov chain of changes in order to sample in a well-defined ensample, e.g. the canonical. Most commonly used is the simple Monte Carlo procedure which generates the thermodynamical energy E and pressure p for N objects in the volume V and at the temperature T. The new positions are simply generated with a Boltzmann probability from the instant positions r^N. The MC technique has demonstrated its usefulness in a series of simulations for different systems in natural science, including simulations of macromolecules. One of its advantages is that it is easy to use for a discreet space and lattice models represent a drastic reduction of computer time and scales in the correct way, giving the qualitative behaviour [4]. But the method also has it limitation which is often overlooked. The time behaviour can in general not be obtained from MC based methods. This is simply due to the fact that it is a Markov procedure for calculating the volume of a phase space. Only in the Brownian limit for heavy obstacles can we determine the self-diffusion from the generated energy-trajectories [6].

Another statistical method, used to simulate macromolecules is the Langevin equation (LE) - and its statistical mechanical founded version the generalised Langevin equation (GLE). In its simple and exact form the GLE for a particle with mass m, moving with the velocity v and exposed to a force f is:

$$m \frac{dv(t)}{dt} = - m\int_o^t dt \; M(t-t') \, v(t') + f(t) \qquad (2)$$

where the memory M(t-t') of previous velocities can be obtained (in principle) from the force moments of the solvent molecules [7]. If the memory is short (Brownian limit), insertion of a delta function in (2) gives the usual Langevin equation (LE).

$$m \frac{dv(t)}{dt} = - \gamma v(t) + f(t) \qquad (3)$$

The deterministic forces f(t) is replaced by stochastic forces and energetically balanced out with the dispersion term $\gamma v(t)$ through the fluctuation - dispersion relaxation. But also this method has its limitation and short-coming. The next step is to separate the forces into two categories. One (dominant) category which is treated explicitly and deterministic and the other which is treated stochastically. A model of this will be a flexible macromolecule consisting of subunits with intramolecular forces and immersed in a stochastic solvent [8]. The short-coming of this method is that it treats the background isotropically. In many situations e.g. dilute colloidal suspension this is an excellent approximation [9], but for conformal changes and phase transitions a more complex model is needed as demonstrated by a simple example in Figure 1 [10]. The figure gives the shape of a tetramer as a function of temperature. The shape is expressed as the mean-square of the ratio between the longest axis l_1 of the chain and the ratio of gyration S. The GLE solution (Δ) agrees well at high temperatures with the exact MD result (•); but at low temperatures the chain remains too open and behaves more like an isolated chain (x). The simple example demonstrates the short-coming of the statistical model, but unfortunately this defect is overlooked in many simulations of collapsed polymers using Langevin dynamics.

The stochastic GLE and LE are, however, very useful in many other contexts. It is due to the fact that it is usually possible to leave out many atomic details so that the models can be extended in space and time. This is demonstrated in the MD-LE model by Kremer, Crest and Carmesin [11]. The equation of motion of a polymer unit is a combination of (1) and (2) for a weakly damped heavy subunit i:

$$r_i(t+h) = 2r_i(t) - r_i(t-h) - \gamma(r_i(t+h) - r_i(t-h))/2h \qquad (4)$$

$$+ \; w_i(t) - \partial U(r^N)/\partial r_i \; (h^2/m_i)$$

where w(t) is a Gaussian white-noise source.

$$<w_i(t)w_j(t')> = \delta_{ij}\ \delta(t-t')\ 6kT\gamma/m \qquad (5)$$

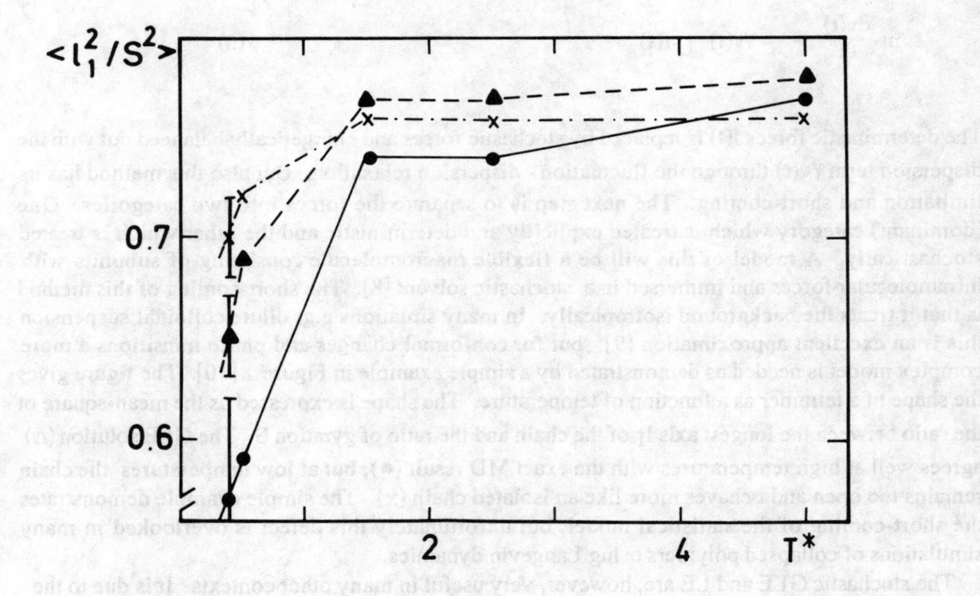

Figure 1. **The asphericity ratio of a simple tetramer as a function of temperature T^*, given by the mean-square of the ratio between the longest axis l_1 of the chain and the ratio of gyration S.**

The friction (viscosity) γ gives the solvent damping of the polymer units, and the friction in (4) and (5) act as a temperature bath. Using Eq. (4) and (5) for a fluid system of a polymer consisting of N = 150 subunits they were able to confirm the reptation dynamics predicted by de Gennes [11].

With this introduction of the simulation technique we shall demonstrate its utility by calculating the dynamical behaviour and the equation of state of some n-alkanes.

3. Simulation of n-alkanes

The n-alkanes are suitable as test examples for computer simulations of chain molecules since they are well documented in the sense that we know the bond-lengths, bond-angles, intramolecular potentials and on a macroscopic scale their thermodynamics. Using n-alkanes it is therefore possible to answer questions like:
How detailed is it necessary to depict a chain molecule in order to get the thermodynamics quantitatively or qualitatively correct? As mentioned the equation of state for n-alkanes to some extent obey a law of corresponding states [1], [2]; a fact used in many empirical equations of state $p(T, \rho)$ for the pressure p as a function of temperature T and density ρ. Figure 2 demonstrates the degree of similarity for $p(T,\rho)$ for three different alkanes, namely propane, pentane and decane. The figure gives the high pressure in compressed fluids at constant temperatures as a function of the relative molar volumes V/V_c, where V_c is the critical molar volume. There are two sets of isotherms, T_1 and T_2, and with the same ratio T_1/T_2 for pentane (full line) and decane (dashed curve). The dash-dotted curve is for propane. As demonstrated in the figure the n-alkanes scale in a simple and universal way with chain length for points of state away from the critical region and this fact makes it possible to demonstrate an agreement or disagreement of a calculated pressure with p_{exp} on a single figure. The question presented at the beginning of this section of how detailed a computer model should be, can now be made more precise and formulated as: Can the data presented in Figure 2 be obtained by a computer simulation and how important are the details like bond-length, intramolecular potentials etc for the pressure in a system of chain molecules?

Computer simulations of alkanes have been performed for a long time [15]. Usually one simplifies the chain by a 'united atom' (UA) model in which the centre of the carbon atoms is taken as nodes or sites for the methylene units. The site units are characterised energetically by a Lennard-Jones potential

$$u(r_{ij}) = 4\,\varepsilon \left[\left(\frac{\sigma_{ij}}{r_{ij}}\right)^{12} - \left(\frac{\sigma_{ij}}{r_{ij}}\right)^{6} \right] \qquad (6)$$

for the interaction between the i'th and the j'th sites, separated a distance r_{ij}. The bond-length and angles are determined experimentally, usually by electron diffraction of gases [16]-[18]. Other intramolecular potential data are taken from [19] or given in [20]. It shall be mentioned that some of the parameters are encumbered with relatively big uncertainties; e.g. the torsions potential and the UA-site potential-parameters. This problem is discussed in Ref. [20]. As a guide for the choice of parameters we have chosen the latest published values!

The molecular dynamics simulations are described in Ref. [20] and [21]. The pressure p as a function of density and temperature is calculated using the virial expression with careful corrections for the long range attraction between the molecules. The results of the MD calculations for the UA-model are given in Figure 3. The figure gives a more detailed graph of the high

pressure density branch of the fluid isotherms for the fluid alkanes, also shown in Figure 2. The signature is the same as in Figure 2 and, the two crosses for the MD-pressure of propane agree perfectly with the experimentally (dash-dotted curve), as already shown in Ref. [20]. Also the MD-pressure of pentane (Δ) at moderate densities agrees well with real pentane (full line). But, the MD-pressure for decane (o) deviate systematically from the experimental isotherm (dashed line). A simple readjustment of some of the parameters will easily compensate for this short-coming. There are, however, no theoretical or experimental justifications for doing so.

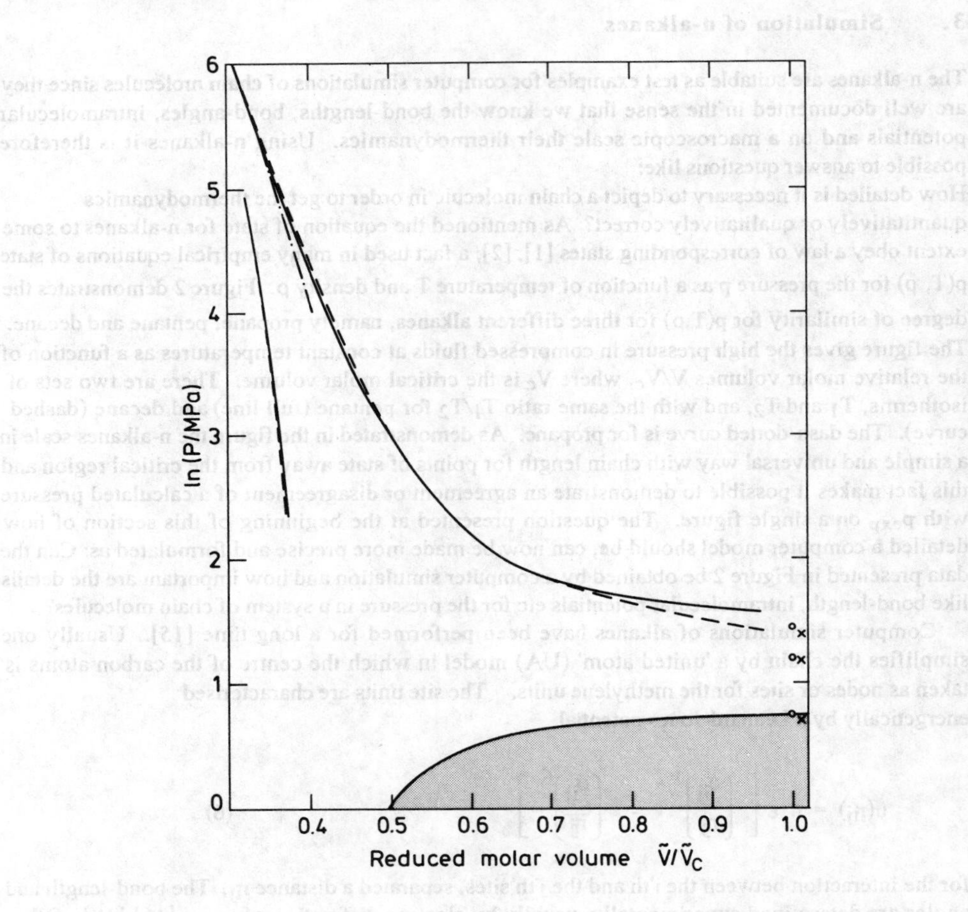

Figure 2 Experimentally obtained pressures of fluid propane, pentane and decane as a function of reduced molar volume V/V_c and for different isotherms. Dash-dotted line: propane at T = 346 K [12]. Full line: pentane at T_1 = 313 K and T_2 = 498 K, respectively, [13], dashed line: decane at T_1 = 423 K and T_2 = 673 K, respectively, [14].

Although the parameters are defected by uncertainties the readjustment necessary for agreement is bigger than the uncertainties and unphysical since we need individual potential parameters for the alkanes and thereby ignoring the law-of-corresponding state behaviour of n-alkanes. So the conclusion must be that one cannot obtain a perfect agreement with experimental data using a united-atom model for n-alkanes. On the other hand the qualitative behaviour of chain molecules can be derived using a much simpler model. e.g. the scaling relation can, as mentioned, be obtained from lattice models and simple bead-spring models for star-polymers scale correctly [22]. This difference between quantity and quality is not trivial and will be discussed in the next section.

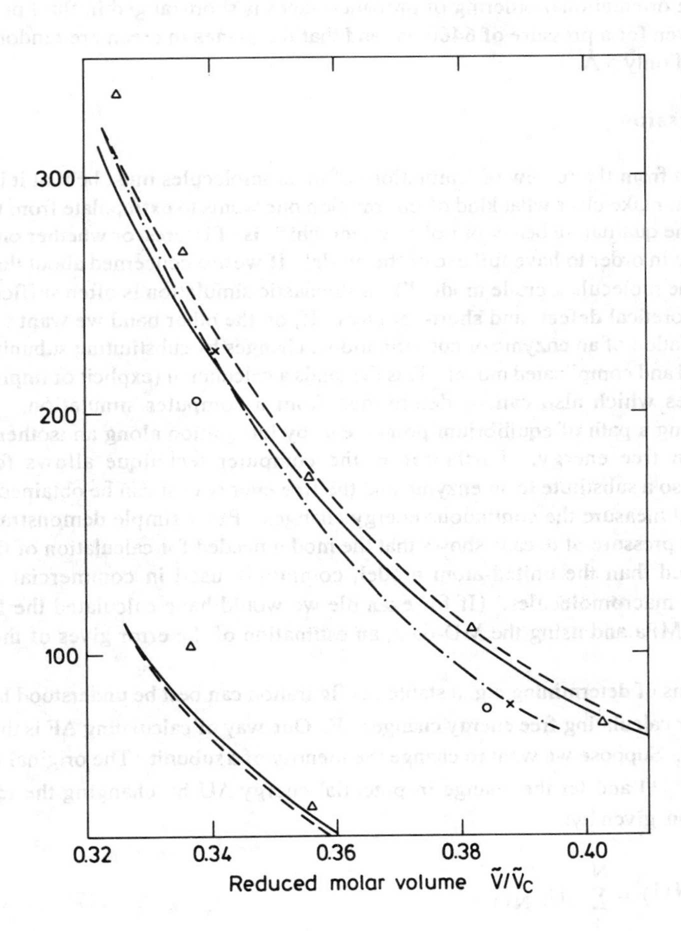

Figure 3 **Detailed graph of the experimentally obtained isotherm from Figure 2 with the same signature, compared with MD-calculated pressures. (x), (Δ) and (o) give the MD-pressures in MPa of propane, pentane and decane**

The real benefit of a good computer model is of course that it offers a detailed insight on the molecular behaviour, an insight that is very difficult or impossible to get from a real experiment. For propane, e.g. we can calculate the molecular (mean) orientation whereas a scattering experiment usually only gives the radial distribution of molecules and only at a lower pressure. The self-diffusion coefficient and diffusion mechanism is another example of commercial interest. We shall end the presentation of MD-simulations of n-alkanes by showing the orientational ordering in propane at a (very) high pressure. The MD-model agrees perfectly with experimental data which gives confidence to the predictions from the model. Figure 4 shows the mean angle $\phi_1(r)$ between pairs of planes as a function of their center-of-mass distance R. The calculations predict that the orientational ordering of propane-planes is short-ranged in fluid propane at room temperature even for a pressure of 6460 bar and that the planes in mean are randomly orientated for distances of only 5 Å.

4. Discussion

The conclusion from the review of simulations of macromolecules must be that it is crucial from the beginning to make clear what kind of information one wants to extrapolate from the model. Whether it is the qualitative behaviour of a system which is of interest or whether one really needs the quantitative in order to have full use of the model. If we are concerned about the sensitivity of an impact in the molecule a crude model like a stochastic simulation is often sufficient despite its (principal) theoretical defects and short-comings. If, on the other hand we want to calculate the stable conformation of an enzyme or conformational changes by substituting subunits etc, we need a more detailed and complicated model. This demands a calculation (explicit or implicit) of the free energy changes which also can be determined from a computer simulation. It is done by integration along a path of equilibrium points, e.g. by integration along an isotherm and from a state of known free energy. Furthermore, the computer technique allows for unphysical manipulations so a substitute to an enzyme and the free energy cost can be obtained by scaling up the subunit and measure the continuous energy changes. But a simple demonstration, given in Figure 3 of the pressure of decane shows that the model needed for calculation of the free energy is more detailed than the united-atom model, commonly used in commercial programs for simulations of macromolecules. (If for example we would have calculated the free energy of decane at 200 MPa and using the MD-data, an estimation of the error gives of the order 1.5 kJ mol^{-1}).

The problems of determining e.g. a stable configuration can best be understood by considering a formalism for calculating free energy changes ΔF. One way of calculating ΔF is the so-called λ-expansion [23]. Suppose we want to change the identity of a subunit. The original total potential energy is $U(r^{N+1})$ and let the change in potential energy ΔU by changing the identity be an analytic function given by:

$$\Delta U\,(r^{N+1}) = \sum_{i}^{N} \Delta U_{i,\,N+1} \qquad (7)$$

The change in energy is then gradually switched on as $\lambda\,\Delta\,U$ by letting λ varying from zero to one. The derivative of energy, F, with respect to λ can be obtained by differentiating the configurational integral Z_λ for the system with energy $U(r^N) + \lambda\Delta U$. It gives:

$$\frac{\partial F}{\partial \lambda} = Z_{\lambda}^{-1} \int e^{-\beta (U(r^N) + \lambda \Delta U)} \Delta U \, dr^{N+1} \tag{8}$$

$$= \langle \Delta U \rangle_{\lambda}$$

and the total change in free energy is

$$\Delta F = \int_{0}^{1} d\lambda \, \langle \Delta U \rangle_{\lambda} \tag{9}$$

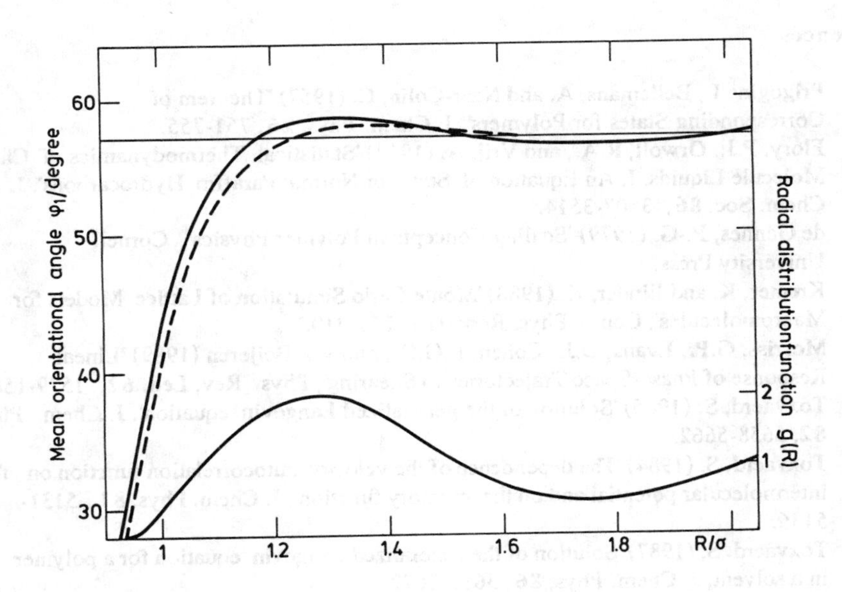

Figure 4 **The mean angle $\phi_1(R)$ between pairs of propane planes as a function of the center-of-mass distance R between the propane molecules. The two angle distribution functions go asymptotically to $\phi_{\infty} = 57.30^{\circ}$. The full line is for a high pressure state (6460 bar), and the dashed line is for a moderate pressure (320 bar). Also shown is the radial distribution function g(R) with the ordinate units to the right in the figure.**

A computer simulation can, in principle calculate ΔF and commercial programs for simulations of macromolecules offer this facility. The calculation, however, is faced with several difficulties. At first ΔF is relatively small compared to the total free energy of the system, and it depends often strongly on the configuration. This means that ΔF fluctuates and necessitates long calculations in order to get a good statistic. Another reason is that the energy function ΔU simply is not good enough. A subunit will typically involve Coulomb forces, and if the solvent is water it might be necessary to consider polarizability changes introduced by the new forces. On the other hand we can hope that some of the errors cancel out if one only considers the relative changes in F by substituting different groups.

To conclude the computer simulation technique has shown its ability in physical-chemical analysis by giving the qualitative behaviour correctly. The present generations of programs are, however, not so detailed and accurate that they allow for calculations of free-energy changes associated with conformational changes and substitutions in systems of macromolecules.

Acknowledgement

A grant No. 11-7785 by the Danish Natural Research Council is gratefully acknowledged.

References

[1] Prigogine I., Bellemans, A. and Naar-Colin, C. (1957) 'Theorem of Corresponding States for Polymers', J. Chem. Phys. **26**, 751-755.

[2] Flory, P.J., Orwoll, R.A., and Vrij, A. (1964) 'Statistical Thermodynamics of Chain Molecule Liquids. I. An Equation of State for Normal Paraffin Hydrocarbons', J. Am. Chem. Soc. **86**, 3507-3514.

[3] de Gennes, P.-G. (1979) 'Scaling Concepts in Polymer Physics', Cornell University Press.

[4] Kremer, K. and Binder, K. (1988) 'Monte Carlo Simulation of Lattice Models for Macromolecules', Comp. Phys. Reports 7, 261-310.

[5] Morriss, G.P., Evans, D.J., Cohen, E.G.D., and van Beijeren (1989) 'Linear Response of Phase-Space Trajectories to Shearing', Phys. Rev. Lett. **62**, 1579-1582.

[6] Toxvaerd, S. (1985) 'Solution of the generalized Langevin equation', J. Chem. Phys. **82**, 5658-5662.

[7] Toxvaerd, S. (1984) 'The dependence of the velocity autocorrelation function on the intermolecular potential and on the memory function', J. Chem. Phys. **81**, 5131-5136.

[8] Toxvaerd, S. (1987) 'Solution of the generalized Langevin equation for a polymer in a solvent, J. Chem. Phys, **86**, 3667-3672.

[9] Xue, W. and Grest, G.S. (1989) 'Brownian dynamics simulations for interacting colloids in the presence of shear flow', Phys. Rev. A **40**, 1709-1712.

[10] Padilla, P. and Toxvaerd, S. (1988) 'Solvent Effect on the Shape of a Tetramer', Mol. Simulation **1**, 399-402.

[11] de Gennes, P.G. (1971) 'Reptation of a Polymer Chain in the Presence of Fixed Obstacles', J. Chem. Phys. **55**, 572-579.

[12] Babb, S.E. Jr. and Robertson, S.L. (1970) 'Isotherms of Ethylene and Propane to 10000 bar*', J. Chem. Phys. **53**, 1097-1099.

[13] Gehrig, M. and Lentz, H. (1979) 'Values of p(V,T) for n-pentane in the range 5 to 250 MPa and 313 to 673 K', J. Chem. Thermodynamics, 291-300.

[14] Gehrig, M. and Lentz, H. (1983) 'Values of p(M_m,T) for n-decane up to 300 MPa and 673 K', J. Chem. Thermodynamics **15**, 1159-1167.

[15] Ryckaert, J.-P. and Bellemans, A, (1975) 'Molecular Dynamics of Liquid n-Butane near its Boiling Point', Chem. Phys. Lett. 30, 123-125.

[16] Iijima, T. (1972) 'Molecular Structure of Propane', Bull. Chem. Soc. Japan **45**, 1291-1294.

[17] Heenan, R.K. and Bartell, L.S. (1983) 'Electron diffraction studies of supersonic jets', J. Chem. Phys. **78**, 1270-1274.

[18] Fitzwater, S. and Bartell, L.S. (1976) 'Vapor-Phase Structure and Conformation of a Long-Chain n-alkane. An Electron Diffraction Study' J. Am. Chem. Soc. **98**, 8338-8344.

[19] Steele, D. (1985) 'An ab initio Investigation of the Torsional Potential Function of n-Butane', J. Chem. Soc. Faraday Trans 2,**81**, 1077-1083.

[20] Toxvaerd, S. (1989) 'Molecular dynamics calculation of the equation of state of liquid propane', J. Chem. Phys, **91**, in press.

[21] Toxvaerd, S. (1988) 'Molecular dynamics of liquid butane', J. Chem. Phys. **89**, 3808-3813.

[22] Grest, G.S., Kremer, K., Milner, S.T., and Witten, T.A., (1989), 'Relaxation of Self-Entangled Many-Atom Star Polymers', Macromolecules **22**, 1904-1910.

[23] Hansen, J.-P. and McDonald, I.R. (1976) 'Theory of Simple Liquids, Academic Press, London.

THE FORM OF COLLOIDAL CRYSTALS FROM SILICA LATICES IN NON-AQUEOUS DISPERSION

D.J.WEDLOCK
Shell Research Ltd
Sittingbourne Research Centre
Sittingbourne, Kent ME9 8AG, U.K.

S.D.LUBETKIN and C.EDSER
Dept of Physical Chemistry
Bristol University
Cantocks Close
Bristol BS8 2LR

and

S. HAWKSWORTH
Dept. of Physics
University of Nottingham
Notts. U.K.

ABSTRACT. A suspension of silica latices of narrow polydispersity, dispersed in pure ethanol was allowed to sediment and consolidate. Under the long range ordering influence of an electrical double layer, an ordered, crystalline phase was formed. Using an acoustic technique (ultrasound velocity scanning) it was shown that the equilibrium state of the crystal was determined by the balance of gravitational compression and electrostatic repulsion between the silica latices, and that an exponential inter-particle potential could be used to describe the resultant spacing.

1. Introduction

Ordering in colloidal suspensions is a subject that has received considerable attention in recent years (1-5). Alder and Hoover (4) demonstrated that freezing and melting phenomena could be predicted in hard sphere colloidal dispersions, at well defined volume fractions using computer simulation. The phenomenon was perhaps most elegantly demonstrated by Pusey (5) using latices of sterically stabilised, essentially monodisperse poly(methylmethacrylate), dispersed in a non-aqueous, continuous phase refractive index matched to the particles. The index matching minimises the van der Waals attractive interactions and allows the quasi hard-sphere potential associated with steric stabilisation to induce ordering or crystallisation at a well defined effective particle volume fraction. The ordered state of the particles showed the characteristic iridescence of such colloidal crystals under white light illumination. In this study we propose to show that the ordered state for a particle with a relatively high buoyant density difference such as a charge stabilised silica latex in ethanol, is subject to gravitational compression effects both in the approach to equilibrium and in the final state of sedimentation equilibrium. The ordering is obtained by the repulsive effect of an unscreened double layer in a continuous phase of ethanol, a

D. M. Bloor and E. Wyn-Jones (eds.),
The Structure, Dynamics and Equilibrium Properties of Colloidal Systems, 741–747.

relatively low dielectric constant (~ one quarter that of water).

2. Experimental

2.1 MATERIALS

The silica samples were prepared by the method of Stober et al. They had an average size of 535 nm and a coefficient of variation of 4 %, determined by scanning electron microscopy.

The sample was cleaned by extensive dialysis, first against distiled water and then against water / ethanol mixtures, with the proportion of ethanol being changed from 0 to 100% in steps of 5 % over a period in excess of fifty days, with the dialysis medium being changed twice daily for the first two weeks and then daily thereafter.

A silica suspension of 16 % m/m silica lattices in ethanol was allowed to sediment to equilibrium in a glass cell, ~240 mm in height, and 32mm square cross section.

2.2 METHODS

2.2.2. *Ultrasound Velocity Scanning* The principles of the experimental technique have been described previously in detail (7,8). Essentially the ultrasound velocity measuring technique involves determination of the time-of-flight of a pulse of ultrasound of an appropriate frequency, in this case 1 M Hz, between two water immersible transducers, external to the cell containing the suspension of particles in ethanol, acoustically linked by the thermostatting water. The time-of-flight data is logged in a vertically resolved manner and transposed to ultrasound velocity from a calibration of internal cell path length as a function of height in combination with the Urick (9) equation, using appropriate dispersed and continuous phase densities and adiabatic compressibilities.

2.2.3. *Zeta Potential Determination.* This was kindly determined by A. Parker of Malvern Instruments, using a Malvern Instruments Zeta Sizer 3, with a standard AZ4 cell appropriate to aqueous and polar liquid systems.

3. Results and Discussion

The inversion of the ultrasound velocity data for such a suspension is controlled by a non-monotonic calibration function, since there is a minimum in the suspension ultrasound velocity as a function of particle volume fraction (8). Fig. 1 shows the literal interpretation of the velocity / volume fraction inversion, during the process of sedimentation as the dotted line. It is noteworthy because the vertical variation in the volume fraction of particles through the ordered phase (visually iridescent) varies linearly, and upon reaching the limit of the iridescent region becomes apparently constant at a break point, equal to the volume fraction which corresponds to the onset of ordering.

At these early stages, before the sediment has consolidated, the concentration gradient through the ordered phase may correspond purely to the extrusion of solvent at different rates through a plug of varying porosity, formed by the sediment slowly consolidating.

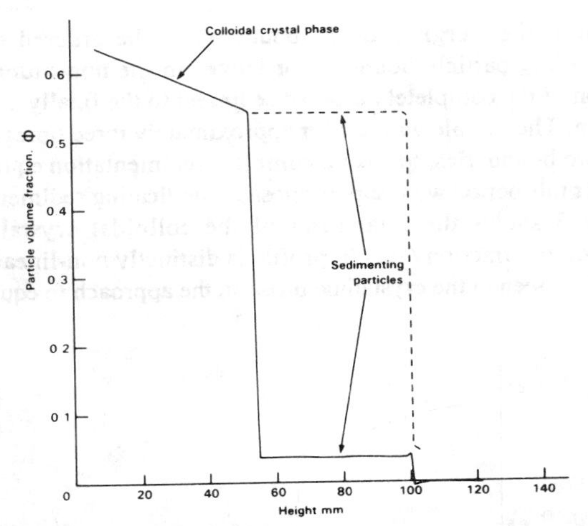

Fig. 1 Particle volume fraction as a function of height in a sedimenting column of silica latices, using alternative root solutions of the Urick equation, above the ordered region.

The breakpoint at the boundary of the colloidal crystalline phase and the sedimenting particles represented by the dotted line, is useful since it images the upper vertical limit of the ordered region in the approach to equilibrium. In reality the point corresponds to a sharp concentration gradient, although not a sharp ultrasound velocity gradient, where the root for the solution of the Urick equation (9) must be switched to obtain the correct concentration in the non-ordered, low particle concentration region. This is shown in the form of the solid line in fig.1, above the crystalline phase.

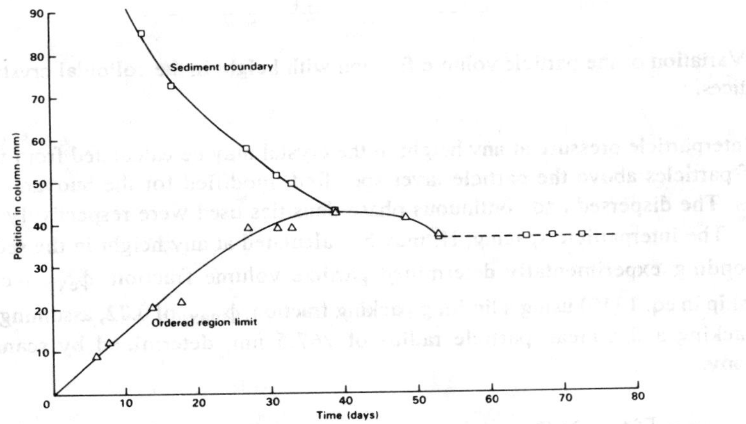

Fig. 2 The sedimenting particle boundary and the ascending limit of the ordered region of the sediment as a function of time.

Fig. 2 shows the form of the merging, of the boundary of the ordered region as it moves upwards, with the descending particle boundary of latices in the non-ordered state, which is followed by the relaxation of the completely ordered sediment to the finally consolidated state of sedimentation equilibrium. The sample was left for approximately three times as long as required to obtain merger of the two boundaries, to ensure complete sedimentation equilibrium. Repeated scans of the sediment over this period were superimposable, indicating sedimentation equilibrium to our satisfaction. Fig. 3 shows the final form of the colloidal crystal at sedimentation equilibrium, where the volume fraction / height profile is distinctly non-linear, compared to the linear concentration gradient seen in the crystalline phase in the approach to equilibrium in fig. 1.

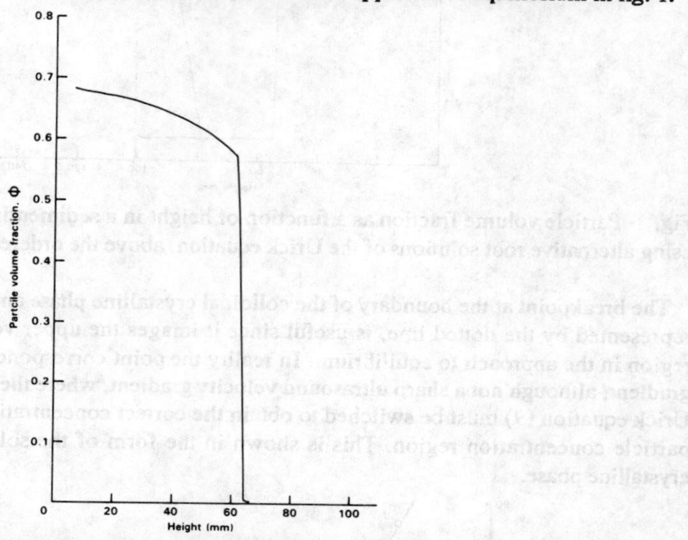

Fig. 3 Variation of the particle volume fraction with height in the colloidal crystal sediment of silica latices.

The interparticle pressure at any height in the crystal may be calculated from the cumulative mass of particles above the particle layer specified, modified for the buoyant density of the particles. The dispersed and continuous phase densities used were respectively 2600 and 796 kg.m^{-3}. The interparticle spacing, H, may be calculated at any height in the sediment with a corresponding experimentally determined particle volume fraction, ϕ_{exp} according to the relationship in eq. 1 (10) using a limiting packing fraction, ϕ_{max}, of 0.72, assuming body-centred cubic packing and a mean particle radius of 267.5 nm, determined by scanning electron microscopy.

$$H = 2\left[\left(\frac{\phi_{max}}{\phi_{exp}}\right)^{1/3} a - a\right] \qquad (1)$$

Fig. 4 shows how calculated interparticle pressure varies with both sediment height and the corresponding calculated interparticle spacing. Fig. 5 is included to show more clearly the exponential nature of the relationship between the interparticle pressure and the interparticle spacing.

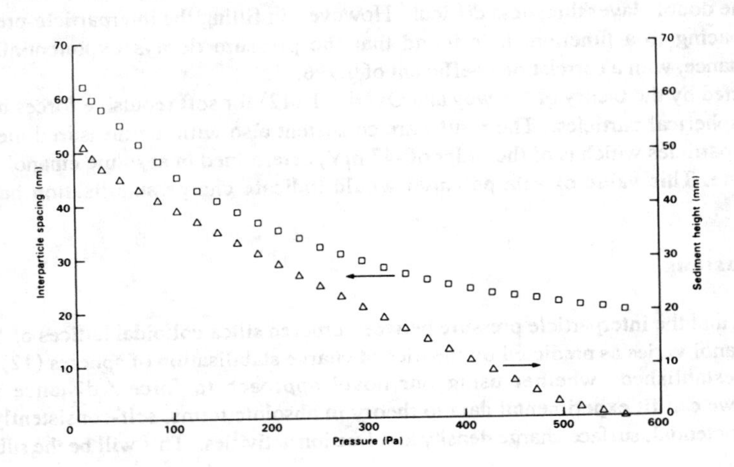

Fig. 4 Interparticle pressure variation with both colloidal crystal height and corresponding interparticle spacing.

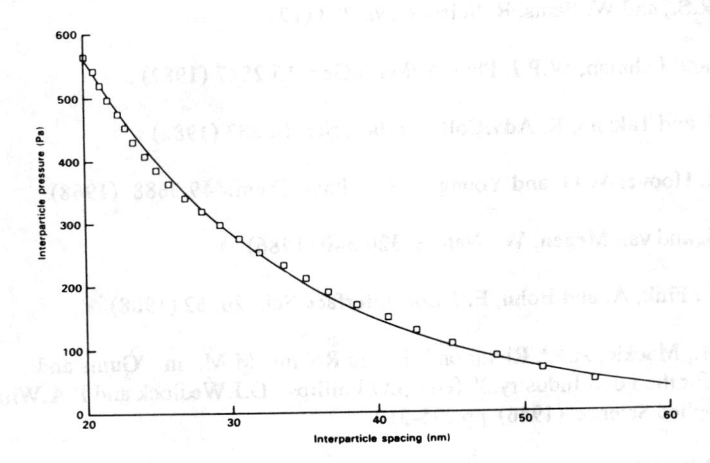

Fig. 5 Interparticle pressure as a function of interparticle spacing, fitted to a single exponential. Correlation coefficient 0.996.

The interparticle pressure corresponds to the osmotic pressure at that point in the sediment, and is a composite of the hard sphere osmotic pressure plus the osmotic pressure due to the free counterions in the double layer (11). The hard sphere osmotic pressure component is less than 0.1% of the total. This study has been performed on silica lattices with no excess simple electrolyte and therefore the total concentration of the counterions is uncertain, which makes estimation of the double layer thickness difficult. However, in fitting the interparticle-pressure / interparticle spacing to a function, it is found that the pressure decays exponentially with interparticle distance, with a correlation coefficient of 0.996.

This is predicted by the theory of Verwey and Overbeek (12) for soft repulsive forces between charged small spherical particles. The results are consistent also with the measured mean zeta potential of the particles which is of the order of -47 mV, determined in absolute ethanol with no added electrolyte. This value of zeta potential would indicate charge stabilisation has to be significant.

4. Conclusions

We have shown that the interparticle pressure between ordered silica colloidal lattices of 535 nm dispersed in ethanol varies as predicted by theories of charge stabilisation of spheres (12), and it remains to be established whether using our novel approach to force / distance profile measurements, we can fit experimental data to theory in absolute terms, self-consistently, with particle surface potential, surface charge density and free ion activities. This will be the subject of a future paper.

5. References

1. Crandall, R.S., and Williams, R. Science 198 293 (1977)

2. Gruner F. and Lehman, W.P.J. Phys.A :Math.Gen 15 2847 (1982)

3. Hachisu, S. and Takano, K. Adv.Coll.Interface Sci. 16 233 (1982)

4. Alder, B.J., Hoover W.G. and Young, D.A. J.Phys.Chem. 49 3688 (1968)

5. Pusey, P.N. and van Megen, W. Nature 320 340 (1986)

6. Strober, W., Fink, A. and Bohn, E. J.Coll.Interface Sci. 26 62 (1968)

7. Howe, A.M., Mackie, A.R., Richmond, P. and Robins, M.M. in 'Gums and Stabilisers for the Food Industry.3' (eds G.O.Phillips, D.J.Wedlock and P.A.Williams) Elsevier Applied Science (1986) pp 295-310.

8. Lubetkin, S.D., Wedlock, D.J. and Edser, C. Colloids and Surfaces in press

9. Urick, R.J. J.Appl.Phys 18 983 (1947)

10. Rohrsetzer, S., Kovacs P., and Nagy, M. Coll.Polym.Sci 264 812 (1986)

11. In 'Intermolecular and Surface Forces' by Israelachvili, J.N., Academic Press (1985) pp 167-173

12. Vervey E.J.W. and Overbeek, J.Th.G. in 'The Theory of Stability of Lyophobic Colloids', Elsevier, Amsterdam, 1948.

CONSOLIDATION OF DEPLETION FLOCCULATED CONCENTRATED SUSPENSIONS

INFLUENCE OF NON-ADSORBING POLYMER CONCENTRATION ON CONSOLIDATION RATE CONSTANTS

D.J.WEDLOCK A.MOMAN AND J.GRIMSEY
Shell Research Ltd
Sittingbourne Research Centre
Sittingbourne, Kent ME9 8AG
Great Britain

ABSTRACT. Observations have been made on the effect of varying the concentration of a non-adsorbing polymer in sterically stabilised concentrated suspensions of organic crystals (a herbicide) using an ultrasound velocity scanning technique. The rate of sediment or floc consolidation is satisfactorily quantitatively interpreted using a first order rate equation and the variation in that rate constant with polymer concentration is described by a single power law for observations spanning the dilute to semi-dilute polymer concentration regime.

Qualitative interpretation of ultrasound velocity scans of the consolidating sediment suggest that the process of depletion flocculation of the dispersed particles occurs at free polymer concentrations at least an order of magnitude less than the dilute /semi-dilute transition regime, the so-called coil overlap concentration in these polydisperse suspensions.

1. Introduction

The gravitational consolidation or compression of stable or flocculated concentrated suspensions has received theoretical consideration (1-5) but there has been relatively little detailed experimental data published, because of the lack of readily available techniques for following such processes in any great detail.

The use of high energy radiations to derive concentration profiles for concentrated suspensions has received some attention (3), but is obviously not a technique that can be used as a matter of routine. The approach of observing boundary movement in gravitational fields has provided much useful information on floc consolidation (4,5) but does not give detailed information on the concentration profiles in the region of perhaps most interest, at the base of the column of particles, where the floc is most extensively compressed. Uniform compression can not be assumed (6). The advent of ultrasound velocity scanning (6-8) has allowed the use of a non-invasive, low energy technique for the investigation of the sedimentation and consolidation processes in opaque colloidal and non- colloidal dispersions, over a range of volume fractions from little more than zero to the close-packed limit.

D. M. Bloor and E. Wyn-Jones (eds.),
The Structure, Dynamics and Equilibrium Properties of Colloidal Systems, 749–757.

In this report, we describe the application of the ultrasound technique to investigating the case of consolidation of a water continuous, polydisperse suspension of organic particles made into an initially colloidally stable state with a non-ionic block co-polymer, and subsequently flocculated by adding a non-adsorbing high molecular weight water soluble polymer in order to moderate sedimentation. This is typical of the approach that may be followed in an industrial formulation process.

In a previous report (9) the flocculation of a similar concentrated suspension by the same water soluble polymer, a cellulose ether, hydroxyethylcellulose, was shown to be due to bridging flocculation of the particles. By changing the dispersant type from an anionic polyelectrolyte / block copolymer that essentially charge stabilises, to a non-ionic dispersant that confers colloidal stability by a steric mechanism, the cellulose ether can be changed from an adsorbing polymer to a non-adsorbing polymer. The non-adsorbing polymer has the capability to cause flocculation of the particles by depletion forces (10) and this floc is a sufficiently self-hindering, connected assembly that close packing of the particles by gravitational sedimentation can be resisted. The dilatant properties of close-packed colloidally stable particles has been well documented (11), and its prevention or retardation is of industrial relevance, enabling control of the flow properties of concentrated suspensions. A full understanding of the consolidation process is important, both theoretically and experimentally.

We show that it is possible to define a rate of consolidation of a floc in the region of importance i.e. at the base of the column, the position where dilatant particulate assemblies are likely to be initially formed, and relate this rate to polymer concentration. The kinetics of particle accumulation are analysed by a first order rate equation, considering the approach of the volume fraction of the dispersed phase at a fixed level in the suspension, to the ultimate achievable packing fraction. This packing fraction is shown to have a value between random packing and close packing of spheres.

The first order rate constant is a continuous function of polymer concentration over a range of polymer concentration from the dilute to semi-dilute regime, that is, either side of the coil overlap concentration, c*. Previous reports (10,12-13) suggest that there is a molecular weight dependence of the propensity of a non-adsorbing polymer to induce depletion flocculation and that further, there is a critical free polymer concentration dependence of the onset of flocculation for a given molecular weight of free polymer. There is also said to be a particle volume fraction dependence of the critical polymer concentration for the onset of flocculation, although this last effect is weak according to Sperry (13) for particle volume fractions below 30%. There have been reports (12) of discontinuities in rheological behaviour of suspensions at this critical polymer (continuous phase) concentration, together with microscopic observations of the onset of flocculation above this concentration. We have different, but not inconsistent findings.

2. Materials and Methods

2.1 NON-ADSORBING POLYMER

Hydroxyethylcellulose (HEC), was supplied by Hercules Ltd. It had a degree of substitution of 2.5 hydroxyethyl groups per glucose residue and a molecular weight of $\sim 1.01 \times 10^5$ determined from intrinsic viscosity measurements. It was added to the suspensions in the form of a pre-dissolved concentrate.

A spin labelled version of the same polymer was prepared as described previously (9) using 2,2',6,6' tetramethyl-piperidino oxyl (TEMPO) , a nitroxyl spin label, covalently linked after cyanogen bromide activation of the polymer in water.

2.2 SUSPENSIONS

The preparation and milling of these suspensions of technical cyanazine (2-[[4-chloro-6-(ethylamino)-1,3,5-triazin-2-yl]amino]-2- methylpropanenitrile) has been described previously (11), except that only samples dispersed with the polyethyleneoxide / polypropyleneoxide block co-polymer were investigated. Samples of the milled dispersions with a volume mean diameter of 2.4 µm were prepared at various added HEC levels. The HEC was added as a pre-dissolved concentrate. Concentrations of HEC are expressed as concentrations in the continuous phase.

2.3 ULTRASOUND VELOCITY SCANNING

The essential instrumental details have been reported previously (8).Ultrasound scans of various samples of a 40 % v/v water continuous suspension of the particles (described under suspensions), were logged for up to ~150 days at 20°C in undisturbed cells of horizontal cross section 32x32 mm.

2.4 ELECTRON SPIN RESONANCE

The spectrometer was a Brucker model ER 200. The absence of adsorbed HEC on the surface of the particles was established by equilibrating the stabilised particles with various levels of spin labelled HEC, centrifuging the suspensions, discarding the supernatant and re- suspending and washing the particles with water pre-saturated with cyanazine. When the supernatant was established to be free of non- adsorbed spin labelled polymer, then a concentrated slurry of suspended particles was examined in the spectrometer, using a short pathlength aqueous type ESR cell (9) in order to minimise dielectric losses.

3. Results and Discussion

3.1 NON-ADSORPTION OF ADDED POLYMER

The electron spin resonance spectra of washed particles that had previously been equilibrated with spin labelled HEC, showed no evidence of any adsorbing polymer (9). It is considered that flocculation of the particles, due to added polymer, would not be by a bridging mechanism in this particular instance but more likely by the depletion mechanism associated with non-adsorbing polymers (13-16).

3.2 SEDIMENTATION OF SUSPENSIONS AND COMPRESSION OF FLOCS

Fig. 1 shows sedimentation in progress for a 40% v/v particulate suspension in water with no added polymer. In the form of ultrasound scans, we can see the process of sedimentation of colloidally stable particles to form a dilatant sediment. A maximum packing fraction of 0.68 is obtained for these particles. This is slightly larger than the random packing limit of 0.62 but is explicable in terms of the size and shape polydispersity of the particles, allowing some interstitial packing.

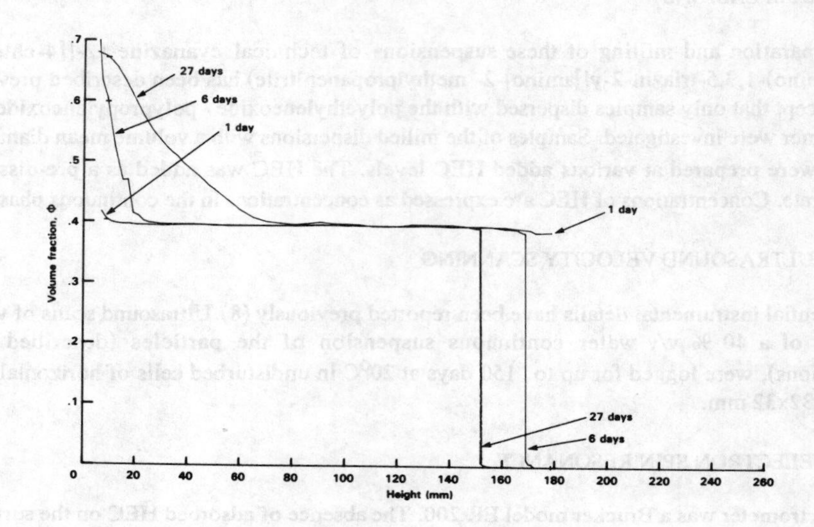

Fig. 1 Sedimentation of a 40% v/v suspension of sterically stabilised cyanazine particles. Volume fraction, ϕ, versus height of column (mm). No polymer added.

The overlapping plateau regions in the centre of the column scans, we feel, demonstrate the colloidally stable nature of the suspension. Influx and efflux of individual sedimenting particles will be balanced and lead to no net change in particle volume fraction in the region in the column centre. There is thus no long range particle connectivity in this system as indicated by the sedimentation profiles.

Fig. 2 on the other hand shows no such overlapping plateau regions, with compression of the total mass of underlying particles occurring from relatively early times. The concentration of added polymer is ~ 0.15 c*, the coil overlap concentration, and hence well within the dilute region of polymer behaviour. c* for this polymer sample was determined to be 12.5 g/l from both intrinsic viscosity and direct measurement of specific viscosity / concentration power law discontinuities.

Interestingly, fig. 2 also shows particle packing at the base of the cell, to the same packing limit as the system containing no added polymer. In fact, for the lowest concentrations of added polymer, some 0.05 x c*, the same overall effect was observed as seen in fig. 2. We might interpret this either as co-existence of flocculated and de-flocculated phases within the suspension, or alternatively formation of a very weak homogeneous floc phase with a sufficiently weak

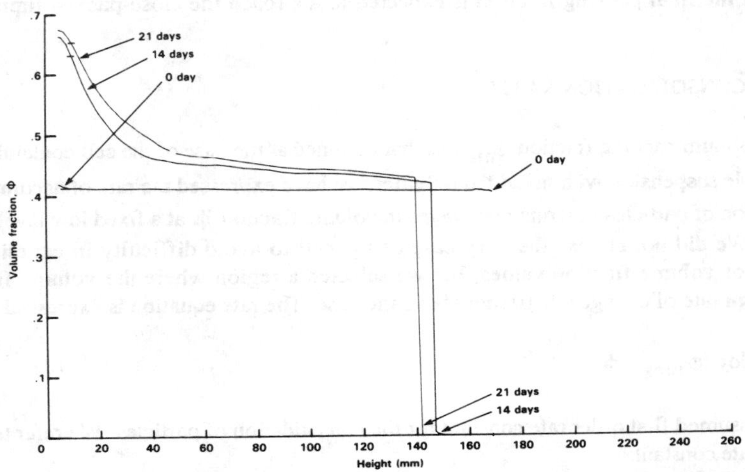

Fig. 2 Sedimentation of a 40% v/v suspension of flocculated cyanazine particles containing 1.9 g/l hydroxyethylcellulose. Volume fraction, ϕ, versus height of column (mm).

network structure (small uniaxial compression modulus), to permit compression of the floc to the packing limit of the particles. Fig. 3 shows consolidation of a flocculated suspension containing 21 g/l HEC, a polymer concentration well above c*. The rate of consolidation is

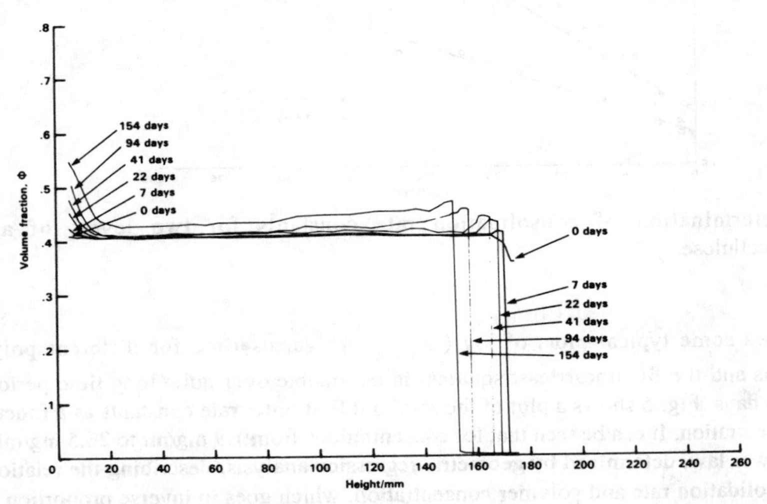

Fig. 3 Consolidation of a column of flocculated suspension of cyanazine, 40% v/v. 21 g/l hydroxyethylcellulose. No dilatant sediment at cell base.Volume fraction, ϕ, as a function of height of column (mm).

attenuated, but the final packing fraction is expected to approach the close-packed limit given sufficient time.

3.3 FLOC CONSOLIDATION RATES

Taking the maximum packing fraction ϕ_{max} as that obtained at the base of the cell containing the colloidally stable suspension with no added polymer, we have expressed the rate of accumulation and consolidation of particles in terms of changes in volume fraction ϕ, at a fixed low level above the cell base. We did not choose the very base of the cell to avoid difficulty in experimental determination of volume fraction values, but we selected a region where the volume fraction undergoes a high rate of change i.e. 10 mm above the base. The rate equation is expressed as,

$$kt = -\log(\phi_{max} - \phi) \tag{1}$$

where k is an assumed first order rate constant for the consolidation of particles. We refer to k as a consolidation rate constant.

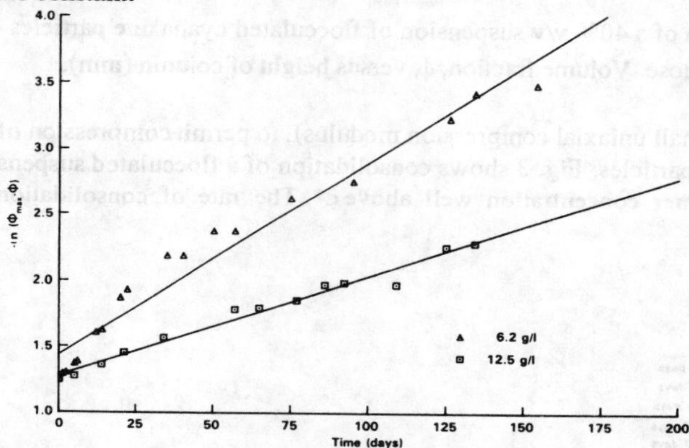

Fig. 4 Determination of consolidation rate constants for two levels of added hydroxyethylcellulose.

Fig.4 shows some typical plots of log (ϕ_{max} - ϕ) versus time for different polymer concentrations and the fit (linear least squares) is reasonable over quite long time periods, in excess of 100 days. Fig. 5 shows a plot of the resultant first order rate constants as a function of polymer concentration. It can be seen that for concentrations from 1.9 mg/ml to 26.5 mg/ml there is a single power law, determined by geometric regression analysis, describing the relationship between consolidation rate and polymer concentration, which goes in inverse proportion to the polymer concentration to the power 3/2. Discontinuities in this power law relationship are only apparent at polymer concentrations below ~1.9 mg/ml, an order of magnitude below c*, where the consolidation rates become constant. Polymer concentrations below 1.9 g/l were not used in the fitting procedure, but are shown on fig. 5 for completeness.

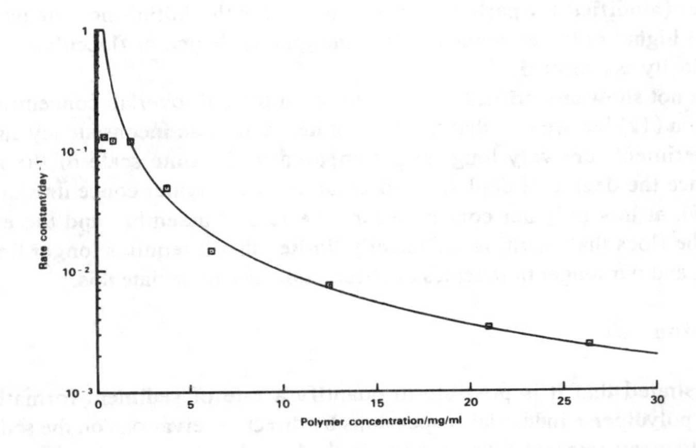

Fig. 5 Consolidation rate constant versus hydroxyethylcellulose concentration. Suspensions were 40% v/v cyanazine suspensions.

The reason that concentrations of polymer below 1.9 mg/ml were not used in the fitting procedure was that observations of the sedimentation velocity of the boundary with the supernatant (tangents to the height / time plot) showed that there was an initial increase in the sedimentation velocity with increasing polymer concentration which maximised and then decreased at concentrations of free polymer ~1.9 mg/ml upwards. The sedimentation velocity was measured as the tangent to the boundary movement line after five days, and the resultant sedimentation velocities as a function of free polymer concentration are shown in fig. 6. The effect

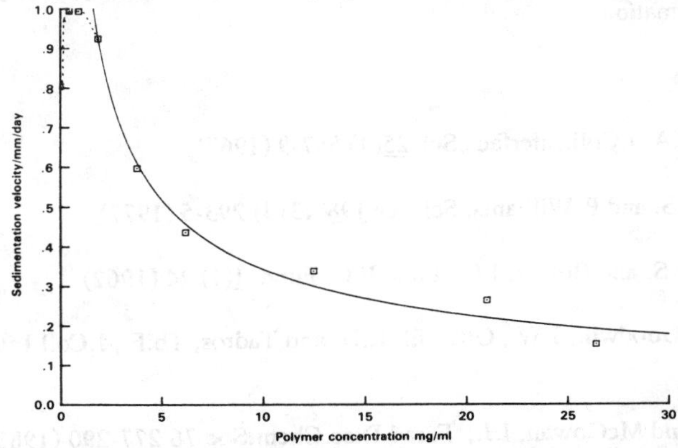

Fig. 6 Sedimentation velocity as a function of added hydroxyethylcellulose, Rate determined after 5 days.

is ascribed to the formation of flocs, at low polymer concentration, with no significant self-hindering connectivity, such that the overall effect is to increase the sedimentation velocity in a Stokesian manner (modified for particle concentration) for the initial increments in polymer concentration. At higher polymer concentrations the greater degree of flocculation reduces the sedimentation velocity as expected.

Our results do not show any critical discontinuities at the coil overlap concentration or any other concentration (12) but we feel that there is not necessarily an inconsistency since the time scale of our experiments are very long term compared to the time scale of the microscopic observations. Since the degree of depletion flocculation is a polymer concentration dependent phenomenon (17), at low polymer concentrations the rate of assembly and the extent of the connectivity of the flocs that result, is sufficiently limited that it requires longer time scales to become manifest, and our longer time scales of observation can appreciate this.

4. Conclusion

We have demonstrated that it is possible to quantify a rate of sediment formation or floc consolidation for polydiperse industrial suspensions by direct observations on the sediment using ultrasound velocity measurements. Observations on the boundary between the suspension and the supernatant continuous phase only give limited insight into the most significant changes. The first order rate constant that we determine for floc consolidation is in inverse proportion to the added non-adsorbing polymer concentration to the power 3/2. The rate of boundary subsidence goes approximately with the inverse of the square root of polymer concentration in the same free polymer concentration range. The functions describing both processes do not appear to have any discontinuities at the transition from dilute to semi-dilute polymer concentration regimes or any other critical concentration.

The above findings mean that it should be possible to predict the likely rate of dilatant sediment formation in long term stored polydisperse suspensions from ultrasound scanning experiments. Direct observation of free liquid formation in concentrated suspensions is probably a poor guide to dilatant sediment formation.

5. References

1. Ratcliffe, G.A. J.Coll.Interface.Sci 25(4) 587-9 (1967)

2. Crandall, R.S. and R.Williams, Science 198(4314) 293-5 (1977)

3. Micheals, A.S. and Bolger, J.C., I and E.C. Fund 1(1) 24 (1962)

4. Buscall, R. Goodwin, J.W., Ottewill, R.H. and Tadros, Th.F., J.Coll.Interface.Sci 85 78 (1982)

5. Buscall, R and McGowan, I.J., Farad.Disc.Chem.Soc 76 277-290 (1983)

6. Wedlock, D.J., Fabris, I.J and Grimsey, J.,Colloids and Surfaces, 43 67-81 (1990).

7. Gunning, P.A., Hibberd, D.J., Howe A.M. and Robbins, M.M., Food Hydrocolloids 2 119 (1988)

8. Lubetkin, S.D., Wedlock D.J. and Edser, C. Colloids and Surfaces in press

9. Wedlock, D.J., ACS Symposium Series (eds B.Cross and H.Scher)no 371 (1988) pp 190-205

10. Vincent, B., Luckham P.F. and Waite, C.F., J.Coll. Interface.Sci 73 (1980)

11. Solid/Liquid Dispersions (ed. Th.F.Tadros) 1987 Academic Press pp 205

12. Tadros, Th.F. and Zsednai, A., Colloids and Surfaces 43 (1990) 95 and 43 (1990) 105

13. Sperry, P.R., Hopfenberg, H.B., and Thomas, N.L.J., Coll.Interface.Sci 82 (1981) 62

14. Asakura. S. and Oosawa,F., J.Chem.Phys., 22 (1954) 1255

15. Vrij, A., Pure Appl.Chem., 48 (1976) 471

16. Heath, D., and Tadros Th.F., Faraday Discuss.Chem.Soc 76 (1983) 203

17. Dickinson, E., J.Coll.Interface.Sci 132 (1989) 274

COLLOIDAL SEMICONDUCTORS

J. K. THOMAS[1]
Department of Chemistry & Biochemistry
University of Notre Dame
Notre Dame, IN 46556

ABSTRACT. Various techniques are described to prepare colloidal semiconductors. The unique properties of these materials, which depend on size, are assessed in terms of semiconductor theory, utilising procedures of photochemistry and spectroscopy.

1. Introduction

Some of the earliest aqueous colloids reported in the literature have a semiconductor constituent.[2] These systems are often metal sulfides, the ease of preparation and their importance in analytical chemistry accounts for their popularity. Throughout the century, a continued interest in the development of the systems has been evident.[3] A sharp increase in interest was reached around 1980 when it was realized that these colloids and suspensions might provide a means of storing solar energy, and of mimicking photosynthetic systems. Concomitant with the last impetus was the realisation that semiconductor materials of small dimensions exhibited altered properties that might be of vital commercial value as well as of academic interest. References 4 to 28 contain many of the published works that reflect on the activity in this area over the past ten years or so.

2. Concepts

The two basic lines of research on colloidal semiconductors, alluded to above, are photoinduced electron transfer, and effects of dimension on the semiconductor properties.

Photoexcitation of semiconductors results in promotion of electrons from the valence band to the conduction band of the material. The result is the formation of holes h^+, and electrons e^-. Recombination of the charged species can give rise to luminescence, which is observed in nearly all semiconductors. Adsorption of molecules e.g., RH or MV^{2+} at the surface may give rise to electron transfer.

$$h^+ + RH \rightarrow RH^+ \rightarrow \text{products} \quad (1)$$

$$e^- + MV^{2+} \rightarrow MV^{+\bullet} \rightarrow \text{products} \quad (2)$$

D. M. Bloor and E. Wyn-Jones (eds.),
The Structure, Dynamics and Equilibrium Properties of Colloidal Systems, 759–771.

The lifetimes of the charged species on the semiconductor tend to be short, and solute diffusion to the particle is too slow to participate in the prior processes. Molecules have to be adsorbed at the semiconductor surface so that charge transfer can take place via reaction 1 or 2. Increasing the surface area, as with colloids, tends to enhance efficiency of charge transfer.

The energy gap between the valence and the conduction bands depends on the dimensions of the particle. This leads to marked changes in the absorption and emission spectra of semiconductor particles as their sizes decrease below about 100 Å. Both calculations and experiments confirm the above suggestion and Table 1 shows calculations giving the separation of ground and first excited states for various spherical particles of CdS.[4a] The marked increase in band gap with decreasing particle size leads to a blue shift on the absorption spectrum, i.e. in general the color of the particles tends to become lighter.

Table 1[(a)]

Number of CdS molecules per particle	radius of spherical particle (Å)[(b)]	Wavelength of Onset of Absorption (nm) of 1st excited states
10	4.9	189
20	6.2	256
50	8.4	353
100	10.6	411
500	18.1	496
1000	22.9	513
∞	∞	522

(a) Taken from reference 4a

(b) Calculated from the 1st column using 5×10^{-23} cm^3 as the volume of a CdS molecule

3. Preparation

For the most part the preparation of semiconductor colloids is achieved by standard techniques of colloid chemistry. Colloids of TiO_2,[4e,5c,6], Fe_2O_3,[7,8] etc. are prepared by hydrolysis of suitable compounds of the metal e.g.

$$TiCl_4 + 2OH^- \rightarrow TiO_2 + 2Cl^- \ 2HCl$$

Sulfides are prepared by direct reaction while avoiding precipitation,[4,5,9,10]

$$Cd^{++} + S^{--} \rightarrow CdS$$

In conventional colloid chemistry the formation of the colloids is controlled so as to produce monodispersed systems. In photochemical procedures large surface areas and hence small particle sizes are essential, and monodispensity is sacrificed to achieve this end. In aqueous solution, stabilisers such as polyvinyl alcohol or hexametaphosphate are used. Alternatively non aqueous solvents such as acetonitrile and methanol are used to decrease Ostwald ripening and hence to maintain small particles.

Surface areas may be increased by producing porous particles[6g], but this is not a common procedure. CdS has been prepared in polymer films [9d,11c,15] a useful procedure for some of the model storage systems. Preparation in Nafion film leads to CdS with properties that are markedly different from those observed in aqueous colloids. The procedure is to exchange or diffuse the metal ion into the polymer matrix followed by immersion in sulfide solution or in H_2S.

Microemulsions have been used [9e,9d,13] as microreactors to produce small particles of CdS. The draw back of such a system is that the particles cannot be removed from the supporting oil medium.

It is possible to form the semiconductor in a confined space such as vesicles[22], porous glass[6], between clay sheets[14], and in zeolites[13,14,3] The procedure again is to exchange the metal ions into the system followed by treatment with H_2S or sulfide solution. One dimensional CdS is formed in porous glass, two dimensional CdS between clay sheets, and small particles in zeolites.

4. Oxide Colloids

Photoexcitation of the colloidal semiconductor leads to hole and electron migration to the surface where subsequent electron transfer reactions with adsorbed molecules eventually leads to stable products. Typical reactions are indicated by processes 1 and 2. The initial e^- transfer at the surface leads to transitory species which can back react,

$$\text{viz.} \quad h^+ + MV^{+\bullet} \rightarrow \text{semiconductor} + MV^{2+} \tag{3}$$

A very necessary feature of the system design is to prevent such processes.

5. TiO_2

One of the earliest examples of photochemical transformation on semiconductors is afforded by the photolysis of organic acids on TiO_2 powders.[11a,11b] The products with acetic acid are CH_4, C_2H_6, CO_2 and H_2, i.e., a type of photo-Kolbe process. Light is absorbed by the TiO_2 producing h^+ and e^- which interact with the adsorbed acid.

$$e^- + H^+ \rightarrow H$$

$$h^+ + CH_3COO^- \rightarrow CH_3COO \rightarrow CH_3 + CO_2$$

$$CH_3 + CH_3 \rightarrow C_2H_6$$

$$CH_3 + H \rightarrow CH_4$$

$$H + H \rightarrow H_2$$

This system provides an example where both species of the photoproduced charged pair, the hole h^+, and the electron e^- enter into reaction with the adsorbed solute. This is not always the case, as with adsorbed I^-. In deaerated solutes little reaction takes place,[5,6] but in the presence of O_2 significant amounts of I_2 is formed. Flash photolysis studies show that the processes are,

$$h^+ + I^- \longrightarrow I \xrightarrow{I^-} I_2^-$$

$$I_2^- + I_2^- \rightarrow I_2 + 2I^-$$

The intermediate I_2^- is readily detected in flash photolysis.

The back reaction $I_2^- + e^- \rightarrow 2I^-$ occurs unless e^- is removed e.g., by O_2,

$$e^- + O_2 \rightarrow O_2^-$$

Many similar inorganic[5f] and organic reactions of particular use in chemical synthesis have been reported[17].

A popular adsorbent in these systems is methyl viologen MV^{2+}. The reduced form $MV^{+\bullet}$, is stable in the absence of O_2 and exhibits an intense blue color. Irradiation of colloidal suspensions of TiO_2 and MV^{2+} can, under the correct conditions, give rise to MV^+.[5,6] Flash photolysis of solutions containing crystalline TiO_2, $(TiO_2)_c$ and MV^{2+} leads to instantaneous formation of MV^+, which is observed by its absorption spectrum with maxima at 395 nm and 605 nm. The products arise by direct electron transfer from excited $(TiO_2)_c$ to adsorbed MV^{2+}. Benzylviologen, a derivative of MC^{2+}, which is not adsorbed on the surface, is not reduced on photoexcitation.

The charge separated products are too short lived to provide any storage of energy. The photoreduced MV^{2+} is still adsorbed on the TiO_2 surface and readily reacts with the hole giving back the starting materials. Photoproduced MV^+ on $(TiO_2)_c$ shows an initial fast decay ($\tau 1/2$ 60 ns) followed by a slower decay with a lifetime of 1 μs; the fraction the initial decay depends on the pH of the medium. In acidic media all $MV^{+\bullet}$ decay rapidly whereas in basic media only about 60% of the signal decays rapidly, the remaining MV^+ being stable. It should be noted that MV^{2+} is adsorbed more strongly on $(TiO_2)_c$ in basic media than in acidic media.

Amorphous $(TiO_2)_a/MV^{2+}$ systems behave quite differently on photo-excitation. MV^{2+} is not adsorbed on amorphous TiO_2 and instantaneous formation of MV^2 is not observed. However, in the presence of electron donors such as polyvinyl alcohol (PVA), or halide ions, a slow growth of MV^+ formation is observed over a period of several μs. In the case of PVA< a permanent reduction of MV^{2+} is observed and reported.[5a] Flash photolysis studies show large yields of MV^+, and rapid back reaction of MV^+ and Cl_2^- yield the starting materials. In 0.1 M $HClO_4$ the reduction of MV^{2+} is not efficient, as electron transfer from ClO_4^- to the TiO_2 hole is not efficient.

The data can be explained as follows:

$$(TiO_2)_a \xrightarrow{h\nu} TiO_2\ (e^-, h^+)$$

$$e^- + MV^{++} \rightarrow MV^+$$

$$h^+ + RCH_2\text{-}OH \longrightarrow R\dot{C}H\text{-}OH + H^+$$

$$MV^{2+} + R\dot{C}H\text{-}OH \longrightarrow MV^+ + RC(=O)H + H^+$$

The photoproduced electrons have a longer lifetime in $(TiO_2)_a$ than on the $(TiO_2)_c$, as illustrated by the slower growth of $MV^{+\bullet}$ on excitation of $(TiO_2)_a/MV^{2+}$.

The photogenerated hole can be intercepted by various inorganic and organic ions, such as $S_2O_8^{2-}$, acetate, and I^-. In the case of $S_2O_8^{2-}$ on $(TiO_2)_c$ photolysis gives a transient spectrum with a maximum at 455 nm which is attributed to the $SO_4^{\bullet-}$ radical. Similar studies with iodide give a transient with a maximum at 390 nm, which has been assigned to the I_2^- radical. These transitory species appear with-in the pulse duration (6×10^{-9} s) even at 10^{-4} M concentrations. It is reasonable to assume that only the ions adsorbed on the surface are reactive. In homogeneous solution the halide anion radicals disproportionate to give neutral molecules and halide ions, and the decay kinetics follow a second-order process. However, the transients generated on the particle surface of $(TiO_2)_c$ colloid in water decay by an apparent first-order process. This indicates that the transients are formed on the surface. Before the radical escapes from the surface of the particle to the solution bulk, they react with the photogenerated electrons on the surface, and the decay appears to be first order.

Iodide ions are adsorbed on amorphous (TiO_2) particles. Flash photolysis of amorphous TiO_2 and iodide results in the instantaneous formation of I_2^- as the adsorbed iodide transfers an electron to the photogenerated hole and the iodine atom combines with the excess iodide to give I_2^- ion. As with crystalline TiO_2, the conduction band electron and I_2^- back react to give the starting materials. Since the electron lives longer on the amorphous TiO_2 as seen from the reaction with MV^{2+}, I_2^- is expected to live longer . The observed half-life of the I_2^- decay is ~35 μs compared to 14 μs for the crystalline TiO_2 particle suspensions.

The surface kinetics of the photoinduced transitory species exhibit complex behaviour due to the variation in colloidal size. A treatment involving a Gaussian distribution in the ln or natural logarithm of the rate constant successfully describes most situations encountered.[18]

The above discussion emphasises the importance of preventing back reaction of the products at the colloid surface. Two unreactive species may be produced such as O_2^- and I_2^- so that products H_2O_2 and I_2 can be formed on dimerisation of the intermediates. Another approach is to use electrostatic effects to influence charged reactants at the surface. This will be illustrated in the CdS section also.

A detailed study of electrolyte and pH effects on TiO_2 has been published.[18b,18c] In these studies it was concluded that pH, which affects the colloid surface charge and hence the thermodynamics of surface e^- transfer, and salt affects, were the two most important features for photoinduced e^- transfer at colloid surfaces. The measurements involved the yield and rate of formation of $MV^{+\bullet}$ on a TiO_2 - MV^{2+} system. The rate of $MV^{+\bullet}$ formation increased markedly over the pH range 2.0 to 11.0. At low pH (~ 3) both Na_2SO_4 and NaCl increase the rate of formation of $MV^{+\bullet}$, the former more than the latter. At high pH (~ 9) both salts decreased the rate of formation of $MV^{+\bullet}$. Studies with a neutral viologen showed that the rate of e^- transfer is more sensitive to electrostatic rather than to thermodynamic factors.

At low pH, TiO_2 colloids possess a positive charge which decreases the efficiency of approach of MV^{2+} to the surface and hence the efficiency of e^- transfer; the opposite effect occurs at high pH where the colloid has a negative charge. This picture also explains the observed salt effects.

In some instances co-catalysts such as Pt are coated onto the TiO_2 colloids, in particular to promote H_2 formed by excess e^- in the particle .[5]

Flash photolysis of $CO_3^{=}$ adsorbed on TiO_2 exhibits instantaneous formation of the carbonate anion radical.[6c] As in other systems, carbonate anion radicals back react with conduction band electrons giving the starting materials, carbonate anion radicals decay faster in aerated solution ($\tau_{1/2}$ = 0.9 ± 0.1 μs) than in deaerated solution ($\tau_{1/2}$ = 2.9 ± 0.2 μs) indicating that oxygen reacts with the photoproducts.

Illumination of aerated or deaerated solutions containing TiO_2 and carbonate (1 M), results in the formation of formaldehyde. As the surface is always saturated with CO_3^{2-} decreasing the concentration of carbonate by 100 fold does not appreciably affect the formation of the final product, and, the formaldehyde yield only decreases at concentrations less that 1 mM carbonate The HCHO yield decreases at long irradiation times, due to secondary reactions of HCHO. Methanol was not observed as one of the products at longer irradiation times and, oxidation of formaldehyde to formate is proposed as a secondary reaction.

$$(TiO_2) \xrightarrow{h\nu} TiO_2\,(e^-, h^+)$$

$$CO_3^{2-} + h^+ \rightarrow CO_3^-; \quad O_2 + e^- \rightarrow O_2^-$$

$$CO_3^- + O_2^- \rightarrow CO_3^{2-} + O_2$$
$$CO_3^- + e^- \rightarrow CO_3^{2-}$$

$$CO_3^- + O_2 \rightarrow CO_3 + O_2^-$$

$$CO_3 \rightarrow CO + O_2$$

$$CO + 2e^- \xrightarrow{2H^+} HCHO$$

$$\text{Net reaction } CO_3^{2-} \xrightarrow[H^+]{e,\ h^+} HCHO + O_2$$

The important aspect of the reaction is that both holes and electrons are used to drive the net chemical reaction, the free energy stored is 138 kcal/mole.

6. Fe_2O_3

Iron oxide colloids are readily prepared in aqueous solution. However, their photochemical reactivity is low, and acidic conditions are needed to adsorb anions such as I^- to the surface and hence to form I_2^- and I_2 on photoexcitation.[7,8,19] To some extent this can be corrected by using co-adsorbed polymers. e.g., polyacrylic acid which enhances bonding of cations e.g., ruthenium tris bipyridyle $Ru(Bipy)^{2+}$, to the surface.[7] Modest modifications of the photochemistry of $Ru(Bipy)^{2+}$ are observed. Long lived radicals e.g., $MV^{+\bullet}$ can reduce the colloid[7,19], which is of interest in the prevention of corrosion in nuclear reactors.

7. CdS

Cadmium sulfide colloids have received much attention from photo-chemists[4,5,9,10,11,13,20,21,22,28], as these systems convert visible light into chemical energy and products.[4,5,9] Hole reactions are not common in these systems, but e^- transfer to adsorbed solutes such as MV^{2+} are readily attained. The quantum yield of $MV^{+\bullet}$ depends on the nature of the surface and on the presence of a sacrificial agent to prevent back reaction and to inject e⁻ into the colloid. A typical agent is ethylene diamine tetracetate, EDTA. The nature of the colloid surface is important as, unlike TiO_2, the sulfide surface is readily chemically altered by pH. In the presence of excess OH^- (high pH) the CdS surface becomes coated with $Cd(OH)_2$ which promotes luminescence of the particles.

With negatively charged stabilisers e.g. sodium lauryl sulfate,NaLS, the quantum yield of $MV^{+\bullet}$ in the CdS-NaLS-MV^{2+} system is small, < 10^{-2}. With a positively charged surfactant as stabiliser e.g. cetyltrimethyl ammonium bromide, CTAB, the yield is even smaller as MV^{2+} is repelled from the surface. However, MV^{2+} exists as a negatively charged complex in the presence of EDTA, and adsorbs at the positively charged surface. Efficient photoinduced e⁻ transfer takes place on visible light excitation of CdS. The $MV^{+\bullet}$ - EDTA complex dissociates and $MV^{+\bullet}$ is repelled from the positive surface. The quantum yield of $MV^{+\bullet}$ can now approach 0.2, Fig. 1. Other anions e.g. hexametaphosphate also produce similar results. This again illustrates the pronounced affect of surface electrostatic charge on the photoinduced charge transfer reactions.

Photochemistry of other semiconductor colloids viz. ZnS, ZnO, CdSe etc. have been reported. Space does not permit a discussion of the reported chemistry, but for the most part the observed events are quite similar to those in TiO_2 and CdS.

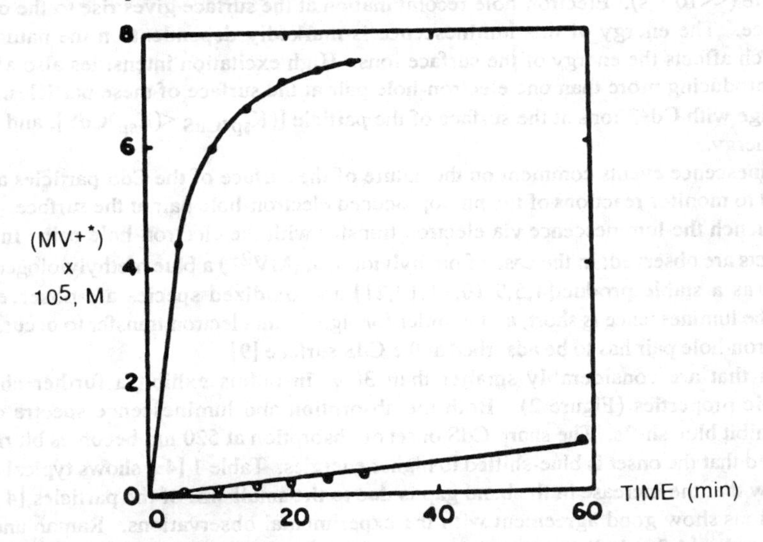

Figure 1. Formation of $MV^{+\bullet}$ as a function of irradiation time in a CdS/CTAB colloid, (O); MV^{2+}/$HEDTA^{3-}$ complex, •. From ref. 9b.

8. Spectral Studies

Small particles of CdS were prepared previously in a glass matrix.[23] It was noted that these materials exhibited blue-shifted absorption and emission spectra. These effects, were later interpreted in terms of the symmetry of the Cd^{2+} environment.[24] These results are similar to those observed in colloid semiconductors.

Semiconductors exhibit emission characteristic of their band gaps (conduction to valence band) at an energy which also corresponds closely to the onset of absorption. For cadmium sulfide the onset of absorption is at about 520 nm, with a green emission at 520 nm on excitation. These properties are altered if the dimensions of the cadmium sulfide are small, as in colloids. It is reported that colloidal CdS (30-Å radius), stabilized by a variety of surface additives, exhibits emission that is red-shifted with respect to that observed with large single crystals of CdS (Figure IIa). The luminescence is short-lived[5,9] ($T < 10^{-9}$ s) and depends on the mode of preparation of the CdS and on the nature of the surface additive. The red shifted luminescence can be associated with sulfur vacancies of the particle. These entities act as electron donors to the holes and produce red emission. The luminescence is altered by surface treatment i.e with different stabilisers such as sodium lauryl sulfate (NaLS) and cetyltrimethylammonium bromide (CTAB), and as markedly red-shifted if small amounts of cupric ion are bound to the surface.[9]

The emission spectrum shows a blue shift from the red to the green at higher excitation intensities. The quenching of the red CdS luminescence by Cu^{2+} to give red-shifted luminescence follows Poisson kinetics[9] as seen in other colloid systems.[25] Such measurements give an estimate of the radii of the particles. In the CdS system a radius of 30 Å was found for several preparations, a result which was confirmed with electron microscopy measurements. Excitation of the colloidal CdS leads to an electron-hole pair which rapidly migrates to the surface of the small particle ($<<10^{-9}$ s). Electron-hole recombination at the surface gives rise to the observed luminescence. The energy of this luminescence is markedly dependent on the nature of the surface which affects the energy of the surface ions. High excitation intensities also affect the surface by producing more than one electron-hole pair at the surface of these particles. Cupric ions exchange with Cd^{2+} ions at the surface of the particle [$(K_{sp})_{CuS} < (K_{sp})CdS$], and trap the excitation energy.

The luminescence events comment on the nature of the surface of the CdS particles and may also be used to monitor reactions of the photoproduced electron-hole pair at the surface. Several additives quench the luminescence via electron transfer with the electron-hole pair. In several cases products are observed; in the case of methylviologen (MV^{2+}) a blue methylviologen (MV^{+}) is observed as a stable product[4,5,9,10,11,19,21] and oxidized species are observed. The lifetime of the luminescence is short, and in order for significant electron transfer to occur, the trap for the electron-hole pair has to be adsorbed at the CdS surface.[9]

Particles that are considerably smaller than 30 Å in radius exhibit a further change in spectroscopic properties (Figure 2). Both the absorption and luminescence spectra of these particles exhibit blue shifts. The sharp CdS onset of absorption at 520 nm becomes blurred, and it is suggested that the onset is blue-shifted to higher energies. Table 1 [4a] shows typical calculations to show that the increase in the band gap is due to the small size of the particles,[4,28] and the calculations show good agreement with the experimental observations. Raman and X-ray diffraction studies[4,20] indicate that at least some of the CdS is crystalline in character; the surface regions, which must now constitute a large proportion of the CdS, must be amorphous, and little can be said about such material. Parts b and c of Figure 2 give spectral details of CdS clusters of 4 to 5 molecules ($W_o = 5$ where $W_o = [H_2O/AOT]$ constructed in the protective environment of reversed micelles.[9e,94,12] These CdS colloids exhibit weak spectral absorption above 410 nm, which increases to a maximum at 330 nm, while the emission is maximum at about 450 nm. Such behaviour is now associated with molecular properties of CdS rather than with semiconductor behaviour. Larger aggregates ($W_o = 10$, - 10 CdS; $W_o = 32$, - 35 CdS) exhibit intermediate behaviour.

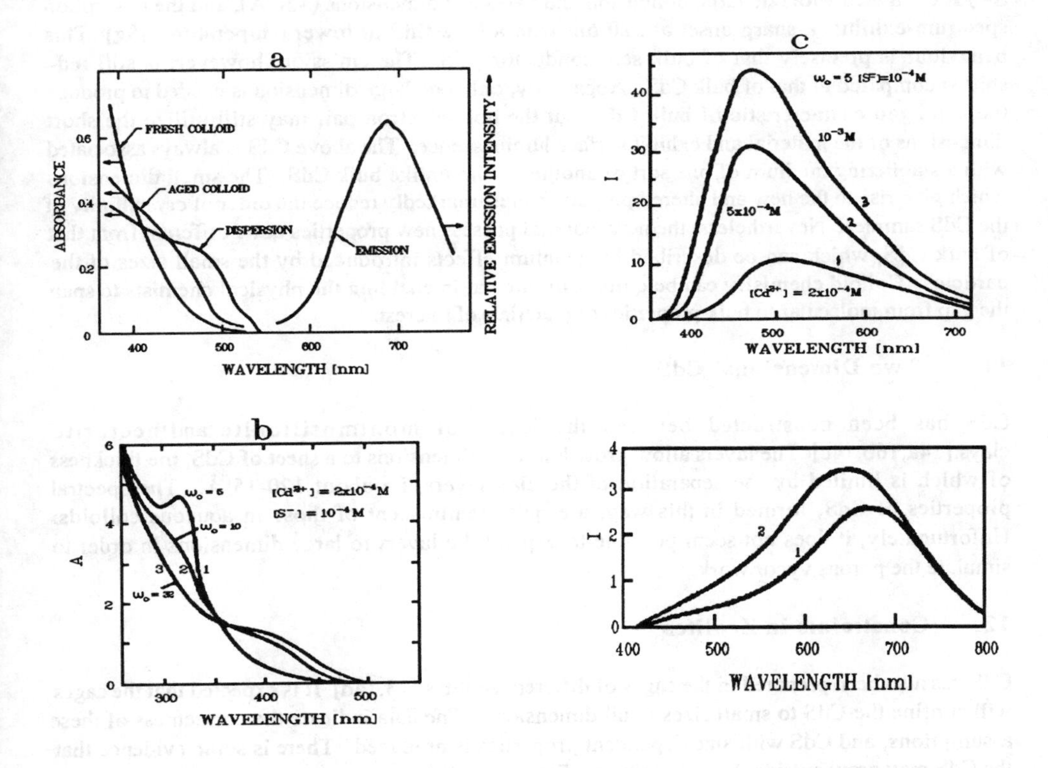

Figure 2. (a) Absorption and emission spectrum of CdS/CTAB colloid, [CdS]= 3 x 10^{-4}M, and a CdS dispersion, [CTAB] = 5 x 10^{-4} M. Excitation wavelength = 400 nm. From ref 19. (b) Absorption spectra of CdS particles in 0.5 M OAT heptane at various ω_o = [H_2O]/[OAT]. From ref 19. (c) Emission spectra of CdS particles in 0.5 M AOT in heptane at ω_o = 5 (a) and ω_o = 32 (b). The numbers 1, 2, and 3 correspond to the shown Cd^{2+} concentrations. From ref 9a.

9. CdS of Varied Dimensions, Shades of Flatland

Solution techniques are convenient for the preparation of small particles where all the dimensions are altered. Colloids may also be formed in "molds" where two or three dimensions may be changed. This type of chemistry in restricted dimensions has all the trappings of the thinking of the earlier part of the century which is so well portrayed in Abbotts "Flatland".[29]

10. One Dimensional CdS

Porous Vycor glass consists of long tubelike pores of 30-Å radius, and CdS may be constructed in this environment. The CdS material trapped in the pores (via reaction of adsorbed Cd^{2+} with S^{2-}) is elongated with one large dimension and two short dimensions (<30 Å), and the absorption spectrum exhibits a sharp onset at 520 nm with a blue shift at lower temperatures.[9g] This behaviour is precisely that of bulk semiconductor CdS. The emission, however, is still red-shifted compared to that of bulk CdS. Apparently, only one large dimension is needed to produce the band gap characteristic of bulk CdS, but the hole-electron pair may still utilize the short dimensions of the material and exhibit surface luminescence. The above CdS is always associated with a stabilizing medium of one sort or another, quite unlike bulk CdS. The small dimensions which give rise to the new and altered properties must markedly reduce the order of crystallinity of the CdS samples. Nevertheless, the new material possess new properties quite different from that of bulk CdS, which can be described by quantum effects introduced by the small sizes of the particles. Colloid chemistry can be quite instrumental in enabling the physical chemists to span the gap from molecular to bulk properties of materials of interest.

11. Two Dimensional CdS

CdS has been constructed between the layers of montmorillonite and hectorite clays.[14a,16b,14c] The layers allow growth in two dimensions to a sheet of CdS, the thickness of which is limited by the separation of the clay layers, i.e about 120-15 Å. The spectral properties of CdS, formed in this way, are quite reminiscent of those in aqueous colloids. Unfortunately, it does not seem possible to expand the layers to large dimensions in order to simulate the porous vycor work.

12. Constraints in Zeolites

CdS has also been prepared in the cages of different zeolites.[13,14d] It is expected that the cages will confine the CdS to small sizes in all dimensions. The data indicate the correctness of these assumptions, and CdS with size dependent properties is produced. There is some evidence that the CdS may grow outside the containing zeolite cage.[13b]

Several semiconductors materials shows the above "size" effects: TiO_2[26], CdS,[4,5,9, 20], Cd_3P_2,[4], ZnS,[4], Fe_2O_3,[7], and PbS.[19]

Conclusion

The two scientific regimes of colloid and photochemistry show a marked interface with semiconductor materials. The techniques of colloid chemistry provide a convenient way of synthesizing important semiconductor materials with unique properties, while the procedures of photochemistry enable fine details of the materials to be assessed and utilised.

References

1. Work supported by the National Science Foundation.

2. Winssinger, Bull Soc. Chim.(1988) 49, 452; Linder & Picton, J. Chem. Soc.(1892) 61, 114 , Picton, J. Chem. Soc. (1978) 61, 137.

3. Matijevic, E., & Scheiner, D., J. Coll. Interface Sci. (1978), 63, 509.

4. (a) Fojtik, A.; Weller, H.; Koch, V.; Henglein, A. Ber. Bunsen-Ges. Phys. Chem. (1984), 88, 969. (b) Weller, H.; Fojtik, A.; Henglein, A. Chem. Phys. Lett. (1985), 117, 484. (c) Baral, S.; Fojtik, A.; Weller, H.; Henglein, A. J. Am. Chem. Soc (1986), 108, 375. (d) Spanhel, L.; Haase, M.; Weller, H.; Henglein, A. J. Am. Chem. Soc. (1987), 109, 5649. (e) Henglein, A., Pure and Applied Chem. (1984), 56, 1215.

5. (a) Ramsden, J. J.; Webber, S. E.; Grätzel, M. J. Phys. Chem. (1985), 89, 1740. (b) Ramsden, J. J.; Grätzel, M. J. Chem. Soc., Faraday Trans. 1 (1984), 80, 919. (c) Duong, H. D.; Ramsden, J.; Grätzel, M. J. Am. Chem. Soc. (1982), 104, 2977. (d) Kalyanasundaram, K.; Borgarello, E.; Duongrong, D.; Grätzel, Angew Chem. Intern. Ed. Engl. (1981) 20 987. (e) Dimitrijevic, N. M.; Li, S.; Grätzel, M. J. Am. Chem. Soc. (1984), 106, 6565. (f) Grätzel, M. p 217 in Photoelectrochemistry, Photocatalysis and Photoreactions Ed. M. Schiavello, Reidel, Holland, (1985).

6. (a) Chandrasekaran, K.; Thomas, J. K. J. Coll. & Interface Sci (1985), 106, 532. (b) (1984), 100, 116. (c) Chem. Phys. Letts. (1983), 99, 7. (d) Chem. Phys. Letts. (1983), 97, 357. (e) J. A. Chem. Soc. (1983), 105, 6383. (f) J. Chem. Soc. Faraday 1. (1984), 80, 1163.

7. Stramel, R.; Thomas, J. K. J. Colloid & Interface Sci. (1986), 110, 121.

8. Kormann, C.; Bahnemann, D. W.; Hoffmann, M. R.; J. Photochem. & Photobiol. A. Chemistry (1989), 48, 161.

9. (a) Thomas, J. K. J. Phys. Chem. (1987), 91, 267. (b) Kuczynski, J.; Thomas, J. K. Langmuir (1985), 1, 158. (c) Kuczynski, J.; Thomas, J. K. J. Phys. Chem. (1983), 87, 5498; (1985), 89, 2720. (d) Kuczynski, J.; Milosavljevic, B. H.; Thomas, J. K. J. Phys. Chem. (1984), 88, 980. (e) Lianos, P.; Thomas, J. K. Chem. Phys. Lett. (1986), 125, 299. (f) Lianos, P.; Thomas, J. K. J. Coll. & Interface Sci. (1987), 117, 505. (g) Kuczynski, J.; Thomas, J. K. J. Phys. Chem. (1985), 89, 2720.

10. Darwent, J. R.; Porter, G. J. Chem. Soc., Chem. Commun. (1981), 145.

11. Kraeutler, B. K.; Bard, A. J. J. Am. Chem. Soc. (1978), 100, 2239. (b) Bard, A. J. Science (1980), 207, 139. (c) (1978), 100, 4317; (1978) 100, 5985.

12. Dannhauser, T.; O'Neil, M.; Johanson, K.; Whitten, D.; McLendon, G. J. Phys. Chem. (1988), 90, 6074.

13. (a) Wang, Z.; Herron, N. J. Phys. Chem. (1987), 91, 257. (b) Herron, N.; Wang, Y.; Eddy, M. M.; Stucky, G. D.; Cox, D. E.; Moller, K.; Blin, T. J. Am. Chem. Soc. (1989), 111, 530.

14. (a) Stramel, R. D.; Nakamura, T.; Thomas, J. K. Chem. Phys. Letts. (1986), 130, 423. (b) J. Chem. Soc. Faraday 1, (1988), 84, 1287. (c) Liu, X.; Thomas, J. K. J. Coll. & Interface Sci. (1989), 129, 476. (d) Langmuir (1989), 5, 58.

15. Micic, O.; Meisel, D. p 227 in "Homogeneous and Heterogeneous Photochemistry" Ed. Pelizzetti, E.; Serpone, N., Reidel Publishing Co. Boston, (1986).

16. (a) Meissner, D.; Memming, R.; Kastening, B. Chem. Phys. Letts. (1983), 96, 34. (b) Meissner, D.; Memming, R.; Li, S.; Yesodharan, S.; Grätzel, M. Ber. Bun. Phys. Chem. (1985), 89, 121.

17. (a) Fox, M. A.; Chen, C. C. J. Am. Chem. Soc. (1982), 103, 6757. (b) Fox, M. A.; Lindig, B.; Chen, C. C. J. Am. Chem. Soc. (1982), 104, 28. (c) Fox, M. A.; Chen, C. C. Tetrahedron Lett. (1983), 24, 547. (d) Fox, M. A.; Chen, C. C.; Younathan, J. N. J. Org. Chem. (1984), 49, 9. (e) Fox, M. A.; Chen, M. J. J. Am. Chem. Soc. (1983), 105, 4497. (f) Kamat, P. V.; Fox, M. A. J. Phys. Chem. (1983), 87, 59. (g) Kamat, P. V.; Fox, M. A. Chem. Phys. Lett. (1983), 104. (h) Fox, M. A. p 363 in "Homogeneous and Halogeneous Photocatalysis". Ed. E. Delizzetti & N. Ssapone Reidel, Holland, (1988).

18. (a) Alberry, W. J.; Bartlett, P. N.; Wiulde, C. P.; Darwent, J. R. J. Am. Chem. Soc. (1985), 107, 6446. (b) Brown, G. T.; Darwent, J. R. Chem. Comm. (1985), 981. (c) Darwent, J. R.; Lepre, A. J. Chem. Soc. Faraday Trans. 2 (1986), 82, 2323.

19. Nozik, A. J.; Williams, F.; Nenadovic, M. T.; Rajh, T.; Micic, O. I. J. Phys. Chem. (1985), 89, 397. (bv) Nedeljkovic, J. M.; Memadovoc, M. T.; Micic, O. I.; Nozik, A J. J. Phys. Chem. (1986), 90, 12. (c) Nozik, A. J.; Williams, F. Nenadovic, M. T.; Rajh, T.; Micic, O. I. J. Phys. Chem. (1985), 89, 397.

20. (a) Brus, L. New J. Chem. (1987), 11, 123. (b) Rosetti, R.; Nakahara, S.; Brus, L. E. J. Chem. Phys. (1983), 79, 1086. (c) Brus, L. E. J. Chem. Phys. (1984), 80, 4403. (d) Rosetti, R.; Ellison, J. L.; Gibson, J. M.; Brus, L. E. J. Chem. Phys. (1984), 80, 4464.

21. (a) Sarkata, T.; Kawai, T. Chem. Phys. Letts. (1981), 80, 341. (b) J. Chem. Soc. Chem. Commun. (1980), 694.

22. Tricot, Y.-M.; Fendler, J. H. J. Am. Chem. Soc. (1984), 106, 7359.

23. Von Georg, T.; Z. Tech. Phys. (1926), 7, 301.

24. Inman, J. K. Mraz, A. M.; Weyl, W. A. p 198 in Solid Luminescent Materials. J. Wiley, NY (1948).

25. Thomas, J. K. Chemistry of Excitation at Interfaces ACS Monograph 181, Washington, DC (1983). Thomas, J. K. J. Phys. Chem. (1987), 91, 267.

26. Anpo, M.; Aikawa, N.; Kubokawa, Y.; Che, M.; Jonis, C.; Giamello, E. J. Phys. Chem. (1985), 89, 5017.

27. Buxton, G. V.; Rhodes, T.; Sellars, R. M. J. Chem. Soc. Faraday. Trans. 1 (1983), 79, 2961.

28. (a) Harbour, J. R.; Wolkow, R.; Hair, M. L. J. Phys. Chem. (1981), 85, 4026. (b) Harbour, J. R.; Tromp, J.; Hair, M. L. Can. J. Chem. (1985), 63, 204.

29. Abbott , E. A., Flatland. Dover reprint, NY 1952; original pub. 1884, London.

OBTENTION AND CHARACTERISATION OF ULTRAFINE MAGNETIC COLLOIDAL PARTICLES IN SOLUTION

M.A.LOPEZ-QUINTELA and J.RIVAS
Dpts. of Physical Chemistry and Applied Physics
University of Santiago de Compostela
E-15706 Santiago de Compostela
Spain

ABSTRACT. Ultrafine NdFeB alloy particles of different sizes have been obtained by chemical reduction of metallic salts by borohydride ions in microemulsions. The amorphous final powders obtained were characterised by X-ray diffractometry and differential thermo analysis. Annealing of the samples leads to the formation of several crystalline phases. Magnetic properties of some representative powders are also reported.

1. Introduction

1.1. MAGNETIC PARTICLES

The use and properties of magnetic materials depend mainly on two factors: 1) on their structural parameters, such as their composition and atomic arrangement. These parameters condition aspects such as the atomic aggregation state (crystal, quasi-crystal, amorphous, etc.), anisotropy, magnetization, magnetoelasticity, etc. and 2) on their microstructural parameters: grain size, porosity, texture in polycrystalline materials, size and shape in particles, etc.

In particular, and limiting ourselves to the field of magnetic particles, their properties, for a specific material, depend among other factors, on their size. In a first approximation we can divide the particles into three categories depending on their size [1] (see Fig.1):

1) Magnetic multidomain particles. In particles with a size in the range of μm or larger, two or more magnetic domains appear.Due to the presence of magnetic walls between the domains, the magnetization can change because of the translation movement of these magnetic walls as well as because of the rotation of the spontaneous magnetization. This permits us to obtain materials with relatively soft magnetic properties and great technological interest (pressed powder cores for low and intermediate frequencies, materials for electromagnetic machinery, etc.).

D. M. Bloor and E. Wyn-Jones (eds.),
The Structure, Dynamics and Equilibrium Properties of Colloidal Systems, 773–788.

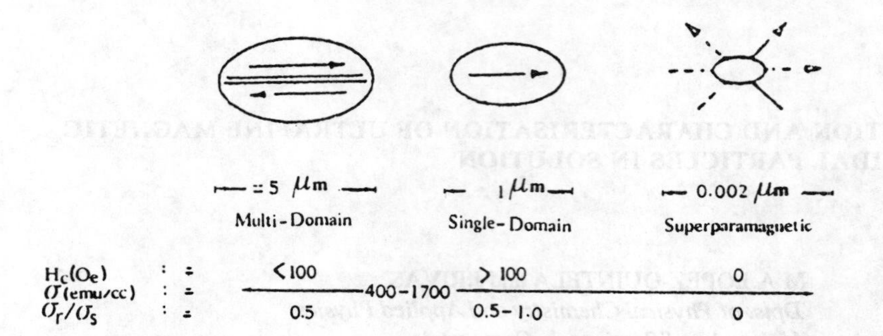

Figure 1. Magnetic behaviour vs particle size. H_c = coercivity field; σ_s = saturation magnetization; σ_r = remanence.

2) Single domain particles. They are particles with a size in the range of μm or less presenting a single magnetic domain. The magnetization can only change due to the rotation of the spontaneous magnetization. This process is energetically more difficult than the translation of the walls generating, this way anisotropies with large values. These materials present hard magnetic properties with a high technological interest, for example in magnetic recording and as pressed powder for the design of permanent magnets.

3) Superparamagnetic particles. They are particles with sizes in the range of nm and therefore fall within the group of particles called ultrafine [2] (size ~ 10 nm). They present special magnetic properties such as superparamagnetic behaviour due to the fact that the size of the particle is so small that they develop a paramagnetic behaviour in a material which is initially ferromagnetic. The scientific importance of these particles is high as they represent a state of matter which is intermediate between the molecular and solid world. Because of their interesting properties, often different from those of solid and molecular material, a specific terminology has been developed for these particles: quasi-solids, quasi-molecules, clusters, etc [3]. Although currently they are not used in industrial systems as much as those mentioned in the previous sections, they are, however, potentially important candidates in the new technologies of the future, such as new catalysts, molecular magnetic separators, ferrofluids, etc.

1.2 STRUCTURE

With respect to the atomic arrangement in magnetic particles we can say that, as a first approximation, there are two fundamental structures: amorphous state and crystalline state. Out of these, the amorphous state has recently achieved a great importance due to the following facts: 1) It is easy to obtain. 2) Its properties -normally different from the crystalline state- open a new range of possibilities for its technological use and 3) as it is a metastable state, other metastable states as well as the crystalline state are easily obtained from it. This widens the range of applications for these materials.

1.3. OBTENTION

For obtaining amorphous magnetic particles we roughly say that there are four basic procedures [4]:

1) Rapid quenching. This metallurgic technique is based on the fast cooling of the material which was fused at high temperatures. In figure 2 is shown a phase diagram for a simple binary

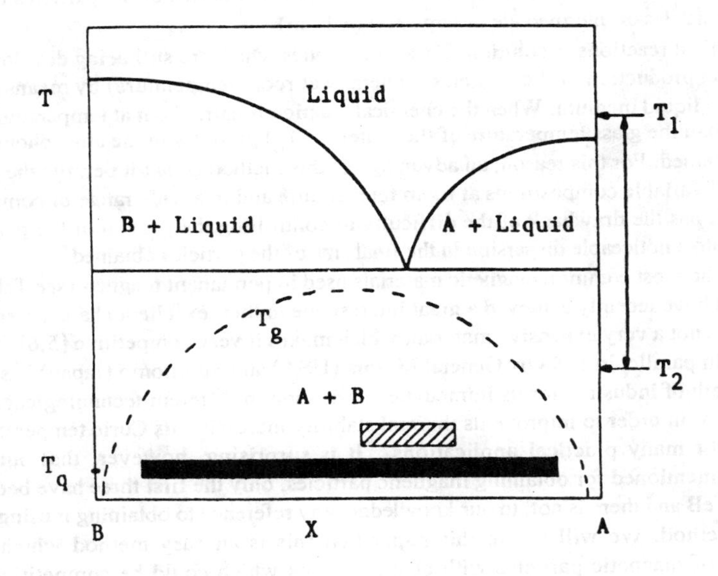

Figure 2. Phase diagram of an AB system. Hatched area denotes compositions of metallic glasses which can be obtained by rapid quenching. Black area shows the corresponding region for the chemical reaction method. T_g = glass temperature. T_q = temperature at which the chemical reaction takes place.

system. To obtain the amorphous material, brusque changes in temperature are needed (T_1 - T_2) for the system to reach a final temperature $T_2 < T_g$ (glass temperature). As we can see in the figure, the leap in temperatures which is needed is larger as we separate from the eutectic point, which in practice limits our obtaining the amorphous material to compositions which are close to those of the eutectic point. This technique is the most used both in investigation and in industry because of its ease.

2) Deposition techniques. They include:

a) Sputtering techniques which consist in hitting the material with positive ions of a rare gas and collecting them by means of deposition on a solid substrate.

b) Thermal evaporation. In contrast with the previous technique, the atoms of the material under study are pulled out thermally in high vacuum.

Although these two techniques are excellent for obtaining fine and ultrafine particles with a controlled size and composition, they are not technologically competitive due to their high production costs.

3) Mechanical alloying. A new technique has been recently developed which consists in previously milling for long periods of time the base products of the material wanted and later heating it. This produces the amorphous material by a solid-state reaction. Although this technique is industrially very promising, it has limits with respect to the size of the particles it obtains. It is limited to particles of intermediate or large size (≥ 1 μm).

4) Chemical reactions in solution. These techniques which are still being developed are based on the direct production of the particles (generally at room temperature) by means of a chemical reaction in a liquid medium. When the chemical reaction is carried out at temperatures (T_q) which are lower than the glass temperature of the material (T_g) particles in the amorphous state can be directly obtained. For this reason, an advantage of this method is that it permits the production of materials of variable compositions at room temperature and in a wide range of compositions (see figure 2). It has the drawback of the difficulty in controlling the growth of the particles, which translates into a noticeable dispersion in the final size of the particles obtained.

Among the most common magnetic materials used in permanent magnets (see Table 1),NdFeB compounds have recently achieved a great interest due to their excellent characteristics and to the fact that it is not a very expensive material, which makes it very competitive [5,6]. This material, developed in parallel in 1983 by General Motors (USA) and Sumitomo (Japan) has attracted the attention both of industry -for its immediate application in different technological areas- and of investigation -in order to improve its thermal stability increasing its Curie temperature, which is very low for many practical applications. It is surprising, however, that out of the four procedures mentioned for obtaining magnetic particles, only the first three have been used in the case of NdFeB and there is not, to our knowledge, any reference to obtaining it using the chemical reaction method. We will see in this paper that this is an easy method which permits the production of magnetic particles with characteristics which could be competitive with those already in existence.

TABLE 1. Optimum magnetic energy products for three generations of permanent magnet alloys[7]

Date	Alloy composition	Magnetic energy product(kJ/m^2)
1935	$Pt_{23}Fe_{77}$	36
1960	$Pt_{23}Co_{77}$	80
1968	$SmCo_5$	160
1975	$(Sm,Pr)Co_5$	184
1977	Sm_2Co_{17}	240
1983	$Nd_{15}B_8Fe_{77}$	290
1985	$Nd_xB_yFe_{(100-x-y)}$	350
1989	$^*Nd_x(RE)_yB_zFe_{(100-x-y-z)}$	>400

*RE = rare earth

Specifically, in this paper we will describe a new procedure developed for the production of NdFeB particles based on the chemical reduction of Nd^{3+} and Fe^{2+} salts by means of $NaBH_4$ or KBH_4 in solution [8,9].

2. Experimental Method

2.1 CHEMICAL REACTION

Until now several procedures have been used for obtaining magnetic particles through chemical reactions. In table 2 we show some of them. It can be observed that these procedures vary from simple precipitations to chemical reductions, including hydrolysis processes, thermal decompositions, etc.

TABLE 2. Some examples of magnetic particles obtained by chemical reaction.

Particles	Method	References
Fe_2O_3	$Fe^{2+}/Fe^{3+} + OH^-$	10, 11
Fe_2O_3	Thermal decomp. of ferric solutions	12
FeC	Thermal decomp. of $Fe(CO)_5$	13
$Fe_{62}B_{38}$	Chemical reduction	14
$Fe_{37}Ni_{28}B_{34}$	Chemical reduction	14
$Fe_{44}Co_{19}B_{37}$	Chemical reduction	14
$Fe_{40}Cr_{16}B_{44}$	Chemical reduction	15
$Fe_{55}Mn_{20}B_{24}$	Chemical reduction	15
$SrFe_{12}O_{19}$	Ferrous hydroxide + Sr^{2+} + oxid. agent(90°C)	16
$SrFe_{12}O_{19}$	Hydrolysis of metalorganic complexes	17
$Fe(OH)_nH_2O$	Precipitation	18

In our case, for obtaining the NdFeB magnetic particles, after different trials, we found that the best procedure consists in the reduction of Fe^{2+} and Nd^{3+} salts using borohydride ions in aqueous solution. The simplified outline of the procedure we followed is shown in figure 3.

During the first trials we carried out, we saw that different parameters influenced the final characteristics of the particles thus obtained. Among these aspects we must highlight the following:

1) Speed of addition of reactives
2) Atmosphere under which the reaction is carried out
3) Concentration of reactives, pH and temperature at which the reaction is carried out.

With respect to the first point, it is well known that the addition of reactives influences the composition and structure of the final particles [19]. In our case, the optimum procedure consisted in adding dropwise a $NaBH_4$ or KBH_4 solution to another one containing the appropriate concentrations of Nd^{3+} and Fe^{2+} salts. In a typical experiment, 100 ml of the reducing agent in excess ($[NaBH_4]$ = 0.156 M) were added in aprox. 15 min to 100 ml of a solution containing the appropriate ratio of Nd^{3+}/Fe^{2+} salts (ex. $[Fe^{2+}]$ = 0.4 M; $[Nd^{3+}]$ = 0.08M).

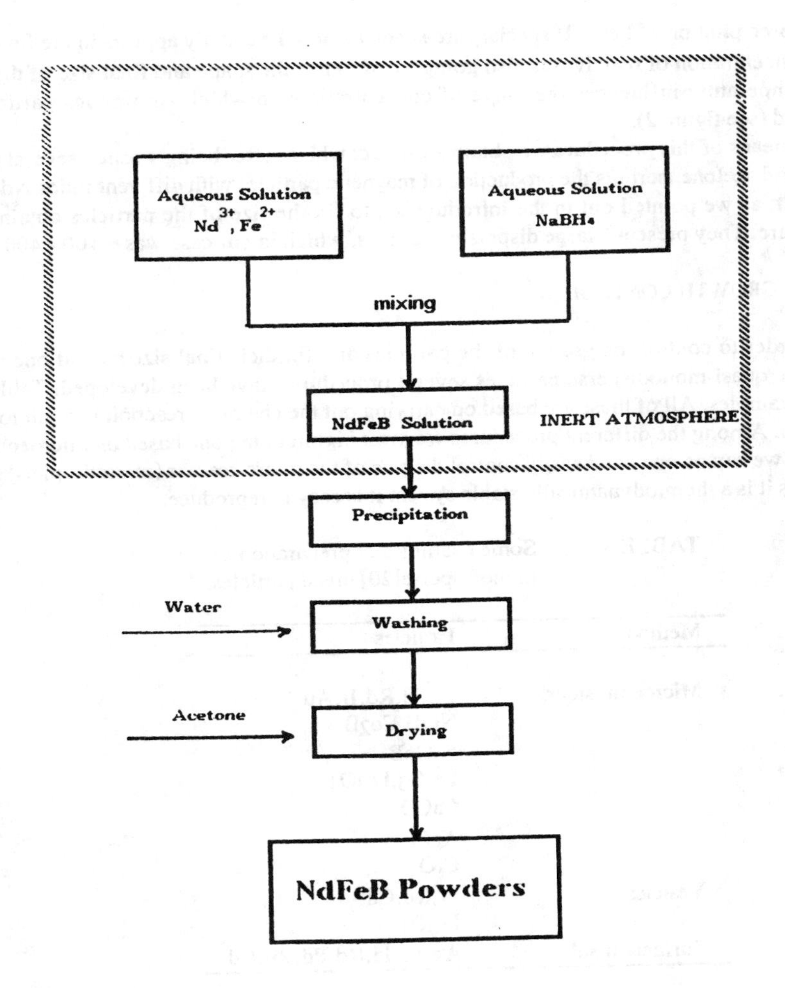

Figure 3. Flow chart showing the preparation method of NdFeB fine particles by chemical reduction.

As the fine particles, due to their large specific surface, oxidize easily, it is necessary to take great care in this point. In our case we have used deoxygenated solutions and the reaction was carried out in a glove box under argon atmosphere.

Other factors which usually affect the speed of a chemical reaction play an important role when obtaining magnetic particles. Thus, the control of the pH is essential in order to avoid/favour the initial precipitation of Fe^{2+}. If special care is not taken α-Fe usually appears in the final product. The concentration of reactives is also going to influence the shape and final size of the particles. The temperature influences the range of concentrations in which amorphous particles can be obtained (see figure 2).

By means of this procedure we obtain a product which, after being washed several times with water and acetone, permits the production of magnetic particles with different ratios Nd/Fe/B. It is difficult, as we pointed out in the introduction, to fix the size of the particles obtained by this procedure. They present a large dispersion in sizes, which in our case was ~ 100 - 400 nm.

2.2 GROWTH CONTROL

In order to control the growth of the particles and fix their final size so that one can obtain ultrafine quasi-monodisperse particles several procedures have been developed. Table 3 shows some examples. All of these are based on carrying out the chemical reaction in a microstructured medium. Among the different procedures we must highlight the one based on microemulsions as in them we can on one hand easily control the size of the micro-reactor (microdroplets) and on the other, as it is a thermodynamically stable system it is easy to reproduce.

TABLE 3. Some examples of preparation of monodisperse[20] metal particles.

Method	Particles
Microemulsions	Pt,Pd,Rd,Ir,Au
	Ni_2B,Co_2B
	NiCoB
	Fe_2O_3,Fe_3O_4
	$CaCO_3$
	AgCl
	CrO_2
Vesicles	Pt,Rd,CdS
	Fe_3O_4
Surfactant sol.	Ag,Pt,Tl,Rd,Pd,Ru,Cd

The size of the droplets in the microemulsion can be modified changing the composition, temperature or pressure of the system. This way, quasi-monodisperse systems of droplets with different sizes in the range 2 - 50 nm can be obtained[21].

For the reasons we have mentioned before and in order to limit the growth of the particles, different experiences for the reduction of Fe^{2+} and Nd^{3+} in the presence of various types of microemulsions were carried out [8,9].

As an example, we now describe a procedure using "water-in-oil" microemulsions made up of heptane/aqueous solution/aerosol-OT (sodium bis-(2-ethyl,hexyl)sulphosuccinate) with a concentration of 0.05M in aerosol-OT (AOT) and a ratio $R = [H_2O]/[AOT] = 22$. In these conditions, the microemulsions are made up by droplets of water with an approximate radius of 4.5 nm. We then prepared a microemulsion containing an aqueous solution of Fe^{2+} and Nd^{3+}. Another microemulsion with the same characteristics containing the reducing agent was added. By means of this procedure we could obtain quasi-monodisperse particles of NdFeB with an approximate diameter of 9 nm.

3. Composition

The composition of the samples obtained was analyzed through chemical analysis, ICP (induced coupled plasma) techniques and X-ray fluorescence. Table 4 shows the ratio of the starting components in the reaction mixture and the composition in the samples obtained.

TABLE 4. Influence of the initial composition of the reaction mixture on the particle composition.

Fe/Nd	
Initial composition	Particle composition
2.0	2.23
4.0	4.17
5.0	5.12
6.2	6.24
7.0	7.10
10.0	9.97

As it was expected, we observe that there is a direct correlation between the initial and final ratios. With respect to the initial concentration of B added, it is always done in excess to guarantee the stoichiometric reduction of the Fe and Nd salts and achieve the introduction of the B into the particles [19]. As an example we tried with the simplest binary alloy of FeB. Thus, mixing a solution of Fe^{2+} with a concentration of 0.05M with one of $NaBH_4$ with a concentration of $0.165M = 2\,[Fe^{2+}] + [B]$ we obtained amorphous magnetic particles with a composition $Fe_{85}B_{15}$(at. %). This is practically the theoretical composition $Fe_{80}B_{20}$.

4. Structure

With respect to the structure of the particles, it was determined by the X-ray diffraction technique (CuK_α radiation). In figure 4 we show two diffractograms representative of the products as they were obtained (amorphous phase) and after annealing the samples through thermal treatments in an Ar atmosphere (crystalline phases).

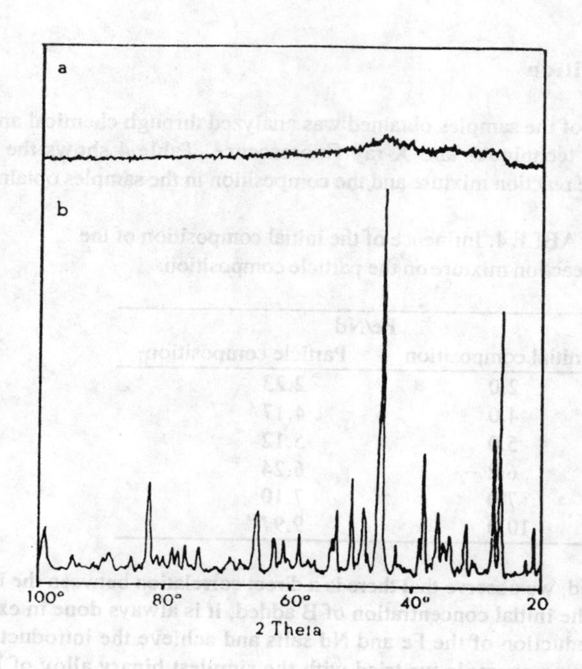

Figure 4. X-ray diffraction patterns of NdFeB powders (a) unannealed samples and (b) after 1h at 800°C heat treatment in argon atmosphere.

We must highlight the complexity of the diffraction spectrum obtained in the crystalline state. This proves that, although they are still magnetic samples, they present a variety of phases, some of them not identified yet.

In order to analyze the amorphous-crystal transition with more detail, we have carried out a study by differential scanning calorimetry (DSC) and differential thermal analysis (DTA) [9]. Figure 5 shows the thermograph obtained by DTA in an inert atmosphere for one representative sample. We can observe that there appears a fusion peak at high temperatures which can be associated with the well known $Nd_2Fe_{14}B$ phase as well as other peaks which we suppose correspond to other unidentified secondary phases. This agrees with the X-ray spectra obtained.

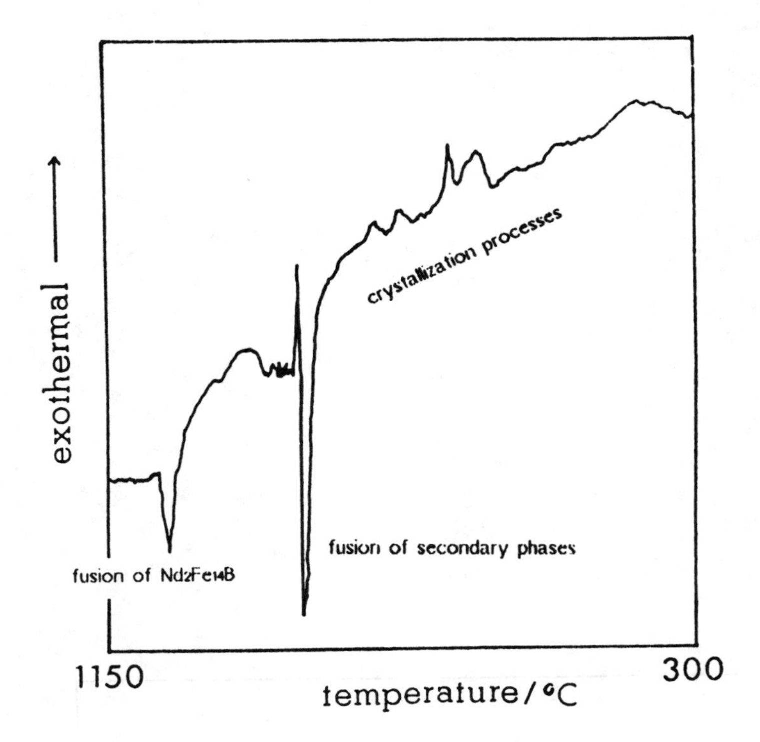

Figure 5. Differential thermal analysis trace of representative NdFeB powders (5K/min heating rate).

As we mentioned in section 2.1, one of the fundamental problems when obtaining the particles is the appearance of αFe. This occurs mainly when the reaction is carried out in a basic medium due to the precipitation of iron hydroxides. In figure 6 we show the X-ray spectra in samples obtained in these conditions and we can see the characteristic peaks of αFe superimposed on the amorphous structure of the samples.

5. Magnetic properties

Finally we will comment the basic magnetic properties of these materials. The magnetization, remanence, coercivity and saturation magnetization curves have been determined in the $2 < T(K) < 300$ temperature range using a SQUID magnetometer on pressed powder samples using a solution of polyvinyl alcohol as bonding, in cylindric shapes with a diameter of 1 mm and a length of 3 mm. In figure 7 we present a typical magnetization curve at low temperatures for samples obtained

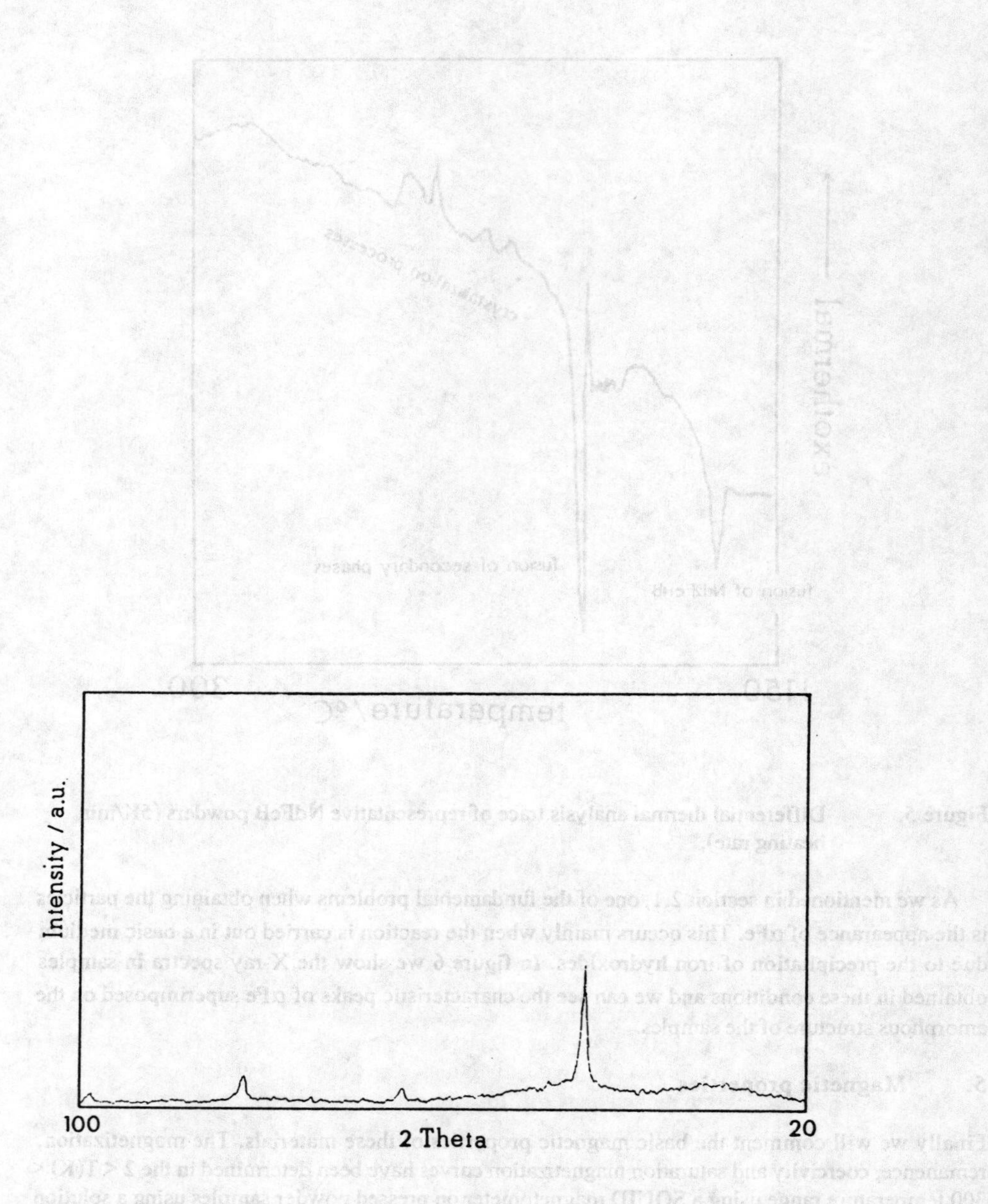

Figure 6. X-ray diffractogram showing the presence of α-Fe.

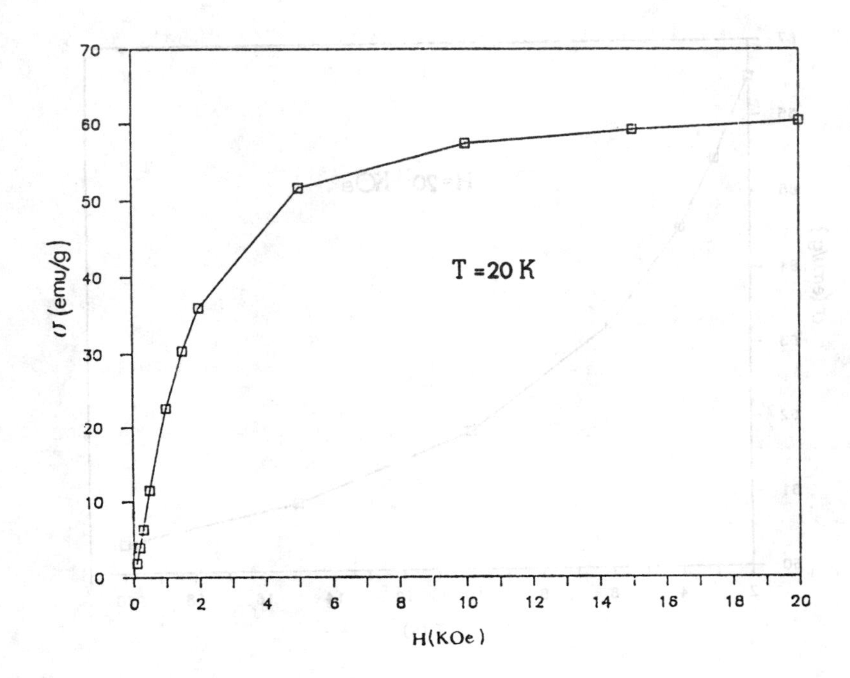

Figure 7. Magnetization curve at low temperature for NdFeB powders.

in aqueous solution. In figure 8 we show the saturation behaviour of the same samples.

We must point out that the saturation magnetization does not follow the $T^{3/2}$ law. This indicates a possible speromagnetic behaviour [22] of the spin system in the amorphous structure of the particles. This behaviour was systematically found in all the samples that have been studied.

The magnetic products obtained are soft, presenting, at room temperature, values of the coercivity field H_c= 170 Oe, of remanence $\sigma_r = 6$ emu/g and of saturation magnetization $\sigma_s = 60$ emu/g which happens to be ≈ 1/3 of the value of $Nd_2Fe_{14}B$ in the crystalline state.

Using cores made of pressed powders we have measured the initial a.c. susceptibility using an a.c. bridge at 1 kHz in the range $80 < T(K) < 300$. We have observed that the susceptibility remains almost constant (= 0.01) in the temperature range under study.

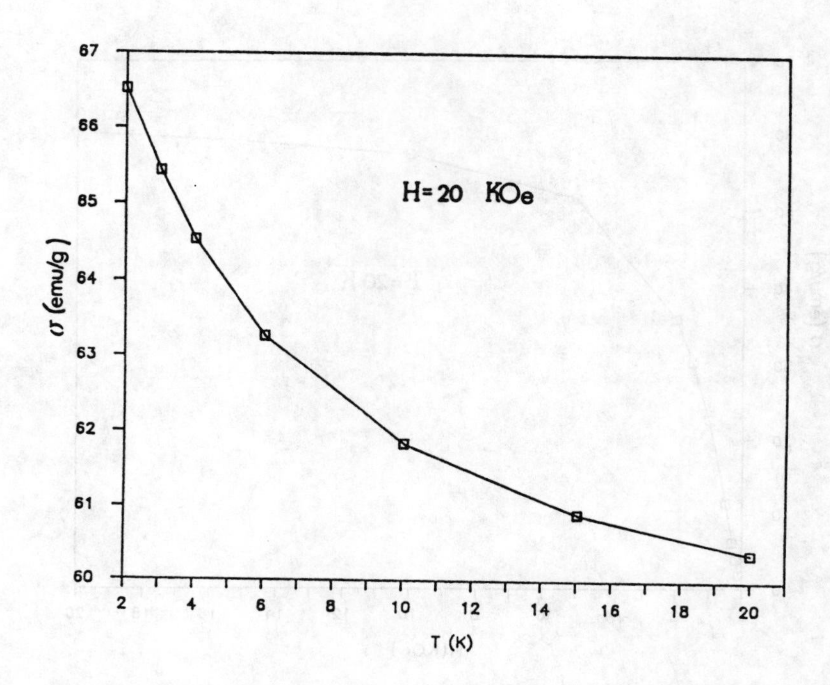

Figure 8. Specific saturation magnetization vs temperature in the range 2 < T(K) < 20 for NdFeB powders. The experimental points do not fit to the $T^{3/2}$ law.

6. Conclusions

Bearing in mind the results obtained, we must say that this is a new and simple procedure for obtaining fine and ultrafine amorphous particles of NdFeB with a technological interest. This method seems to be competitive with the traditional methods as, on one hand it has a great technological interest within magnetic materials, an on the other it has an evident scientific interest as demonstrated by the preliminary results we present. Experiments such as neutron diffraction, synchrotron, Mössbauer, amorphous-crystal transitions and high temperature magnetic behaviour, among others, are being carried out in order to study these magnetic particles.

Finally, it is necessary to point out that the experiments we mention with respect to the properties of the NdFeB particles have centered on one particular method for obtaining them. The results corresponding to variations of this method (for example: obtaining the particles in presence of electric and magnetic fields; different ways for adding the reactives; presence of other metallic salts in the initial reaction mixture; etc.) as well as those corresponding to other chemical methods will be the object of future publications.

Acknowledgements

This research was supported by the Spanish CICYT No. MAT89-0425-C03 and in part of the European Communities contract No. SC1-000058. We thank CEAM Group for many stimulating suggestions.

7. References

1. Bate, G. (1986) 'Particulate recording materials', Proc. IEEE, 74, 1513-1525.
2. Hayashi, C. (1987) 'Ultrafine particles', Phys.Today, december, 44-51.
3. Davenas, J. and Rabette, P.M. (eds.) (1986) 'Contribution of cluster physics to materials science and technology', Martinus Nijhoff Publishers, Dordrecht.
4. Moorjani, K. and Coey, J.M.D. (1984) 'Magnetic Glasses', Elsevier, Amsterdam.
5. Croat, J.J. (1989) 'Current status of rapidly solidified Nd-Fe-B permanent magnets', IEEE Trans. Magn., 25, 3550-3554.
6. Higuchi, A. and Hirosawa, S. (1989) 'Sintered Nd-Fe-B permanent magnetis', IEEE Trans. Magn., 25, 3555-3560.
7. Hadfield, D. (1985) 'Sintered rare-earth transition metal permanent magnets and their applications', in I.V. Mitchell (ed.), Nd-Fe permanent magnets: their present and future applications, Elsevier, London, pp. 229-234.
8. López Quintela, M.A., Rivas, J. and Quibén, J. (1988) 'Process to obtain fine magnetic Nd-Fe-B particles of various sizes', Spanish Patent 2009404.
9. López Quintela, M.A., Rivas, J. and Quibén, J. (1989) 'Preparation of ultrafine Nd-Fe-B magnetic particles by chemical reduction in microemulsions', Proc. European Conf. Adv. Mat. Proc., Aachen, FRG, Nov. 22-24.
10. Bennetch, L.M., Greiner, H.S., Hancock, K.R. and Hoffman, M. (1975) 'Magnetic impulse record member', U.S. Patent 3904540
11. Matsumoto, S., Koga, T., Fukai, K. and Shinya, S. (1980) 'Production of acicular ferric oxide', U.S. Patent 4202871.
12. Van Diepen, A.M. and Popma, T.J.A.(1976) 'Mössbauer effet and magnetic properties of an amorphous Fe_2O_3',, J. Physique, 37, C6-755
13. Van Wonterghem, J., Morup, S., Charles, S.W., Wells, S. and Villadsen, J. (1985) 'Formation of a metallic glass by thermal decomposition of $Fe(CO)_5$', Phys. Rev. Lett., 55, 410-413.
14. Van Wonterghem, J., Morup, S., Koch, C.J.W., Charles, S.W. and Wells, S. (1986) 'Formation of ultrafine amorphous alloy particles by reduction in aqueous solution', 322, 622-623.
15. Inoue, A., Saida, J. and Masumoto, T. (1988) 'Formation of ultra-fine amorphous powders in Fe-M-B (M=transition metal) systems by chemical reduction method and their thermal and magnetic properties', Metall. Trans. A, 19, 2315-2318.
16. Fan, X. and Matijevic E. (1988) 'Preparation of uniform colloidal strontium ferrite

particles', J. Am. Ceram. Soc., 71, C60-C62.

17. Haneda, K., Miyakawa, C. and Goto, K. (1987) 'Preparation of small particles of $SrFe_{12}O_{19}$ with high coercivity by hydrolysis of metal-organic complexes', IEEE Trans. Magn., 23, 3134-3136.
18. Okamoto, S. and Sekizawa, H. (1979) 'Magnetic properties of amorphous ferric hydroxide gels', J. Physique, 40, C2-137.
19. Kim, S.G. and Brock, J.R. (1987) 'Growth of ferromagnetic particles from cation reduction by borohydride ions', J. Colloid Interf. Sci., 116, 431-443.
20. Nagy, J.B., Derouane, E.G., Gourgue, A., Lufimpadio, N., Ravet, I. and Verfaillie, J.P., (1986) 'Physico-Chemical characterization of microemulsions: Preparation of mono-disperse colloidal metal boride particles', Proc. Sixth Int. Symp. Surf. Sol., New Delhi, India, Aug. 18-22.
21. Shah, D.O. (ed.) (1985) 'Macro- and microemulsions', American Chemical Society, Washington D.C.
22. Coey, J.M.D., (1985) ' in 'Amorphous Antiferromagnetism', in D. Adler, H. Fritzsche and S.R. Ovshinsly (eds.), Physics of Disordered Materials, Plenum Press, N.Y., pp. 729-738.

NON-EQUILIBRIUM DOUBLE LAYERS IN CONNECTION WITH COLLOID STABILITY

J. LYKLEMA
Department of Physical and Colloid Chemistry
Dreijenplein 6
6703 HB Wageningen
The Netherlands

ABSTRACT. **A dynamic interpretation is presented for the interaction between pairs of colloidal particles. In this approach, the transient disequilibration of double layers during encounters is considered. Principles and some elaborations are given. It is concluded that inclusion of double layer dynamics significantly retards the approach of particles, and hence reduces the rate of coagulation. One of the qualitatively new features is that a relation with surface conductivity is established.**

1. Introduction to the problem

When two electrically charged colloidal particles approach each other, their electrical double layers overlap, and when the charges have the same sign, this overlap leads to repulsion. For so-called electrostatic colloids (lyophobic systems, deriving their stability against coagulation solely from this electrical repulsion) the theory for this has been worked out in the DLVO theory after Deryagin and Landau [1] and Verwey and Overbeek [2] who developed it independently during the Second World War.

The DLVO theory is basically an *equilibrium* theory. At any distance of approach h the Gibbs energy of the interacting pair of particles $G(h)$ is calculated, assuming the ions in the interacting double layers to be distributed according to Poisson-Boltzmann statistics. In doing so, equilibrium is tacitly implied. In physical parlance, the interaction forces are conservative and $G(h)$ is the isothermal reversible work done by these forces.

In the present paper, we shall subject the issue of equilibrium to further study. Basically, the question is to what extent double layers have the time to attain their equilibrium state during the brief period that two particles interact during, say, a Brownian encounter. If there are deviations from equilibrium, that may or may not be transient, the next question is whether or not these departures affect colloid stability. Obviously, the time t now becomes an additional variable and instead of $G(h)$ we have to consider $G(h, t)$.

There is another recurrent issue in interpreting colloid stability, namely the question whether interaction takes place at *constant* (surface) *charge* σ^o or at *constant* (surface) *potential* Ψ^o. On closer inspection it becomes clear that solving this problem also comes down to finding the departure from equilibrium during collisions. Let us discuss this in more detail.

In their monography, Verwey and Overbeek [2] discussed the AgI-aqueous solution interface, which at that time was one of the best studied model colloids. For this system σ^o is due to adsorption of Ag^+ and I^- ions, called *potential-determining* (p.d.) or *charge-determining* (c.d.)

D. M. Bloor and E. Wyn-Jones (eds.),
The Structure, Dynamics and Equilibrium Properties of Colloidal Systems, 789–799.

ions. In formula,

$$\sigma^o = F\left(\Gamma_{Ag^+} - \Gamma_{I^-}\right) = F\left(\Gamma_{AgNO_3} - \Gamma_{KI}\right) \quad (1)$$

where Γ is the surface concentration of the species mentioned and F the Faraday constant. The second equality of (1) refers to the actually measured excesses: double layers being electroneutral as a whole, only the adsorption of electroneutral species is measurable. The surface potential was assumed to obey Nernst's law, a statement that was later confirmed by, for instance, measurements with AgI electrodes. Hence,

$$\psi^o = -\frac{RT}{F}\left(pAg - pAg^o\right) \quad (2)$$

where $pAg = -\log c_{Ag^+}$ and pAg^o is the value of pAg at the point of zero charge (p.z.c.). According to (2), ψ^o is counted with respect to the (thermodynamically inaccessible) value of the surface potential at the p.z.c.

Equations (1) and (2) are typical for *reversible* interfaces, that are interfaces where σ^o, and hence ψ^o, is due to equilibrium adsorption. The counterpart is the mercury-solution interface, which under not extreme conditions is *polarizable*, meaning that no (Faradaic) current flows across the interface. In that case, ψ^o can (apart from a constant) be controlled by an external source; it is an independent variable, and the charge on the surface is a result of the applied potential, just the other way around as in the AgI case, where the surface potential is the result of spontaneous accumulation of charge due to exchange with the solution.

It follows from electrostatics that, when the two surfaces are approaching each other, the (integral) capacitance σ^o/Ψ^o of each of the double layers decreases. Physically this can be understood in two ways

(i) when the potentials are fixed, σ^o goes down: with the two surfaces nearby each other, new charges give rise to a larger increase of Ψ^o than when the surfaces were isolated. Consequently, a lower σ^o suffices to obtain the same ψ^o.

(ii) when the charges are fixed, ψ^o rises for the same reason.

In both cases, work has to be done to bring the surfaces together, or, in other words, repulsion ensues. It is recognized that cases (i) and (ii) stand for interaction at constant potential and at constant charge, respectively, and it is also realized that there may be intermediate situations in which both σ^o and ψ^o change to some extent.

Verwey and Overbeek, considering AgI a typical colloid, chose to elaborate their theory for constant ψ^o (Deryagin and Landau [1] did the same, without giving the present issue much attention). They argued that ψ^o would be fixed because of equilibrium with the solution, as determined by [2]. Hence, during encounter, σ^o should decrease and the implication is that this decrease is fast enough for equilibrium to persist. The reduction of σ^o can be achieved either by desorption of p.d. ions or by adsorption of counterions from the bulk, to neutralize the surface charge. Whatever the mechanism, this discharge process must be fast compared with the approach of the particles.

Systems that would be typical representatives of the "constant charge" category are clay mineral sols. A substantial fraction of the particle charge is due to isomorphic substitution of multivalent cations in the interior of the particles by cations of lower valency, imparting a negative volume charge. Diffusion of solids inside a solid is a relatively slow process and this charge cannot move during the rapid encounters of two clay platelets. Hence the charge is "frozen" and repulsion arises upon interaction because ψ^o increases.

The conclusion of this analysis is that the distinction between interaction at constant σ^o or at constant ψ^o is virtually determined by the ratio between the time scale of transport of certain ions and the time scale of encounters between the two particles.

It is obvious that this ratio can vary widely, not only between different systems, but also between investigations done at different time scales. For example, it is quite imaginable that certain ionic transport processes, needed to reduce σ^o cannot be completed during a brief Brownian encounter, whereas they would come to completion under shelf stability conditions.

For all these reasons, the study of double layer disequilibration as a function of time, as taking place under the influence of an external field, is an important and interesting task we shall henceforth call such phenomena *double layer dynamics*.

2. Double layer dynamics and the kinetics of coagulation

By "kinetics of coagulation" one usually understands the set of processes leading to the formation of particle aggregates. Very generally, the study of kinetics considers the processes involved in bringing the particles together (diffusion, sometimes also shear) taking into account the interaction of (Gibbs) energy between the particles (the resilience they have against approach). Double layer dynamics must be incorporated in the models that have been derived for these kinetics.

Historically, the study of the kinetics has been initiated by Smoluchowski [3]. His theory was fairly primitive, but it must be added that it was developed more than two decades before DLVO theory came to maturation. Smoluchowski only considered unstable colloids in which each encounter led to definitive aggregation. The particles move by diffusion, hence the force $F(t)$, acting on them, has a stochastic nature. No other forces were considered. For rapid coagulation, the Smoluchowski theory is still the basic model. When the insights into the interaction grew, it became possible to also include forces due to particle interaction. Fuchs [4] introduced this idea, which was later extended and experimentally tested by Reerink and Overbeek [5]. We shall call this force F_{DLVO}. It consists of an electrical (usually repulsive) and a London-Van der Waals part (usually attractive) part. We disregard steric contributions due to polymers. A variety of equations are available for F_{DLVO}, each applying to certain experimental conditions.

Another type of force, called the hydrodynamic force, F_{hydr}, was introduced by Spielman [6] and, independently, by Honig and Roebersen [7], following an earlier suggestion by Deryagin. Basically this elaboration incorporates the hydrodynamic hindering that moving objects experience from each other when they are in each others neighbourhood. Mathematically this can be expressed by writing the diffusing coefficient D as $D(h)$; D decreases with decreasing h. The reduction already sets in at distances further than twice the double layer thickness ($h > 2\kappa^{-1}$), but the deviations become particularly significant for short h, where F_{DLVO} also has to be accounted for.

Theory for all of this has been worked out by the above authors.

3 Problems to solve

Incorporation of double layer dynamics into the theory for the kinetics of coagulation involves at least three main problems that we shall now describe, together with our approach taken to solve them.

The first issue is: "how to deal with the force balance"? Generally, there are several forces involved. In addition to $F(t)$, F_{DLVO} and F_{hydr}, already mentioned, we now need F_{dyn}, for which an expression has to be found. In addition there is a friction force: if the particle moves at constant velocity v under the influence of a sum of forces F,

$$F = fv \tag{3}$$

where f is a geometry-dependent *friction coefficient*. For a sphere of radius a, according to Stokes, we have the familiar formula

$$f = 6\pi\eta a \tag{4}$$

where η is the viscosity of the medium.

As upon approach the particles may accelerate or decelerate, inertia forces have to be accounted for. According to Newton

$$F_{in} = m\frac{dv}{dt} \tag{5}$$

where m is the mass of the sphere.

Also according to Newton, the sum of all these forces is zero. If we explicit the various contributions, we obtain a special case of the Langevin *equation*. It is obvious that this equation is not easily solved because the forces are not independent. For instance, it is appreciated that when friction, hydrodynamic or inertia contributions reduce the rate of approach, F_{dyn} will also diminish, because now more time is available for the double layers to come to equilibrium.

The solution to this problem is not to consider the dynamics in terms of forces but in terms of the various friction coefficients, the argument being that these *are* additive. In a well-known paper, Chandrasekhar postulated this [8], he proved the correctness a posteriori. Unlike forces, friction coefficients are scalars. Their dimensions are N s m^{-1}, and the product fv^2, (dimensions N m s^{-1} = J s^{-1}) is the energy dissipation per unit of time. Also for the stochastic part of the transport a friction coefficient can be introduced. Einstein proved that

$$D = kT/f \tag{6}$$

which for spheres gives $D = kT/6\pi\eta a$, a well-known expression. The product fv^2 must now be replaced by $f\langle v\rangle^2$ which is zero; diffusion does not lead to energy dissipation.

The second problem is: "what double layer model do we want?" Several decharging mechanisms can be imagined: desorption of c.d. ions into the solution, followed by their transport, transport of these ions towards the interior of the particle, tangential outflow of such ions parallel to the surface, compensation of σ^0 by ion pair formation with counterions coming from the solution etc. It will depend on the nature of the particles and their geometries which of these mechanisms may become operative. For example, transport of charges through a particle occurs only if this particle has a certain conductivity and lateral charge flow does not play a significant role in interaction between infinitely large flat plates. The various flows may occur in four places, the bulk of the particle, the surface of the particle, the Stern layer and the diffuse part of the double layer. As these flows may be coupled and are determined by the transient field and by diffusion, an intricate picture emerges.

As a first step we shall use a simplified double layer model in which only two types of ions are considered, bound (b) and free (f). The bound ions (i) (identical to the surface ions) are trapped in a Lennard-Jones type of potential energy well of depth $w_i = \Delta_{ads}G_i^o/RT$ and width Δ. Here, $\Delta_{ads}G_i^o$ is the standard molar Gibbs energy of adsorption of i. The well forms a channel for the surface charges through which they can move laterally. All remaining ions are free and their distribution at equilibrium is governed by Poisson-Boltzmann statistics. More specifically, we shall consider the familiar situation that the solution contains only two electrolytes, one containing the c.d. ions and the other indifferent, the latter being present in excess so that it controls the double layer thickness κ^{-1}. Equations for this model have been derived by Martynov [9].

Problem number three is "how does a double layer look like under non-equilibrium conditions?". In our approach this will be treated as a *perturbation* phenomenon, meaning that only relatively small deviations from equilibrium will be considered. According to this for σ^o and ψ^o we write

$$\sigma^o = \sigma^o(eq) + \Delta\sigma^o \tag{7}$$

and

$$y^o = y^o(eq) + \Delta y^o \tag{8}$$

respectively, where we have introduced the dimensionless potential $y = F\psi/RT$. Typically, the perturbation terms $\Delta\sigma^o$ and y^o are functions of distance h and time t whereas $\sigma^o(eq)$ and $y^o(eq)$ are generally functions of h only with the limiting cases $\sigma^o = \sigma^o(eq)$ = constant (interaction at constant charge) and $y^o = y^o(eq)$ = constant (interaction at constant potential). We shall take a more general approach in which $y^o = y^o(h,t)$, $y^o(eq) = y^o(eq,h)$ and $\Delta y^o = \Delta y^o(h,t)$.

For the relation between surface charge and potential under non-equilibrium conditions we write

$$\sigma^o = \sigma^o(eq)e^{-\Delta y^o}e^{-\Phi} \tag{9}$$

This equation states that σ^o differs from its equilibrium value for two reasons: (1) the potential is not at its equilibrium value but somewhat out of it (accounting for the factor exp $-\Delta y^o$) and (2) the double layer does not have its equilibrium distribution but deviates somewhat from it. The latter deviation is written in terms of a *perturbation potential* Φ, a dimensionless quantity introduced by O'Brien and White [10]. At the very onset Φ is not known, but in the next step, when the transport equation is solved, it can be explicited and thereafter eliminated, assuming certain limiting behaviour, depending on the geometry. More specifically, the current, or charge flux J_b is due to diffusion (driving force grad $\sigma^o = \nabla\sigma^o$) and conduction (driving force grad $y^o = \nabla y^o$), so that we arrive at this special form of the *Nernst-Planck equation*

$$FJ_b = -D_b\nabla\sigma^o - D_b\sigma^o\nabla y^o \tag{10}$$

which, using (9), can be simplified to

$$FJ_b = D_b\sigma^o\nabla\Phi \approx D_b\sigma^o(eq)\nabla\Phi \tag{11}$$

where the (double) integration can be carried out if two boundary values or expressions for Φ are available, which will depend on geometry. One of these conditions is usually a symmetry consideration, for instance for two spheres $d\Phi/dz = 0$ at the points of closest approach, if z is normal to h. The second condition is derived from the fact that far away from the interaction region $\Phi \rightarrow 0$. Hence, in principle enough information is available to solve the *Nernst-Planck equation* (10).

4. Solutions of the equation of flow

So far, solutions have been obtained for the interaction between two discs [11] and between two identical spheres [12], see fig. 1 for an explanation of the geometries. The case of spheres is more realistic. However, as it leads to very involved mathematical equations, we shall give some equations below for discs, which are easier. The final application holds again for the spherical case.

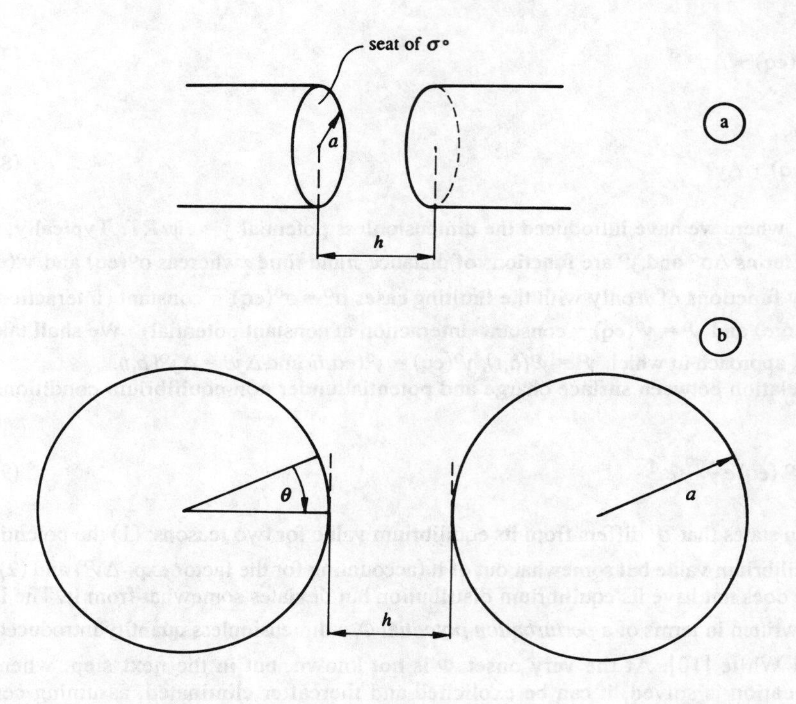

Figure 1. Explanation of the symbols for the interaction between two flat discs (a) and two spheres (b).

For the case of two discs, integration of (10) or (11) leads to

$$\Delta y^o = -\frac{1}{8}\frac{\partial y^o\,(eq)}{\partial t}\frac{a^2 - r^2}{D_b} \tag{12}$$

$$\frac{dy^o\,(eq)}{dt} = \frac{dy^o\,(eq)}{dh}\frac{dh}{dt} \tag{13}$$

where r indicates the radial direction. The physical picture is a bit artificial because Φ is assumed to decrease gradually from its maximum at $r = 0$ to $r = a$, beyond which it is suddenly zero. Nevertheless, (12) is interesting and useful because it relates a dynamic property (Δy^o) to the time dependency of an *equilibrium* property ($\partial y^o(eq)/\partial t$), which, according to (13) can be translated into the product of the change of $y^o(eq)$ with distance and the rate of approach (dh/dt). In this way, the dynamic problem is reduced to a static one.

For spheres the solution is different. In this geometry it is expedient to distinguish between an *interaction area* (low θ, see fig. 1b), where the surfaces experience each other's presence and a *transport area* (high θ) where the excess charge is transported towards the backsides of the particles. The equations are more complicated but the principle remains that the transient disequilibration is reduced to the product of the rate and static properties.

As we have chosen for the Martynov double layer model [9], we must rederive the DLVO expressions for $y^o(h)$. These derivations are straightforward, although somewhat laborious. For the disc model the result is

$$\frac{dy^o(eq)}{dh} \approx \frac{\Delta e^{-\Delta_{ads}G^o/RT}}{h^2} \tag{14}$$

5. Introduction of the disjoining pressure

In DLVO theory the interaction between two colloidal particles is usually written as a *disjoining pressure* $\Pi(h)$, viz., the force per unit area. For spherical particles, the total force can be given. A number of equations for $\Pi(h)$ exist in the literature which are valid for a variety of conditions with respect to κ, σ^o, y^o, h etc. Of course, all of these are static expressions, i.e. σ^o is $\sigma^o(eq)$ and $y^o = y^o(eq)$. If we want to find the dynamic contribution to interaction, we must derive the corresponding expressions for non-equilibrated double layers.

Consider the case of strong double layer overlap ($\kappa h << 1$). For that case, under equilibrium

$$\pi(h)(eq) = \left(\frac{RT}{F}\right)^2 \frac{2\varepsilon_0\varepsilon\kappa^2}{\pi} \frac{\tanh\left[y^o(eq)\right]/2}{\sinh^2(\kappa h)} \tag{15}$$

The choice of the particular expression for the disjoining pressure is not critical. For the static case of two discs, Π is a function of h only. Under dynamic conditions this is no longer true: because of the radial outflow of charge, the conditions change with r, so that for each ring between r and $r + dr$, $\Pi(h,r)$ has a different value. The total force, which, according to (3), is related to the rate of approach (= dh/dt) is obtained by integration of $\Pi(h,r)$ with respect to r.

$$F_{dyn} = f_{dyn}\frac{dh}{dt} = 2\pi\int_0^a r\Pi(h,r)dr \tag{16}$$

We obtain $\Pi(h,r)$ from $\Pi(h)(eq)$ by (again) a perturbation procedure: in (15) we replace $y^o(eq)$ by $y^o = y^o(eq) + \Delta y^o$ develop $\tanh^2[y^o(eq) + \Delta y^o]$ into a Taylor series of which the first two terms are retained. Subtracting the equilibrium part, we are left with the increment in Π due to dynamics. Details can be found in ref. 11. The outcome is

$$\Pi_{dyn}(r,h) = 2\left(\frac{RT}{F}\right)^2 \frac{\varepsilon_o\varepsilon\kappa^2}{\pi} \frac{\sinh\left[y^o(eq)/2\right]\Delta y^o}{\sinh^2(\kappa h)\cosh^3\left[y^o(eq)/2\right]} \quad (17)$$

As Δy^o is known (see (12) – (14)), this can be written in terms of the Martynov double layer model parameters. The integration can then be carried out and from (16) the following formula is eventually obtained for the dynamic friction coefficient

$$f_{dyn} = \left(\frac{RT}{F}\right)^2 \frac{\varepsilon_o\varepsilon\kappa^2 a^4 \Delta e^{-\Delta_{ads}G^o/RT}}{8D_b h^2} \frac{\sinh\left[y^o(eq)/2\right]\Delta y^o}{\sinh^2(\kappa h)\cosh^3\left[y^o(eq)/2\right]} \quad (18)$$

This materially completes the quintessence of the theory, but (18) is not yet in a handy form to check its relevance to colloid stability. In fact, the corresponding equation for spheres is even more complex. In the next section we shall consider some applications.

6. Double layer dynamics and colloid stability

We shall discuss two approaches to estimate the impact of double layer dynamics on colloid interaction.

In the first, we compare f_{dyn} with the hydrodynamic (viscous) friction coefficient

$$f_R = \frac{3}{2}\frac{\pi\eta a^4}{h^3} \quad (19)$$

that two discs experience upon approach. The subscript R refers to Reynolds, who derived (19). Equation (18) can be simplified by replacing sin(κh) by κh, which for our situation (κh<< 1) is a good approximation. Furthermore, Overbeek [13] showed that

$$\left(\frac{RT}{F}\right)^2 \frac{\varepsilon_o\varepsilon}{6\pi\eta D_f} \approx 0.2 \quad (20)$$

for aqueous solutions, where D_f is the diffusion coeffient for free charge-determining ions. With this in mind

$$\frac{f_{dyn}}{f_R} \approx 0.2 \frac{\Delta e^{-\Delta_{ads}G^o/RT}}{h} \frac{D_f}{D_b} \frac{\sinh\left[y^o(eq)/2\right]}{\cosh^3\left[y^o(eq)/2\right]} \quad (21)$$

Equation [21] can be used to judge the impact of dynamics semi-quantitatively. The quotient Δ/h is of order of unity but < 1; D_f/D_b is of order of unity but > 1; the exponent may be of the order of 10^2 and the ratio of the hyperbolic functions is about 0.3–0.4 under realistic experimental conditions. It is concluded that for discs f_{dyn} and f_R are of the same magnitude, implying that there is a clear contribution of dynamics to stability, although the effect is not so strong as to modify stability qualitatively.

In a second approach, double layer dynamics is incorporated in the rate of coagulation. Recall that the *stability ratio W*, a measurable quantity, is defined as the ratio between the rate of fast coagulation and the actual rate. The higher *W*, the more stable the sol is. For interacting spheres, *W* can be related to the various friction coefficients and the interaction Gibbs energy, which in this case is usually written as $V(s)$, where s is the reduced distance between the centres of the spheres, defined as $s = (h + 2a)/a$ (see fig. 1b); s is dimensionless. Extending the pertaining Fuchs [4] and Reerink and Overbeek [5] equation,

$$W = 2 \int_{2}^{\infty} \frac{f_{hydr} + f_{dyn}}{6\pi\eta a} e^{V(s)/kT} \frac{ds}{s^2} \tag{22}$$

where f_{hydr} is the hydrodynamic friction coefficient belonging to F_{hydr} and $6\pi\eta a$ is the Stokes viscous resistance, given in (4). If $V(s)$ has a pronounced maximum (which is always the case unless for very unstable sols), the integration is dominated by the exponential factor and one may write

$$W = \frac{f_{hydr} + f_{dyn}}{f_{visc}} 2 \int_{2}^{\infty} e^{V(s)/kT} \frac{ds}{s^2} \tag{23}$$

The advantage is that the integral is now again entirely classical. The quotient of friction coefficients has to be taken at $h = h_{max}$, the distance where $V(s)$ has its maximum. For f_{visc} we have (4), for f_{hydr} we write [6,7]

$$f = \frac{3\pi\eta a^2}{h_{max}} \tag{24}$$

and for spheres f_{dyn} is a very involved function of D_f/D_b, κa, κh, Δ, the indifferent electrolyte concentration and $\Delta_{ads}G^o$. Details will be published elsewhere [12].

Figure 2 gives an illustration for $y^o = 0.5$ and a certain combination of values for Δ, $\Delta_{ads}G^o/RT$ and the ratio between the concentrations of charge determining and indifferent electrolytes. From this figure it is seen that for $D_f/D_b \rightarrow 1$ the dynamic contribution is negligible as compared with the hydrodynamic correction, but for higher values of this quotient, the dynamic correction becomes progressively important. For very high values of D_f/D_b the rate of coagulation can be reduced by orders of magnitude, but under those conditions the underlying assumptions become worse, although at present it is difficult to say by how much.

Both illustrations indicate that the incorporation of dynamic effects leads to significant quantitative deviations from the classical static behaviour.

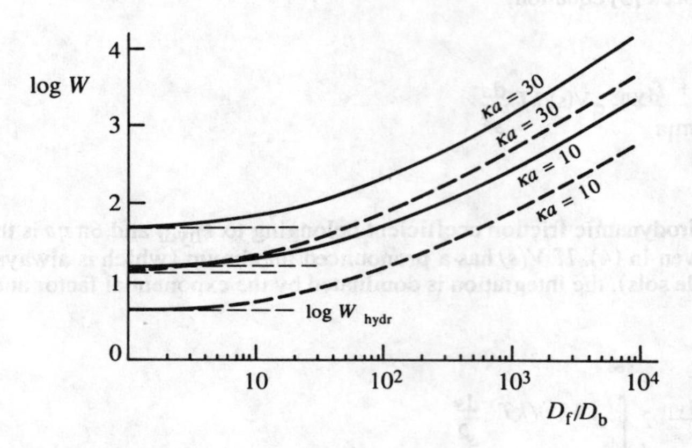

Figure 2. Increase of the stability ratio due to double layer dynamics.
Conditions: $y^0 = 0.5$: $\kappa \Delta_{q_b} \exp(-\Delta_{ads} G^0/RT)/2 q_{tot} = 1$
Drawn curves $\kappa h_m = 0.3$; dashed curves, $\kappa h_m = 1$.
For $D_f \rightarrow D_b$ the dynamic contribution to the stability ratio vanishes and W approaches $W_{(hydr)}$

7. Conclusion

Incorporation of transient disequilibration of electrical double layers does not only lead to an improvement in the theory for the kinetics of coagulation. Non-equilibrium double layers are of wider interest, they play for instance basic roles in the dielectric response of colloidal systems in applied electric fields and the theory is readily modified to describe the influence of ad- or desorbing organic molecules on particle interaction.

Of particular interest is the relation to electrokinetics. Static theories usually incorporate electrokinetics by using the electrokinetic potential ζ as an effective surface potential, or rather the outer Helmholtz potential. In the dynamic interpretation the tangential flow of charge (in the present model in terms of D_b) plays an essential role and this flow is closely coupled to the surface conductance.

In conclusion, double layer dynamics improves existing theories quantitatively but it also introduces qualitatively new features into these theories.

References

1. B.V. Derjaguin (= Deryagin), L. Landau, Acta Physicochim URSS *10*(1939) 333.
2. J.Th.G. Overbeek, E.J.W. Verwey. Theory of the Stability of Lyophobic Colloids. Elsevier, Amsterdam 1948.
3. M. von Smoluchowski, Physik. Z. *17*(1916) 557, 585; Z. physik. Chem. *92*(1917) 129.
4. N.A. Fuchs, Z. Phys. *89*(1934) 736.
5. H. Reerink, J.Th.G. Overbeek, Discuss. Faraday Soc. *18*(1954) 74.
6. L.A. Spielman, J. Colloid Interface Sci. *33*(1970) 562.
7. E.P. Honig, G.J. Roebersen and P.H. Wiersema, J. Colloid Interface Sci. *36*(1971) 97.
8. S. Chandrasekhar. Rev. Modern Physics *15*(1943) 1.

9. G.A. Martynov, Elektrokhimiya *15*(1979), 494 (transl. 418).
10. R.W. O'Brien, L. White, J. Chem. Soc., Faraday Trans. (II) *74*(1978) 1607.
11. S.S. Dukhin, J. Lyklema, Langmuir *3*(1987) 94.
12. S.S. Dukhin, J. Lyklema, Faraday Discuss. Chem. Soc. *90*(1990) submitted.
13. J.Th.G. Overbeek, Kolloid-Beih. *54*(1943) 287.

RECENT THEORIES ON THE ELECTRIC DOUBLE LAYER

R.D. GROOT
Unilever Research Laboratorium
PO Box 114, 3130 AC Vlaardingen
The Netherlands

ABSTRACT. The classical theory of double layers always predicts a repulsion between equally charged colloidal particles. It appears that if modern theories on the liquid state are applied to the double layer problem, this classical result is obtained only at low surface charge, low salt concentration and high dielectric permittivity. Beyond these limits the new theory predicts a significant lateral ordering of the electrolyte adjacent to the particle surface. This lateral ordering leads to a net electrostatic attraction between the particles, and to spontaneous charge inversion of an isolated particle.

1. Introduction

In the classical (DLVO) theory of colloidal interactions, an electrically charged particle is surrounded by a layer of counterions that screen its charge. When two particles of opposite charge approach each other, these double layers start to overlap. As a consequence, the ion concentration between the particles is increased. This concentration increase directly leads to a pressure increase between the particles - the ions are described by the ideal gas model - and hence to an effective repulsive force between the particles. Over the last decade, this phenomenon has attracted renewed attention. Firstly, computer simulations have been carried out to test the theory on details of the distribution of the ionic clouds. Furthermore, a new type of theory on the liquid state has been developed that goes far beyond the ideal gas law description of liquids. Using these theories, an accurate description of the thermodynamics can be given, even in highly non-ideal systems. These theoretical developments now have been applied to describe the electric double layer.

Both the computer simulations and the liquid state theory have indicated that something quite unexpected happens as the particle surface charge becomes sufficiently high, and if divalent ions are present. It appears that the particles may attract each other at some distance, due to electrostatic interactions. Furthermore, a single double layer may show one or more oscillations in the electrostatic potential as a function of the distance from the particle surface. These predictions are so qualitatively different from the classical picture, that decisive experiments had to be carried out to determine which of the two descriptions is the better. At present, accumulating experimental evidence tends to confirm the predictions of the new theory. To explain these non-classical phenomena, it is essential first to introduce the theory of homogeneous liquids. This will be done in the next section. In the third section, this theory will be applied to the thermodynamic description of electrolytes in a double layer, and the results of the theory will be discussed in connection with the experimental results described in section four.

D. M. Bloor and E. Wyn-Jones (eds.),
The Structure, Dynamics and Equilibrium Properties of Colloidal Systems, 801–812.

2. Homogeneous electrolytes

In 1923 Debye and Hückel published a theory on electrolytes that has determined our view on the problem since then [1]. Their description considered point ions that interact by a purely Coulombic interaction. Their consideration rests on two essential points. Firstly, the mean electrostatic field at the point $\vec{r}$ is considered, given the fact that a positive ion is kept fixed at the origin. This field in general has two contributions: one from the ion itself, and one from a diffuse cloud of other ions that surrounds this first ion. Hence

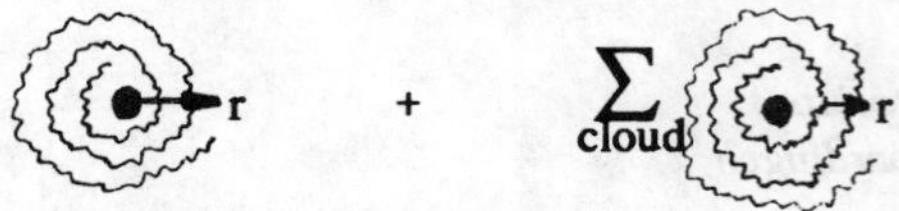

which can be described mathematically as

$$\phi(r) = eV_{direct}(r) + \int \rho(r') V_{direct}(r - r')\, dr' \tag{1}$$

where V_{direct} is the bare Coulomb interaction, and $\rho(r')$ is the local charge density of the cloud.

To determine this charge density, this equation alone is not sufficient, we also have to relate it directly to the electrostatic potential at that position. This second step is always to some extent approximate, whereas equation (1) is exact. As an approximation, we relate the charge density to the field by a Boltzmann distribution, and furthermore we shall linearize the exponential functions:

$$\begin{aligned} \rho(r') &= ec_+(r') - ec_-(r') \\ &\approx ec[\exp - e\phi(r')/kT - \exp e\phi(r')/kT] \\ &\approx -2ce^2\phi(r')/kT \end{aligned} \tag{2}$$

On combining equations (1) and (2), we find

$$\phi(r) = eV_{direct}(r) - \frac{2ce^2}{kT}\phi * V_{direct}(r) \tag{3}$$

where the * stands for a spatial convolution. To solve this equation, we use the fact that the Fourier transform of a convolution equals the product of the Fourier transforms of the separate functions. This means that the difficult integral equation (3) in real space, becomes a simple algebraic equation in Fourier space. By inserting the Fourier transform of the direct Coulomb potential, $V_{direct}(r) = 1/4\pi\varepsilon r \rightarrow \tilde{V}_{direct}(k) = 1/\varepsilon k^2$, and transforming both sides of equation (3), we obtain

$$\tilde{\phi}(k) = \frac{e}{\varepsilon}\frac{1}{k^2} - \frac{2ce^2}{\varepsilon kT}\tilde{\phi}(k)\frac{1}{k^2} \Rightarrow \tilde{\phi}(k) = \frac{e}{\varepsilon}\frac{1}{k^2 + \kappa_D^2} \tag{4}$$

where $\kappa_D = \sqrt{2ce^2/\varepsilon kT}$ is the Debye inverse screening length. The last step towards our description of the ionic cloud around an ion, is the back transformation of the electrostatic field to real space. We thus find the field in real space upon performing the following integration:

$$\phi(r) = \frac{1}{2\pi^2 r} \int_0^\infty k \sin kr\, \tilde{\phi}(k)\, dk$$

$$= \frac{e}{4\pi^2 \varepsilon r} \frac{1}{i} \int_{-\infty}^{\infty} \frac{k e^{ikr}}{k^2 + \kappa_D^2} dk$$

$$= \frac{e}{\varepsilon} \frac{\exp - \kappa_D r}{4\pi r} \qquad (5)$$

where a simple contour integration has been used in the last step.

The physical content of these formulas is as follows. Any positive ion attracts negative charge from its surrounding. This counter charge forms a negative cloud around the ion. At very large distances, one simultaneously observes the ion and its counter charge, and these exactly cancel. Therefore the electrostatic field does not decay as a 1/r function, but much faster: it decays as an exponential. Note however that the total amount of charge in the ionic cloud equals unity, hence on the average only *one* negative charge is "bound" to the positive ion.

Now if the ions are all of finite size, this picture cannot be correct at high salt concentration. On increasing the salt concentration, the screening length is reduced in proportion to the inverse square root of the concentration. At a certain concentration, the screening length becomes smaller than the ionic size, and hence the field of the first ion has decayed before the end of its counter ion is reached. This implies that behind the first layer of negative counter charge around a positive ion, the positive cloud of the counter charge must show up. The first one who predicted these oscillating layers of charge was Kirkwood (1936), who considered ions of finite size [2]. To mimic the fact that a certain region around an ion is excluded, he replaced the direct potential by an effective direct interaction. This effective interaction equals the Coulomb interaction outside the hard core distance, and it is put equal to zero inside the hard core. Hence

$$V_{direct} = \frac{1}{4\pi\varepsilon r} \rightarrow V_{eff} = \begin{cases} 0 & \text{if } r < d \\ \dfrac{1}{4\pi\varepsilon r} & \text{if } r > d \end{cases}$$

which leads to the following field equation in Fourier space.

$$\tilde{\phi}(k) = \frac{e}{\varepsilon}\frac{\cos kd}{k^2} - \frac{2ce^2}{\varepsilon kT}\tilde{\phi}(k)\cdot\frac{\cos kd}{k^2}$$

The solution of this equation corresponds in real space to

$$\phi(r) \approx \frac{e}{\varepsilon}\frac{\exp(-\kappa r)\cos(\omega r)}{4\pi r} \tag{6}$$

where κ and ω are the screening and wavenumber respectively. For low salt concentration $\omega = 0$, and a Debye-Hückel behaviour is found. At the concentration where $\kappa_D d = 1.03$, and where d is the ion diameter, the oscillatory behaviour sets in. Figure 1 shows the region where the ionic cloud oscillations occur in the Kirkwood theory.

In fact what we have calculated above, is the ionic charge density near an arbitrary but fixed ion. As in this approximation the charge density is taken to be proportional to the field (see

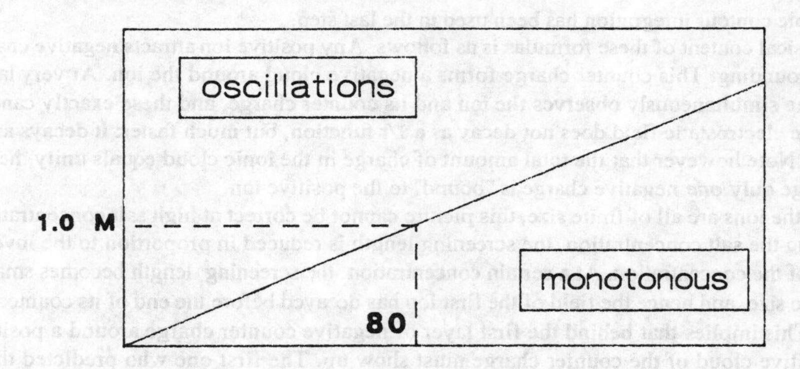

Figure 1. Boundary of monotonous/oscillatory potentials in the Kirkwood theory.

equation 2), the field should vanish inside the hard core diameter. Instead, we have put the effective interaction at zero in an *ad hoc* manner. It can be shown that we can satisfy the more physical requirement $\rho(r) = 0$ if $r < d$, by inserting a linear function in the effective interaction, inside the hard core. Thus

$$V_{eff} = \begin{cases} a + br & \text{if } r < d \\ \dfrac{1}{4\pi\varepsilon r} & \text{if } r > d \end{cases}$$

where a and b are functions of the salt concentration. This theory is called the *Mean Spherical Approximation* (MSA), and it was introduced by Waisman and Lebowitz in 1970 [3]. The thermodynamic results of this theory, such as activity coefficients, are very accurate, and this theory also predicts a transition to oscillatory charge distributions. According to the MSA this transition takes place at $\kappa_D d = 1.23$.

3. Inhomogeneous Electrolytes

To determine the distribution of ions near a charged surface, we shall introduce a *local* thermodynamic description of the electrolyte. For this purpose, we consider an infinitely large, hard planar wall that has a given surface charge density. Adjacent to this wall we divide space up into parallel bins. This division has no physical consequences, it is only a means to set up the calculations. At the end of the day, the "administrative" bins will be chosen at a vanishing thickness, and a continuous liquid description is recovered. The number of ion centres in each bin, divided by the bin volume will now be used as a *local* ionic concentration c(r). The whole theory now focuses on determining this c(r) as a function of the distance from the wall. The exact answer as to how the ions are divided over the bins follows from the requirement that in thermodynamic equilibrium the electrochemical potential of any ion type in a specific bin should equal the chemical potential of that ion type in all other bins. Hence the chemical potential of all species must be constant throughout the system. To determine the ionic distributions, we must therefore know how the chemical potential in a specific bin is determined by the ion concentrations in this bin and in all other bins. In general the chemical potential consists of two major parts, an entropic part and an energetic part. The configurational entropy leads to the contribution

$$\mu^{conf}(r) = kT \ln c(r) \tag{7}$$

to the chemical potential in the bin at the position r. The other parts of the chemical potential are due to energetic interactions with the fixed surface charge, and due to interactions with the other ions. This interaction part equals

$$\mu^{int}(r) = V_{surface}(r) + \sum \int c(r')V_{eff}(r - r')d^3r' \tag{8}$$

where the sum runs over all ion types, and where the integration in fact is approximated by a summation over the bins. It should be noted that equation 1 and equation 8 are very analogous.

The differences are that in the latter $V_{surface}$ represents the electrostatic interaction with a two-dimensional charged sheet, and that V_{eff} again must contain a correction inside the hard core that describes the excluded volume effects. Thus V_{eff} contributes both to the energy and to the (free volume) entropy. However, as appeared in the case of homogeneous electrolytes, this effective interaction in general depends on the ionic concentration. In the case of inhomogeneous electrolyte distributions, it is therefore not so easy to define this effective interaction.

Two approximate theories have been published from which this effective ion-ion interaction can be obtained. One theory is the *Anisotropic Hypernetted Chain Theory*, [4-6] and the other is the *Density Functional Theory* [7-9]. In the latter theory, the effective interaction in the inhomogeneous electrolyte is *assumed* to equal the effective interaction of a homogeneous electrolyte of concentration $\bar{c}(r)$ where this reference concentration is chosen differently at each point. The concentration $\bar{c}(r)$ is determined as the concentration of the best fitting homogeneous electrolyte that describes the thermodynamics of the inhomogeneous electrolyte locally. In practice this means that $\bar{c}$ is a weighted average of the local concentration c, where the weight function follows from an *objective* criterion. Once the effective interaction is known, the concentrations $c_i = c(r_i)$ can be found from the diffusion equilibrium. If we approximate the liquid by *n* bins, equations (7), (8) and the equilibrium condition lead to the *n* equations

$$\mu^{conf}(c_i) + \mu^{int}(c_1,...,c_n;i) = \mu \quad \text{for } i = 1...n$$

from which the *n* unknown concentrations $c_1...c_n$ can be solved numerically. On taking the limit $n \rightarrow \infty$ while keeping the system volume fixed, we obtain the desired continuous description.

In the Anisotropic Hypernetted Chain Theory (HNC), the effective interaction is determined without reference to a local coarse-grained concentration. Instead, two equations are supplied for the density-density distribution function between all pairs of points r and r'. One relation results from an integral equation similar to equation (1), which describes that the electrostatic interactions of the bare ion and its counterion cloud are coupled. The other relation basically is the Boltzmann factor (equation 2) that relates the ionic cloud to the local potential change at r', which results from inserting an ion at position r. In the HNC theory of homogeneous liquids, this Boltzmann factor contains a combination of the bare Coulombic interaction, the effective interaction as defined above, and the local density-density distribution function, but the principle remains the same. The special point in the inhomogeneous HNC theory is the fact that these relations are applied to a large number of combinations of points near the charged surface. Because the point r can be varied on the line perpendicular to the surface, and the point r' can be varied in the plane perpendicular to the surface, one typically has to solve some 10^6 coupled non-linear equations to obtain the concentration profiles in the inhomogeneous HNC theory. In contrast, in the density functional theory one typically must solve a few 10^2 equations. Therefore it is relatively easy to describe multi-component systems by the density functional theory provided that a reliable (analytical) theory exists on the corresponding homogeneous system.

To calculate the electric double layer using one of the theories mentioned above, it is necessary to define a mathematical model of the electrolyte. One model that is frequently used because of its simplicity, is the so-called Restricted Primitive Model (RPM). This model describes a two-component system of equally sized, charged hard spheres. To mimic the solvent, the spheres are embedded in a medium of fixed dielectric permitivity. The medium itself is structureless, it only introduces a reduction of the electrostatic interaction by a factor ε_r. This simple picture of the solvent has led to the description "primitive". Using the density functional theory, one can introduce another model of electrolyte solutions with only little extra computational effort. In this model, the solvent is described by the introduction of *uncharged* hard spheres, whereas the electrostatic interaction is taken equal to the interaction in the RPM. In this way, the effects that are caused by the excluded volume of the solvent, i.e. the liquid structuring, can be investigated. The model should by no means be interpreted as a realistic model for aqueous solutions, it is no more than a tool to gain some insight into the importance of the repulsive interactions with the solvent.

4. Results and Discussion

The theory described in the former section has been applied to several double layer problems. Firstly, one may study the ion distribution in a single double layer as described by the restricted primitive model. Next, one can introduce the hard sphere solvent, and study its influence on observables like the electrostatic potential at contact. Another problem that was studied is the electrostatic interaction between two charged surfaces. In addition, experiments and computer simulations have been carried out to check the theoretical predictions. Generally it is found that the classical DLVO theory of double layers gives a very reasonable description of the phenomena. However, on increasing the particle surface charge, and on decreasing the dielectric permitivity (which is equal to increasing the ionic valency) some results are obtained that are at variance with the classical model. These results will be discussed in conjunction with the available experimental data.

The first result to be mentioned is the fact that the ion distribution tends towards a layered-like structure on increasing the surface charge. This phenomenon is illustrated by the curves of figure (2) that show the coion and counterion distributions adjacent to a charged surface[8]. The bulk concentration is chosen at 1 *mol/l*, and the ion size is taken equal to d = 4.25 Å, corresponding to the value that is taken in the simulation literature [10-12]. At a surface charge density of $\sigma^* = \sigma d^2/e = 0.4$, a small shoulder starts to appear in the counterion concentration profile which grows into a second layer of counterions at higher surface charge densities. This second layer is also observed in the corresponding simulations [10,11]. The build up of such a second layer is entirely caused by the non-electrostatic repulsion between the ions. As the surface attracts counterions, the average salt concentration adjacent to the surface is increased. This concentration increase in turn would lead to a large increase in the chemical potential of the counterions due to the loss of free volume entropy, if all ions would be placed in the first layer. Instead, the double layer is broadened: the energy loss associated with this solution is compensated for by the gain in entropy of free volume.

As a consequence, the counter charge is divided over a wider range and the resulting electrostatic potential at ion-surface contact is increased relative to the classical theory. The contact potential is shown in figure (3), together with the simulation results of reference 11. In the density functional theory that was used, no fit parameters are used to fix the theory to the simulations. It is possible to improve the fit to the simulation data by adjusting the way to determine the reference concentration [7], but this approach does not lead to a better understanding of the processes, and the theory loses much of its predictive value.

Another new point is a possible explanation of the phenomenon of charge inversion. This phenomenon was observed as early as 1954 by Strauss and coworkers [13]. In electrophoresis experiments on polyelectrolytes at high salt strength (from 0.6 M KBr or 0.5 M $MgCl_2$ the polyacid was observed to move *towards* the cathode. This charge inversion has also been found in simulation studies in cylindrical symmetry, using the RPM to model a divalent electrolyte [12], and again the liquid state theory is able to describe this phenomenon quantitatively. Plischke and Henderson used the anisotropic hypernetted chain theory to calculate the double layer of a 2-2 electrolyte in flat plate symmetry [5], and found good agreement with simulation data. However, also with monovalent electrolyte this phenomenon was found, both in flat plate symmetry [8] and in cylindrical symmetry [9] on introduction of a simple hard sphere description of the solvent. The results of these calculations are presented in figure 4, where the concentration profiles of coions, counterions and solvent are plotted, and in figure 5, where the electrostatic potential is shown as a function of the distance from the surface, for various surface charges.

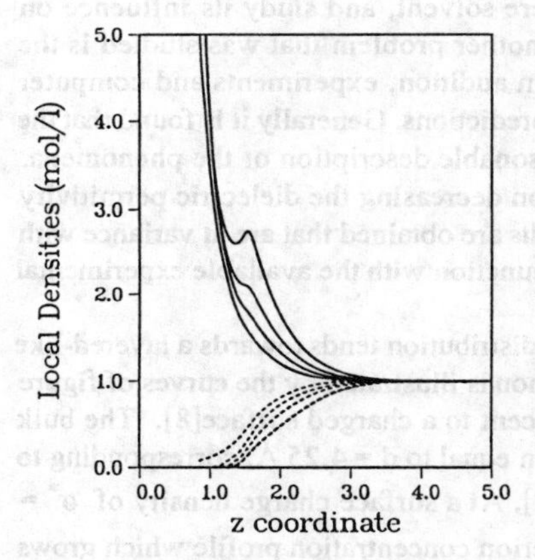

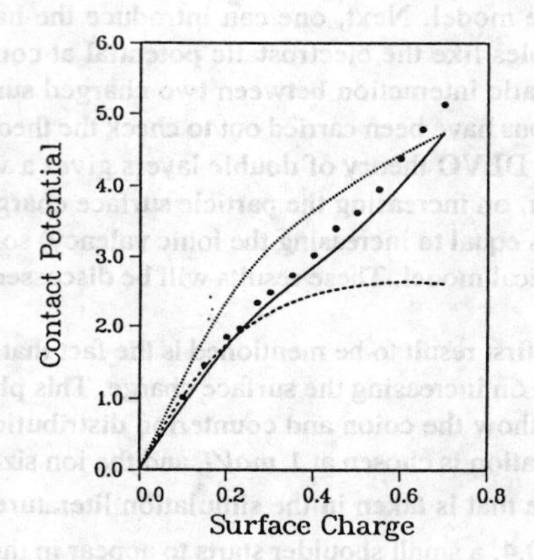

Figure 2 (left) The counterion and coion concentration as a function of the distance from the surface, for the surface charges $\sigma^* = 0.25, 0.42, 0.55, 0.70$. Figure 3. (right) The dimensionless contact potential as a function of the surface charge, in the Gouy-Chapman theory, the *homogeneous* HNC theory and the density functional theory (dotted, dashed and full curves) compared to the simulation results (dots). The figures are taken from reference 8.

The explanation of the charge inversion phenomenon is quite straight forward in the light of the density functional theory. From the theory of homogeneous electrolytes, we know that the electrostatic potential around a single ion tends to oscillate when the salt concentration exceeds some critical value (see equation 6 and figure 1). A similar behaviour holds for the the potential near a charged surface in the first instance, but there is one major difference. The surface tends to attract counterions, and generally the total salt concentration is increased near the surface. This indicates that, especially at high surface charges, the salt concentration adjacent to the surface may be increased beyond the critical concentration where oscillations start to occur, whereas it is below this concentration in the bulk. As a result, the potential starts off oscillating, but before the first oscillation has finished, the salt concentration has decayed already to a value below the critical concentration. Thus the potential has the "wrong" sign at large distances.

A more intuitive explanation can be given from a microscopic picture. Suppose a negatively charged surface is covered with a single layer of counterions, that precisely cancels its charge. This system behaves as a capacitor, the electrostatic field is present only between the surface and the counterion layer. Hence if another positive ion is brought towards the layer from infinity, it *will not* have any electrostatic interaction with the capacitor as a whole, until it starts to push away the other positive ions. Thereby, a hole in the double layer is created and the bare negative surface

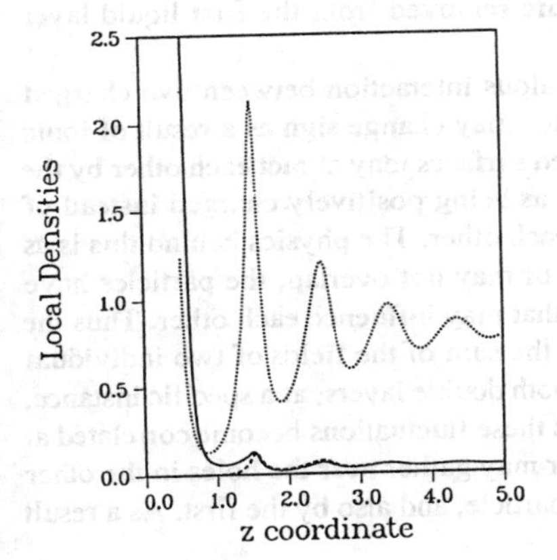

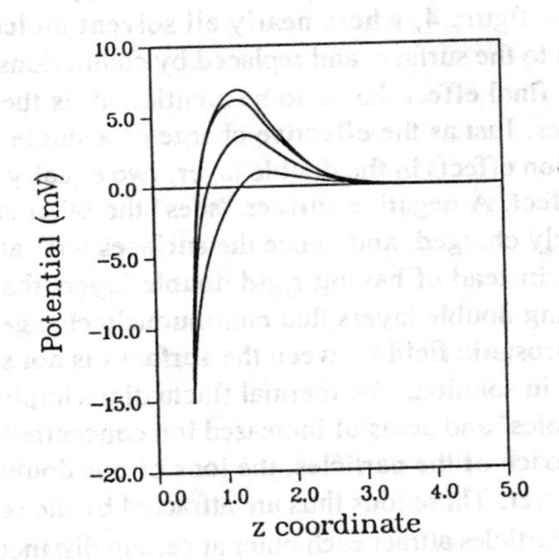

Figure 4. (left) The ionic concentrations and the solvent density profile as a function of the distance from the surface, for the surface changes $\sigma^* = 0.25$. Figure 5. (right) The electrostatic potential adjacent to a negatively charged surface, for $\sigma^* = 0.25, 0.42, 0.55, 0.70$. The figures are taken from reference 8.

becomes exposed. The positive ion is then attracted by the negative surface. If a negative ion would try to penetrate the double layer, it would also find a negative surface at the end of the hole it has created, and hence it would be repelled from the double layer. On inserting more counterions in the first layer than are necessary to compensate for the surface charge, an excess of positive charge arises, which must be compensated for by a third layer of negative charge. We thus have two capacitors in series.

The introduction of an excess charge δQ in the first counterion layer leads to an energy decrease proportional to δQ resulting from the rearrangements in this layer, and it leads to an energy increase proportional to $(\delta Q)^2$ that is necessary to charge the second capacitor. Hence in the first instance the energy is lowered. The entropy, however, is also reduced as a result of the reduced free volume per ion in the first double layer. A balance between these two effects determines whether a charge inversion will occur or not. If we wish to favour the charge inversion, we can choose between two options. One is to increase the energy gain of these correlation effects, i.e. we can increase the ionic charge. Indeed, inversion is found in simulations on divalent electrolytes [5,12]. Another option is to reduce the entropy barrier by introducing an explicit solvent molecule into the model. In that case high volume fractions of molecules occur both adjacent to the surface and far away from the surface. The free volume per ion at the surface in that case equals the free volume per ion in the bulk. The system now can simply replace solvent molecules from the interfacial layer by counterions. The price to pay is only the (ideal) entropy of mixing, which is much lower than the entropy loss associated with packing effects. This is indeed shown in figure 4, where nearly all solvent molecules are removed from the first liquid layer adjacent to the surface, and replaced by counterions.

The final effect that is to be mentioned, is the anomalous interaction between two charged interfaces. Just as the effective charge of a single interface may change sign as a result of ionic correlation effects in the double layer, two equally charged surfaces may attract each other by the same effect. A negative surface "sees" the other surface as being positively charged instead of negatively charged, and hence the surfaces will attract each other. The physics behind this is as follows: instead of having rigid double layers that may or may not overlap, the particles have fluctuating double layers that continuously change, and that may influence each other. Thus the net electrostatic field between the surfaces is not simply the sum of the fields of two individual particles in solution. The thermal fluctuations imply that both double layers, at a specific instance, show "holes" and areas of increased ion concentration. As these fluctuations become correlated at the approach of the particles, the ions of one double layer may gather near the holes in the other double layer. These ions thus are attracted by the second particle, and also by the first. As a result the two particles attract each other at certain distances.

The question when this picture applies to a real colloidal system, again is governed by a balance between the fluctuation effects, and the entropy effects that tend to oppose the former. From the charge inversion in the RPM which only appears when divalent ions are present, one might guess that two charged colloid particles only attract if a divalent electrolyte is used. Indeed this was found by Kjellander and coworkers [14,15] who solved the anisotropic hypernetted chain equations for the restricted primitive model in the geometry of two equally charged surfaces. For divalent electrolyte they found an osmotic pressure as a function of the interface distance that very much looks like the Van der Waals pV-diagram, if the temperature is replaced by the inverse surface charge density (see figure 2 of reference 15). At low surface charge density, they found a continuously decreasing pressure with increasing distance. At one specific surface charge density, the pressure shows a horizontal deflection point, similar to that of a critical isotherm in a pV-diagram. For larger surface charges the pressure shows a minimum and may become negative.

The resulting attractive interaction between the equally charged surfaces is much larger than the usual London dispersion forces, though the effects are similar. The dispersion forces have their origin in the correlation of the quantum fluctuation of the local polarizabilities, whereas the present effect results from the correlation of the thermal fluctuations of the ionic clouds. The presence of this effect successfully explains the swelling properties of mineral clays [15]. The distance between the clay lamellae was measured by X-ray diffraction, for Ca-clays and Na-clays. It appears that the clays do swell in the presence of monovalent salt, whereas they remain in a stable state in the presence of divalent ions. Kjellander and coworkers also reported on the direct measurement of the force between two mica surfaces, using an advanced experimental technique [14]. These measurements show an oscillatory force between the surfaces. However, at a distance between 0.7 nm and 1.7 nm one oscillation clearly is missing. From this observation it must be concluded that the surfaces attract each other in the given range (only repulsive forces can be detected by this technique). The existence of this force minimum is in line with the anisotropic hypernetted chain calculations on the RPM. A similar result has been obtained by Guldbrand and coworkers, who studied the RPM by the Monte Carlo simulation technique [16]. These simulations also indicate the existence of a negative minimum in the double layer force with divalent ions. The fact that no fast oscillations in the force were predicted, is a consequence of the use of the RPM. If the electrolyte would be embedded in a more realistic solvent, the latter would tend to build up layers adjacent to the surfaces (see figure 4). The overlap of these layers gives rise to an oscillating force that superimposes on the electrostatic force.

5. Conclusions

The classical double layer theory is valid only within a finite range of physical parameters. For divalent ions and high surface charge, it leads to qualitatively wrong predictions, because it does not incorporate ionic correlations in the double layers. When these effects become important, one should also be careful in using the RPM to model the electrolyte, since this model cannot cope with entropy effects in connection with the solvent. The latter is important because the charge inversion and double layer attraction result from a subtle balance between energetic and entropic effects.

6. References

1. P. Debye and E. Hückel, Z. Phys. **24**, 185 (1923).
2. J.G. Kirkwood, Chem. Rev. **19**, 275 (1936).
3. E. Waisman and J.L. Lebowitz, J. Chem. Phys. **52**, 4307 (1970).
4. R. Kjellander and S. Marčelja J. Chem. Phys. **82**, 2122 (1985).
5. M. Plischke and D. Henderson, J. Chem. Phys. **88**, 2712 (1988).
6. D. Henderson and M. Plischke, J. Phys. Chem. **92**, 7177 (1988).
7. T. Alts, P. Nielaba, B.D'Aguanno and F. Forstmann, Chem. Phys.**111**, 223 (1987).
8. R.D. Groot, Phys. Rev. A **37**, 3456 (1988).

9. R.D. Groot and J.P. van der Eerden, J. Electroanal. Chem. **247**, 73 (1988).

10. G.M. Torrie and J.P. Valleau, J. Chem. Phys. **73**, 5807 (1980).

11. P. Ballone, G. Pastore and M.P. Tosi, J. Chem. Phys. **85**, 2943 (1986).

12. V. Vlachy and A.D.J. Haymet, J. Chem. Phys. **84**, 5874 (1986).

13. U.P. Strauss, N.L. Gershfield and H. Spiera, J. Am. Chem. Soc. **76**, 5909 (1954).

14. R. Kjellander, S. Marčelja , R.M. Pashley and J.P. Quirk, J. Phys. Chem. **92**, 6489 (1988).

15. R. Kjellander, S. Marčelja , and J.P. Quirk, J. Coll. Int. Sc. **126**, 194 (1988).

16. L. Guldbrand, B. Jönsson, H. Wennerström and P. Linse, J. Chem. Phys. **80**, 2221 (1984).

APPLICATION OF DOUBLE LAYER THEORY TO MODERATELY COMPLEX SYSTEMS

D.G. HALL,
Unilever Research,
Port Sunlight Laboratory,
Quarry Road East,
Bebington,
Wirral,
Merseyside.
L63 3JW

ABSTRACT. According to the Grahame model, the electrical double layer is divided into a diffuse region which conforms to the Poisson-Boltzmann equation and an inner region which is treated as one or more electrical condensers in series. The aim of this talk is to present an alternative thermodynamic approach, based on a theorem derived some years ago by the author. This alternative approach utilises the concept of an outer Stern potential (ψ_d) at the plane separating inner and diffuse regions, but does not rely on any detailed model of the inner regions. Electrostatic concepts of dubious significance, such as the surface (or wall) potential, ψ_o, and inner region capacity are thereby avoided. Various applications of the alternative approach are discussed. These include tests for specific adsorption, interactions between overlapping double layers, a demonstration that the constant potential boundary condition is not a lower bound for interactions between identical parallel plates, and the relationship of less than constant potential interactions to two dimensional phase separation.

1. Introduction

Recent theoretical work on the electrical double layer [1,2] has tended to focus increasingly on the application to the primitive model of statistical mechanical methods such as the mean spherical approximation (MSA) [3,4] and the hypernetted chain approximation (HNC)[3,5] which were originally developed to describe bulk liquids and bulk solutions. The results of this work agree fairly well with simulation studies [6,7] and have lead to the interesting predication that in some situations interactions between identical overlapping double layers can be attractive [8,9]. At present however, it is not altogether obvious how to use this modern work in the interpretation of experimental data. Consequently experimentalists tend to rely on more traditional procedures which use the Poisson-Boltzmann equation to describe the diffuse region and the Stern-Grahame model [10] to describe the inner regions. The limitations of these procedures have been discussed previously by the author [11,14] who has forwarded an alternative approach based on thermodynamic arguments which is less model dependent and more general [11] but which has yet to be widely used despite some favourable publicity [15,16]. The aims of this paper are to review this alternative approach; to compare and contrast it with the better known Stern-Grahame model [10] and to highlight some more recent developments [17,18]. The emphasis is on doublelayers formed at the non metallic solid/solution interface. Double layers of this kind can form in a variety of ways [13]. For example the ions of crystals such as $BaSO_4$ may exhibit different tendencies to

D. M. Bloor and E. Wyn-Jones (eds.),
The Structure, Dynamics and Equilibrium Properties of Colloidal Systems, 813–829.

dissolve. Fixed groups on the surface such as COOH or OH may lose or gain protons from solution. Surfactant ions may adsorb from solution. The material presented below is sufficiently general that it encompasses all of these situations.

2. Double Layer Model and Experimental Methods

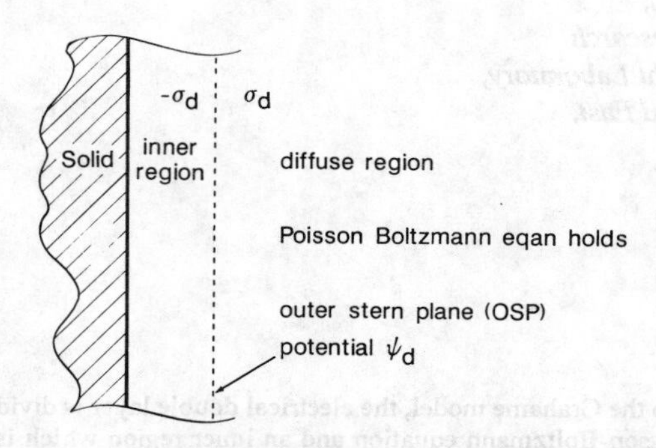

Figure 1 Schematic representation of double layer model.

The model used is depicted schematically in Figure 1. According to this model the double layer is regarded as consisting of inner and diffuse regions. The boundary between these regions is referred to as the Outer Stern Plane (OSP) [13,26] and it is supposed that the ion distribution in the diffuse region conforms to the Poisson-Boltzmann equation. Consequently the electrical potential at the OSP relative to the bulk solution, ψ_d, is related to the charge per unit area to the left of the OSP by the well known equation

$$\sigma_d^2 = \frac{\varepsilon kT}{2\pi} \sum_i n_i^b [\exp - \frac{\nu_i e \psi_d}{kT} - 1] \tag{1}$$

where the n_i^b, are bulk number densities and ν denotes ionic valency including the sign and where we have used unrationalised units.

For the present the inner regions are regarded as a black box whose state is determined by the nature of the solid concerned and the bulk solution in contact with it. According to this model we may divide the Gibbs surface excess of a species i into contributions from inner and diffuse regions by writing [18b]

$$\Gamma_i = \Gamma_i^\sigma + \Gamma_i^d \tag{2}$$

where, when the Poisson-Boltzmann equation applies, the Γ_i^d are given by [19]

$$\Gamma_i^d = n_i^b \int_0^{\psi_d} \frac{\left[\exp - \frac{v_i e \psi}{kT} - 1\right] d\psi}{\left[\frac{8\pi kT}{\varepsilon} \sum_i n_i^b \left\{\exp - \frac{v_i e \psi}{kT} - 1\right\}\right]^{1/2}} \tag{3}$$

Given ψ_d it is straightforward to obtain the Γ_i^d from equation 3 numerically or, when all ions are z valent, analytically.

Two experimental methods used to study the double layers formed by solids are potentimetric titration and electrokinetic measurements. Essentially potentimetric titration methods measure the adsorption (Γ_1) of a potential determining species as a function of its concentration in solution in the presence of background electrolyte [20]. For example, we may take a suspension of AgI, add KI and measure the bulk I^- concentration using an I^- selective electrode. The difference between the amount of I^- added and the amount which enters the bulk solution is the change in the amount adsorbed by the suspension. Electrokinetic measurements such as the determination of electrophoretic mobilities and streaming potentials lead to a quantity known as the zeta potential ζ which is the potential at the so called plane of shear relative to the bulk solution [16]. Although there is no a-priori reason for doing so it is common to regard ζ as a measure of ψ_d. In the absence of contrary evidence this a-postiori identification is adopted below. It is also worth mentioning that estimates of ζ are not based on the Poisson-Boltzmann equation when the conditions for the Helmholtz-Smoluchowsi equation to apply are met [21]. However, when the Poisson-Boltzmann equation does apply then given ψ_d we may estimate σ_d *via* equation 1 and the Γ_i^d *via* equation 2. The situation where ζ and hence $\sigma_d = 0$ is referred to as the isoelectric point (I.E.P.) whereas that in which $\Gamma_1 = 0$ is known as the zero point of charge (ZPC).

In principle at least potentiometric titration studies may be used to obtain quantitative information about the adsorption of species whose uptake cannot be measured directly. For example, if we have a single potential determining species 1 and supporting electrolyte consisting of counterions 2 and coions 3 it may be shown thermodynamically that

$$\left(\frac{\partial \Gamma_3}{\partial \Gamma_1}\right)_{\theta_3} = - \left(\frac{\partial \theta_1}{\partial \theta_3}\right)_{\Gamma_1} \tag{4}$$

where

$$\theta_1 = \bar{\mu}_1 - \frac{\nu_1}{\nu_2}\bar{\mu}_2$$

$$\theta_3 = \bar{\mu} - \frac{\nu_3}{\nu_2}\bar{\mu}_2 \qquad (5a,b)$$

and where $\bar{\mu}$ denotes electrochemical potential [20,22,23]. θ_1 and θ_3 are the chemical potentials of the electrically neutral salts formed by species 1 and 2 and by species 3 and 2 respectively. The right hand side of equation 4 can be obtained from potentiometric titration data. On integration between two states I and II we find from 4 that

$$\Gamma_3^{II} - \Gamma_3^{I} = -\int_{I}^{II} \left(\frac{\partial\theta_1}{\partial\theta_3}\right)_{\Gamma_1} d\Gamma_1 \qquad (6)$$

where the integration is performed at constant θ_3. Equation 6 gives changes in Γ_3 at a given θ_3. To get changes in Γ_3 with θ_3 and absolute values of Γ_3 we need to know how Γ_3 depends on θ_3 at some fixed Γ_1. This may be done as follows. For some systems it is found that at a particular activity a_1 of the potential determining species potentiometric titration curves at different concentrations of background electrolyte give a common intersection point [20,23]. If it is also found that $\zeta = 0$ at the same a_1, this indicates that $\Gamma_1 = 0$ and that there is no specific adsorption of supporting electrolyte ions. If this state is taken as state I we may use equation 6 to get Γ_3. Γ_2 is then obtainable from the electrical neutrality condition

$$\nu_1\Gamma_1 + \nu_2\Gamma_2 + \nu_3\Gamma_3 = 0 \qquad (7)$$

From electrokinetic measurements we may find σ_d and then determine Γ_2^d and Γ_3^d from equation 3. Hence from a combined electrokinetic and potentiometric titration study on the same system we may in principle obtain Γ_1, Γ_2^σ, Γ_3^σ, Γ_2^d and Γ_3^d at different bulk concentrations of the potential determining species and supporting electrolyte. In practice a variety of difficulties may arise namely

a) It may not be possible to perform both sets of measurements.

b) To get Γ_1 from potentiometric titration studies a surface area is required.

c) The procedure of obtaining Γ_3 from equation 6 is not very accurate.

Consequently it is worthwhile to consider how one should proceed when only one of the techniques is available. We now turn to this issue by discussing a useful theorem and its implications.

3. A Useful Theorem and its Implications

a) STATEMENT OF THE THEOREM

The theorem may be stated as follows [11c,13,14] dY given by

$$dY = \sum_i \Gamma_i^\sigma d\mu_i \tag{8}$$

is an exact differential where

$$\mu_i = \mu_i^b - \nu_i e \psi_d \tag{9}$$

and where μ_i^b

$$\mu_i^b = \mu_i^\ominus (T,p) + kT \ln n_i^b \tag{10}$$

Equations 8-10 apply both to isolated and interacting double layers when the diffuse region conforms to the Poisson-Boltzmann equation.

For a surface with a single potential determining species in the presence of supporting electrolyte there are three $d\mu_i$ on the right hand side of equation 8. This is one more than can be varied independently at equilibrium. However, if we can argue that Γ_i^σ for one species, typically an indifferent coion, is zero than equation 8 becomes quite useful.

For many systems this restriction is not seriously limiting. In some instances it may be overcome by adding species which are not specifically adsorbed. Also if a species is not specifically adsorbed at the ZPC or IEP it seems reasonable to suppose that this will remain the case when it is a coion. However, problems can be expected to arise with porous solids such as silica because in this case bulk electrolyte may penetrate the particle interior.

The precise physical significance of Y in the present context is not important. For an isolated surface it may be regarded as the total surface tension minus the contribution thereto from the diffuse part of the double layer [24]. The importance of the theorem is that it enables inner and diffuse regions to be discussed separately in a general but straightforward way. The value of equation 8 is best illustrated by considering some applications.

b) TESTS FOR SPECIFIC ADSORPTION OF SUPPORTING ELECTROLYTE [11c,13,14]

When there is no specific adsorption of supporting electrolyte Γ_2^σ and Γ_3^σ are both zero $\nu_1 e \Gamma_1^\sigma = -\sigma_d$ and equation 8 takes the simple form.

$$dY = \Gamma_1^\sigma d\mu_1 \tag{11}$$

This equation shows that at constant T and p Γ_1^{σ} depends only on μ_1 and vice versa. It follows that a graph of ψ_d vs $\frac{kT}{\nu_1 e} \ln n_1^b$ at constant σ_d should be linear with a Nernstian slope and that graphs of σ_d vs $[kT \ln n_1^b - \nu_1 e \psi_d]$ obtained at different ionic strengths should superimpose. We refer to the former graph as an Nernst plot and the latter graph as a congruence plot [13,14]. These plots are readily applied to electrokinetic data by measuring ψ_d and calculating σ_d using equation 1. The reverse procedure applies to the interpretation of potentiometric titration data. One measures σ_d and calculates ψ_d. A series of Nernst plots for a nylon sol are shown in Figure 2 and a congruence plot for the same sol is shown in Figure 3. These results show that there is no significant specific adsorption of supporting electrolyte ions by nylon under the conditions studied [13,14]. Figures 4 and 5 show corresponding results for a silver iodide sol. It is clear from both graphs that there is specific adsorption of supporting electrolyte at higher surface charges and ionic strengths and that the Nernst plot in particular shows this very clearly [13,14]. Figure 6 shows the results of attempting a congruence plot for silica. As expected equation 11 does not apply in this case and this is clear from the plot [13,14].

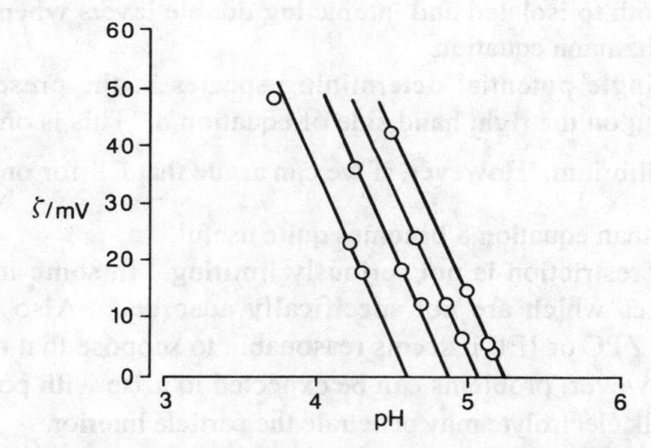

Figure 2. >>Nernst plot<< for nylon sol, ζ(mv) vs pH at constant σ_d, from L–R, σ_d=0.4, 0.3, 0.2, 0.1 $\mu C/cm^2$ respectively.

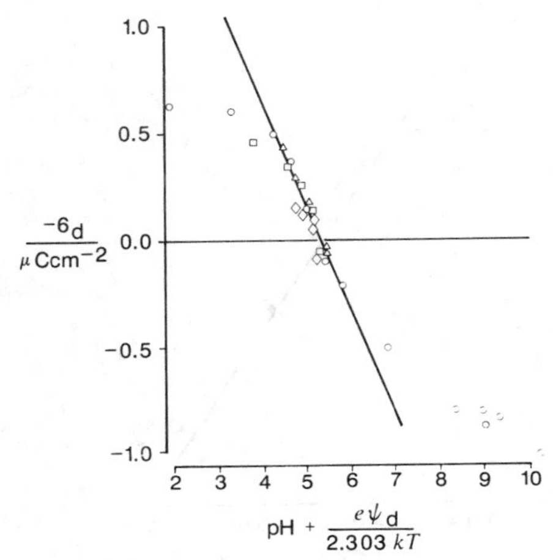

Figure 3. >>Congruence plots<< for nylon sol, $-\sigma_d$ vs $[\mathrm{pH} + e\psi_d/2.303\,kT]$ at various ionic strengths, ○ = 10^{-3} mpl, △ = 10^{-2} mpl, □ = 5×10^{-2} mpl, ◇ = 10^{-1} mpl

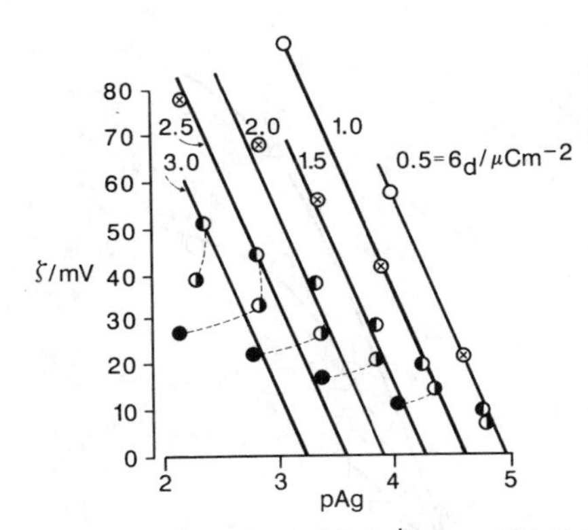

Figure 4. >>Nernst plot<< for AgI Sol, ζmv vs pAg at const σ_d, from L–R, σ_d = 3.0, 2.5, 2.0, 1.5, 0.5, μCm/cm^2 respectively, Ionic strengths O = 10^{-3}, ⊗ = 10^{-2}, ◐ = 5×10^{-2}, ◑ = 10^{-1}, ● = 1.5×10^{-1} mpl respectively.

A further application of the plot is the estimation of surface areas from potentiometric titration data [14]. If a common intersection point is obtained at the ZPC it may be possible to obtain an area which leads to congruence behaviour near the ZPC. This procedure is described in further detail elsewhere [14].

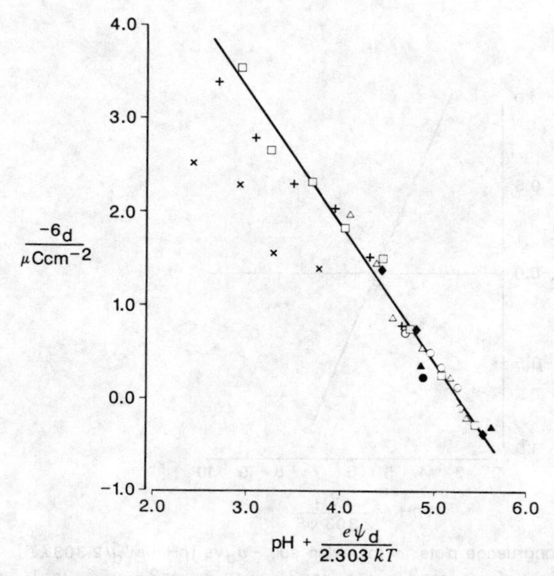

Figure 5 >> Congruence plots << for AgI sol., $-\sigma_d$ vs [pAg + $e\psi_d/2.303kT$] at various ionic strengths, ○ $=10^{-3}$, △ $=10^{-3}$, □ $=5\times10^{-3}$, + $=10^{-1}$, × $=1.5\times10^{-1}$, ◆ $=10^{-1}$, mpl respectively ●, ▲, ◆ denote titration data[10].

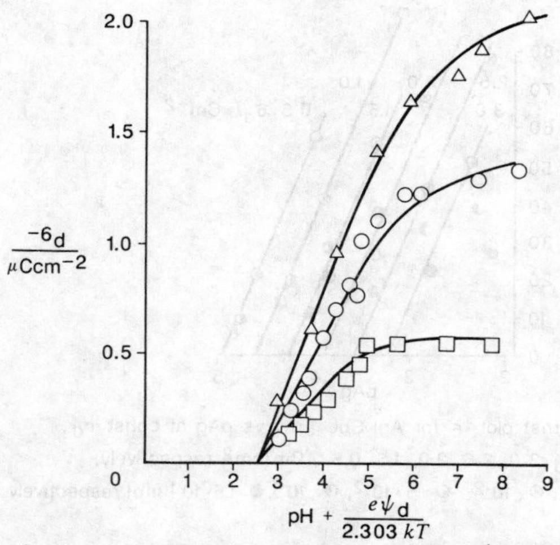

Figure 6. >>Congruence plots<< for silica, $-\sigma_d$ vs [pH + $e\psi_d/2.303kT$] at various strengths, □ $=10^{-4}$, ○ $=10^{-3}$, △ $=10^{-2}$mpl respectively.

c) SPECIFIC ADSORPTION OF COUNTERIONS FROM ELECTROKINETIC DATA [11,13]

When counterions but not coions are specifically adsorbed equation 8 becomes

$$dY = \Gamma_1^{\sigma} d\mu_1 + \Gamma_2^{\sigma} d\mu_2 \tag{12}$$

and may be rewritten as

$$dY = -\frac{\sigma_d}{\nu_1 e} d\mu_1 - \Gamma_2^{\sigma} \frac{\nu_2}{\nu_1} d\theta_1 \tag{13}$$

where $\sigma_d = -\nu_1 e \Gamma_1^{\sigma} - \nu_2 e \Gamma_2^{\sigma}$

and $\theta_1 = \mu_1 - \dfrac{\nu_1}{\nu_2} \mu_2$

on cross differentiation equation 13 gives

$$\frac{1}{\nu_1 e}\left(\frac{\partial \sigma_d}{\partial \theta_1}\right)_{\mu_1} = \frac{\nu_2}{\nu_1}\left(\frac{\partial \Gamma_2^{\sigma}}{\partial \mu_1}\right)_{\theta_1} \tag{14}$$

which in turn may be integrated to give

$$\Gamma_2^{\sigma II} - \Gamma_2^{\sigma I} = \int_I^{II}\left(\frac{\partial \Gamma_2^{\sigma}}{\partial \mu_1}\right)_{\theta_1} d\mu_1 = \frac{1}{\nu_2 e}\int_I^{II}\left(\frac{\partial \sigma_d}{\partial \theta_1}\right)_{\mu_1} d\mu_1 \tag{15}$$

where the path of integration is at constant θ_1. Equation 15 enables us to obtain absolute values of Γ_2^{σ} and hence Γ_1^{σ} if $\Gamma_3^{\sigma} = 0$ and $\Gamma_2^{\sigma} = 0$ when $\sigma_d = 0$.

d) SEVERAL POTENTIAL DETERMINING SPECIES [14,18b]

Suppose that there are two potential determining species i and j and that we can measure Γ_i and Γ_j individually from potentiometric titration studies. If there is no specific adsorption of other species then graphs of σ_d vs $\ln a_i$ at different levels of supporting electrolyte will have a common intersection point when $\sigma_d = 0$ if a_j is varied in such a way that either $\nu_i d\ln a_j = \nu_j d\ln a_i$ or a_j is held constant. In the former case σ_d vs $kT \ln a_i - \nu_i e \psi_d$ where $\sigma_d = - \nu_i \Gamma_i^{\sigma} - \nu_j e \Gamma_j^{\sigma}$ should show congruence behaviour and $kT \ln a_i$ vs $\nu_i e \psi_d$ should show Nernstein behaviour [18b]. However, neither a congruence plot nor a Nernst plot should result if a_i is varied at constant a_j. Thus, if there is a hidden potential determining species in the system whose bulk activity is constant it is possible that a common intersection point will be found at the IEP but that congruence will not be found near the IEP.

It follows from the above argument that if there are several species i which may be specifically absorbed we may determine which are and which are not by varying all the a_i in accordance with equation 16 and relaxing this condition for each species in turn [18b]. If when this is done the Nernst plot and the congruence plot are maintained the species concerned is not specifically adsorbed. However, to apply this procedure we must have a counterion and a coion species which are known not to be specifically adsorbed.

The adsorption of uncharged species presents no new problems and we may use electrokinetic studies to determine changes in the Γ_i for such species by applying the cross differentiation procedure already described.

e. INTERACTIONS BETWEEN OVERLAPPING DOUBLE LAYERS [11a→c,12,13]

Consider two charged surfaces whose OSPs are separated by a distance x. When x is varied at equilibrium σ_d and ψ_d for the two surfaces can be expected to change. Consequently to calculate interactions by assuming that either σ_d or ψ_d is constant is strictly incorrect. A more correct calculation can be made if for both surfaces graphs of σ_d vs ψ_d are available which include the values of these quantities that actually occur as x is varied [11b]. When the above theorem applies this information may be obtained empirically from studies of the isolated surfaces concerned.

To show how this can be done consider a variation of x with equilibrium otherwise maintained. Since the electrochemical potentials remain constant it follows that changes in the quantities μ_i are given by

$$\Delta\mu_i = - \nu_i \Delta\psi \qquad (17)$$

where $\Delta\psi$ refers to the change in electrical potential at the Outer Stern Plane. It is apparent from equation 17 that $\Delta\mu_i/\nu_i$ is the same for all specifically adsorbed or potential determining species found on both sides of the OSP.

Consider now an isolated plate and suppose that we vary the bulk solution in such away that $kT \Delta \ln n_i^b/\nu_i$ is the same for all i found on both sides of the OSP. In practice this can only be done without violating bulk solution electroneutrality if there is at least one species, typically an indifferent coion, which does not occur in significant amounts in the inner regions. We now have for the isolated plate

$$\frac{\Delta\mu_i}{\nu_i} = \frac{kT\,\Delta\ln n_i^b}{\nu_i} - e\,\Delta\psi_d \qquad (18)$$

However, if the $\Delta\mu_i$ in equation 18 are the same as those in 17 it follows that

$$\Delta\psi = \Delta\psi_d - \frac{kT}{\nu_i e}\,\Delta\ln n_i^b \qquad (19)$$

which shows how changes in ψ_d on the isolated plate are related to changes in ψ on the interacting plate when the $\Delta\mu_i$ are the same. However, according to our theorem if the μ_i in the two cases are the same then so also are the Γ_i^σ and therefore so also is σ_d. In other words the σ_d which corresponds to $\Delta\psi_d$ for the isolated plate, and which may be obtained from electrokinetic studies, is the same as the σ_d which corresponds to $\Delta\psi$ for the interacting plate. Hence we may obtain the charge vs potential curve required to calculate the interactions.

As an example consider two negative silver iodide surfaces with KNO_3 as the supporting electrolyte. As we have already noted some specific adsorption of K^+ appears to occur at higher ionic strengths and surface charges. For this system the appropriate experiment on the isolated surfaces is to vary the solution composition keeping the ion product $[K^+]\,[I^-]$ constant and to determine the zeta potential. This variation is possible because the indifferent nitrate ion concentration can be varied to preserve electrical neutrality.

4. Upper and Lower Bounds on Interactions between Identical Double Layers [17,18a]

It is easy to demonstrate that the constant charge boundary condition is an upper bound on double layer interactions at electrochemical equilibrium [11b]. For identical plates it might also be thought that the constant potential condition is a lower bound and that interactions should fall between these limits [25]. However, this need not be so. To show why, the simplest case to consider is a surface with but a single potential determining species at which there is no specific adsorption of supporting electrolyte. For this case it can be shown that

if $\left(\frac{\partial\mu}{\partial\Gamma}\right) > 0$ interactions fall between constant charge and constant potential

if $\left(\frac{\partial\mu}{\partial\Gamma}\right) = 0$ interactions occur at constant potential

if $\left(\frac{\partial\mu}{\partial\Gamma}\right) < 0$ interactions are below constant potential

Since Γ always decreases with decreasing x an increase in μ with decreasing x must be accompanied by a decrease in ψ. At first sight it appears that a negative $(\partial\mu/\partial\Gamma)$ violates the conditions for thermodynamic stability. However, all that can be deduced from the necessary conditions for interfacial stability is that

$$\left(\frac{\partial\mu}{\partial\Gamma}\right) + \nu e\left(\frac{\partial\psi}{\partial\Gamma}\right) \geq 0 \tag{20}$$

Since $\nu e\left(\frac{\partial\psi}{\partial\Gamma}\right)$ must be positive this does not rule out the possibility that $\left(\frac{\partial\mu}{\partial\Gamma}\right)$ may be negative.

Whether or not this state of affairs occurs in practice can be checked by plotting ψ_d vs $\log_{10} n^b$ at a constant concentration of supporting electrolyte. A sub-Nernstian slope denotes interactions between constant charge and constant potential. A Nernstian slope denotes interactions at constant potential. A super-Nernstian slope denotes interactions at less than constant potential.

It is straightforward to show that $\nu e\,(\partial\psi_d/\partial\Gamma)$ decreases on addition of electrolyte at constant Γ. Hence if $(\partial\mu/\partial\Gamma)$ is negative equation 20 will eventually be breached and phase separation into two coexistent surface phases should occur [17,18a].

The more realistic situation where there are several potential determining species or where supporting electrolyte is specifically adsorbed can be dealt with using similar methods provided that there is at least one species which is not specifically adsorbed. However, when there is only one such species it will not be possible to vary the concentration of supporting electrolyte keeping all Γ_i^σ constant.

The type of system which might show interactions at less than constant potential is a surface at which an ionic surfactant forms hemi-micelles or other adsorbed aggregates at low ionic strengths.

5. Bulk Solution Non-Ideality [18,24]

One of the drawbacks of the Poisson-Boltzmann equation is that ideal behaviour of the bulk solution is assumed. This assumption is not inherent in the above theorem. Nor is it required to obtain zeta potentials when electrophoretic mobilities are given by the Helmholtz-Smoluchowski equation [21]. The main difficulty of allowing for solution non-ideality is that the relationship between charge density and electrical potential is no longer known. This relationship is a characteristic property of the bulk electrolyte and must either be calculated from a theory which allows for non-ideality or determined experimentally. The experimental determination is possible in principle from a combination of electrokinetic and potentiometric titration studies on the same system. The procedure may be applied even when one species of the supporting electrolyte is specifically adsorbed [24]. However, it is necessary to assume that reliable single ionic activities can be obtained from cells with liquid junctions.

6. The Stern-Grahame Model

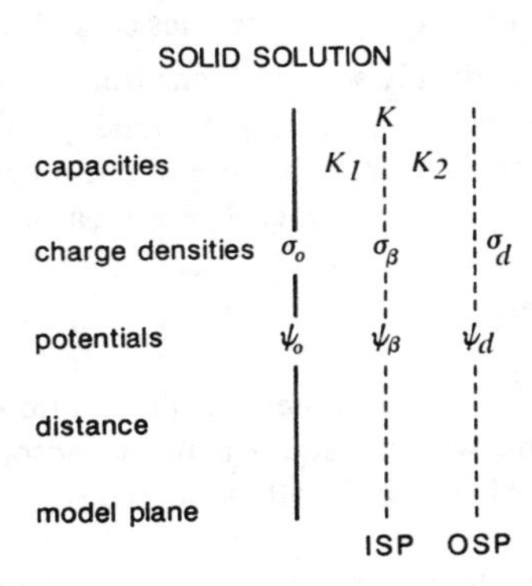

Figure 7. Schematic representation of the Stern--Grahame double layer model.

The conventional approach used to interpret double layer behaviour is the Stern-Grahame model [10,13,26]. The key quantities that form the basis of this model are shown in Figure 7. The inner part of the double layer with integral capacity K is subdivided into two regions having integral capacities K_1 and K_2 by the Inner Stern Plane (I.S.P.) which is taken to be the adsorption plane of specifically adsorbed ions from the supporting electrolyte. ψ_o, ψ_β and ψ_d respectively refer to electrical potentials relative to the bulk solution at the wall the ISP and the OSP. σ_o, σ_β and σ_d respectively refer to the corresponding charges. The nine quantities referred to above are related by the following equations

$$\sigma_d^2 = \frac{\varepsilon kT}{2\pi} \sum_i n_i^b \left(\exp - \frac{v_i e \psi_d}{kT} - 1 \right) \qquad (20)$$

$$\frac{1}{K} = \frac{1}{K_1} + \frac{1}{K_2} \qquad (21)$$

$$\psi_o - \psi_\beta = \sigma_o / K_1$$

$$\psi_\beta - \psi_d = \sigma_\beta / K_2 \qquad (22a,b)$$

$$\sigma_o + \sigma_\beta + \sigma_d = 0 \qquad (23)$$

There are also expressions for the chemical potentials of potential determining and specifically adsorbed species. These expressions are assumed to take the form

$$\mu_i^b = \mu_i^\sigma = f(T,p,\Gamma_i^\sigma) + \nu_i e \psi \tag{24}$$

where ψ refers to the electrical potential seen by the species concerned. Thus $\psi = \psi_o$ for potential determining ions, and $\psi = \psi_\beta$ for specifically adsorbed ions from the supporting electrolyte.

Equations 20-24 provide seven equations relating the nine quantities referred to above. It is clear from them that applying the model even to the case of one potential determining species and one specifically adsorbed species is far from simple. There are also other drawbacks.

1) K_1 and K_2 may depend on σ_o and σ_β.

2) It is not easily extended to more complex systems without introducing further complications. For example it can be argued that we may require a distinct adsorption plane and an associated capacitance for every species which is speficially adsorbed.

3) There is too much emphasis on electrostatic concepts. This tends to disguise the underlying thermodynamic nature of the equilibria involved. The quantities ψ_o, ψ_β, K_1 and K_2 are inaccessible to direct experiment.

4) The model is not required to calculate double layer interactions.

Clearly the approach described above provides a better basis for interpreting experimental data.

In defence of the Stern-Grahame model it can be argued that when all ionic double layer parameters cited above are known the model has some predictive capability. The same can also be said for the approach described above when the dependence of μ_1^σ and μ_2^σ on Γ_1^σ and Γ_2^σ are known. A further defence of the Stern-Grahame model is that some of the expressions for chemical potentials which naturally stem from it agree well with experimental observations. However, even this evidence turns out to be less strong on close examination than it appears to be at first sight. To illustrate why this is the case the simple example of a surface such as AgI is considered in the situation that specific adsorption of supporting electrolytes is absent [11d,13]. For this kind of system it is usually argued that ψ_o is given by the Nernst equation

$$e\psi_o = f(T,p) - kT \ln n_\Gamma^b \tag{25}$$

Hence

$$e(\psi_o - \psi_d) = f(T,p) - [kT \ln n_\Gamma^b + e\psi_d] \tag{26}$$

However, according to the Stern Grahame model [10,26]

$$e(\psi_o - \psi_d) = \frac{e^2\Gamma_\Gamma}{K} \tag{27}$$

where K is the integral capacitance of the inner regions. It follows from equations 5, 26 and 27 that a graph of $[kT \ln n^b_{I^-} + e\psi_d]$ vs Γ_{I^-} should be linear and Figure 5 shows this is indeed the case over a substantial range of conditions. In this case at least therefore the Stern-Grahame model appears to be vindicated.

We may however argue alternatively as follows [11d,13]. By application of the Gibbs adsorption equation we know that when $\Gamma_{I^-} = 0$ the interfacial tension γ of the AgI solution interface has a maximum analogous to the electrocapillary maximum observed for mercury. Let γ_o denote this maximum value and let $\pi = \gamma_o - \gamma$. When Γ_{I^-} is small it is reasonable to write

$$\pi = B(\Gamma_{I^-})^2 \tag{28}$$

where B is independent of Γ_{I^-} but depends on the concentration of supporting electrolyte. From the Gibbs absorption equation we also have

$$d\pi = \Gamma_{I^-} kT\, d\ln n^b_{I^-} = \Gamma_{I^-} [d\mu_{I^-} + ed\psi_d] \tag{29}$$

By equating $d\pi$ as given by equation 28 with $d\pi$ as given by equation 29 we find that

$$\mu_{I^-} - e\psi_d = f(T,p) + 2B\Gamma_{I^-} \tag{30}$$

Since at constant T and p μ_{I^-} depends only on Γ_{I^-} it follows that $2B\Gamma_{I^-} + e\psi_d$ does not depend on the concentration of supporting electrolyte. Now when Γ_{I^-} is small

$$\psi_d = -\frac{4\pi e\Gamma_{I^-}}{\kappa\varepsilon} \tag{31}$$

where κ denotes the Debye screening length and ε denotes dielectric constant. It follows that

$$B = B_o(T,p) + \frac{2\pi e^2}{K\varepsilon} \tag{32}$$

and that

$$\mu_{I^-} = f(T,p) + 2B_o\Gamma_{I^-} \tag{33}$$

so that $(kT \ln n^b_{I^-} + e\psi_d)$ vs Γ_{I^-} should be linear. Thus on the basis of thermodynamic arguments the same result is obtained as predicted by the Stern-Grahame model. In view of this all we can conclude is that equations 26 and 27 conform to expectations based on more general thermodynamic grounds. The agreement of such expressions with experiment is of course to be expected. Consequently such agreement cannot be taken as supporting evidence for the assumptions of the model [11d,13].

7. Conclusions

The approach described above shows clearly that thermodynamic methods can be used to advantage in the interpretation of experimental studies of electrical double layers formed at the solid solution interface. The approach compares favourable with more traditional methods based on the Stern-Grahame model. Also much of the predictive capability of the Stern-Grahame model can be obtained on the basis of more general thermodynamic considerations.

These conclusions should not be taken to imply that modelling is redundant in this area. There remains a clear need to understand why isotherms are obtained by a applying the above thermodynamic methods behave as they do. There is also a clear need to relate thermodynamic data with structural information obtained from spectroscopic studies and to obtain a predictive capability of equilibrium and kinetic properties. The role of modelling in achieving these aims is crucial. The existence of a sound thermodynamic base is vital for such models to be developed efficiently.

References

1. S.L. Carnie and G.M. Torrie, *Advances in Chemical Physics*, 1984, 56, 141.

2. R.D. Groot, Proceedings of this meeting.

3. D. Henderson and L. Blum, *J. Chem. Phys*, 1978, 69, 5441.

4. S.L. Carnie, D.Y.C. Chen, D.J. Mitchell and B.W. Ninham, *J. Chem. Phys*, 1982, 77, 5150.

5. D. Henderson, L. Blum, and W.R. Smith, *Chem. Phys. Lett*, 1979, 63, 381.

6. G.M. Torrie and J.P. Valleau, *J. Chem. Phys*, 1980, 73, 5807.

7. G.M. Torrie and J.P. Valleau, *J. Chem. Phys*, 1982, 86, 3251.

8. L. Gulbrand, B. Jonsson, H. Wennerstrom and P. Linse, *J. Chem. Phys*, 1984, 80, 2221.

9. R. Kjellander and S. Marcelja, *Chem. Phys. Letters*, 1986, 127, 402.

10. D.C. Grahame, *Chem. Rev*, 1947, 41, 441.

11. D.G. Hall, *J. Chem. Soc., Faraday Trans. II*, a) 1975, 71, 937, b) 1977, 73, 101, c) 1978, 74, 1757, d) 1980, 76, 1254.

12. D.G. Hall and M.J. Sculley, *J. Chem. Soc., Faraday Trans II*, 1977, 73, 869.

13. D.G. Hall and H.M. Rendall and A.L. Smith, *Croatica Chim. Acta*, 1980, 53, 147.

14. D.G. Hall and H.M. Rendall, *J. Chem. Soc., Faraday I*, 1980, 76, 2575.

15. D.B. Hough and H.M. Rendall, C6 of Adsorption from Solution at the Solid-Liquid

Interface Ed. G.D. Parfitt and C.H. Rochester Academic Press, London 1983.

16. R.J. Hunter Foundations of Colloid Science Vol. 2, Clarendon Press, Oxford 1989.

17. D.G. Hall, *J. Colloid Interface Sci*, 1985, 108, 411.

18. D.G. Hall, *J. Chem. Soc., Faraday Trans. I* a) 1988, 84, 2215, b) 1988, 84, 2227.

19. J.Th.G Overbeek, *Prog. Biophys., Biophys. Chem.*, 1956, 6, 57.

20. B.H. Bisterbosch and J. Lyklema, *Advances in Colloid and Interface Sci*, 1978, 9, 147.

21. P.C. Hiemenz, Principles of Colloid and Surface Chemistry Ch. 11 Marcel Dekker Inc., New York, 1977.

22. R. Parsons, 2nd Int. Congress, Surface active substances, London, 1956, proc. vol. 3, Butterworths, London, 1957, p. 38.

23. J. Lyklema, *J. Electroanal. Chem*, 1972, 37, 53.

24. D.G. Hall, to be published.

25. D.Y.C. Chan and D.J. Mitchell, *J. Colloid and Interface Sci*, 1983, 95, 193.

26. A.L. Smith, *J. Colloid and Interface Sci*, 1976, 55, 525.

WETTING PHENOMENA

A.M. CAZABAT
Physique de la matière condensée
Collège de France
15, place Marcellin-Berthelot
75231 Paris cedex 05 -France-
Tel.: (33) 1-44271080

ABSTRACT. The spreading rate of a liquid on a solid surface usually depends on many uncontrolled parameters. Present theories are relevant for simple model systems, i.e. pure liquids spreading on smooth surfaces and are in very good agreement with experimental findings. An important result of these analyses is that in most cases the macroscopic, hydrodynamic behaviour of the liquid can be treated independently of the phenomena occurring at microscopic scales.

This paper presents a review of the main results obtained in these simple models. Then we discuss briefly how they can be adapted to describe real situations, where surface roughness and adsorption or contamination processes play a role.

1. Introduction

The spreading of liquids on solid surfaces plays a major role in many phenomena like : adhesion, coating, painting, printing,imbibition of porous media... For practical applications, the understanding (and, possibly, control) of the basic processes is needed.

When a liquid front advances on a solid surface, the description must be performed at different length scales :

- the macroscopic part of the liquid corresponds to thicknesses larger than, say, one micrometer. It has the properties of the bulk liquid and its motion is described by usual hydrodynamics.
- at smaller thicknesses, the film no longer has the bulk properties . It is the microscopic part where, again, two domains must be distinguished.
- the mesoscopic part (1) corresponds to thicknesses much larger than the molecular size. Here, the film can still be described by the hydrodynamics of the continuous media, but its intensive properties are, now, thickness-dependent.
- finally, the molecular part cannot be described by a continuous theory.

The aim of theoretical studies was first to account for the dynamics of the macroscopic part (2-3). Although the experimental behaviour was satisfactorily described, some intriguing paradoxes remained unsolved. A further step was to investigate the influence of the microscopic part on the macroscopic dynamics. Some authors put emphasis on the mesoscopic part (4-6); other ones, on the molecular part (7-8).

The next step was to study the microscopic dynamics itself, which has been done in the

D. M. Bloor and E. Wyn-Jones (eds.),
The Structure, Dynamics and Equilibrium Properties of Colloidal Systems, 831–855.

mesoscopic range (5-6).

What is left now is the dynamics of the molecular part. It is by no means an academic problem : as a matter of fact, the first molecular layer often has an abnormal behaviour. It is able to advance on the solid much faster than the macroscopic front (9-10). This is of importance in coating and adhesion, for example.

In the following, we shall first introduce the parameters relevant for the macroscopic (2.1) and microscopic(2.2) description of the liquid front. Next(3), the main theoretical macroscopic analyses will be presented and compared with experimental findings in partial (3.1) and complete (3.2) wetting,for pure liquids spreading on smooth, clean surfaces.

In many practical cases, these requirements are not met. The liquid is not really pure, the solid surface always has a certain degree of roughness, and time dependent processes as adsorption and contamination interfere with the spreading. These effects are discussed in section 4, mostly in a semi-empirical way, as no theory is presently available.

The next sections are devoted to the microscopic dynamics. Thin films, i.e. mesoscopic range, are treated in section 5, ultrathin films, i.e. molecular range find their place in section 6.

2. Relevant parameters for the analysis of spreading

2.1 THE MACROSCOPIC SCALE

At the macroscopic scale, relevant parameters for describing a gas/liquid/solid system are the interfacial tensions:

solid-liquid	γ_{SL}
solid-gas	γ_{SG}
liquid-gas	$\gamma_{LG} \equiv \gamma$

and, in dynamical situations, the viscosity η of the liquid (viscous losses in gas are usually negligible).

If a three phase contact line exists, the liquid wedge is characterized by the contact angle θ and, in non equilibrium case, the mean velocity U.

2.1.1 *Partial wetting, complete wetting.* Partial wetting corresponds to equilibrium situations with non zero contact angle θ_e. A balance of the horizontal components of capillary forces at the contact line is expressed by the Young equation :

$$\gamma_{SG} - \gamma\cos\theta_e - \gamma_{SL} = 0$$

which can be written as:

$$\cos\theta_e = 1 + \frac{S}{\gamma}$$

$$S = \gamma_{SG} - \gamma - \gamma_{SL} < 0$$

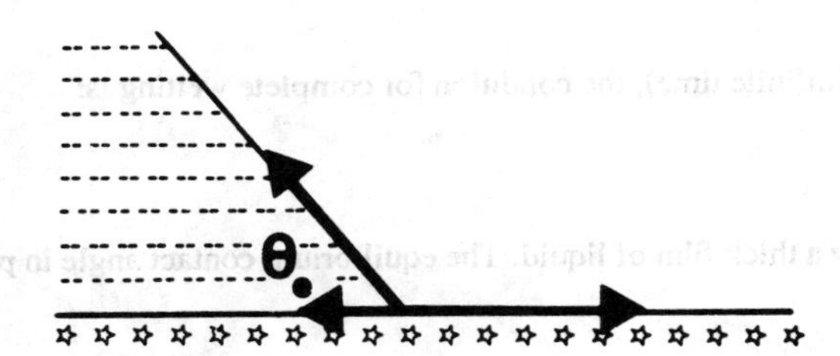

Fig.1 Balance of forces at the contact line. θ_e is the equilibrium contact angle.

In complete wetting, the equilibrium situation corresponds to the presence on the solid of a macroscopic liquid film :

$$\gamma_{SG} = \gamma + \gamma_{SL} \quad \text{so that} \quad S = 0$$

This equilibrium situation is achieved at infinite time. At finite time, the instantaneous contact angle $\theta(t)$ and the mean velocity $U(t)$ are used to describe the behaviour of the liquid front.

A relevant parameter is no longer the equilibrium spreading coefficient S (or final spreading coefficient (11-12)) but , at least at short times, the initial spreading coefficient S_0 (12-13):

$$S_0 = \gamma_{SO} - \gamma - \gamma_{SL}$$

where γ_{SO} is the solid vacuum interfacial tension. Using S_0 in the equations means that no transfer of molecules through the gaseous phase takes place during the time of the experiment. S_0 is strictly positive for complete wetting.

The unbalanced capillary driving force for spreading is then (per unit length):

$$F_c = \gamma_{SO} - \gamma\cos\theta - \gamma_{SL}$$
$$F_c = S_o + \gamma\,(1 - \cos\theta) \qquad [1]$$

2.1.2 *Wetting Criteria*. At short times, the condition for complete wetting is:

$$S_o > 0$$

In partial wetting, the contact angle will be given by:

$$\cos\theta_o = 1 + \frac{S_o}{\gamma} \qquad S_o < 0$$

At thermodynamic equilibrium (infinite time), the condition for complete wetting is:

$$S = 0$$

It means that the solid is covered by a thick film of liquid. The equilibrium contact angle in partial wetting is given by:

$$\cos\theta_e = 1 + \frac{S}{\gamma} \qquad S < 0$$

2.2 THIN LIQUID FILMS

2.2.1 *Thin Liquid Films* do not possess the properties of the bulk. This is conveniently accounted for by introducing the thickness dependent disjoining pressure (14-15):

$$\Pi(Z) = \frac{-1}{V_o}\left[\mu_F(Z) - \mu_B\right] \qquad [2]$$

Here, Z is the film thickness, V_o the molecular volume and μ_F and μ_B the chemical potentials of a molecule in the film and in the bulk, respectively.

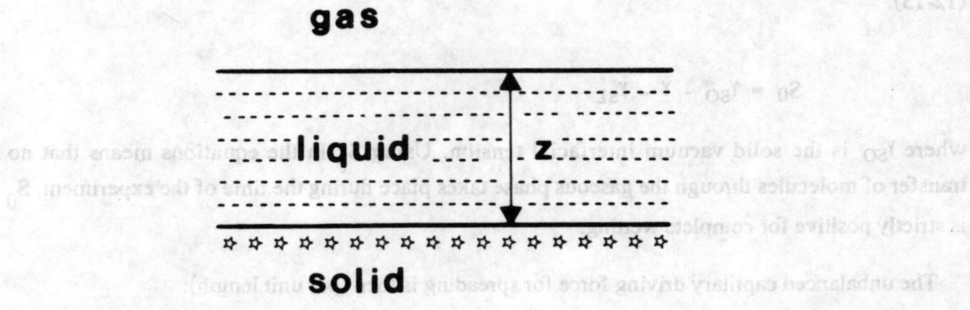

Fig.2 The film geometry.

Positive values of Π ($\mu_F < \mu_B$) mean repulsive interactions between solid/liquid and liquid/gas interfaces. They tend to thicken the film ("disjoining" effect,fig.2).

In the mesoscopic part, Π has a smooth behaviour, usually dominated by Van der Waals (and possibly coulombic) interactions(fig.3).

For pure Van der Waals forces:([3])

$$\Pi(Z) = \frac{A}{6\pi Z^3} \qquad 20\ \text{Å} \leq Z \leq 200\ \text{Å} \qquad [3a]$$

$$\Pi(Z) = \frac{B}{Z^4} \qquad Z > 200\ \text{Å} \qquad [3b]$$

Here, A and B are effective Hamaker constants which can be calculated (in principle !) if the dielectric constants of the media are known in a wide frequency range (14,16-17).

In the molecular part, no general behaviour can be described. Often, an oscillatory behaviour of $\Pi(Z)$ at short distances reflects some layering of the film (17-19) induced by the solid walls.

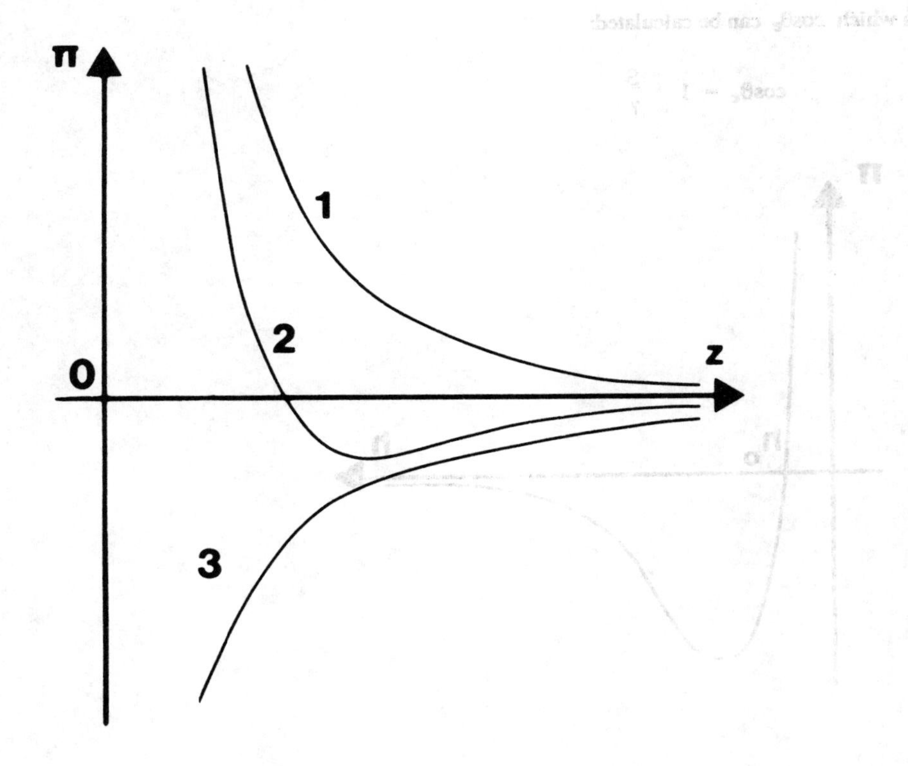

Fig.3 Schematic behaviour of the disjoining pressure $\Pi(Z)$

2.2.2 ***If the Disjoining Pressure* is known, the wetting properties of the system can be deduced.** **For example:**

$$S_o = \int_o^\infty \Pi(Z)dZ$$

In partial wetting, at short times

$$\cos\theta_o = 1 + \frac{S_o}{\gamma}$$

At long times, S differs from S_o if a (necessary thin) film has been adsorbed on the solid. Let h_0 be the thickness of this film (fig.4):

$$S = \int_{h_0}^\infty \Pi(Z)dZ$$

from which $\cos\theta_e$ can be calculated:

$$\cos\theta_e = 1 + \frac{S}{\gamma}$$

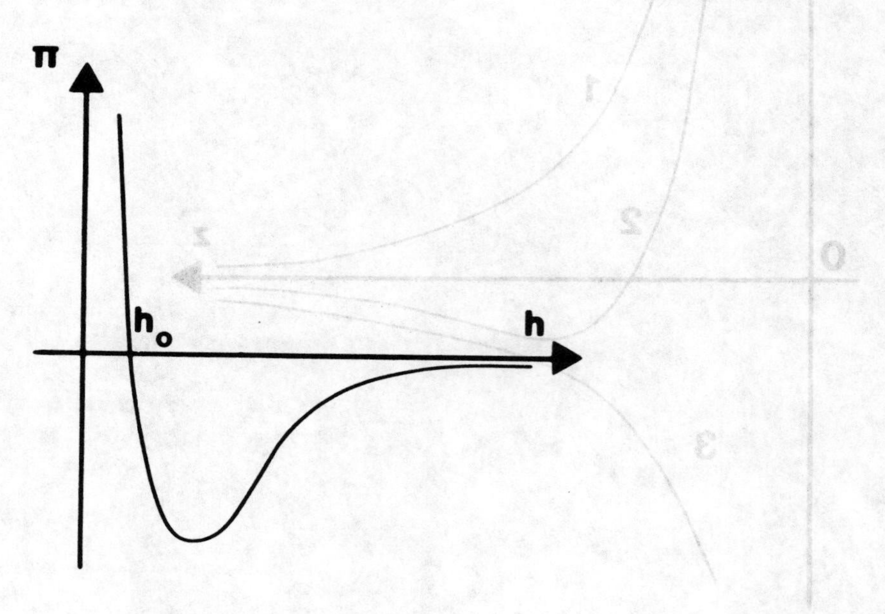

Fig.4 **Non wetting situation with a thin film of thickness h_0.**

3. Spreading dynamics at the Macroscopic Scale

In the following, we shall restrict ourselves to short times, i.e., when the transport of molecules through the gaseous phase is negligible.

A convenient situation for discussing the dynamics is the stationary case of a liquid front advancing with constant mean velocity U on the solid.

For this stationary case, the driving force F_C just balances the friction force. Various theories have been proposed for describing the problem both in wetting and non wetting situations. They differ mostly by the expression of the friction term.

3.1 PARTIAL WETTING

In forced flow experiments, the liquid is forced to advance on the solid with a constant velocity U. The contact angle θ differs from the static one θ_0. The unbalanced force can be written:

$$F_C = \gamma\,(\cos\theta_0 - \cos\theta) \qquad \theta > \theta_0$$

The friction force is the sum of the viscous terms in the macroscopic and mesoscopic part of the front (the hydrodynamical terms) and of adsorption-desorption effects at the molecular scale (3)(6-9).

In partial wetting, the mesoscopic part is totally negligible (6). If θ is large enough, the viscous terms will be small. Several authors (8) neglect them and assume that adsorption-desorption processes give the main contribution to friction. The viscosity of the liquid appears only through the diffusion process of the molecules toward the contact line (7). The result can be written as (20):

$$U = \frac{K\lambda}{\eta}\,\sinh\left[\frac{\gamma}{2nkT}\,(\cos\theta_0 - \cos\theta)\right] \qquad [5]$$

where n is the density of adsorption sites on the surface and λ the distance between them ($\lambda \sim n^{-1/2}$). K depends on the energy of the activation process.

Experimental studies agree well with this formula (7,8), (20-22) for θ-θ_0 larger than, say, 20°. Below this value, the poor accuracy of forced flow experiments does not allow to conclude (fig.5).

At small θ and θ_0 values, the viscous friction becomes significant. The corresponding term must be included in the equation (23). If it overcomes the adsorption-desorption effect, we obtain a completely hydrodynamic description of the problem (21):

$$U = C\left(\frac{\gamma}{\eta}\right)(\cos\theta_0 - \cos\theta)^{3/2} \qquad [6]$$

Where C is a numerical constant.

Amusingly enough, the comparison with experiments is again very satisfactory, over the whole range of angles ! This means that the accuracy and reproducibility of the forced flow experiment does not permit discrimination between theories. Moreover, both equations [5] and [6] give

unrealistic results $(\cos\theta > 1)$ at large U; they have a restricted validity.

Let us note that equation [6] can be written introducing the capillary number

$$Ca = \frac{\eta U}{\gamma}$$

$$Ca = C\,(\cos\theta_0 - \cos\theta)^{3/2} \qquad [7]$$

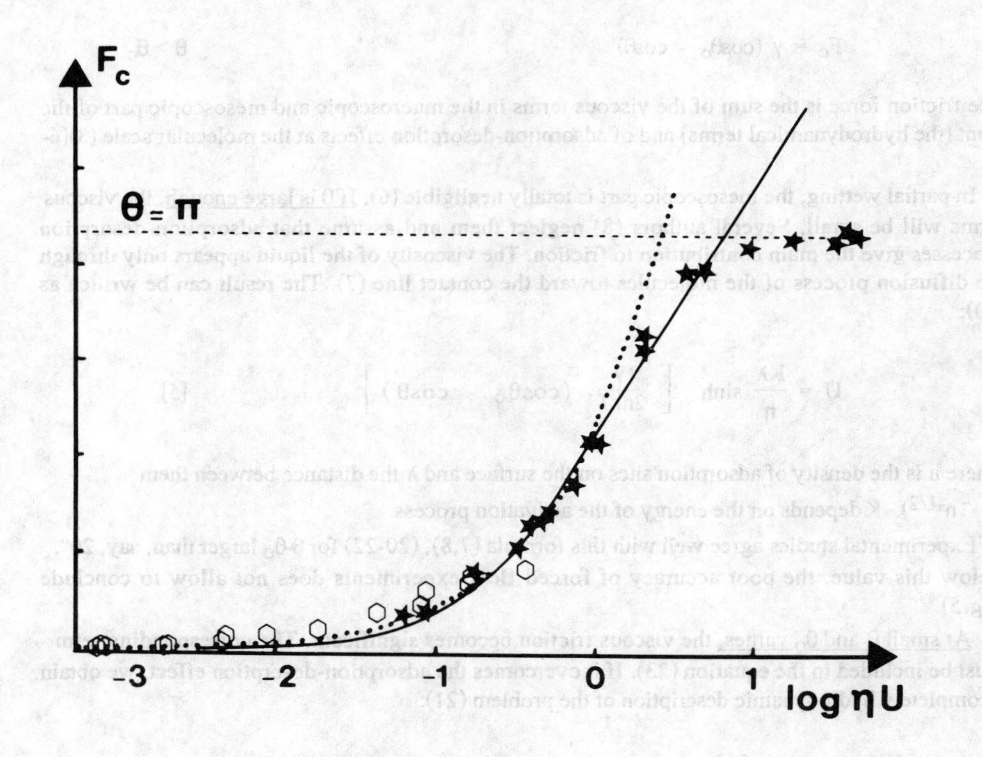

Fig.5 Driving force F_C as a function of velocity U . Stars and hexagons: Experiments (Hoffmann 75). Full line (-):adsorption desorption theory,equ.[5].Points (·) hydrodynamic theory, equ.[6].

Such a reduced form cannot be obtained with [5], except in the low Ca range :

$$Ca \approx \frac{K\lambda}{2nkT}(\cos\theta_o - \cos\theta) \qquad \theta \rightarrow \theta_o \qquad [8]$$

Complete theories, including both viscous and adsorption-desorption effects, have been recently proposed (23). In this case, a new, implicit equation is obtained:

$$U = \frac{K\lambda}{\eta} \sinh \left[\frac{\gamma(\cos\theta o - \cos\theta) + \beta C_a^{2/3}}{2nkT} \right]$$

3.2 COMPLETE WETTING

In complete wetting, the unbalanced driving force can be written [1]:

$$F_c = S_o + \gamma (1 - \cos\theta)$$

It is not just the limit of [4] for θ_0=0 because of the contribution of the initial spreading coefficient S_o, which is strictly positive.

In fact, many authors have ignored this contribution, although it is obviously the dominant term in F_c (2-3), (7-8), (20-23). The corresponding formulae are obtained by putting θ_0=0 in the preceding equations [5-8]. The agreement with experiments is very good.

The reason of this surprising situation has been explained by de Gennes (6): the S_0 term controls the microscopic part of the liquid front, the macroscopic part being actually driven by $\gamma(1-\cos\theta)$. Thus we can forget about S_o when macroscopic dynamics is studied.

At large θ, where the microscopic part is just of molecular size, the conclusion is not so clear. But here, equation [5] becomes:

$$U \approx \frac{K\lambda}{\gamma} \exp\left(\frac{F_c}{2nkT}\right)$$

Inserting S_o in F_c or not just changes the activation energy, i.e. the constant K.

Let us discuss further the opposite case: $\theta \rightarrow 0$ Here, equation [7] becomes (2-3),(6):

$$Ca = D\,\theta^3 \qquad [9]$$

with $D \approx 10^{-2}$ .Equation [8] becomes (8,20):

$$Ca = \left(\frac{K\lambda}{4nkT}\right)\theta^2 \qquad [10]$$

Small values of θ are conveniently studied by investigating spreading drops. It is a non stationary problem where the radius of the wetted spot R(t) for small drops of volume Ω obeys:

$$R(t) \cong \Omega^{3/10}\left(\frac{\gamma t}{\eta}\right)^{1/10} \qquad [11]$$

in the hydrodynamic theory (Eq.[9]) and

$$R(t) \sim t^{1/7}$$

in the activation theory(Eq.[10]).
Here, precise experiments are available which unambiguously lead to (fig.6)

$$R(t) \sim t^{1/10}$$

Thus the hydrodynamic theory (6) is the correct one at low angles.

Note: We have ignored the gravity. It is not possible for large drops($R \geq$ 2-3 mm). In this case, the spreading is accelerated (25-27) and different power laws are observed:

$$R(t) \cong \Omega^{3/8}\left(\frac{\rho g t}{\eta}\right)^{1/8}$$

Here, g is the gravitational acceleration and ρ is the liquid density.

4. Macroscopic dynamics in non ideal situations

Surface roughness and contamination are always present in practical cases and their effect must be investigated.

Many authors have studied the effect of controlled roughness on the equilibrium contact angle value (28-31). However, dynamic studies are rare (27).

In the same way, work has been done on the spreading of mixtures (32-33), but the influence of contamination has not been systematically analysed.

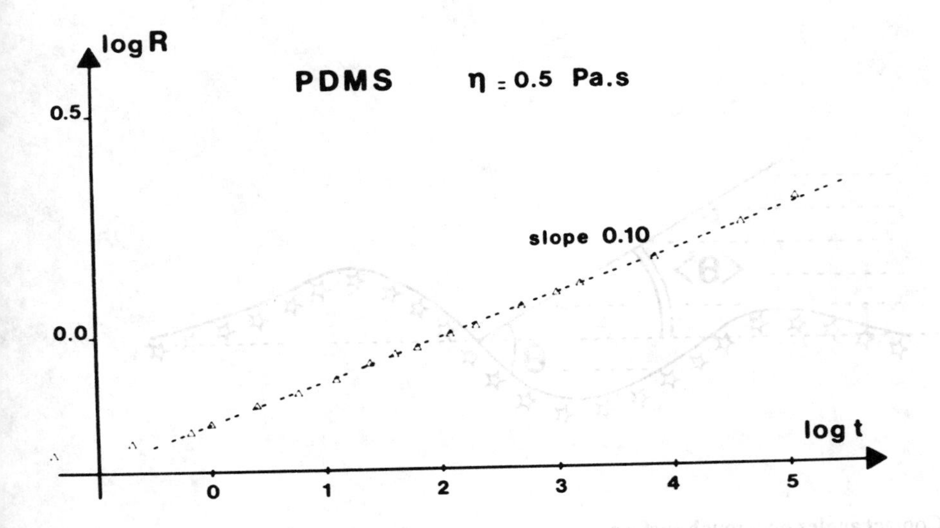

Fig.6 Normal spreading in complete wetting case: Log R versus Log t. (R unit 0.5 mm, t unit mn)

4.1 ROUGHNESS

Evidence of roughness (or chemical heterogeneity) of the solid is given by the occurrence of stick-slip phenomena during the advance of the liquid front (34-38).

These effects play little role in forced flow experiments at large velocity, where the macroscopic part of the driving force

$$F_{c(macro)} = \gamma\ (\cos\theta_o\ -\ \cos\theta)$$

overcomes their influence.

It is no longer the case at low U. Here, the behaviour of the liquid front depends not only on the roughness amplitude, but also on the roughness profile. We shall propose an introductory discussion of the problem in terms of a very simple model. At the current time, no elaborated models are available.

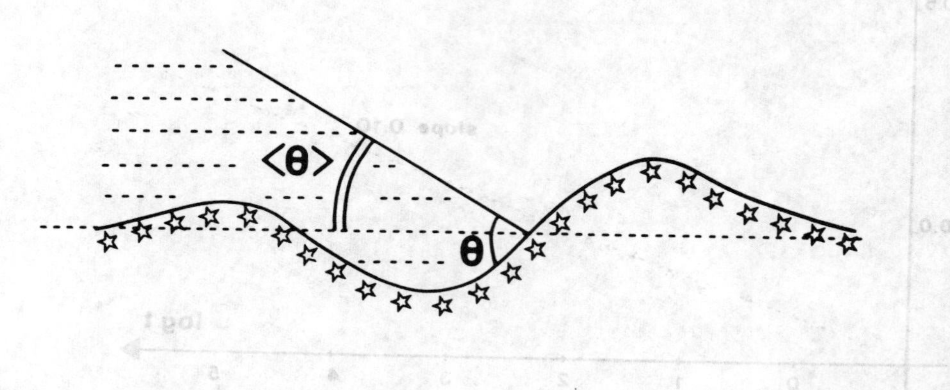

Fig.7 Contact angles on a rough surface.

4.1.1 *"Easy" channels (34).* The average slope of a rough, horizontal surface is zero. To first order, one would expect the averaged effect to be weak (fig.7)

$$\langle F_c \rangle = \gamma\ (\cos\theta_0 - \langle \cos\theta \rangle)$$

$$F_{c(smooth)} = \gamma\ (\cos\theta_0 - \cos \langle \theta \rangle)$$

In fact this analysis does not take into account the role of preferential easy paths on the surface: if the main part of the liquid finds its way through such channels, only a limited range of slopes will be explored by the advancing front: the real $\langle \cos\theta \rangle$ average has no relation with $\cos\langle\theta\rangle$.

What characterizes the roughness is the density of connected easy paths, and obviously their depth and profile.

This approach gives evidence of the similarity between roughness effects in spreading dynamics and the spontaneous imbibition of 3-dimensional porous media:

4.1.2 *Imbibition of Porous Media.* The spontaneous imbibition of a porous medium connected to a reservoir has been treated by Washburn (39). The medium is considered as an ensemble of parallel pipes of radius r (fig.8)

Spontaneous imbibition occurs if $P_1 < P_R$, i.e. if $\theta < 0.5\pi$. The corresponding driving force per unit length of contact line is $\gamma\cos\theta$.

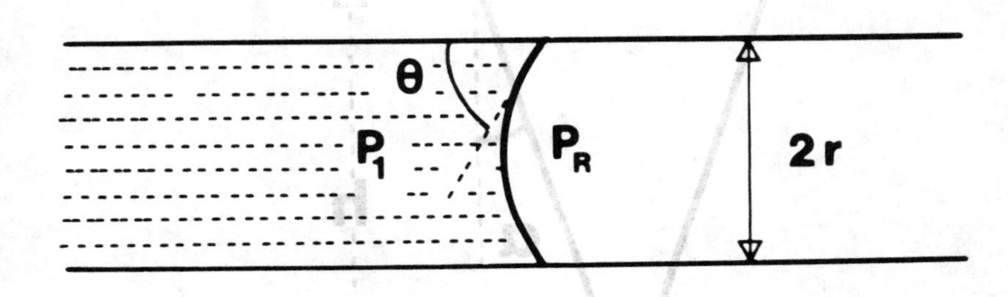

Fig.8 The pipe geometry.

Balancing it again viscous friction (a Poiseuille-flow is assumed in the pipe) leads to:

$$L(t) = \sqrt{\frac{\gamma r\cos\theta}{2\eta}}\ \sqrt{t}$$

The velocities are rather low and θ is taken identical to θ_o.

Models with rectangular pipes [40] or averaged values of dimensions and "tortuosity" can be developed [41]

4.1.3 *Channels on Surfaces.* Let us suppose that channels with triangular cross section exist on the surface (Fig.9). The Laplace pressure is approximately:

$$p = \frac{\gamma}{r}\cos(\theta_o + \alpha)$$

The liquid will fill the channel if:

$$\theta_o + \alpha < \frac{\pi}{2}$$

Here, the geometry is more complex than with pipes, but at long times a Washburn-like equation holds [40,42]. One expects:

$$L(t) \sim \sqrt{\frac{\gamma r \cos(\theta_o + \alpha)}{\eta}} \sqrt{t}$$

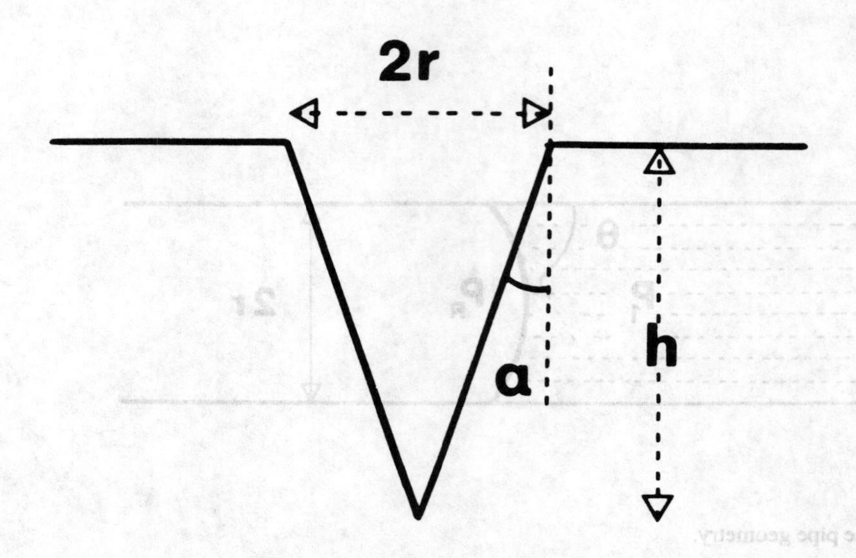

Fig.9 Cross section of a "triangular" channel.

For a network of interconnected channels, a reasonable result will be:

$$L^2(t) = \Phi \frac{\gamma t}{\eta} < r\cos(\theta_o + \alpha) >$$

the average being taken only on positive values of the cosine. Φ is a empirical coefficient which accounts for the fraction of efficiently connected channels.

4.1.4 *The Spreading of Drops.* Spontaneous spreading of drops on random rough surfaces (27) is satisfactorily described by the previous model:

At short times, the central part of the drop acts as a reservoir. A rim develops, the length of which increases like the square root of the time (27,42).

At longer times, the reservoir is empty. If volume conservation is taken into account, the variation of the total radius of the wetted spot is expected to be $R \approx t^{1/4}$. The agreement with experiment is again satisfactory (27).

An important point is that the spreading condition is not $\theta_0 = 0$, but $\alpha + \theta_o < 0.5\,\pi$. Non wetting liquid may spread spontaneously on rough surfaces (42).

4.1.5 *Forced Flow Geometries.* Spontaneous filling of easy channels will be of importance at low velocities. Let us propose the following (crude) model for the surface: The driving force is supposed to be the sum of the "smooth" term:

$$\gamma\,(\cos\theta_0 - \cos\theta)$$

and an extra term accounting for the influence of the connected channels:

$$\varepsilon\,\gamma < \cos\,(\theta_0 + \alpha) >$$

If the connectivity of the easy paths is low, ε is close to zero. The opposite case corresponds to grooves parallel to the direction of motion ($\varepsilon \le 1$).

$$F_c \approx \gamma\,[\,\cos\theta_0 - \cos\theta\; + \varepsilon < \cos\,(\theta_0 + \alpha) >]$$

at large U, the "smooth" term is dominant; at low U ($\theta_0 \neq \theta$), one gets, for ($\theta_0 \neq 0$):

$$F_c \approx \gamma\,[(\theta - \theta_0)\sin\theta_0\; + \varepsilon < \cos\,(\theta_0 + \alpha) >]$$

and, for $\theta_0 = 0$ (complete wetting):

$$F_c \approx \gamma\left[\frac{\theta^2}{2} + \varepsilon\; < \cos\,(\theta_0 + \alpha) > \right]$$

According to the value of the parameter ε, the influence of the roughness is observable or not.

It might be of interest to check this model against experiments on controlled rough surfaces where α and ε could be known in an independent way.

4.2 CONTAMINATION

Contamination results in a change of the interfacial tensions. The viscosity of the liquid may also be affected.

An initial wetting situation may become non wetting after some time, and conversely. However, the more striking effects occur when a spatial gradient of the liquid-gas interfacial tension γ is produced.

Let us consider a liquid film of thickness h, and call x the spatial coordinate (fig.10). The interfacial tension gradient produces a stress which is associated to a surface velocity:

$$V_s = \frac{h}{\eta}\,\frac{d\gamma}{dx}$$

The liquid is driven towards the large γ regions.It is the Marangoni effect (43); a well known manifestation of it are the tears of wine.

Gradients in γ may arise from surface contamination, if the liquid is able to solubilize adsorbed species, or atmospheric contamination (44). Obviously, low velocities U and long times are critical. The spontaneous spreading of drops is especially sensitive.

Usually, an excess of contaminant molecules occurs at the wedge of the liquid. According to

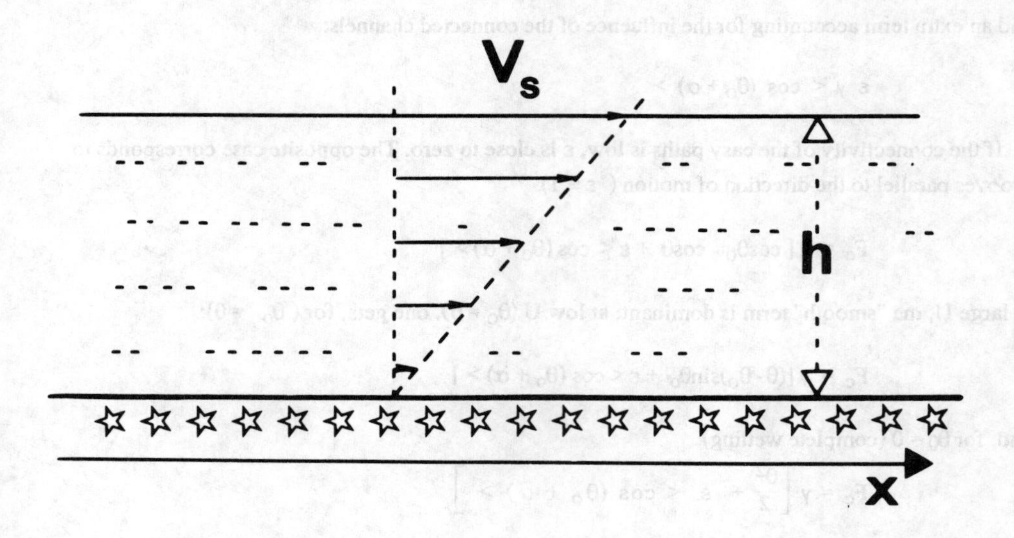

Fig.10 Couette velocity profile in the presence of an interfacial tension gradient.

the sign of the difference $\gamma_{CONT} - \gamma$ the spreading can be accelerated ($\gamma_{CONT} > \gamma$), or stopped ($\gamma_{CONT} < \gamma$). The drop may even retract. These effects are rapidly dominant over spontaneous spreading, which is a slow phenomenon (because of the weakness of the driving force $F_c \approx \theta^2$ at low angles). Examples of spreading under contamination are given in Figure 11.

No theoretical approach is presently available. Spreading in presence of insoluble surfactant layers, which is a much simpler case, has been treated in the stationary case (forced flow). (45).

We must keep in mind that only macroscopic scales have been considered in this part. It is clear that roughness and contamination will also interfere with microscopic spreading. Here the situation is even worse because the laws for "normal" spreading are not precisely known. Comparison between theory and experiment requires a careful and critical examination of the relevance of both calculations and measurements !

5. The Dynamics of Thin Films : Mesoscopic Scale

5.1 THEORETICAL ANALYSIS

Thin films are described by hydrodynamic equations. In the following, we shall assume θ to be small, in order to have a well developed mesoscopic film at the edge of the macroscopic front.

Only the complete wetting case and non volatile liquids are considered.

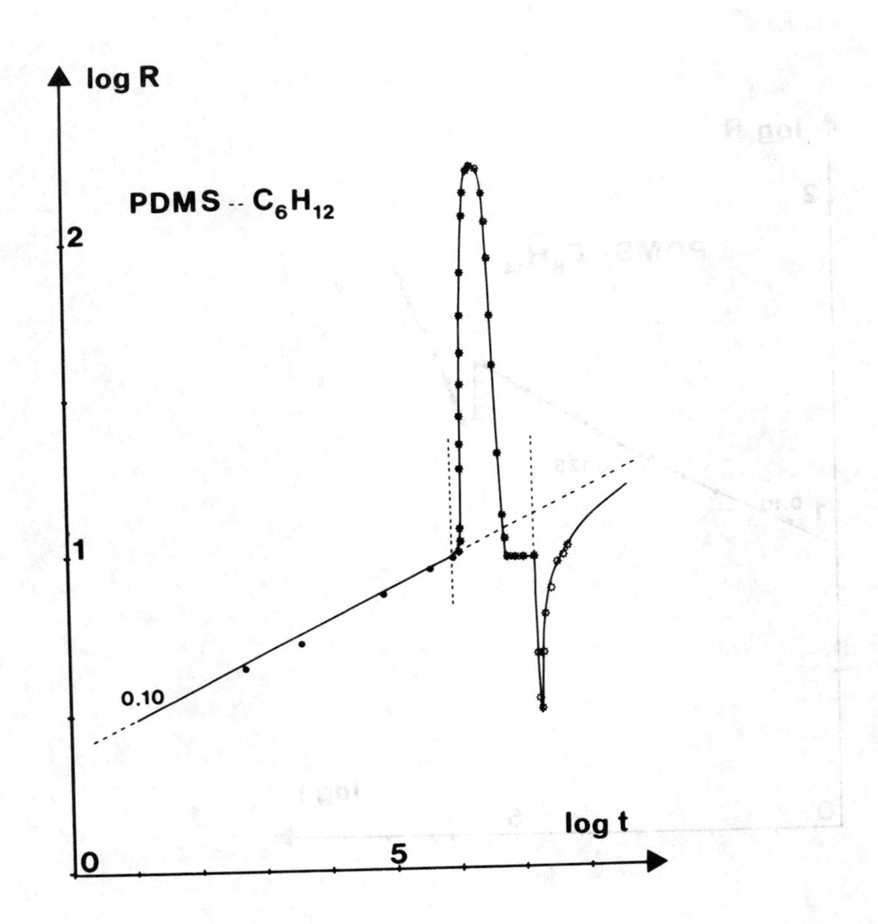

Fig.11a Spreading under contamination by cyclohexane
* Full circles : no contamination.
* Full suns : contamination present.
* Half full suns: contamination present, saturation effects (a).
* Open suns : no contamination (a).

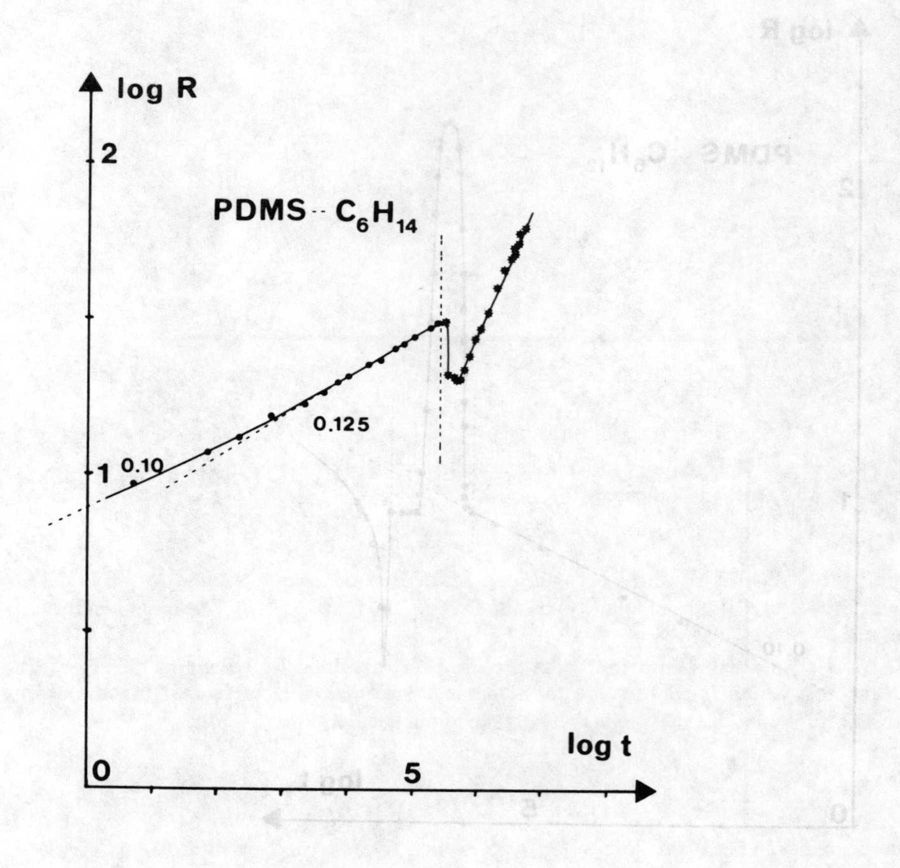

Fig.11b Spreading under contamination by hexane
* Full circles : no contamination.
* Full suns : contamination present.

The velocity field in these conditions is a Poiseuille flow (6) (fig.12):

$$V(Z) = V_s\ (2hZ - Z^2)$$

$$V_{max} = V_s = \frac{3}{2} U$$

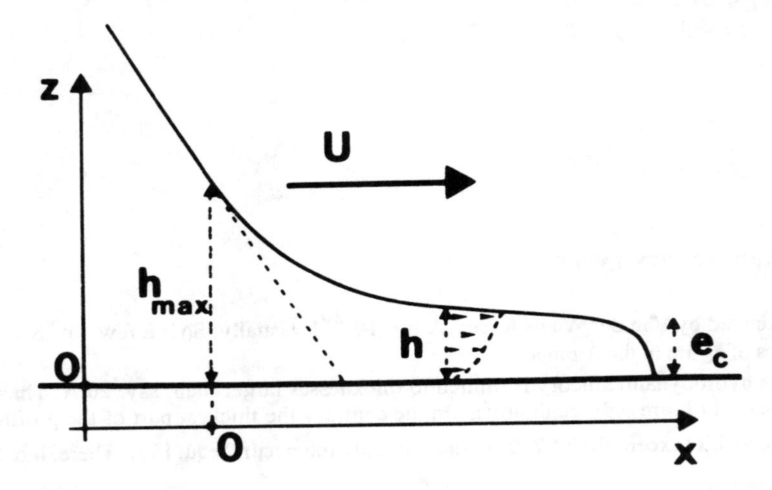

Fig.12 Microscopic (schematic) profile of an advancing liquid front. Complete wetting case: Poiseuille velocity profile is indicated.

The film profile h(x) is obtained by inserting, in the hydrodynamic equations, the disjoining pressure term. Details of the calculations can be found elsewhere (6),(46). The main results are as follows:

* There is a finite minimum thickness for the film, obtained by balancing the spreading coefficient S_o, which is the driving term, against the thickening effect of the disjoining pressure. The minimum thickness e_c is given by the implicit equation:

$$S_o = e_c\ \Pi(e_c) + \int_{e_c}^{\infty} \Pi(h)dh \qquad [12]$$

For non retarded Van der Waals forces ([3a]), an explicit formula can be written (6):

$$e_c = \frac{1}{2} \sqrt{\frac{A}{\pi S_o}}$$

* The film profile scales like 1/x between e_c and a maximum value at the crossover to macroscopic edge

$$h_{max} \approx \frac{1}{\theta} \sqrt{\frac{A}{6\pi\gamma}}$$

* The length of the film is then

$$L \approx \frac{1}{3\eta U} \sqrt{\frac{AS_o}{\pi}} \approx \frac{1}{3\gamma Ca} \sqrt{\frac{AS_o}{\pi}}$$

with

$$Ca = D\ \theta^3$$

5.2 COMPARISON WITH EXPERIMENTS

For liquids dominated by Van der Waals forces, A is $\approx 10^{-20}$ J. Usually, S_0 is a few 10^{-2} $N.m^{-1}$. Calculated values of e_c are in the Å range.
The validity of a hydrodynamic theory is limited to thicknesses larger than, say, 20 Å. Thus the predictions for e_c and L are only qualitative. On the contrary the thickest part of the profile is expected to scale as 1/x (except for h >200 Å where retardation occurs, equ. [3b]. There, h α $1/\sqrt{x}$ (47)).

Experiments (47-48) agree with this prediction. However, the significance of such an agreement is weak: curves are fitted over less than one decade by an hyperbola whose asymptote is unknown...

5.3 DIFFUSIVE FILMS

Let us add a few words about cases where the macroscopic edge is static (U = 0 , capillary rise) or absent (tiny droplets).

Here, no sationary situation is obtained. A hydrodynamic treatment can still be performed. The diffusive film is found to obey a pseudo-diffusive equation: (49)

$$\frac{\partial h}{\partial t} - \frac{\partial}{\partial x}\left[D(h) \frac{\partial h}{\partial x}\right] = 0$$

with

$$D(h) = -\frac{h^3}{3\eta} \frac{d\Pi}{dh}$$

* For Van der Waals fluids: [3a]

$$D(h) = \frac{A}{6\pi\eta h}$$

The profile scales like $1/x^2$ and is truncated at $h = e_C$.

The predictions are not easily checked by experiments : the remnant motion of the meniscus (which reaches its equilibrium position only at infinite time in the capillary rise geometry) is enough for the profile to scale like $1/x$ in its thicker part (50). Then, we are outside the validity range of the theory.

6. The Dynamics of Ultrathin Films : Molecular Scale

At the present time, no theory is available for the description of ultrathin films. By analogy with thin film description, we expect the film to be longer and thinner at large S_o, and diffusion-like behaviour to occur if there is no moving macroscopic front.

We also expect specific features due to the steep variations of $\Pi(z)$ in this range. Moreover, 2-dimensional molecular diffusion might play a role.

6.1 EXPERIMENTAL OBSERVATIONS

In this range of thicknesses ($h \leq 20$ Å) and for dynamic studies, ellipsometry is the most convenient technique.

Beaglehole (50-51) was the first one to obtain ellipsometers with high spatial and thickness resolution, and fast response. The use of acousto-optic modulators at high frequency (50 kHz) is the basic improvement of his setup.

Our ellipsometer belongs to this class, with a thickness resolution of 0.1 Å for a spot size of 200 µm x1mm and acquisition rate of 1 point per millisecond (52). Such a thickness resolution can be reached by using a very smooth substrate, with high contrast index against the liquid, and by reducing the contamination by a flow of dry nitrogen.

The substrates are silicon wafers (optical index ≈ 3.8) whose residual roughness (a few Å) plays no role in the thickness measurements because we measure the film thickness by comparison of data with and without the film. The requirement is the stability of the baseline which is of the order of 0.2 Å per week under nitrogen flow.

With this setup, we have studied diffusive ultrathin films obtained in capillary rise geometry (9) or with tiny droplets (53-54).

The liquids are:

- light polydimethylsiloxane (PDMS) with a molecular weight of 2400, $\eta = 20 \times 10^{-3}$ Pa.s, a worm-like molecule of diameter about 7 Å (55),
- squalane: a branched saturated alcane $C_{30}H_{62}$ with $\eta = 21 \times 10^{-3}$ Pa.s (SQ),
- tetrakis(2-ethylhexoxy)silane: a star-like molecule with a central silicon atom and four aliphatic arms, with molecular weight of 545, $\eta = 6.8 \times 10^{-3}$ Pa.s (TK).

* The more striking observation is the relatively rapid growth of a monomolecular layer of molecules in front of the main film. The thickness of the monolayer is the molecular diameter, which means for example that the PDMS molecules are lying flat on the surface (9,53,54).

The length of the layer is found to obey a diffusion law:

$$L(t) = \sqrt{D_1 t}$$

$$D_l(TK) = 1.6 \times 10^{-10}\ m^2s^{-1}$$

$$D_l(PDMS) = 2.8 \times 10^{-10}\ m^2s^{-1}$$

*** The edge of the layer, or the droplet itself at very long times, smooths out by a 2-dimensional diffusion of isolated molecules on the surface. The width of the density probability of molecules obeys:**

$$a(t) = \sqrt{D_m t}$$

with

$$D_m(PDMS) \approx 6 \times 10^{-11}\ m^2s^{-1}$$

$$D_m(SQ) = 7 \times 10^{-11}\ m^2s^{-1}$$

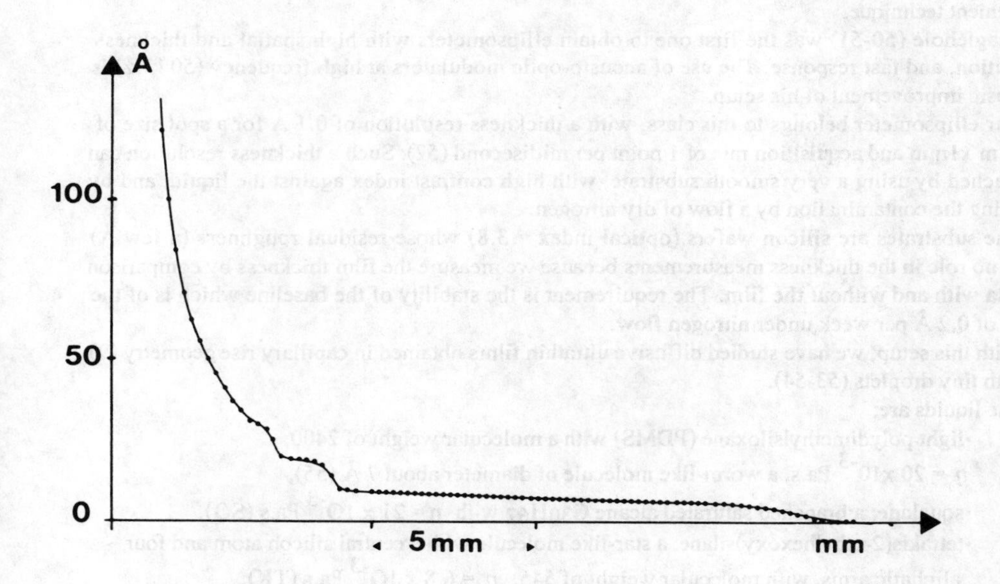

Fig.13 Ellipsometric profile of a TK film in capillary rise geometry:
x axis : film length (vertical coordinate), y axis : film thickness
- 270 hours after contact -.

*** With PDMS and TK, several steps are visible at the liquid edge, corresponding to successive layers of molecules. A typical capillary rise profile of TK is given in figure 13.**

6.2 DISCUSSION

Obvious analogies exist between thin and ultrathin film dynamics. However, specific behaviours of ultrathin films, i.e. layering and growth of the first layer, deserve further discussion.

6.2.1 *Layering* Step-like adsorption isotherms are known for long (56-58). An adsorption isotherm is an experimental determination of the variation $h(\Pi)$. The disjoining pressure Π is directly related to the chemical potential μ_F which can be changed in various ways (14). For volatile liquids, the vapour pressure is used as the control parameter.

Theoretical models account satisfactorily for these effects, which are a direct consequence of liquid-liquid and liquid-wall interactions (18-19),(59-60). Layering in equilibrium situations is well understood.

What part of our "dynamic" layering can be explained by static properties has to be discussed.

6.2.2 *First Layer Dynamics.* The dynamics of the first layer is conveniently described in the frame-work of monomolecular films. Here,no Poiseuille flow can be written, and the basic diffusion equation is somewhat different (10) :

Let c be the number of film molecules per unit area of the solid, the film "thickness" is just:

$$h = C V_o$$

The diffusion equation is :

$$\frac{\partial C}{\partial t} - \frac{\partial}{\partial x}\left[D(C)\frac{\partial C}{\partial x}\right] = 0$$

with

$$D(C) = -\frac{V_o C}{\alpha}\frac{d\Pi}{dC}$$

or

$$\frac{\partial h}{\partial t} - \frac{\partial}{\partial x}\left[D(h)\frac{\partial h}{\partial x}\right] = 0$$

with

$$D(h) = -\frac{V_o h}{\alpha}\frac{d\Pi}{dh}$$

α is a friction coefficient.

Steep variations of Π in the molecular range lead to high values of D. Reasonable Π-shapes allow to explain the specific behaviour of the first monolayer (fig.13)

7. Acknowledgements

Fruitful discussions with P.G.de Gennes, J.Lyklema, M.A.Cohen Stuart and J.Scheutjens are gratefully acknowledged. Figure 13 is due to the courtesy of N. Fraysse and F. Heslot.

8. References

(1) de Gennes P.G.: Lectures - Collège de France- 1988.
(2) Fritz G. :Z. Ang. Phys. 19 , 374 (1965).
(3) Tanner L. :J. Phys.D. 12 , 1473 (1979).
(4) Summ B.D., Yushchenko V.S., Shchukin E.D. :Colloids and surfaces 27 43 (1987)
(5) Teletzke G.F. : Thesis- University of Minnesota-(1983) Teletzke G.F., Scriven L.E., Davis H.T. :J. Chem. Phys. 77, 5794 (1982); ibid. 78, 1431 (1983).
(6) de Gennes P.G. : Rev. Mod. Physics 57, 828 (1985).
(7) Cherry B.W. , Holmes C.M.: J. Coll. Int. Sci. 29,174 (1969).
(8) Blake T.D. , Haynes J.M. : J. Coll. Int. Sci. 30, 421 (1969).
(9) Heslot F., Cazabat A.M., Fraysse N. J of Phys. Cond. Matter - Liquids 1, 5793 (1989) .
(10) Cazabat A.M. : CRAS, Paris 310 II 107 (1990).
(11) Adamson A.W. :"Physical chemistry of surfaces" 4th Edn. - Wiley N.Y. (1982).
(12) Blake T.D. in "Surfactants" - Acad. Press London (1984) p.221.
(13) Cooper W.A. , Nuttall W.A.: J. Agr. Sci. 7, 219 (1915).
(14) Deryaguin B.V. , Churaev N.V. , Muller V.M.: "Surface forces" (1987) - Consultant Bureau - N.Y. and London - ,and references.
(15) De Feijter J.A. in "Thin liquid films" - Ivanov I.B. Editor - :Surfactant Science Series 29, M. Dekker p.1.
(16) Mahanty J. , Ninham B.W. :"Dispersion forces" - Ac. Press - (1976).
(17) Israelachvili J.N. :"Intermolecular and surfaces forces" - Ac. Press - (1985)
(18) Bassignana I.C. , Larher Y. : Surface science 147, 48 (1985).
(19) Ball P.C. , Evans R.: J. Chem. Phys. 89, 4412 (1988).
(20) Blake T.D.: unpublished lecture (1980); Hoffmann R.L: J. Coll. Int. Sci. 94, 470 (1983).
(21) Hoffmann R.L. : J. Coll. Int. Sci. 50, 228 (1975).
(22) Ström G.: Thesis - Royal Institute of technology - (1988) Stockholm , Sweden.
(23) Petrov J.G. , Radoev B.P.: Colloid Polymer Sci. 259, 753 (1981); Blake T.D. :A.I. Ch. E "International Symp. on the mechanics of thin film coating" (1988).
(24) de Gennes P.G.: Coll. Pol. Sci. 264, 463 (1986).
(25) Lopez J. ,Miller C. ,Ruckenstein E.: J. Coll. Int. Sci. 56, 460 (1976).
(26) Joanny J.F. : Thesis - University of Paris - (1985).
(27) Cazabat A.M. ,Cohen Stuart M.A.: J. Phys. Chem. 90, 5845 (1986).
(28) Johnson R.E. ,Dettre R.H. in "Contact angle, wettability and adhesion" - (Gould R.F. Editor) Adv.in Chemistry Series 43,p.112, p.135 (1964)
(29) Oliver J.F. ,Huh C. Mason S.G.: Colloids and surfaces 1, 79 (1980).
(30) Wenzel R.N. :Ind. Eng.Chem. 28, 988 (1936).
(31) Good R.J. : J. Am. Chem. Soc. 74, 5041 (1952).
(32) Marmur A. , Lelah M.D.: J. Coll. Int. Sci. 78,262 (1980).
(33) Pesach D. , Marmur A. : Langmuir 3, 519 (1987).
(34) Bayramli E. ,Van de Ven T.G.M. , Mason S.G. : Colloids and surfaces 3, 131 (1981); id. : Canadian J. of Chemistry 59, 1954-1962 (1981).
(35) Dussan E.B. : Ann. Rev. Fluid Mech. 11, 371 (1979).
(36) Haines W.G. : J. Agric. Sci. 20, 97 (1930).
(37) Blake T.D. ,Ruschak K.J.: Nature 282, 489 (1979).
(38) Cohen Stuart M.A. ,Cazabat A.M.: Prog. Coll. Polymer Sci. 74, 64 (1987) ; Princen H.M., Cazabat A.M., Cohen Stuart M.A., Heslot F. ,Nicolet S.: J. Coll. Int. Sci. 126,

84 (1988).
(39) Washburn E.D. : Phys. Rev. 17, 374 (1921).
(40) Lenormand R., Zarcone C. :59th Ann. Tech. Cong. and Exhibition So. of Petroleum Eng. - Houston - (1984).
(41) Lyne M.B. : J. of pulp and paper research 2, 141 (1989).
(42) Bouillault A., Cazabat A.M., Cohen Stuart M.A. : C.R.A.S.-Paris- 303, 525 (1986).
(43) Marangoni C. : Nuovo Cimento 5, 239 (1871).
(44) Carles P. , Cazabat A.M.: Coll. and Surfaces; to appear in Progress in Coll. and Polymer Science, 41 97 (1989).
(45) Joanny J.F. :J. Coll. Int. Sci. 128, 407 (1989).
(46) Joanny J.F. : Physicochemical Hydr. 9, 183 , B.Spalding Ed. (1987)- and ref.(26) .
(47) Beaglehole D. : J. Phys. Chemistry 93, 893 (1989) and previous (1984) unpublished results.
(48) Leger L., Erman M., Guinet A.M., Ausserre D., Strazielle G., Benattar J.J., Rieutord F.,Daillant J. Bosio L.:Rev. Phys. Appl. 23, 1047 (1988).
(49) Joanny J.F., de Gennes P.G. : J. Phys. -Paris- 47,121 (1986)
(50) Beaglehole D. : Physica 100B, 163 (1980).
(51) Beaglehole D. :in press in: Rev. Sci. Inst.
(52) Beaglehole D., Heslot F., Cazabat A.M. :to appear in: Proceedings of the E.P.S. meeting "Hydrodynamics of dispersed media" (1988).
(53) Heslot F., Cazabat A.M., Levinson P. :Phys. Rev. Lett. 62, 1286 (1989).
(54) Heslot F., Fraysse N., Cazabat A.M.: Nature 338, 640 (1989).
(55) Horn R.G.,Israelachvili J.N.: Macromolecules 21, 2836 (1988).
(56) Drir M., Nham H.S., Hess G.B.: Phys. Rev. B 33, 5145 (1986).
(57) Gilquin B.:Thesis - University of Nancy - (1979).
(58) Drir M., Hess G.B.:Phys. Rev. B 33, 4758 (1986).
(59) Ebner C.: Phys. Rev. A 23, 1925 (1981).
(60) Saam W.: Surf. Sci. 125, 253 (1983)
(61) Lekner J. : Physica - Amsterdam - 113A, 506 (1982)

THERMODYNAMICS OF ADSORPTION FROM DILUTE SOLUTIONS

D.G. HALL,
Unilever Research,
Port Sunlight Laboratory,
Quarry Road East,
Bebington,
Wirral,
Merseyside.
L63 3JW

ABSTRACT. A brief description of the Gibbs adsorption equation and developments since are forwarded, with emphasis given to situations where it makes sense to talk about an adsorbed state. Ionic and nonionic solutes and mixed solvents are considered. Specific adsorption is treated explicitly and non specific adsorption is treated implicitly. The application of bulk solution thermodynamic methods to treat the specific adsorption of several solutes is described. Generalised Clausius-Clapeyron equations for integral and differential heats of adsorption are derived. The physical significance of these heats in terms of well defined transfer processes is explained. Partial molar areas are defined, the treatment of ideal and non ideal mixing are discussed. Competitive adsorption, the dependence of low interfacial tensions on temperature, adsorption maxima, the effects of adsorption on interparticle interactions and on reaction rates at interfaces are all reviewed briefly in general thermodynamic terms.

1. Introduction

The thermodynamics of interfaces rests securely on the work of Gibbs published over 100 years ago [1]. This work is elegant, general and exact but is also somewhat formal and it is not always obvious how to apply it in a way which displays clearly the physical issues involved. Much of the work subsequent to Gibbs has been concerned with overcoming this problem [2-9]. At a formal level this work has essentially three aims. The first is to derive rigourous thermodynamic expressions which relate the results of different kinds of experiments such as calorimetry and the temperature dependence of adsorption equilibria. The second is to provide methods for estimating quantities of interest at a molecular level from macroscopic experiments. The third is to provide a basis for the development of more approximate theories with predictive capability. All three issues are addressed in this paper which is concerned with solute adsorption from dilute solutions.

2. Brief Derivation of the Gibbs Adsorption Equation for a Planar Interface

Consider two coexisting bulk phases α and β consisting of c independent components $i = 1 \rightarrow c$ and separated by a planar interface. Let the system of interest depicted in Fig. 1 be a cylinder of length l and cross sectional area A with one end in bulk phase α, the other in bulk phase β and

D. M. Bloor and E. Wyn-Jones (eds.),
The Structure, Dynamics and Equilibrium Properties of Colloidal Systems, 857–878.

with both ends parallel to the interface. The state of this system is completely determined by the temperature T, the pressure p of the bulk phases, the amounts N_i of the components i and the area A. Hence we may write the following fundamental equation

$$dG = - SdT + Vdp + \sum_i \mu_i \, dN_i + \sigma dA \tag{1}$$

where G, S and V respectively are the Gibbs free energy, the entropy and the volume of the system, μ denotes chemical potential and σ denotes surface tension.

Since G is a linear homogeneous function of the N_i and A we have

$$G = \sum_i \mu_i N_i + \sigma A \tag{2}$$

It follows from 1 and 2 that

$$0 = - SdT + Vdp - \sum_i N_i \, d\mu_i - A d\sigma \tag{3}$$

and on dividing by A we obtain

$$d\sigma = - \frac{S}{A} \, dT + l dp - \sum_i \frac{N_i}{A} \, d\mu_i \tag{4}$$

where

$$l = V/A$$

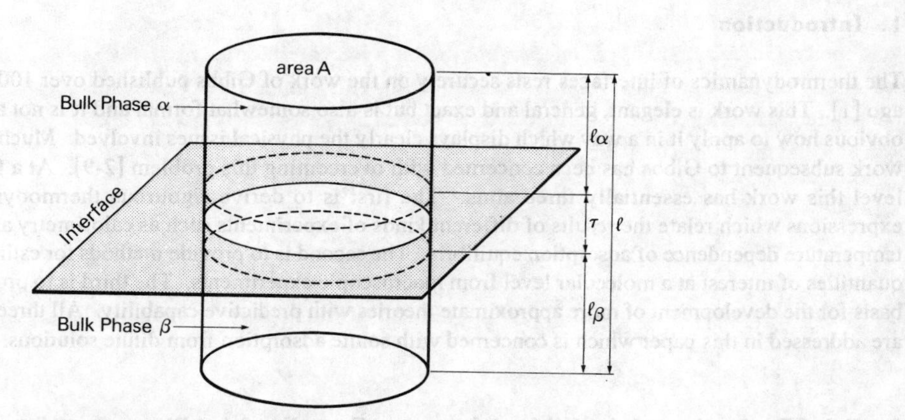

Figure 1. Schematic representation of system under consideration.

The Gibbs Duhem equation for the two bulk phases may be written as

$$dp = s_\alpha\, dT + \sum_i n_i^\alpha\, d\mu_i$$

$$dp = s_\beta\, dT + \sum_i n_i^\beta\, d\mu_i \qquad (5a,b)$$

where s denotes entropy per unit volume and n denotes number density.

Multiplying equations 5a and 5b by l_α and l_β respectively, and subtracting their sum from equation 4 we obtain.

$$d\sigma = -s^\sigma\, dT + \tau dp - \sum_i \Gamma_i\, d\mu_i \qquad (6)$$

where

$$S^\sigma = \frac{S}{A} - l_\alpha s_\alpha - l_\beta\, s_\beta$$

$$\tau = l - l_\alpha - l_\beta$$

$$\Gamma_i = \frac{N_i}{A} - l_\alpha n_i^\alpha - l_\beta n_i^\beta \qquad (7a\text{-}c)$$

Equation 6 is generally true for any choices of l_α and l_β but because of equations 5a, b only c of the c+2 variables T, p and the μ_i are independently variable at equilibrium. However, l_α and l_β may be chosen so that any two members of the set S^σ, τ, Γ_i^σ are zero. For the gas solution interface a common choice of l_α and l_β is that for which τ and $\Gamma_{solvent}$ are both zero. This is the well known Gibbs zero convention [1]. For the oil water interface an alternative is to choose l_α and l_β so that Γ_{oil} and Γ_{water} are both zero [5,6]. This choice has the advantages that the independent variables T, p and solute chemical potentials are a particularly convenient set to work with and that the Γ_i thus defined given by

$$\Gamma_i = -\left(\frac{\partial\sigma}{\partial\mu_i}\right)_{T,p,\mu_j} \qquad (8)$$

are measurable experimentally. Other choices of l_α and l_β correspond to different sets of independent variables.

Clearly the quantities of S^σ, τ and the Γ_i depend on the choice of l_α and l_β. However the Γ_i of solutes which adsorb strongly from dilute solution are insensitive to different reasonable choices of these quantities and the Γ_i of solutes which are confined to the interface do not depend at all on such a choice.

3. Dilute Solutions and Surface phases

When a solute is strongly adsorbed from dilute solution the adsorbed material can be expected to have a markedly different environment from solute in either bulk phase. Consequently it makes sense to think in terms of an adsorbed state. This leads to the concept of a surface phase which at equilibrium coexists with the bulk phases on either side of it and in turn suggests that it may be useful to apply the methods of bulk phase thermodynamics. However the analogy is not perfect because unlike bulk phases, interfaces are non autonomous [10]. In other words their properties are determined not only by their contents but also by the nature of the bulk phases they separate. Indeed what should be regarded as the contents of the surface phase is itself controversial.

Consider for example the gas solution interface of a dilute solution of a nonionic surfactant. According to equation 6 at constant T and p the surface excess Γ depends only on the surfacetension σ. This is analogous to the relationship between number density and pressure for a single component fluid and suggests that we may regard the surface phase as consisting of a single component [11,12]. Alternatively it has been argued that the surface phase should be regarded as a binary mixture of surfactant and water [13-15]. Both viewpoints have their adherents and to some extent it is a matter of taste which is preferable. The view adopted in this paper is that the surface phase should be regarded as consisting only of specifically adsorbed material by which we mean material whose environment differs distinctly from that in either bulk phase. It is arguable that only under these circumstances does it make sense to talk about transferring material from a bulk state to an absorbed state.

In some cases this may involve splitting the total Gibbs surface excess Γ_i into separate contributions from the surface phase and the bulk solution by writing [9,16]

$$\Gamma_i = \Gamma_i^\sigma + \Gamma_i^b \tag{9}$$

where Γ_i^σ refers to the amount present in the distinct environment referred to above. It is also useful sometimes to regard adsorbed and non-adsorbed members of the same species as separate species whose chemical potentials happen to be equal at equilibrium. Indeed such a viewpoint is adopted naturally when discussing the kinetics of adsorption. These considerations lead us to re-write equation 6 in the following form

$$d\sigma = -S^\sigma dT + \tau dp - \sum_i \Gamma_i^\sigma d\mu_i^\sigma - \sum \Gamma_i^b d\mu_i^b - \sum_r \Gamma_r d\mu_r \tag{10}$$

where i refers to specifically adsorbed species and r refers to species which are not specifically adsorbed. Examples of such species include supporting electrolyte ions, which in some cases may be negatively adsorbed [5,17] and cosolvents such as methanol in methanol water mixtures where a gradation in the environment of methanol in moving from the surface to the bulk might be expected.

4. Clausius-Clapeyron equations and partial molar quantities

To keep matters simple the discussion in this section is restricted to the adsorption of dilute uncharged species which are present in only one of the bulk phases and for which the Γ_i^b can be expected to be sufficiently small that the term $\sum_i \Gamma_i^b d\mu_i^b$ is negligible. It is also supposed that each of the two bulk phases contain but a single solvent component. Alternative methods for discussing the adsorption of charged species for which the Γ_i^b may not be negligible are considered in section 7.

The appropriate form of equation 10 we require is

$$d\sigma = -S^\sigma dT + \tau dp - \sum_i \Gamma_i^\sigma d\mu_i^\sigma \tag{11}$$

An equivalent expression is

$$d\left(\frac{\sigma}{T}\right) = -\frac{h^\sigma}{T^2} dT + \tau dp - \sum_i \Gamma_i^\sigma d\left(\frac{\mu_i^\sigma}{T}\right) \tag{12}$$

which has the advantage that we may work directly with enthalpy rather than entropy. It is apparent from equation 12 that $\frac{\mu_k^\sigma}{T}$ may be regarded as a function of T, p and the Γ_i^σ. Hence we may write

$$d\left(\frac{\mu_k^\sigma}{T}\right) = -\frac{h_k^\sigma}{T^2} dT + \frac{v_k^\sigma}{T} dp + \frac{1}{T}\sum_i \left(\frac{\partial \mu_k^\sigma}{\partial \Gamma_i}\right)_{\Gamma_j} d\Gamma_i \tag{13}$$

where at constant T and p

$$h_k^\sigma = \left(\frac{\partial h^\sigma}{\partial \Gamma_k}\right)_{\Gamma_j} \tag{14a,b}$$

$$v_k^\sigma = \left(\frac{\partial \tau}{\partial \Gamma_k}\right)_{\Gamma_j}$$

If component k behaves ideally in bulk solution so that

$$\mu_k^b = \mu_k^{\ominus}(T,p) + RT \ln C_k \tag{15}$$

then

$$d\left[\frac{\mu_k^b}{T}\right] = -\frac{h_k^b}{T^2}\, dT + \frac{v_k^b}{T}\, dp + R d\ln C_k \tag{16}$$

where h_k^b, and v_k^b, respectively are the partial molar enthalpy and volume of species k. For variations at equilibrium

$$d\left(\mu_k^\sigma/T\right) = d\left(\mu_k^b/T\right)$$

hence we obtain from equations 13 and 15

$$0 = -\frac{\left(h_k^\sigma - h_k^b\right)}{T^2}\, dT + \frac{\left(v_k^\sigma - v_k^b\right)}{T}\, dp + \frac{1}{T}\sum_i \left[\frac{\partial \mu_k^\sigma}{\partial \Gamma_i}\right]_{T,\mu,\Gamma j} d\Gamma_i - R d\ln C_k \tag{17}$$

which in turn gives the Clausius-Clapeyron equations

$$\left(h_k^\sigma - h_k^b\right) = -RT^2 \left(\frac{\partial \ln C_k}{\partial T}\right)_{p,\Gamma_i}$$

$$\left(v_k^\sigma - v_k^b\right) = RT \left(\frac{\partial \ln C_k}{\partial p}\right)_{T,\Gamma_i} \tag{18a,b}$$

The LHS of equation 18a is the enthalpy of transferring a k molecule from the bulk solution to the interface under the condition that the Γ^σ of all other components are kept fixed. $\left(v_k^\sigma - v_k^b\right)$ is the corresponding volume change and equations 18a,b show how these quantities can be found from the effects of T and p on the adsorption behaviour. Both $\left(h_k^\sigma - h_k^b\right)$ and $\left(v_k^\sigma - v_k^b\right)$ are differential quantities of adsorption. Also $\left(h_k^\sigma - h_k^b\right)$ is analogous to the isosteric heat encountered in gas adsorption.

The differential quantities h_k^σ and v_k^σ are not analogous to the partial molar quantities used widely in bulk solution thermodynamics. To obtain quantities which are, we may proceed as follows. Let σ_o be the surface tension in the absence of absorbing solutes and let $(\sigma_o - \sigma) = \pi$. In analogy with equation 12 we have

$$d\left(\frac{\sigma_o}{T}\right) = -\frac{h_o^\sigma}{T^2}\,dT + \frac{\tau_o}{T}\,dp \tag{19}$$

Subtracting equation 12 from equation 19 we obtain

$$d\frac{\pi}{T} = \frac{\left[h^\sigma - h_o^\sigma\right]}{T^2}\,dT - \frac{(\tau - \tau_o)}{T}\,dp + \sum_i \Gamma_i^\sigma\, d\left(\mu_i^\sigma/T\right) \tag{20}$$

Let

$$H^\sigma = A\left[h^\sigma - h_o^\sigma\right]$$

$$V^\sigma = A\left(\tau - \tau_o\right) \tag{21a-c}$$

$$N_i^\sigma = A\,\Gamma_i^\sigma$$

and let

$$G^\sigma = \sum_i \mu_i^\sigma N_i^\sigma \tag{22}$$

Multiplying equation 20 by A and adding the result to the differential of equation 22 divided by T, we obtain

$$d\left(G^\sigma/T\right) = -\frac{H_\sigma}{T^2}\,dT + \frac{V^\sigma}{T}\,dp + \sum_i \frac{\mu_i^\sigma}{T}\,dN_i^\sigma + A\,d(\pi/T) \tag{23}$$

We may now define partial molar enthalpies, volumes and areas by writing

$$h_i^* = \left(\frac{\partial H^\sigma}{\partial N_i^\sigma}\right)_{T,p,\pi,N_j^\sigma} = -T^2\left(\frac{\partial \mu_i^\sigma/T}{\partial T}\right)_{p,\pi/T,N_i^\sigma} = -T^2\left(\frac{\partial \mu_i^\sigma/T}{\partial T}\right)_{p,\pi/T,x_i^\sigma}$$

$$v_i^* = \left(\frac{\partial V^\sigma}{\partial N_i^\sigma}\right)_{T,p,\pi,N_j^\sigma} = \left(\frac{\partial \mu_i^\sigma}{\partial p}\right)_{T,\pi,N_i^\sigma} = \left(\frac{\partial \mu_i^\sigma}{\partial p}\right)_{T,\pi,x_i^\sigma} \tag{24a-c}$$

$$\omega_i = \left(\partial A/\partial N_i^\sigma\right)_{T,p,\pi,N_j^\sigma} = \left(\partial \mu_i^\sigma/\partial \pi\right)_{T,p,N_i^\sigma} = \left(\partial \mu_i^\sigma/\partial \pi\right)_{T,p,x_i^\sigma}$$

where x_k^σ given by $x_k^\sigma = N_k^\sigma / \sum_i N_i^\sigma$ is the surface mole fraction of adsorbed k and constant x_i^σ denotes constant surface composition. Also

$$\frac{H^\sigma}{A} = \sum_i \Gamma_i^\sigma h_i^*$$

$$\frac{V^\sigma}{A} = \sum_i \Gamma_i^\sigma v_i^* \tag{25a-c}$$

$$1 = \sum_i \omega_i \Gamma_i^\sigma$$

The inclusion of h_o^σ and τ_o in the definitions of H^σ and V^σ as given by equations 21a,b are necessary if the h_i^* and v_i^* are all to remain finite in the limit that all $\Gamma_i^\sigma \rightarrow 0$.

It follows from equations 24a-c that we may write

$$d\left(\frac{\mu_k^\sigma}{T}\right) = -\frac{h_k^*}{T^2}\,dT + \frac{v_k^*}{T}\,dp + \frac{1}{T}\sum_i \left(\frac{\partial \mu_k^\sigma}{\partial x_i^\sigma}\right)_{T,p,x_j^\sigma} dx_i^\sigma + \omega_k\, d(\pi/T) \tag{26}$$

which together with equation 16 gives

$$\left(h_k^* - h_k^b\right) = -RT^2 \left(\frac{\partial \ln C_k}{\partial T}\right)_{p,x_i^\sigma,\pi/T}$$

$$\left(v_k^* - v_k^b\right) = RT \left(\frac{\partial \ln C_k}{\partial p}\right)_{T,x_i^\sigma,\pi/T} \tag{27a,b}$$

$\left(h_k^* - h_k^b\right)$ is the enthalpy change accompanying the transfer of one mole of k from bulk to surface keeping all N_i^σ constant but allowing A to vary. It is the generalisation to multicomponent surface phases of the integral heat of adsorption. $\left(h^\sigma - h_o^\sigma - \sum_i \Gamma_i h_i^b\right)$ is the enthalpy change of forming the interface of interest from bare oil-water interface by absorbing material from the bulk solution. It is the quantity typically measured in calorimetric studies of adsorption heats. It is related to differential heats by the expression

$$\left(h^{\sigma} - h_o^{\sigma}\right) - \sum_i \Gamma_i^{\sigma} h_i^{b} = \int_o^{\Gamma_i} \sum_i \left(h_i^{\sigma} - h_i^{b}\right) d\Gamma_i \tag{28}$$

where the integral on the RHS is independent of the path of integration and may for convenience be taken as constant x_i^{σ}.

Almost identical results are obtained for mixed solvents [9] provided that the solvent composition is maintained constant. Also like equations 18a,b equations 27a,b show how the heats and volumes of adsorption may be obtained from the temperature dependence of adsorption equilibria.

The basis of the work in this section is given in papers by Hill [3] and Everett [4].

5. Competitive adsorption

Consider a solution containing two adsorbing species, typically a surfactant s and a polymer p. A question of interest is how the adsorption of polymer Γ_p depends on the polymer concentration C_p when the ratio C_s/C_p is kept constant where C_s denotes surfactant concentration. For simplicity we will suppose that the polymer does not self associate in solution and that the amount of surfactant bound to the polymer is always negligible compared with C_s.

At constant T and p we may write the Gibbs adsorption isotherm as

$$d\pi = \Gamma_p\, d\mu_p + \Gamma_s\, d\mu_s \tag{29}$$

which gives on cross differentiation

$$\left(\frac{\partial \Gamma_p}{\partial \mu_s}\right)_{\mu_p} = \left(\frac{\partial \Gamma_s}{\partial \mu_p}\right)_{\mu_s} = \left(\frac{\partial \Gamma_s}{\partial \Gamma_p}\right)_{\mu_s} \left(\frac{\partial \Gamma_p}{\partial \mu_p}\right)_{\mu_s} \tag{30}$$

Since Γ_p is a function of μ_s and μ_p

$$d\Gamma_p = \left(\frac{\partial \Gamma_p}{\partial \mu_s}\right)_{\mu_p} d\mu_s + \left(\frac{\partial \Gamma_p}{\partial \mu_p}\right)_{\mu_s} d\mu_p \tag{31}$$

and it follows that

$$d\Gamma_p = \left(\frac{\partial \Gamma_p}{\partial \mu_p}\right)_{\mu_s} \left[d\mu_p + \left(\frac{\partial \Gamma_s}{\partial \Gamma_p}\right)_{\mu_s} d\mu_s \right] \tag{32}$$

From bulk solution thermodynamics it is straightforward to show that in the present case

$$d\mu_p = -\left(\frac{\partial C_s}{\partial C_p}\right)_{\mu_s} d\mu_s + RT\, d\ln C_p \tag{33}$$

which together with equation 32 gives

$$\left(\frac{\partial \Gamma_p}{\partial \ln C_p}\right)_{C_s/C_p} = \left(\frac{\partial \Gamma_p}{\partial \mu_p}\right)_{\mu_s}\left[RT + \left(\left(\frac{\partial \Gamma_s}{\partial \Gamma_p}\right)_{\mu_s} - \left(\frac{\partial C_s}{\partial C_p}\right)_{\mu_s}\right)\left(\frac{\partial \mu_s}{\partial \ln C_s}\right)_{C_s/C_p}\right] \tag{34}$$

In the particular case that $d\mu_s = RT\, d\ln C_s$ equation 34 simplifies further to give

$$\left(\frac{\partial \Gamma_p}{\partial \ln C_p}\right)_{C_s/C_p} = RT\left(\frac{\partial \Gamma_p}{\partial \mu_p}\right)_{\mu_s}\left[1 + \left(\frac{\partial \Gamma_s}{\partial \Gamma_p}\right)_{\mu_s} - \left(\frac{\partial C_s}{\partial C_p}\right)_{\mu_s}\right] \tag{35}$$

We note that $(\partial\Gamma_p/\partial\mu_p)_{\mu_s}$ is almost certainly positive, $(\partial\Gamma_s/\partial\Gamma_p)_{\mu_s}$ is the increase in surfactant adsorption which accompanies the adsorption of a p molecule and may be regarded as the amount of adsorbed surfactant associated with an adsorbed p molecule where association may be used in a positive or negative sense. Hence at a crowded interface $(\partial\Gamma_s/\partial\Gamma_p)_{\mu_s}$ can be expected to be negative because absorption of a p molecule can be expected to displace surfactant from the surface. Moreover if the p molecule is large $(\partial\Gamma_s/\partial\Gamma_p)_{\mu_s}$ may substantially less than -1. $(\partial C_s/\partial C_p)_{\mu_s}$ may be regarded as the amount of surfactant associated with a p molecule in bulk solution. If surfactant associates strongly with the polymer this quantity may be quite large and positive but is unlikely to be large and negative, especially in dilute solutions. We conclude therefore that $(\partial\Gamma_p/\partial \ln C)_{Cs/C_p}$ is likely to be quite large and negative is especially as monolayer coverage is approached. This conclusion is in accord with similar ones drawn from simulations [18] and calculations based on models [19,20]. Since the above analysis rests on classical thermodynamics it is arguably more powerful. To conclude this section it is perhaps worth noting that in the critical micellar region $^1/_{RT}\left(\frac{\partial \mu_s}{\partial \ln C_s}\right)$ changes typically from one to a very small value. Hence it is quite possible that the right hand side of equation 34 changes sign at the onset of micellisation. Also the above arguments apply in an open system situation where amounts in the bulk solution far exceed amounts adsorbed.

6. Dependence of Low Interfacial Tensions on Temperature

In a series of fine papers Aveyard and coworkers have made a careful and systematic study of systems exhibiting ultra low oil/water interfacial tensions [21-24]. For several systems they have reported minima in interfacial tension as a function of temperature. Two other features of this work are that no noticeable changes in surface tension with increasing surfactant concentration were noted and that the temperature studies were done in closed systems using the spinning drop technique. On the basis of a thermodynamic argument Aveyard and coworkers concluded that at the minimum in surface tension the entropies of surfactant aggregation and surfactant adsorption are equal [21,23]. In this section an alternative analysis is presented which leads to modifications of this conclusion.

For simplicity we suppose that there is but one solute, a nonionic surfactant. Let N_o N_w and

N_d determine the total amounts of oil, water and surfactant in the system and let C_o and C_w denote the surfactant concentrations in the oil and water. According to equation 1

$$\left[\frac{\partial \sigma}{\partial T}\right]_{p,N_i,A} = -\left(\frac{\partial S}{\partial A}\right)_{T,p,N_i} \tag{36}$$

which shows clearly that there is a minimum in surface tension versus temperature when the entropy changes associated with the formation of new surface is zero and when the heat capacity change on forming new surface is negative. It is convenient to regard this process as the formation of bare oil/water interface followed by adsorption of surfactant from the bulk. However, when surfactant is present in both bulk phases each phase loses surfactant to the interface. Hence

$$\left(\frac{\partial S}{\partial A}\right)_{T,p,N_i} = S_o^\sigma + \Gamma s^* - f_w s_w - f_o s_o \tag{37}$$

where S_o^σ is the excess entropy per unit area of the bare oil/water interface. s^* is the integral partial molar entropy of adsorbed surfactant given by $S^* = (S^\sigma - S_o^\sigma)/\Gamma$, s_w and s_o are tbe bulk partial molar entropies in oil and water and f_o and f_w are the respective fractions of adsorbed surfactant drawn from the oil phase and the water phase. These quantities are given by [25]

$$f_o = (1 - f_w) = \frac{N_o(\partial C_o/\partial \mu)}{N_o\left(\frac{\partial C_o}{\partial \mu}\right) + N_w\left(\frac{\partial C_w}{\partial \mu}\right)} \tag{38}$$

and after some simple algebra we obtain

$$\left[\frac{\partial \sigma}{\partial T}\right]_{p,N_i} = -\left[S_o^\sigma + \Gamma(s^* - s_w) - \frac{\Gamma(h_o - h_w)\, N_o(\partial C_o/\partial \mu)}{T\left(N_o\left(\frac{\partial C_o}{\partial \mu}\right) + N_w\left(\frac{\partial C_w}{\partial \mu}\right)\right)}\right] \tag{39}$$

where $(h_o - h_w) = T(s_o - s_w)$

Aveyard and coworkers concluded that

$$\left[\frac{\partial \sigma}{\partial T}\right] = -(S^\sigma - \Gamma s_w) \tag{40}$$

and then identified the righthandside with the adsorption entropy of Γ moles of micellar surfactant. However, this interpretation of $(S^\sigma - \Gamma s_w)$ ignores the contribution of S_o^σ. Aveyard and coworkers also ignored the final term of equation 39. This term makes a positive contribution to $(\partial\sigma/\partial T)$ if $h_o - h_w$ is positive and this is necessarily true if increasing T favours transfer from water to the oil phase.

In a spinning drop experiment N_o/N_w is typically of order 10^{-3}. Consequently the final term on the righthandside of equation 39 will be negligible unless $(\partial C_o/\partial \mu) >> (\partial C_w/\partial \mu)$. Now in a micellar solution of non interacting micelles.[26]

$$\frac{\partial C}{\partial \mu} = m + n^2 C_m \,\tilde{}\, nC \qquad (41)$$

where m, C_m and n respectively denote the monomer concentration, the concentration of micelles and the aggregation number. Since the interfacial tensions in the systems of interest are of order 10^{-2} dynes per cm and change hardly at all with increasing surfactant concentration this suggests that aggregates containing of order 10^3 or more surfactant molecules are present somewhere in the system.

If such aggregates are present in the oil phase rather than the aqueous phase a value of $(\partial C_o/\partial \mu)/(\partial C_w/\partial \mu)$ of order 10 - 100 may not be unreasonable. If in addition $C_o >> C_w$ then it is possible that the final term of equation 39 is significant. Minima in σ versus T are often accompanied by a transfer of surfactant from water to oil. This finding has been explained in terms of changes in surfactant head group size and aggregate curvature with temperature [21]. The above argument indicates that surfactant transfer from water to oil may be responsible for the observed minima rather than merely coincident therewith. A similar analysis can be given of minima in σ induced by salt addition.

7. Ionic Surfactant plus Salt [9]

So far the discussion has been confined to solutions of uncharged species. When ionic species are present we define the quantities θ_i by writing

$$\theta_i = \left[\bar{\mu}_i - \frac{\nu_i}{\nu_c} \bar{\mu}_c \right]$$

where $\bar{\mu}$ denotes the electrochemical potential, ν denotes ionic valency including the sign and where species c is a reference species present in the bulk solution and preferably not specifically adsorbed. The θ_i thus defined may be used in the above sections in exactly the same way as the μ_i all quantities such as heats and volumes of adsorption retain their significance.

In this section the adsorption of an ionic surfactant from an electrolyte solution with a common counterion is considered. For simplicity it is supposed that the surfactant is absent from one of the bulk phases. Subscripts 1, 2 and 3 respectively refer to surfactant ions, counterions and coions. In this context the electrolyte solution may be regarded as a mixed solvent and Γ_i^b may not be negligible. To eliminate the term $\Gamma_i^b \, d\left(\theta_i/T\right)$ in the equation analogous to equation 10 we assume that the activity coefficients of surfactant ions and coions are equal and denote both by γ. For submicellar solutions electrical neutrality gives

$$C_1 + C_3 = C_2 = C \text{ and it follows that } \gamma = \gamma(C).$$

We also make the reasonable assumptions that

$$\frac{\Gamma_1^b}{C_1} = \frac{\Gamma_3}{C_3} = \frac{\Gamma_c}{C} \tag{42}$$

and that $\Gamma_3 = 0$ when $\Gamma_1^\sigma = 0$. From now on we denote Γ_1^σ by Γ. By equating the chemical potentials of adsorbed and bulk surfactant we obtain the following equations

$$0 = -\left(\frac{h_1^\sigma - h_1^b}{T^2}\right) dT + \frac{v_1^\sigma - v_1^b}{T} dp + \frac{1}{T}\left(\frac{\partial \theta_1^\sigma}{\partial \Gamma}\right)_{T,p,C_3} d\Gamma$$

$$- R\, d\ln C_1 - R\left[1 - 2\left(\frac{\partial \Gamma_c}{\partial \Gamma}\right)_{T,p,C_2} - (2 - 2\left(\frac{\partial \Gamma_c}{\partial \Gamma}\right)_{T,p,C_2}\left(\frac{\partial \ln\gamma}{\partial \ln C_2}\right)_{T,p})\right] d\ln C \tag{43}$$

$$0 = -\frac{h_1^* - h_1^b}{T^2} dT + \frac{v_1^* - v_1^b}{T} dp + \frac{1}{\Gamma} d\,(\pi/T)$$

$$- R\, d\ln c_1 - R\left[1 - \frac{2\Gamma_c}{\Gamma} - \left(2 - \frac{2\Gamma_c}{\Gamma}\right)\left(\frac{\partial \ln\gamma}{\partial \ln C_2}\right)_{T,p}\right] \tag{44}$$

Let

$$J^\sigma = -\left(\frac{\partial \ln C_1}{\partial \ln C_2}\right)_{T,p,\Gamma} \tag{45a}$$

$$J^* = -\left(\frac{\partial \ln C_1}{\partial \ln C_2}\right)_{T,p,\pi} \tag{45b}$$

we find from equations 43 and 44 that

$$RT\ \Gamma = \left(\frac{\partial \pi}{\partial \ln C_1}\right)_{T,p,C_2}$$

$$h_1^\sigma - h_1^b = -RT^2\left(\frac{\partial \ln c_1}{\partial T}\right)_{p,\Gamma_1} - RT^2 J^\sigma \left(\frac{\partial \ln C_2}{\partial T}\right)_{p,\Gamma_1} \qquad (46a\text{-}c)$$

$$h_1^* - h_1^b = -RT^2\left(\frac{\partial \ln C_1}{\partial T}\right)_{p,\pi/T} - RT^2 J^* \left(\frac{\partial \ln C_2}{\partial T}\right)_{p,\pi/T}$$

We note that Γ is the amount of surfactant present in the monolayer and includes no contribution from underlying diffuse double layer. Moreover Γ_1^b is by no means always negligible. According to diffuse double layer theory when $C_3 = 0$ $\Gamma_1^b/\Gamma \to -1/2$ in the limit that $\Gamma \to 0$. However, it turns out that as Γ approaches its limiting value Γ_1^b/Γ is much less significant [15]. In this case we have approximately $\pi = \pi(\theta_1)$ and $\Gamma = \Gamma(\theta_1)$. $\left(h_1^\sigma - h_1^b\right)$ is the enthalpy of transferring a surfactant ion from bulk solution to the interface keeping the total area A constant whilst allowing solvent and supporting electrolyte to equilibrate. $\left(h_1^* - h_1^b\right)$ is the corresponding integral quantity where the transfer takes place at constant π. Finally we note that according to equation 46a-c no knowledge of activity coefficients is required to obtain Γ or the heats and volumes of surfactant adsorption from the dependence of π on T, p and solution composition.

8. Adsorption Maxima [16]

Much of the above discussion has made use of the analogy between surface and bulk phases. However, there are important differences which arise from the fact that, in contrast to bulk phases, surface phases are non autonomous. This is well illustrated by the issue of adsorption maxima [16,27-29].

For binary bulk mixtures it is well known that $(\partial\mu_{solute}/\partial C_{solute}) >, O$ is necessary for stability with respect to separation into two coexistent phases. However, $(\partial\mu_{solute}/\partial\Gamma) >, O$ is not a necessary condition for the stability of surface phases. This in itself is not surprising. $(\partial\mu_{solute}/\partial\Gamma) \geq O$ for electrolytes at the air water interface which are negatively adsorbed and it is clear that $(\partial\mu_{solute}/\partial\Gamma)$ must change sign for any binary mixture which gives a surface tension minimum. What is surprising is that $(\partial\Gamma_m/\partial\theta_{solute})$ may be negative where Γ_m is the amount present in the monolayer. This implies that the solute number density need not increase monotonically with increasing bulk concentration everywhere in the system.

A practical consequence of the discussion in ref. 16 is that some adsorption maxima which have hitherto been regarded with suspicion, [29] despite reasonable precautions to ensure purity of materials, may in fact be genuine.

9. Ideal and Non-ideal Mixed Monolayers

So far we have been concerned with the development of thermodynamic arguments which are useful in the interpretation of data and in which any approximations made have been minimal. However, thermodynamics also provides a sound basis for developing non exact models with predictive capability. Examples of such models for bulk systems are theories of ideal and regular solutions. In this section we review a similar approach for surfaces. Following Goodrich [30] and Garrett [31] we define an ideal mixed monolayer of two surfactants i and j as one in which the surfactant chemical potentials μ_i and μ_j are given by

$$\mu_i = \mu_i^o(T,p,\pi) + RT \ln x_i$$
$$\mu_j = \mu_j^o(T,p,\pi) + RT \ln x_j \qquad (47a,b)$$

where μ_i^o is the chemical potential of i in a monolayer of pure i which has the same surface tension as the mixed monolayer of interest at the same T and p and where $x_i = (1 - x_j) = \Gamma_i(\Gamma_i + \Gamma_j)$

It is straightforward to deduce from equation 47 that

$$a_i = a_i^o x_i \qquad (48)$$

and that

$$\frac{a_i}{a_i^o} + \frac{a_j}{a_j^o} = 1 \qquad (49)$$

where a_i is the activity of i in the solution of interest and a_i^o is the activity corresponding to μ_i^o. It is also straightforward to show that

$$\frac{x_i}{\Gamma_i^o} + \frac{x_j}{\Gamma_j^o} = \frac{1}{\Gamma_i + \Gamma_j} \qquad (50)$$

where Γ_i^o is the value of Γ_i for a solution of pure i with chemical potential μ_i^o and activity a_i^o.

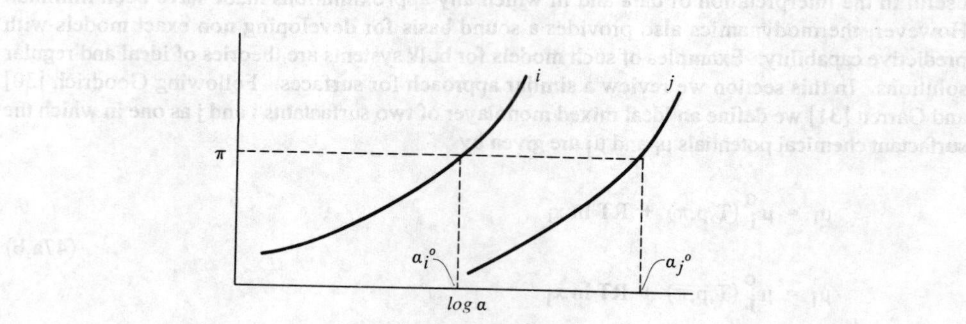

Figure 2. Illustration of method for determining a_i^o and a_j^o from experimental surface pressure data.

Equations 48-50 enable the behaviour of mixtures to be calculated from that of the pure components if we know the a_i of the mixture. To do this take the two π vs a_i graphs for the pure surfactants and find the π for which equation 49 is satisfied. This gives us a_i^o and a_j^o which in turn leads to Γ_i^o and Γ_j^o and to x_i. The procedure is illustrated in Fig (2). $(\Gamma_i + \Gamma_j)$ may now be found from equation 50 and this in turn on multiplication by x_i gives Γ_i. Thus given a_i and a_j we may calculate π, Γ_i and Γ_j for the mixture. As is shown in Figs. 3 and 4 this procedure works well for mixtures of $C_{12}E_3$ and $C_{12}E_6$ even though the dependence of π on Γ for these two pure actives differ considerably.

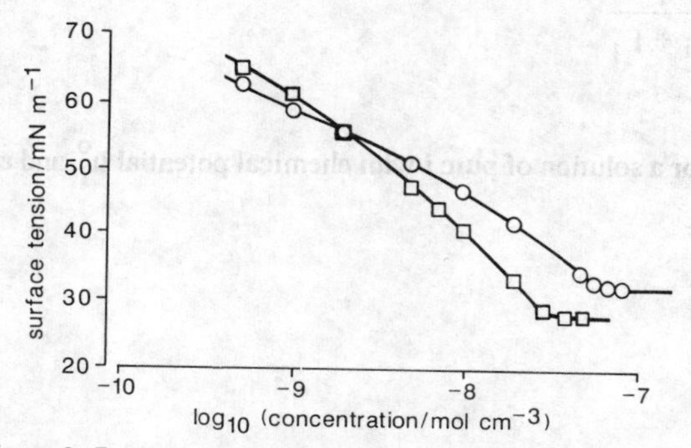

Figure 3. Equilibrium surface tensions of noninic surfactant solutions.
□ $C_{12}E_3$; ○ $C_{12}E_6$

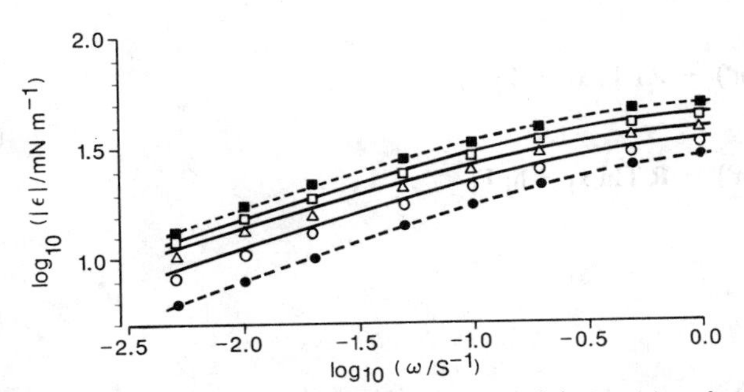

Fig.4 Frequency dependence of dynamic dilational modulus |ε| of noninic surfactant mixtures. ■ 10^{-8} mol cm^{-3} pure $C_{12}E_3$; □ 7.5×10^{-9} mol cm^{-3} $C_{12}E_3$ + 2.5×10^{-9} mol cm^{-3} $C_{12}E_6$; △ 5.0×10^{-9} mol cm^{-3} $C_{12}E_3$ + 5.0×10^{-9} mol cm^{-3} $C_{12}E_6$; ○ 2.5×10^{-9} mol cm^{-3} $C_{12}E_3$ + 7.5×10^{-9} mol cm^{-3} $C_{12}E_6$; ● 10^{-8} mol cm^{-3} pure $C_{12}E_6$.
Full drawn lines, theory for mixed system.

Rosen and coworkers [32,33] have attempted to extend this approach by writing in analogy with regular solution theory

$$\mu_i = \mu_i^o(T,p,\pi) + RT\ln x_i + l_{ij}\, x_j^2$$

$$\mu_j = \mu_j^o(T,p,\pi) + RT\ln x_j + l_{ij}\, x_i^2 \qquad (51a,b)$$

where l_{ij} is a parameter which allows for non ideal mixing. Although they turn out to be quite useful in some practical situations equations 51a,b suffer from the drawback that l_{ij} should be zero in the limit that $\pi \to 0$. In other words l_{ij} is not fixed at a constant T and p but depends in addition on π. Also, as a colleague has pointed out, [34] equations 51a,b cannot accommodate changes in partial molar areas with surface composition at constant π.

To obtain a description of non ideal mixing which uses a single non ideality parameter that is independent of π we proceed as follows. Suppose that two monolayers at the same surface pressure π' are mixed in such away that their total area is unchanged. If they mix ideally the final pressure π is equal to π' but this is unlikely otherwise. To allow for non ideal mixing we suppose that

$$\pi - \pi' = l_{ij} \Gamma_i \Gamma_j \tag{52}$$

where l_{ij} depends on T and p. It is now fairly straightforward to show that the chemical potentials μ_i and μ_j are given by

$$\mu_i = \mu_i^o (T,p,\pi') + RT \ln x_i + l_{ij} \Gamma_j$$

$$\mu_j = \mu_j^o (T,p,\pi') + R T \ln x_j + l_{ij} \Gamma_i \tag{53a,b}$$

that

$$\frac{\Gamma_i}{\Gamma_i^o} + \frac{\Gamma_j}{\Gamma_j^o} = 1 \tag{54}$$

and that

$$a_i = a_i^o \, x_i \exp (l_{ij} \Gamma_j / RT)$$

$$a_j = a_j^o \, x_j \exp (l_{ij} \Gamma_i / RT) \tag{55a,b}$$

where Γ_i^o, Γ_j^o, a_i^o and a_j^o refer to the pure monolayers of i and j at T, p and π'

Given a_i and a_j for a mixture, isotherms for the pure components and a value of l_{ij} equations 52-55 may be used to calculate Γ_i, Γ_j and π. The procedure is less straightforward than in the ideal case. It is however straightforward to calculate a_i, a_j and π given Γ_i and Γ_j. To do this one uses equation 54 to get the π' corresponding to Γ_i^o and Γ_j^o. This then gives a_i^o and a_j^o. Equations 55a,b then give a_i and a_j. By adjusting the input values of Γ_i and Γ_j in accordance with an appropriate iteration procedure the a_i and a_j of the solution of interest should emerge eventually. The determination of l_{ij} experimentally may be accomplished by fitting an experimental value of π for some appropriate mixture.

10. Interparticle Interactions

The effects of adsorption on interparticle interactions may be described thermodynamically by the following equation[35]

$$d\Delta G = -\Delta S \, dT + \Delta V dp - \sum_i \Delta\Gamma_i \, d\mu_i + X dx \tag{56}$$

where the free energy of interaction ΔG is given by

$$\Delta G = G(x) - G(\infty) = \int_{\infty}^{x} X dx \tag{57}$$

X denotes the force between the particles +ve for attraction. x denotes the interparticle separation and where the integration is performed at constant T, p and bulk solution composition. ΔS is an appropriate entropy term.

$$\Delta\Gamma_i = \Gamma_i(x) - \Gamma_i(\infty) \tag{58}$$

where the Γ_i are relative adsorptions with respect to solvent (o) which is not included in the summation.

$$\Delta V = \frac{N_o(x) - N_o(\infty)}{n_o} \tag{59}$$

where N_o denotes the total amount of solvent in the system and n_o is the bulk number density. The quantities ΔG, ΔS, ΔV and ΔN_i all refer to changes in an open system of fixed volume.

From equation 56 it follows that

$$\left(\frac{\partial X}{\partial \mu_i}\right)_{T,p,\mu_j,x} = -\left(\frac{\partial \Gamma_i}{\partial x}\right)_{T,p,\mu_i} \tag{60}$$

This equation shows that the effect of additives on the interparticle force are related to changes in amounts adsorbed with interparticle separation. In particular an additive will increase the attraction between the particles if its adsorption increases with decreasing x [35,36]. This is in accord with Le Chatelier's principle. Depletion flocculation, [37] where a negative adsorption becomes less negative with decreasing x, is a special case of the thermodynamic approach as is the DLVO theory of double layer interactions [38,39].

11. Reactions at Interfaces

Consider the reaction

A + B → transition state → products

according to transition state theory we may write [40]

$$\text{Excess surface rate} = k\Gamma^{\neq} \tag{61}$$

where $\Gamma^{\neq}$ is the surface excess of activated complexes, assumed for simplicity to be positive, and k is a quantity which is taken to be insensitive to changes in the reaction environment. When the equilibrium hypothesis is valid $\Gamma^{\neq}$ may be handled thermodynamically in the same way as the Γ for any stable species present in very small quantities. After procedures described elsewhere [40] we obtain for variations at constant T the expression

$$RT \, d\ln\Gamma^{\neq} = RT \, d\ln \text{(excess surface rate)} =$$

$$\left[1 + \left(\frac{\partial \Gamma_A}{\partial \Gamma^{\neq}}\right)_{\mu_i,\mu_A,\mu_B}\right] d\mu_A + \left[1 + \left(\frac{\partial \Gamma_b}{\partial \Gamma^{\neq}}\right)_{\mu_i,\mu_A,\mu_B}\right] d\mu_B + \sum_i \left(\partial \Gamma_i / \partial \Gamma^{\neq}\right)_{\mu_i, \mu_A, \mu_B} d\mu_i \quad (62)$$

The derivatives on the right hand side of equation (62) have exactly the same significance as $\left(\partial \Gamma_s / \partial \Gamma_p\right)_{\mu_s}$ in section 5 and may be regarded as the amount of the species concerned associated with an adsorbed activated complex. In this context association like adsorption may be used in a positive or negative sense.

12. Conclusions

It is hoped that the above sections have convinced the reader that the aims set out in the introduction can be met in practice. The Clausius-Clapeyron equations in sections (4) and (7) relate calorimetric heats to the temperature dependence of adsorption isotherms thus providing useful relationships between measurable quantities. The differential and integral heats and volumes of adsorption refer to well defined quantities associated with the transfer of molecules from bulk to the adsorbed state whilst allowing the rest of the system to reequilibrate. Even if not well understood such quantities can be discussed in terms of molecular interactions. The adsorption of cosolvents has been treated implicitly so that awkward quantities such as the partial molar enthalpy of negatively adsorbed material are avoided. It has been shown how ideal and non ideal mixing in monolayers can be defined and used as a basis for predicting surface properties of surfactant mixtures given adsorption isotherms for the pure components and a non ideal mixing parameter for each surfactant pair. Finally it has been shown briefly how thermodynamic methods can be used to discuss the affects of additives on interparticle interactions and on reactions at interfaces.

References

1. J.W. Gibbs, Scientific Papers, Vol. 1, Dover Inc., N.Y. 1961.

2. E.A. Guggenheim, *Trans Faraday Soc.*, 1940, **36**, 397.

3. T.L. Hill, *J. Chem. Phys.*, 1949, **17**, 520, 1950, **18**, 246.

4. D.H. Everett, *Trans. Faraday Soc.*, 1950, **46**, 453, 942, 957.

5. R. Parsons, *Canadian J. Chem.*, 1959, **37**, 308.

6. R.S. Hansen, *J. Phys. Chem*, 1962, **66**, 410.

7. G. Schay, *J. Colloid and Interface Sci.*, 1973, **42**, 478.

8. K. Motomura *J. Colloid and Interface Sci*, 1974, **48**, 307.

9. D.G. Hall in Adsorption from Solution, Ed, R.H. Ottewill, C.H. Rochester and A.L. Smith, Academic Press, 1983, p. 31.

10. R. Defay, I. Prigogine, A. Bellemans and D.H. Everett, Surface Tension and Adsorption, Longmans, 1966.

11. N.K. Adam, The Physics and Chemistry of Surfaces Oxford University Press, London and New York, 1941.

12. J.T. Davies and E.K. Rideal, Interfacial Phenomena, Academic Press, New York, 1961.

13. J.A.V. Butler, *Proc. Roy. Soc. A*, 1932, **135**, 348.

14. E.H. Lucassen-Reynders, *Prog. Surface and Membrane Sci.*, 1976, **3**, 253.

15. E.H. Lucassen-Reynders, Ch1 of Anionic Surfactants, Surfactants Science Series Vol. 11, Marcel Dekker Inc., 1981.

16. D.G. Hall, *J. Chem. Soc., Faraday I*, 1980, **76**, 386.

17. H.D. Hurwitz, *J. Electroanal. Chem.*, 1965, **10**, 35.

18. E. Dickinson and S.R. Euston, *Mol. Phys.*, 1989, **68**, 407.

19. M.A. Cohen-Stuart, G.J. Fleer and J.M.H.M. Scheujtens, *J. Colloid and Interface Sci.*, 1984, **97**, 515.

20. J.A. De Feijter, J. Benjamins and M. Tamboer, *Colloids and Surfaces,* 1987, **27**, 413.

21. R. Aveyard, B.P. Binks, T.A. Lawless and J. Mead, *J. Chem. Soc., Faraday Trans. I*, 1985, **81**, 2155.

22. R. Aveyard, B.P. Binks, and J. Mead, *J. Chem. Soc., Faraday Trans I*, 1985, **81**, 2169, 1986, **82**, 1755, 1987, **83**, 2347.

23. R. Aveyard and T.A. Lawless, *J. Chem. Soc., Faraday Trans. I*, 1986, **82**, 2951.

24. R. Aveyard, B.P. Binks, T.A. Lawless and J. Mead, *Canadian J. Chem.*, 1988, **66**, 3031.

25. D.G. Hall, unpublished work.

26. D.G. Hall and B.A. Pethica, Ch. 16 of Nonionic Surfactants Ed. M. Schick, Marcel Dekker Inc., New York, 1967.

27. F.H. Sexsmith and H.J. White, *J. Colloid Sci.*, 1959, **14**, 630.

28. M.L. Corrin, E.L. Lind, A. Roginsky and W.D. Harkins, *J. Colloid Sci.*, 1949, **4**, 485.

29. A. Fava and H. Eyring, *J. Phys. Chem.*, 1956, **60**, 890.

30. F.C. Goodrich, Proc. 2nd Intern Congress of Surface Activity, 1957, **1**, 85, Butterworths, U.K.

31. P.R. Garrett, *J. Chem. Soc., Faraday Trans. 1*, 1976, **72**, 2174.

32. M.J. Rosen and X.G. Hua, *J. Colloid and Interface Sci.*, 1982, **86**, 164.

33. X.G. Hua and M.J. Rosen, *J. Colloid and Interface Sci.*, 1982, **90**, 212.

34. P.R. Garrett, personal communication.

35. D.G. Hall, *J. Chem. Soc., Faraday Trans. 2*, 1972, **68**, 2169.

36. S.G. Ash, D.H. Everett and C.J. Radke, *J. Chem. Soc.*, Faraday 2, 1973, **69**, 1256.

37. G.J. Fleer, J.H.M.H. Scheujtens and B. Vincent, ACS Symposium Series, 1984, **240**, 245.

38. B.V. Derjaguin and L. Landeau, *Acta Phys. Chem.*, 1941, **14**, 633.

39. E.J.W. Verwey and J.Th.G. Overbeek, Theory of the Stability of Lyophobic Colloids Elsevier, Amsterdam, 1948.

40. D.G. Hall, *J. Chem. Soc., Faraday Trans. I*, 1989, **85**, 1881.

LIST OF PARTICIPANTS

M Aara, Kjemisk Institut, Universitet I Bergen, Allegt 41-N-5007, Bergen, Norway
D. Andreu, Departamento de Quimica Fisica, Facultad de Quimica, Universidad de Santiago, Spain
M. Aston, BP Research Centre, Chersey Road, Sunbury on Thames, Middlesex, UK
P. Atkinson, School of Chemical Science, University of East Anglia, Norwich, UK
Z. Attay, Bogazici Universitesi, Fen-Edebiyat Fakultesi, PK 2 Bebek, Istanbul, Turkey
D. Attwood, University of Manchester, Oxford Road, Manchester, UK
W.D. Bauer, Fritz-Haber Institut de Max Planck Gesellschaft, Faradayweg 4-6, 1000 Berlin 33, Germany
C.J. Bergh, Unilever Research Laboratories, Vlaardigen, The Netherlandss
J. Bibette, Centre de Recherche Paul Pascal, CNRS, Chateau Brivazac, F-33600 Pessac, France
T. Bleasdale, University of Salford, Department of Chemistry & Applied Chemistry, The Crescent, Salford, UK
D.M. Bloor, University of Salford, Department of Chemistry & Applied Chemistry, The Crescent, Salford, UK
A. Bommarius, 25 Ames Street, Cambridge, Massachusetts, USA
Bonekamp, Netherlands Energy Research Foundation, Westerduinweg 3, P O Box 1, 1755 ZG Petten, The Netherlands
S. Bucci, Eniricerche, via Maritano 26, 20097 S. Donato, Milanese (Milano), (M1), Italy
H. Burrows, Universidade de Coimbra, Departamento de Quimica, P-3049 Coimbra, Cedex, Portugal
R. Buscall, ICI Corporate Colloid Science Group, P O Box 1, The Heath, Runcorn, Cheshire, UK
M. Catavro, University of East Anglia, School of Chemical Sciences, Norwich, UK
A.M. Cazabat, A.M,College du France, Physique de la Matiere Condensee, 11 Place Marcelin-Berthelot, 75321, France
R. Clark, University of East Anglia, Norwich, UK
J.H. Clint, Colloid Science Branch, B P Research Centre, BP Research Ltd, Chertsey Road, Sunbury-on-Thames, U.K.
M.A. Cohen-Stuart, Department of Physical and Colloid Science, Landbouwuniversiteit, The Netherlands
S. Costa, Centro de Quimica Estrutural, Complexo I, Instituto Superior Tecnico, 1096 Lisbon Codex, Portugal
T. Crowley, University of Salford, Department of Chemistry & Applied Chemistry, The Crescent, Salford, UK
J.M.R. D'Oliveira, Centro de Quimica Fisica Molecular, Complexo Interdisciplinar, Instituto Superior Tecnico, Portugal
E. Dandy, University of Salford, Department of Chemistry & Applied Chemistry, The Crescent, Salford, UK
V. Degiorgio, Dipartimento di Elettronica, Universita di Pavia, 27100 Pavia, via Abbniatagrasso 209, Italy
A. Delgado, Departamento de Fisica Aplicada, Facultad de Ciencias, Universidad de Granada, 18071 Granada, Spain
E. Dickinson, Department of Food Science, University of Leeds, Leeds, UK
J. Eastoe, Department of Chemistry, University of East Anglia, Norwich,UK
E.K. Ersland, University of Bergen, Norway

S.R. Euston, Department of Food Science, University of Leeds, Leeds, UK
I. Fabris, Silsoe College, Silsoe, Bedford, UK
P.D.I. Fletcher, School of Chemistry, University of Hull, Cottingham Road, Hull, UK
S. Frenzel, Christian-Albrechts-Universitat Zu Kiel, Institut fur Angewandte Physik, de Universitat Kiel, Germany
S.E. Friberg, Clarkson University, Department of Chemistry, Cora & Bayard Science Center, Potsdam, USA
J.C. Gee, Chemistry Department, University of Tennessee, Knoxville, TN 37996, USA
H. Gharibi, University of Salford, Department of Chemistry & Applied Chemistry, The Crescent, Salford, UK
J.W. Goodwin, Department of Physical Chemistry, University of Bristol, Cantock's Close, Bristol, UK
R.D. Groot, Unilever Research Laboratorium, Oliver van Noortlaan 120, postbus 114, The Netherlands
D.G. Hall, Unilever Research, Port Sunlight Laboratories, Quarry Road East, Bebington, Wirral, UK
Y. Hanif, University of Salford, Department of Chemistry & Applied Chemistry, The Crescent, Salford, UK
K. Hatchman, University of Salford, Department of Chemistry & Applied Chemistry, The Crescent, Salford,
T.A. Hatton, Department of Chemical Engineering, Massachusetts Institute of Technology, Cambridge, USA
C. Herrmann, Universitat Bielefeld, Fakultat fur Chemie, Postfach 8640, 4800 Bielefeld 1, Germany
H. Hoffmann, Physikalische Chemie I, Universitat Bayreuth, Universitatsstrasse 30, Postfach 3008, Germany
R. Hoffmann, University of Bielefeld, West Germany.
H.E.J. Hofland, Division of Pharm. Technology, University of Leiden, P O Box 9502, Leiden, The Netherlands
H. Hoiland, Kjemisk Institut, Universitetet I Bergen, Allegt 41-N-5007, Bergen, Norway
J. Holzwarth, Department of Chemistry, Freie University, 1000 Berlin 45, Gartnerstr 2d, Germany
A. Holzwarth, Fritz-Haber Institut der Max Planck Gesellschaft, Faradayweg 406, D-1000 Berlin, 33-Dahlem, Germany
C. Hughes, Department of Pure and Applied Physics, The Queen's University of Belfast, Belfast, UK
M. Huser, Lehrstruhl fur Physikalische Chemie II, Templergraben 59, 5100 Aachen, Germany
L. Johansson, Drug Delivery Research, Pharmaceutical R & D, AB Hassle, S-431 83 Molndal, Sweden
M. Kahlweit, Max-Planck-Institut fur Physikalische Chemie, 3400 Gottingen-Nikolousberg, Germany
J. Lang, CNRS, Institut Charles Sadron, (CRM-EAHP), 6 Rue Boussingault, 67083 Strasbourg Cedex, France
T. Mehrian, Physical and Colloid Chemistry Dept, Dreijenplein 6, 6703 HB, Wageningen Agricultural Centro, The Netherlands
P. Lianos, University of Patras, School of Engineering, 26000 Patras, Greece
B. Lindman, Chemical Center, University of Lund, Physical Chemistry 1, P O Box 124, S-221 00 Lund,

J.E. Lofroth, Drug Delivery Research, Pharmaceutical R & D, AB Hassle, S-431 83 Molndal, Sweden
M.A. Lopez-Quintela, Universidad de Santiago de Compostela, Facultad de Quimica Fisica, Spain
H. Lyklema, Department of Physical & Colloid Chemistry, Landbouwuniversiteit Wageningen, Dreijenplein 6, The Netherlands
A. Malliaris, National Research Center, Demokritos, Athens 153 10, Greece
M. Manabe, Department of Industrial Chemistry, Niihama National College of Tech., Yagumo 7-1, Niihama, Japan
G.H. Markx, Department of Biological Sciences, University College of Wales, Aberystwyth, Dyfed, N. Wales, U.K.
E. McCoo, Department of Pure & Applied Physics, Queen's University of Belfast, Belfast, UK
J.F.R.S. Martins, Centro de Quimica Estrutural, Complexo I, Instituto Superior Tecnico, Av. Rovisco Pais, 1096, Portugal
K. Matsui, Kanto Gakuin University, Mabie Memorial School, Mutsuura, Kanazawa-Ku, Yokohoma 236, Japan
V. Maurino, Departamento Chimica Analitica, University of Torino, via Giuria 5, 10125 Torino, Italy
M. Miguel, Chemistry Department, Coimbra University, P-3049 Coimbra Codex, Portugal
C. Minero, Dipartimento Chimica Analitica, University of Torino, via Giuria 5, 10125 Torino, Italy
W.G. Morley, Unilever Research, Colworth House, Sharnbrook, Bedfordshire, UK
Y. Moroi, Department of Chemistry, Faculty of Science, Kyushu University, Hakozaki, Higashi-ku, Japan
E. Morris, Silsoe College, Silsoe, Bedford, UK
H. Mwakibete, University of Dar-es-Salaam, Tanzania
G. Nezzal, Institute de Chimie Industrielle, Universite des Sciences et de la Technologie, B.P. 32 El Alia, Algeria
R.F. Pasternack, Chairman, Swarthmore College, Swarthmore, PA 19081, USA
E. Pelan, Unilever Research, Colworth Laboratory, Sharnbrook, Bedfordshire, UK
E. Pelizetti, Dipartimento di Chimica Analitica, Universita di Torino, via Pietro Giuria, 10125 Torino, Italy
E. Perez-Benito, Departamento de Quimica Fisica, Quimica Analitica E Ingenieria Quimica, Spain
M. Pietralla, Abtellung due Experimentelle Physik, Universitat Ulm, D-7900 Ulm, den Oberer Esselsberg, Germany
M.P. Pileni, Universite Pierre et Marie Curie, Structure Et Reactivite Aux Interfaces, France
T.J.T. Pinheiro, Chemistry Department, Coimbra University, p.3049 Coimbra Codex, Portugal
R. Rajagopalan, Dept. of Chemical Engineering, Cullen College of Engineering, University of Houston, Houston, Texas
J.D.F. Ramsay, UK Atomic Energy Authority, Chemistry Division, Building 429, Harwell, Oxfordshire, UK
B.H. Robinson, School of Chemical Sciences, University of East Anglia, Norwich, UK
A. Sanderson, University of Salford, Department of Chemistry & Applied Chemistry, The Crescent, Salford,.
R. Seiders, Biological Sciences Branch, Department of the Army, United States Army Material Command, UK
M.A. Thomason, University of Salford, Department of Chemistry & Applied Chemistry, The Crescent, Salford, UK

H.K. Schatzel, Christian-Albrechts-Universitat Zu Kiel, Institut fur Angewandte Physik, Olshausenstr 40, Germany
P. Schrimpf, Fritz-Haber Institut de Max Planck Gesellschaft, Faradayweg 4-6, D-1000 Berlin, 33-Dahlem, Germany
K. Shirahama, Department of Chemistry, Faculty of Science and Eng., Saga University, Saga 840, Japan
M. Sjoberg, Institute for Surface Chemistry, Box 5607, S-114 86 Stockholm, Sweden
R. Skurtveit, Kjemisk Institut, Universitetet I Bergen, Allegt 41-N-5007, Bergen, Norway
A. Sobotta, Inst. fur Angewandte Physik, Christian-Albrechts Universitat Zu Kiel, Leibnizstr 11, Olshausenstr , Germany
A. Sputtek, Helmholtz-Institut fur Biomedizinische Technik, an der RWTH Aachen, Pauwelstrasse 30, Germany
J.Q. Solla, Universidad de Santiago de Compostela, Facultad de Quimica Fisica, Departamento de Quimica Fisica, Valiente, Departamento de Quimica Fisica, Universidad de Alcala, Campus Universitario etc, Spain
A. Souto, Departamento de Quimica, Universidade de Coimbra, P-3048 Coimbra Codex, Portugal
H.N. Stein, Technische Universiteit Eindhoven, Laboratory of Colloid Chemistry, Postbus 513, The Netherlands
M.B. Suhaimi, Chemistry Department, Faculty of Science & Environmental Studies, University Pertanian
J.K. Thomas, University of Notre Dame, College of Science, Notre Dame, Indiana 46556, USA
G.J.T. Tiddy, Unilever Research, Port Sunlight Laboratories, Quarry Road East, Bebington, Wirral, UK
B.A. Timimi, Unilever Research, Port Sunlight Laboratories, Quarry Road East, Bebington, Wirral, UK
T. Towey, School of Chemical Sciences, University of East Anglia, Norwich, UK
S. Toxvaerd,Kemisk Laboratorium III, H.C. Orsted Institut, University of Copenhagan, Universitetsparken 5, Denmark
C. Tripp, Xerox Research Centre of Canada, 2660 Speakman Drive, Mississauga, Ontario, Canada
T. Valis, National Hellenic Research Foundation, Biological Research Center, 48 Vas Constantinou Av, Athens,
G. Van Aken, State University Utrecht, Van't Hoff Laboratory, Padualaan 8, 3508 TB Utrecht, The Netherlands
J. Vecer, Department of Biophysics, Paterson Institute for Cancer Research, Christie Hospital, UK
L.G.J. Verhoeven, Unilever Research, Colworth House, Sharnbrook, Bedfordshire, UK
S. Wall, Department of Physical Chemistry, Chalmers University of Tech., and University of Goteberg,
J. Wates, Akzo Chemicals Ltd, Research Centre, Hollingworth Road, Littleborough, lancashire, UK
P. Waugh, University of Salford, Department of Chemistry & Applied Chemistry, The Crescent, Salford, UK
D. Wedlock, Shell Research Ltd, Sittingbourne Research Centre, Sittingbourne, Kent, UK
I. Winter, Universidad de Santiago de Compostela, Facultad de Quimica Fisica, Departamento de Quimica, Portugal
E. Wyn-Jones, Department of Chemistry and Applied Chemistry, University of Salford, Salford M5 4WT, UK

X. Lekkerkerker, Rijikuniversitiet te Utrecht, Van't Hoff Laboratorium, Postbus 80,501, 3508 TB
Utrecht, The Netherlands
Ye, The University of Hull, School of Chemistry, Hull, UK

INDEX OF SUBJECTS